Automotive Electricity, Electronics, and Computer Controls

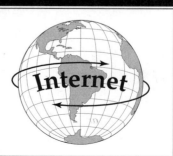

Automotive Electricity, Electronics, and Computer Controls

Barry Hollembeak

Delmar Publishers

an International Thomson Publishing company I(T)P®

Albany • Bonn • Boston • Cincinnati • Detroit • London • Madrid
Melbourne • Mexico City • New York • Pacific Grove • Paris • San Francisco
Singapore • Tokyo • Toronto • Washington

Notice to the Reader

Publisher does not warrant or guarantee any of the products described herein or perform any independent analysis in connection with any of the product information contained herein. Publisher does not assume, and expressly disclaims, any obligation to obtain and include information other than that provided to it by the manufacturer.

The reader is expressly warned to consider and adopt all safety precautions that might be indicated by the activities herein and to avoid all potential hazards. By following the instructions contained herein, the reader willingly assumes all risks in connection with such instructions.

The publisher makes no representation or warranties of any kind, including but not limited to, the warranties of fitness for particular purpose or merchantability, nor are any such representations implied with respect to the material set forth herein, and the publisher takes no responsibility with respect to such material. The publisher shall not be liable for any special, consequential, or exemplary damages resulting, in whole or in part, from the readers' use of, or reliance upon, this material.

Delmar Staff

Publisher: Alar Elken
Acquisitions Editor: Vernon R. Anthony
Project Editor: Megeen Mulholland
Production Coordinator: Karen Smith
Art and Design Coordinator: Michele Canfield, Cheri Plasse
Editorial Assistent: Betsy Hough

Copyright © 1999
By Delmar Publishers
an International Thomson Publishing company I(T)P®

The ITP logo is a trademark under license.

Printed in the United States of America

For more information, contact:

Delmar Publishers
3 Columbia Circle, Box 15015
Albany, New York 12212-5015

International Thomson Editores
Seneca 53
Colonia Polanco
11560 Mexico D F Mexico

International Thomson Publishing Europe
Berkshire House
168-173 High Holborn
London WC1V 7AA
United Kingdom

International Thomson Publishing GmbH
Königswinterer Strasse 418
53227 Bonn
Germany

Nelson ITP, Australia
102 Dodds Street
South Melbourne
Victoria, 3205 Australia

International Thomson Publishing Asia
221 Albert Street
#15-01 Albert Complex
Singapore 189969

Nelson Canada
1120 Birchmont Road
Scarborough, Ontario
M1K 5G4, Canada

International Thomson Publishing Japan
Hirakawa-cho Kyowa Building, 3F
2-2-1 Hirakawa-cho
Chiyoda-ku,
Tokyo 102 Japan

International Thomson Publishing France
Tour Maine-Montparnasse
33 Avenue du Maine
75755 Paris Cedex 15, France

ITE Spain/Paraninfo
Calle Magallanes, 25
28015-Madrid, Espana

5 6 7 8 9 10 XXX 08 07 06 05

Library of Congress Cataloging-in Publication Data
Hollembeak, Barry.
    Automotive electricity, electronics, and computer controls / Barry Hollembeak
        p.   cm.
    Includes index.
    ISBN 0–8273–6566–7 (alk. paper)
    1. Automobiles—Electric equipment.   2. Automobiles—Electronic equipment.
    3. Automobiles—Motors—Computer control systems.
    I. Title.
TL272.H6223   1998
629.25'4--dc21

97–42032
CIP

Contents

Chapter 1
Basic Electrical Theories

Objective 1

Introduction 1

Basics of Electron Flow 1

Electricity Defined 3

Voltage Drop...6

Photo Sequence 1—Performing
a Voltage Drop Test 7

Types of Current...9

Electrical Circuits 9

Capacitance 14

Semiconductors 15

Magnetism Principles.................................16

Theory of Induction 17

EMI Suppression 18

Summary 19

Terms-To-Know 20

Review Questions.......................................20

ASE Style Review Questions 21

Chapter 2
Electrical and
Electronic Components

Objective 23

Introduction 23

Electrical Components 23

Electronic Components 28

Circuit Protection Devices 35

Summary 39

Terms-To-Know 40

Review Questions.......................................40

ASE Style Review Questions 41

Chapter 3
Circuit Problems and
Electrical Test Equipment

Objective 43

Introduction 43

Circuit Defects 43

Test Equipment...45

Summary 56

Terms-To-Know 57

Review Questions.......................................57

ASE Style Review Questions 58

Chapter 4
Basic Electrical
Troubleshooting and Service

Objective 59

Introduction 59

Testing Circuit Protection Devices.............60

Testing and Replacing Electrical Components............62

Testing for Circuit Defects 66

Scan Tester Features 70

Digital Storage Oscilloscope (DSO) 72

Summary 76

Terms-To-Know 77

ASE Style Review Questions 77

Chapter 5
Wiring and Circuit Diagrams

Objective 79

Introduction 79

Electrical Troubleshooting 79

Automotive Wiring.....................................80

Wiring Diagrams 85

Reading Wiring Diagrams 90
Summary 97
Terms-To-Know 97
Review Questions 98
ASE Style Review Questions 98

Chapter 6
Wiring Repair

Objective 101
Introduction 101
Copper Wire Repairs 101
Repairing Aluminum Wire 103
Photo Sequence 2—Soldering Copper Wire 104
Splicing Twisted/Shielded Wires 106
Replacing Fusible Links 107
Repairing Connector Terminals 107
Summary 111
Terms-To-Know 111
ASE Style Review Questions 111

Chapter 7
Automotive Batteries

Objective 113
Introduction 113
Conventional Batteries 114
Maintenance-Free Batteries 117
Hybrid Batteries 119
Recombination Batteries 119
Battery Terminals 120
Battery Ratings 120
Battery Cables 121
Battery Hold-Downs...................................121
Battery Failure 121
Summary 122
Terms-To-Know 123
Review Questions.......................................123
ASE Style Review Questions 124

Chapter 8
Battery Diagnosis and Service

Objective 125
Introduction 125
General Precautions 126

Battery Inspection......................................126
Charging the Battery 127
Battery Test Series 129
Battery Removal and Cleaning....................133
Battery Drain Test 134
Photo Sequence 3—Removing the Battery 135
Jumping the Battery 136
Summary 137
Terms-To-Know 138
ASE Style Review Questions 138

Chapter 9
Direct Current Motors
and the Starting System

Objective 141
Introduction 141
Motor Principles 142
Starter Drives 145
Cranking Motor Circuits 147
Starter Control Circuit Components.............147
Cranking Motor Designs 150
Summary 153
Terms-To-Know 153
Review Questions.......................................154
ASE Style Review Questions 155

Chapter 10
Starting System
Diagnosis and Service

Objective 157
Introduction 157
Starting System Service Cautions 157
Starting System Troubleshooting 158
Testing the Starting System 158
Starter Motor Removal 164
Starter Motor Disassembly.........................165
Starter Motor Component Tests 166
Photo Sequence 4—Typical Procedure
for Delco Remy Starter Disassembly 167
Starter Reassembly 169
Summary 171
Terms-To-Know 172
ASE Style Review Questions......................172

Chapter 11
Charging Systems

Objective 175
Introduction 175
Principle of Operation 176
AC Generator Circuits 183
AC Generator Operation Overview 183
Regulation 184
Charging Indicators 189
Summary 191
Terms-To-Know 192
Review Questions..192
ASE Style Review Questions 193

Chapter 12
Charging System
Testing and Service

Objective 195
Introduction 195
Charging System Service Cautions 196
AC Generator Noises 197
Charging System Troubleshooting..............197
AC Generator Removal and Replacement 207
Bench Testing the AC Generator 207
AC Generator Disassembly 209
AC Generator Component Testing 210
Photo Sequence 5—Typical
Procedure for IAR Alternator Disassembly 212
AC Generator Reassembly 215
Diode Pattern Testing 215
Summary 218
Terms-To-Know 218
ASE Style Review Questions 218

Chapter 13
Lighting Circuits

Objective 221
Introduction 221
Lamps..222
Headlights...222
Headlight Switches..224
Dimmer Switches 225
Headlight Circuits..225

Concealed Headlights...226
Flash to Pass 228
Exterior Lights 228
Interior Lights..236
Lighting System Complexity 238
Summary 240
Terms-To-Know 240
Review Questions...240
ASE Style Review Questions 241

Chapter 14
Lighting Circuits
Repair and Diagnosis

Objective 243
Introduction 243
Headlights...244
Headlight Switch Testing and Replacement..............250
Dimmer Switch Testing and Replacement 251
Photo Sequence 6—Removal
of the Multifunction Switch 254
Taillight Assemblies 255
Interior Lights..258
Summary 260
Terms-To-Know 261
ASE Style Review Questions 261

Chapter 15
Conventional Analog
Instrumentation, Indicator
Lights, and Warning Devices

Objective 263
Introduction 263
Speedometers 263
Odometers 264
Tachometers 265
Gauges 265
Gauge Sending Units 269
Warning Lamps 270
Audible Warning Systems 273
Summary 274
Terms-To-Know 274
Review Questions...274
ASE Style Review Questions 275

Chapter 16
Conventional Instrument Cluster Diagnosis and Repair

Objective 277
Introduction 277
Instrument Panel and Printed Circuit Removal 277
Speedometers 279
Tachometers 283
Gauges 283
Single Gauge Failure 283
Multiple Gauge Failure 284
Gauge Sending Units 285
Warning Lamps 285
Photo Sequence 7—Bench Testing
the Fuel Level Sender Unit........................286
Testing Charging Indicator Lamp Circuits 287
Audible Warning Systems 289
Summary 291
ASE Style Review Questions 291

Chapter 17
Electrical Accessories

Objective 293
Introduction 293
Horns 293
Horn Switches 294
Horn Circuits 294
Windshield Wipers 295
Wiper Motors 295
Blower Motor Circuit 301
Electric Defoggers 302
Power Mirrors...303
Automatic Rear View Mirror 303
Electrochromic Mirrors 305
Power Windows 305
Power Seats 307
Power Door Locks 307
Summary...311
Terms-To-Know 311
Review Questions......................................312
ASE Style Review Questions 312

Chapter 18
Electrical Accessories Diagnosis and Repair

Objective 315
Introduction 315
Horn Diagnosis...315
Wiper System Service 317
Windshield Washer System Service 322
Blower Motor Service 323
Electric Defogger Diagnosis and Service....325
Power Window Diagnosis 326
Power Seat Diagnosis................................327
Power Door Lock Diagnosis 327
Summary 329
ASE Style Review Questions 329

Chapter 19
Basic Distributor and Electronic Ignition Systems

Objective 331
Introduction 331
Basic Circuitry 332
Ignition Components 333
Ignition Timing...337
Spark Timing Systems 340
Electronic or Solid-State Ignition..............345
Computer-Controlled Electronic Ignition 349
Summary 350
Terms-To-Know 350
Review Questions......................................350
ASE Style Review Questions 351

Chapter 20
Basic Ignition System Diagnosis and Service

Objective 353
Introduction 353
Logical Troubleshooting 353
Visual Inspection 354
Ignition System Diagnosis 358
Oscilloscope Testing.................................360
Individual Component Testing 366

Spark Plug Service 373

Setting Ignition Timing 375

Summary 378

Terms-To-Know 380

ASE Style Review Questions 380

Chapter 21
Introduction to the Computer

Objective 383

Introduction 383

Analog and Digital Principles 384

Central Processing Unit 386

Computer Memory 387

Information Processing................................389

Inputs 392

Outputs 399

Multiplexing 400

Summary 402

Terms-To-Know 403

Review Questions...403

ASE Style Review Questions 404

Chapter 22
Body Computer
System Diagnosis

Objective 407

Introduction 407

Electronic Service Precautions.............409

Trouble Codes 409

Entering Diagnostics 410

Testing Actuators 413

Testing Sensors...414

Testing Hall-effect Sensors......................416

PROM Replacement.......................................417

Summary 417

Photo Sequence 8—Typical
Procedure for Replacing the PROM...........418

Terms-To-Know 419

ASE Style Review Questions 420

Chapter 23
Computer-Controlled
Ignition Systems

Objective 421

Introduction 421

Common Sensors 422

Computer-Controlled
Ignition System Operation 425

Electronic Ignition System Operation 427

Examples of EI Systems.............................430

Summary 432

Terms-To-Know 432

Review Questions...433

ASE Style Review Questions 433

Chapter 24
Computer-Controlled
Ignition System Diagnosis

Objective 435

Introduction 435

Isolating Computerized
Engine Control Problems 436

EI System Service 441

Photo Sequence 9—Removing
and Replacing Various DIS Components ...442

Testing Hall-effect Sensors......................444

Computer-Controlled DI Ignition
System Service and Diagnosis 445

Electronic Ignition (EI)
System Diagnosis and Service 447

Diagnosis and Service of Ford EEC
Low Data and High Data Rate EI Systems 449

General Motors Electronic
EI System Service and Diagnosis............452

Summary 457

Terms-To-Know 458

ASE Style Review Questions 458

Chapter 25
Electronic Fuel Control

Objective 461

Introduction 461

Input Sensors 462

Outputs 473
Open Loop vs Closed Loop 475
Computer Air/Fuel Ratio Strategy 475
Modes of Operation 477
Adaptive Memory....................................478
Fuel Pump Circuits.................................480
Pressure Regulators 482
Throttle Body Injection Systems 485
Port Fuel Injection Systems 486
Typical Domestic
Sequential Fuel Injection System 488
Typical Domestic
Multipoint Fuel Injection System.............491
Central Port Injection 493
Electronic Continuous
Injection System (CIS-E) 495
Summary 497
Terms-To-Know 498
Review Questions...................................498
ASE Style Review Questions 499

Chapter 26
Electronic Fuel Injection Diagnosis and Service

Objective 501
Introduction 501
Disconnecting Battery Cables 502
Service Precautions 503
Preliminary Inspection 503
Locating Service Information...................503
Fuel Pressure Testing 505
Causes of Low Fuel Pump Pressure 508
Fuel Pump Volume Testing 508
Diagnosis of Computer
Voltage Supply and Ground Wires 508
Input Sensor Diagnosis and Service........509
Photo Sequence 10—MAP Sensor Testing 516
Exhaust Gas Recirculation Valve
Position Sensor Diagnosis 519
Injector Testing......................................520
Cold Start Injector Diagnosis 523
Minimum Idle Speed Adjustment 524
Idle Air Control Motor Testing 525
Flash Code Retrieval 526

Scan Tester Diagnosis 530
Ford Breakout Box Testing 532
Fuel Injection Data Recording Interpretation............532
Diagnosing Repeated Component Failures 536
Diagnosing Multiple Component Failures 538
Summary 538
Terms-To-Know 539
ASE Style Review Questions 539

Chapter 27
Onboard Diagnostics Second Generation (OBD II)

Objective 541
Introduction 541
OBD I vs OBD II 542
OBD II Component Requirements 543
Chrysler OBD II 551
General Motors OBD II 557
Monitors 561
Ford OBD II 568
Monitors 569
Summary 574
Terms-To-Know 574
Review Questions...................................576
ASE Style Review Questions 576

Chapter 28
OBD II Diagnosis

Objective 577
Introduction 577
OBD I vs OBD II Diagnosis....................578
Summary 580
ASE Style Review Questions 580

Chapter 29
Advanced Lighting Circuits and Electronic Instrumentation

Objective 583
Introduction 583
Computer-Controlled Concealed Headlights 583
Automatic Headlight Dimming 586
Automatic On/Off with Time Delay..........587
Daytime Running Lamps 589

Illuminated Entry Systems 592

Instrument Panel Dimming 592

Fiber Optics 592

Lamp Outage Indicators 593

Electronic Instrumentation Introduction 596

Head-up Display..609

Voice Warning Systems 610

Travel Information Systems 611

Summary 614

Terms-To-Know 614

Review Questions..614

ASE Style Review Questions 615

Chapter 30
Advanced Lighting Systems and Electronic Instrumentation Diagnosis and Repair

Objective 617

Introduction 617

Computer-Controlled Concealed
Headlight Diagnosis 618

Automatic Headlight System Diagnosis...................620

Photo Sequence 11—Typical Procedure
for Replacing a Photocell Assembly 623

Illuminated Entry System Diagnosis 624

Fiber Optics Diagnosis 627

Diagnosing and Servicing
Electronic Instrumentation 627

Electronic Speedometers and Odometers.................636

Electronic Gauges..639

Trip Computers..640

Head-Up Display (HUD) Diagnosis....................641

Summary 642

Terms-To-Know 643

ASE Style Review Questions 643

Chapter 31
Electronic Climate Control

Objective 645

Introduction 645

Introduction to Semiautomatic and
Electronic Automatic Temperature Control 645

Electronic Automatic Temperature Control 652

Summary 661

Terms-To-Know 661

Review Questions..662

ASE Style Review Questions 663

Chapter 32
Diagnosis of Electronic Climate Control Systems

Objective 665

Introduction 665

Preliminary Inspection 665

Semiautomatic Temperature
Control System Diagnosis 666

Electronic Automatic Temperature
Control System Diagnosis 670

Photo Sequence 12—Typical Procedure
for Performing a Scan Tester Diagnosis of an
Automatic Temperature Control System 681

Summary 681

ASE Style Review Questions 682

Chapter 33
Vehicle Accessories

Objective 685

Introduction 685

Introduction to Electronic
Cruise Control Systems 685

Memory Seats..691

Electronic Sunroof Concepts 692

Anti-theft Systems 695

Automatic Door Locks 700

Keyless Entry 702

Electronic Heated Windshield 704

Intelligent Windshield Wipers 706

Vehicle Audio Entertainment Systems.....................707

Summary 708

Terms-To-Know 711

Review Questions..711

ASE Style Review Questions 712

Chapter 34
Diagnosing and Servicing Vehicle Accessories

Objective 715

Introduction 715

Diagnosis and Service of
Electronic Cruise Control Systems 715

Self-Diagnostics 716

Diagnosing Systems without Trouble Codes 723

Photo Sequence 13—Typical Procedure
for Replacing the Servo Assembly 727

Memory Seat Diagnosis 728

Electronic Sunroof Diagnosis 730

Anti-theft System Troubleshooting 730

Automatic Door Lock
System Troubleshooting 734

Keyless Entry Diagnosis 737

Electronically Heated Windshield Service 739

Radio-Stereo Sound Systems 742

Summary 744

Terms-To-Know 744

ASE Style Review Questions 745

Chapter 35
Passive Restraint Systems

Objective 747

Introduction 747

Passive Seatbelt Systems 747

Air Bag Systems 749

Air Bag Deployment 755

General Motors' SIR 755

Summary 760

Terms-To-Know 761

Review Questions 761

ASE Style Review Questions 762

Chapter 36
Diagnosing Passive Restraint Systems

Objective 765

Introduction 765

Automatic Seatbelt Service 765

Air Bag Safety and Service Warnings 771

General Motors SIR Diagnostics 772

Ford Air Bag System Flash Code Diagnosis 775

Toyota Air Bag System Flash Code Diagnosis 776

Honda Air Bag System Voltmeter Diagnosis 776

Inspection After an Accident 777

Clean-up Procedure After Deployment 778

Component Replacement 778

Photo Sequence 14—Typical Procedure
for Removing the Air Bag Module 779

Summary 781

ASE Style Review Questions 781

Glossary
.... 783

Index
.... 803

Basic Electrical Theories

Objective

Upon completion and review of this chapter, you should be able to:

❑ Explain the theories and laws of electricity.

❑ Describe the difference between insulators, conductors, and semiconductors.

❑ Define voltage, current, and resistance.

❑ Define and use Ohm's law correctly.

❑ Explain the basic concepts of capacitance.

❑ Explain the difference between AC and DC currents.

❑ Define and illustrate series, parallel, and series-parallel circuits and the electrical laws that govern them.

❑ Explain the basic theory of semiconductors.

❑ Explain the theory of electromagnetism.

❑ Explain the principles of induction.

Introduction

The electrical systems used in today's vehicles can be very complicated. However, through an understanding of the principles and laws governing electrical circuits, technicians can simplify their job of diagnosing electrical problems. In this chapter you will learn the laws that dictate electrical behavior, how circuits operate, the difference between types of circuits, and how to apply Ohm's law to each type of circuit. You will also learn the basic theories of semiconductor construction. Because magnetism and electricity are closely related, a study of electromagnetism and induction is included in this chapter.

Basics of Electron Flow

Because electricity is an energy form that cannot be seen, some technicians regard the vehicle's electrical system as being more complicated than it is. These technicians approach the vehicle's electrical system with some reluctance. It is important for technicians to understand that electrical behavior is confined to definite laws that produce predictable results and effects. To facilitate the understanding of the laws of electricity, a short study of atoms is presented.

Atomic Structure

An **atom** is the smallest part of a chemical element that still has all the characteristics of that element. An atom is constructed of a fixed arrangement of **electrons** in orbit around a **nucleus**, much like planets orbiting the sun (Figure 1-1). The nucleus is made up of positive charged particles called **protons** and particles that have no charge called **neutrons**. These two types of particles are tightly bound together. The electrons are free to move within their orbits at fixed distances around the nucleus. The attraction between the negative electrons and the positive protons causes the electrons to orbit the

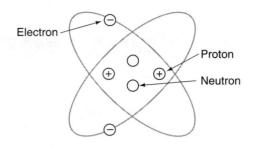

Figure 1-1 Basic construction of an atom.

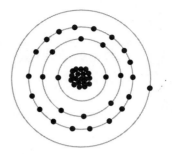

Figure 1-3 Basic structure of a copper atom.

nucleus. All of the electrons surrounding the nucleus are negatively charged, so they repel each other if they get too close. The electrons attempt to stay as far away from each other as possible without leaving their orbits.

Atoms attempt to have the same number of electrons as there are protons in the nucleus. This makes the atom **balanced** (Figure 1-2). To remain balanced, an atom will shed an electron or attract an electron from another atom. A specific number of electrons are in each of the electron orbit paths. The orbit closest to the nucleus has room for two electrons; the second orbit holds up to 8 electrons; the third holds up to 18; and the fourth and fifth hold up to 32 each. The number of orbits depends on the number of electrons the atom has. A copper atom contains 29 electrons, 2 in the first orbit, 8 in the second, 18 in the third orbit, and 1 in the fourth (Figure 1-3). The outer orbit, or shell as it is sometimes called, is referred to as the **valence ring**. This is the orbit we care about in our study of electricity.

In studying the laws of electricity, the only concern is with the electrons that are in the valence ring. Since an atom seeks to be balanced, an atom that has several electrons missing in its valence ring will attempt to gain other electrons from neighboring atoms. Also, if the atom has an excess amount of electrons in its valence ring it will try to pass them on to neighboring atoms.

Conductors and Insulators

Some substances can shed or attract electrons very easily. The ease with which the atom accomplishes this depends upon the number of electrons needed in the valence ring to balance the atom. If an atom sheds electrons easily, it is a good **conductor**. A conductor is a material that supports the flow of electricity. If the atom does not shed its electrons easily, the substance is a good **insulator**. Insulators are materials that do not support the flow of electricity.

The difference between the atomic structures of conductors and insulators is how tight the electrons in the valence ring are held together by the nucleus. The outermost electrons in a conductor's atoms are loosely held together by the nucleus while the atoms in an insulator substance hold the electrons very tightly. For example, because a copper atom has 29 electrons and 29 protons arranged as shown in Figure 1-3, there is only one electron in the valence ring. For the valence ring to be completely filled it would require 32 electrons. Since there is only one electron, it is loosely tied to the atom and can be easily removed, making it a good conductor.

Copper, silver, gold, and other good conductors of electricity have only one to three electrons in their valence ring. These atoms can be made to give up the electrons in their valence ring with little effort.

Since electricity is the movement of electrons from one atom to another, atoms that have one to three electrons in their valence ring support electricity. They allow the electron to easily move from the valence ring of one atom to the valence ring of another atom. Therefore, if we have a wire made of millions of copper atoms, we have a good conductor of electricity. To have electricity, we simply need to add one electron to one of the copper atoms. That atom will shed the electron it had to another atom, which will shed its original electron to another, and so on. As the electrons move from atom to atom, a force is released. This force is what we use to light lamps, run motors, and so on. As long as we keep the electrons moving in the conductor, we have electricity.

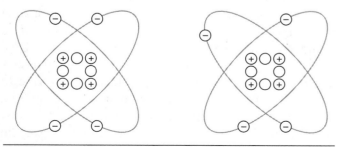

Figure 1-2 If the number of electrons and protons in an atom are the same, the atom is balanced.

The electrons in the atoms of a conductor can be freed from their outer orbits by forces such as heat, friction, light, pressure, chemical reaction, and magnetism. When electrons are moved out of their orbit, they form an electrical **current** under proper conditions. Insulators are required to control the routing of the flow of electricity.

A BIT OF HISTORY

Electricity was discovered by the Greeks over 2,500 years ago. They noticed that when amber was rubbed with other materials it was charged with an unknown force that had the power to attract objects, such as dried leaves and feathers. The Greeks called amber "elektron." The word electric is derived from this word and means "to be like amber."

The atoms of materials that make good insulators have five to eight electrons in their valence ring. The electrons are held tightly around the atom's nucleus and they can not be moved easily. Materials such as some plastics, glass, and rubber make good insulators. Insulators are used to prevent electron flow or to contain it within a conductor. Insulating material covers the outside of most conductors to keep the moving electrons within the conductor.

In summary, the number of electrons in the valence ring determines whether an atom is a good conductor or insulator. Some atoms are not good insulators or conductors; these are called semiconductors. Basically, the number of atoms in the valence ring determines the type of material:

1. Three or fewer electrons - conductor
2. Five or more electrons - insulator
3. Four electrons - semiconductor

CAUTION *The human body is a conductor of electricity. When performing service on electrical systems, be aware that electrical shock is possible. Although the shock is usually harmless, the reaction to the shock can cause injury. Observe all safety rules associated with electricity. Never wear jewelry when servicing or testing the electrical system.*

WARNING *Any broken, frayed or damaged insulation material requires replacement or repair to the conductor. Exposed conductors can result in a safety hazard and circuit component damage.*

Electricity Defined

Electricity is the movement of free electrons through a conductor (Figure 1-4). It is important to note that random movement of electrons is not electric current; the electrons must move in the same direction. The negatively charged electrons are attracted to the positively charged protons. Since there are excess electrons on the other end of the conductor, there are many electrons being attracted to the protons. This attraction acts to push the electrons toward the protons. This push is normally called **electrical pressure**. The amount of electrical pressure is determined by the number of electrons that are attracted to protons. The electrical pressure or **electromotive force (EMF)** attempts to push an electron out of its orbit and toward the excess protons. If an electron is freed from its orbit, the atom acquires a positive charge because it now has one more proton than it has electrons. The unbalanced atom or **ion** attempts to return to its balanced state so it will attract electrons from the orbit of other balanced atoms. This starts a chain reaction as one atom captures an electron and another releases an electron. As this action continues to occur, electrons will flow through the conductor. A stream of free electrons forms and an electrical current is started. This does not mean a single electron travels the length of the insulator; it means the overall effect is electrons moving in one direction. All this happens at the speed of light at 186,000 miles per second (299,000 kilometers per second). The strength of the electron flow is dependant on the **potential** difference or voltage.

The three elements of electricity are voltage, current, and resistance. How these three elements interrelate governs the behavior of electricity. Once the technician comprehends the laws that govern electricity, understanding

Conductor

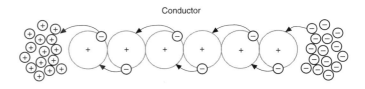

Figure 1-4 As electrons flow in one direction from one atom to another, an electrical current is developed.

the function and operation of the various automotive electrical systems is an easier task. This knowledge will assist the technician in diagnosis and repair of automotive electrical systems.

Voltage

Voltage can be defined as an electrical pressure (Figure 1-5) and is the electromotive force (EMF) that causes the movement of the electrons in a conductor. In Figure 1-4, voltage is the force of attraction between the positive and negative charges. An electrical **pressure difference** is created when there is a mass of electrons at one point in the circuit, and a lack of electrons at another point in the circuit. In the automobile, the battery or generator is used to apply the electrical pressure.

The amount of pressure applied to a circuit is stated in the number of volts. If a voltmeter is connected across the terminals of an automobile battery it may indicate 12.6 volts. This is actually indicating that there is a difference in potential of 12.6 volts. There is 12.6 volts of electrical pressure between the two battery terminals. Remember, the term voltage is the difference or electrical potential that exists between the negative and positive sides of a circuit.

In a circuit that has current flowing, voltage will exist between any two points in that circuit (Figure 1-6). The only time voltage does not exist is when the potential drops to zero. In Figure 1-6 the voltage potential between points A and C, and B and C is 12.6 volts. However, between points A and B the pressure difference is zero and the voltmeter will indicate 0 volts.

Current

Current can be defined as the rate of electron flow (Figure 1-7) and is measured in **amperes**. Current is a measurement of the electrons passing any given point in the circuit in one second. Because the flow of electrons is at nearly the speed of light, it would be impossible to physically see electron flow. However, the rate of elec-

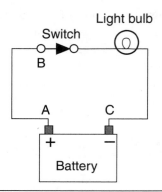

Figure 1-6 A simplified light circuit illustrating voltage potential.

tron flow can be measured. Current will increase as pressure or voltage is increased, provided circuit resistance remains constant.

A BIT OF HISTORY

The ampere is named after André Ampère, who in the late 1700s worked with magnetism and current flow to develop some foundations for understanding the behavior of electricity.

An electrical current will continue to flow through a conductor as long as the electromotive force is acting on the conductor's atoms and electrons. If a potential exists in the conductor, with a build-up of excess electrons at the end of the conductor the farthest from the EMF, and there is a lack of electrons at the EMF side, current will flow. The effect is called **electron drift** and accounts for the method in which electrons flow through a conductor.

An electrical current can be formed by the following forces: friction, chemical, heat, pressure, and magnetism. Whenever electrons flow or drift in mass an electrical

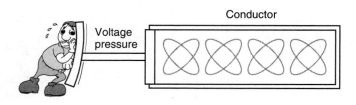

Figure 1-5 Voltage is the pressure that causes the electrons to move.

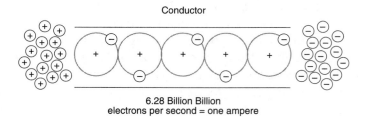

Figure 1-7 The rate of electron flow is called current and is measured in amperes.

current is formed. There are eight laws that regulate this electrical behavior:

1. Electrons repel each other.

2. Like charges repel each other.

3. Unlike charges attract each other.

4. Electrons flow in a conductor only when affected by an electromotive force.

5. A voltage difference is created in the conductor when an EMF is acting on the conductor.

6. Electrons flow only when a voltage difference exists between the two points in a conductor.

7. Current tends to flow to ground in an electrical circuit.

8. **Ground** is defined as the common negative connection of the electrical system and is the point of lowest voltage. The symbol for ground is ≡. The ground circuit used in most automotive systems is through the vehicle chassis and/or engine block.

So far we have described current as the movement of electrons through a conductor. Electrons move because of a potential difference. This describes one of the common theories about current flow. The **electron theory** states that since electrons are negatively charged, current flows from the most negative to the most positive point within an electrical circuit. In other words, current flows from negative to positive. This theory is widely accepted by the electronic industry.

Another current flow theory is called the **conventional theory**. This states that current flows from positive to negative. The basic idea behind this theory is simply that although electrons move toward the protons, the energy or force that is released as the electrons move begins at the point where the first electron moved to the most positive charge. As electrons continue to move in one direction, the released energy moves in the opposite direction. This theory is the oldest theory and serves as the basis for most electrical diagrams.

Trying to make sense of it all may be difficult for you. It is also difficult for scientists and engineers. In fact, another theory has been developed to explain the mysteries of current flow. This theory is called the **hole-flow theory** and is actually based on both electron theory and the conventional theory.

As a technician, you will find references to all of these theories. Fortunately, it really doesn't matter as long as you know what current flow is and what affects it. From this understanding you will be able to figure out how the circuit basically works, how to test it, and how to repair it. In this text, we will present current flow as moving from positive to negative and electron flow as

moving from negative to positive. Remember that current flow is the result of the movement of electrons, regardless of the theory.

Resistance

The third component in electricity is **resistance**. Resistance is the opposition to current flow and is measured in **ohms** (Ω). The size, type, length, and temperature of the material used as a conductor will determine its resistance. Devices that use electricity to operate, such as motors and lights, have a greater amount of resistance than the conductor.

A complete electrical **circuit** consists of: (1) a power source, (2) a load or resistance unit, and (3) conductors. Resistance (load) is required to change electrical energy to light, heat, or movement. There is resistance in any working device of a circuit, such as a lamp, motor, relay, or other load component.

There are five basic characteristics that determine the amount of resistance in any part of a circuit:

1. The atomic structure of the material: The fewer the number of electrons in the outer valence ring, the higher the resistance of the conductor.

2. The length of the conductor: The longer the conductor, the higher the resistance.

3. The diameter of the conductor: The smaller the cross-sectional area of the conductor, the higher the resistance.

4. Temperature: Basically, as the temperature of the conductor increases so does its resistance.

5. Physical condition of the conductor: If the conductor is damaged by nicks or cuts, the resistance will increase because the conductor's diameter is decreased by these.

There may be unwanted resistance in a circuit. This could be in the form of corrosion, loose connection, or a broken conductor. In these instances the resistance may cause the load component to operate at reduced efficiency or to not operate at all.

It does not matter if the resistance is from the load component or from a corroded connection, there are certain principles that dictate its impact in the circuit:

1. Voltage always drops as current flows through the resistance.

2. An increase in resistance causes a decrease in current.

3. All resistances change the electrical energy into heat energy to some extent.

Voltage Drop

Voltage drop occurs when current flows through a load device or resistance. Voltage drop is the amount of electrical pressure lost or converted as it pushes current flow through a resistance. After a resistance, the voltage is lower than it was before the resistance.

There must be a voltage present for current to flow through a resistor. Kirchhoff's law basically states that the sum of the voltage drops in an electrical circuit will always equal source voltage. In other words, all of the source's voltage is used by the circuit.

Voltage drop can be measured by using a voltmeter (Figure 1-8). With current flowing through a circuit, the voltmeter may be connected in parallel over the resistor, wire, or component to measure voltage drop. The voltmeter indicates the amount of voltage potential between two points in the circuit. The voltmeter reading indicates the difference between the amount of voltage available to the resistor and the amount of voltage after the resistor.

Gustav Kirchhoff was a German scientist who in the 1800s discovered two facts about the characteristics of electricity. One is called his voltage law, which states, "The sum of the voltage drops across all resistances in a circuit must equal the voltage of the source." His law on current states, "The sum of the currents flowing into any point in a circuit equals the sum of the currents flowing out of the same point." These laws describe what happens when electricity is applied to a load. Voltage drops while current remains constant; current does not drop.

Measuring Voltage Drop Using a Voltmeter

Many times a technician will be required to perform a voltage drop test to determine if excess resistance is in the circuit. This test is usually easier to perform than measuring resistance with an ohmmeter. Follow Photo Sequence 2 as a guide in performing a voltage drop test on the headlight circuit of a vehicle.

It is also possible to use the voltmeter to test for excessive resistance in the ground side of the circuit. Connect the red voltmeter test lead to the ground side terminal of the headlight socket and the black voltmeter test lead to a good ground. Turn on the headlights and observe the voltmeter reading. The reading should be close to zero volts.

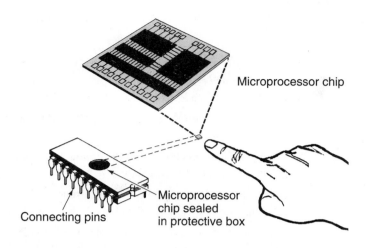

Figure 1-8 Using a voltmeter to determine the voltage drop between two points in the circuit.

After completing the voltage drop test, compare the test results with specifications to determine if a fault exists.

Ohm's Law

Understanding **Ohm's law** is the key to understanding how electrical circuits work. The law states it takes one volt of electrical pressure to push one ampere of electrical current through one ohm of electrical resistance. This law can be expressed mathematically as:

$$1 \text{ Volt} = 1 \text{ Ampere} \times 1 \text{ Ohm}$$

This formula is most often expressed as: $E = I \times R$, where E stands for EMF or electrical pressure (voltage), I stands for intensity of electron flow (current), and R represents resistance. This formula is often used to find the value of one electrical characteristic when the other two are known. As an example: If we have 2 amps of current and 6 ohms of resistance in a circuit, we must have 12 volts of electrical pressure:

$$E = 2 \text{ Amps} \times 6 \text{ Ohms} \quad E = 2 \times 6 \quad E = 12 \text{ volts}$$

If the voltage and resistance of a circuit is known but not the current, Ohm's law can be used to determine the amount of amperage. Since $E = I \times R$, I would equal E divided by R. For example, if the voltage is 12 and the resistance is 6 ohms, the amount of current is calculated in this way:

$$\frac{E}{R} = \frac{12}{6} = 2 \text{ amperes}$$

The same logic is used to calculate resistance when voltage and current are known. $R = E/I$. One easy way to remember the formulas of Ohm's law is to draw a circle

Performing a Voltage Drop Test

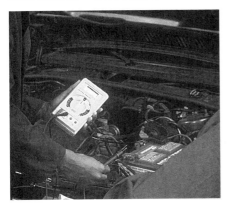

P1-1 Tools required to perform this task: voltmeter and fender covers.

P1-2 Set the voltmeter on its lowest DC volt scale.

P1-3 To test the voltage drop of the entire system, connect the red (positive) voltmeter test lead to the battery positive (+) terminal.

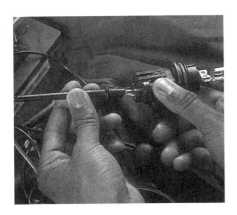

P1-4 Connect the black (negative) voltmeter test lead to the low beam terminal of the headlight socket. Make sure you are connected to the input side of the headlight.

P1-5 Turn on the headlights (low beam) and observe the voltmeter readings. The voltmeter will indicate the amount of voltage that is dropped between the battery and the headlight.

and divide it into three parts as shown in Figure 1-9. Simply cover the unknown value you want to calculate and the formula is exposed.

To show how easy this works, consider the 12-volt circuit in Figure 1-10. This circuit contains a 3-ohm light bulb. To determine the current in the circuit cover the I in the circle to expose the formula $I = E/R$. Then plug in the numbers, $I = 12/3$. Therefore the circuit current is 4 amperes.

A BIT OF HISTORY:

Georg S. Ohm was a German scientist in the 1800s who discovered that all electrical quantities are proportional to each other and therefore have a mathematical relationship.

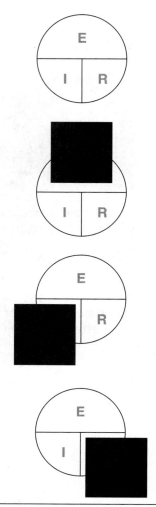

Figure 1-9 The mathematical formula for Ohm's law using a circle to help understand the different formulas that can be derived from it. To identify which formula to use, cover the unknown value. The exposed formula is the one to use to calculate the unknown.

Ohm's Law is the basic law of electricity and it states that the amount of current in an electric circuit is inversely proportional to the resistance of the circuit, and is directly proportional to the voltage in the circuit. For example, if the resistance decreases and the voltage

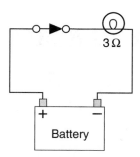

Figure 1-10 Simplified light circuit with 3 ohms of resistance in the lamp.

remains constant, the amperage will increase. If the resistance stays the same and the voltage increases, the amperage will also increase.

For example, refer to Figure 1-11: on the left side is a 12-volt circuit with a 3-ohm light bulb. This circuit will have 4 amps of current flowing through it. If a 1-ohm resistor is added to the same circuit (as shown to the right in Figure 1-11), total circuit resistance is now 4 ohms. Because of the increased resistance, current dropped to 3 amps. The light bulb will be powered by less current and will be less bright than it was before the additional resistance.

Another point to consider is voltage drop. Before adding the 1-ohm resistor, the source voltage (12 volts) was dropped by the light bulb. With the additional resistance, the voltage drop of the light bulb decreased to 9 volts. The remaining 3 volts are dropped by the 1-ohm resistor. This can be proven by using Ohm's law. When the circuit current was 4 amps, the light bulb had 3 ohms of resistance. To find the voltage drop multiply the current by the resistance:

$$E = I \times R \quad \text{or} \quad E = 4 \times 3 \quad \text{or} \quad E = 12$$

When the extra resistor was added to the circuit, the light bulb still had 3 ohms of resistance, but the current in the circuit decreased to 3 amps. Again voltage drop can be determined by multiplying the current by the resistance:

$$E = I \times R \quad \text{or} \quad E = 3 \times 3 \quad \text{or} \quad E = 9$$

The voltage drop of the additional resistor is calculated in the same way: $E = I \times R$ or $E = 3$ volts. The total voltage drop of the circuit is the same for both circuits; however, the voltage drop at the light bulb changed. This also would cause the light bulb to be dimmer.

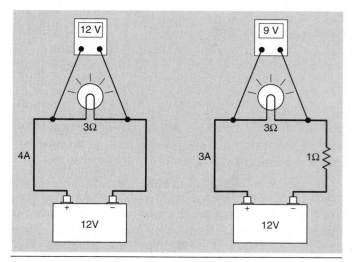

Figure 1-11 The light circuit in Figure 1-9 shown with normal circuit values and with added resistance in series.

Electrical Power

Electrical power is another term used to describe electrical activity. Power is expressed in **watts** and is a measure of the total electrical work being performed per unit of time. A watt is equal to one volt multiplied by one ampere. There is another mathematical formula that expresses the relationship between voltage, current, and power. It is simply: $P = E \times I$. **Power** measurements are measurements of the rate at which electricity is doing work. It is possible to convert horsepower ratings to electrical power rating using the conversion factor: 1 horsepower equals 746 watts.

The best examples of power are light bulbs. Household light bulbs are sold by wattage. A 100-watt bulb is brighter and uses more electricity than a 60-watt bulb. Seldom do technicians worry about wattage when working on cars. However, an understanding of electrical power will help in the understanding of electrical circuits.

Referring back to Figure 1-11. The light bulb in the circuit on the left had a 12-volt drop at 4 amps of current. We can calculate the power used by the bulb by multiplying the voltage and the current:

$$P = E \times I \quad \text{or} \quad P = 12 \times 4 \quad \text{or} \quad P = 48$$

The bulb produced 48 watts of power. When the resistor was added to the circuit the bulb dropped 9 volts at 3 amps of current. The power of the bulb is calculated in the same way as before:

$$P = E \times I \quad \text{or} \quad P = 9 \times 3 \quad \text{or} \quad P = 27$$

This bulb produced 27 watts of power, a little more than half of the original. It would be almost half as bright. The key to understanding what happened is to remember the light bulb didn't change; the circuit changed.

Types of Current

There are two classifications of electrical current flow: direct current (DC) and alternating current (AC). The type of current flow is determined by the direction it flows and by the type of voltage that drives them.

Direct Current

Direct current (DC) can only be produced by a battery and has a current that is the same throughout the circuit and flows in the same direction (Figure 1-12). Voltage and current are constant if the switch is turned on or off. Most of the electrically controlled units in the automobile require direct current.

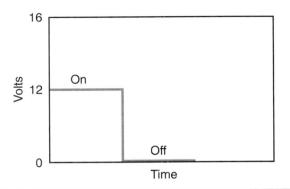

Figure 1-12 Direct current flows in the same direction and is the same throughout the circuit.

Alternating Current

Alternating current (AC) is produced anytime a conductor moves through a magnetic field. In an alternating current circuit, voltage and current do not remain constant. Alternating current changes directions from positive to negative. The voltage in an AC circuit starts at zero and rises to a positive value. Then it falls back to zero and goes to a negative value. Finally it returns to zero (Figure 1-13). The AC voltage shown in Figure 1-13 is called a **sine wave**. One **cycle** is completed when the voltage has gone positive, returned to zero, gone negative, and returned to zero.

Electrical Circuits

The electrical term **continuity** refers to the circuit being continuous. For current to flow, the electrons must have a continuous path from the source voltage to the load component and back to the source. A simple automotive circuit is made up of three parts:

1. Battery (power source).

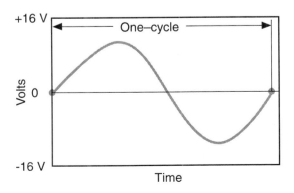

Figure 1-13 Alternating current does not remain constant.

2. Wires (conductors).

3. Load (light, motor, etc.).

The basic circuit shown includes a switch to turn the circuit on and off, a protection device (fuse), and a load. In this instance, with the switch closed, current flows from the positive terminal of the battery through the light and returns to the negative terminal of the battery. To have a complete circuit the switch must be closed or turned on. The effect of opening and closing the switch to control electrical flow would be the same if the switch was installed on the ground side of the light.

The portion of the circuit from the positive side of the source to the load component is called the **insulated side** or "hot" side of the circuit. The portion of the circuit that is from the load component to the negative side of the source is called the **ground side** of the circuit.

There are basically three different types of electrical circuits: (1) the series circuit, (2) the parallel circuit, and (3) the series-parallel circuit.

Series Circuit

A **series circuit** consists of one or more resistors (or loads) with only one path for current to flow (Figure 1-14). If any of the components in the circuit fails, the entire circuit will not function. All of the current that comes from the positive side of the battery must pass through each resistor, then back to the negative side of the battery.

The total resistance of a series circuit is calculated by simply adding the resistances together. For example, a series circuit with three light bulbs: one bulb has 2 ohms of resistance, and the other two have 1 ohm each. The total resistance of this circuit is 2 + 1 + 1 or 4 ohms. In a series circuit, if additional resistance is added to the circuit, total circuit resistance increases and amperage draw decreases.

The characteristics of a series circuit are:

1. The total resistance is the sum of all resistances.

2. The current through each resistor is the same.

3. The current is the same throughout the circuit.

4. The voltage drop across each resistor will be different if the resistor values are different.

5. The sum of the voltage drop of each resistor equals the source voltage.

Parallel Circuit

In a **parallel circuit** each path of current flow has separate resistances that operate either independently or in conjunction with each other (depending on circuit design). In a parallel circuit, current can flow through more than one parallel **leg** at a time (Figure 1-15). In this type of circuit, failure of a component in one parallel leg does not affect the components in other legs of the circuit. The legs of a parallel circuit are also called parallel **branches** or **shunt circuits**.

Total resistance in a parallel circuit is always less than the lowest individual resistance because current has more than one path to follow. If more parallel resistors are added, more circuits are added, and the total resistance will decrease and total amperage draw increases. If all resistances in the parallel circuit are equal, use the following formula to determine total resistance:

$$RT = \frac{\text{The value of one resistor}}{\text{the number of legs}}$$

For example, if a parallel circuit had three legs and each leg had a 12-ohm resistor, total circuit resistance would be :

$$RT = 120 \div 3 = 40 \text{ ohms}$$

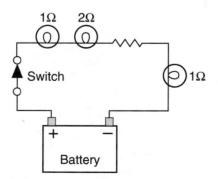

Figure 1-14 An example of a series circuit.

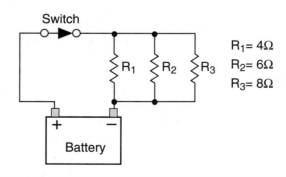

Figure 1-15 In a parallel circuit, current can flow through more than one parallel leg at a time.

The total resistance of a parallel circuit with two legs or two paths for current flow is calculated by using this formula:

$$RT = \frac{R1 \times R2}{R1 + R2}$$

If the value of R1 in Figure 1-16 is 3 ohms and R2 has a value of 6 ohms, the total resistance can be found:

$$RT = \frac{R1 \times R2}{R1 + R2} \quad \text{or} \quad \frac{3 \times 6}{3 + 6} \quad \text{or} \quad RT = \frac{18}{9} \quad \text{or} \quad RT = 2$$

Based on this calculation, we can determine that the total circuit current is 6 amps (12 volts divided by 2 ohms). Using basic Ohm's law and a basic understanding of electricity, other factors of the circuit can quickly be determined.

Each leg of the circuit has 12 volts applied to it; therefore, each leg must drop 12 volts. So the voltage drop across R1 is 12 volts, and the voltage drop across R2 is also 12 volts. Using the voltage drops, current flow through each leg can be determined. To calculate current in a parallel circuit, each shunt branch is treated as an individual circuit. Applied voltage is the same to all branches. To determine the branch current simply divide the source voltage by the shunt branch resistance:

$$I = E \div R$$

Since R1 has 3 ohms and drops 12 volts, the current through it must be 4 amps. R2 has 6 ohms and drops 12 volts and its current is 2 amps. The total current flow through the circuit is 4 + 2 or 6 amps.

The total resistance of a circuit with more than two legs can be calculated with the following formula:

$$RT = \frac{1}{\dfrac{1}{R1} + \dfrac{1}{R2} + \dfrac{1}{R3} + \cdots \dfrac{1}{Rn}}$$

For example, with three parallel branches with resistance values of 4Ω, 6Ω, and 8Ω the total resistance is figured as follows:

$$RT = \frac{1}{\dfrac{1}{4} + \dfrac{1}{6} + \dfrac{1}{8}} = \frac{1}{\dfrac{6}{24} + \dfrac{4}{24} + \dfrac{3}{24}} = \frac{1}{\dfrac{13}{24}} = \frac{24}{13} = 1.85\Omega$$

Often it is much easier to calculate total resistance of a parallel circuit by using total current. Begin by finding the current through each leg of the parallel circuit; then add them together to find total current. Use basic Ohm's law to calculate the total resistance:

$$R = E \div I \text{ or } R = 12 \div 6.5 \text{ amps or } R = 1.85 \text{ ohms}$$

The characteristics of a parallel circuit are:

1. The voltage applied to each parallel leg is the same.
2. The voltage dropped across each parallel leg will be the same; however, if the leg contains more than one resistor, the voltage drop across each of them will depend on the resistance of each resistor in that leg.
3. The total resistance of a parallel circuit will always be less than the resistance of any of its legs.
4. The current flow through the legs will be different if the resistance is different.
5. The sum of the current in each leg equals the total current of the parallel circuit.

Series-Parallel Circuits

The **series-parallel circuit** has some loads that are in series with each other and some that are in parallel (Figure 1-17). To calculate the total resistance in this type of circuit, calculate the **equivalent series loads** of the parallel branches first. The equivalent series load, or **equivalent resistance**, is the total resistance of a parallel circuit and is equivalent to the resistance of a single load in series with the voltage source. Next, calculate the series resistance and add it to the equivalent series load. For example, if the parallel portion of the circuit has two

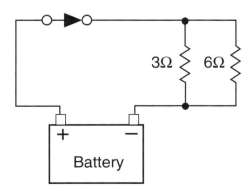

Figure 1-16 Parallel circuit with two resistors.

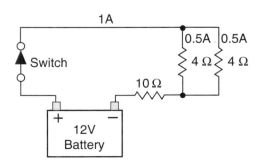

Figure 1-17 A series-parallel circuit with known resistance values.

branches with 4Ω resistance each, and the series portion has a single load of 10Ω, use the following method to calculate the equivalent resistance of the parallel circuit:

$$RT = \frac{\text{Value of one resistor}}{\text{Total number of resistors}} = \frac{4}{2} = 2 \text{ ohms}$$

Then add this equivalent resistance to the actual series resistance to find the total resistance of the circuit.

$$2 \text{ ohms} + 10 \text{ ohms} = 12 \text{ ohms}$$

With the total resistance now known, total circuit current can be calculated. Because the source voltage is 12 volts, 12 is divided by 12 ohms:

$$I = E \div R \quad \text{or} \quad I = 12 \div 12 \quad \text{or} \quad I = 1 \text{ amp}$$

The current flow through each parallel leg is calculated by using the resistance of each leg and voltage drop across that leg. To do this, first find the voltage drops. Since all 12 volts are dropped by the circuit, we know that some are dropped by the parallel circuit and the rest by the resistor in series. We also know that the circuit current is 1 amp, that the equivalent resistance value of the parallel circuit is 2 ohms, and the resistance of the series resistor is 10. Using Ohm's law, voltage drop of the parallel portion of the circuit can be calculated:

$$E = I \times R \quad \text{or} \quad E = 1 \times 2 \quad \text{or} \quad E = 2$$

Two volts are dropped by the parallel circuit. This means 2 volts are dropped by each of the 4 ohm resistors. Using this voltage drop, it is possible to calculate current flow through each parallel leg:

$$I = E \div R \quad \text{or} \quad I = 2 \div 4 \quad \text{or} \quad I = 0.5 \text{ amps}$$

Since the resistance on each leg is the same, each leg has 0.5 amps through it. The sum of the amperages in each leg will equal the current of the whole circuit (0.5 + 0.5 = 1 ampere).

It is important to realize that the actual or measured values of current, voltage, and resistance may be somewhat different than the calculated values. The change is caused by the effects of heat on the resistances. As the voltage pushes current through a resistor, the resistor heats up. The resistor changes the electrical energy into heat energy. This heat may cause the resistance to increase or decrease depending on the material it is made of. The best example of a resistance changing electrical energy into heat energy is a light bulb. A light bulb gives off light because the conductor inside the bulb heats up and glows when current flows through it.

Technicians will seldom have the need to calculate the values in an electrical circuit. The primary importance of being able to use Ohm's law is to understand the relationship between voltage, amperage, and resistance. Technicians use electrical meters to measure current, voltage, and resistance. When a measured value is not within specifications, you should be able to explain why. Ohm's law is used to do that.

Using Ohm's Law

Most automotive electrical systems are wired in parallel. Actually the system is made up of a number of series circuits wired in parallel. This allows each electrical component to work independently of the others. When one component is turned on or off, the operation of the other components should not be affected.

In a 12-volt circuit with one 3-ohm light bulb the switch controls the operation of the light bulb. When the switch is closed, current flows and the bulb is lit. Four amps will flow through the circuit and the bulb.

$$I = E \div R \quad \text{or} \quad I = 12 \div 3 \quad \text{or} \quad I = 4 \text{ amps}$$

Now take the same circuit with a 6-ohm light bulb added in parallel to the 3-ohm light bulb. With the switch for the new bulb closed, 2 amps will flow through that bulb. The 3-ohm bulb is still receiving 12 volts and has 4 amps flowing through it; it will operate in the same way and with the same brightness as it did before adding the 6-ohm light bulb. Since total circuit resistance decreased, current flow increased. Total current draw is now 6 amperes (4 + 2 = 6):

Leg #1 $I = E \div R \quad \text{or} \quad I = 12 \div 3 \quad \text{or} \quad I = 4 \text{ amps}$

Leg #2 $I = E \div R \quad \text{or} \quad I = 12 \div 6 \quad \text{or} \quad I = 2 \text{ amps}$

If the switch to the 3-ohm bulb is opened, the 6-ohm bulb works in the same way and with the same brightness as it did before we opened the switch. In this case two things happened: the 3-ohm bulb no longer is lit, and the circuit current dropped 2 amps.

Use the same circuit just described except add a 1-ohm light bulb and switch in parallel to the circuit. With the switch for the new bulb closed, 12 amps will flow through that leg of the circuit. The other bulbs are working in the same way and with the same brightness as before. Again, total circuit resistance decreases so the total circuit current increases. Total current is now 18 amperes:

Leg #1 $I = E \div R \quad \text{or} \quad I = 12 \div 3 \quad \text{or} \quad I = 4 \text{ amps}$

Leg #2 $I = E \div R \quad \text{or} \quad I = 12 \div 6 \quad \text{or} \quad I = 2 \text{ amps}$

Leg #3 $I = E \div R \quad \text{or} \quad I = 12 \div 1 \quad \text{or} \quad I = 12 \text{ amps}$

Total current = 4 + 2 + 12 or 18 amps

When the switch for any of these bulbs is opened or closed, the bulbs either turn off or on, and the total resistance and current through the circuit changes. Notice as more parallel legs are added, total circuit current goes up. There is a commonly used statement, "Current always takes the path of least resistance." This statement is not totally correct. As illustrated in the previous circuits, current flows to all of the bulbs regardless of the bulbs' resistance. The resistances with lower values will draw higher currents, but all of the resistances will receive the current they allow. To be more accurate the statement should be, "Larger amounts of current will flow through lower resistances." This is very important to remember when diagnosing electrical problems.

From Ohm's law, we know that when resistance decreases, current increases. If a 0.6-ohm light bulb is installed in place of the 3-ohm bulb, the other bulbs will work in the same way and with the same intensity as they did before. However, 20 amps of current will flow through the 0.6-ohm bulb. This will raise our circuit current to 34 amps. Lowering the resistance on the one leg of the parallel circuit greatly increases the current through the circuit. This high current may damage the circuit or components. It is possible the high current can cause wires to burn. In this case the wires that would burn are the wires that would carry the 34 amps or the 20 amps to the bulb, not the wires to the other bulbs.

Leg #1 $I = E \div R$ or $I = 12 \div 0.6$ or $I = 20$ amps

Leg #2 $I = E \div R$ or $I = 12 \div 6$ or $I = 2$ amps

Leg #3 $I = E \div R$ or $I = 12 \div 1$ or $I = 12$ amps

Total current = 20 + 2 + 12 or 34 amps

Let's see what happens when resistance is added to one of the parallel legs. An increase in resistance should cause a decrease in current. In Figure 1-18, a 1-ohm resistor was added after the 1-ohm light bulb. This resistor is in series with the light bulb, and the total resistance of that leg is now 2 ohms. The current through that leg is now 6 amps. Again, the other bulbs were not affected by the change. This changes the total circuit current, which now drops to 12 amps. The added resistance lowered total circuit current and changed the way the 1-ohm bulb works. This bulb will now drop only 6 volts. The remaining 6 volts will be dropped by the added resistor. The 1-ohm bulb will be much dimmer than before; its power rating dropped from 144 watts to 36 watts. Additional resistance causes the bulb to be dimmer. The bulb itself wasn't changed, only the resistance of that leg changed. The dimness is caused by the circuit, not the bulb.

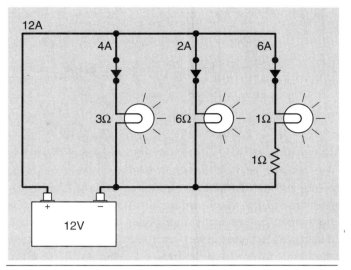

Figure 1-18 A series circuit contained in a leg of a parallel circuit.

Leg #1 $I = E \div R$ or $I = 12 \div 3$ or $I = 4$ amps

Leg #2 $I = E \div R$ or $I = 12 \div 6$ or $I = 2$ amps

Leg #3 $I = E \div R$ or $I = 12 \div 1+1$ or
$I = 12 \div 2$ or $I = 6$ amps

Total current = 4 + 2 + 6 or 12 amps

Now let's see what happens when we add a resistance that is common to all of the parallel legs. In Figure 1-19, a 0.333 ohm resistor was added (0.333 was chosen to keep the math simple!) to the negative connection at the battery. This will cause the circuit's current to decrease; it will also change the operation of the bulbs

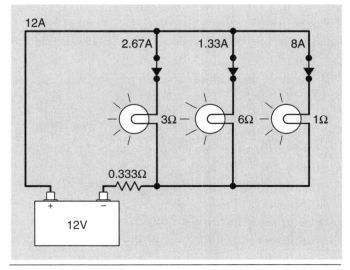

Figure 1-19 A resistor in series with a parallel circuit.

in the circuit. The total resistance of the bulbs in parallel is 0.667 ohms.

$$RT = \frac{1}{\frac{1}{3} + \frac{1}{6} + \frac{1}{1}} \quad \text{or} \quad \frac{1}{0.333 + 0.167 + 1} = \frac{1}{1.5} = 0.667$$

The total resistance of the circuit is 1 ohm (0.667 + 0.333), which means the circuit current is now 12 amps. Because there will be a voltage drop across the 0.333 ohm resistor, each of the parallel legs will have less than source voltage. To find the amount of voltage dropped by the parallel circuit we multiply the amperage by the resistance. Twelve amps multiplied by 0.667 equal 8. So 8 volts will be dropped by the parallel circuit; the remaining 4 volts will be dropped by the 0.333 resistor. The amount of current through each leg can be calculated by taking the voltage drop and dividing it by the resistance of the leg.

Leg #1 $I = E \div R$ or $I = 8 \div 3$ or $I = 2.667$ amps

Leg #2 $I = E \div R$ or $I = 8 \div 6$ or $I = 1.333$ amps

Leg #3 $I = E \div R$ or $I = 8 \div 1$ or $I = 8$ amps

Total circuit current = 2.667 + 1.333 + 8 or 12 amps

The added resistance affected the operation of all the bulbs, because it was added to a point that was common to all of the bulbs. All of the bulbs would be dimmer, and circuit current would be lower.

Capacitance

Some automotive electrical systems will use a **capacitor** or **condenser** to store electrical charges (Figure 1-20). **Capacitance** is the ability of two conducting surfaces to store voltage. The two surfaces must be separated by an insulator. A capacitor does not consume any power. All of the voltage stored in the capacitor is returned to the circuit when the capacitor discharges. Because the capacitor stores voltage, it will also absorb voltage changes in the circuit. By providing for this storage of voltage, damaging voltage spikes can be controlled. They are also used to reduce radio noise.

A capacitor is made by wrapping two conductor strips around an insulating strip. The insulating strip, or **dielectric** prevents the plates from coming in contact while keeping them very close to each other. The dielectric can be made of some insulator material such as ceramic, glass, paper, plastic, or even the air between the two plates. A capacitor blocks direct current. A small amount of current enters the capacitor and charges it.

Most capacitors are connected in parallel across the circuit (Figure 1-21). Capacitors operate on the principle that opposite charges attract each other, and there is a potential voltage between any two oppositely charged points. When the switch is closed, the protons at the positive battery terminal will attract some of the electrons on one plate of the capacitor away from the area near the dielectric material. As a result, the atoms of the positive plate are unbalanced because there are more protons than electrons in the atom. This plate now has a positive charge because of the shortage of electrons (Figure 1-22). The positive charge of this plate will attract electrons on the other plate. The dielectric keeps the electrons on the negative plate from crossing over to the positive plate, resulting in a storage of electrons on the negative plate. The movement of electrons to the negative plate and away from the positive plate is an electrical current.

Current will flow "through" the capacitor until the voltage charges across the capacitor and the battery are equalized. Current flow through a capacitor is only the effect of the electron movement onto the negative plate and away from the positive plate. Electrons do not actually pass through the capacitor from one plate to another. The charges on the plates do not move through

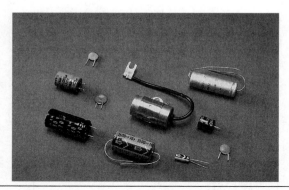

Figure 1-20 Capacitors that can be found in automotive electrical circuits.

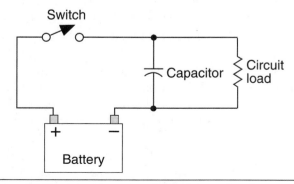

Figure 1-21 A capacitor connected to a circuit.

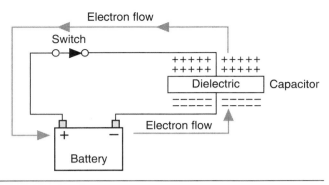

Figure 1-22 The positive plate sheds its electrons.

the **electrostatic field** (the field that is between the two oppositely charged plates). They are stored on the plates as **static electricity** (electricity that is not in motion).

When the charges across the capacitor and battery are equalized, there is no potential difference and no more current will flow "through" the capacitor. Current will now flow through the load components in the circuit.

When the switch is opened, current flow from the battery through the resistor is stopped. However, the capacitor has a storage of electrons on its negative plate. Because the negative plate of the capacitor is connected to the positive plate through the resistor, the capacitor acts as a battery. The capacitor will discharge the electrons through the resistor until the atoms of the positive plate and negative plate return to a balanced state (Figure 1-23).

In the event a high voltage spike occurs in the circuit, the capacitor will absorb the additional voltage before it is able to damage the circuit components. A capacitor can also be used to stop current flow quickly when a circuit is opened (such as in the ignition system). It can also store a high voltage charge and then discharge it when a circuit needs the voltage (such as in some air bag systems).

Capacitors are rated in units called **farads**. A one-farad capacitor connected to a one-volt source will store

$6.28 \times 10_{18}$ electrons. A farad is a large unit, and most commonly used capacitors are rated in picofarad (a trillionth of a farad) or microfarads (a millionth of a farad).

Semiconductors

As discussed earlier, electrical materials are classified as conductors, insulators, or **semiconductors**. Semiconductors are crystal materials that can act as both a conductor or an insulator. Semiconductors include diodes, transistors, and silicon-controlled rectifiers. These semiconductors are often called solid-state devices because they are constructed of a solid material. The most common materials used in the construction of semiconductors are silicon or germanium. Both of these materials are classified as **crystals**. A crystal is the term used to describe a material that has a definite atom structure.

Silicon and germanium have four electrons in their outer orbits. Because of their crystal-type structure, each atom shares an electron with four other atoms (Figure 1-24). As a result of this **covalent bonding** (atoms sharing electrons with other atoms), each atom will have eight electrons in its outer orbit. All the orbits are filled and there are no free electrons, thus the material (as a category of matter) falls somewhere between conductor and insulator.

Perfect crystals are not used for manufacturing semiconductors. They are **doped** with impurity atoms. This doping adds a small percentage of another element to the crystal. The doping element can be arsenic, antimony, phosphorous, boron, aluminum, or gallium.

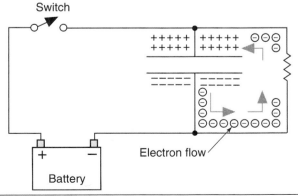

Figure 1-23 Current flow with the switch open and the capacitor discharging.

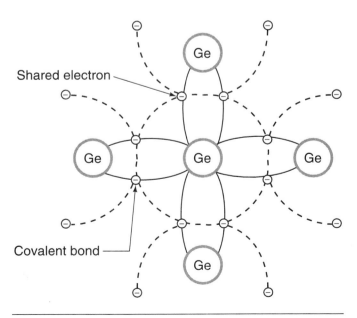

Figure 1-24 Crystal structure of germanium.

If the crystal is doped by using arsenic, antimony, or phosphorous the result is a material with free electrons. Materials such as arsenic have five electrons, which leaves one electron left over. This doped material becomes negatively charged. Under the influence of an EMF, it will support current flow. Negatively charged materials are referred to as **N-type materials**.

If boron, aluminum, or gallium are added to the crystal, a **P-type material** is produced. Materials like boron have three electrons in their outermost orbit. Because there is one fewer electron, there is an absence of an electron that produces a hole and becomes positively charged.

By putting N-type and P-type materials together in a certain order, solid-state components are built that can be used for switching devices, voltage regulators, electrical control, and so on.

Magnetism Principles

Magnetism is an atomic force that can attract or repel through space, air, or solid matter. Magnetism is the force that is used to produce most of the electrical power in the world. It is also the force used to create the electricity to recharge a vehicle's battery, make a starter work, and produce signals for various operating systems. A magnet is a material that attracts iron, steel, and a few other materials. Because magnetism is closely related to electricity, many of the laws that govern electricity also govern magnetism.

There are two types of magnets used on automobiles, **permanent magnets** and **electromagnets**. Permanent magnets are magnets that do not require any force or power to keep their magnetic field. Electromagnets depend on electrical current flow to produce and, in most cases, keep their magnetic field.

Magnets

All magnets have polarity. A magnet that is allowed to hang free will align itself north and south. The end facing north is called the **north seeking pole**, and the end facing south is called the **south seeking pole**. Like poles will repel each other, and unlike poles will attract each other. These principles are shown in Figure 1-25. The magnetic attraction is the strongest at the poles.

A strong magnet produces many **lines of force**, and a weak magnet produces fewer lines of force. Invisible lines of force leave the magnet at the north pole and enter again at the south pole. While inside the magnet, the lines of force travel from the south pole to the north pole. The concentration of the magnetic lines of force is called

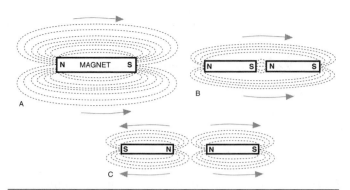

Figure 1-25 Magnetic principles: (A) all magnets have poles, (B) unlike poles attract each other, and (C) like poles repel.

magnetic flux density.

The **field of force** (or magnetic field) is all the space, outside the magnet, that contains lines of magnetic force. Magnetic lines of force penetrate all substances; there is no known insulation against magnetic lines of force. The lines of force may be deflected only by other magnetic materials or by another magnetic field.

Electromagnetism

Electromagnetism is the use of electricity to create a magnetic field. Whenever an electrical current flows through a conductor, a magnetic field is formed around the conductor. The number of lines of force, and the strength of the magnetic field produced, will be in direct proportion to the amount of current flow.

The direction of the lines of force is determined by the **right-hand rule**. Using the conventional theory of current flow being from positive to negative, the right hand is used to grasp the wire with the thumb pointing in the direction of current flow. The fingers will point in the direction of the magnetic lines of force.

André Marie Ampère noted that current flowing in the same direction through two nearby wires will cause the wires to be attracted to one another. Also, he observed that if current flow in one of the wires is

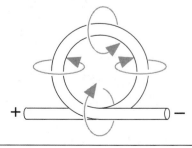

Figure 1-26 Looping the conductor increases the magnetic field.

reversed, the wires will repel one another. In addition, he found that if a wire is coiled with current flowing through the wire, the same magnetic field that surrounds a straight wire combines to form one larger magnetic field (Figure 1-26). This magnetic field has true north and south poles. Looping the wire doubles the flux density where the wire is running parallel to itself.

The north pole can be determined in the coil by use of the right-hand rule. Grasp the coil with the fingers pointing in the direction of current flow (+ to -) and the thumb will point toward the north pole.

As more loops are added, the fields from each loop will join and increase the flux density (Figure 1-27). To make the magnetic field even stronger, an iron core can be placed in the center of the coil. The soft iron core is a material that has high permeability and provides an excellent conductor for the magnetic field that travels through the center of the wire coil. Permeability is the term used to indicate the magnetic conductivity of a substance compared with the conductivity of air. The greater the **permeability**, the greater the magnetic conductivity and the easier a substance can be magnetized. **Reluctance** is the term used to indicate a material's resistance to the passage of flux lines. The strength of an electromagnetic coil is affected by the following factors (Figure 1-28):

1. The amount of current flowing through the wire.
2. The number of windings or turns.
3. The size, length, and type of core material.
4. The direction and angle at which the lines of force are cut.

The strength of the magnetic field is measured in ampere-turns:

ampere-turns = amperes × number of turns

The magnetic field strength is measured by multiplying the current flow in amperes through a coil by the num-

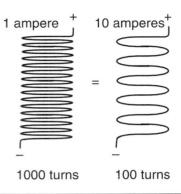

Figure 1-28 Magnetic field strength is determined by the amount of amperage and the number of coils.

ber of complete turns of wire in the coil. For example, in the illustration (2-28), a 1,000 turn coil with 1 ampere of current would have a field strength of 1,000 ampere-turns. This coil would have the same field strength as a coil with 100 turns and 10 amperes of current.

Theory of Induction

Electricity can be produced by **magnetic induction**. **Induction** is the magnetic process of producing a current flow in a wire without any actual contact to the wire. Magnetic induction occurs when a conductor is moved through the magnetic lines of force (Figure 1-29) or when a magnetic field is moved across a conductor. A difference of potential is set up between the ends of the conductor and a voltage is induced. This voltage exists only when the magnetic field or the conductor is in motion.

The **induced voltage** can be increased by either increasing the speed at which the magnetic lines of force cut the conductor, or by increasing the number of conductors that are cut. It is this principle that is behind the

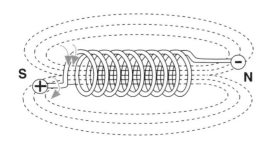

Figure 1-27 Adding more loops of wire increases the magnetic flux density.

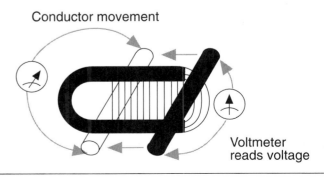

Figure 1-29 Moving a conductor through a magnetic field induces an electrical potential difference.

operation of all ignition systems, starter motors, charging systems, and relays.

A common induction device is the ignition coil. As the current increases, the coil will reach a point where the magnetic strength eventually levels off, and where current will no longer increase as it passes through the coil. The magnetic lines of force, which represent stored energy, will collapse when the applied voltage is removed. When the lines of force collapse, the magnetic energy is returned back to the wire as electrical energy.

If voltage is induced in the wires of a coil when current is first connected or disconnected, it is called **self-induction** (also referred to as counter EMF (CEMF) or as a **voltage spike**). The resulting current is in the opposite direction of the applied current and tends to reduce the magnetic force. Self-induction is governed by Lenz's law, which states that an induced current flows in a direction opposite the magnetic field that produced it.

Self-induction is generally not wanted in automotive circuits. For example, when a switch is opened, self-induction tends to continue to supply current in the same direction as the original current because as the magnetic field collapses, it induces voltage in the wire. According to Lenz's law, voltage induced in a conductor tends to oppose a change in current flow. Self-induction can cause an electrical arc to occur across an opened switch. The arcing may momentarily bypass the switch and allow the circuit that was turned off to operate for a short period of time. The arcing will also burn the contacts of the switch.

Self-induction is commonly found in electrical components that contain a coil or an electric motor. To help reduce the arc across contacts, a capacitor or clamping diode may be connected to the circuit. The capacitor will absorb the high voltage arcs and prevent arcing across the contacts. **Diodes** are semiconductors that allow current flow in only one direction. A clamping diode can be connected in parallel to the coil and will prevent current flow from the self-induction coil to the switch.

Magnetic induction is also the basis for a generator and many of the sensors on today's vehicles. In a generator, a magnetic field rotates inside a set of conductors. As the magnetic field crosses the wires, a voltage is induced. The amount of voltage induced by this action depends on the speed of the rotating field, the strength of the field, and the number of conductors the field cuts through.

Magnetic sensors are used to measure speeds, such as engine, vehicle, and shaft speeds. These sensors typically use a permanent magnet. Rotational speed is determined by the passing of blades or teeth in and out of the magnetic field. As a tooth moves in and out of the magnetic field, the strength of the magnetic field is changed and a voltage signal is induced. This signal is sent to a control device, where it is interpreted.

EMI Suppression

As manufacturers began to increase the number of electronic components and systems in their vehicles, the problem of **electro-magnetic induction (EMI)** had to be controlled. Electromagnetic Induction is an undesirable creation of electromagnetism whenever current is switched on and off.

The low power integrated circuits used on modern vehicles are sensitive to the signals produced as a result of EMI. EMI is produced as current in a conductor is turned on and off. EMI is also caused by static electricity that is created by friction. The friction is a result of the tires and their contact with the road, or from fan belts contacting the pulleys.

EMI can disrupt the vehicle's computer systems by inducing false messages in the computer. The computer requires messages to be sent over circuits in order to communicate with other computers, sensors, and actuators. If any of these signals are disrupted, the engine and/or accessories may turn off.

EMI can be suppressed by any one of the following methods:

1. Adding a resistance to the conductors. This is usually done to high-voltage systems such as the secondary circuit of the ignition system.

2. Connecting a capacitor in parallel and a **choke** coil in series with the circuit. A choke is an inductor in series with a circuit.

3. Shielding the conductor or load components with a metal or metal impregnated plastic.

4. Increasing the number of paths to ground by using designated ground circuits. This provides a clear path to ground that is very low in resistance.

5. Adding a clamping diode in parallel to the component.

6. Adding an isolation diode in series to the component.

Summary

❏ An atom is constructed of a complex arrangement of electrons in orbit around a nucleus. If the number of electrons and protons are equal, the atom is balanced or neutral.

❏ A conductor allows electricity to easily flow through it.

❏ An insulator does not allow electricity to easily flow through it.

❏ Electricity is the movement of electrons from atom to atom. In order for the electrons to move in the same direction, an electromotive force (EMF) must be applied to the circuit.

❏ The electron theory defines electron flow as motion from negative to positive.

❏ The conventional theory of current flow states that current flows from a positive point to a less positive point.

❏ Voltage is defined as an electrical pressure and is the difference between the positive and negative charges.

❏ Current is defined as the rate of electron flow and is measured in amperes. Amperage is the amount of electrons passing any given point in the circuit in one second.

❏ Resistance is defined as opposition to current flow and is measured in ohms (Ω).

❏ Ohm's law defines the relationship between current, voltage, and resistance. It is the basic law of electricity and states that the amount of current in an electric circuit is inversely proportional to the resistance of the circuit, and is directly proportional to the voltage in the circuit.

❏ Wattage represents the measure of power (P) used in a circuit. Wattage is measured by using the power formula, which defines the relationship between amperage, voltage, and wattage.

❏ Capacitance is the ability of two conducting surfaces to store voltage.

❏ Direct current results from a constant voltage and a current that flows in one direction.

❏ In an alternating current circuit, voltage and current do not remain constant. AC current changes direction from positive to negative and negative to positive.

❏ For current to flow, the electrons must have a complete path from the source voltage to the load component and back to the source.

❏ The series circuit provides a single path for current flow from the electrical source through all the circuit's components, and back to the source.

❏ A parallel circuit provides two or more paths for current to flow.

❏ A series-parallel circuit is a combination of the series and parallel circuits.

❏ The equivalent series load is the total resistance of a parallel circuit plus the resistance of the load in series with the voltage source.

❏ Voltage drop is caused by a resistance in the circuit that reduces the electrical pressure available after the resistance.

❏ Kirchhoff's voltage law states that the total voltage drop in an electrical circuit will always equal the available voltage at the source.

Terms To Know

Alternating current (AC)	Electromotive force (EMF)	Ohms
Amperes	Electron drift	Ohm's law
Atom	Electrons	Parallel circuit
Balanced	Electron theory	Permanent magnets
Branches	Electrostatic field	Permeability
Capacitance	Equivalent resistance	Potential
Capacitor	Equivalent series loads	Power
Choke	Farads	Pressure difference
Circuit	Field of force	Protons
Condenser	Ground	P-type material
Conductor	Ground side	Reluctance
Continuity	Hole-flow theory	Resistance
Conventional theory	Induced voltage	Right-hand rule
Covalent bonding	Induction	Self-induction
Crystals	Insulated side	Semiconductors
Current	Insulator	Series circuit
Cycle	Ion	Series-parallel circuit
Dielectric	Leg	Shunt circuits
Diodes	Lines of force	Sine wave
Direct current (DC)	Magnetic flux density	South seeking pole
Doped	Magnetic induction	Static electricity
Electrical pressure	Magnetism	Valence ring
Electricity	Neutrons	Voltage
Electro-magnetic induction (EMI)	North seeking pole	Voltage drop
Electromagnetism	N-type materials	Voltage spike
Electromagnets	Nucleus	Watts

Review Questions

Short Answer Essays

1. List and define the three elements of electricity.
2. Describe the use of Ohm's law.
3. List and describe the three types of circuits.
4. Explain the principle of electromagnetism.
5. Describe the principle of induction.
6. Describe the basics of electron flow.
7. Define the two types of electrical current.
8. Describe the difference between insulators, conductors, and semiconductors.
9. Explain the basic concepts of capacitance.
10. What does the measurement of "Watt" represent?

Fill-in-the-Blanks

1. _____ are negatively charged particles. The nucleus contains positively charged particles called _____ and particles that have no charge called _____ .

2. A _____ allows electricity to easily flow through it. An _____ does not allow electricity to easily flow through it.

3. For the electrons to move in the same direction, there must be an _____ applied.

4. The _____ _____ of current flow states that current flows from a positive point to a less positive point.

5. Resistance is defined as _____ to current flow and is measured in _____ .

6. _____ is the ability of two conducting surfaces to store voltage.

7. Kirchhoff's voltage law states that the _____ _____ _____ in an electrical circuit will always _____ available voltage at the source.

8. The _____ of all the resistors in series is the total resistance of that series circuit.

9. _____ is defined as an electrical pressure.

10. _____ is defined as the rate of electron flow.

ASE Style Review Questions

1. The methods that can be used to form an electrical current are being discussed:
   Technician A says electricity can be generated by magnetic induction.
   Technician B says electricity can be produced by a battery.
   Who is correct?
   a. A only
   b. B only
   c. Both A and B
   d. Neither A nor B

2. Circuit resistance is being discussed:
   Technician A says in a series circuit total resistance is figured by adding together all of the resistances in the circuit.
   Technician B says in a parallel circuit the total resistance is less than the lowest resistor.
   Who is correct?
   a. A only
   b. B only
   c. Both A and B
   d. Neither A nor B

3. While discussing voltage drop:
   Technician A says all of the voltage from the source is dropped in a complete circuit.
   Technician B says corrosion is not a contributor to voltage drop.
   Who is correct?
   a. A only
   b. B only
   c. Both A and B
   d. Neither A nor B

4. Technician A says voltage is the electrical pressure that causes electrons to move. Technician B says voltage will exist between any two points in that circuit unless the potential drops to zero. Who is correct?
   a. A only
   b. B only
   c. Both A and B
   d. Neither A nor B

5. Technician A says wattage is a measure of the total electrical work being performed. Technician B says the power formula is expressed as $P \times R = I$. Who is correct?
   a. A only
   b. B only
   c. Both A and B
   d. Neither A nor B

6. Technician A says a capacitor consumes electrical power. Technician B says a capacitor induces voltage. Who is correct?
   a. A only
   b. B only
   c. Both A and B
   d. Neither A nor B

7. Types of electrical currents are being discussed:
   Technician A says alternating current is produced from a voltage and current that remain constant and flow in the same direction.
   Technician B says direct current changes directions from positive to negative.
   Who is correct?
   a. A only
   b. B only
   c. Both A and B
   d. Neither A nor B

8. The principles of induction are being discussed:
   Technician A says induction is the magnetic process of producing a current flow in a wire without any actual contact to the wire.
   Technician B says induced voltage only exists when the magnetic field or the conductor is in motion.
   Who is correct?
   a. A only
   b. B only
   c. Both A and B
   d. Neither A nor B

9. Technician A says if the resistance increases and the voltage remains constant, the amperage will increase. Technician B says Ohm's law can be stated as $I = E \times R$. Who is correct?
   a. A only
   b. B only
   c. Both A and B
   d. Neither A nor B

10. Technician A says two 4-ohm resistors in parallel have an equivalent resistance of 2 ohms.
    Technician B says two 4-ohm resistors in series have an equivalent resistance of 8 ohms. Who is correct?
    a. A only
    b. B only
    c. Both A and B
    d. Neither A nor B

2 Electrical and Electronic Components

Objective

Upon completion of this chapter, you will be able to:

- ❏ Describe the purpose and operation of a switch.
- ❏ Explain the difference between normally open (NO) and normally closed (NC) switches.
- ❏ Describe the purpose and operation of a relay.
- ❏ Describe the difference between a rheostat and a potentiometer.
- ❏ Describe the construction, purpose and operation of a diode.
- ❏ Describe the use and construction of different types of diodes.
- ❏ Describe the purpose and function of different types of transistors.
- ❏ Explain the purpose of circuit protection devices.
- ❏ List the most common types of circuit protection devices.

Introduction

In this chapter you will be introduced to common electrical and electronic components. These components include circuit protection devices, switches, relays, variable resistors, diodes, and different forms of transistors. Today's technician must comprehend the operation of these components and how they affect electrical system operation to be able to accurately and quickly diagnose many electrical failures. More specialized components will be introduced in later chapters.

Electrical Components

Electrical circuits require different components depending on the type of work they do and how they are to perform it. A light may be wired directly to the battery, but it will remain on until the battery drains. A switch will provide for control of the light circuit, however if dimming of the light is also required, the switch is not the only component needed.

There are several electrical components that may be incorporated into a circuit to achieve the desired results from the system. These components include switches, relays, buzzers, and various types of resistors.

Switches

A **switch** is the most common means of providing control of electrical current flow to an accessory (Figure 2-1). A switch can control the on/off operation of a circuit or direct the flow of current through various circuits. The contacts inside the switch assembly carry the current when they are closed. When they are open, current flow is stopped.

A **normally open (NO)** switch will not allow current flow when it is in its rest position. The contacts are open until they are acted upon by an outside force that closes them to complete the circuit. A **normally closed (NC)** switch will allow current flow when it is in its rest position. The contacts are closed until they are acted upon by an outside force that opens them to stop current flow.

The simplest type of switch is the single-pole, single-throw (SPST) switch (Figure 2-2). The term throw refers

Figure 2-1 Common types of switches used in the automotive electrical system.

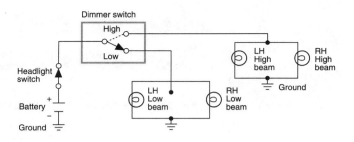

Figure 2-3 A simplified schematic of a headlight system illustrating the SPDT dimmer switch.

to the number of output circuits. The term **pole** refers to the number of input circuits. This switch controls the on/off operation of a single circuit. The most common type of SPST switch design is the hinged-pawl. The **pawl** acts as the contact and changes position as directed to open or close the circuit. The dotted lines used in the symbol indicate the movement of the switch pawl from one position to the other.

Some SPST switches are designed to be a **momentary contact switch**. This switch usually has a spring that holds the contacts open until an outside force is applied that closes them. The horn button on most vehicles is of this design.

Some electrical systems may require the use of a **single-pole, double-throw switch (SPDT)**. The dimmer switch used in the headlight system is usually a SPDT switch. This switch has one input circuit with two output circuits. Depending on the position of the contacts, voltage is applied to the high beam circuit or to the low beam circuit (Figure 2-3).

The most complex switch is the **ganged switch** used for ignition switches. Ganged means that the wipers all move together. In Figure 2-4, the five wipers are all ganged together and will move together. Battery voltage

is applied to the switch from the starter relay terminal. When the ignition key is turned to the start position, all wipers move to the "S" position. Wipers D and E will complete the circuit to ground to test the instrument panel warning lamps. Wiper B provides battery voltage to the ignition coil. Wiper C supplies battery voltage to the starter relay and the ignition module. Wiper A has no output. The **make-before-break wiper** prevents any break in voltage to the ignition coil when the switch is moved from the S position to the R position.

Once the engine starts, the wipers are moved to the Run position. Wipers D and E are moved out of contact with any output terminals. Wiper A supplies battery voltage to the comfort controls and turn signals, wiper B supplies battery voltage to the ignition coil and other accessories, and wiper C supplies battery voltage to other accessories. The jumper wire between terminals A and R of wiper C indicate that those accessories listed can be operated with the ignition switch in RUN or ACC position.

Mercury switches are used by many vehicle manufacturers to detect motion. This switch uses a capsule that is partially filled with mercury and has two electrical

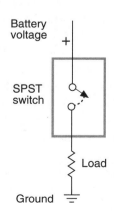

Figure 2-2 A simplified illustration of a SPST switch.

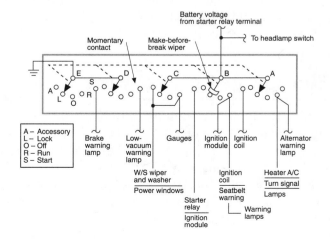

Figure 2-4 Illustration of an ignition switch.

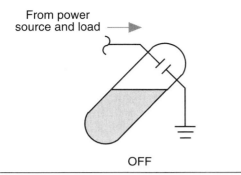

Figure 2-5 A mercury switch in the open position. The mercury does not contact the points.

contacts located at one end. If the switch is constructed as a normally open switch, the contacts are located above the mercury level (Figure 2-5). Since mercury is an excellent conductor of electricity, if the capsule is moved so the mercury touches both of the electrical contacts, the circuit is completed (Figure 2-6). This type of switch is used to illuminate the engine compartment when the hood is opened. While the hood is shut the capsule is tilted in such a position that the mercury is not able to complete the circuit. Once the hood is opened, the capsule tilts with the hood and the mercury completes the circuit and the light turns on.

Relays

Some circuits utilize electromagnetic switches called **relays** (Figure 2-7). A relay is a device that uses low current to control a high current circuit. The coil in the relay has a very high resistance, thus it will draw very low current. This low current is used to produce a magnetic field that will close the contacts. Normally open relays have their points closed by the electromagnetic field. Normally closed relays have their points opened by the mag-

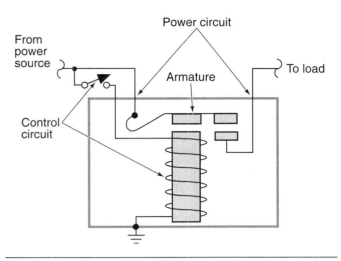

Figure 2-7 A relay uses electrical current to create a magnetic field to draw the contact point closed.

netic field. The contacts are designed to carry the high current required to operate the load component. When current is applied to the coil, the contacts close and heavy battery current flows to the load component that is being controlled.

Figure 2-8 shows a relay application in a horn circuit. Battery voltage is applied to the coil. Since the horn button is a normally open type switch, the current flow to ground is open. Pushing the horn button will complete the circuit, allowing current flow through the coil. The coil develops a magnetic field, which closes the contacts. With the contacts closed, battery voltage is applied to the horn (which is grounded). Used in this manner, the horn relay becomes a control of the high current necessary to blow the horn. The control circuit may be wired with very thin wire since it will have low current flowing through it. The control unit may have only 1/4 ampere flowing through it, while the horn may require 24 or more amperes.

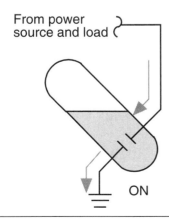

Figure 2-6 When the mercury switch is tilted, the mercury contacts the points and closes the circuit.

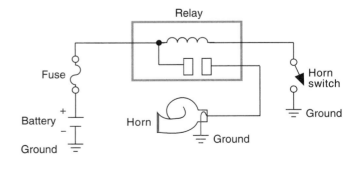

Figure 2-8 A relay can be used in the horn circuit to reduce the size of the conductors required to be installed in the steering column.

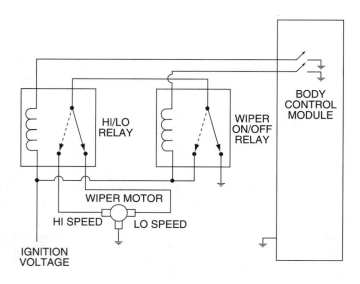

Figure 2-9 Using a relay as a diverter to control hi/lo wiper operation. The Hi/Lo relay diverts current to the different brushes of the wiper motor.

Relays can also be used as a circuit diverter (Figure 2-9). In this example the HI/LO wiper relay will direct current flow to either the high speed brush or low speed brush of the wiper motor to control wiper speeds.

ISO Relays

ISO relays conform to the specification of the International Standards Organization (ISO) for common size and terminal patterns (Figure 2-10). The terminals are identified as 30, 87A, 87, 86, and 85. Terminal 30 is usually connected to battery voltage. This source voltage can be either switched (on or off by some type of switch) or connected directly to the battery. Terminal 87A is connected to terminal 30 when the relay is de-energized. Terminal 87 is connected to terminal 30 when the relay is energized. Terminal 86 is connected to battery voltage (switched or unswitched) to supply current to the electromagnet. Finally, terminal 85 provides ground for the electromagnet. Once again the ground can be switched or unswitched.

Solenoids

A **solenoid** is an electromagnetic device and operates in the same way as a relay; however, a solenoid uses a movable iron core. Solenoids can do mechanical work such as switching electrical, vacuum, and liquid circuits. The iron core inside the coil of the solenoid is spring loaded. When current flows through the coil, the magnetic field created around the coil attracts the core and moves it into the coil. To do work, the core is attached to a mechanical linkage, which causes something to move. When current flow through the coil stops, the spring pushes the core back to its original position. Some power door locks use solenoids to work the locking devices. Solenoids may also switch a circuit on or off, in addition to causing a mechanical action. Such is the case with some starter solenoids. These devices move the starter gear in and out of mesh with the flywheel. At the same time they complete the circuit from the battery to the ignition circuit. Both of these actions are necessary to start an engine.

Buzzers

A **buzzer** is similar in construction to a relay except for the internal wiring (Figure 2-11). A buzzer, or sound

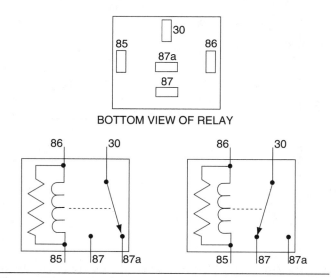

Figure 2-10 ISO relay terminal identification.

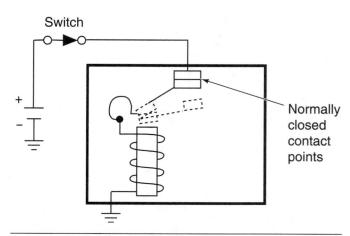

Figure 2-11 A buzzer reacts to the current flow to open and close rapidly, creating a noise.

generator, is sometimes used to warn the driver of possible safety hazards by emitting an audio signal (such as seat belt not buckled).

The coil is supplied current through the normally closed contact points. When voltage is applied to the buzzer, current flows through the contact points to the coil. When the coil is energized, the contact arm is attracted to the magnetic field. As soon as the contact arm is pulled down the current flow to the coil is opened, and the magnetic field is dissipated. The contact arm then closes again, and the circuit to the coil is closed. This opening and closing action occurs very rapidly and it is this movement that generates the vibrating signal.

Resistors

All circuits require resistance in order to operate. If the resistance is there to perform a useful function it is referred to as the **load device**. However resistance can also be used to control current flow and as sensing devices for computer systems. There are several types of resistors that may be used within a circuit. These include: fixed resistors, stepped resistors, and variable resistors.

Fixed Resistors

Fixed resistors are usually made of carbon or oxidized metal (Figure 2-12). These resistors have a set resistance value and are used to limit the amount of current flow in a circuit. The resistance value can be determined by the color bands on the protective shell (Figure 2-13). Usually there are four or five color bands. If there are four bands, the first two are the digit bands, the third is the "multiplier", and the fourth is the tolerance. On a resistor with five bands, the first three are digit bands.

For example, if the resistor has four color bands of yellow, black, brown, and gold the resistance value is determined as follows:

The first color band (yellow) gives the first digit value of 4.

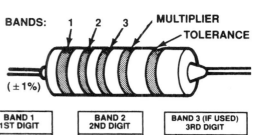

| BANDS: | 1 | 2 | 3 | MULTIPLIER |
| | | | | TOLERANCE |

(±1%)

BAND 1 1ST DIGIT		BAND 2 2ND DIGIT		BAND 3 (IF USED) 3RD DIGIT	
COLOR	DIGIT	COLOR	DIGIT	COLOR	DIGIT
BLACK	0	BLACK	0	BLACK	0
BROWN	1	BROWN	1	BROWN	1
RED	2	RED	2	RED	2
ORANGE	3	ORANGE	3	ORANGE	3
YELLOW	4	YELLOW	4	YELLOW	4
GREEN	5	GREEN	5	GREEN	5
BLUE	6	BLUE	6	BLUE	6
VIOLET	7	VIOLET	7	VIOLET	7
GRAY	8	GRAY	8	GRAY	8
WHITE	9	WHITE	9	WHITE	9

MULTIPLIER	
COLOR	MULTIPLIER
BLACK	1
BROWN	10
RED	100
ORANGE	1,000
YELLOW	10,000
GREEN	100,000
BLUE	1,000,000
SILVER	0.01
GOLD	0.1

RESISTANCE TOLERANCE	
COLOR	TOLERANCE
SLIVER	± 10%
GOLD	± 5%
BROWN	± 1%

Figure 2-13 Resistor color code chart (Courtesy of Chrysler Corporation).

The second color band (black) gives the second digit value of 0.

The digit value is now 40. Multiply this by the value of the third band. In this case brown has a value of 10 so the resistor should have 400 ohms of resistance (40 × 10 = 400).

The last band gives the tolerance. Gold equals a tolerance range of +/- 5%.

Stepped Resistors

A **stepped resistor** has two or more fixed resistor values. The stepped resistor can have an integral switch or have a switch wired in series. A stepped resistor is commonly used to control electrical motor speeds (Figure 2-14). By changing the position of the switch, resistance is increased or decreased within the circuit. If the switch is set to a low resistance, then higher current flows to the motor and its speed is increased. If the switch is placed in the low speed position, additional resistance is added to the circuit and less current flows to the motor, causing it to operate at a reduced speed.

Another use of the stepped resistor is to convert digital to analog signals in the computer circuit. This is accomplished by converting the on/off digital signals into a continuously variable analog signal.

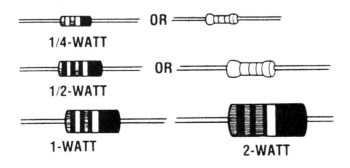

1/4-WATT   OR

1/2-WATT   OR

1-WATT   2-WATT

Figure 2-12 Fixed resistors (Courtesy of Chrysler Corporation).

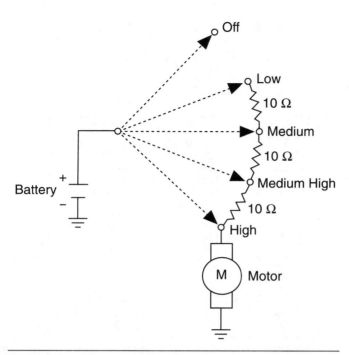

Figure 2-14 The stepped resistor used to control motor speeds. The total resistance is 30Ω in the low position, 20Ω in the medium position, 10Ω in the medium-high position, and 0Ω in the high position.

Variable Resistors

A **variable resistor** provides for an infinite number of resistance values within a range. The most common types of variable resistors are **rheostats** and **potentiometers**. A rheostat is a two terminal variable resistor used to regulate the strength of an electrical current. One terminal is connected to the fixed end of a resistor and a second terminal is connected to a movable contact called a **wiper** (Figure 2-15). By changing the position of the wiper on the resistor, the amount of resistance can be increased or decreased. The most common use of the rheostat is in the instrument panel lighting switch. As the switch knob is turned, the instrument lights dim or brighten depending on the resistance value.

A potentiometer is a three wire variable resistor that acts as a voltage divider to produce a continuously variable output signal proportional to a mechanical position. When a potentiometer is installed into a circuit, one terminal is connected to a power source at one end of the resistor. The second wire is connected to the opposite end of the resistor and is the ground return path. The third wire is connected to the wiper contact (Figure 2-16). The wiper senses a variable voltage drop as it is moved over the resistor. Since the current always flows through the same amount of resistance, the total voltage drop measured by the potentiometer is very stable. For this reason the potentiometer is a common type of input sensor for the vehicle's on-board computers.

Electronic Components

Because a semiconductor material can operate as both a conductor and an insulator, it is very useful as a switching device. How a semiconductor material works depends on the way current flows, or tries to flow, through it.

Diodes

As introduced in Chapter 2, a diode is an electrical one-way check valve that will allow current to flow in one direction only. A diode is the simplest semiconductor device. It is formed by joining P-type semiconductor material with N-type material. The N (negative) side of a diode is called the cathode and the P (positive) side, the anode (Figure 2-17). The point where the cathode and anode join together is called the PN junction. When a

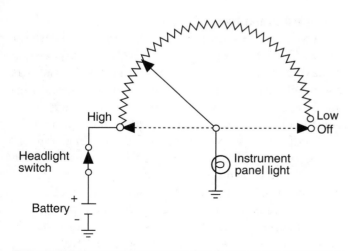

Figure 2-15 A rheostat can be used to control the brightness of a lamp.

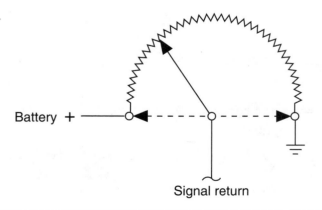

Figure 2-16 A potentiometer is used to send a signal voltage from the wiper.

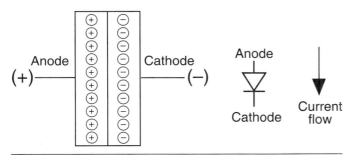

Figure 2-17 A diode and its symbol.

diode is made, the positive holes from the P region and the negative charges from the N region are drawn toward the junction. Some charges cross over and combine with opposite charges from the other side. When the charges cross over, the two halves are no longer balanced and the diode builds up a network of internal charges opposite to the charges at the PN junction. The internal EMF between the opposite charges limits the further diffusion of charges across the junction.

When the diode is incorporated within a circuit and a voltage is applied, the internal characteristics change. If the diode is **forward-biased** there will be current flow (Figure 2-18). Forward-bias means a positive voltage is applied to the P-type material and negative voltage to the N-type material. In this state, the negative region will push electrons across the barrier as the positive region pushes holes across. When forward-biased, the diode acts as a conductor.

If the diode is **reverse-biased**, positive voltage is applied to the N-type material and negative voltage is applied to the P-type material. In this case there will be no current flow (Figure 2-19). The negative region will attract the positive holes away from the junction and the positive region will attract electrons away. This makes the diode act as an insulator.

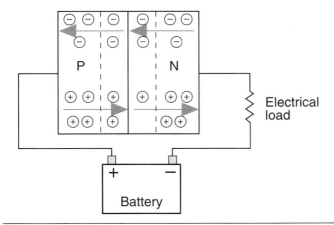

Figure 2-18 Forward-biased voltage causes current flow.

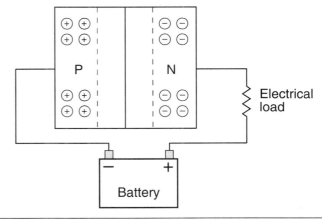

Figure 2-19 Reverse-bias voltage prevents current flow.

Because the diode is a semiconductor there will always be a voltage drop across it.

Zener Diodes

As stated, if a diode is reverse-biased it will not conduct current. However, if the reverse voltage is increased, a voltage level will be reached at which the diode will conduct in the reverse direction. This voltage is called **zener voltage**. Reverse current can destroy a simple PN-type diode, but the diode can be doped with materials that will withstand reverse current.

A **zener diode** is designed to operate in the breakdown region. At the point that zener voltage is reached, a large current flows in reverse bias. This prevents the voltage from climbing any higher. This makes the zener diode an excellent component for regulating voltage. If the zener diode is rated at 15 volts, it will not conduct in the reverse direction when voltage is below 15 volts. At 15 volts it will conduct and the voltage will not increase over 15 volts.

Figure 2-20 illustrates a simplified circuit that has a zener diode in it to provide a constant voltage level to the

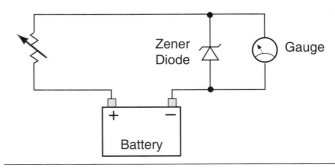

Figure 2-20 Simplified instrument gauge circuit that uses a zener diode to maintain a constant voltage to the gauge.

instrument gauge. In this example, the zener diode is connected in series with the resistor and in parallel to the gauge. If the voltage to the gauge must be limited to 7 volts, the zener diode used would be rated at 7 volts. Since the zener diode maintains a constant voltage drop, and the total voltage drop in a series circuit must equal the amount of source voltage, voltage that is greater than the zener voltage must be dropped over the resistor. Even though source voltage may vary (as a normal result of the charging system) causing different currents to flow through the resistor and zener diode, the voltage dropped by the zener diode remains the same.

When system voltage reaches 7 volts the zener breaks down. At this point the zener diode conducts reverse current, causing an additional voltage drop across the resistor. The amount of voltage to the instrument gauge will remain at 7 volts since the zener diode "makes" the resistor drop the additional voltage to maintain this limit.

Light-Emitting Diodes

A **light-emitting diode (LED)** is similar in operation to the diode, except the LED emits light when it is forward-biased. The light-emitting diode has a small lens built into it so light can be seen when current flows through the diode (Figure 2-21). When the LED is forward-biased, the holes and electrons combine and current is allowed to flow through it. The energy generated is released in the form of light. It is the material used to make the LED that will determine the color of the light emitted.

Normally an LED requires 1.5 to 2.2 volts to light. The light from an LED is not heat energy as is the case with other lights; it is electrical energy. Because of this, LEDs last longer than light bulbs.

Photo Diodes

A **photo diode** also allows current to flow in one direction only. However the direction of current flow is opposite a standard diode. Reverse current flow only occurs when the diode receives a specific amount of light. These types of diodes can be used in automatic headlight systems.

Clamping Diodes

Whenever the current flow through a coil (such as used in a relay or solenoid) is discontinued, a voltage surge or spike is produced. This surge results from the collapsing of the magnetic field around the coil. The movement of the field across the windings induces a very high voltage spike, which can damage electronic components as it flows through the system. In some circuits a capacitor can be used as a shock absorber to prevent component damage from this surge. In today's complex electronic systems, a **clamping diode** may be used to prevent the voltage spike. By installing a clamping diode in parallel across the coil, a bypass is provided for the electrons during the time the circuit is open (Figure 2-22). A clamping diode is nothing more than a standard diode, the term "clamping" refers to its function.

An example of the use of clamping diodes is on some air conditioning compressor clutches. Since the clutch operates by electromagnetism, opening of the clutch coil produces a voltage spike. If this voltage spike were left unchecked it could damage the vehicle's on-board computers. The installation of the clamping diode prevents the voltage spike from reaching the computers.

Relays may also be equipped with a clamping diode. However some use a resistor to dissipate the voltage spike. The two types of relays are not interchangeable.

Transistors

The word **transistor** is a combination of two words, transfer and resist. The transistor is used to control current flow in the circuit. It can be used to allow a predetermined amount of current flow or to resist this flow.

A transistor is a three layer semiconductor (Figure 2-23). It is used as a very fast switching device. Transistors are made by combining P-type and N-type materials in

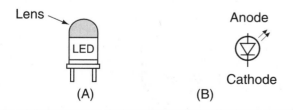

Lens

LED

(A)

Anode

Cathode

(B)

Figure 2-21 (A) A light emitting diode uses a lens to emit the generated light. (B) Symbol for LED.

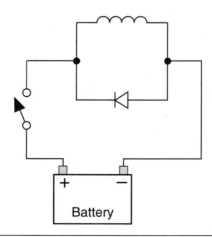

Battery

+   –

Figure 2-22 A clamping diode in parallel to a coil prevents voltage spikes when the switch is opened.

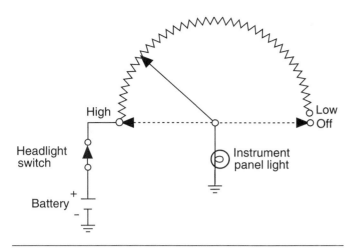

Figure 2-23 Transistors that are used in automotive applications.

groups of three. The two possible combinations are NPN (Figure 2-24) and PNP (Figure 2-25).

WARNING *The PNP-type transistor is the most commonly used transistor in automotive electronics. A PNP transistor cannot be replaced with an NPN transistor.*

The three layers of the transistor are designated as **emitter**, **collector**, and **base**. The emitter is the outside layer of the forward-biased diode and has the same polarity as the circuit side to which it is applied. The arrow on the transistor symbol refers to the emitter lead. The arrow points in the direction of positive current flow and to the N-type material. The collector is the outside layer of the reverse-biased diode. The base is the shared middle layer. Each of these different layers have their own lead for connecting to different parts of the circuit. In effect a transistor is two diodes that share a common center layer. When a transistor is connected to the circuit,

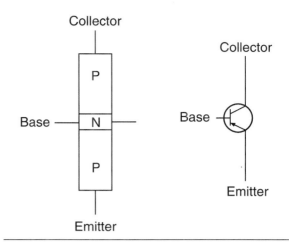

Figure 2-25 A PNP transistor and its symbol.

the emitter-base junction will be forward-biased and the collector-base junction will be reverse-biased.

In the NPN transistor, the emitter conducts current flow to the collector when the base is forward-biased. The transistor cannot conduct unless the voltage applied to the base leg exceeds the emitter voltage by approximately 0.7 volt. This means both the base and collector must be positive with respect to the emitter. With less than 0.7 volt applied to the base leg (compared to the voltage at the emitter), the transistor acts as an opened switch. When the voltage difference is greater than 0.7 volt at the base, compared to the emitter voltage, the transistor acts as a closed switch (Figure 2-26).

When an NPN transistor is used in a circuit, it normally has a reverse-bias applied to the base-collector junction. If the emitter-base junction is also reverse-biased, no current will flow through the transistor (Figure 2-27). If the emitter-base junction is forward-biased (Figure 2-28), current flows from the emitter to the base.

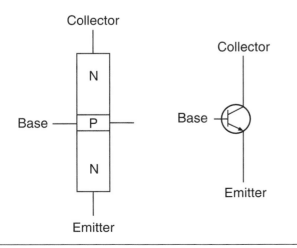

Figure 2-24 A NPN transistor and its symbol.

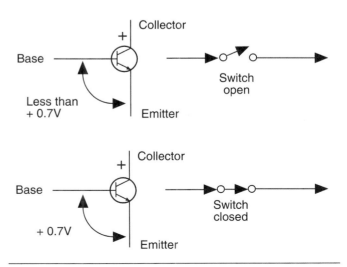

Figure 2-26 NPN transistor action.

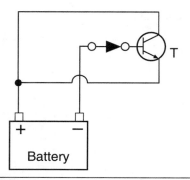

Figure 2-27 NPN transistor with reverse-biased voltage applied to the base. No current flow.

The base is a thin layer and a positive voltage is applied to the collector and electrons flow from the emitter to the collector.

In the PNP transistor, current will flow from the emitter to the collector when the base leg is forward-biased with a voltage that is more negative than that at the emitter (Figure 2-29). For current to flow through the emitter to the collector, both the base and the collector must be negative in respect to the emitter.

Since current can be controlled through a transistor, this component can be used as a very fast electrical switch. It is also possible to control the amount of current flow through the collector. This is because the output current is proportional to the amount of current through the base leg.

A transistor has three operating conditions:

1) **Cutoff:** when reverse-bias voltage is applied to the base leg of the transistor. In this condition the transistor is not conducting and no current will flow.

2) **Conduction:** bias voltage difference between the base and the emitter has increased to the point that the transistor is switched on. In this condition the transistor is conducting. Output current is proportional to that of the current through the base.

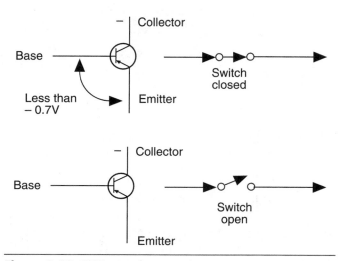

Figure 2-29 PNP transistor action.

3) **Saturation:** this is the point where forward-bias voltage to the base leg is at a maximum. With bias voltage at the high limits, output current is also at its maximum.

These types of transistors are called **bipolar** because they have three layers of silicon, with two of them being the same. Another type of transistor is the **field-effect transistor (FET)**. The FET's leads are listed as source, drain, and gate. The source supplies the electrons and is similar to the emitter in the bipolar transistor. The drain collects the current and is similar to the collector. The gate creates the electrostatic field that allows electron flow from the source to the drain. It is similar to the base.

The FET transistor does not require bias voltage, only a voltage needs to be applied to the gate terminal to get electron flow from the source to the drain. The source and drain are constructed of the same type of doped material. They can be either N-type or P-type materials. The source and drain are separated by a thin layer of either N-type or P-type material.

Using Figure 2-30, if the source is held at 0 voltage and 6 volts are applied to the drain, no current will flow

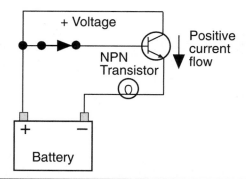

Figure 2-28 NPN transistor with forward-biased voltage applied to the base. Current flows.

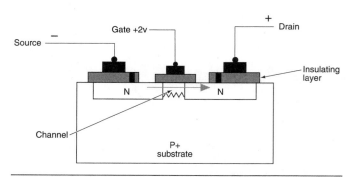

Figure 2-30 A FET uses a positive voltage to the gate terminal to create a capacitive field to allow electron flow.

between the two. However, if a lower positive voltage to the gate is applied, the gate forms a capacitive field between the channel and itself. The voltage of the capacitive field attracts electrons from the source and current will flow through the channel to the higher positive voltage of the drain.

This type of FET is called an **enhancement-type FET** because the field effect improves current flow from the source to the drain. This operation is similar to that of a normally open switch. A **depletion-type FET** is like a normally closed switch, whereas the field effect cuts off current flow from the source to the drain.

The most harmful things to a transistor are excess voltage and heat. If you have a problem with a module that stops operating after the engine is warm, heat the module slowly using a hair dryer to test the module. Two principle uses for transistors are as switches and amplifiers. The average transistor should last 100 years unless it receives some abuse.

WARNING *Some forms of FETs have a very thin insulation layer between the gate and channel that static electricity from your hands is able to burn through. Be careful not to touch the connector pins of the computer or the integrated circuits inside of the computer.*

Transistor Amplifiers

A transistor can be used in an amplifier circuit to amplify the voltage. This is useful when using a very small voltage for sensing computer inputs, but needing to boost that voltage to operate an accessory (Figure 2-31). The small signal voltage that is applied to the base leg of a transistor may look like that of Figure 2-32a. However, the corresponding signal through the collector may be like that shown in Figure 2-32b. Three things happen in an amplified circuit:

1) The amplified voltage at the collector is greater than that of the base voltage.

2) The input current increases.

3) The pattern has been inverted.

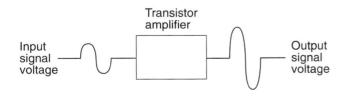

Figure 2-31 A simplified amplifier circuit.

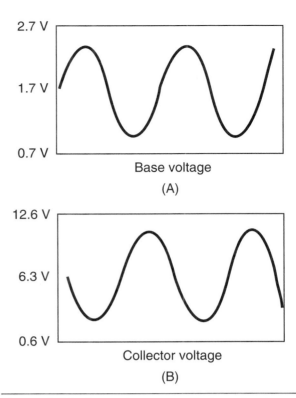

Figure 2-32 The voltage applied to the base (A) is amplified and inverted through the collector (B).

Some amplifier circuits use a **Darlington pair**, which is two transistors that are connected together (Figure 3-33). The first transistor in a Darlington pair is used as a preamplifier to produce a large current to operate the second transistor. The second transistor is isolated from the control circuit and is the final amplifier. The second transistor boosts the current to the amount required to operate the load component. The Darlington pair is utilized by most control modules used in electronic ignition systems.

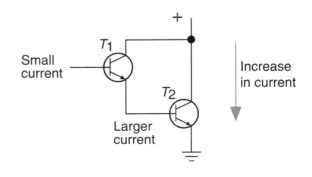

Figure 2-33 A Darlington pair used to amplify current. T1 acts as a preamplifier that creates a larger base current for T2, which is the final amplifier that creates a larger current.

Phototransistors

A **phototransistor** is a transistor that is sensitive to light (Figure 2-34). In a phototransistor, a small lens is used to focus incoming light onto the sensitive portion of the transistor. When light strikes the transistor, holes and free electrons are formed. These increase current flow through the transistor according to the amount of light. The stronger the light intensity, the more current will flow. This type of phototransistor is often used in automatic headlight dimming circuits.

Thyristors

A **thyristor** is a semiconductor switching device composed of alternating N and P layers. It can also be used to rectify current from AC to DC. The most common type of thyristor used in automotive applications is the **silicon controlled rectifier (SCR)**. Like the transistor, the SCR has three legs. However, it consists of four regions arranged PNPN (Figure 2-35). The three legs of the SCR are called the anode (or P-terminal), the cathode (or N-terminal), and the gate (one of the center regions).

The SCR requires only a trigger pulse (not a continuous current) applied to the gate to become conductive. Current will continue to flow through the anode and cathode as long as the voltage remains high enough or until the gate voltage is reversed.

The SCR can be connected into a circuit either in the forward or reverse direction. Using Figure 2-35 of a forward direction connection, the P-type anode is connected to the positive side of the circuit and the N-type cathode is connected to the negative side. The center PN junction blocks current flow through the anode and cathode.

Once a positive voltage pulse is applied to the gate, the SCR turns on. Even if the positive voltage pulse is removed, the SCR will continue to conduct. If a negative voltage pulse is applied to the gate, the SCR will no longer conduct.

Figure 2-34 Phototransistor.

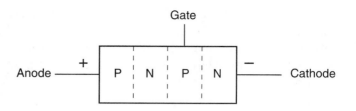

Figure 2-35 A forward direction SCR.

The SCR will also block any reverse current from flowing from the cathode to the anode. Since current can flow only in one direction through the SCR, it can rectify AC current to DC current.

Integrated Circuits

An **integrated circuit** is a complex circuit of thousands of transistors, diodes, resistors, capacitors, and other electronic devices that are formed onto a tiny silicon chip (Figure 2-36). As many as 30,000 transistors can be placed on a chip that is 1/4 inch (6.35 mm) square. Integrated circuits are constructed by photographically reproducing circuit patterns on to a silicon wafer. The process begins with a large scale drawing of the circuit. This drawing can be room size. Photographs of the circuit drawing are reduced until they are the actual size of the circuit. The reduced photographs are used as a mask. Conductive P-type and N-type materials, along with insulating materials, are deposited onto the silicon wafer. The mask is placed over the wafer and selectively exposes the portion of material to be etched away or the portions requiring selective deposition. The entire process of creating an integrated circuit chip takes over 100 separate steps. Out of a single wafer 4 inches in diameter, thousands of integrated circuits can be produced. The integrated circuit is also called an **IC chip.**

The small size of the integrated chip has made it possible for the vehicle manufacturers to add several computer controlled systems to the vehicle without taking much space. Also a single computer is capable of performing several functions.

WARNING *Integrated circuits can be damaged by static electricity. Use caution when working with these circuits. There are anti-static straps available for the technician to wear to reduce the possibility of destroying the integrated circuit.*

WARNING *Do not connect or disconnect an I/C to the circuit with the power on. The arc that is produced may damage the chip.*

WARNING *Do not test an IC chip with an ohm-meter.*

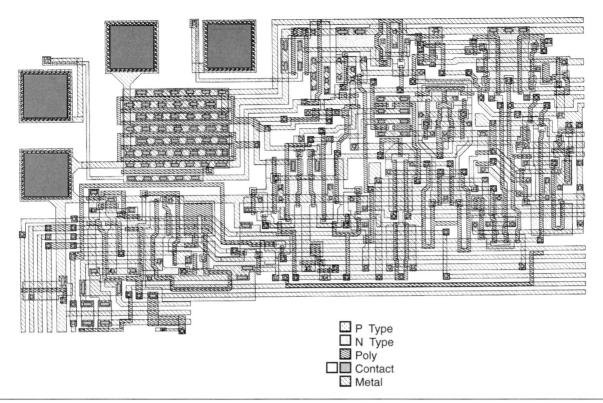

P Type
N Type
Poly
Contact
Metal

Figure 2-36 An enlarged illustration of an integrated circuit with thousands of transistors, diodes, resistors, and capacitors. Actual size can be less than 1/4 inch square.

Circuit Protection Devices

Most automotive electrical circuits are protected from current flow that would exceed the capacity of the conductors and/or the load components. Excessive current flow results from a decrease in circuit resistance. Circuit resistance will decrease when too many components are connected in parallel or when a component or wire becomes shorted. A **short** is an undesirable, low resistance path for current flow. To prevent damage to the components and conductors, these circuits use some form of a protection device. The protection device is designed to "turn off" the system that it protects. This is done by creating an open (like turning off a switch) to prevent a complete circuit. If an **overload** (excessive current flow in a circuit) occurs, the protection device opens, preventing current from flowing in the circuit.

Fuses

The most common circuit protector is the in-line **fuse** (Figure 2-37). A fuse contains a metal strip that will melt when the current flowing through it exceeds its rating. The thickness of the metal strip determines the rating of the fuse. When the metal strip melts, excessive current is indicated. The cause of the excessive current must be found and repaired; then a new fuse of the same rating should be installed. The most commonly used automotive fuses are rated from 4 to 30 amps.

The fuses are generally installed in a central **fuse box**. Fuse box is the term used to identify the central location of the fuses contained within a single holding fixture. The most common location of the fuse box is

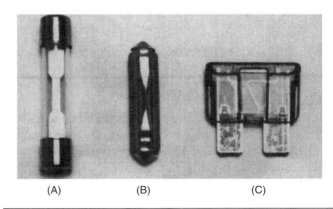

(A)   (B)   (C)

Figure 2-37 Three types of commonly used fuses. (A) glass cartridge, (B) ceramic, and (C) blade (or minifuse).

under the instrument panel. However, it can be located behind kick panels, in the glove box, and electrical junction box on the fender well. Fuse identification and specifications are usually labeled on the fuse box or on the fuse box cover. Of course, this information can also be found in the vehicle's owner's manual and the service manual.

The in-line fuse is connected in series with the circuit that it is protecting. Battery voltage is applied to the main bus bar which is connected to one end of the fuse, with the other end of the fuse connected to the circuit. The bus bar is a common electrical connection to which all of the fuses are attached. The bus bar is connected to battery voltage.

There are three basic types of fuses: glass or ceramic fuses, blade-type fuses, and bullet or cartridge fuses. Glass and ceramic fuses are found mostly on older vehicles. Sometimes, however, you can find them in a special holder connected in series with a circuit. Glass fuses are small glass cylinders with metal caps. The metal strip connects the two caps. The rating of the fuse is normally marked on one of the caps.

Blade-type fuses are flat plastic units and are available in three different physical sizes: mini, standard, and maxi (Figure 2-38). The plastic housing is formed around two male blade-type connectors. The metal strip

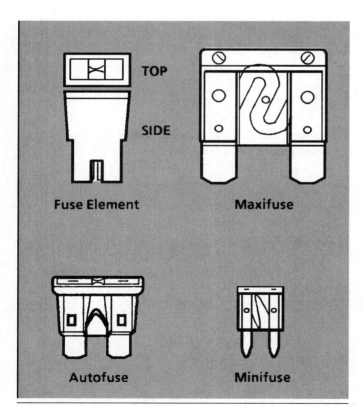

Figure 2-38 Common blade-type fuses. (Courtesy of General Motors Corporation, Service Technology Group)

connects these connectors inside the plastic housing. The rating of these fuses is on top of the plastic housing and the plastic is color coded (Figure 2-39).

Cartridge-type fuses are used in many older European vehicles. These fuses are made of plastic or a ceramic material. They have pointed ends and the metal strip rounds from end to end. This type of fuse is much like a glass fuse except the metal strip is not enclosed.

In a parallel circuit, a single fuse can be used before the first branch (Figure 2-40) or an individual fuse can be used with each branch (Figure 2-41).

When adding accessories to the vehicle, the correct fuse rating must be selected. Use Watt's law to determine the correct fuse rating to use (watts ÷ volts = amperes). The fuse selected should be slightly higher rated than the actual current draw to allow for current surges (5 to 10 %).

WARNING *Fuses are rated by amperage and voltage. Never install a larger rated fuse into a circuit than the one that was designed by the manufacturer. Doing so may damage or destroy the circuit.*

Fusible Links

A vehicle may use one or several **fusible links** to provide protection for the main power wires before they are divided into smaller circuits at the fuse box. A fusible link is made of meltable material with a special heat resistant insulation. When there is an overload in the circuit, the link melts and opens the circuit. To properly test a fusible link, use an ohmmeter or continuity tester. The fusible links are usually located at a main connection near the battery (Figure 2-42). The current capacity of a fusible link is determined by its size. A fusible link is usually four wire sizes smaller (four numbers larger) than the circuit it protects. The smaller the wire the larger its number. A circuit using 14-gauge wire requires an 18-gauge fusible link for protection. A "blown" fusible link is usually identified by bubbling of the insulator material around the link.

CAUTION *Do not replace a fusible link with a resistor wire or vise-versa.*

Maxi-fuses

In place of fusible links, many manufacturers use a **maxi-fuse**. A maxi-fuse looks similar to blade-type fuses except it is larger and has a higher amperage capacity. They are also referred to as **cartridge fuses**. By using

AUTOFUSE

CURRENT RATING	COLOR
3	VIOLET
5	TAN
7.5	BROWN
10	RED
15	BLUE
20	YELLOW
25	NATURAL
30	GREEN

MAXIFUSE

CURRENT RATING	COLOR
20	YELLOW
30	GREEN
40	AMBER
50	RED
60	BLUE
70	BROWN
80	NATURAL

MINIFUSE

CURRENT RATING	COLOR
5	TAN
7.5	BROWN
10	RED
15	BLUE
20	YELLOW
25	WHITE
30	GREEN

Figure 2-39 Fuses are rated by the amperage they can carry. (Courtesy of General Motors Corporation, Service Technology Group)

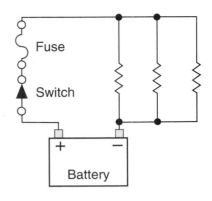

Figure 2-40 One fuse to protect the entire parallel circuit.

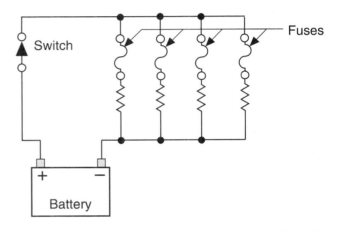

Figure 2-41 Fuses used to protect each branch of a parallel circuit.

maxi-fuses, the manufacturers are able to break down the electrical system into smaller circuits. If a fusible link burns out, many of the vehicle's electrical systems may be affected. By breaking down the electrical system into smaller circuits, and installing maxi-fuses, the consequence of a circuit defect will not be as severe as it would have been with a fusible link. In place of a single fusible link, there may be many maxi-fuses, depending on how the circuits are divided. This makes the techni-

Figure 2-42 Fusible links located near the battery.

cian's job of diagnosing a faulty circuit much easier. Maxi-fuses are used because they are less likely to cause an under-hood fire when there is an overload in the circuit. If the fusible link burned in two, it is possible that the "hot" side of the fuse could come into contact with the vehicle frame and the wire catch on fire.

Circuit Breakers

A circuit that is susceptible to an overload on a routine basis is usually protected by a **circuit breaker**. A circuit breaker uses a **bimetallic strip** that reacts to excessive current (Figure 2-43). A bimetallic strip consists of two different types of metals. One strip will react more quickly to heat than the other, causing the strip to flex in proportion to the amount of current flow. When an overload or circuit defect occurs that causes an excessive amount of current draw, the current flowing through the bimetallic strip causes it to heat. As the strip heats, it bends and opens the contacts. Once the contacts are opened current can no longer flow. With no current flowing, the strip cools and closes again. If the high-current cause is still in the circuit, the breaker will open again. The circuit breaker will continue to cycle open and closed as long as the overload is in the circuit. This type of circuit breaker is self-resetting or a "cycled" circuit breaker. Some circuit breakers require manually resetting by pressing a button while others must be removed from the power to reset (Figure 2-44).

SERVICE TIP *A circuit breaker, fitted with alligator clips, is useful to bypass a fuse that keeps blowing. The circuit breaker will keep the current flowing through the circuit so that it may be checked for the cause of high current draw while protecting the circuit. (Figure 2-45).*

An example of the use of a circuit breaker is in the power window circuit. Since the window is susceptible to jams, due to ice build-up on the window, a current overload is possible. If this should occur, the circuit breaker will heat up and open the circuit before the window motor is damaged. If the operator continues to attempt to operate the power window, the circuit breaker will cycle open and close until the cause of the jam is removed.

PTCs as Circuit Protection Devices

Automotive engineers are faced with conflicting needs to provide reliable circuit protection against shorts to ground or other overload conditions yet at the same

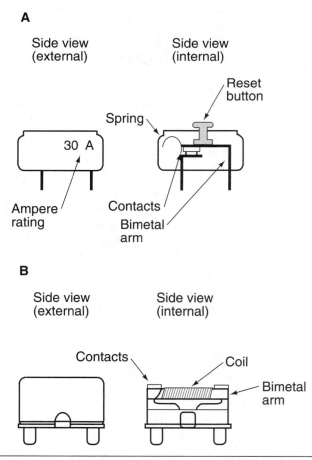

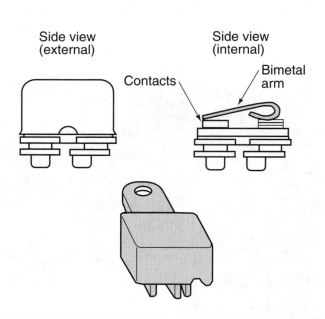

Figure 2-43 The circuit breaker uses a bimetallic strip that opens if current draw is excessive.

Figure 2-44 Non-cycling circuit breakers. (A) Manual reset type circuit breaker and (B) circuit breaker that requires being removed from the power to reset.

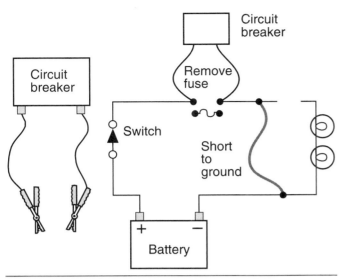

Figure 2-45 A circuit breaker fitted with alligator clips used to diagnose electrical problems.

time reduce vehicle weight and cost. Traditionally fuses are used to protect multiple circuits. However, this results in large, heavy, and complex wiring assemblies. The use of polymer, **positive temperature coefficient (PTC)** resistors provides a means of meeting these needs. A PTC resistor increases in resistance as temperature increases. Because of its design, a PTC resistor has the ability to trip (increase resistance to the point that it becomes the load device in the circuit) during an over-current condition and reset after the fault is no longer present.

Conductive polymers consist of specially formulated plastics and various conductive materials. At normal temperatures the plastic materials form a crystalline structure. The structure provides a low resistance conductive chain. The resistance is so low that it does not affect the operation of the circuit. However, if the current flow increases above the trip threshold, the additional heat causes the crystalline structure to change to an amorphous state. In this condition the conductive paths separate, causing a rapid increase in the resistance of the PTC. The increased resistance reduces the current flow to a safe level.

Summary

❑ A switch can control the on/off operation of a circuit or direct the flow of current through various circuits.

❑ A relay is a device that uses low current to control a high current circuit.

❑ A buzzer is sometimes used to warn the driver of possible safety hazards by emitting an audio signal.

❑ A fixed resistor has a set resistance value.

❑ A stepped resistor has two or more fixed resistor values.

❑ A variable resistor provides for an infinite number of resistance values within a range.

❑ A rheostat is a two terminal variable resistor used to regulate the strength of an electrical current.

❑ A potentiometer is a three wire variable resistor that acts as a voltage divider to produce a continuously variable output signal proportional to a mechanical position.

❑ A diode is an electrical one-way check valve that will allow current to flow in one direction only.

❑ A transistor is a three layer semiconductor that is commonly used as a very fast switching device.

❑ An integrated circuit is a complex circuit of thousands of transistors, diodes, resistors, capacitors, and other electronic devices that are formed onto a tiny silicon chip.

❑ The protection device is designed to "turn off" the system it protects. This is done by creating an open (like turning off a switch) to prevent a complete circuit.

Terms To Know

Base	Fuse box	Positive temperature coefficient (PTC)
Bimetallic strip	Fusible links	Potentiometers
Bipolar	Ganged switch	Relays
Bus bar	IC chip	Reverse-biased
Buzzer	Integrated circuit	Rheostats
Cartridge fuses	ISO relays	Saturation
Circuit breaker	Light-emitting diode (LED)	Short
Clamping diode	Load device	Silicon-controlled rectifier (SCR)
Collector	Make-before-break wiper	Single-pole, double-throw switch (SPDT)
Conduction	Maxi-fuse	
Cutoff	Mercury switches	Solenoid
Darlington pair	Momentary contact switch	Stepped resistor
Depletion-type FET	Normally open (NO) switch	Switch
Emitter	Normally closed (NC) switch	Thyristor
Enhancement-type FET	Overload	Transistor
Field-effect transistor (FET)	Pawl	Variable resistor
Fixed resistors	Photo diode	Wiper
Forward-biased	Phototransistor	Zener diode
Fuse	Pole	Zener voltage

Review Questions

Short Answer Essays

1. What is the purpose of a relay?
2. Explain the difference between normally open (NO) and normally closed (NC) switches.
3. Describe the difference between a rheostat and a potentiometer.
4. What is the function of a diode?
5. List the different types of diodes.
6. What is the purpose of transistors?
7. Explain the operation of a bipolar transistor.
8. Describe the purpose of a zener diode.
9. Explain the purpose of circuit protection device.
10. List the most common types of circuit protection devices.

Fill-In-The-Blanks

1. Never install a larger rated _____ into a circuit than the one that was designed by the manufacturer.

2. A _____ can control the on/off operation of a circuit or direct the flow of current through various circuits.

3. A normally _____ switch will not allow current flow when it is in its rest position. A normally _____ switch will allow current flow when it is in its rest position.

4. An _____ _____ is a complex circuit of thousands of transistors, diodes, resistors, capacitors, and other electronic devices that are formed onto a tiny silicon chip.

5. A _____ is used in electronic circuits as a very fast switching device.

6. A _____ is an electrical one-way check valve that will allow current to flow in one direction only.

7. A _____ is an electrical-mechanical device that uses low current to control a high current circuit.

8. A _____ _____ provides for an infinite number of resistance values within a range. A _____ is a two terminal variable resistor used to regulate the strength of an electrical current.

9. The _____ requires only a trigger pulse applied to the gate to become conductive.

10. A PTC resistor _____ in resistance as temperature increases.

ASE Style Review Questions BCC

1. While discussing protection devices: Technician A says a fuse automatically resets after the cause of the overload is repaired. Technician B says the protection device creates an open when an overload occurs. Who is correct?
   a. A only
   b. B only
   c. Both A and B
   d. Neither A nor B

2. While discussing circuit components: Technician A says a switch can control the on/off operation of a circuit or direct the flow of current through various circuits. Technician B says a relay controls high current with low current. Who is correct?
   a. A only
   b. B only
   c. Both A and B
   d. Neither A nor B

3. While discussing diodes: Technician A says the zener diode can be used for regulating voltage. Technician B says the installation of a clamping diode across a coil provides a bypass for the electrons when the circuit is opened suddenly. Who is correct?
   a. A only
   b. B only
   c. Both A and B
   d. Neither A nor B

4. While discussing the use of transistors: Technician A says a transistor can be used to control the switching on/off of a circuit. Technician B says a transistor can be used to amplify voltage. Who is correct?
   a. A only
   b. B only
   c. Both A and B
   d. Neither A nor B

5. Circuit protection devices are being discussed. Technician A says a "blown" fuse is identified by bubbling of the insulator material around the link. Technician B say a "blown" fusible link is identified by a burned through metal wire in the capsule. Who is correct?
   a. A only
   b. B only
   c. Both A and B
   d. Neither A nor B

6. Transistor operation is being discussed. Technician A says the emitter of an NPN transistor is usually connected to the ground circuit. Technician B says in an NPN transistor, a more negative voltage at the base than what is on the emitter is needed to turn it on. Who is correct?
   a. A only
   b. B only
   c. Both A and B
   d. Neither A nor B

7. Diodes are being discussed. Technician A says a diode connected in a circuit so it does not conduct current is forward-biased. Technician B says a diode provides high speed switching. Who is correct?
   a. A only
   b. B only
   c. Both A and B
   d. Neither A nor B

8. Variable resistors are being discussed. Technician A says rheostats have three wires and are used as a voltage divider. Technician B says a rheostat provides only three different resistance values. Who is correct?
   a. A only
   b. B only
   c. Both A and B
   d. Neither A nor B

9. Technician A says an integrated circuit is a complex circuit of thousands of electronic devices formed onto a tiny silicon chip. Technician B says an integrated circuit can be tested with an ohmmeter. Who is correct?
   a. A only
   b. B only
   c. Both A and B
   d. Neither

10. Technician A says a solenoid is an electromagnetic device that operates much like a relay. Technician B says resistance used to perform a useful function in a circuit is referred to as the load device. Who is correct?
    a. A only
    b. B only
    c. Both A and B
    d. Neither A nor B

Circuit Problems and Electrical Test Equipment

Objective

Upon completion of this chapter, you will be able to:

- ❏ Define circuit defects including opens, shorts, grounds, and excessive voltage drops.
- ❏ Explain the effects that each type of circuit defect will have on the operation of the electrical system.
- ❏ Explain the proper use of analog volt/amp/ohmmeters.
- ❏ Explain the proper use of digital volt/amp/ohmmeters.
- ❏ Describe when to use the different types of multimeters.
- ❏ Explain the proper use of digital storage oscilloscope.
- ❏ Explain the proper use of a test light.
- ❏ Explain the proper use of a logic probe.

Introduction

To be able to properly diagnose electrical components and circuits, you must be able to use many different types of electrical test equipment. In addition, you must have an understanding of the different types of circuit defects. In this chapter, you will learn when and how to use the most common types of test equipment. You will also learn which test instrument is best to use to identify the cause of the various types of electrical problems.

Circuit Defects

All electrical problems can be classified as being one of three types of problems: an open, short, or high resistance. Each one of these will cause a component to operate incorrectly or not at all. Understanding what each of these problems will do to a circuit is the key to proper diagnosis of any electrical problem.

Open

An **open circuit** is simply a break in the circuit (Figure 3-1). An open is caused by turning a switch off, a break in a wire, a burned out light bulb, a disconnected wire or connector, or anything that opens the circuit. When a circuit is open, current does not flow and the component does not work. Because there is no current flow, there are no voltage drops in the circuit. Source voltage is available everywhere in the circuit up to the point at which it is open. Source voltage is even available after a load, if the open is after that point.

Opens caused by a blown fuse will still cause the circuit not to operate, but the cause of the problem is the excessive current that blew the fuse. Nearly all other opens are caused by a break in the continuity of the circuit. These breaks can occur anywhere in the circuit.

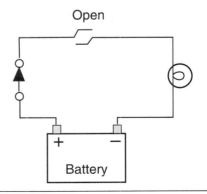

Figure 3-1 An open circuit stops all current flow.

Shorts

A **shorted circuit** is a circuit that allows current to bypass part of the normal path. Shorts cause an increase in current flow. This increased current flow can burn wires or components.

An example of a shorted circuit could be found in a faulty coil. The windings within a coil are insulated from each other, however, if this insulation breaks down a copper-to-copper contact is made between the turns. Since part of the windings will be bypassed, this reduces the number of windings in the coil through which current will flow. This results in the effectiveness of the coil being reduced. Also, since the current bypasses a portion of the normal circuit resistance, current flow is increased and excess heat can be generated.

Another example of a shorted circuit is if the insulation of two adjacent wires breaks down, and allows a copper-to-copper contact (Figure 3-2). If the short is between point A and B, light 1 would be on all the time. If the short was between point B and C, both lights would illuminate when either switch was closed.

A third type of short is a **grounded circuit**. A grounded circuit is a condition that allows current to flow in an unintentional path to ground (Figure 3-3). To see what happens in a circuit that has a short to ground refer to Figure 3-4. If normal resistance of the two bulbs is 3 ohms and 6 ohms, since they are in parallel total circuit resistance is 2 ohms. The short makes a path from the power side of one bulb to the return path to the battery. The short creates a low resistance path. If the low resistance path has a resistance value of 0.001 ohms, it is possible to calculate what would happen to the current in this circuit. The short becomes another leg in the parallel circuit. Since the total resistance of a parallel circuit is always lower than the lowest resistance, we know the

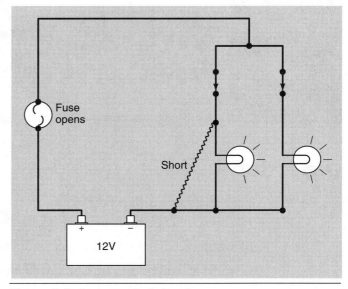

Figure 3-3 A grounded circuit.

total resistance of the circuit is now less than 0.001 ohms. Using Ohm's law we can calculate the current flow through the circuit.

$$I = E/R \quad \text{or} \quad I = 12/.001 \quad \text{or} \quad I = 12{,}000 \text{ Amps}$$

Needless to say, it would take a large wire to carry that kind of amperage. The 10-amp fuse would melt quickly when the short occurred. This would protect the wires and light bulbs.

SERVICE TIP *A grounded circuit can be checked by removing the fuse and connecting a test lamp in series across the fuse connections. If the lamp lights, the circuit is grounded. Any load that is completely bypassed by having a short to ground causes the voltage drop across the load to be zero.*

High Resistance

High resistance problems occur when there is unwanted resistance in the circuit. The high resistance can come from a loose connection, corroded connection, wrong size wires, and so on. Since the resistance becomes an additional load in the circuit the effect is that the load component, with reduced voltage and current applied, operates with reduced efficiency. An example would be a taillight circuit with a load component (light bulb) that is rated at 50 watts. To be fully effective this bulb must draw 4.2 amperes at 12 volts ($I = P \div E$). This means a full 12 volts should be applied to the bulb. If resistance is present at other points in the circuit, some of the 12 volts will be dropped. With less voltage (and cur-

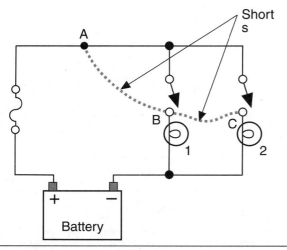

Figure 3-2 A short circuit can be a copper-to-copper contact between two adjacent wires.

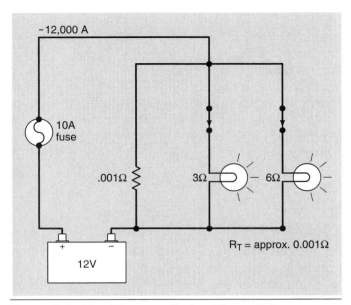

Figure 3-4 Ohm's law applied to Figure 3-3.

rent) being available to the light bulb, the bulb will illuminate with less intensity.

High resistance may appear on either the insulated or grounded return side of a circuit. It may also be present in both at the same time. To test for voltage drop, the circuit must be active (current flowing). The source voltage must be as specified before voltage drop readings can be valid. Whenever voltage drop is suspected, both sides of the circuit must be checked.

Test Equipment

Since electricity is an invisible force, the proper use of test tools will permit the technician to "see" the flow of electrons. Knowing what is being looked at and being able to interpret various meter types will assist in electrical system diagnosis. To diagnose and repair electrical circuits correctly, a number of common tools and instruments are used. The most common tools are jumper wires, test lights, voltmeters, ammeters, and ohmmeters.

Jumper Wires

One of the simplest types of test equipment is the **jumper wire**. A jumper wire is simply a wire with an alligator clip on each end. Connecting one end of the jumper wire to battery positive will provide an excellent 12-volt power supply for testing a component. Jumper wires can be used to check the load components by bypassing switches, conductors, and connections in the circuit. Jumper wires can also be used to provide the ground to test that portion of the circuit. Jumper wires can never be used to jump the load of the circuit.

To protect the circuit being tested, it is recommended the jumper wire be fitted with an in-line fuse holder. This will allow the quick changing of fuses to correctly protect the circuit. Using a fused jumper wire will help prevent damage to the circuit if the jumper wire is connected improperly.

WARNING *Never connect a jumper wire across the terminals of the battery. The battery could explode, causing series injury.*

Test Lights

A **test light** is used when the technician needs to "look" for electrical power in the circuit. The test light handle is transparent and contains a light bulb. A sharp probe extends from one end of the handle while a ground wire with a clamp extends from the other end (Figure 3-5). If the circuit is operating properly, clamping the lead of the test light to ground and probing the insulated side of the circuit, the lamp should light (Figure 3-6).

A test light is limited in that it does not display how much voltage is at the point of the circuit being tested. However, by understanding the effects of voltage drop the technician will be able to interpret the brightness of the test light and relate the results to that which would be expected in a good circuit. If the lamp is connected after a voltage drop, the lamp will light dimly. Connecting the test lamp before the voltage drop should light the lamp brightly. The light should not illuminate at all if it is probing for voltage after the last resistance.

Another type of circuit tester is the self-powered continuity tester (Figure 3-7). The continuity tester has an

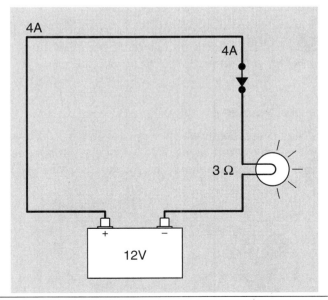

Figure 3-5 A typical test light used to probe for voltage in a circuit.

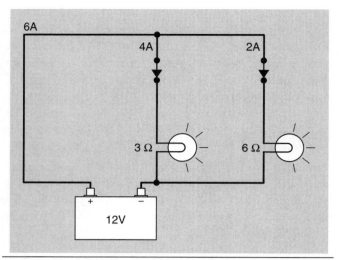

Figure 3-6 If voltage is present, the test light will illuminate.

internal battery that powers a light bulb if the circuit is complete between the probe and the ground wire. With the power in the circuit turned off or disconnected, the ground clip is connected to the ground terminal of the load component. By probing the feed wire, the light will illuminate if the circuit is complete (has continuity). If there is an open in the circuit, the lamp will not illuminate. Never use a self-powered test light to test the air bag system. The battery in the tester can cause the air bag to deploy.

> **WARNING** It is not recommended that a test light be used to probe for power in a computer controlled circuit. The increased draw of the test light may damage the system components.

> **WARNING** Do not connect a self-powered test light to a circuit that is powered. Doing so will damage the test light.

Logic Probes

Many computer controlled systems use a pulsed voltage to transmit messages or to operate a component. A standard or self-powered test light should not be used to

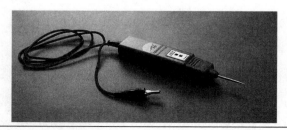

Figure 3-7 Typical self-powered continuity tester.

test these circuits since they may damage the computer. However a **logic probe** (Figure 3-8) can be used. A logic probe looks something like a test light except it contains three different colored LEDs. The red LED will light if there is high voltage at the point in the circuit being tested. The green LED will light to indicate the presence of low voltage. The yellow LED lights to indicate the presence of voltage pulses. If the voltage is a pulsed voltage from a high level to a low level, the yellow LED will be on and the red and green LEDs will cycle indicating the change in voltage.

Multimeters

A **multimeter** is an electrical test meter capable of measuring voltage, resistance, and amperage. In addition, some types of multimeters are designed to test diodes, measure frequency, duty cycle, temperature, and rotation speed. Multimeters are available in analog (swing needle) and digital display.

Analog Meters

Analog meters use a sweeping needle and a scale to display test values (Figure 3-9). All analog meters use a D'Arsonval movement. A D'Arsonval movement is a small coil of wire mounted in the center of a permanent horseshoe-type magnet. A pointer or needle is mounted to the coil (Figure 3-10). When taking a measurement, current flows through the coil and creates a magnetic field around the coil. The coil rotates within the permanent magnet as its magnetic field interacts with the magnetic field of the permanent magnet. The amount of rotation is determined by the strength of the magnetic field around the coil. The direction of the rotation is determined by the direction of the current flow through the coil. Since the needle moves with the coil, it reflects the amount of coil movement and its direction. To give accurate readings, analog meters must be calibrated. This

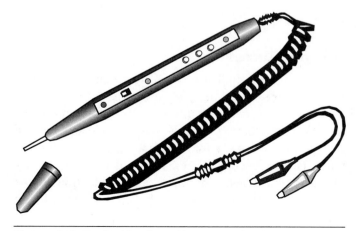

Figure 3-8 Typical logic probe.

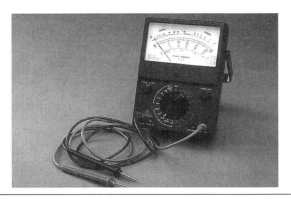

Figure 3-9 An analog meter.

is done through a zero adjustment knob or adjusting screw in the meter. These adjustments move the coil to its base point.

There are some service procedures that specify that an analog meter be used instead of a digital meter. Always follow the manufacturer's recommendations while performing diagnostic and/or service routines.

Digital Meters

With modern vehicles incorporating computer controlled systems, the need for **digital multimeters (DMM)** is required (Figure 3-11). Computer systems have integrated circuits that operate on very low amounts of current. Analog meters will download computer circuits and burn out the IC chips since they allow a large amount of current to flow through the circuit. On the other hand, most digital multimeters have very high input resistance (**impedance**) which prevents the meter from drawing current when connected to a circuit. Most DMMs have at least 10 megohms (10 million ohms) of impedance. This reduces the risk of damaging computer circuits and components. DMMs are also referred to as **DVOMs (digital volt/ohmmeter).**

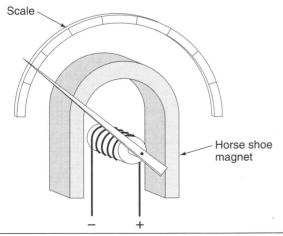

Figure 3-10 D'Arsonval movement is the basis for the movement of an analog meter.

WARNING *Not all DMMs are rated at 10 megohms of impedance. Be sure of the meter you are using to prevent electronic component damage.*

Digital meters rely on electronic circuitry to measure electrical values. The measurements are displayed with LEDs or on a liquid crystal display (LCD). Digital meters tend to give more accurate readings and are certainly much easier to read. Rather than reading a scale at the point where the needle lines up, digital meters simply display the measurement in a numerical value. This also eliminates the almost certain error caused by viewing an analog meter at an angle.

Voltmeter

A **voltmeter** is one of the most used meters in the shop. A voltmeter is used to read the pressure behind the flow of electrons. A voltmeter has two leads: a red positive lead and a black negative lead. The red lead should be connected to the positive side of the circuit or component. The black should be connected to ground or to the negative side of the component. A voltmeter is connected in parallel with a circuit and reads directly in volts (Figure 3-12). The voltmeter hookup should be in parallel to the circuit that is being measured to prevent the high resistance meter from disrupting the circuit.

Figure 3-13 shows how to check for voltage in a closed circuit. The voltage at point A is 12 volts positive. There is a drop of 6 volts over the 1Ω resistor and the reading is 6 volts positive at point B. The remaining voltage drops in the motor load and the voltmeter reads 0 at point C, indicating normal motor circuit operation.

Figure 3-11 Digital Multimeter. (Courtesy of TIF Instruments, Inc.)

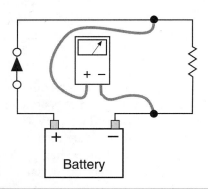

Figure 3-12 Connecting a voltmeter in parallel to the circuit.

When reading voltage in the same circuit that has an open (Figure 3-14), 12 volts will be indicated at any point ahead of the open. This is indicated at points A, B, and C, but not through X. Since the circuit is open and there is no electrical flow, there is no voltage drop across a resistor or load.

Voltage drop is the amount of electrical energy converted to another form of energy. For example, to make a lamp light, electrical energy is converted to heat energy. It is the heat that makes the lamp light. To measure voltage drop across each load it must be determined what point is the most positive and what point is the most negative in the circuit. A point in the circuit can be either positive or negative, depending on what is being measured. Referring to Figure 3-15, if voltage drop across R_1 is being measured, the positive meter lead is placed at point A and the negative lead at point B. The voltmeter will measure the difference in volts between these points. However, to measure the voltage drop across R_2, the polarity of point B is positive and point C is negative because point B is the most negative for R_1, yet it is the most positive point for R_2. To measure voltage drop across R_3, point C is positive and point D is negative.

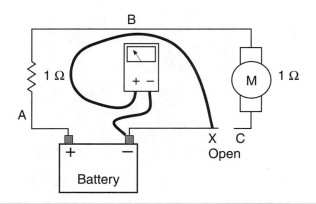

Figure 3-14 Checking voltage in an open circuit.

The positive lead of the voltmeter should be placed as close as possible to the positive side of the battery in the circuit.

There are two methods used to display AC voltage: **root mean square (RMS)** and **average responding**. RMS meters convert the AC signal to a comparable DC voltage signal. Average responding meters display the average voltage peak. If the AC voltage signal is a true sine wave, both methods would display the same value. However, most automotive sensors do not produce a pure sine wave signal. A sine wave is a waveform that shows voltage changing polarity from positive to negative. The technician must know how the different meters will display the AC voltage reading under these circumstances.

Ohmmeter

An **ohmmeter** will measure resistance and continuity. The ohmmeter is powered by an internal battery, thus the power to the circuit being tested must be disconnected. By connecting the ohmmeter leads in parallel to the portion of the circuit being tested, an open or excessive resistance can be detected (Figure 3-16). The meter sends a

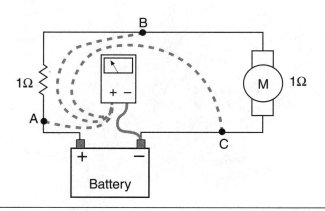

Figure 3-13 Checking voltage in a closed circuit.

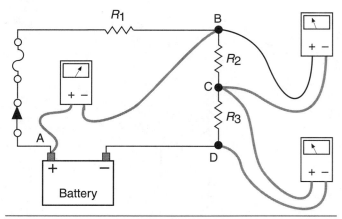

Figure 3-15 Using a voltmeter to measure voltage drops.

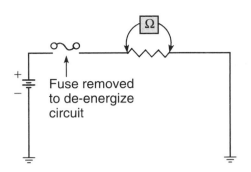

Figure 3-16 Measuring resistance with an ohmmeter. The meter is connected in parallel with component being tested after power is removed from the circuit.

current through the component and determines the amount of resistance based on the voltage drop across the load. The meter reads from zero to infinity (∞). A reading of zero means there is no resistance in the circuit. This may indicate the presence of a short in a component that requires resistance. For example, a coil winding should have a high resistance value, a zero ohms reading would indicate the coil windings are being bypassed. If the meter indicates an infinity reading, this means the resistance is higher than the meter can read on the selected scale. If an infinity reading is obtained on the highest scale this usually indicates the circuit has an open.

Most analog ohmmeters use a multiplier to figure higher resistances. The multi-position switch on the front of the meter indicates four ranges. These ranges are usually labeled R × 1, R × 10, R × 100, and R × 1K. The reading on the ohmmeter scale must be multiplied by the value indicated by the range to get the actual resistance. For example, if the reading is 22Ω (resistance) and the switch is on R × 1, the resistance is 22Ω. However, if the same meter reading of 22Ω is obtained on the R × 100 scale the reading is 2,200Ω (22 × 100). If the reading of 22Ω is obtained on the R x 1K scale, then the reading is actually 22,000Ω (22 × 1,000). The ranges are selected based on the component being tested. If the resistance is unknown, start on the lowest scale and work up to the higher scales. This will prevent misinterpreting infinity readings.

The test chart shown in Figure 3-17 illustrates the readings that may be expected from an ohmmeter or voltmeter under different conditions. It is important to become familiar with these examples in order to analyze circuits.

WARNING *Since the ohmmeter is self-powered, never use an ohmmeter on a powered circuit.*

Ammeter

An **ammeter** measures current flow in a circuit. The ammeter must be connected in series with the circuit being tested (Figure 3-18). To make a series connection, open the circuit at a convenient location (such as a relay or fuse) and connect the ammeter leads across the open, observing polarity. Do not connect an ammeter in parallel with the circuit as this can cause damage to the circuit and/or the meter.

Most hand-held multimeters have a 10 ampere protection device. This is the highest amount of current flow the meter can read. When using the ammeter, start on a high scale and work down to obtain the most accurate readings.

It is easier to test current flow using an ammeter equipped with an induction type clamp-on pickup (Figure 3-19). The pickup is clamped around the wire or cable being tested. The ammeter determines the amount of amperage by the strength of the magnetic field created by the current flowing through the wire.

Using Figure 3-18, assume the circuit was designed to draw 5 amperes and is protected by a 6 ampere fuse. If the fuse constantly blows, this indicates the presence of a short to ground or other low resistance fault. Mathematically, each load in the circuit should draw 1.25 amperes (5 ÷ 4 = 1.25). To locate the short begin by removing all of the lights in the circuit from their sockets. Connect the ammeter into the series portion of the circuit. Connect a fused jumper wire (or self-resetting circuit breaker) in

Type of Defect	Test Unit	Expected Results
Open	Ohmmeter	∞ infinite resistance between conductor ends
	Test light	No light after open
	Voltmeter	∅ volts at end of conductor after the open
Ground	Ohmmeter	∅ resistance to ground
	Test light	Lights if connected across fuse
	Voltmeter	Generally not used to test for ground
Short	Ohmmeter	Lower than specified resistance through load component
		∅ resistance to adjacent conductor ∞ infinite resistance to ground
	Test light	Light will illuminate on both conductors
	Voltmeter	A voltage will be read on both conductors

Figure 3-17 Circuit test chart.

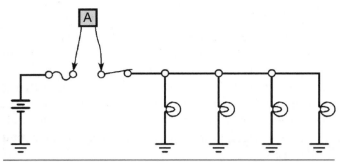

Figure 3-18 Measuring current flow with an ammeter. The meter must be connected in series with the circuit.

place of the circuit protection device. The fuse in the jumper wire should be rated at no more than 50% higher capacity than the specified fuse used in the circuit. Next, close the switch and read the ammeter. With all of the loads disconnected the ammeter should read zero amperes. If there is any reading, there is a short to ground.

If zero amperes was measured, reconnect each light in sequence. The ammeter reading should increase 1.25 amperes with each bulb. If the reading increases more than 1.25 amperes whenever a bulb is installed, the problem is in that leg of the circuit.

CAUTION *Always use an ammeter that can handle the expected current, since excessive current can damage the meter.*

Additional DOM Functions

Some multimeters feature additional functions besides measuring AC or DC voltage amperage and ohms. Some

Figure 3-19 Ammeter with inductive pickup.

meters have a MIN MAX function that displays the minimum, maximum, and average values received by the meter during the time the test was being recorded. This feature is valuable when checking sensors, output commands, or circuits for electrical **noise.** Noise is usually the result of **radio frequency interference (RFI).** RFI is an unwanted voltage signal that rides on a signal. The noise causes slight increases and decreases in the voltage signal to or from the computer. Another definition of noise is an AC signal riding on a DC voltage. The computer may attempt to react to the minute changes in the signal as a result of the noise. This means the computer is responding to the noise and not the voltage signal, resulting in incorrect component operation.

Also, some multimeters may have the capabilities to measure duty cycle, pulse width, and frequency. **Duty cycle** is the measurement of the amount of on time as compared to the time of a cycle (Figure 3-20). A cycle is one set of changes in a signal that repeats itself several times. The duty cycle is displayed in a percentage. For example, a 60% duty cycle means the device is on 60% of the time and off 40% of the time of one cycle.

Pulse width is similar to duty cycle except it is a measurement of the time the device is turned on within a cycle (Figure 3-21). Pulse width is usually measured in milliseconds.

Frequency is a measure of the number of cycles that occur in one second (Figure 3-22). The higher the frequency, the more cycles that occur in a second. Frequency is measured in **hertz.** If the cycle occurs once per second the frequency is one hertz. If the cycle is 300 times per second then the frequency is 300 hertz.

Reading The DOM

With the increased use of the digital volt/ohmmeter (DOM), it is important for the technician to be able to accurately read the meter. By becoming proficient in the use of the DOM, technicians will have confidence in their conclusions and recommended repairs. This will eliminate the replacement of parts that were not faulty.

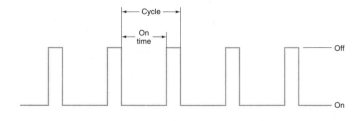

Figure 3-20 Duty cycle.

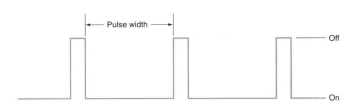

Figure 3-21 Pulse width.

PREFIX	SYMBOL	RELATION TO BASIC UNIT
Mega	M	1,000,000
Kilo	K	1,000
Milli	m	0.001 or $\frac{1}{1000}$
Micro	μ	.000001 or $\frac{1}{1000000}$
Nano	n	.000000001
Pico	p	.000000000001

Figure 3-23 Common prefixes used on digital multimeters.

There are deviations between the different DOM manufacturers as to the way the display is presented, but most follow the method described here.

Multimeters have either an "auto range" feature, in which the appropriate scale is automatically selected by the meter, or they must be manually set to a particular scale. Either way, to designate particular ranges and readings meters display a prefix before the reading or range. Meters use the prefix because they cannot display long numbers. Most digital meters are capable of displaying 3 1/2 digits (1999). Values such as 20200Ω cannot be displayed as a whole number. As a result, scales are expressed as a multiple of tens or use the prefix k, M, m, and μ (Figure 3-23). The prefix K stands for kilo and represents 1,000 units. For example, a reading of 10K equals 10,000. An M stands for mega and represents 1,000,000 units. A reading of 10M would represent 10,000,000. An m stands for milli and represents 0.001 of a unit. A reading of 10m would be 0.010. The symbol stands for micro and represents 0.000001 of a unit. In this case a reading of 125.0 μ would represent 0.0000125.

If the display has no prefix before the unit being measured (V, A, Ω) the reading displayed is read directly. For example, if the reading was 1.243V the actual voltage value is 1.243. However, if there is a prefix displayed then the decimal point will need to be floated to determine actual readings. If the prefix is M (mega) then the decimal is floated six places to the right. For example, a reading display of 2.50MΩ is actually 2, 500,000 ohms. A reading display of .250MΩ is actually 250,000 ohms.

A prefix of K (kilo) means the decimal point needs to be moved three places to the right. For example, a display reading of 56.4KΩ is actually 56,400 ohms. A reading of 1.264KΩ is actually 1,264 ohms.

If the prefix is m (milli) then the decimal is floated three places to the left. For example, a reading of 25.4mA is representing 0.025 amperes. A display of 165.0mA is actually 0.165 amperes.

Finally, if the prefix is a μ (micro) the decimal is floated six points to the left. A reading displayed as 125.3μA would represent 0.0001253 amperes while a reading of 4.6μA is actually 0.0000046 amperes.

When using the ohmmeter function of the DOM, make sure power to the circuit being tested is turned off. Before taking measurements, be sure to calibrate the meter. This is done by holding the two test leads together. Most DVOMs will self-calibrate while others will need to be adjusted by turning a knob until the meter reads zero. Connect the DOM to parallel to the portion of the circuit being tested. If continuity is good, the DOM will read zero or close to zero even on the low scale. If the continuity is very poor, the DOM will display an infinite reading. This reading is usually shown as a blinking "1.000", a blinking "1", or an "OL".

If the meter is not auto-ranging, the meter will have a knob or buttons that are selected to display values in different scales. The most common scale gradations for the ohmmeter function are 200, 2K, 20K, 200K, 2M, and 20M. It is important the correct scale be used to obtain an accurate reading. When the DOM is connected to a circuit to test its resistance, begin with the lowest scale and scale up until the infinite indicator is no longer displayed. This is true unless the lowest scale does not produce the infinite reading, then stay on the lowest scale. Also, if the DOM indicates an infinite reading on the highest scale then there is an open in the circuit. This method of scaling up will put the DOM to the lowest scale on which it is still capable of reading the resistance of the component or circuit. This will assure the highest possible degree of accuracy is attained.

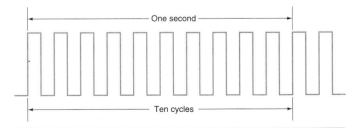

Figure 3-22 Frequency signal.

The numbers that represent the scale reading are the potential number of units that can be read on that particular scale, before the infinite indicator is displayed. Any letter suffix represents the multiplying factor. On the 200 scale there is no multiplying factor and the meter is capable of reading up to 200 ohms before the infinite indicator is displayed. On the 2K scale the multiplying factor is 1,000. This scale is capable of reading 2,000 ohms before an infinite indicator is shown. The 20K scale is capable of reading up to 20,000 ohms before the infinite indicator is shown. The 20M scale has a multiplier factor of 1,000,000 and is capable of reading up to 20,000,000 ohms of resistance before the infinite indicator is shown.

If the resistance reading is on the 200 scale, simply read the display as it appears. If the display shows 154, the resistance is 154 ohms. If the display shows 104.4 ohms, the resistance is 104.4 ohms. If the resistance value of the component or circuit is over 200 ohms the display would show the infinite indicator. If the infinite indicator is shown, then scale up to the next range.

If the meter is set on the 2K scale, the resistance is read by placing the decimal point three places to the right. For example, if the resistance value is 450 ohms the display will read .450Ω. The first space would be blank since the value is less than 1,000 (Figure 3-24). Moving the decimal point on the display three digits to the right provides the correct resistance value (450.0 ohms). If the resistance value was 1,230 ohms, this would be displayed as 1.230Ω (Figure 3-25). The first space is not left blank since the value is greater than 1,000. Again, by moving the decimal point to the right three spaces the resistance value is determined. If the resistance was over 2,000 ohms the display would show the infinite indicator and the technician would go to the next highest scale.

If the DOM is on the 20K scale, always move the decimal point three places to the right, whether all three

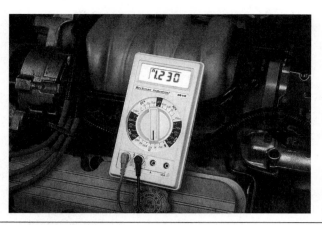

Figure 3-25 DOM displaying resistance on the 2K scale. Since the resistance is greater than 1,000 ohms, the first space is filled.

places are actually shown on the display or not. For example, if the display showed 15.00Ω this would indicate 15,000 ohms of resistance. A reading of 50 ohms would be indicated by .05Ω.

The 200K scale is read the same way as the 20K scale. Move the decimal point three places to the right. If the display shows 150.0Ω the resistance value is 150,000 ohms.

On the 20M scale always move the decimal point six places to the right. A display of 19.00Ω would actually be 19,000,000 ohms, while a display of 1.456 would be 1,456,000 ohms. If the amount of resistance is higher than 20M ohms the display will show the infinite indicator. This usually means that there is an open in the circuit.

Voltage readings on the non-auto-ranging DOM is very simple. The scales are usually 2V, 20V and 200V. On the 2V scale the display would show up to 2.000 volts. On the 20V scale the display will be up to 20.00 volts and on the 200V scale the display will be up to 200.0 volts. Be sure that the meter is on the appropriate DC or AC voltage to test the circuit.

Oscilloscope

The **oscilloscope** is very useful in diagnosing many electrical problems quickly and accurately. Digital and analog voltmeters do not react fast enough to read systems that cycle quickly. The oscilloscope may be considered as a very fast reacting voltmeter that reads and displays voltages. The scope allows the technician to view voltage over time. These voltage readings appear as a voltage trace on the oscilloscope screen (Figure 3-26). Some smaller oscilloscopes use liquid crystal displays (LCD). However most larger screens are cathode ray tube (CRT), which is very similar to the picture tube in a television set. High voltage from an internal source is

Figure 3-24 DOM displaying resistance on the 2K scale. Since the resistance is less than 1,000 ohms, the first space is blank.

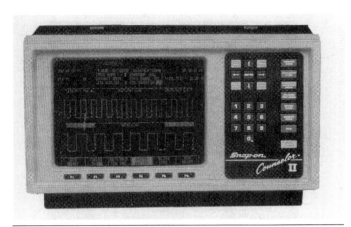

Figure 3-26 Oscilloscope. (Courtesy of Snap-on Tools Company, Copyright Owner)

supplied to an electron gun in the back of the CRT when the oscilloscope is turned on. This electron gun emits a continual beam of electrons against the front of the CRT. The external leads on the oscilloscope are connected to deflection plates above and below, and on each side of the electron beam. When a voltage signal is supplied from the external leads to the deflection plates, the electron beam is distorted and strikes the front of the screen in different locations to indicate the voltage signal from the external leads.

An upward movement of the voltage trace on an oscilloscope screen indicates an increase in voltage, and a downward movement of this trace represents a decrease in voltage. As the voltage trace moves across an oscilloscope screen, it represents a specific length of time. Most oscilloscopes of this type are referred to as analog scopes or real-time scopes. This means the voltage activity is displayed without any delay.

Oscilloscopes are usually incorporated within a large diagnostic machine. However, smaller portable units are available. Today most technicians use a variation of the oscilloscope called a **lab scope**. The screen of a lab scope is divided into small divisions of time and voltage (Figure 3-27). The division of the screen creates a grid pattern. Time is represented by the horizontal movement of the waveform. Voltage is measured with the vertical position of the waveform. Since the scope displays voltage over time, the waveform moves from left to right. The value of the divisions can be adjusted to improve the view of the waveform. For example, the vertical scale can be adjusted so each division represents 0.5 volts and the horizontal scale can be adjusted so each division equals 0.005 (5 milliseconds). This allows the technician to view small changes in voltage that occur in a very short period of time. The grid serves as a reference for measurements.

Since a scope displays actual voltage, it will display any electrical noise or disturbances accompanying the voltage signal (Figure 3-28). Electrical **glitches** are momentary changes in the voltage signal. This may be the result of momentary shorts to ground, shorts to power, or opens in the circuit. By observing a voltage signal while wiggling a wiring harness, any looseness can be detected by a change in the voltage signal.

Most lab scopes are digital scopes, commonly referred to as **digital storage oscilloscopes (DSO)**. The digital scope converts the voltage signal into digital information and stores it into its memory. This allows the technician to capture the signal for closer analysis. Most DSOs have a sampling rate of one million samples per second. This means the signal displayed is not real time. However, the delay is very slight.

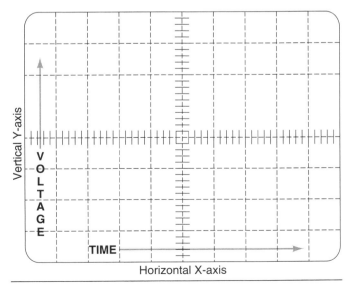

Figure 3-27 Grids on a scope screen serve as a time and voltage reference.

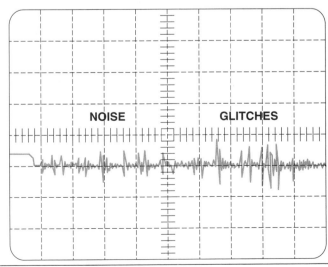

Figure 3-28 RFI noise and glitches showing as voltage signals.

Most scopes are **dual trace scopes**. This allows the technician to view two patterns at the same time (Figure 3-29). This is helpful to monitor the cause and effect of a sensor or compare a good pattern to the one being sampled.

Scope Waveforms

The change of voltage over time is displayed on the **waveform**. Any change in the amplitude of the trace indicates a change in voltage. If the trace is a straight horizontal line, the voltage is constant (Figure 3-30). A diagonal line up or down represents a gradual increase or decrease in voltage. A vertical line indicates a sudden change in voltage values, such as turning on or off the circuit.

Scopes can display AC and DC voltage. A normal AC signal changes its polarity and amplitude over a period of time. The waveform created by AC voltage is typically a sine wave (Figure 3-31). One complete sine wave shows the voltage moving from zero to its positive peak, then moving down through zero to its negative peak, and returning to zero. If the rise and fall from positive and negative is the same, the wave is said to be **sinusoidal**. If the rise and fall are different, the wave is nonsinusoidal.

DC voltage waveforms may appear as a straight line or as a line showing a change of voltage over time. Sometimes a DC voltage waveform will appear as a **square wave**. A square wave is identified by having straight vertical lines with flat horizontal lines (Figure 3-32). This type of waveform represents a voltage that is being turned on and off.

Scope Controls

Scope controls will vary depending on the manufacturer and model of the scope. However, most will provide controls for intensity, vertical (Y-axis) adjustment,

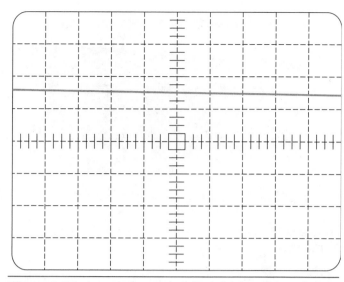

Figure 3-30 A waveform showing a constant voltage.

horizontal (X-axis) adjustment, and trigger adjustments. The vertical adjustment controls the voltage displayed. The voltage setting is the voltage that will be shown per division (Figure 3-33). For example, if the scope is set on 0.5 (500 milli) volts, a 5 volt signal would require 10 divisions. If the scope is set on 1 volt, a 5 volt signal would require 5 divisions.

The horizontal adjustment controls the time of the trace (Figure 3-34). The time is measure per division. If the division time is set at .4 seconds and there are ten divisions across the screen then total screen time is four seconds (.4 × 10 = 4.0).

Trigger controls tell the scope when to begin a trace across the screen. Setting the trigger is required when

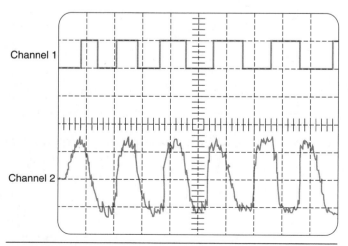

Figure 3-29 A dual trace scope can show two patterns at the same time.

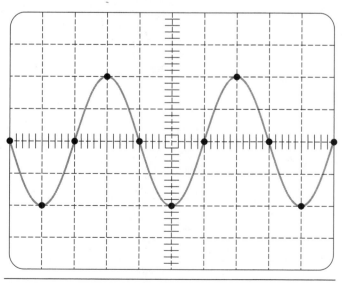

Figure 3-31 An AC voltage sine wave.

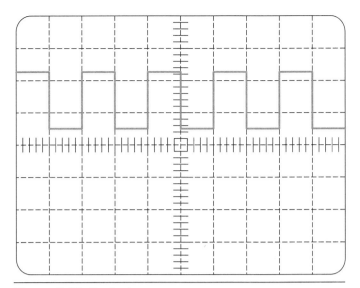

Figure 3-32 Typical square wave.

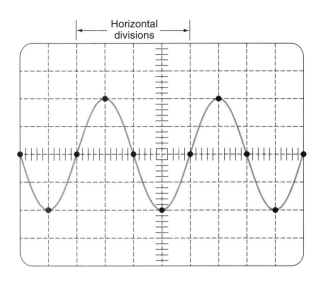

Figure 3-34 Horizontal divisions represent time.

attempting to observe the timing of something. Proper triggering will allow the trace to repeatedly begin and end at the same points on the screen. The trigger can usually be set for NORM or AUTO. In the NORM position, no trace will be displayed until a voltage signal occurs within the set time base. The AUTO setting will display the trace regardless of the time base.

Slope and level controls are used to define the actual trigger voltage. The slope switch determines whether the trace will begin on a rising or falling voltage signal (Figure 3-36). The level control determines where the time base will be triggered according to a certain point on the slope.

A trigger source switch tells the scope which input signal to trigger on. This can be Channel 1, Channel 2, line voltage, or an external signal. External signal trig-

gering is useful when desiring to observe a trace of a component that may be affected by the operation of another component. An example would be observing fuel injector activity when changes in throttle position are made. The external trigger would be voltage change at the throttle position sensor. The display trace would be the cycling of a fuel injector.

Scan Testers

Scan testers are used to test automotive computer systems. These testers retrieve fault codes from the onboard computer memory and display these codes in the digital reading on the tester. The scan tester performs

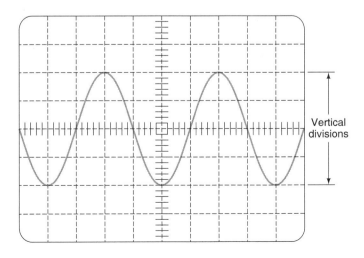

Figure 3-33 Vertical divisions represent voltage.

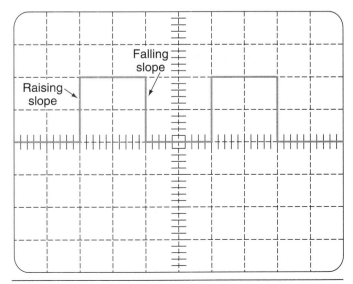

Figure 3-35 A trigger can be set to start the trace with a rise or fall of voltage.

Figure 3-36 Scan tester with various modules. (Courtesy of Snap-on Tools Company, Copyright Owner)

many other diagnostic functions depending on the year and make of vehicle. Most scan testers have removable modules that are updated each year. These modules are designed to test the computer systems on various makes of vehicles. For example, some scan testers have a 3-in-1 module which tests the computer systems on Chrysler, Ford, and General Motors vehicles. A 10-in-1 module is also available to diagnose computer systems on vehicles imported by 10 different manufacturers. These modules plug into the scan tester (Figure 3-36).

Scan testers have the capability to test many onboard computer systems such as engine computers, antilock brake computers, air bag computers, and suspension computers, depending on the year and make of vehicle, and the type of scan tester. In many cases, the technician must select the computer system to be tested with the scan tester after the tester is connected to the vehicle.

The scan tester is connected to specific diagnostic connectors on various vehicles. Some manufacturers have one diagnostic connector, and they connect the data wire from each onboard computer to a specific terminal in this connector. Other vehicle manufacturers have several different diagnostic connectors on each vehicle, and each of these connectors may be connected to one or more onboard computers. A set of connectors is supplied with the scan tester to allow tester connection to various diagnostic connectors on different vehicles.

The scan tester must be programmed for the model year, make of vehicle, and type of engine. With some scan testers, this selection is made by pressing the appropriate buttons on the tester, as directed by the digital tester display. On other scan testers, the appropriate memory card must be installed in the tester for the vehicle being tested. Some scan testers have a built-in printer to print test results, while other scan testers may be connected to an external printer.

As automotive computer systems become more complex, the diagnostic capabilities of scan testers continue to expand. Many scan testers now have the capability to store, or "freeze," data into the tester during a road test, and then play back this data when the vehicle is returned to the shop.

Some scan testers now display diagnostic information based on the fault code in the computer memory. Service bulletins published by the scan tester manufacturer may be indexed by the tester after the vehicle information is entered in the tester. Other scan testers will display sensor specifications for the vehicle being tested.

Summary

❏ An open circuit is a circuit in which there is a break in continuity.

❏ A shorted circuit is a circuit that allows current to bypass part of the normal path.

❏ A grounded circuit is a condition that allows current to return to ground before it has reached the intended load component.

❏ Conventional 12-V or self-powered test lights should not be used to test computer circuits. High-impedance test lights should be used for this purpose.

❏ Multimeters read AC volts, DC volts, milliamperes, amperes, and ohms on various scales.

❏ Only high-impedance digital multimeters should be used to test computer system components.

❏ An analog oscilloscope contains a cathode ray tube (CRT), which provides a voltage trace.

❏ Scan testers retrieve fault codes from the computer memory and perform many other diagnostic functions.

Terms-To-Know

Ammeter	Grounded circuit	Oscilloscope
Analog meters	Hertz	Pulse width
Average responding	High resistance	Radio frequency interference (RFI)
D'Arsonval movement	Impedance	Root mean square (RMS)
Digital multimeters (DMM)	Jumper wire	Scan testers
Digital storage oscilloscope (DSO)	Lab scope	Shorted circuit
Dual trace scopes	Logic probe	Sinusoidal
Duty cycle	Multimeter	Square wave
DVOMs (Digital Volt/Ohmmeters)	Noise	Test light
Frequency	Ohmmeter	Voltmeter
Glitches	Open circuit	Waveform

Review Questions

Short Answer Essay

1. Describe the purpose of an oscilloscope.
2. What is the ohmmeter used to test for?
3. Define what is meant by opens, shorts, and high resistance and explain the effects that each type of circuit defect will have on the operation of the electrical system.
4. Describe how to use a DOM to measure voltage.
5. Explain the proper use of a test light.
6. Explain the proper use of a logic probe.
7. List and define the most common prefixes used on auto-ranging DVOMs.
8. Explain the process of performing a voltage drop test.
9. Explain the difference between analog and digital multimeters.
10. Describe the proper method of connecting an ammeter to a circuit.

Fill-In-The-Blanks

1. When continuity is good, the DOM will read close to _____, even on the low scale. If the continuity is very poor, the DOM will display an _____ reading.
2. All electrical problems can be classified as being one of three types of problems: an _____, _____, or _____ _____.
3. A _____ _____ is a circuit that allows current to bypass part of the normal path.
4. Jumper wires can never be used to jump the _____ of the circuit.
5. _____ meters use a sweeping needle and a scale to display test values.
6. A voltmeter is used to read the _____ behind the flow of electrons.
7. An _____ measures current flow in a circuit.
8. _____ _____ is the measurement of the amount of on time as compared to the time of a cycle. _____ _____ is a measurement of the time the device is turned on within a cycle.
9. The change of voltage over time is displayed on the _____ shown on the oscilloscope.
10. _____ _____ are used to test automotive computer systems and to retrieve fault codes from the onboard computer memory.

ASE Style Review Questions

1. The use of an ammeter is being discussed:
Technician A says the ammeter is used to measure current flow. Technician B says the ammeter must be connected in parallel to the circuit being tested. Who is correct?
   a. A only
   b. B only
   c. Both A and B
   d. Neither A nor B

2. A DOM is being used to measure current flow. The meter is displaying 85.5mA. Technician A says this represents 0.0855 amperes. Technician B says the decimal point needs to be moved six points to the left. Who is correct?
   a. A only
   b. B only
   c. Both A and B
   d. Neither A nor B

3. While discussing circuit defects: Technician A says an open means there is continuity in the circuit. Technician B says a short bypasses a portion of the circuit. Who is correct?
   a. A only
   b. B only
   c. Both A and B
   d. Neither A nor B

4. Circuit defects are being discussed. Technician A says added resistance can cause a lamp in a parallel circuit to burn brighter than normal. Technician B says excessive resistance may appear on either the insulated or grounded return side of a circuit. Who is correct?
   a. A only
   b. B only
   c. Both A and B
   d. Neither A nor B

5. Technician A says a 10% duty cycle indicates that the load device is turned on most of the time. Technician B says the pulse width is measured in degrees. Who is correct?
   a. A only
   b. B only
   c. Both A and B
   d. Neither A nor B

6. A vehicle is being tested for a draw against the battery with the ignition switch in the OFF position. The specifications state the draw should be between 10 and 30 milliamps. The voltmeter reads .251 volts. Technician A says this draw is within the specification range. Technician B says the draw is too high. Who is correct?
   a. A only
   b. B only
   c. Both A and B
   d. Neither A nor B

7. While discussing circuit testers: Technician A says a conventional 12-V test light may be used to diagnose automotive computer circuits. Technician B says a self-powered test light may be used to diagnose an air bag circuit. Who is correct?
   a. A only
   b. B only
   c. Both A and B
   d. Neither A nor B

8. Technician A says an analog voltmeter that is connected in parallel to the load device will indicate the voltage drop across the device. Technician B says that an ohmmeter reading of 0.003 when connected in parallel to the coil of an A/C compressor indicates the coil is shorted. Who is correct?
   a. A only
   b. B only
   c. Both A and B
   d. Neither A nor B

9. While discussing oscilloscopes: Technician A says the upward voltage traces on an oscilloscope screen indicate a specific length of time. Technician B says the cathode ray tube (CRT) in an oscilloscope is like a very fast reacting voltmeter. Who is correct?
   a. A only
   b. B only
   c. Both A and B
   d. Neither A nor B

10. Technician A says voltage drop testing with a voltmeter is used to determine if there is excessive resistance in the circuit. Technician B says voltage drop testing will determine the wattage rating of the load device. Who is correct?
   a. A only
   b. B only
   c. Both A and B
   d. Neither A nor B

4 Basic Electrical Troubleshooting and Service

Objective

Upon completion and review of this chapter, you should be able to:

❑ Describe how different electrical problems cause changes in an electrical circuit.

❑ Diagnose and repair circuit protection devices.

❑ Test switches with a variety of test instruments.

❑ Test relays and relay circuits for proper operation.

❑ Identify and test fixed and variable resistors with a lab scope, voltmeter, or ohmmeter.

❑ Diagnose diodes for opens, shorts, and other defects.

❑ Locate and repair opens in a circuit.

❑ Locate and repair shorts in a circuit.

❑ Locate and repair the cause of unwanted high resistance in a circuit.

Introduction

Troubleshooting electrical problems involves the same tools and methods, regardless of which circuit has the problem. All electrical circuits must have voltage, current, and resistance. Testing for the presence of these, measuring them, and comparing your measurements to specifications is the key to effective diagnosis. To do this you must have a solid understanding of these basic electrical properties.

Troubleshooting electrical problems involves using meters, test lights, and jumper wires to determine if any part of the circuit is open or shorted, or if there is unwanted resistance. To troubleshoot a problem, always begin by verifying the customer's complaint. Then operate the system and others, to get a complete understanding of the problem. Often there are other problems, which are not as evident or bothersome to the customer, that will provide helpful information for diagnostics.

Obtain the correct wiring diagram for the car and study the circuit that is affected. From the diagram you should be able to identify testing points and probable problem areas. Then test and use logic to identify the cause of the problem.

An ammeter and a voltmeter connected to a circuit at the different locations shown in Figure 4-1 should give readings as indicated when there are no problems in the circuit. An open exists whenever there is not a complete path for current flow. If there is an open anywhere in the circuit, the ammeter will read zero current. If the open is in the 1-ohm resistor, a voltmeter connected from C to ground will read zero. However, if the resistor is open and the voltmeter is connected to points B and C, the reading will be 12 volts. The reason is that the battery, ammeter, voltmeter, 2-ohm resistor, and 3-ohm resistor are all connected together to form a series circuit. Because of the open in the circuit, there is only current flow in the circuit through the meter, not the rest of the

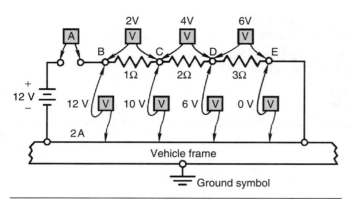

Figure 4-1 A basic circuit being tested with an ammeter and a voltmeter.

circuit. This current flow is very low because the meter has such high resistance. Therefore, the voltmeter will show a reading of 12 volts, indicating no voltage drop across the resistors.

To help you understand this concept, look at what happens if the 2-ohm resistor is open instead of the 1-ohm resistor. A voltmeter connected from point C to ground would indicate 12 volts. The 1-ohm resistor in series in the high resistance of the voltmeter would have little effect on the circuit. If an open should occur between point E and ground, a voltmeter connected from points B, C, D, or E to ground would read 12 volts. A voltmeter connected across any one of the resistors, from B to C, C to D, or D to E, would also read zero volts, because there will be no voltage drops if there is no current flow.

A short would be indicated by excessive current and/or abnormal voltage drops. These examples illustrate how a voltmeter and ammeter may be used to check for problems in a circuit. An ohmmeter also may be used to measure the values of each component and compare these measurements to specifications. If there is no continuity across a part, it is open. If there is more resistance than specified, there is high internal resistance. If there is less resistance than specified, the part is shorted.

WARNING *Any broken, frayed, or damaged insulation material requires replacement or repair to the wire. Exposed conductor material from damaged insulation can result in a safety hazard and damage to circuit components.*

CAUTION *Because the human body is a conductor of electricity, observe all safety rules when working with electricity.*

Testing Circuit Protection Devices

WARNING *Fuses and other protection devices normally do not wear out. They fail because something went wrong. Never replace a fuse or fusible link, or reset a circuit breaker, without finding out why they failed.*

A protection device is designed to "turn off" the system whenever excessive current or an overload occurs. There are three basic types of fuses in automotive use: cartridge, blade, and ceramic. The cartridge fuse is found on most older domestic cars and a few imports. To check this type of fuse, look for a break in the internal metal strip. Discoloration of the glass cover or glue bubbling around the metal caps is an indication of overheating. Late-model domestic vehicles, and many imports, use blade or spade fuses. To check the fuse, pull it from the fuse panel and look at the fuse element through the transparent plastic housing. Look for internal breaks and discoloration. The ceramic fuse is used on many European imports. To check this type of fuse, look for a break in the contact strip on the outside of the fuse. All types of fuses can be checked with an ohmmeter or test light. If the fuse is good, there will be continuity through it.

Fuses are rated by the current at which they are designed to blow. A three-letter code is used to indicate the type and size of fuses. Blade fuses have codes ATC or ATO. All glass SFE fuses have the same diameter, but the length varies with the current rating. Ceramic fuses are available in two sizes, code GBF (small) and the more common code GBC (large). The amperage rating is also embossed on the insulator. Codes such as AGA, AGW, and AGC indicate the length and diameter of the fuse. Fuse lengths in each of these series is the same, but the current rating can vary. The code and the current rating is usually stamped on the end cap.

SERVICE TIP *To calculate the correct fuse rating, use Watt's law: watts divided by volts equals amperes. For example, if you are installing a 55-watt pair of fog lights, divide 55 by the battery voltage (12 volts) to find out how much current the circuit has to carry. Since 55 ÷ 12 = 4.58, the current is approximately 5 amperes. To allow for current surges, the correct in-line fuse should be rated slightly higher than the normal current flow. In this case, an 8-ampere fuse would do the job.*

Fuse links are used in circuits where limiting the maximum current is not extremely critical. They are often installed in the positive battery lead to the ignition switch and other circuits that have power with the key off. Fuse link wire is covered with a special insulation that bubbles when it overheats, indicating that the link has melted. If the insulation appears good, pull lightly on the wire. If the link stretches, the wire has melted. Of course, when it is hard to determine if the fuse link is burned out, check for continuity through the link with a test light or ohmmeter.

To replace a fuse link, cut the protected wire where it is connected to the fuse link. Then, tightly crimp or solder a new fusible link of the same rating as the original link. Since the insulation on the manufacturer's fuse links is flameproof, never fabricate a fuse link from ordinary wire because the insulation may not be flameproof.

WARNING *Always disconnect the battery ground cable prior to servicing any fuse link.*

Many late-model vehicles use maxi-fuses instead of fusible links. Maxi-fuses look and operate like two-prong, blade or spade fuses, except they are much larger and can handle more current. (Typically, a maxi-fuse is four to five times larger).

Maxi-fuses are easier to inspect and replace than fuse links. To check a maxi-fuse, look at the fuse element through the transparent plastic housing. If there is a break in the element, the maxi-fuse has blown. To replace it, pull it from its fuse box or panel. Always replace a blown maxi-fuse with a new one having the same ampere rating.

Some circuits are protected by circuit breakers. Like fuses, they are rated in amperes. There are two types of circuit breakers: cycling or those that must be manually reset.

In the cycling type, the bimetal arm will begin to cool once the current to it is stopped. Once it returns to its original shape, the contacts are closed and power is restored. If the current is still too high, the cycle of breaking the circuit will be repeated.

Two types of non-cycling or resettable breakers are used. One is reset by removing the power from the circuit. There is a coil wrapped around a bimetal arm. When there is excessive current the contacts open and a small current passes through the coil. This current through the coil is not enough to operate a load, but it does heat up both the coil and the bimetal arm. This keeps the arm in the open position until power is removed. The other type is reset by depressing a reset

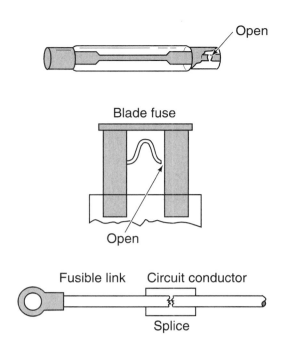

Figure 4-2 A fuse can have a hidden fault that cannot be seen by the technician.

button. A spring pushes the bimetal arm down and holds the contacts together. When an over current condition exists and the bimetal arm heats up, the bimetal arm bends enough to overcome the spring and the contacts snap open. The contacts stay open until the reset button is pushed, which snaps the contacts together again.

A visual inspection of a fuse or fusible link will not always determine if it is blown (Figure 4-2). To accurately test a circuit protection device, use an ohmmeter, voltmeter, or test light. With the fuse or circuit breaker removed from the vehicle, connect the ohmmeter's test leads across the protection device's terminals (Figure 4-3). On its lowest scale, the ohmmeter should read 0 to 1

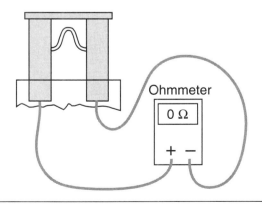

Figure 4-3 Using an ohmmeter to test the fuse. A good fuse will have zero resistance.

ohms. If it reads infinite, the protection device is open. Test a fusible link in the same way (Figure 4-4). Before connecting the ohmmeter across the fusible link, make sure there is no current flow through the circuit. To be safe, disconnect the negative cable of the battery.

To test a circuit protection device with a voltmeter, check for available voltage at both terminals of the unit (Figure 4-5). If the device is good, voltage will be present on both sides. A test light can be used in place of a voltmeter.

SERVICE TIP *Before using a test light it is good practice to check the tester's lamp. To do this, simply connect the test light across the battery. The light should come on.*

Measuring voltage drop across a fuse or other circuit protection device will tell you more about its condition than just if it is open. If a fuse, a fuse link, or circuit breaker is in good condition, a voltage drop of zero will be measured. If 12 volts is read, the fuse is open. Any reading between zero and 12 volts indicates some voltage drop. If there is voltage drop across the fuse, it has resistance and should be replaced. Make sure you check the fuse holder for resistance as well.

WARNING *Fuses are rated by amperage. Never install a larger rated fuse into a circuit than the one that was designed by the manufacturer. Doing so may damage or destroy the circuit. Also do not replace a fusible link with a resistor wire or vice versa.*

WARNING *Do not use an unfused jumper wire to bypass the protection device. Circuit damage may result.*

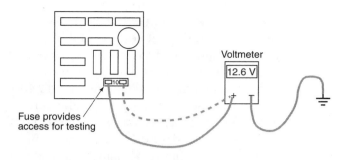

Figure 4-5 Voltmeter test of a circuit protection device. Battery voltage should be present on both sides.

Testing and Replacing Electrical Components

All electrical components can fail. Testing them is the best way of determining if they are good or bad. For the most part, the proper way for checking electrical components is determined by what the component is supposed to do. If we think about what something is supposed to do and how it does it, we can figure out how to test it. Often, removing the component and testing it on a bench is the best way to check it.

Switches

The easiest method of testing a normally open (NO) switch is to use a jumper wire to bypass the switch (Figure 4-6). If the circuit operates with the switch bypassed, the switch is defective. Voltage drop across switches should also be checked. Ideally when the switch is closed there should be no voltage drop. Any voltage drop indicates resistance, and the switch should be replaced.

A voltmeter or test light can be used to check for voltage on both sides of the switch. A faulty NO switch would have voltage present at the input side of the switch but not on the output side when in the ON position (Figure 4-7).

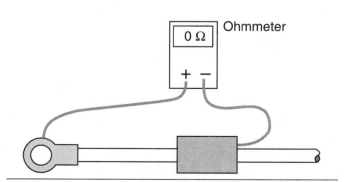

Figure 4-4 Testing a fusible link with an ohmmeter.

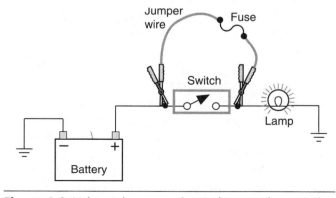

Figure 4-6 Using a jumper wire to bypass the switch.

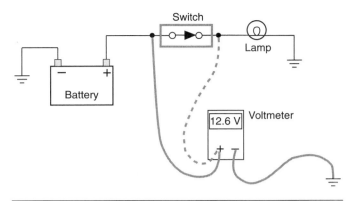

Figure 4-7 Using a voltmeter to test a NO switch.

WARNING *Use a jumper wire to bypass non-resistive portions of the circuit. Do not use the jumper wire to bypass the load component. The high current will damage the circuit.*

If the switch is removed it can be tested with an ohmmeter. With the switch contacts opened there should be no continuity between the terminals (Figure 4-8). When the contacts are closed, there should be zero resistance through the switch contacts. On complex ganged-type switches, the technician should consult the service manual for a continuity diagram (Figure 4-9). If there is no continuity chart, use the wiring diagram to make your own chart.

Relays

The relay can be checked using a jumper wire, voltmeter, ohmmeter, or test light. If the terminals are easily accessible, the jumper wire and test light may be the fastest method. Check the wiring diagram for the relay

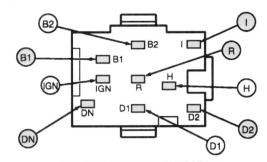

Connector End Views

HEADLAMP SWITCH CONNECTOR K21352-A

Pin Number	Circuit	Circuit Function
B1	38 (BK/O)	Power Supply to Battery
B2	195 (T/W)	Tail Lamp Switch Feed
I	19 (LB/R)	Instrument Panel Lamp Feed
IGN	—	Not Used
R	14 (BR)	Headlamp Switch to Tail Lamp and Side Marker Lamps
H	15 (R/Y)	Headlamp Dimmer Switch Feed
DN	—	Not Used
D1	54 (LG/Y)	Interior Lamp Switch Feed
D2	706 (GY)	Battery Saver Door Switch Feed

Figure 4-9 Headlight switch continuity diagram. (Reprinted with permission of Ford Motor Company)

being tested to determine if the control is through an insulated or ground switch. Use the illustration (Figure 4-10) as a guide to test a ground-switch-controlled relay. Follow these steps:

1. Use a voltmeter to check for available voltage to the battery side of the relay coil (terminal A). If voltage is not present at this point, the fault is in the circuit from the battery to the relay. Also, check for voltage at terminal D. If there is zero voltage here, there is an open in the circuit between the relay and the battery. If voltage is present at both terminals, continue testing.

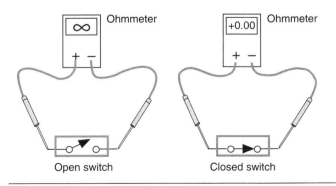

Figure 4-8 The continuity through a switch can be checked with an ohmmeter. With the switch closed there should be zero resistance. With the switch open, there should be infinite resistance.

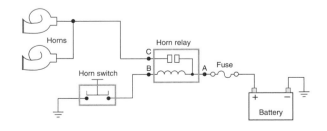

Figure 4-10 A relay with a ground control switch.

2. Probe for voltage at control terminal B. If voltage is not present at this terminal, then the fault is in the relay coil. If voltage is present, continue testing.

3. Use a jumper wire to connect terminal B to a good ground. If the horn sounds, the fault is in the control circuit from terminal B to the horn switch ground. If the horn does not sound, continue testing.

4. Connect the jumper wire from the battery positive to terminal C. If the horn did not sound, there is a fault in the circuit from the relay to the horn ground. If the horn sounded, the fault is in the relay.

If the relay is controlled by the computer, it is not recommended that a test light be used. The test light may draw more current than the circuit is designed to carry and damage the computer. Refer to the illustration (Figure 4-11) for procedures using a voltmeter to test a relay. Use a digital volt-ohm meter (DOM) set on the 20-V DC scale as follows:

1. Connect the negative voltmeter test lead to a good ground.

2. Connect the positive voltmeter test lead to the output wire (terminal B). Turn on the ignition switch. If no voltage is present at this terminal, go to step 3. If the voltmeter reads 12.0 volts or higher, turn off the control circuit. The voltmeter should then read 0 volts. If it does, then the relay is good. If the voltmeter still reads any voltage, the relay is not opening and needs to be replaced.

3. Connect the positive voltmeter test lead to the power input terminal (terminal A). The voltmeter should indicate at least 12.0 volts. If below this value, the circuit from the battery to the relay is faulty. If the voltage value is correct, continue testing.

4. Connect the positive voltmeter test lead to the control circuit terminal (terminal C). The voltage should read 10.5 volts or higher. If not, check the circuit from the battery to the relay (including the ignition switch). If the voltage is 10.5 volts or higher, continue testing.

5. Connect the positive voltmeter test lead to the relay ground terminal (terminal D). If more than 1 volt is indicated on the meter, there is a poor ground connection.

NOTE *It may be necessary to switch the DOM to the 2-V scale. If the reading is less than 1 volt, replace the relay.*

WARNING *It is not recommended that a test light be used to probe for power in a computer-controlled circuit. The increased draw of the test light may damage the system components.*

If the relay terminals are not accessible, remove the relay from its holding fixture and bench test it. Use an ohmmeter to test for continuity between the relay coil terminals (Figure 4-12). If the meter indicates an infinite reading, replace the relay. However, some relays have a resistor wired in parallel to the relay coil for spark suppression. If the coil is open, the ohmmeter would indicated the value of the resistor. For this reason always refer to the service manual for the proper specifications. If there is the correct amount of resistance in the coil, use a pair of jumper wires to energize the coil (Figure 4-13). Check for continuity through the relay contacts. If the meter indicates an infinite reading, the relay is defective. If there is continuity, the relay is good and the circuits will have to be checked.

Be sure to check your service manual for resistance specifications and compare the relay to them. It is easy to check for an open coil; however, a shorted coil will also prevent the relay from working. Low resistance across a coil would indicate that it is shorted. Too low of resis-

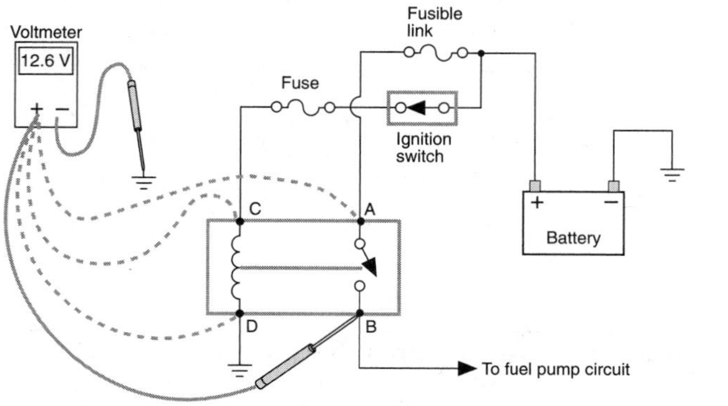

Figure 4-11 Testing relay operation with a voltmeter.

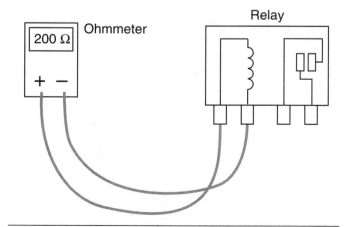

Figure 4-12 Testing the relay coil continuity.

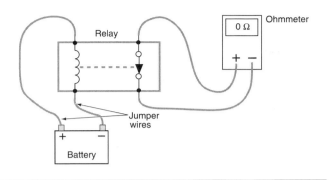

Figure 4-13 Bench testing the relay.

tance may also damage the transistors and/or driver circuits because of the excessive current that would result.

SERVICE TIP *The above procedures can be used to test the relay to determine the type of fault it has. However, the easiest way to test a relay is to substitute it with a known good relay of the same type. If the circuit operates with the substitute relay, the old relay is the faulty component.*

Testing Stepped Resistors

Stepped resistors are commonly used in heater blower motor circuits to control blower speed. The best method of testing a stepped resistor is to use an ohmmeter. To obtain accurate test results it is a good practice to remove the resistor from the circuit. Connect the ohmmeter leads to the two ends of the resistor (Figure 4-14). Compare the results with manufacturer's specifications. Be sure to place the ohmmeter on the correct scale to read the anticipated amount of resistance.

A stepped resistor can also be checked with a voltmeter. By measuring the voltage after each part of the resistor block and comparing the readings to specifications, you can tell if the resistor is good or not.

Testing Variable Resistors

As with the stepped resistor, the best method of testing a variable resistor is with an ohmmeter. However, it is possible to use an oscilloscope, voltmeter, or test light.

To test a rheostat, locate the input and output terminals and connect the test leads to them. Rotate the resistor knob slowly while observing the ohmmeter. The resistance value should remain within the specification limits and change in a smooth and constant manner. If the resistance values are out of limits or the resistance value jumps as the knob is turned, replace the rheostat.

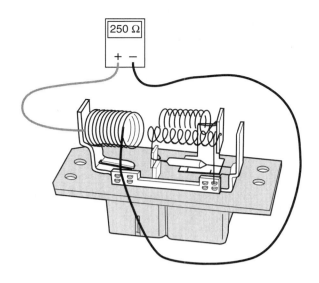

Figure 4-14 Ohmmeter testing of a stepped resistor. (Reprinted with permission of Ford Motor Company)

If an oscilloscope or a voltmeter is used, the readings should be smooth and consistent. Use Ohm's law to determine the specification limits. A test light should change in brightness as the knob is turned; the rheostat is defective if the light blinks at any point.

To test a potentiometer, connect the ohmmeter test leads to terminals A and C (Figure 4-15). Check the results with specifications. Next connect the ohmmeter test leads to terminals A and B (Figure 4-16). Check the resistance at the stop and observe the ohmmeter as the wiper is moved to the other stop. The resistance values should be within specification and smooth and constant.

A voltmeter can be used in the same manner. However, jumper wires will have to be used to gain access to the test points (Figure 4-17). Because potentiometers are primarily used in computer-controlled circuits, it is not recommended that a test light be used.

WARNING *Do not pierce the insulation to test the potentiometer. These circuits usually operate on 5 to 9 volts. Piercing the insulation may break some of the wire strands, resulting in a voltage drop that will give errant information to the computer. Even if the conductor is not broken, moisture can enter and cause corrosion.*

Testing Diodes

Regardless of the bias of the diode, it should allow current flow in one direction only. To test a diode, use an ohmmeter. Connect the meter's leads across the diode

Figure 4-15 Using an ohmmeter to test the continuity between terminals A and C.

(Figure 4-18). Observe the reading on the meter. Then reverse the meter's leads and observe the reading on the meter. The resistance in one direction should be very high or infinite and in the other direction, the resistance should be close to zero. If any other readings are observed, the diode is bad. A diode that has low resistance in both directions is shorted. A diode that has high resistance or an infinite reading in both directions is open.

You may run into problems when checking a diode with a high impedance digital ohmmeter. Since many diodes won't allow current flow through them unless the voltage is at least 0.6 volts, a digital meter may not be able to forward bias the diode. This will result in readings that indicate the diode is open, when in fact it may not be. Because of this problem, many multimeters are equipped with a diode testing feature. This feature allows for increased voltage at the test leads. Some meters will display the voltage required to forward bias the diode. If the diode is open, the meter will display "OL" or another reading to indicate infinity or out-of-range. Some meters during diode check will make a beeping noise when there is continuity.

Diodes may also be tested with a voltmeter. Using the same logic as when testing with an ohmmeter, test the voltage drop across the diode. The meter should read low

Figure 4-16 Testing continuity between terminal A and B while the wiper is being moved.

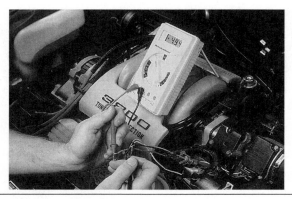

Figure 4-17 Use jumper wires connecting the wire connector to the sensor so that voltmeter readings can be obtained.

voltage in one direction and near source voltage in the other direction. Most automotive diodes will drop 500 to 650m volts.

Testing for Circuit Defects

Electrical circuits may develop an open, a short, a ground, or excessive resistance that will cause the circuit to operate improperly.

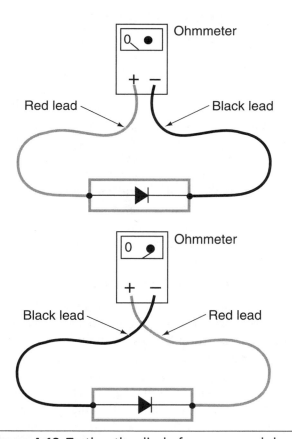

Figure 4-18 Testing the diode for opens and shorts.

Testing for Opens

It is possible to test for opens using a voltmeter, test light, self-powered test light, ohmmeter, or a jumper wire. The test equipment used will depend on the circuit being tested and the accessibility of the components.

The technician must determine the correct operation of the circuit before attempting to determine what is wrong. The illustration (Figure 4-19) shows the voltmeter readings that should be obtained in a properly operating parallel circuit.

The easiest method of testing a circuit is to start at the most accessible place and work from there. If the load component is easily accessible, test for voltage at the input to the load (Figure 4-20). Use the following procedure for locating the open:

1. Check for voltage at point A. If voltage is 10.5 volts or higher, check the ground side (point B). If voltage is less than 1 volt, the load component is faulty. If more than 1 volt is present, there is excessive resistance or an open in the ground circuit. If the voltage at point A was less than 10.5 volts, continue testing.

2. Work toward the battery. Test all connections for voltage. If voltage is present at a connection, then the open is between that connection and the previously tested location (Figure 4-21). Use a jumper wire to bypass that section to confirm the location of the open.

3. If battery voltage is present at point B, the open is in the ground circuit. Use a jumper wire to connect the ground circuit. Then retest the component.

In more complex circuits the open may have very different results (Figure 4-22). In a normally operating circuit the voltmeter readings would be as indicated in the illustration. If an open occurs in the ground side of the circuit, the circuit converts to a series circuit (Figure 4-23). This is a form of feedback that results in lamps coming on that are not intended to. Normal voltage is applied

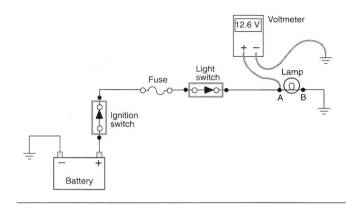

Figure 4-20 Testing for voltage to locate the open.

to lamp 3, but lamps 1, 2, and 4 are in series and will illuminate dimmer than normal. The voltmeter will read 12 volts at the locations illustrated in Figure 4-23. However, the voltmeter will not indicate 0 volts on the ground side of bulb 1.

Testing for Shorts

Locating a copper-to-copper short can be one of the most difficult tasks for a technician. If the short is within a component, the component will operate at less than optimum or not at all. An ohmmeter can be used to check the resistance of the component. If there is a short, then the amount of resistance will be lower than specified. If specifications for the component are not available, it may be necessary to replace the component with a known good unit. Do this only after it has been determined that the insulated and ground side circuits are in good condition.

If the short is between circuits, the result will be components operating when not intended (Figure 4-24). Visually check the wiring for signs of burned insulation and melted conductors that will indicate a short. Also check common connectors that are shared by the two

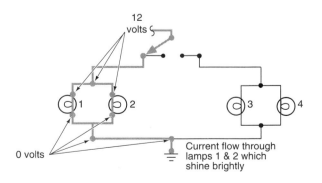

Figure 4-19 Voltmeter readings that would be expected in a properly operating parallel circuit. (Courtesy of Chrysler Corporation)

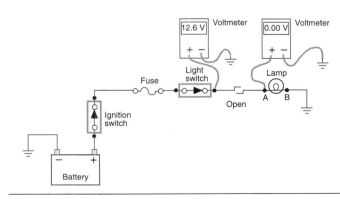

Figure 4-21 Locating the open.

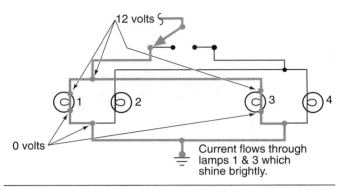

Figure 4-22 Properly operating complex parallel circuit. (Courtesy of Chrysler Corporation)

affected circuits. Corrosion can form between two terminals of the connector and result in the short.

If the visual inspection does not isolate the cause of the copper-to-copper short, remove one of the fuses for the affected circuits. (If the affected circuits share a common fuse, remove it.) Install a buzzer that has been fitted with terminals across the fuse holder terminals (Figure 4-25). Activate the circuit that the buzzer is connected to. In Figure 4-24, if the buzzer is connected to fuse B, then switch 1 would be turned on. Disconnect the loads that are supposed to be activated by this switch (lamp 1). Disconnect the wire connectors in the circuit from the load back to the switch. If the buzzer stops when a connector is disconnected, the short is in that portion of the circuit.

Testing for a Short to Ground

A fuse that blows as soon as it is installed indicates a short to ground. If the circuit is unfused, the insulation and conductor will melt. Not all shorts will blow the fuse, however. If the short to ground is on the ground side of the load component but before a grounding switch, the component will not turn off (Figure 4-26). If the short to ground is after the load and grounding switch (if applicable), circuit operation will not be affected.

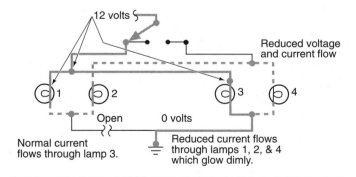

Figure 4-23 An open in the ground circuit can convert the circuit to a series circuit. (Courtesy of Chrysler Corporation)

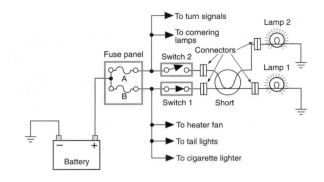

Figure 4-24 A copper-to-copper short between two circuits.

To confirm that the circuit has a ground before the load, remove the fuse and connect a test light in series across the fuse connections. If the lamp lights, the circuit has a short to ground.

WARNING *Use a circuit breaker that is rated between 25 and 30 amperes. The use of a circuit breaker rated too high will damage the circuit.*

It is difficult to test for shorts to ground with a test light or voltmeter because the fuse blows before any testing can be conducted. To prevent this, connect a cycling circuit breaker that is fitted with alligator clips across the fuse holder (Figure 4-27). The circuit breaker will continue to cycle open and closed, allowing the technician to test for voltage.

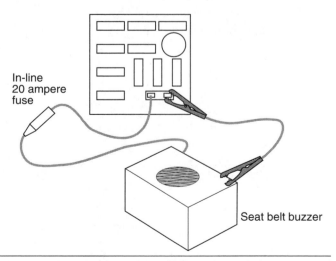

Figure 4-25 The buzzer will sound until the cause of the short is found.

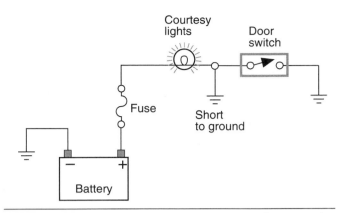

Figure 4-26 A ground in this location will cause the lamp to remain on.

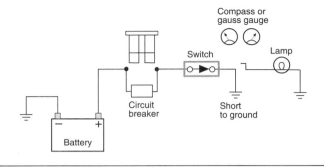

Figure 4-28 The gauge needle will fluctuate over the portion of the circuit that has current flowing through it. Once the short to ground has been passed, the needle will stop fluctuating.

Testing for shorts may be complicated if there are several circuits protected by a single fuse and if the ground is located in a section of wire that is not accessible. There are a couple of methods that can be used to locate the fault.

One method is to connect a test light, in series with a cycling circuit breaker, across the fuse holder. While observing the test light, disconnect individual circuits one at a time until the light goes out. The fault is in the circuit that was disconnected when the light went off.

A second method is to use a **Gauss gauge** or a compass to locate the short to ground. The gauge or compass works on the principle that a magnetic field is developed around a conductor that is carrying current. With a cycling circuit breaker bypassing the blown fuse, trace the path of the circuit with the gauge or compass. The needle will fluctuate as long as the gauge is over the conductor. The needle will stop fluctuating when the point of the short to ground is passed (Figure 4-30). This method will work even through the vehicle's trim. It will be necessary to follow all of the circuits protected by the fuse. Consult the wiring diagram for this information.

Testing for Voltage Drop

Voltage drop, when considered as a defect, defines the portion of applied voltage used up in other points of

the circuit other than by the load component. It is a resistance in the circuit that reduces the amount of electrical pressure available beyond the resistance. Excessive voltage drop may appear on either the insulated or ground return side of a circuit. To test for voltage drop, the circuit must be active (current flowing). The source voltage must be as specified before voltage drop readings can be valid. Whenever voltage drop is suspected, both sides of the circuit must be checked.

Excessive voltage drop caused by high resistance can be identified by dim or flickering lamps, inoperative load components, or slower than normal electrical motor speeds. Excessive resistance will not cause the fuse to blow.

To perform a voltage drop test on any circuit, the positive voltmeter lead must be connected to the most positive portion of the circuit. Refer back to Photo Sequence 2 (in Chapter 2) for the procedure to conduct a voltage drop test. Consult the service manual for the maximum amount of voltage drop allowed.

When testing the ground side of the circuit, the ground connection terminal of the load component is the most positive location and the battery negative post is the most negative (Figure 4-29). Usually more than 0.1 volt indicates excessive resistance in the ground circuit.

According to many manuals, the maximum allowable voltage drop for an entire circuit, except for the drop across the load, is 10% of the source voltage. Although 1.2 volts is the maximum acceptable amount, it is still too much. Many good technicians use 0.5 volt as the maximum allowable drop. However, there should be no more than 0.1 volt dropped across any one wire or connector. This is the most important specification to consider and remember.

It is possible to calculate voltage drop by testing for available voltage. Use Ohm's law to determine the correct amount of voltage drop that should be across a component. Test for available voltage on both sides of the

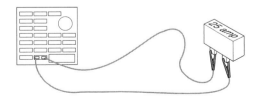

Figure 4-27 Use a circuit breaker to protect the circuit while checking for the ground.

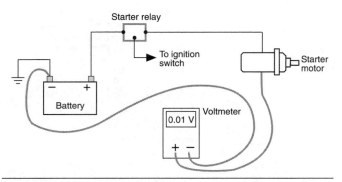

Figure 4-29 Testing for voltage drop on the ground side of the starter motor circuit. Notice the voltmeter connections.

load component (Figure 4-30). Subtract the available voltage readings to obtain the amount of voltage drop across the component.

Scan Tester Features

Scan testers display data and **diagnostic trouble codes (DTCs)** on computer systems and perform many other diagnostic functions. On many vehicles, scan testers have the capability to diagnose various computer systems such as engine, transmission, antilock brake system (ABS), suspension, and air bag. Scan testers vary depending on the manufacturer, but many scan testers have the following features:

1. Display window — displays data and messages to the technician. Messages are displayed from left to right. Most scan testers display at least four readings on the display at the same time.

2. Memory cartridge — plugs into the scan tester (Figure 4-31). These memory cartridges are designed for specific vehicles and electronic systems. For example, a different cartridge may be required for the transmission computer and the engine computer. Most scan tester manufacturers supply memory cartridges for domestic and imported vehicles.

3. Power cord — connected from the scan tester to the battery terminals or cigarette lighter socket.

4. Adapter cord — plugs into the scan tester and connects to the data link connector (DLC) on the vehicle (Figure 4-32). A special adapter cord is supplied with the tester for the diagnostic connector on each make of vehicle.

5. Serial interface — optional devices, such as a printer, terminal, or personal computer, may be connected to this terminal.

6. Keypad — allows the technician to enter data and reply to tester messages.

Typical keys on a scan tester are (Figure 4-33):

1. Numbered keys, digits 0 through 9

2. Horizontal or vertical arrow keys, allow the technician to move back and forward through test modes and menus

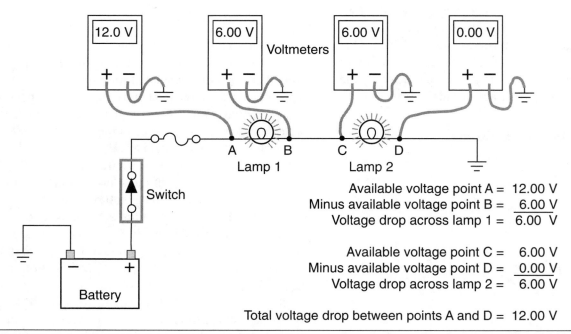

Available voltage point A = 12.00 V
Minus available voltage point B = 6.00 V
Voltage drop across lamp 1 = 6.00 V

Available voltage point C = 6.00 V
Minus available voltage point D = 0.00 V
Voltage drop across lamp 2 = 6.00 V

Total voltage drop between points A and D = 12.00 V

Figure 4-30 Using available voltage to calculate voltage drop over a component. This method is used if the circuit is too long to test with standard test leads.

Figure 4-31 Scan tester memory cartridge. (Courtesy of OTC-SPX Corp. Aftermarket Tool & Equipment Group)

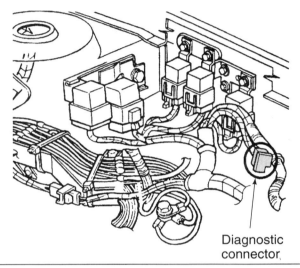

Diagnostic connector

Figure 4-32 The scan tester connects to the diagnostic link connector (DLC). (Courtesy of Chrysler Corporation)

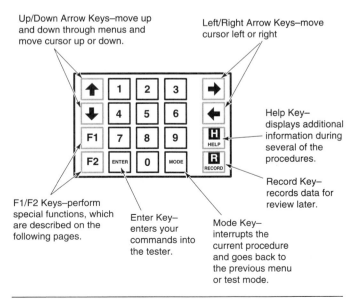

Numbered Keys–digits 0 through 9.

Up/Down Arrow Keys–move up and down through menus and move cursor up or down.

Left/Right Arrow Keys–move cursor left or right

Help Key–displays additional information during several of the procedures.

Record Key–records data for review later.

F1/F2 Keys–perform special functions, which are described on the following pages.

Enter Key–enters your commands into the tester.

Mode Key–interrupts the current procedure and goes back to the previous menu or test mode.

Figure 4-33 Scan tester features. (Courtesy of OTC-SPX Corp. Aftermarket Tool & Equipment Group)

3. ENTER keys, enter information into the tester

4. EXIT MODE keys, allow the technician to interrupt the current procedure and go back to the previous modes.

5. F1 and F2 keys, allow the technician to perform special functions described in the scan tester manufacturer's manuals

6. HELP key, allows the technician to obtain additional diagnostic information from the scan tester software

7. YES AND NO keys, allow the technician to select or reject specific procedures

8. RECORD key, allows the technician to record data in the scan tester memory for future reference.

WARNING *The tester manufacturer's and vehicle manufacturer's recommended scan tester diagnostic procedures must be followed while diagnosing computer systems. Improper test procedures may result in scan tester damage and computer system damage.*

CAUTION *Always keep scan tester leads away from rotating parts such as belts and fan blades. Personal injury or property damage may result if scan tester leads become tangled in rotating parts.*

Scan testers are now have OBD II capabilities. OBD II is discussed in a later chapter. Some previous scan testers may require a special adapter and the proper cables to provide OBD II capabilities. The advantages offered by the use of a scan tool include:

1. A scan tester provides quick access to data from various onboard computers. Some vehicles have several diagnostic link connectors (DLCs) to which the scan tester must be connected to access data from a specific computer. For example, some vehicles have separate DLCs to access the powertrain control module (PCM) and antilock brake system (ABS) computer data. Many vehicles that have onboard diagnostic (OBD II) systems have a central DLC and data links from the various onboard computers to this DLC. Accessing this computer data greatly reduces diagnostic time.

2. A wide variety of modules are available for many scan testers. These modules allow the same scan tester to display data from many vehicles including imported vehicles. Some scan tester modules access service bulletin information related to engine and transmission problems. This information is available in a book published by the scan tester manufacturer.

3. The vehicle may be driven on a road test with the scan tester connected to the DLC. This allows the technician to observe computer data during various operating conditions when a specific problem may occur. Most scan testers have a snap shot capability which freezes computer data into the scan tester memory for a specific period of time. This data may be played back after the technician returns the car to the shop.

4. Most scan testers may be connected to a printer and a copy of the scan tester data may be printed. This allows improved communication between the customer, service writer, and the technician.

5. A scan tester may be connected to a personal computer (PC). This connection allows data to be transferred from the scan tester to the PC. This data may be saved and recalled at a future time. With a computer modem, this information may be transferred to an off-site diagnostic center for analysis.

The disadvantages of a scan tool include:

1. Some vehicles do not have data wires from the computer to the DLC. On these applications, the scan tester cannot display computer data. Some import vehicle manufacturers provide a computer data stream which their dealers can access, but this data information is not made available to the aftermarket.

2. The scan tester only displays computer data; therefore, it is not very useful in diagnosing certain problems. For example, the scan tester does not display fuel pressure.

3. The scan tester displays the data received by the computer. This data indicates a problem in a specific area. For example, the data from an engine coolant temperature (ECT) sensor may be incorrect. The technician still has to perform diagnostic tests with a volt-ohmmeter to determine if the defect is in the ECT sensor, connecting wires to the PCM, or the PCM.

4. When displaying output data in some cases, the scan tester only displays the command issued by the computer to the output actuator. Defective wires or actuators may still cause the actuator to be inoperative.

Digital Storage Oscilloscope (DSO)

The greatest advantage of a DSO is the speed at which it samples electrical signals. Mechanical switching speed of switches and relays is measured in thousands of a second or milliseconds. Electrical/electronic switching speed is measured in millionths of a second or microseconds. Radio frequency interference (RFI) is measured in billionths of a second. The DSO operates at 25 million

samples per second. Another method of expressing this operating speed is to say the DSO is capable of sampling a signal in 40 billionths of a second. DSO sampling speed is at least 47,000 times faster than other automotive testers such as other engine analyzers.

This sampling speed allows the DSO to provide an extremely accurate, expanded display of input sensors and output actuators compared to multimeters or other analog scopes. This increased sampling speed allows the DSO to display glitches or momentary defects in input sensors and output actuators. The extremely fast sampling speed of the DSO allows this scope to display a graph of input sensor and output actuator operation. Some DSOs have the capability to display two voltage traces across the screen (Figure 4-34). Other DSOs, such as the Simu-Tech, display six voltage traces simultaneously.

DSO Screen

WARNING *While diagnosing computer systems, always place test equipment such as DSOs or scan testers in a secure position where they will not fall on the floor or into rotating components. Severe meter damage may occur if the DSO or scan tester is dropped.*

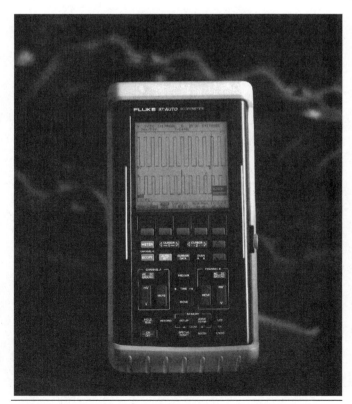

Figure 4-34 Digital storage oscilloscope. (Reproduced with permission of Fluke Corporation)

On the DSO screen, voltage is displayed vertically. Voltage change is shown as a vertical movement. Vertical grids on the screen provide a voltage measurement, and the voltage level between the grids is adjustable (Figure 4-35). The technician must know the voltage in the circuit being tested and then select the voltage per division on the DSO that provides the most detail with the signal remaining on the screen.

Horizontal movement on the screen represents time. The milliseconds per division on the horizontal grid are adjustable with the time button on the DSO. Each time a DSO samples a voltage signal, it displays a dot on the screen. The DSO then connects the dots to provide a waveform. When a faster signal is being read, a shorter time base should be selected on the DSO.

If the time base selected is too long and the voltage too high, the waveform is too small to read. Conversely, if the time base is too short and the voltage scale too low, the waveform is too large for the screen. The technician must select the proper time base for the voltage signal being measured so the waveform is displayed on the screen.

Peak, Average, and Root Mean Square Related to an AC Voltage Waveform

The term **peak** represents the highest point in one cycle of an AC voltage waveform. When both the highest and lowest peaks are considered in an AC voltage waveform, the term **peak-to-peak voltage** is the total voltage measured between these peaks. For example, an AC voltage waveform with a 60-V peak would have a 120-V peak-to-peak. The average voltage in an AC voltage waveform is calculated by multiplying 0.637 × peak voltage. The average voltage on a 60-V peak would be 0.637 × 60 = 38.2 V.

In many cases, root mean square (RMS) is used to describe AC voltage. For example, if one cycle of an AC

voltage waveform from a 120-V household electrical outlet is divided into four parts at 90 degree intervals, the instantaneous voltage and current are recorded for each degree in a 90 degree interval and then averaged. The square root of the average may be calculated by multiplying 0.7071 × the peak voltage. The peak voltage for the average 120-V household outlet is about 170 V at 60 hertz (Hz). Therefore, 0.7071 × 170 = 120.207 RMS.

Selecting DSO Voltage and Time Base

To display a waveform from a 120-V household electrical outlet, round off the peak voltage of 170V to 200V. There are eight vertical voltage divisions on the DSO screen with four divisions above and below the centerline. Select 50V per division to display the high and low peaks on the waveform. Assuming the 0-V position in the waveform is positioned in the center of the screen, the 50-V per division selection provides 200V above and below the screen centerline to display the 170-V peak voltage above and below the centerline. If the volts per division setting is increased, the peaks appear shorter on the screen.

In a 60-Hz AC voltage, one cycle occurs in approximately 18 milliseconds (ms). To display one complete AC cycle requires about 20ms. The average DSO has 10 horizontal divisions. Since 20 ÷ 10 = 2ms per division, this time base selection displays one AC voltage waveform. If 4ms per division is selected, 2 AC voltage waveforms are displayed. Increasing the time base displays more AC voltage cycles on the screen. Conversely, decreasing the time base displays fewer AC voltage cycles on the screen, and the waveform appears expanded.

Each time a DSO takes a voltage sample, it displays a dot on the screen, and then connects these dots to display a waveform. When the ms time base is too low, the waveform is expanded horizontally and a reduced number of dots are used in the waveform display. This may result in an altered and incomplete waveform display. Ideally one to three cycles should be displayed on the screen for the best display.

When the DEFAULT button is pressed on some DSOs, a baseline volts per division and ms per division is automatically selected internally. If the volts per division and ms per division selected by the technician are incorrect for the voltage signal being tested, this DEFAULT mode baseline should provide settings to display a waveform on the screen. Then the volts per division and ms per division may be adjusted to provide the desired display.

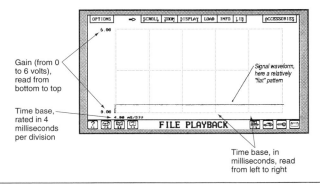

Figure 4-35 DSO vertical and horizontal screen grids. (Courtesy of EDGE Diagnostic Systems)

Trigger and Trigger Slope

The **trigger** selection tells the DSO when to begin displaying a waveform. Until the DSO has a trigger level, it doesn't know when to begin the waveform display. When testing an input sensor that operates in a 0-V to 5-V range, select a trigger level of one-half this range.

Trigger slope informs the DSO whether the voltage signal is moving upward or downward when it crosses the trigger level. When a negative trigger slope is selected, the voltage signal is moving downward when it crosses the trigger level. Selecting a positive trigger level results in an upward voltage signal trace when it crosses the trigger level. A marker on the left side of the screen indicates the 0-V or ground voltage position. This marker may be moved with the DSO controls (Figure 4-36). A second marker at the top of the screen indicates the trigger location. Since the control buttons on DSOs vary, the technician must spend some time to become familiar with a particular DSO.

Types of Voltage Signals

An **analog signal** is a varying voltage within a specific range over a period of time (Figure 4-37). A throttle position sensor (TPS) produces an analog voltage signal each time the throttle is opened (Figure 4-39).

A **digital signal** is either on or off. It may be described as one that is always high or low. The leading edge of a digital signal represents an increasing voltage and while the trailing edge of the signal represents a decreasing voltage. If the component is turned on by insulated side switching, the line across the top of the leading and trailing edge signals represents the length of component on time, which is called pulse width (Figure 4-39).

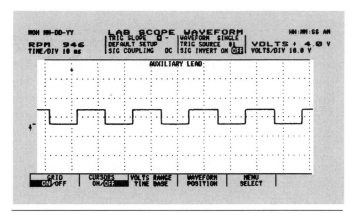

Figure 4-36 Ground level reference marker on the left side of the screen. (Courtesy of Snap-on Tools Company, Copyright Owner)

The computer measures the distance between the leading edge of a digital signal and leading edge of the next signal to determine the frequency of the waveform. The distance between the leading edge of one digital signal and the leading edge of the next digital signal is referred to as one cycle (Figure 4-40). The computer counts the number of cycles over a period of time to establish the frequency. For example, if 92 cycles are occurring per second, the frequency is 92 hertz (Hz).

The relationship between the on time and off time in a digital signal is called duty cycle. For example, if the component on time and off time are equal, the component has a 50% duty cycle. When the component has a 90% duty cycle, the component is on for 90% of the time in a cycle and off for 10% of the time in a cycle. The computer controls some outputs, such as a carburetor mixture control solenoid, by varying the pulse width while the frequency remains constant. This type of com-

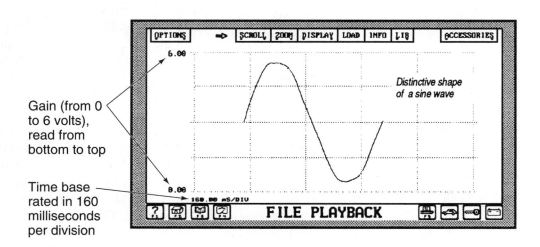

Gain (from 0 to 6 volts), read from bottom to top

Time base rated in 160 milliseconds per division

Figure 4-37 Analog voltage signal. (Courtesy of EDGE Diagnostic Systems)

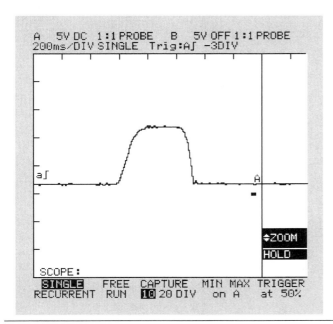

Figure 4-38 TPS analog voltage signal. (Reprinted with permission of Fluke Corporation)

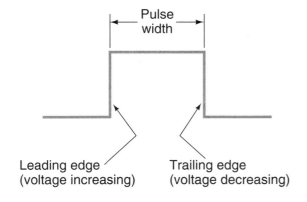

Figure 4-39 The component on time in a cycle is called pulse width. (Courtesy of EDGE Diagnostic Systems)

puter control is referred to as **pulse width modulation (PWM)**. Other outputs, such as fuel injectors, are controlled by varying the frequency and the pulse width.

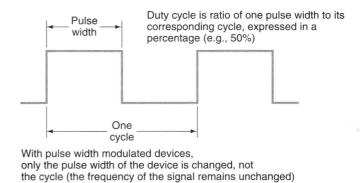

With pulse width modulated devices, only the pulse width of the device is changed, not the cycle (the frequency of the signal remains unchanged)

Figure 4-40 The distance between leading edges is one cycle. (Courtesy of EDGE Diagnostic Systems)

User-Friendly DSOs

DSOs with simplified, user-friendly controls have recently been introduced to the automotive service industry. In these DSOs the auto-range function automatically selects the proper voltage and time base for the signal being received. The technician may use the DSO controls to turn off the auto-range function, and manually select the voltage range and time base.

When the menu key is pressed, various menus are displayed and the vertical arrow keys allow the technician to scroll through the menu to select a specific test. Digital readings are displayed on the screen with most waveforms. For example, minimum, average, and maximum millivolt (mV) readings are provided with an O_2 sensor waveform. The DSO automatically adjusts for zirconia or titania O_2 sensors. This DSO has multimeter and ignition waveform capabilities.

Some user-friendly DSOs, such as the OTC Vision, have a removable application module to help prevent scope obsolescence. A software program card plugs into the bottom of the DSO. This DSO also sets the voltage range and time base automatically for the signal being tested. Four voltage waveforms or six multimeter functions may be displayed on the DSO screen.

Summary

❑ Troubleshooting electrical problems involves using meters, test lights, and jumper wires to determine if any part of the circuit is open or shorted, or if there is unwanted resistance.

❑ To check a cartridge type fuse, look for a break in the internal metal strip.

❑ Fuse link wire is covered with a special insulation that bubbles when it overheats, indicating that the link has melted.

❑ A visual inspection of a fuse or fusible link will not always determine if it is blown. To accurately test a circuit protection device, use an ohmmeter, voltmeter, or test light.

❑ The easiest method of testing a normally open (NO) switch is to use a jumper wire to bypass the switch.

❑ The best method of testing a stepped resistor is to use an ohmmeter. A stepped resistor can also be checked with a voltmeter.

❑ The best method of testing a variable resistor is with an ohmmeter. However, it is possible to use an oscilloscope, voltmeter, or test light.

❑ To test a diode, use an ohmmeter. Connect the meter's leads across the diode. The resistance in one direction should be very high or infinite and in the other direction, the resistance should be close to zero. Many multimeters are equipped with a diode testing feature.

❑ It is possible to test for opens using a voltmeter, test light, self-powered test light, ohmmeter, or a jumper wire. The test equipment used will depend on the circuit being tested and the accessibility of the components.

❑ A fuse that blows as soon as it is installed indicates a short to ground.

❑ Not all shorts will blow the fuse. If the short to ground is on the ground side of the load component but before a grounding switch, the component will not turn off. If the short to ground is after the load and grounding switch (if applicable), circuit operation will not be affected.

❑ Voltage drop, when considered as a defect, defines the portion of applied voltage used up in other points of the circuit other than by the load component. It is a resistance in the circuit that reduces the amount of electrical pressure available beyond the resistance. Excessive voltage drop may appear on either the insulated or ground return side of a circuit.

❑ To perform a voltage drop test on any circuit, the positive voltmeter lead must be connected to the most positive portion of the circuit.

❑ Scan testers display data and diagnostic trouble codes (DTCs) on computer systems and perform many other diagnostic functions.

❑ The greatest advantage of a DSO is the speed at which it samples electrical signals. High sampling speed allows the DSO to provide an extremely accurate, expanded display of input sensors and output actuators compared to multimeters or other analog scopes.

❑ An analog signal is a varying voltage within a specific range over a period of time.

❑ A digital signal is either on or off. It may be described as one that is always high or low.

❑ The computer measures the distance between the leading edge of a digital signal and leading edge of the next signal to determine the frequency of the waveform.

❑ The relationship between the on time and off time in a digital signal is called duty cycle.

❑ The computer controls some outputs by varying the pulse width while the frequency remains constant. This type of computer control is referred to as pulse width modulation (PWM).

Terms To Know

Analog signal	Gauss gauge	Pulse width modulation (PWM)
Diagnostic trouble codes (DTCs)	Peak	Trigger
Digital signal	Peak-to-peak voltage	Trigger slope

ASE Style Review Questions

1. Circuit defects are being discussed. Technician A says that an open can only be on the ground side of the circuit. Technician B says excessive circuit resistance can be the result of a poor electrical connection. Who is correct?
   a. A only
   b. B only
   c. Both A and B
   d. Neither A nor B

2. Testing the fuse is being discussed. Technician A says that sometimes a visual inspection of a fuse or fusible link does reveal an open. Technician B says to use a jumper wire to bypass the fuse to test the circuit. Who is correct?
   a. A only
   b. B only
   c. Both A and B
   d. Neither A nor B

3. Testing of a switch is being discussed. Technician A says that a switch can be tested with a voltmeter. Technician B says to use an ohmmeter to test a switch. Who is correct?
   a. A only
   b. B only
   c. Both A and B
   d. Neither A nor B

4. Technician A says that a relay can only be tested while it is connected to the circuit. Technician B says that a stepped resistor should be tested disconnected from the circuit. Who is correct?
   a. A only
   b. B only
   c. Both A and B
   d. Neither A nor B

5. Technician A says it is best to use a test light to test the operation of a potentiometer. Technician B says it is best to use a test light to test a rheostat. Who is correct?
   a. A only
   b. B only
   c. Both A and B
   d. Neither A nor B

6. Voltage drop testing is being discussed. Technician A says that it is possible to calculate voltage drop by testing for available voltage on both sides of the component. Technician B says that excessive voltage drop can be on either side of the circuit. Who is correct?
   a. A only
   b. B only
   c. Both A and B
   d. Neither A nor B

7. The results of copper-to-copper shorts is being discussed. Technician A says if there is a short in an electrical motor then the amount of resistance will be higher than specified. Technician B says a short between circuits can result in both circuits operating by closing one switch. Who is correct?
   a. A only
   b. B only
   c. Both A and B
   d. Neither A nor B

8. The fuse for an A/C blower motor blows shortly after it is installed. Technician A says this can be caused by excessive resistance on the ground circuit. Technician B says the blower motor may be binding internally. Who is correct?
   a. A only
   b. B only
   c. Both A and B
   d. Neither A nor B

9. Technician A says if a test light is connected across the fuse holder and it lights, there is a short to ground in the circuit. Technician B says the magnetic field around a conductor is not strong enough to affect a compass through the vehicle trim. Who is correct?
   a. A only
   b. B only
   c. Both A and B
   d. Neither A nor B

10. Testing of a potentiometer is being discussed. Technician A says a voltmeter can be used to test the potentiometer. Technician B says that the wires can be pierced to test for voltage. Who is correct?
    a. A only
    b. B only
    c. Both A and B
    d. Neither A nor B

5 Wiring and Circuit Diagrams

Objective

Upon completion and review of this chapter, you should be able to:

❏ Demonstrate the proper troubleshooting process.

❏ Explain when single-stranded or multi-stranded wire should be used.

❏ Explain the use of resistive wires in a circuit.

❏ Describe the construction of spark plug wires.

❏ Explain how wire size is determined by the American Wire Gauge (AWG) and metric methods.

❏ Describe how to determine the correct wire gauge to be used in a circuit.

❏ Explain how temperature affects resistance and wire size selection.

❏ Explain the purpose and use of printed circuits.

❏ Explain why wiring harnesses are used and how they are constructed.

❏ Explain the purpose of wiring diagrams.

❏ Identify the common electrical symbols that are used.

❏ Explain the purpose of the component locator.

Introduction

Today's vehicles have a vast amount of electrical wiring that, if laid end to end, can stretch for half a mile. Today's technician must be proficient at reading wiring diagrams to be able to sort through this great maze of wires. Trying to locate the cause of an electrical problem can be quite difficult if you do not have a good understanding of wiring systems and diagrams.

In this chapter you will learn how wiring harnesses are made, how to read the wiring diagram, how to interpret the symbols used, and how terminals are used. This will reduce the amount of confusion you may experience when repairing an electrical circuit. It is also important to understand how to determine the correct type and size of wire to carry the anticipated amount of current. It is possible to cause an electrical problem by simply using the wrong gauge size of wire. A technician must understand the three factors that cause resistance in a wire: length, diameter, and temperature, to perform repairs correctly.

Electrical Troubleshooting

Many electrical repairs will involve the replacement or repairing of a damaged conductor. To locate the problem area, today's technician must be capable of reading and understanding electrical schematics. Once it is determined that the battery is operating correctly, the schematic should always be the starting point in troubleshooting an electrical system. By using the schematic, the technician is able to understand how the circuit should work. This is essential before attempting to figure out why it does not work. Troubleshooting is the diagnostic procedure of locating and identifying the cause of the fault. It is a step-by-step process of elimination by use of cause and effect.

The process of troubleshooting an electrical problem is as follows:

1. Confirm the problem. Perform a check of the system to gain an understanding of what is wrong. If the faulty system is monitored by the onboard computer, enter diagnostics to retrieve any trouble codes.

2. Study the electrical schematic. This will indicate any shared circuits. Trying to operate the shared circuits will help direct the technician to the problem area. If the shared circuits operate correctly, the problem is isolated to the wiring or components of the problem system. If the shared circuits do not operate, the problem is usually in the power or ground circuit.

3. Locate and repair the fault. By narrowing down the possible causes and taking measurements as required, the fault is located. Before replacing any components, check the ground and power leads. If these are good, then the component is bad.

4. Test the repair. Repeat a check of the system to confirm that it is operating properly.

Automotive Wiring

Primary wiring is the term used for conductors that carry low voltage. The insulation of primary wires is usually thin. **Secondary wiring** refers to wires used to carry high voltage, such as ignition spark plug wires. Secondary wires have extra thick insulation.

Most of the primary wiring conductors used in the automobile are made of several strands of copper wire wound together and covered with a polyvinyl chloride (PVC) insulation (Figure 5-1). Copper has low resistance and can be connected to easily by using crimping connectors or soldered connections. Other types of conductor materials used in automobiles include silver, gold, aluminum, and tin-plated brass.

WARNING *A solid copper wire may be used in low-voltage, low-current circuits where flexibility is not required. Do not use solid wire where high voltage, high current, or flexibility is required, unless solid wire was used by the manufacturer.*

Stranded wire is used because it is very flexible, and has less resistance than solid wire. Electrons tend to flow on the outside surface of conductors. Because there is

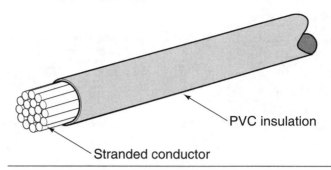

Figure 5-1 Stranded primary wire.

more surface area exposed in a stranded wire (each strand has its own surface), there is less resistance in the stranded wire than in the solid wire. The PVC insulation is used because it can withstand temperature extremes and corrosion. PVC insulation is also capable of withstanding battery acid, antifreeze, and gasoline. The insulation protects the wire from shorting to ground and from corrosion.

A **ballast resistor** is used to protect the ignition primary circuit from excessive voltage. It reduces the current flow through the coil's primary windings and provides a stable voltage to the coil. The resistance value of most ballast resistors is usually between 0.8 and 1.2 ohms.

Some automobiles use a **resistance wire** in the ignition system instead of a ballast resistor. The resistance wire is built with a certain amount of resistance per foot. It is most commonly used to limit the amount of current flow through the primary windings of the ignition coil. This wire is called the ballast resistor wire and is located between the ignition switch and the ignition coil (Figure 5-2) in the ignition "RUN" circuit.

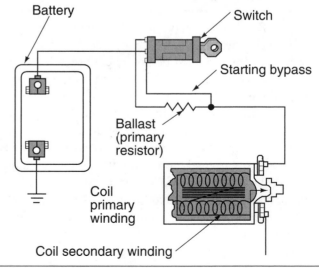

Figure 5-2 Ballast resistor used in some ignition primary wiring circuits.

Spark plug wires are also resistance wires. The resistance lowers the current flow through the wires. By keeping current flow low, the magnetic field created around the wires is kept to a minimum. The magnetic field needs to be controlled because it causes radio interference. The result of this interference is noise on the vehicle's radio and all nearby radios and televisions. The noise can interfere with emergency broadcasts and the radios of emergency vehicles. Because of this concern, all ignition systems are designed to minimize radio interference; most do so with resistance-type spark plug wires. Spark plug wires are targeted because they carry high voltage pulses. The lower current flow has no adverse effect on the firing of the spark plug. Spark plug wires are sometimes referred to as **television-radio-suppression (TVRS) cables**.

Most spark plug wire conductors are made of nylon, rayon, fiberglass, or aramid thread impregnated with carbon. This core is surrounded by rubber (Figure 5-3). The carbon-impregnated core provides sufficient resistance to reduce RFI, yet does not affect engine operation. As the spark plug wires wear because of age and temperature changes, the resistance in the wire will change. Most plug wires have a resistance value of 3,000Ω to 6,000Ω per foot. However, some have between 6,000Ω and 12,000Ω. The accepted value when testing is 10,000Ω per foot as a general specification.

Because the high voltage within the plug wires can create electromagnetic induction, proper wire routing is important to eliminate the possibility of **cross-fire**. Cross-fire is the electromagnetic induction spark that can be transmitted in another wire close to the wire carrying the current. To prevent cross-fire, the plug wires must be installed in the proper separator. Any two parallel wires next to each other in the firing order should be positioned as far away from each other as possible (Figure 5-4). When induction cross-fire occurs, no spark is jumped from one wire to the other. The spark is the result of induction from another field. Cross-fire induction is most common in two parallel wires that fire one after the other in the firing order.

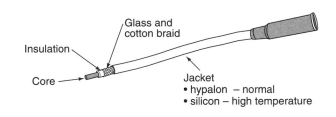

Figure 5-3 Typical spark plug wire.

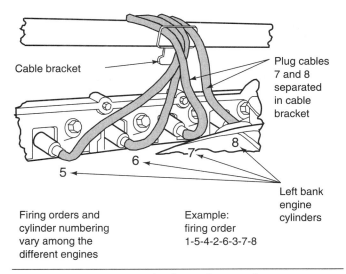

Figure 5-4 Proper spark plug wire routing to prevent cross-fire. (Reprinted with permission of Ford Motor Company)

Wire Sizes

An additional amount of consideration must be given for some margin of safety when selecting wire size. There are three major factors that determine the proper size of wire to be used:

1. The wire must have a large enough diameter, for the length required, to carry the necessary current for the load components in the circuit to operate properly.

2. The wire must be able to withstand the anticipated vibration.

3. The wire must be able to withstand the anticipated amount of heat exposure.

Wire size is based on the diameter of the conductor. The larger the diameter, the less the resistance. There are two common size standards used to designate wire size: **American wire gauge (AWG)** and metric.

The AWG standard assigns a number to the wire based on its diameter (gauge). The higher the number, the smaller the wire diameter. For example, 20-gauge wire is smaller in diameter than 10-gauge wire. Most electrical systems in the automobile use 14-, 16-, or 18-gauge wire. Most battery cables are 2-, 4-, or 6-gauge cable.

Both wire diameter and wire length affect resistance. Sixteen-gauge wire is capable of conducting 20 amperes for 10 feet with minimal voltage drop. However, if the current is to be carried for 15 feet, 14-gauge wire would be required. If 20 amperes were required to be carried for 20 feet, then 12-gauge wire would be required. The additional wire size is needed to prevent voltage drops in the wire. The illustration (Figure 5-5) lists the wire size required to carry a given amount of current for different lengths.

Total Approximate Circuit Amperes	Wire Gauge (for Length in Feet)								
12 V	3	5	7	10	15	20	25	30	40
1.0	18	18	18	18	18	18	18	18	18
1.5	18	18	18	18	18	18	18	18	18
2	18	18	18	18	18	18	18	18	18
3	18	18	18	18	18	18	18	18	18
4	18	18	18	18	18	18	18	16	16
5	18	18	18	18	18	18	18	16	16
6	18	18	18	18	18	18	16	16	16
7	18	18	18	18	18	18	16	16	14
8	18	18	18	18	18	16	16	16	14
10	18	18	18	18	16	16	16	14	12
11	18	18	18	18	16	16	14	14	12
12	18	18	18	18	16	16	14	14	12
15	18	18	18	18	14	14	12	12	12
18	18	18	16	16	14	14	12	12	10
20	18	18	16	16	14	12	10	10	10
22	18	18	16	16	12	12	10	10	10
24	18	18	16	16	12	12	10	10	10
30	18	16	16	14	10	10	10	10	10
40	18	16	14	12	10	10	8	8	6
50	16	14	12	12	10	10	8	8	6
100	12	12	10	10	6	6	4	4	4
150	10	10	8	8	4	4	2	2	2
200	10	8	8	6	4	4	2	2	1

Note: 18 AWG as indicated above this line could be 20 AWG electrically. 18 AWG is recommended for mechanical strength.

Figure 5-5 The distance the current must be carried is a factor in determining the correct wire gauge to use.

Another factor to wire resistance is temperature. An increase in temperature creates a similar increase in resistance. A wire may have a known resistance of 0.03 ohms per 10 feet at 70° F. When exposed to temperatures of 170° F, the resistance may increase to 0.04 ohms per 10 feet. Wires that are to be installed in areas that experience high temperatures, as in the engine compartment, must be of a size such that the increased resistance will not affect the operation of the load component. Also, the insulation of the wire must be capable of withstanding the high temperatures.

In the metric system, wire size is determined by the cross-sectional area of the wire. Metric wire size is expressed in square millimeters (mm²). In this system the smaller the number, the smaller the wire conductor. The approximate equivalent wire size of metric to AWG is shown (Figure 5-6).

Terminals and Connectors

To perform the function of connecting the wires from the voltage source to the load component reliably, terminal connections are used. Today's vehicles can have as

Metric Size (mm^2)	AWG (Gauge) Size	Ampere Capacity
0.5	20	4
0.8	18	6
1.0	16	8
2.0	14	15
3.0	12	20
5.0	10	30
8.0	8	40
13.0	6	50
19.0	4	60

Figure 5-6 Approximate AWG to metric equivalence.

many as 500 separate circuit connections. The terminals used to make these connections must be able to perform with very low voltage drop. A loose or corroded connection can cause an unwanted voltage drop that results in poor operation of the load component. For example, a connector used in a light circuit that has as little as 10% voltage drop (1.2V) may result in a 30% loss of lighting efficiency.

Terminals can be either crimped or soldered to the conductor. The terminal makes the electrical connection and it must be capable of withstanding the stress of normal vibration. The illustration (Figure 5-7) shows several different types of terminals used in the automotive electrical system. In addition, the following connectors are used on the automobile:

1. **Molded connector:** These connectors usually have one to six wires that are molded into a one-piece component (Figure 5-8).

2. **Multiple-wire hard-shell connector:** These connectors usually have a hard plastic shell that holds the connecting terminals of separate wires (Figure 5-9). The wire terminals can be removed from the shell to be repaired.

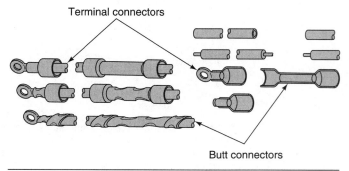

Terminal connectors

Butt connectors

Figure 5-7 Primary wire terminals used in automotive applications.

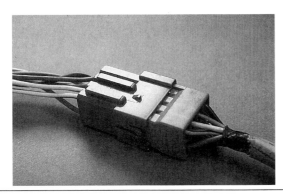

Figure 5-8 Multiple circuit wire molded connector.

3. **Bulkhead connectors:** These connectors are used when several wires must pass through the bulkhead (Figure 5-10).
4. **Weather-Pack Connectors:** These connectors have rubber seals on the terminal ends and on the covers of the connector half (Figure 5-11). These connectors are used on computer circuits to protect the circuit from corrosion, which may result in a voltage drop.
5. **Metri-Pack Connectors:** These are like the weather-pack connectors, but do not have the seal on the cover half (Figure 5-12).
6. Heat shrink covered butt connectors are recommended for air bag applications by some manufacturers. Other manufacturers allow NO repairs to the circuitry, while still others require silver-soldered connections.

To reduce the number of connectors in the electrical system, a **common connection** can be used. If there are several electrical components that are physically close to each other, a single common connection or splice eliminates using a separate connector for each wire.

Printed Circuits

Most instrument panels use **printed circuit** boards as circuit conductors. A printed circuit is made of a thin phenolic or fiberglass board that copper (or some other conductive material) has been deposited on. Portions of the conductive metal are then etched or eaten away by acid. The remaining strips of conductors provide the circuit path for the instrument panel lights, warning lights, indicator lights, and gauges of the instrument panel (Figure 5-13). The printed circuit board is attached to the back of the instrument panel housing. An edge connector joins the printed circuit board to the vehicle wiring harness.

Whenever it is necessary to perform repairs on or around the printed circuit board, it is important to follow these precautions:

1. When replacing light bulbs, be careful not to cut or tear the surface of the printed circuit board.

2. Do not touch the surface of the printed circuit with your fingers. The acid present in normal body oils can damage the surface.
3. If the printed circuit board needs to be cleaned, use a commercial cleaning solution designed for electrical use. If this solution is not available, it is possible to clean the board by lightly rubbing the surface with an eraser.

Wiring Harness

Most manufacturers use **wiring harnesses** to reduce the amount of loose wires hanging under the hood or dash of an automobile. The wiring harness is an assembled group of wires that branch out to the various electrical components. The wiring harness provides for a safe path for the wires of the vehicle's lighting, engine, and accessory components. The wiring harness is made by grouping insulated wires and wrapping them together. The wires are bundled into separate harness assemblies that are joined together by connector plugs. The multiple-pin connector plug may have more than 60 individual wire terminals.

There are several complex wiring harnesses in a vehicle, in addition to several simpler harnesses. The engine compartment harness and the under dash harness are examples of a complex harness (Figure 5-14). A complex harness serves many circuits. Lighting circuits usually use a more simple harness (Figure 5-15). The simple harness services only a few circuits. Some individual circuit wires may branch out of a complex harness to other areas of the vehicle.

Most wiring harnesses use a flexible tubing to provide for quick wire installation (Figure 5-16). The tubing has a seam that can be opened to accommodate the installation or removal of wires from the harness. The seam will close once the wires are installed, and will remain closed even if the tubing is bent.

Wiring Protective Devices

Often overlooked, but very important to the electrical system, are proper **wire protection devices** (Figure 5-19). These devices prevent damage to the wiring by maintaining proper wire routing and retention. Special clips, retainers, straps, and supplementary insulators provide additional protection to the conductor over what the insulation itself is capable of providing. Whenever the technician must remove one of these devices to perform a repair, it is important that the device be reinstalled to prevent additional electrical problems.

Whenever it is necessary to install additional electrical accessories, try to support the primary wire in at least

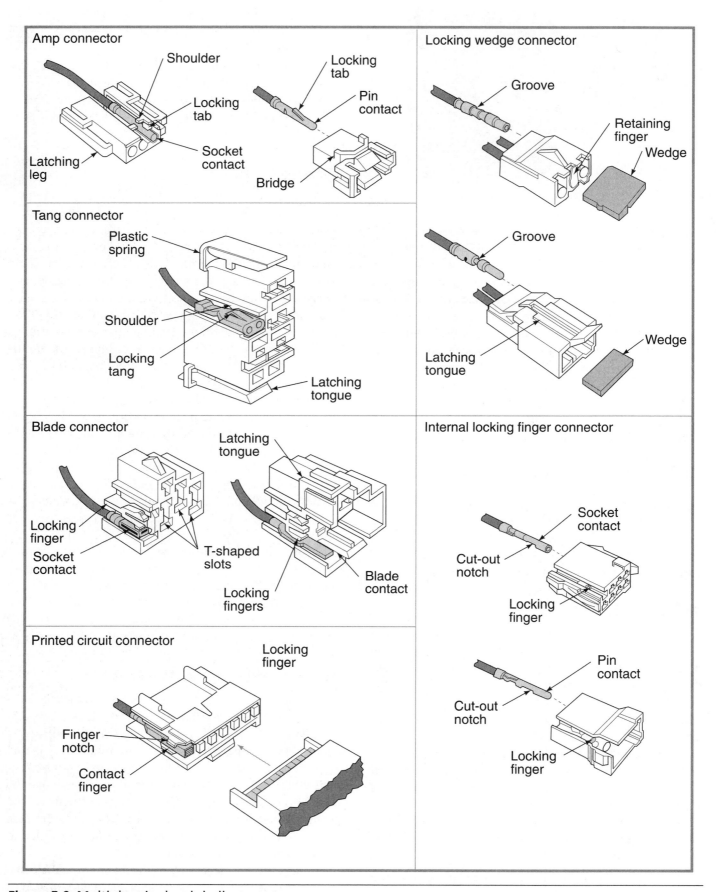

Figure 5-9 Multiple-wire hard shell connectors.

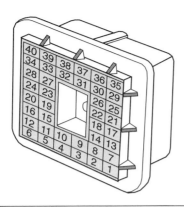

Figure 5-10 Bulkhead connector. (Courtesy of Chrysler Corporation)

Figure 5-12 Metri-pack connector.

Figure 5-11 Weather-pack connector is used to prevent connector corrosion.

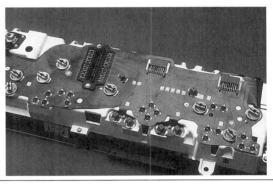

Figure 5-13 Printed circuits eliminate bulky wires behind the instrument panel.

1-foot intervals. If the wire must be routed through the frame or body, use rubber grommets to protect the wire.

WARNING *Do not use metal clamps to secure wires to the frame or body of the vehicle. The metal clamp may cut through the insulation and cause a short to ground. Use plastic clips in place of metal.*

Wiring Diagrams

One of the most important tools for diagnosing and repairing electrical problems is a **wiring diagram**. These diagrams identify the wires and connectors from each circuit on a vehicle. They also show where different circuits are interconnected, where they receive their power, where the ground is located, and the colors of the different wires. All of this information is critical to proper diagnosis of electrical problems. Some wiring diagrams also give additional information that helps you understand how a circuit operates and how to identify certain

components (Figure 5-18). Wiring diagrams do not explain how the circuit works; this is where your knowledge of electricity comes in handy.

A wiring diagram can show the wiring of the entire vehicle or a single circuit (Figure 5-19). These single circuit diagrams are also called block diagrams. In both types, the wire colors and the connectors are shown. Wiring diagrams of the entire vehicle tend to look more complex and threatening than block diagrams. However, once you simplify the diagram to only those wires, connectors, and components that belong to an individual circuit, they become less complex and more valuable.

Wiring diagrams show the wires, connections to switches and other components, and the type of connector used throughout the circuit. Total vehicle wiring diagrams are normally spread out over many pages of a service manual. Some are displayed on a single large sheet of paper that folds out of the manual. A system wiring diagram is actually a portion of the total vehicle diagram. The system and all related circuitry are shown on a single page. System diagrams are often easier to use than vehicle diagrams simply because there is less information to sort through.

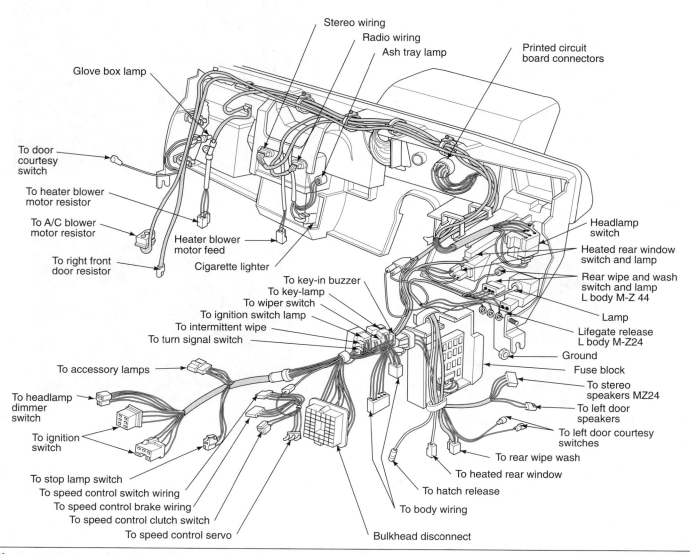

Stereo wiring
Radio wiring
Ash tray lamp
Printed circuit board connectors
Glove box lamp
To door courtesy switch
To heater blower motor resistor
To A/C blower motor resistor
Heater blower motor feed
To right front door resistor
Cigarette lighter
Headlamp switch
Heated rear window switch and lamp
Rear wipe and wash switch and lamp L body M-Z 44
To key-in buzzer
To key-lamp
To wiper switch
To ignition switch lamp
To intermittent wipe
To turn signal switch
Lamp
Lifegate release L body M-Z24
Ground
Fuse block
To accessory lamps
To headlamp dimmer switch
To ignition switch
To stereo speakers MZ24
To left door speakers
To left door courtesy switches
To rear wipe wash
To heated rear window
To stop lamp switch
To speed control switch wiring
To speed control brake wiring
To speed control clutch switch
To speed control servo
To hatch release
To body wiring
Bulkhead disconnect

Figure 5-14 Complex wiring harness. (Courtesy of Chrysler Corporation)

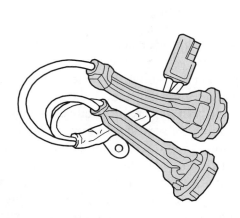

Figure 5-15 Simple wiring harness. (Courtesy of Chrysler Corporation)

Figure 5-16 Flexible tubing used to make wiring harnesses.

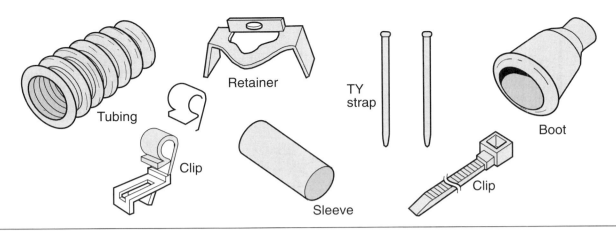

Figure 5-17 Typical wire protection devices. (Courtesy of Chrysler Corporation)

Remember that electrical circuits need a complete path in order to work. A wiring diagram shows the insulated side of the circuit and the point of ground. Also, when lines (or wires) cross on a wiring diagram, this does not mean they connect. If wires are connected, there will be a connector or a dot at the point where they cross. Most wiring diagrams do not show the location of the wires, connectors, or components in the vehicle. Some have location reference numbers displayed by the wires. After studying the wiring diagram you will know what you are looking for. Then you move to the car to find it.

In addition to entire vehicle and system specific wiring diagrams, there are other diagrams that may be used to diagnose electricity problems. An **electrical schematic** shows how the circuit is connected. It does not show the colors of the wires or their routing. Schematics are what have been used so far in this book. They display a working model of the circuit. These are especially handy when trying to understand how a circuit works. Schematics are typically used to show the internal circuitry of a component or to simplify a wiring diagram. One of the troubleshooting techniques used by good electrical technicians is to simplify a wiring diagram into a schematic.

Installation diagrams show where and how electrical components and wiring harnesses are installed in the vehicle. These are helpful when trying to locate where a particular wire or component may be in the car. These diagrams also may show how the component or wiring harness is attached to the vehicle (Figure 5-20).

Electrical Symbols

Most wiring diagrams do not show an actual drawing of the components. Rather they use **symbols** to represent the components. Often the symbol displays the basic operation of the component. Many different symbols have been used in wiring diagrams through the years. Figure 5-21 shows some of the commonly used symbols. Recently most manufacturers have begun to use some new style symbols (Figure 5-22). The reason for the change is due to most manufacturers going to electronic media instead of paper forms of service manuals. Some of the older symbols are not distinguishable when viewed on a monitor. You need to be familiar with all of the symbols; however, you don't need to memorize all of the variations. Wiring diagram manuals include a "legend" that helps you interpret the symbols.

A BIT OF HISTORY

The service manuals for early automobiles were hand drawn and labeled. They also had drawings of the actual components. As more and more electrical components were added to cars, this became impractical. Soon schematic symbols replaced the component drawings.

Color Codes and Circuit Numbering

Nearly all of the wires in an automobile are covered with colored insulation. These colors are used to identify wires and electrical circuits. The color of the wires is marked on a wiring diagram. Some wiring diagrams also include circuit numbers. These numbers, or letters and numbers, help identify a specific circuit. Both types of coding makes it easier to diagnose electrical problems. Unfortunately, not all manufacturers use the same method of wire identification.

Figure 5-23 shows common color codes and their abbreviations. Most wiring diagrams list the appropriate

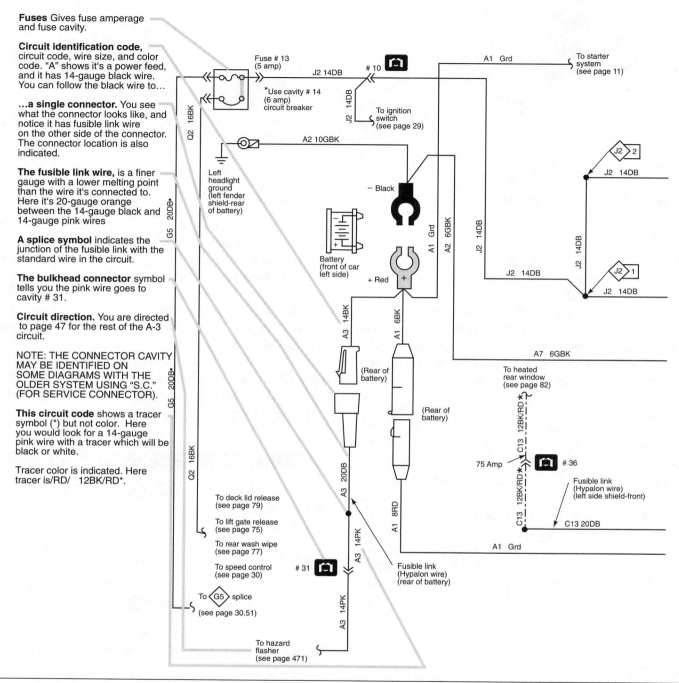

Fuses Gives fuse amperage and fuse cavity.

Circuit identification code, circuit code, wire size, and color code. "A" shows it's a power feed, and it has 14-gauge black wire. You can follow the black wire to…

…a single connector. You see what the connector looks like, and notice it has fusible link wire on the other side of the connector. The connector location is also indicated.

The fusible link wire, is a finer gauge with a lower melting point than the wire it's connected to. Here it's 20-gauge orange between the 14-gauge black and 14-gauge pink wires

A splice symbol indicates the junction of the fusible link with the standard wire in the circuit.

The bulkhead connector symbol tells you the pink wire goes to cavity # 31.

Circuit direction. You are directed to page 47 for the rest of the A-3 circuit.

NOTE: THE CONNECTOR CAVITY MAY BE IDENTIFIED ON SOME DIAGRAMS WITH THE OLDER SYSTEM USING "S.C." (FOR SERVICE CONNECTOR).

This circuit code shows a tracer symbol (*) but not color. Here you would look for a 14-gauge pink wire with a tracer which will be black or white.

Tracer color is indicated. Here tracer is/RD/ 12BK/RD*.

Figure 5-18 Wiring diagrams provide the technician with necessary information to accurately diagnose the electrical systems. (Courtesy of Chrysler Corporation)

color coding used by the manufacturer. Make sure you understand what color the code is referring to before looking for a wire.

In most color codes, the first group of letters designates the base color of the insulation. If a second group of letters is used, it indicates the color of the tracer. For example, a wire designated as WH/BLK would have a white base color with a black tracer.

Ford uses four methods of color coding its wires (Figure 5-24):

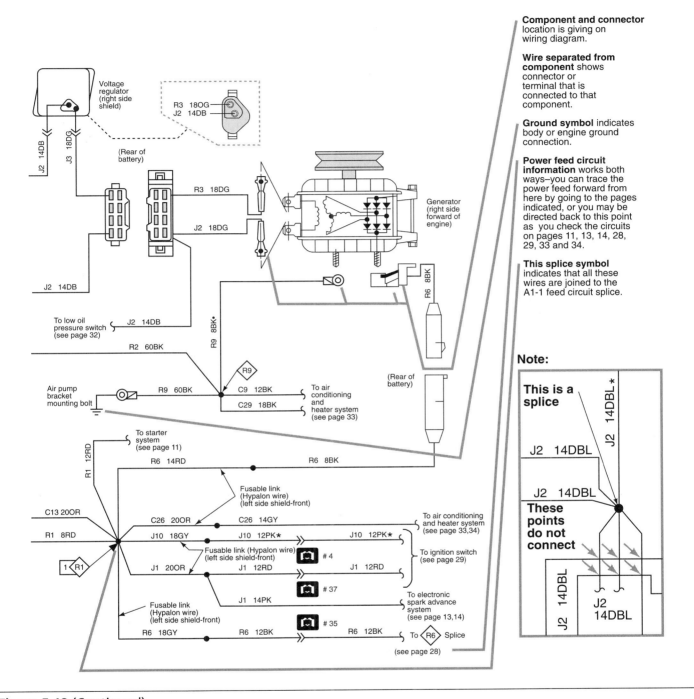

Figure 5-18 (Continued)

1. Solid color.
2. Base color with a stripe (tracer).
3. Base color with hash marks.
4. Base color with dots.

Chrysler uses a numbering method to designate the circuits on the wiring diagram (Figure 5-25). The circuit identification, wire gauge, and color of the wire are included in the wire number. Chrysler identifies the main circuits by using a main circuit identification code

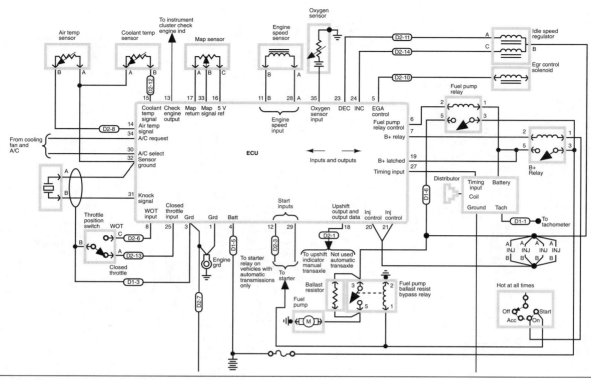

Figure 5-19 Wiring diagram that illustrates only one specific circuit for easier reference. This is also known as a block diagram (Courtesy of Chrysler Corporation)

that corresponds to the first letter in the wire number (Figure 5-26).

General Motors uses numbers that include the wire gauge in metric millimeters, the wire color, the circuit number, splice number, and ground identification (Figure 5-27). In this example, the circuit is designated as 120, the wire size is 0.8mm2, the insulation color is black, the splice is numbered S114, and the ground is designated as G117. Most manufacturers also number connectors and terminals for identification.

The Society of Automotive Engineers (SAE) is attempting to standardize the circuit diagrams used by the various manufacturers. The system that is developed may be similar to the DIN used by import manufacturers. DIN assigns certain color codes to a particular circuit as follows:

- Red wires are used for direct battery-powered circuits and also ignition-powered circuits.
- Black wires are also powered circuits controlled by switches or relays.
- Brown wires are usually the grounds.
- Green wires are used for ignition primary circuits.

A combination of wire colors is used to identify sub-circuits. The base color still identifies the circuit's basic purpose. In addition to standardized color coding, DIN attempts to standardize terminal identification and circuit numbering.

Component Locators

The wiring diagrams in most service manuals may not indicate the exact physical location of the components of the circuit. Another shop manual called a **component locator** is used to find where a component is installed in the vehicle. The component locator may use both drawings and text to lead the technician to the desired component.

Many electrical components may be hidden behind kick panels, dash boards, fender wells, and under seats. The use of a component locator will save the technician time in finding the suspected defective unit.

Reading Wiring Diagrams

When attempting to locate a possible cause for system malfunction, it is important to have the correct wiring diagram for the vehicle being worked on. There may be a different diagram for each model and even for the same models equipped with different options. Also,

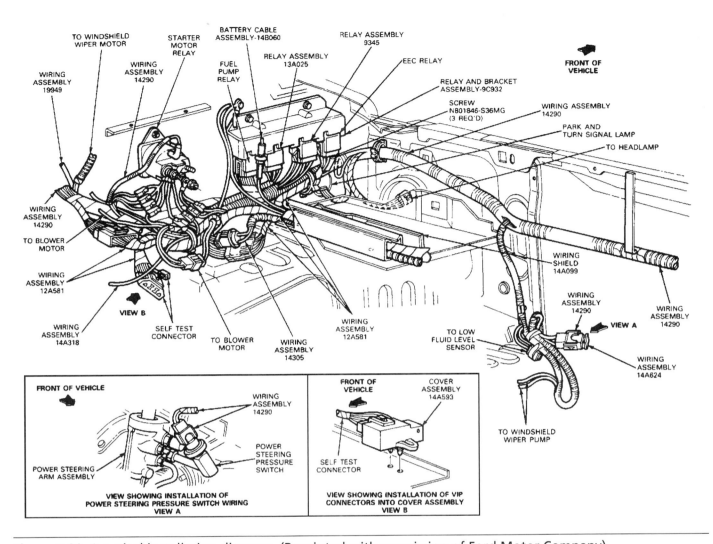

Figure 5-20 A typical installation diagram. (Reprinted with permission of Ford Motor Company)

diagrams may differ between two- and four-door models. In some cases it may be necessary to use the date of manufacture and/or the **vehicle identification number (VIN)** number to determine the correct diagram to use. The VIN is assigned to a vehicle for identification purposes. The identification plate is usually located on the crowl, next to the upper instrument panel. It is visible from the outside of the vehicle (Figure 5-28).

Next, study the method used to identify circuits and color codes. Usually this information is provided in the service manual. Also become familiar with the electrical symbols used by the manufacturer.

Before you begin to try to use a wiring diagram, you must first have an idea of what you are looking for. It may be a particular component, circuit, or connector. The best way to start the process is by identifying the component or one of the components that doesn't work correctly. Then look in the index for the wiring diagram and find where that component is shown.

The electrical section in most service manuals breaks down the electrical system of the automobile into individual circuits. This approach makes it easier to find a particular component. Of course, you still need to use the index to find the page the component and its circuit is on.

If the service manual uses a total vehicle wiring diagram, finding the component may be a little trickier. Wiring diagrams are usually indexed by grids. The diagram is marked into equal sections like a street map. The wiring diagram's index will list a letter and number for each major component and many different connection points. If the wiring diagram is not indexed, you can locate the component by relating its general location in the vehicle to a general location on the wiring diagram. Most system diagrams are drawn so the front of the car is on the left of the diagram.

Once you have found the component or part of the circuit you were looking for, identify all of the components, connectors, and wires that are related to that com-

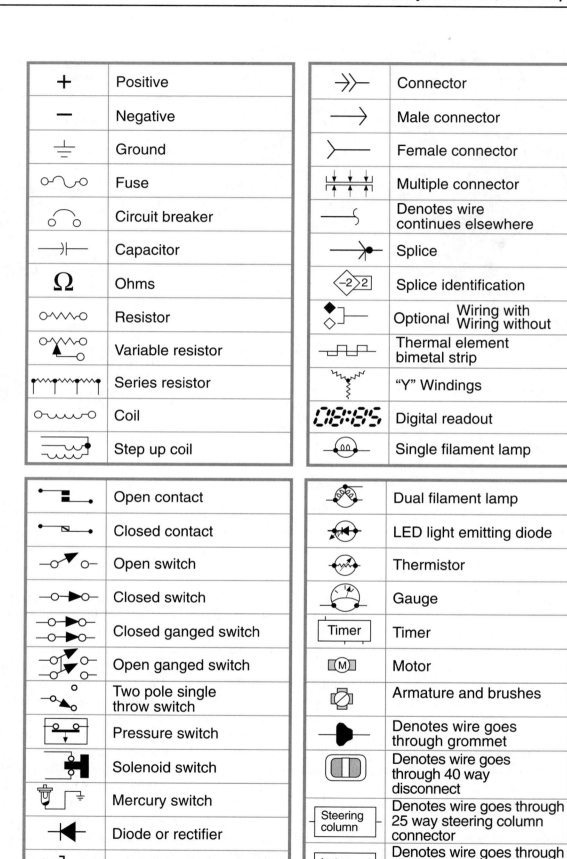

+	Positive		≫—	Connector
—	Negative		—→	Male connector
⏚	Ground		>—	Female connector
Fuse	Fuse			Multiple connector
Circuit breaker	Circuit breaker		—ʃ	Denotes wire continues elsewhere
—)|—	Capacitor		—⟶•	Splice
Ω	Ohms		⟨-2⟩2	Splice identification
Resistor	Resistor			Optional Wiring with / Wiring without
Variable resistor	Variable resistor			Thermal element bimetal strip
Series resistor	Series resistor			"Y" Windings
Coil	Coil		*08:85*	Digital readout
Step up coil	Step up coil			Single filament lamp

Open contact	Open contact			Dual filament lamp
Closed contact	Closed contact			LED light emitting diode
Open switch	Open switch			Thermistor
Closed switch	Closed switch			Gauge
Closed ganged switch	Closed ganged switch		Timer	Timer
Open ganged switch	Open ganged switch		M	Motor
Two pole single throw switch	Two pole single throw switch			Armature and brushes
Pressure switch	Pressure switch			Denotes wire goes through grommet
Solenoid switch	Solenoid switch			Denotes wire goes through 40 way disconnect
Mercury switch	Mercury switch		Steering column	Denotes wire goes through 25 way steering column connector
Diode or rectifier	Diode or rectifier		Instrument panel	Denotes wire goes through 25 way instrument panel connector
Bi directional zener diode	Bi directional zener diode			

Figure 5-21 Common electrical symbols used in wiring diagrams. (Courtesy of Chrysler Corporation)

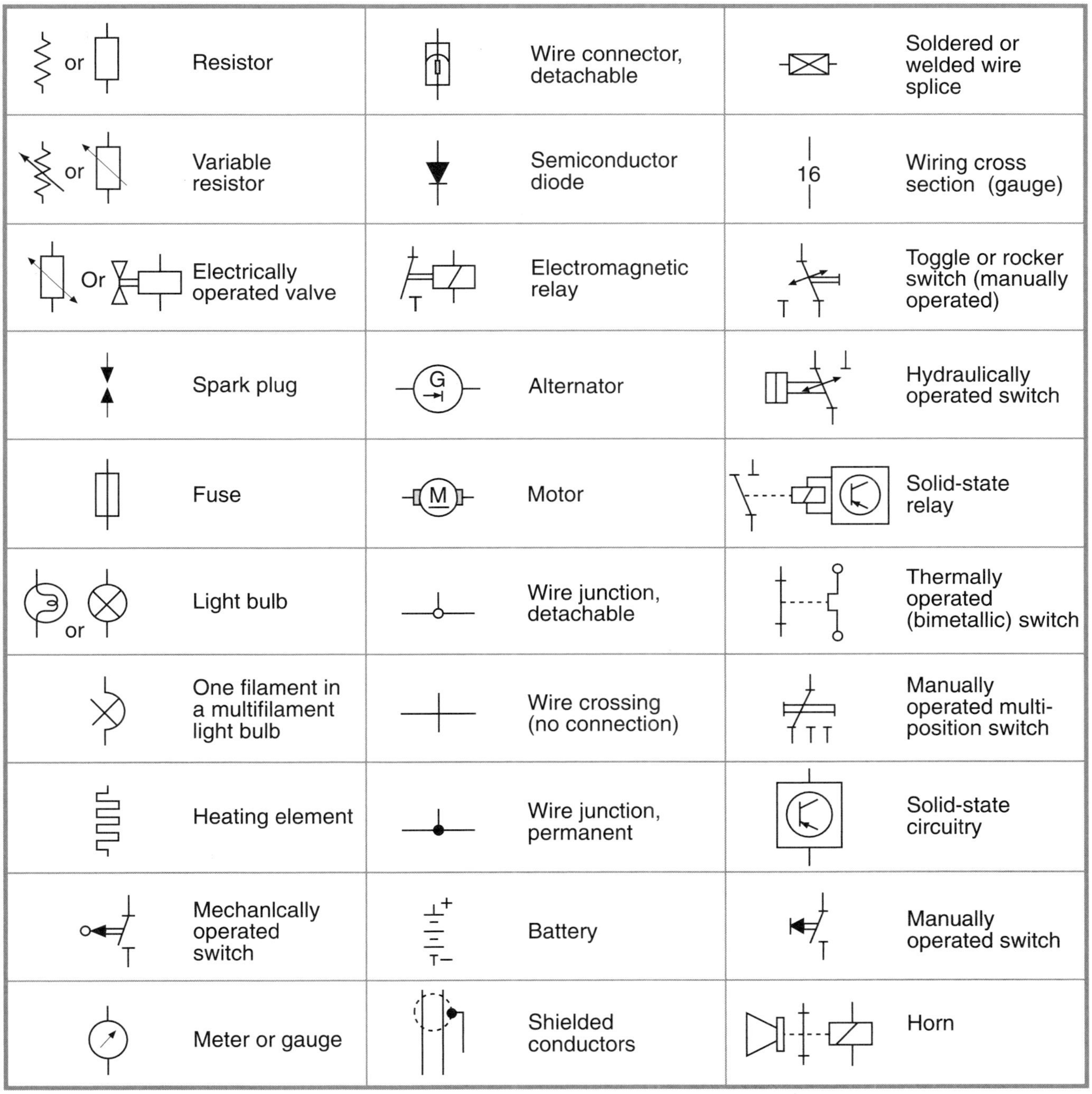

⌇ or ▯	Resistor	▯	Wire connector, detachable	⊠	Soldered or welded wire splice
⌇ or ▱	Variable resistor	▼	Semiconductor diode	16	Wiring cross section (gauge)
▯ Or ▷◁	Electrically operated valve	▯	Electromagnetic relay		Toggle or rocker switch (manually operated)
▼	Spark plug	Ⓖ	Alternator		Hydraulically operated switch
▯	Fuse	Ⓜ	Motor	Ⓚ	Solid-state relay
⊖ ⊗ or	Light bulb	—∘—	Wire junction, detachable		Thermally operated (bimetallic) switch
⊗	One filament in a multifilament light bulb	—+—	Wire crossing (no connection)		Manually operated multi-position switch
⊟	Heating element	—●—	Wire junction, permanent	Ⓚ	Solid-state circuitry
⚬⟋	Mechanically operated switch	⊣⊢ + −	Battery		Manually operated switch
⊙	Meter or gauge		Shielded conductors	▷ ⊡	Horn

Figure 5-22 Electrical symbols used in more recent manuals.

Color	Abbreviations		
Aluminum	AL		
Black	BLK	BK	B
Blue (Dark)	BLU DK	DB	DK BLU
Blue	BLU	B	L
Blue (Light)	BLU LT	LB	LT BLU
Brown	BRN	BR	BN
Glazed	GLZ	GL	
Gray	GRA	GR	G
Green (Green)	GRN DK	DG	DK GRN
Green (Light)	GRN LT	LG	LT GRN
Maroon	MAR	M	
Natural	NAT	N	
Orange	ORN	O	ORG
Pink	PNK	PK	P
Purple	PPL	PR	
Red	RED	R	RD
Tan	TAN	T	TN
Violet	VLT	V	
White	WHT	W	WH
Yellow	YEL	Y	YL

Figure 6-26 Common color codes used in automotive applications.

Figure 5-23 Common color codes used in automotive applications.

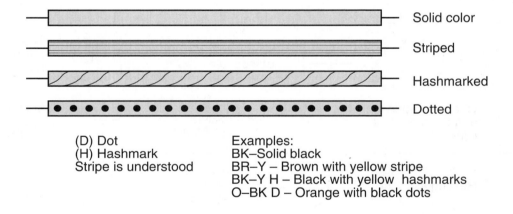

(D) Dot
(H) Hashmark
Stripe is understood

Examples:
BK–Solid black
BR–Y – Brown with yellow stripe
BK–Y H – Black with yellow hashmarks
O–BK D – Orange with black dots

Figure 5-24 Four methods that Ford uses to color code their wires. (Reprinted with permission of Ford Motor Company)

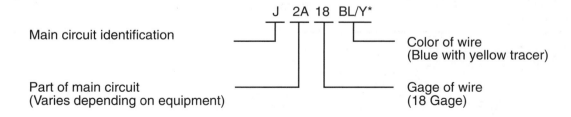

J 2A 18 BL/Y*

Main circuit identification

Part of main circuit
(Varies depending on equipment)

Color of wire
(Blue with yellow tracer)

Gage of wire
(18 Gage)

Figure 5-25 Chrysler wiring code identification. (Courtesy of Chrysler Corporation)

A	BATTERY FEED (i.e. Fuselink Feeds, Starter Feeds, Starter Relay)	N	ESA Module Electronic Circuits
B	Brakes	O	Not used as a circuit designator
C	Climate Control (A/C/ Heater, E.B.L. and Heated Mirror Related Circuits)	P	Power Options (Battery Feed) i.e. Seats, Doors Locks, Mirrors, Deck Lid Release, etc.)
D	Diagnostic Circuits	Q	Power Options (Ignition Feed) (i.e. Windows, Power Top, Power Sun Roof, etc.)
E	Dimming Illumination Circuits		
F	Fused Circuits (Non-Dedicated Multi-System Feeds)	R	Passive Restraint
G	Monitoring Circuits (Gages, Clocks, Warning Devices)	S	Suspension and Steering Circuits
		T	Transmission/Transaxle, Differential, Transfer Case and Starter System Circuits
H	**OPEN**		
L	Not used as a circuit designator	U	**OPEN**
J	**OPEN**	V	Speed Control and Wash Wipe Circuit
K	Engine Logic Module Control Circuits	W	**OPEN**
L	Exterior Lighting Circuits	X	Sound System (i.e. Radio and Horn)
M	Interior Lighting Circuits (Dome, Courtesy Lamps, Cargo Lamps)	Y	**OPEN**
		Z	Grounds (B−)

Figure 5-26 Chrysler circuit identification codes. (Courtesy of Chrysler Corporation)

ponent. This is done by tracing through the circuit, starting at the component. Tracing does not mean taking a pencil and marking on the wiring diagram. Tracing means taking your pencil and drawing out the circuit on another piece of paper. It doesn't have to be pretty to work; it just needs to be accurate. Tracing may also mean taking your finger and following the wires to where they lead. In order for tracing to have any value, you need to identify the power source for the component and/or for the circuit, all related loads (sometimes this involves tracing the circuit back through other pages of the wiring diagram), and the ground connection for the component and for all of the related loads.

After you have traced the circuit, study it and make sure you know how the circuit is supposed to work. Then describe the problem you are hoping to solve. Ask yourself what could cause this? Limit your answers to those items in your traced wiring diagram. Also limit your answers to the description of the problem. It is wise to make a list of all probable causes of the problem; then number them according to probability. For example, if no dashlights come on, it is possible that all of the bulbs are burned out. However, it is not as probable as a blown fuse. After you have listed the probable causes in order of probability, look at the wiring diagram to identify how you can quickly test to find out if each is the cause. Diagnostics is made easier as your knowledge of electricity grows. It also becomes easier with a good understanding of how the circuit works.

Figure 5-29 is a schematic for a blower control circuit of a heater system. By tracing through the circuit, the technician should be able to determine the correct opera-

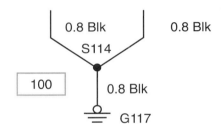

Figure 5-27 GM's method of circuit and wire identification.

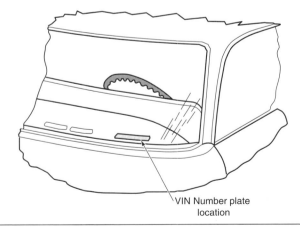

Figure 5-28 VIN plate location.(Courtesy of General Motors Corporation, Service Technology Group)

tion of the system. It cannot be overemphasized that the technician should not attempt to figure out why the circuit is not working until it is understood how it is supposed to work.

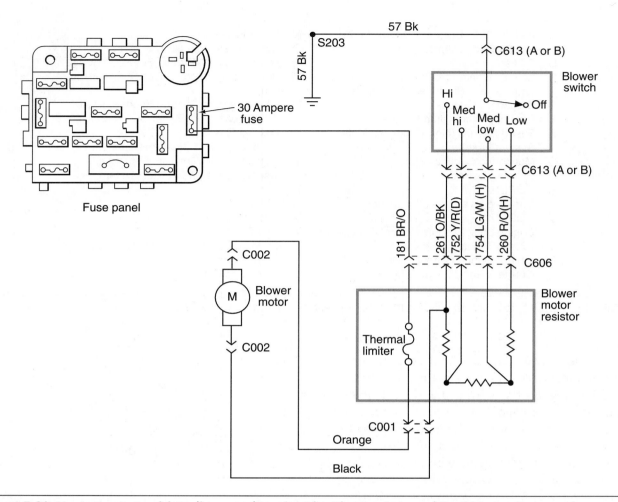

Figure 5-29 Heater system wiring diagram. (Reprinted with permission of Ford Motor Company)

In this circuit, battery voltage is applied to the fuse box where the 30-ampere fuse supplies the battery voltage to the motor through circuit number 181. Notice that circuit 181 connects to the blower motor resistor block through connector C-606. When it leaves the resistor block (after flowing through the thermal limiter) it goes through connector C-001 and attaches to the motor through connector C-002. Details of these connectors are shown with the schematic.

The circuit then leaves the motor and enters the resistor block through the same connectors. There is a splice in the resistor block that connects this wire to a series of resistors. Connector C-606 directs the various circuits out of the resistor block to the blower switch through connector C-613. If the switch is placed in the OFF position, the circuit to ground is opened and the blower motor should not operate. If the switch is placed in the LOW position, the current will flow through all of the resistors and the switch through circuit 260. The switch completes the circuit to ground. As the switch is placed

in different speed positions, the amount of resistance and the circuit number change. The motor speed should increase as the amount of resistance decreases.

If the customer's concern is that the heater motor does not work at all, in any speed position, consider the following possible causes:

1. Open fuse.
2. Open in the lead from the battery to the fuse box.
3. Bad ground connection after the switch.
4. Inoperative blower motor.
5. Open in circuit 181.
6. Open in the orange wire between connectors C-001 and C-002.
7. Open thermal limiter.
8. Open in the black wire between connectors C-002 and C-001.
9. Disconnected, damaged, or corroded connector C-606 or C-613.

However, if the customer says that the motor will not operate in the low position only, the problem is limited to two possibilities:

1. The third resistor in series is open.
2. An open in circuit 260 from the resistor block to the blower switch.

Once the potential problem areas are determined, use the color code to locate the exact wires. Test the leads for expected voltages at different locations in the circuit. By understanding the way the system is supposed to work, the problem of determining where to look for the problems is simplified. Practice in reading wiring schematics is the only good teacher.

Summary

❏ Most of the primary wiring conductors used in the automobile are made of several strands of copper wire wound together and covered with a polyvinyl chloride (PVC) insulation.

❏ Stranded wire is used because there is more surface area exposed in a stranded wire, resulting in less resistance in the stranded wire than in the solid wire.

❏ There are three major factors that determine the proper size of wire to be used: (1) The wire must be large enough in diameter for the length required to carry the necessary current for the load components in the circuit to operate properly, (2) the wire must be able to withstand the anticipated vibration, and (3) the wire must be able to withstand the anticipated amount of heat exposure.

❏ Wire size is based on the diameter of the conductor.

❏ Factors that affect the resistance of the wire include the conductor material, wire diameter, wire length, and temperature.

❏ Terminals can be either crimped or soldered to the conductor. The terminal makes the electrical connection and it must be capable of withstanding the stress of normal vibration.

❏ Printed circuit boards are used to simplify the wiring of the circuits they operate. A printed circuit is made of a thin phenolic or fiberglass board that copper (or some other conductive material) has been deposited on.

❏ A wire harness is an assembled group of wires that branch out to the various electrical components.

❏ A wiring diagram shows a representation of actual electrical or electronic components and the wiring of the vehicle's electrical systems.

❏ In place of actual pictures, a variety of electrical symbols are used to represent the components in the wiring diagram.

❏ Color codes and circuit numbers are used to make tracing wires easier.

❏ A component locator is used to determine the exact location of several of the electrical components.

Terms To Know

American wire gauge (AWG)	Metri-pack connectors	Symbols
Ballast resistor	Molded connector	Television-radio-suppression (TVRS) cables
Bulkhead connectors	Multiple-wire hard-shell connector	
Common connection	Primary wiring	Vehicle identification number (VIN)
Component locator	Printed circuit	Weather-pack connectors
Cross-fire	Resistance wire	Wire protection devices
Electrical schematic	Secondary wiring	Wiring diagram
Installation diagrams	Stranded wire	Wiring harnesses

Review Questions

Short Answer Essays

1. Explain the purpose of wiring diagrams.
2. Explain how wire size is determined by the American Wire Gauge (AWG) and metric methods.
3. Explain the purpose and use of printed circuits.
4. Explain the purpose of the component locator.
5. Explain when single-stranded or multi-stranded wire should be used.
6. Explain how temperature affects resistance and wire size selection.
7. List the three major factors that determine the proper size of wire to be used.
8. List and describe the different types of terminal connectors used in the automotive electrical system.
9. What is the difference between a complex and a simple wiring harness?
10. Describe the methods the three domestic automobile manufacturers use for wiring code identification.

Fill-in-the-Blanks

1. There is _____ resistance in the stranded wire than in the solid wire.
2. _____ _____ is the electromagnetic induction spark that can be transmitted in another wire that is close to the wire carrying the current.
3. Wire size is based on the _____ of the conductor.
4. In the AWG standard, the _____ the number, the smaller the wire _____.
5. An increase in temperature creates a _____ in resistance.
6. _____ connectors are used when several wires must pass through the bulkhead.
7. _____ _____ _____ are used to prevent damage to the wiring by maintaining proper wire routing and retention.
8. A wiring diagram is an electrical schematic that shows a _____ of actual electrical or electronic components (by use of symbols) and the _____of the vehicle's electrical systems.
9. In most color codes, the first group of letters designates the _____ _____ of the insulation. The second group of letters indicates the color of the _____.
10. A _____ _____ is used to determine the exact location of several of the electrical components.

ASE Style Review Questions

1. Automotive wiring is being discussed. Technician A says that most of the primary wiring conductors are made of several strands of copper wire wound together and covered with an insulation. Technician B says that other types of conductor materials used in automobiles include silver, gold, aluminum, and tin plated- brass. Who is correct?
   a. A only
   b. B only
   c. Both A and B
   d. Neither A nor B

2. Stranded wire use is being discussed. Technician A says that there is less surface area exposed in a stranded wire for electron flow. Technician B says that there is more resistance in the stranded wire than in the solid wire. Who is correct?
   a. A only
   b. B only
   c. Both A and B
   d. Neither A nor B

3. Spark plug wires are being discussed. Technician A says that RFI is controlled by using resistances in the conductor of the spark plug wire. Technician B says that current standards require that all ignition systems control RFI. Who is correct?
   a. A only
   b. B only
   c. Both A and B
   d. Neither A nor B

4. Spark plug wire installation is being discussed. Technician A says that there is little that can be done to correct cross-fire. Technician B says that the spark plug wires must be installed in the proper separator and any two parallel wires next to each other in the firing order should be positioned as far away from each other as possible to prevent cross-fire. Who is correct?
   a. A only
   b. B only
   c. Both A and B
   d. Neither A nor B

5. The selection of the proper size of wire to be used is being discussed. Technician A says that the wire must be large enough, for the length required, to carry the amount of current necessary for the load components in the circuit to operate properly. Technician B says that temperature has such a little effect on resistance that it is not a factor in wire size selection. Who is correct?
   a. A only
   b. B only
   c. Both A and B
   d. Neither A nor B

6. Terminal connectors are being discussed. Technician A says that terminal connections can not corrode. Technician B says that the terminals can be either crimped or soldered to the conductor. Who is correct?
   a. A only
   b. B only
   c. Both A and B
   d. Neither A nor B

7. Wire routing is being discussed. Technician A says that whenever it is necessary to install additional electrical accessories, support the primary wire in at least 10-foot intervals. Technician B says that if the wire must be routed through the frame or body, use metal clips to protect the wire. Who is correct?
   a. A only
   b. B only
   c. Both A and B
   d. Neither A nor B

8. Printed circuit boards are being discussed. Technician A says that printed circuit boards are used to simplify the wiring of the circuits they operate. Technician B says that care must be taken not to touch the board with your bare hands. Who is correct?
   a. A only
   b. B only
   c. Both A and B
   d. Neither A nor B

9. Wiring harnesses are being discussed. Technician A says a wire harness is an assembled group of wires that branch out to the various electrical components. Technician B says that most under-hood harnesses are simple harnesses. Who is correct?
   a. A only
   b. B only
   c. Both A and B
   d. Neither A nor B

10. Wiring diagrams are being discussed. Technician A says wiring diagrams give the exact location of the electrical components. Technician B says wiring diagrams will indicate what circuits are interconnected, where circuits receive their voltage source, and what colors of wires are used in the circuit. Who is correct?
   a. A only
   b. B only
   c. Both A and B
   d. Neither A nor B

6 Wiring Repair

Objective

Upon completion and review of this chapter, you should be able to:

❑ Perform repairs to copper wire using solderless connections.

❑ Solder splices to copper wire.

❑ Repair aluminum wire according to manufacturer's requirements.

❑ Repair twisted/shielded wire.

❑ Replace fusible links.

❑ Repair and/or replace the terminals of a hard-shell connector.

❑ Repair and/or replace the terminals of weather-pack and metri-pack connectors.

Introduction

Not all electrical repairs involve removing and replacing a faulty component. Many times the cause of the malfunction is a damaged conductor. The technician must make a repair to the circuit that will not increase the resistance. It should also be a permanent repair. There are many methods to repair a damaged wire. The type of repair used will depend on factors such as:

1. Type of repair required
2. Ease of access to the damaged area
3. Type of conductor
4. Size of wire
5. Circuit requirements
6. Manufacturer's recommendations

The most common methods of wire repair include wrapping damaged insulation with electrical tape or tubing, crimping the connections with solderless connectors, and soldering splices.

Copper Wire Repairs

Copper wire is the most commonly used primary wire in the automobile. The insulation may break down, or the wire may break due to stress or excessive motion. The wire may also be damaged due to excessive current flow through the wire. Any of these conditions require that the wire be repaired.

WARNING *Repair of air bag wiring must be done to manufacturer's specifications.*

In some instances it may be necessary to bypass a length of wire that is not accessible. In this case, cut the wire before it enters the inaccessible portion and at the other end where it leaves the area. Install a replacement wire and reroute it to the load component. Be sure to protect the wire by using straps, hangers, and grommets as needed.

The two most common methods of splicing copper wire are with solderless connectors or by soldering. **Crimping** means to bend, or deform by pinching a connector so the wire connection is securely held in place. Crimping of **solderless connections** is an acceptable means of **splicing** wires that are not subjected to weather elements, dirt, corrosion, or excessive movement. Solderless connections are hollow metal tubes covered with insulating plastic. They can be butt connectors or terminal ends (Figure 6-1). Splice is the term used to mean joining of one or more electrical conductors at a single point. Do the following to make a splice using solderless connections:

1. Use the correct size of stripping opening on the crimping tool to remove enough insulation to allow the wire to completely penetrate the connector. The crimping tool has different areas for performing several functions (Figure 6-2). This tool will cut the wire, strip the insulation, and crimp the connector.

2. Place the wire into the connector and crimp the connector (Figure 6-3). To get a proper crimp, place the open area of the connector facing toward the anvil. Be sure the wire is compressed under the crimp.

3. Insert the stripped end of the other wire into the connector and crimp in the same manner.

4. Use electrical tape or **heat shrink tubing** to provide additional protection. Heat shrink tubing is plastic tubing that shrinks in diameter when exposed to heat.

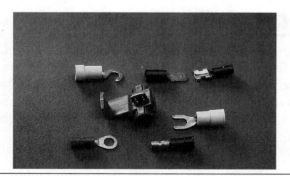

Figure 6-1 Types of solderless connectors and terminals commonly used for wire repair.

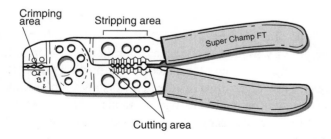

Figure 6-2 A typical crimping tool used for making electrical repairs.

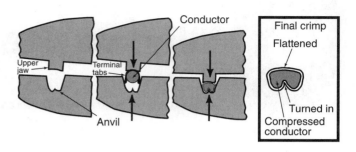

Figure 6-3 Properly crimping a connector.

Another type of crimping connector is the **tap splice connector**. This type of connector allows for adding an additional circuit to an existing feed wire without stripping the wires (Figure 6-4). Although tap connectors make connecting wires easy, these should not be used to provide power to critical components. Also make sure the fuse of the circuit being tapped into has a great enough capacity before adding the circuit. Tap connectors add a circuit in parallel with another circuit. This typically causes circuit resistance to decrease and circuit amperage to increase.

Soldering is the process of using heat and solder (a mixture of metals with a low melting point) to make a splice or connection. Soldering is the best way to splice copper wires. Solder is an alloy of tin and lead. It is melted over a splice to hold the wire ends together. Soldering may be a splicing procedure, but it is also an art that takes much practice. Photo Sequence 3 goes through the soldering process when using a **splice clip**. A splice clip is a special connector used along with solder to assure a good connection. The splice clip is different from a solderless connection in that it does not have insulation. The hole is provided for applying solder (Figure 6-5). If a splice clip is not used, the wire ends should be braided together tightly. Then the splice should be heated with the soldering gun. It is important to note that when soldering, the solder should melt by the heat of the wire splice, not the heat of the soldering tool. Always use rosin core solder when making electrical repairs. Acid core solder is used for other purposes than electrical repairs and can cause the wire to corrode, which would lead to high resistance.

WARNING *Before cutting into a wire to make a splice, look for other splices or connections first. Never have two or more splices within 1.5 inches (40mm) of each other. Also always use wire of the same size or larger than the wire being replaced.*

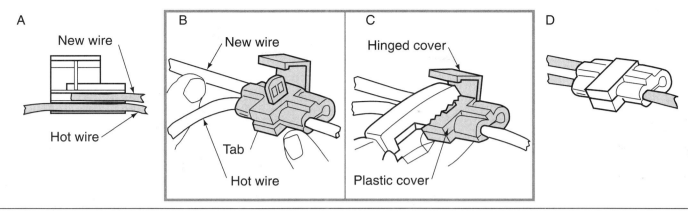

Figure 6-4 Using the tap connector to splice in another wire: (A) place wires in position in the connector, (B) close the connector around the wires, (C) use pliers to force the tab into the conductors, (D) close the hinged cover.

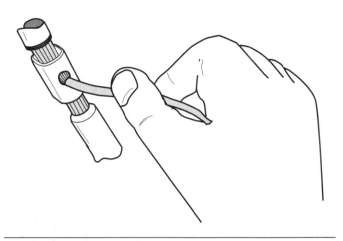

Figure 6-5 Splice clip.(Courtesy of General Motors Corporation, Service Technology)

An alternate method of soldering wires together is to use wire joints in place of splice clips. Remove about one inch of the insulation from the wires. Join the wires using one of the methods illustrated (Figure 6-6). Heat the twisted connection with the soldering iron. Apply the solder to the strands of wire. Do not apply the solder directly to the soldering iron. The solder should melt and flow evenly among all of the wire strands (Figure 6-7). Insulate the splice with electrical tape or heat shrinking tube.

Repairing Aluminum Wire

WARNING *Attempting to solder aluminum wire will damage the conductor.*

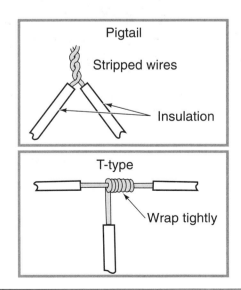

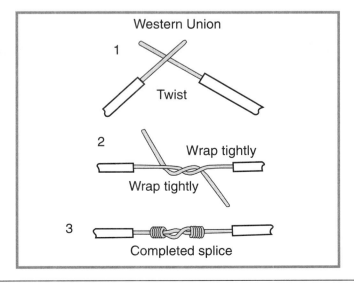

Figure 6-6 Methods for joining wires together.

Soldering Copper Wire

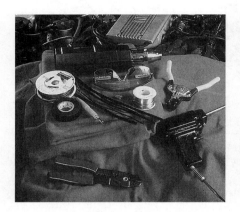

P2-1 Tools required to solder copper wire: 100 watt soldering iron, 60/40 rosin core solder, crimping tool, splice clip, heat shrink tube, heating gun, safety glasses, sewing seam ripper, electrical tape, and fender covers.

P2-2 Disconnect the fuse that powers the circuit being repaired. Note: If the circuit is not protected by a fuse, then disconnect the battery.

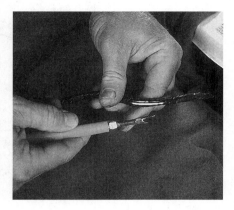

P2-3 If the wiring harness is taped, use a seam ripper to open the wiring harness.

P2-4 Cut out the damaged wire using the wire cutters on the crimping tool.

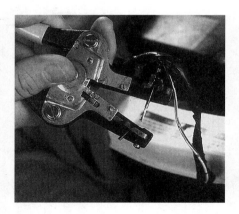

P2-5 Using the correct size stripper, remove about 1/2 inch of the insulation from both wires. Be careful not to nick or cut any of the wires.

P2-6 Determine the correct gauge and length of replacement wire. Using the correct size stripper, remove 1/2 inch of insulation from each end of the replacement wire.

2 Soldering Copper Wire

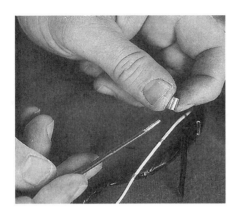

P2-7 Select the proper size splice clip to hold the splice.

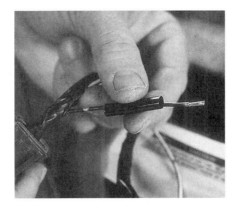

P2-8 Place the correct length and size of heat shrink tube over the two ends of the wire. Slide the tube far enough away so it is not exposed to the heat of the soldering iron.

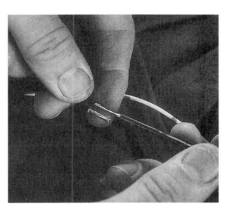

P2-9 Overlap the two splice ends and hold in place with thumb and forefinger.

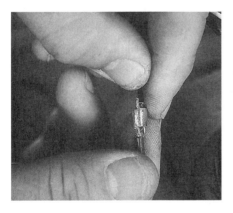

P2-10 Center the splice clip around the wires making sure the wires extend beyond the splice clip in both directions.

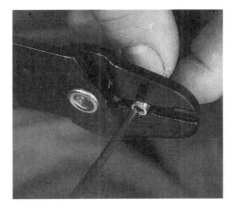

P2-11 Crimp the splice clip firmly in place. Crimp the clip at both ends of the repair.

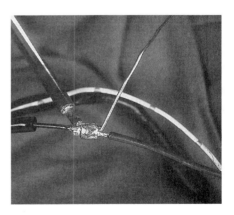

P2-12 Heat the splice clip with the soldering iron while applying solder to the opening in the back of the clip. Do not apply solder to the iron, the iron should be 180 degrees away from the opening of the clip. Apply only enough solder to make a good connection. The solder should travel through the wire.

Soldering Copper Wire

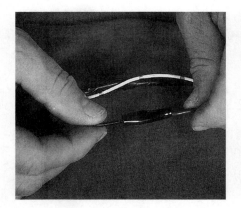

P2-13 After the solder cools, slide the heat shrink tube over the splice.

P2-14 Heat the tube with the hot air gun until it shrinks around the splice. Do not overheat the tube. Repeat for all other splices to complete the repair.

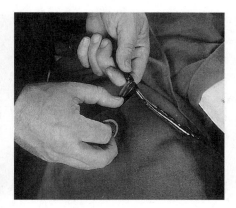

P2-15 Retape the wiring harness.

General Motors has used single-stranded aluminum wire in limited applications where no flexing of the wire is expected. This wire usually has a thick plastic insulator and is placed in a brown harness.

After cutting away the damaged wire, strip all wire of the last 1/4 inch (6mm) of insulation. Be careful not to nick or damage the conductor. Apply a generous coating of petroleum jelly to the wire and connector. The petroleum jelly will prevent corrosion from developing in the core.

Crimp the connector in the usual manner. Insulate the splice with electrical tape or heat shrink tube.

Splicing Twisted/Shielded Wires

Twisted/shielded wire is used in computer circuits. It protects the circuit from electrical noise that would interfere with the operation of the computer controls (Figure 6-8). These wires may carry as low as 0.1 ampere of current. It is important that the splice made in these wires does not have any resistance. The added resistance may give false signals to the computer or actuator. Do the following to splice this type of wire:

1. Locate and cut out the damaged section of wire.

2. Being careful not to cut into the mylar tape or to cut the drain wire, remove about 1 inch of the outer jacket from the ends of the cable.

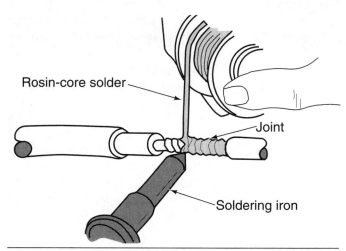

Figure 6-7 When soldering, apply the solder to the joint not to the tip of the soldering iron.

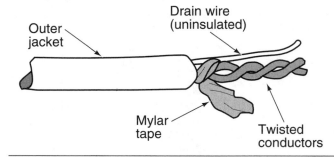

Figure 6-8 Twisted/shielded wire used in computer circuits. (Courtesy of General Motors Corporation, Service Technology Group)

3. Unwrap the mylar tape. Do not remove the tape from the cable (Figure 6-9).

4. Untwist the conductors and remove the insulation from the ends.

5. Use a splice clip to connect the two wires and solder the splice.

6. Wrap the conductors with the mylar tape. Do not wrap the drain wire in the tape.

7. Splice the drain wire and solder the connection.

WARNING *When repairing the wires, stagger the splice connections. This will prevent shorts.*

8. Wrap the drain wire around the conductors and the mylar tape (Figure 6-10).

9. Use electrical tape or heat shrink tube to insulate the cable (Figure 6-11).

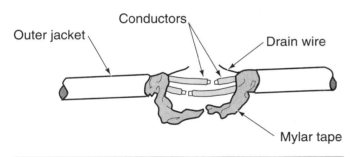

Figure 6-9 Unwrap the mylar tape to expose the conductors before attempting to repair the wire. (Courtesy of General Motors Corporation)

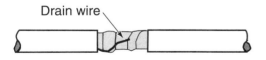

Figure 6-10 The drain wire should be wrapped around the outside of the mylar tape. (Courtesy of General Motors Corporation, Service Technology Group)

Figure 6-11 The completed repair. (Courtesy of General Motors Corporation, Service Technology Group)

Replacing Fusible Links

Not all fusible link open circuits are detectable by visual inspection only. Test for battery voltage on both sides of the fusible link to confirm its condition. If the fusible link must be replaced, it is cut out of the circuit and a new fusible link is crimped or soldered into place.

There are two types of insulation used on fusible links: Hypalon and Silicone/GXL. Hypalon can be used to replace either type of link. However, do not use Silicone/GXL to replace Hypalon. To identify the type of insulation, cut the blown link's insulation back. The insulation of the Hypalon link is a solid color all the way through. The insulation of Silicon/GXL will have a white inner core.

WARNING *Disconnect the battery negative cable before performing any repairs to the fusible link.*

When cutting off the damaged fusible link from the feed wire, cut it beyond the splice (Figure 6-12). When making the repair link, do not use a fusible link that is cut longer than 9 inches (228 mm). This length will not provide sufficient overload protection. Splice in the repair link by crimping or soldering.

If the damaged fusible link feeds two harness wires, use two fusible links. Splice one link to each of the harness wires (Figure 6-13).

WARNING *Use the fusible link gauge required by the manufacturer.*

CUSTOMER CARE *If the fusible link or any of the fuses are blown, it is important to locate the cause. Fuses do not wear out. If they blow it is due to an overload of current in the circuit. By using a cause-and-effect diagnosis approach, the fault can be identified and repaired the first time the vehicle is in the shop.*

Repairing Connector Terminals

The connector terminal is subjected to needed repairs due to abuse, improper disconnecting procedures, and exposure potential to the elements. The method of repair depends on the type of connector.

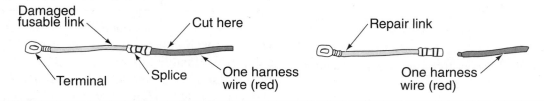

Figure 6-12 Replacing a fusible link. (Courtesy of General Motors Corporation, Service Technology Group)

Figure 6-13 Replacing fusible link that feeds two circuits. (Courtesy of General Motors Corporation, Service Technology Group)

Molded Connectors

Molded connectors usually have one to six wires that are molded into a one piece component. Connectors of this design cannot be separated for repairs. If the connector is damaged, it must be cut off and a new connector spliced in.

Hard-Shell Connectors

CAUTION *Do not place your fingers or body next to the connector. If excessive force is needed to depress the tang, the pick may be pushed out the back and cause injury.*

Hard-shell connectors have a hard plastic shell that holds the connecting terminals of the separate wires. Hard-shell connectors usually provide a means of removing the terminals for repair. Use a pick, or the special tool, to depress the locking tang of the connector (Figure 6-14). Pull the lead back far enough to release the locking tang from the connector. Remove the pick, then pull the lead completely out of the connector. Make the repair to the terminal using the same procedures for repairing copper or aluminum wire.

Reform the terminal locking tang to assure a good lock in the connector (Figure 6-15). Use the pick to bend the lock tang back into its original shape. Insert the lead into the back of the connector. A noticeable "catch" should be felt when the lead is halfway through the connector. Gently push back and forth on the lead to confirm that the terminal is locked in place.

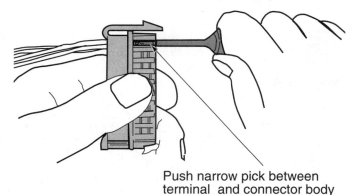

Push narrow pick between terminal and connector body

Figure 6-14 Depress the locking tang to remove the terminal from the connector. (Courtesy of General Motors Corporation, Service Technology Group)

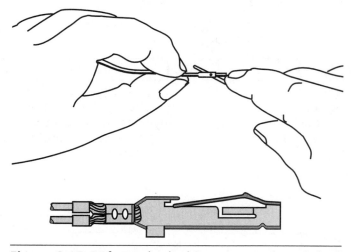

Figure 6-15 Reform the locking tang to its original position before inserting the terminal back into the connector. (Courtesy of General Motors Corporation, Service Technology Group)

Repairing Weather-Pack Connectors

Weather-pack connectors have rubber seals on the terminal ends and on the covers of the connector half to protect the circuit from corrosion. The terminals of the weather-pack connector are secured by a hinged secondary lock or a plastic terminal retainer. To perform repairs, first disconnect the two halves by pulling up on the primary lock while pulling the two halves apart (Figure 6-16). Unlock the secondary locks and swing them open (Figure 6-17).

Depress the terminal locking tangs using the special weather-pack tool. Push the cylinder of the tool into the terminal cavity from the front until it stops (Figure 6-18). Pull the tool out, then gently pull the lead out of the back of the connector (Figure 6-19).

The terminal is either a male or female connector (Figure 6-20). Use the correct terminal for the repair. Feed the wire through the seal and connect the repair lead in the normal manner of crimping and soldering (Figure 6-21). Reform the terminal lock tang by bending it back into its original position (Figure 6-22).

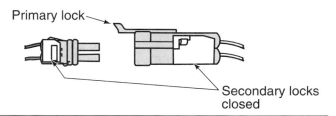

Figure 6-16 The weather-pack connector has two locks. Use the primary lock to separate the halves. (Courtesy of General Motors Corporation, Service Technology Group)

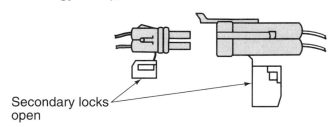

Figure 6-17 Unlock the secondary lock to remove the terminals from the connector (Courtesy of General Motors Corporation, Service Technology Group)

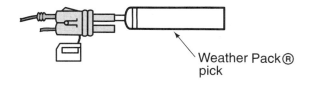

Figure 6-18 Use the recommended special tool to unlock the tang on the terminal. (Courtesy of General Motors Corporation, Service Technology Group)

Figure 6-19 After the lock tang has been depressed, remove the lead from the back of the connector (Courtesy of General Motors Corporation, Service Technology Group)

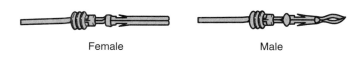

Figure 6-20 Male and female connectors. (Courtesy of General Motors Corporation)

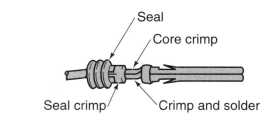

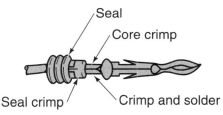

Figure 6-21 Crimp and solder the terminal to the lead. (Courtesy of General Motors Corporation, Service Technology Group)

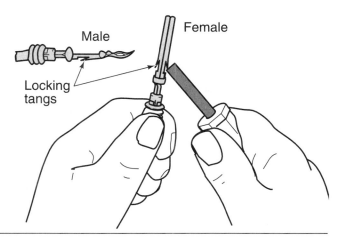

Figure 6-22 Reforming the locking tangs of the terminal. (Courtesy of General Motors Corporation, Service Technology Group)

Insert the lead from the back of the connector until a noticeable "catch" is felt. Gently push and pull on the lead to confirm that it is locked to the connector. Close the secondary locks and reconnect the connector halves.

Repairing Metri-Pack Connectors

Metri-pack connectors do not have the seal on the cover half like weather-pack connectors. There are two types of metri-pack connectors: **pull to seat** and **push to seat**. These terms depict the method used to install the terminals into the connector. The push-to-seat terminal removal is illustrated (Figures 7-23 and 7-24).

The pull-to-seat terminal is removed by inserting a pick into the connector and under the lock tang (Figure 6-25). Gently pull back on the lead while prying up on the lock tang. When the lock tang is free of the tab in the connector, push the lead through the front of the connector.

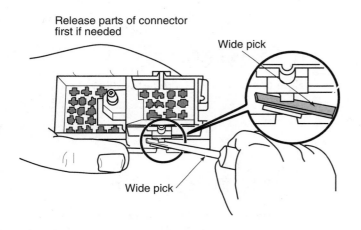

Figure 6-23 Use a wide pick to unlock the male terminal locking tang. (Courtesy of General Motors Corporation, Service Technology Group)

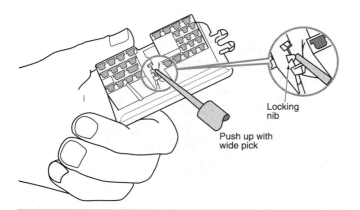

Figure 6-24 Use a wide pick to unlock the nib of the terminal retainer for female terminals. (Courtesy of General Motors Corporation, Service Technology Group)

To make the repairs to the terminal, insert the stripped wire through the seal and the connector body (Figure 6-26). Crimp and solder the terminal to the wire. Pull the wire lead and terminal back into the connector body until the terminal is locked (Figure 6-27).

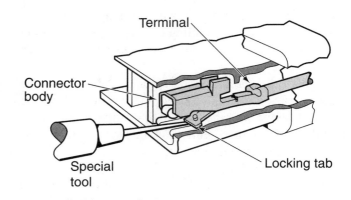

Figure 6-25 Pull up on the lock tang to release the terminal from the connector. (Courtesy of General Motors Corporation, Service Technology Group)

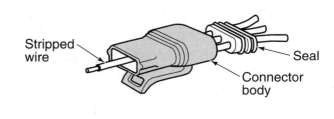

Figure 6-26 The wire lead must be installed into the seal and connector before attaching the terminal. (Courtesy of General Motors Corporation, Service Technology Group)

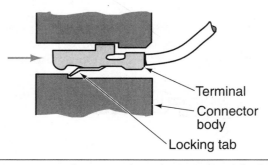

Figure 6-27 Make sure that the terminal locks into the connector body. (Courtesy of General Motors Corporation, Service Technology Group)

Summary

❏ The most common methods of wire repair include wrapping damaged insulation with electrical tape or tubing, crimping the connections with solderless connectors, and soldering splices.

❏ The two most common methods of splicing copper wire are with solderless connectors or by soldering.

❏ Crimping means to bend, or deform by pinching a connector so the wire connection is securely held in place.

❏ Soldering is the process of using heat and solder (a mixture of metals with a low melting point) to make a splice or connection.

❏ Always use rosin core solder when making electrical repairs. Acid core solder is used for other purposes than electrical repairs and can cause the wire to corrode, which would lead to high resistance.

❏ Twisted/shielded wire is used in computer circuits. These wires may carry as low as 0.1 ampere of current. It is important that the splice made in these wires does not have any resistance.

❏ Molded connectors cannot be separated for repairs. If the connector is damaged, it must be cut off and a new connector spliced in.

❏ Hard-shell connectors usually provide a means of removing the terminals for repair. Use a pick, or the special tool, to depress the locking tang of the connector.

❏ There are two types of metri-pack connectors: pull to seat and push to seat. These terms depict the method used to install the terminals into the connector.

Terms to Know

Crimping	Push to seat	Splicing
Heat shrink tubing	Solderless connections	Splice clip
Pull to seat	Soldering	Tap splice connector

ASE Style Review Questions

1. Splicing copper wire is being discussed. Technician A says it is acceptable to use solderless connections. Technician B says acid core solder should not be used on copper wires. Who is correct?
   a. A only
   b. B only
   c. Both A and B
   d. Neither A nor B

2. Technician A says when replacing a fusible link, the gauge size of the replacement link can be decreased but never increased. Technician B says a 14-gauge fusible link can be replaced with an equivalent length of 14-gauge stranded wire. Who is correct?
   a. A only
   b. B only
   c. Both A and B
   d. Neither A nor B

3. The fuse for the parking lights is open. Technician A says find the cause for the blown fuse. Technician B says the fuse probably wore out due to age. Who is correct?
   a. A only
   b. B only
   c. Both A and B
   d. Neither A nor B

4. Repairs to a twisted/shielded wire are being discussed. Technician A says a twisted/shielded wire carries high current. Technician B says because a twisted/shielded wire carries low current, any repairs to the wire must not increase the resistance of the circuit. Who is correct?
   a. A only
   b. B only
   c. Both A and B
   d. Neither A nor B

5. Replacement of fusible links is being discussed. Technician A says not all open fusible links are detectable by visual inspection. Technician B says to test for battery voltage on both sides of the fusible link to confirm its condition. Who is correct?
   a. A only
   b. B only
   c. Both A and B
   d. Neither A nor B

6. Technician A says troubleshooting is the diagnostic procedure of locating and identifying the cause of the fault. Technician B says troubleshooting is a step-by-step process of elimination by use of cause and effect. Who is correct?
   a. A only
   b. B only
   c. Both A and B
   d. Neither A nor B

7. Technician A says a replacement fusible link should be at least 9 inches long. Technician B says if the damaged fusible link feeds two harness wires, use one replacement fusible link. Who is correct?
   a. A only
   b. B only
   c. Both A and B
   d. Neither A nor B

8. Repairing connectors is being discussed. Technician A says molded connectors are a one-piece design and cannot be separated for repairs. Technician B says to replace the seals when repairing the weather-pack connector. Who is correct?
   a. A only
   b. B only
   c. Both A and B
   d. Neither A nor B

9. Technician A says all metri-pack connectors use male terminals. Technician B says the connectors of the metri-pack are called pull to seat or push to seat to depict the method used to install the terminals into the connector. Who is correct?
   a. A only
   b. B only
   c. Both A and B
   d. Neither A nor B

10. Technician A says acid core solder should be used whenever copper wires are to be soldered. Technician B says solderless connectors should not be used if a weather-resistant connection is desired. Who is correct?
    a. A only
    b. B only
    c. Both A and B
    d. Neither A nor B

7 Automotive Batteries

Objective

Upon completion and review of this chapter, you should be able to:

❏ Explain the purposes of the battery.

❏ Describe the construction of conventional, maintenance-free, hybrid, and recombination batteries.

❏ Define the main elements of the battery.

❏ Explain the chemical action that occurs to produce current in a battery.

❏ Explain the chemical reaction that occurs in the battery during cycling.

❏ Describe the differences, advantages, and disadvantages between different types of batteries.

❏ Describe the different types of battery terminals used.

❏ Describe the methods used to rate batteries.

❏ Determine the correct battery to be installed into a vehicle.

❏ Explain the effects of temperature on battery performance.

❏ Describe the different loads or demands placed upon a battery during different operating conditions.

❏ Explain the major reasons for battery failure.

❏ Define battery-related terms such as deep cycle, electrolyte solution, and gassing.

Introduction

An automotive battery is an **electrochemical** device capable of producing electrical energy. When the battery is connected to an external load, such as a starter motor, an energy conversion occurs that results in an electrical current flowing through the circuit. The battery does not store electricity as electrons. The battery stores energy in chemical form. Electrical energy is produced in the battery by the chemical reaction that occurs between two dissimilar plates that are immersed in an electrolyte solution. The automotive battery produces direct current (DC) electricity that flows in only one direction.

When discharging the battery (current flowing from the battery), the battery changes chemical energy into electrical energy. It is through this change that the battery releases stored energy. During charging (current flowing through the battery from the charging system), electrical energy is converted into chemical energy. As a result, the battery can store energy until it is needed.

The automotive battery has several important functions, including:

1. It operates the starting motor, ignition system, electronic fuel injection, and other electrical devices for the engine during cranking and starting.

2. It supplies all the electrical power for the vehicle accessories whenever the engine is not running or when the vehicle's charging system is not working.

3. It furnishes current for a limited time whenever electrical demands exceed charging system output.

4. It acts as a stabilizer of voltage for the entire automotive electrical system.

5. It stores energy for extended periods of time.

The largest demand placed on the battery occurs when it must supply current to operate the starter motor.

The amperage requirements of a starter motor may be over several hundred amperes. This requirement is also affected by temperatures, engine size, and engine condition.

After the engine is started, the vehicle's charging system works to recharge the battery and to provide the current to run the electrical systems. Most AC generators have a maximum output of 60 to 90 amperes. However, many luxury vehicles with extensive electrical accessories may be equipped with generators rated above 110 amperes. This is usually enough to operate all of the vehicle's electrical systems and meet the demands of these systems. However, under some conditions (such as engine running at idle) generator output is below its maximum rating. If there are enough electrical accessories turned on during this time (heater, wipers, headlights, and radio) the demand may exceed the AC generator output. The total demand may be 20 to 30 amperes. During this time the battery must supply the additional current.

Even with the ignition switch turned off, there are electrical demands placed on the battery. Clocks, memory seats, engine computer memory, body computer memory, and electronic sound system memory are all examples of **key-off loads**. The total current draw of key-off loads is usually less than 30 milliamperes. Key off loads are referred to as parasitic loads.

In the event that the vehicle's charging system fails, the battery must supply all of the current necessary to run the vehicle. Most batteries will supply 25 amperes for approximately 120 minutes before discharging low enough to cause the engine to stop running.

The amount of electrical energy that a battery is capable of producing depends on the size, weight, active area of the plates, and the amount of sulfuric acid in the electrolyte solution. In this chapter you will study the design and operation of different types of batteries currently used in automobiles. These include conventional batteries, maintenance-free batteries, hybrid batteries, and recombination batteries.

Conventional Batteries

The conventional battery is constructed of seven basic components:

1. Positive plates
2. Negative plates
3. Separators
4. Case
5. Plate straps
6. Electrolyte
7. Terminals

The difference between "3-year" and "5-year" batteries is the amount of **material expanders** used in the construction of the plates and the number of plates used to build a cell. Material expanders are fillers that can be used in place of the active materials. They are used to keep the manufacturing costs low.

A **plate**, either positive or negative, starts with a frame structure called a **grid**. The grid has horizontal and vertical grid bars that intersect at right angles. An active material made from ground lead alloys (usually antimony) oxide, acid, and material expanders is pressed into the grid in paste form. About 5% to 6% antimony is added to increase the strength of the grid. The positive plate is given a "forming charge" that converts the lead oxide paste into lead peroxide. The lead peroxide is composed of small grains of particles. This gives the plate a high degree of porosity, allowing the electrolyte to penetrate the plate. The negative plate is given a "forming charge" that converts the paste into sponge lead.

The negative and positive plates are arranged alternately in each **cell** element (Figure 7-1). Each cell element can consist of 9 to 13 plates. Usually the negative plate groups contain one more plate than the positive plate groups to help equalize the chemical activity. The positive and negative plates are insulated from each other by **separators** made of microporous materials. Many batteries have envelope-type separators that retain active materials near the plate. The construction of the element is completed when all of the positive plates are connected to each other and all of the negative plates are connected to each other. The connection of the plates is by **plate straps** (Figure 7-2).

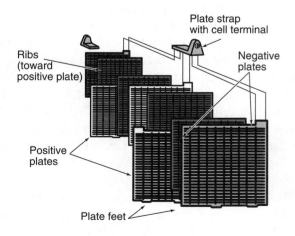

Figure 7-1 A battery cell consists of alternate positive and negative plates. (Reprinted with permission of Ford Motor Company)

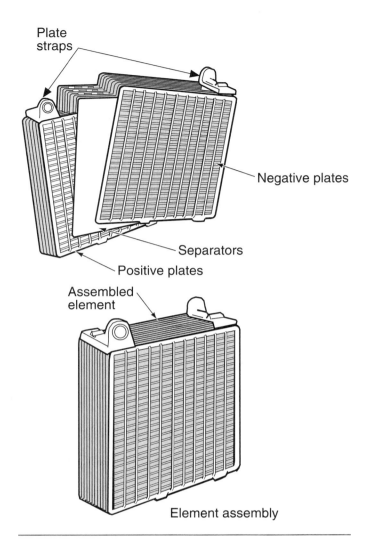

Figure 7-2 Construction of a battery element. (Reprinted with permission of Ford Motor Company)

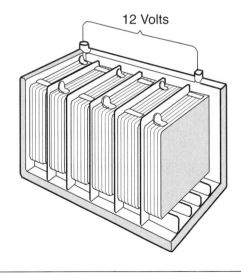

Figure 7-3 The 12-volt battery consists of six 2-volt cells that are wired in series. (Reprinted with permission of Ford Motor Company)

A typical 12-volt automotive battery is made up of six cells connected in series (Figure 7-3). This means the positive side of a cell element is connected to the negative side of the next cell element. This is repeated throughout all six cells. By connecting the cells in series the current capacity of the cell and cell voltage remain the same. The six cells produce 2.1 volts each. Wiring the cells in series produces the 12.6 volts required by the automotive electrical system. The plate straps provide a positive cell connection and a negative cell connection. The cell connection may be one of three types: through the partition, over the partition, or external (Figure 7-4). The cell elements are submerged in a cell case filled with **electrolyte solution**. The electrolyte solution consists of 64% water and 36% sulfuric acid, by weight. Electrolyte is both conductive and reactive.

CAUTION *Electrolyte is very corrosive. It can cause severe injury if it comes in contact with skin and/or eyes. If you come into contact with battery acid, wash immediately with a water and baking soda solution (baking soda will neutralize the acid). If you get the acid in your eyes, flush them immediately with cool water or a commercial eye wash. Then seek medical treatment immediately. If you swallow any electrolyte, do not induce vomiting. Seek medical treatment immediately.*

WARNING *Do not add electrolyte to the battery (after the initial fill) if the cells are low. Add only distilled water. Distilled water has the minerals removed. The minerals in regular tap water can cause the battery cells to short out because they are conductive.*

The battery case is made of polypropylene, hard rubber, and plastic base materials. The battery case must be capable of withstanding temperature extremes, vibration, and acid absorption. The cell elements sit on raised supports in the bottom of the case. By raising the cells, chambers are formed at the bottom of the case that trap the sediment that flakes off the plates. If the sediment was not contained in these chambers, it could cause a conductive connection across the plates and short the cell. The case is fitted with a one-piece cover.

Because the conventional battery releases hydrogen gas when it is being charged, the case cover will have vents. The vents are located in the cell caps of a conventional battery (Figure 7-5).

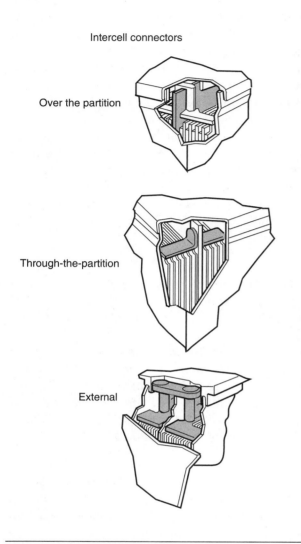

Intercell connectors

Over the partition

Through-the-partition

External

Figure 7-4 The cell elements can be connected using one of three intercell connection methods. (Reprinted with permission of Ford Motor Company)

Figure 7-5 The vents of a conventional battery allow the release of gases.

CAUTION *The hydrogen gases produced from a charging battery are very explosive. Exploding batteries are responsible for over 15,000 injuries per year that are severe enough to require hospital treatment. Do not smoke, have any open flames, or cause any sparks near the battery. Also, do not lay tools on the battery. They may short across the terminals causing the battery to explode. Always wear eye protection and proper clothing when working near the battery. Also, because most jewelry is an excellent conductor of electricity, do not wear it when performing work on or near the battery.*

WARNING *A battery that has been rapidly discharged will create hydrogen gas. Do not attach jumper cables to a weak battery if starting the vehicle has been attempted. Wait for at least 10 minutes before connecting the jumper cable and attempting to start the vehicle.*

Chemical Action

Activation of the battery is through the addition of electrolyte. This solution causes the chemical actions to take place between the lead peroxide of the positive plates and the sponge lead of the negative plates. The electrolyte is also the carrier that moves electric current between the positive and negative plates through the separators.

The automotive battery has a fully charged specific gravity of 1.265 corrected to 80°F. Therefore, a specific gravity of 1.265 for electrolyte means it is 1.265 times heavier than an equal volume of water. As the battery discharges, the specific gravity of the electrolyte decreases.

Specific gravity is the weight of a given volume of a liquid divided by the weight of the same volume of water. Water, therefore, has a specific gravity of 1.000. The electrolyte, being heavier than water, has a specific gravity of more than 1.000. The electrolyte of a fully charged battery has a specific gravity of about 1.265. This decreases as the battery discharges because the electrolyte becomes more like water. The specific gravity of a battery can give you an indication of how charged a battery is:

Fully charged:	1.265 specific gravity
75% charged:	1.225 specific gravity
50% charged:	1.190 specific gravity
25% charged:	1.155 specific gravity
Discharged:	1.120 or lower specific gravity

These specific gravity values may vary slightly according to the design of the battery. However, regardless of the design, the specific gravity of the electrolyte in all batteries will decrease as the battery discharges. Temperature of the electrolyte will also affect its specific gravity. All specific gravity specifications are based on a standard temperature of 80°F. When the temperature is above that temperature, the specific gravity is lower. When the temperature is below that standard, the specific gravity increases. Therefore all specific gravity measurements must be corrected for temperature. A general rule to follow is to add 0.004 for every 10 degrees above 80 degrees and subtract 0.004 for every 10 degrees below 80 degrees.

In operation, the battery is being partially discharged and then recharged. This represents an actual reversing of the chemical action that takes place within the battery. The constant cycling of the charge and discharge modes slowly wears away the active materials on the cell plates. This action eventually causes the battery plates to sulfate. The battery must be replaced once the sulfation of the plates has reached the point that there is insufficient active plate area.

In the charged state, the positive plate material is essentially pure lead peroxide, PbO_2. The active material of the negative plates is spongy lead, Pb. The electrolyte is a solution of sulfuric acid, H_2SO_4, and water. The voltage of the cell depends on the chemical difference between the active materials.

The illustration (Figure 7-6) shows what happens to the plates and electrolyte during discharge. The lead (Pb) from the positive plate combines with sulfate (SO_4) from the acid, forming lead sulfate ($PbSO_4$). While this is occurring, oxygen (O_2) in the active material of the positive plate joins with the hydrogen from the electrolyte, forming water (H_2O). This water dilutes the acid concentration.

A similar reaction is occurring in the negative plate. Lead (Pb) is combining with sulfate (SO_4) forming lead sulfate (PbSO4). The result of discharging is changing the positive plate from lead dioxide into lead sulfate and changing the negative plate into lead sulfate. Discharging a cell makes the positive and negative plates the same. Once they are the same, the cell is discharged.

The charge cycle is exactly the opposite (Figure 7-7). The lead sulfate ($PbSO_4$) in both plates is split into its original forms of lead (Pb) and sulfate (SO_4). The water in the electrolyte splits into hydrogen and oxygen. The hydrogen (H_2) combines with the sulfate to become sulfuric acid again (H_2SO_4). The oxygen combines with the positive plate to form the lead peroxide. This now puts the plates and the electrolyte back in their original form and the cell is charged.

If electrolyte and dirt are allowed to accumulate on the top of the battery case it may create a conductive path between the positive and negative terminals. This can result in a constant discharge on the battery.

Maintenance-Free Batteries

Many batteries are referred to as **maintenance-free**. This means there is no provision for the addition of water to the cells since the battery is sealed. The maintenance-free battery contains cell plates made of a slightly different compound. The plate grids contain calcium, cadmium, or strontium to reduce **gassing** (the conversion of the solution into hydrogen and oxygen gas) and self-discharge. The antimony used in conventional batteries is not used in maintenance-free batteries because it increases the breakdown of water into hydrogen and oxygen and because of its low resistance to overcharging. The use of calcium, cadmium, or strontium reduces the amount of vaporization that takes place during normal operation. The grid may be constructed with additional supports to increase its strength and to provide a shorter path, with less resistance, for the current to flow to the top tab.

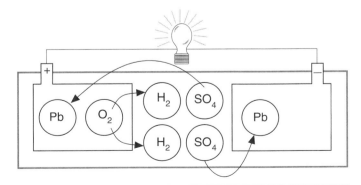

Figure 7-6 Chemical action that occurs inside of the battery during the discharge cycle.

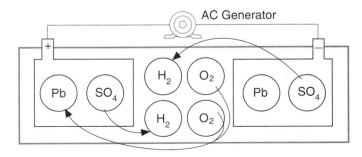

Figure 7-7 Chemical action inside of the battery during the charge cycle.

Each plate is wrapped and sealed on three sides by an envelope design separator. The envelope is made from microporous plastic. By enclosing the plate in an envelope, the plate is insulated and reduces the shedding of the active material from the plate.

The battery is sealed except for a small vent so the electrolyte and vapors cannot escape. An expansion or condensation chamber allows the water to condense and drain back into the cells. Because the water cannot escape from the battery, it is not necessary to add water to the battery on a periodic basis. Containing the vapors also reduces the possibility of corrosion and discharge through the surface because of electrolyte on the surface of the battery. Vapors only leave the case when the pressure inside the battery is greater than atmospheric pressure.

Some maintenance-free batteries have a built-in **hydrometer** that shows the state of charge (Figure 7-8). A hydrometer checks the specific gravity of the electrolyte to determine the battery's state of charge. If the dot that is at the bottom of the hydrometer is green, then the battery is fully charged (more than 65% charged). If the dot is black, the battery state of charge is low. If the battery does not have a built-in hydrometer it cannot be tested with a hydrometer because the battery is sealed.

WARNING *If the dot is yellow or clear, do not attempt to recharge the battery. A yellow or clear eye means the electrolyte is low; the battery must be replaced.*

Many manufacturers have revised the maintenance-free battery to a **"low maintenance battery,"** in that the caps are removable for testing and electrolyte level checks. Also the grid construction contains about 3.4% antimony. To decrease the distance and resistance of the path that current flows in the grid and to increase its strength, the horizontal and vertical grid bars do not intersect at right angles (Figure 7-9).

The advantages of maintenance-free batteries over conventional batteries include:

1. A larger reserve of electrolyte above the plates.
2. Increased resistance to overcharging.
3. Longer shelf life (approximately 18 months).
4. Ability to be shipped with electrolyte installed, reducing the possibility of accidents and injury to the technician.
5. Higher cold cranking amps rating.

The major disadvantages of the maintenance-free battery include:

1. **Grid growth** when the battery is exposed to high temperatures. Grid growth refers to the grid growing little metallic fingers that extend through the separators and short out the plates.
2. Inability to withstand **deep cycling** (discharging the battery to a very low state of charge before recharging it).
3. Low reserve capacity.
4. Faster discharge by parasitic loads.
5. Shorter life expectancy.

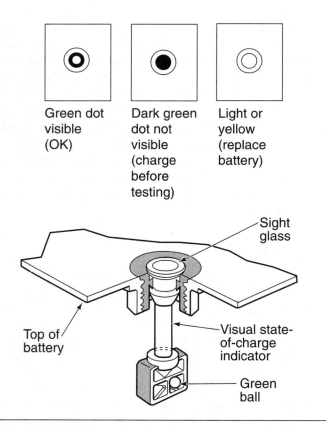

Green dot visible (OK)

Dark green dot not visible (charge before testing)

Light or yellow (replace battery)

Sight glass

Top of battery

Visual state-of-charge indicator

Green ball

Figure 7-8 Built-in hydrometer gives indication of battery condition. (Courtesy of Chrysler Corporation)

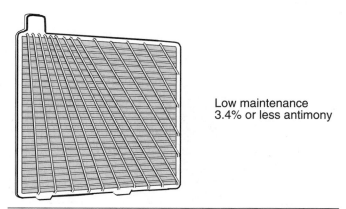

Low maintenance 3.4% or less antimony

Figure 7-9 Low maintenance battery grid with vertical grid bars intersecting at an angle. (Reprinted with permission of Ford Motor Company)

Buick first introduced the storage battery as standard equipment in 1906.

Hybrid Batteries

The **hybrid battery** combines the advantages of the low maintenance and maintenance-free batteries. The hybrid battery can withstand six deep cycles and still retain 100% of its original reserve capacity. The grid construction of the hybrid battery consists of approximately 2.75% antimony alloy on the positive plates and a calcium alloy on the negative plates. This allows the battery to withstand deep cycling while retaining reserve capacity for improved cranking performance. Also, the use of antimony alloys reduces grid growth and corrosion. The lead calcium has less gassing than conventional batteries.

Grid construction differs from other batteries in that the plates have a lug located near the center of the grid. In addition, the vertical and horizontal grid bars are arranged in a radial pattern (Figure 7-10). By locating the lug near the center of the grid and using the radial design, the current has less resistance and a shorter path to follow to the lug). This means the battery is capable of providing more current at a faster rate.

The separators used are constructed of glass with a resin coating. The glass separators offer low electrical resistance with high resistance to chemical contamination. This type of construction provides for increased cranking performance and battery life.

Recombination Batteries

One of the most recent variations of the automobile battery is the **recombination battery** (Figure 7-11). The recombination battery does not use a liquid electrolyte. Instead, it uses separators that hold a gel-type material. The separators are placed between the grids and have very low electrical resistance. Because of this design, output voltage and current are higher than in conventional batteries. The extra amount of available voltage (approximately 0.6V) assists in cold weather starting. Also, gassing is virtually eliminated.

The following are some other safety features and advantages of the recombination battery:

1. Contains no liquid electrolyte. If the case is cracked, no electrolyte will spill.
2. Can be installed in any position, including upside down.
3. Is corrosion free.
4. Has very low maintenance because there is no electrolyte loss.
5. Can last as much as four times longer than conventional batteries.
6. Can withstand deep cycling without damage.
7. Can be rated over 800 cold cranking amperes.

WARNING *The recombination battery requires slightly different testing procedures. Refer to the manufacturer's manual before attempting to test the battery.*

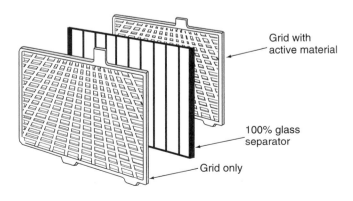

Figure 7-10 Hybrid grid and separator construction. (Reprinted with permission of Ford Motor Company)

Grid with active material

100% glass separator

Grid only

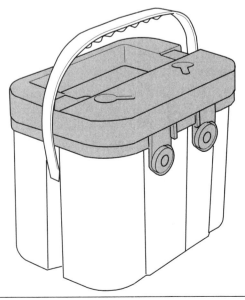

Figure 7-11 The recombination battery is one of the most recent advances in the automotive battery. (Courtesy of Optima Batteries, Inc.)

Battery Terminals

All automotive batteries have two **terminals** to provide a means of connecting the battery plates to the vehicle's electrical system. One terminal is a positive connection, the other is a negative connection. The battery terminals extend through the cover or the side of the battery case. The following are the most common types of battery terminals (Figure 7-12):

1. Post or top terminals: Used on most automotive batteries. The positive post will be larger than the negative post to prevent connecting the battery in reverse polarity.

2. Side terminals: Positioned in the side of the container near the top. These terminals are threaded and require a special bolt to connect the cables. Polarity identification is by positive and negative symbols.

3. L terminals: Used on specialty batteries and some imports.

Battery Ratings

Battery capacity ratings are established by the Battery Council International (BCI) in conjunction with the Society of Automotive Engineers (SAE). Battery cell voltage depends on the types of materials used in the construction of the battery. Current capacity depends on several factors:

1. The size of the cell plates. The larger the surface area of the plates, the more chemical action that can occur. This means a greater current is produced.

Battery terminals

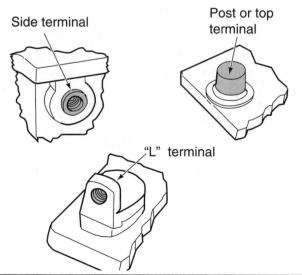

Figure 7-12 The most common types of automotive battery terminals. (Reprinted with permission of Ford Motor Company)

2. The weight of the positive and negative plate active materials.

3. The weight of the sulfuric acid in the electrolyte solution.

The battery's current capacity rating is an indication of its ability to deliver cranking power to the starter motor and of its ability to provide reserve power to the electrical system. The commonly used current capacity ratings are explained in the following sections.

Ampere-Hour Rating

The **ampere-hour rating** is the amount of steady current that a fully charged battery can supply for 20 hours at 80°F (26.78°C) without the cell voltage falling below 1.75 volts. For example, if a battery can be discharged for 20 hours at a rate of 4.0 amperes before its terminal voltage reads 10.5 volts, it would be rated at 80 ampere-hours.

Cold Cranking Amps Rating

Cold cranking amps (CCA) rating is the most common method of rating automotive batteries. It is determined by the load, in amperes, a battery is able to deliver for 30 seconds at 0°F (217.7°C) without terminal voltage falling below 7.2 volts for a 12-volt battery. The cold cranking rating is given in total amperage and is identified as 300 CCA, 400 CCA, 500 CCA, and so on. Some batteries are rated as high as 1,100 CCA.

Reserve Capacity Rating

The **reserve capacity** rating is determined by the length of time, in minutes, that a fully charged battery can be discharged at 25 amperes before battery voltage drops below 10.5 volts. This rating gives an indication of how long the vehicle can be driven, with the headlights on, if the charging system should fail.

Watt-Hour Rating

The starter motor converts the electrical power supplied by the battery into mechanical power, so some battery manufacturers rate their batteries using watt-hour rating. The watt-hour rating of the battery is determined at 0°F (217.7°C) because the battery's capability to deliver wattage varies with temperature. Watt-hour rating is determined by calculating the ampere-hour rating of the battery times the battery voltage.

Battery Size Selection

Some of the aspects that determine the battery rating required for a vehicle include engine size, engine type, climatic conditions, vehicle options, and so on. The requirement for electrical energy to crank the engine increases as the temperature decreases. Battery power drops drastically as temperatures drop below freezing (Figure 7-13). The engine also becomes harder to crank due to the tendency of oils to thicken when cold, which results in increased friction.

As a general rule, it takes 1 ampere of cold cranking power per cubic inch of engine displacement. Therefore, a 200-cubic inch displacement (CID) engine should be fitted with a battery of at least 200 CCA. To convert this into metric, it takes 1 amp of cold cranking power for every 16 cm^3 of engine displacement. A 1.6-liter engine should require at least a battery rated at 100 CCA. This rule may not apply to vehicles that have several electrical accessories. The best method of determining the correct battery is to refer to the manufacturer's specifications.

The battery that is selected should fit the battery holding fixture and the hold-down must be able to be installed. It is also important that the height of the battery not allow the terminals to short across the vehicle hood when it is shut. BCI group numbers are used to indicate the physical size and other features of the batter. This group number does not indicate the current capacity of the battery.

Battery Cables

Battery cables must be of a sufficient capacity to carry the current required to meet all demands (Figure 7-14). Normal 12-volt cable size is usually 4 or 6 gauge. Various forms of clamps and terminals are used to assure a good electrical connection at each end of the cable. Connections must be clean and tight to prevent arcing, corrosion, and high voltage resistance.

The positive cable is usually red (but not always) and the negative cable is usually black. The positive cable will fasten to the starter solenoid or relay. The negative cable fastens to ground on the engine block. Some manufacturers use a negative cable with no insulation. Sometimes the negative battery cable may have a body grounding wire to help assure that the vehicle body is properly grounded.

Temperature	% of Cranking Power
80°F (26.7°C)	100
32°F (0°C)	65
0°F (−17.8°C)	40

Figure 7-13 Temperatures affect the cranking power of the battery.

Battery Hold-Downs

WARNING *Connecting the battery cables in reverse polarity can damage many of the vehicle's computer systems and generators.*

All batteries must be secured in the vehicle to prevent damage and the possibility of shorting across the terminals if it tips. Normal vibrations cause the plates to shed their active materials. **Hold-downs** reduce the amount of vibration and help increase the life of the battery.

A BIT OF HISTORY

The storage battery on early automobiles was mounted under the car. It wasn't until 1937 that the battery was located under the hood for better accessibility. Today with the increased use of maintenance-free batteries, some manufacturers have "buried" the battery again. For example, to access the battery on some vehicles you must remove the left front wheel and work through the wheel well.

Battery Failure

Whenever battery failure occurs, first perform some simple visual inspections. Check the case for cracks, check the electrolyte level in each cell (if possible), and check the terminals for corrosion. The sulfuric acid that vents out with the battery gases attacks the battery terminals and battery cables. As the sulfuric acid reacts with the lead and copper, deposits of lead sulfate and copper sulfate are created (Figure 7-15). These deposits are resistive to electron flow and limit the amount of current that can be supplied to the electrical and starting systems. If the deposits are bad enough, the resistance can increase to a level that prevents the starter from cranking the engine.

One of the most common causes of early battery failure is overcharging. If the charging system is supplying a voltage level over 15.5 volts, the plates may become warped. Warping of the plates results from the excess heat that is generated as a result of overcharging. Overcharging also causes the active material to disintegrate and shed off of the plates.

Figure 7-14 The battery cable is designed to carry the high current required to start the engine and supply the vehicle's electrical systems.

Figure 7-15 Corroded battery terminals reduce the efficiency of the battery.

If the charging system does not produce enough current to keep the battery charged, the lead sulfate that is not converted back to H_2SO_4 can become crystalized on the plates. If this happens, the sulfate is difficult to remove and the battery will resist recharging. This results in battery **sulfation**, which permanently damages the battery.

Vibration is another common reason for battery failure. If the battery is not secure, the plates will shed the active material as a result of excessive vibration. If enough material is shed, the sediment at the bottom of

the battery can create an electrical connection between the plates. The shorted cell will not produce voltage, resulting in a battery that will have only 10.5 volts across the terminals. With this reduced amount of voltage, the starter usually will not be capable of starting the engine. To prevent this problem, make sure that proper hold-down fixtures are used.

During normal battery operation, the active materials on the plates will shed. The negative plate also becomes soft. Both of these events will reduce the effectiveness of the battery.

Summary

❑ An automotive battery is an electrochemical device that provides for and stores electrical energy.

❑ Electrical energy is produced in the battery by the chemical reaction that occurs between two dissimilar plates that are immersed in an electrolyte solution.

❑ An automotive battery has the following important functions:

1. It operates the starting motor, ignition system, electronic fuel injection, and other electrical devices for the engine during cranking and starting.

2. It supplies all the electrical power for the vehicle accessories whenever the engine is not running or at low idle.

3. It furnishes current for a limited time whenever electrical demands exceed charging system output.

4. It acts as a stabilizer of voltage for the entire automotive electrical system.

5. It stores energy for extended periods of time.

❑ Electrical loads that are still placed on the battery when the ignition switch is in the OFF position are called key-off or parasitic loads.

❑ The conventional battery is constructed of seven basic components:

1. Positive plates	3. Separators	5. Plate straps	7. Terminals
2. Negative plates	4. Case	6. Electrolyte	

❑ Electrolyte solution used in automotive batteries consists of 64% water and 36% sulfuric acid by weight.

❑ The electrolyte solution causes the chemical actions to take place between the lead dioxide of the positive plates and the sponge lead of the negative plates. The electrolyte is also the carrier that moves electric current between the positive and negative plates through the separators.

❑ Grid growth is a condition where the grid grows little metallic fingers that extend through the separators and short out the plates.

❑ Deep cycling is discharging the battery almost completely before recharging it.

❑ The recombination battery uses separators that hold a gel-type material in place of liquid electrolyte.

❑ The most common methods of battery rating are cold cranking, reserve capacity, ampere-hour, and watt-hour.

Terms-To-Know

Ampere-hour rating	Hold-downs	Plate straps
Cell	Hybrid battery	Recombination battery
Cold cranking amps (CCA)	Hydrometer	Reserve capacity
Deep cycling	Key-off loads	Separators
Electrochemical	Low maintenance battery	Specific gravity
Electrolyte solution	Maintenance-free	Sulfation
Gassing	Material expanders	Terminals
Grid	Parasitic loads	
Grid growth	Plate	

Review Questions

Short Answer Essays

1. Explain the purposes of the battery.

2. Describe how a technician can determine the correct battery to be installed into a vehicle.

3. Describe the methods used to rate batteries.

4. Describe the different types of battery terminals used.

5. Explain the effects that temperature has on battery performance.

6. Describe the different loads or demands that are placed upon a battery during different operating conditions.

7. List and describe the seven main elements of the conventional battery.

8. What are the major reasons that a battery fails?

9. List at least three safety concerns associated with working on or near the battery.

10. Describe the difference in construction of the hybrid battery as compared to the conventional battery.

Fill-in-the-Blanks

1. An automotive battery is an _____ device that provides for and stores _____ energy.

2. When discharging the battery, it changes _____ energy into _____ energy.

3. The assembly of the positive plates, negative plates, and separators is called the _____ or _____.

4. The electrolyte solution used in automotive batteries consists of _____% water and _____% sulfuric acid.

5. A fully charged automotive battery has a specific gravity of _____ corrected to 80°F.

6. _____ _____ is a condition where the grid grows little metallic fingers that extend through the separators and shorts out the plates.

7. The _____ _____ rating indicates the battery's ability to deliver a specified amount of current to start an engine at low ambient temperatures.

8. The electrolyte solution causes the chemical actions to take place between the lead peroxide of the _____ plates and the _____ _____ of the _____ plates.

9. Some of the aspects that determine the battery rating required for a vehicle include engine _____ , engine _____ , _____ conditions, and vehicle _____ .

10. Electrical loads that are still present when the ignition switch is in the OFF position are called _____ loads.

ASE Style Review Questions

1. Technician A says the battery provides electricity by releasing free electrons. Technician B says the battery stores energy in chemical form. Who is correct?
   a. A only
   b. B only
   c. Both A and B
   d. Neither A nor B

2. Technician A says the largest demand on the battery is when it must supply current to operate the starter motor. Technician B says the current requirements of a starter motor may be over one hundred amperes. Who is correct?
   a. A only
   b. B only
   c. Both A and B
   d. Neither A nor B

3. Technician A says even with the ignition switch turned off, there are electrical demands placed on the battery. Technician B says after the engine is started, the vehicle's charging system works to recharge the battery and to provide the current to run the electrical systems. Who is correct?
   a. A only
   b. B only
   c. Both A and B
   d. Neither A nor B

4. The current capacity rating of the battery is being discussed. Technician A says the amount of electrical energy that a battery is capable of producing depends on the size, weight, and active area of the plates. Technician B says the current capacity rating of the battery depends on the types of materials used in the construction of the battery. Who is correct?
   a. A only
   b. B only
   c. Both A and B
   d. Neither A nor B

5. The construction of the battery is being discussed. Technician A says the 12-volt battery consists of positive and negative plates connected in parallel. Technician B says the 12-volt battery consists of six 2-volt cells wired in series. Who is correct?
   a. A only
   b. B only
   c. Both A and B
   d. Neither A nor B

6. Maintenance of a conventional battery is being discussed. Technician A says electrolyte should be added to the battery if the cells are low. Technician B says only distilled water should be used in batteries. Who is correct?
   a. A only
   b. B only
   c. Both A and B
   d. Neither A nor B

7. Battery terminology is being discussed. Technician A says grid growth is a condition where the grid grows little metallic fingers that extend through the separators and short out the plates. Technician B says deep cycling is discharging the battery almost completely before recharging it. Who is correct?
   a. A only
   b. B only
   c. Both A and B
   d. Neither A nor B

8. Battery rating methods are being discussed. Technician A says the ampere-hour is determined by the load in amperes a battery is able to deliver for 30 seconds at 0°F (217.7°C) without terminal voltage falling below 7.2 volts for a 12-volt battery. Technician B says the cold cranking rating is the amount of steady current that a fully charged battery can supply for 20 hours at 80°F (26.7°C) without battery voltage falling below 10.5 volts. Who is correct?
   a. A only
   b. B only
   c. Both A and B
   d. Neither A nor B

9. The hybrid battery is being discussed. Technician A says the hybrid battery can withstand six deep cycles and still retain 100% of its original reserve capacity. Technician B says the grid construction of the hybrid battery consists of approximately 2.75% antimony alloy on the positive plates and a calcium alloy on the negative plates. Who is correct?
   a. A only
   b. B only
   c. Both A and B
   d. Neither A nor B

10. Technician A says connecting the battery cables in reverse polarity can damage many of the vehicle's computer systems. Technician B says the battery must be secured in the vehicle to prevent internal damage and the possibility of shorting across the terminals if it tips. Who is correct?
    a. A only
    b. B only
    c. Both A and B
    d. Neither A nor B

8 Battery Diagnosis and Service

Objective

Upon completion and review of this chapter, you should be able to:

❏ Demonstrate all safety precautions and rules associated with servicing the battery.

❏ Perform a visual inspection of the battery, cables, and terminals.

❏ Test a conventional battery's specific gravity.

❏ Perform an open circuit test and accurately interpret the results.

❏ Test the capacity of the battery to deliver both current and voltage and accurately interpret the results.

❏ Perform a three-minute charge test to determine if the battery is sulfated.

❏ Correctly slow and fast charge a battery, in or out of the vehicle.

❏ Describe the differences between slow and fast charging and when either method should be used.

❏ Jump start a vehicle by use of a booster battery and jumper cables.

❏ Perform a battery drain test and accurately determine the causes of battery drains.

❏ Perform a battery leakage test and determine the needed corrections.

❏ Do a battery terminal test and accurately interpret the results.

❏ Remove, clean, and reinstall the battery properly.

Introduction

A discharged or weak battery can affect more than just the starting of the engine. The battery is the heart of the electrical system of the vehicle. Because of its importance, the battery should be checked whenever the vehicle is brought into the shop for service. A battery test series will show the state of charge and output voltage of the battery, which determines if it is good, in need of recharging, or must be replaced.

There are many different manufacturers of battery test equipment. A popular tester is the Sun Electrical Corporation's VAT-40 (Figure 8-1). In this chapter some of the procedures call for the use of a VAT-40. However, the procedures for the VAT-40 are very similar to other testers. Some testers are computer-based and conduct the

tests automatically after a particular test is selected (Figure 8-2). Always follow the procedures provided by the equipment manufacturer.

Figure 8-1 A Sun VAT-40 battery, starting, and charging system tester.

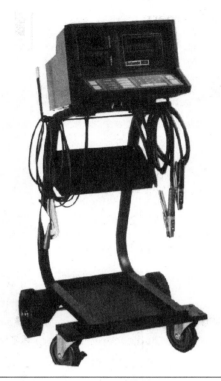

Figure 8-2 A computer-based alternator, regulator, battery, and starter tester. (Reprinted with the permission Ford Motor Company)

General Precautions

Before attempting to do any type of work on or around the battery, the technician must be aware of certain precautions. To avoid personal injury or property damage, take the following precautions:

1. Battery acid is very corrosive. Do not allow it to come in contact with skin, eyes, or clothing. If battery acid gets into your eyes, rinse them thoroughly with clean water and receive immediate medical attention. If battery acid comes in contact with skin, wash with clean water. Baking soda added to the water will help to neutralize the acid. If the acid is swallowed, drink large quantities of water or milk followed by milk of magnesia and a beaten egg or vegetable oil.

2. When making connections to a battery, be careful to observe polarity, positive to positive and negative to negative.

3. When disconnecting battery cables, always disconnect the negative (ground) cable first.

4. When connecting battery cables, always connect the negative cable last.

5. Avoid any arcing or open flames near a battery. The vapors produced by the battery cycling are very explosive. Do not smoke around a battery.

6. Follow manufacturer's instructions when charging a battery. Charge the battery in a well-ventilated area. Do not connect or disconnect the charger leads while the charger is turned on.

7. Do not add additional electrolyte to the battery if it is low. Add only distilled water.

8. Do not wear any jewelry or watches while servicing the battery. These items are excellent conductors of electricity. They can cause severe burns if current flows through them by accidental contact with the battery positive terminal and ground.

9. Never lay tools across the battery. They may come into contact with both terminals, shorting out the battery and causing it to explode.

10. Wear safety glasses or face shield when servicing the battery.

11. If the battery's electrolyte is frozen, allow it to defrost before doing any service or testing of the battery. While it is defrosting, look for leaks in the case. Leakage means the battery is cracked and should be replaced.

Battery Inspection

Before performing any electrical tests, the battery should be inspected, along with the cables and terminals. The complete visual inspection of the battery should include the following items:

1. Battery date code: This provides information as to the age of the battery.

2. Condition of battery case: Check for dirt, grease, and electrolyte condensation. Any of these contaminants can create an electrical path between the terminals and cause the battery to drain. Also check for damaged or missing vent caps and cracks in the case. A cracked or buckled case could be caused by excessive tightening of the hold-down fixture, excessive under-hood temperatures, buckled plates from extended undercharged conditions, freezing, or excessive charge rate.

3. Electrolyte level, color, and odor: If necessary, add distilled water to fill to 1/2 inch above the top of the plates. After adding water, charge the battery before any tests are performed. Discoloration of electrolyte and the presence of a "rotten egg" odor indicate an excessive charge rate, excessive deep cycling, impurities in the electrolyte solution, or an old battery.

4. Condition of battery cables and terminals: Check for corrosion, broken clamps, frayed cables, and loose terminals.

5. Battery abuse: This includes the use of bungee cords and 2 X 4s for hold-down fixtures, too small of a battery rating for the application, and obvious neglect to periodic maintenance. In addition, inspect the terminals for indications that they have been hit by a hammer and for improper cable removal procedures. Finally, check for proper cable length.

6. Battery tray and hold-down fixture: Check for proper tightness. Also check for signs of acid corrosion of the tray and hold-down unit. Replace as needed.

7. If the battery has a built-in hydrometer, check its color indicator.

SERVICE TIP *Grid growth can cause the battery plate to short out the cell. If there is normal electrolyte level in all cells but one, that cell is probably shorted and the electrolyte has been converted to hydrogen gas.*

Charging the Battery

CAUTION *There are many safety precautions associated with charging the battery. The hydrogen gases produced by a charging battery are very explosive. Exploding batteries are responsible for over 15,000 injuries per year that are severe enough to require hospital treatment. Keep sparks, flames, and lighted cigarettes away from the battery. Also, do not use the battery to lay tools on. They may short across the terminals and result in the battery exploding. Always wear eye protection and proper clothing when working near the battery. Also, most jewelry is an excellent conductor of electricity. Do not wear any jewelry when performing work on or near the battery. Do not remove the vent caps while charging. Do not connect or disconnect the charger leads while the charger is turned on.*

WARNING *Before charging a battery that has been in cold weather, check the electrolyte for ice crystals. Do not attempt to charge a frozen battery. Forcing current through a frozen battery may cause it to explode. Allow it to warm at room temperature for a few hours before charging.*

Charging the battery is the process of passing an electric current through the battery in an opposite direction than during discharge. If the battery needs to be recharged, the safest method is to remove the battery from the vehicle. The battery can be charged in the vehi-

cle, however. If the battery is to be charged in the vehicle, it is important to protect any vehicle computers by removing the negative battery cable.

WARNING *If the battery is to be removed from the vehicle, disconnect the negative battery cable first. Lift the battery out with a lift strap (Figure 8-3).*

When connecting the charger to the battery, make sure the charger is turned off. Connect the cable leads to the battery terminals, observing polarity. Attempting to charge the battery while the cables are reversed will result in battery damage. For this reason, many battery chargers have a warning system to alert the technician that the cables are connected in reverse polarity. Rotate

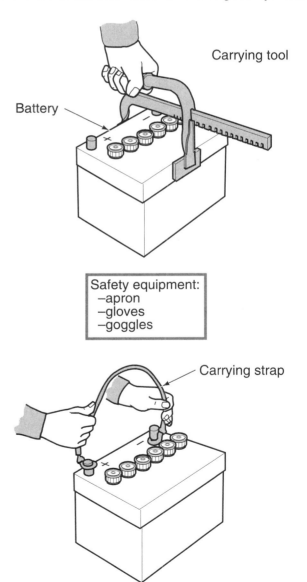

Figure 8-3 Always use a battery lift strap or battery carrier to lift the battery.

the clamps slightly on the terminals to assure a good connection.

Depending on the requirements and amount of time available, the battery can be either slow or fast charged. Each method of charging has its advantages and disadvantages.

Slow Charging

Slow charging recharges the battery at a rate between 3 and 15 amperes for a long period of time. Slow charging the battery has two advantages: It is the only way to restore the battery to a fully charged state, and it minimizes the chances of overcharging the battery. Slow charging the battery causes the lead sulfate on the plates to convert to lead peroxide and sponge lead throughout the thickness of the plate.

Fast Charging

Fast charging the battery will bring the state of charge up high enough to crank the engine. Since fast charging uses high current for a short period of time, it is unable to recharge the battery as effectively as slow charging. Fast charging the battery converts only the lead sulfate on the outside of the plates. The conversion does not go through the plates. After the battery has been fast charged to a point that it will crank the engine, it should then be slow charged to a full state.

WARNING *Fast charging the battery requires that the battery be monitored at all times and the charging time must be controlled. Do not fast charge a battery for longer than two hours. Excessive fast charging can damage the battery. Do not allow the voltage of a 12-volt battery to exceed 15.5 volts. Also don't allow the temperature to rise above 125°F.*

Charge Rate

The charge rate required to recharge a battery depends on several factors:

1. Battery capacity. High capacity batteries require longer charging time.
2. State of charge.
3. Battery temperature.
4. Battery condition.

Slow chargers are the easiest on the battery. However, they require a long period of time to recharge the battery. The basic rule of thumb for slow charging the battery is one ampere for each positive plate in one cell. Use Figure 8-4 to determine the rate of charge according to the reserve capacity of the battery.

SERVICE TIP *If a battery is severely discharged and will not take a slow charge, connect a good battery in parallel (with jumper cables). Fast charge for 30 minutes; then disconnect the good battery and slow charge the bad battery.*

Slow charging of the battery may not always be practical due to the time involved. In these cases fast charging is the only alternative. To determine the fast charge current rate, use the illustration (Figure 8-5).

An alternative method is to connect a voltmeter across the battery terminals while it is charging. If the voltmeter reads fewer than 15 volts, the charging rate is low enough. If the voltmeter reads over 15 volts, reduce the charging rate until voltage reads below 15 volts. Keeping the voltage at 15 volts will insure the quickest charge, and a safe rate for the battery.

WARNING *To prevent damage to the AC generator and computers, disconnect the negative battery cable before fast charging the battery in the vehicle.*

WARNING *Fast charging at rates over 30 amperes for longer than two hours can result in permanent battery damage.*

There are three methods of determining if the battery is fully charged:

1. Specific gravity holds at 1.264 or higher after the battery is stabilized.

Battery Capacity (Reserve Minutes)	Slow Charge
80 minutes or less	10 hrs. @ 5 amperes 5 hrs. @ 10 amperes
Above 80 to 125 minutes	15 hrs. @ 5 amperes 7.5 hrs. @ 10 amperes
Above 125 to 170 minutes	20 hrs. @ 5 amperes 10 hrs. @ 10 amperes
Above 170 to 250 minutes	30 hrs. @ 5 amperes 15 hrs. @ 10 amperes

Courtesy of Battery Council International

Figure 8-4 Table showing the rate and time of slow charging a battery according to reserve capacity.

BATTERY HIGH-RATE CHARGE TIME SCHEDULE					
SPECIFIC GRAVITY READING	CHARGE RATE AMPERES	BATTERY CAPACITY—AMPERE HOURS			
		45	55	70	85
Above 1.225	5	★	★	★	★
1.200–1.225	35	30 min.	35 min.	45 min.	55 min.
1.175–1.200	35	40 min.	50 min.	60 min.	75 min.
1.150–1.175	35	50 min.	65 min.	80 min.	105 min.
1.125–1.150	35	65 min.	80 min.	100 min.	125 min.

★ Charge at 5-ampere rate until specific gravity reaches 1.250 @ 80°F.

Figure 8-5 Charging rates based on state of charge and battery capacity rating. Electrolyte temperatures should not exceed 125°F during charging.

2. An open circuit voltage test indicates 12.68 or higher after the battery has been stabilized.

3. The ammeter on the battery charger falls to approximately 3 amperes or less and remains at that level for one hour.

A BIT OF HISTORY

Lead acid batteries date back to 1859. Alexander Graham Bell used a primitive battery to make his first local call in 1876. Once it was learned how to recharge the lead acid batteries, they were installed into the automobile. These old-style batteries could not hold a charge very well. It was believed that placing the battery on a concrete floor made them discharge faster. Although this fable has no truth to it, the idea has hung on for years.

Battery Test Series

When the battery and cables have been completely inspected and any problems have been corrected, the battery is ready to be tested further. For the tests to be accurate, the battery must be fully charged.

Battery Terminal Test

The **battery terminal test** checks for poor electrical connections between the battery cables and terminals. This simple test will establish whether or not the terminal connection is good. It is good practice to perform this test anytime the battery cable is disconnected and recon-

nected to the terminals. By performing this test, comebacks, due to loose or faulty connections, can be reduced.

Connect the negative voltmeter test lead to the cable clamp and connect the positive meter lead to the battery terminal (Figure 8-6). Disable the ignition system to prevent the vehicle from starting. This may be done by removing the ignition coil secondary wire from the distributor cap and putting it to ground (Figure 8-7). On vehicles with high energy ignition (HEI) remove the battery lead to the distributor (Figure 8-8). Do not ground this lead.

WARNING *Always refer to the manufacturer's service manual for the correct procedure for disabling the ignition system.*

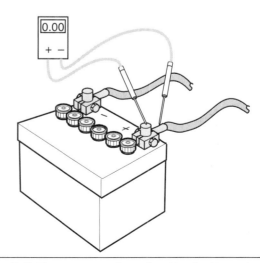

Figure 8-6 Test connections for the battery terminal test.

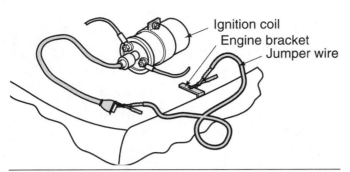

Figure 8-7 Grounding the coil's secondary cable to prevent the engine from starting and to protect the coil.

Crank the engine and observe the voltmeter reading. If the voltmeter shows over 0.3 volt, there is a high resistance at the cable connection. Remove the battery cable using the terminal puller (Figure 8-9). Clean the cable ends and battery terminals (Figure 8-10).

Battery Leakage Test

Battery drain can be caused by a dirty battery. The dirt can actually allow current flow over the battery case. This current flow can drain a battery as quickly as leaving a light on. A **battery leakage test** is conducted to see if current is flowing across the battery case. To perform a battery leakage test, set a voltmeter to a scale around 12 volts DC. Connect the negative test lead to the negative

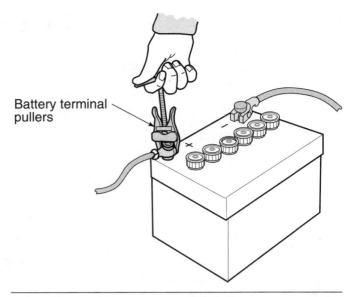

Figure 8-9 Use battery terminal pullers to remove the cable end from the terminal. Do not pry the clamp off.

terminal of the battery. Move the red test lead across the top and sides of the battery case (Figure 8-11). If the meter reads voltage, a current path from the negative terminal of the battery to its positive terminal is being completed through the dirt. Keep in mind you should not measure voltage anywhere on the case of the battery. If voltage is present, remove the battery and use a baking soda and water mixture to clean the case (Figure 8-12). When cleaning the battery, don't allow the baking soda and water solution to enter its cells. After the case is clean, rinse it off with clean water.

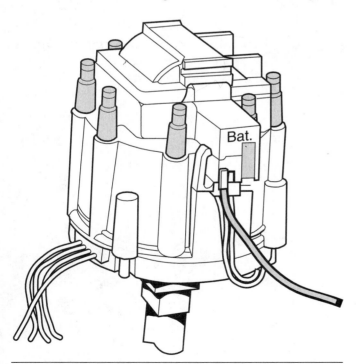

Figure 8-8 To disable the HEI ignition system, disconnect the BAT wire from the coil connector. Similar disconnections are done to disable EI systems.

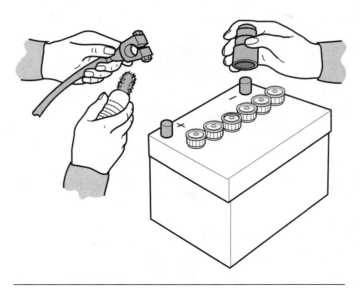

Figure 8-10 The terminal cleaning tool is used to clean the clamp and terminal.

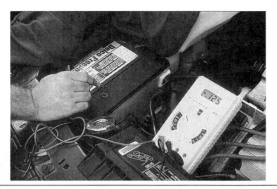

Figure 8-11 Performing the battery leakage test.

Figure 8-12 If the leakage test indicates any voltage, clean the top of the battery with baking soda and water.

State of Charge Test

Measuring the **state of charge** is a check of the battery's electrolyte and plates. It can be determined by testing the specific gravity of the electrolyte using a hydrometer (Figure 8-13).

Follow these steps to test the battery's state of charge:

1. Remove all battery vent caps.

2. Check the electrolyte level. It must be high enough to withdraw the correct amount of solution into the hydrometer.

3. Squeeze the bulb and place the pickup tube into the electrolyte of a cell.

4. Slowly release the bulb. Draw in enough solution until the float is freely suspended in the barrel. Hold the hydrometer in a vertical position.

The float rises and the specific gravity is read where the float scale intersects the top of the solution (Figure 8-14). The reading must also be compensated for temperatures.

Test Results

As a battery becomes discharged, its electrolyte has a larger percentage of water. Thus, a discharged battery's electrolyte will have a lower specific gravity number than that of a fully charged battery.

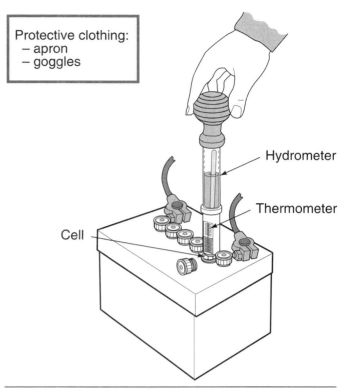

Figure 8-13 Drawing electrolyte into the hydrometer. (Courtesy of Chrysler Corporation)

A fully charged battery will have a hydrometer reading of at least 1.265. Remember, the specific gravity is also influenced by the temperature of the electrolyte and the readings must be corrected to the temperature. If the

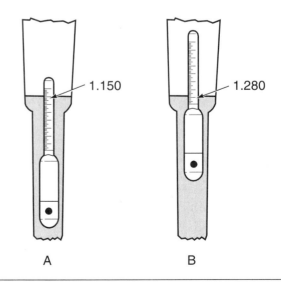

Figure 8-14 The specific gravity of the electrolyte is read at the point where the electrolyte intersects the float. (A) shows a low reading. (B) shows a high reading.

corrected hydrometer reading is below 1.265 the battery needs recharging, or it may be defective.

A defective battery can be determined with a hydrometer by checking every cell. If the specific gravity has a 0.050 point variation between the highest and lowest cell readings, the battery is defective and should be replaced (Figure 8-15). When all the cells have an equal gravity, even if all are low, the battery can usually be regenerated by recharging.

Specific gravity tests should not be used as the sole determinant of battery condition. When the specific gravity of all the cells is good or bad, the voltage of the battery must be considered before coming to a conclusion about the battery's condition. A battery with low specific gravity and acceptable voltage is normally only discharged, perhaps due to a charging system problem. However, a battery with good specific gravity readings but low voltage readings is always bad and needs to be replaced.

Open Circuit Voltage Test

The **open circuit voltage test** is used to determine the battery's state of charge. It is used when a hydrometer is not available or cannot be used. To obtain accurate test results, the battery must be **stabilized** (surface charge removed). If the battery has just been recharged, perform the capacity test, then wait at least 10 minutes to allow battery voltage to stabilize. Connect a voltmeter across the battery terminals, observing polarity (Figure 8-16). Measure the open circuit voltage. Take the reading to the 1/10 volt.

To analyze the open circuit voltage test results, consider that a battery at a temperature of 80°F, in good condition, should show about 12.4 volts (Figure 8-17). If the state of charge is 75% or more, the battery is considered "charged."

Figure 8-16 Open circuit voltage test using a DOM.

Capacity Test

The **capacity test** provides a realistic determination of the battery's condition by checking its ability to perform when loaded. For this test to be accurate, the battery must pass the state of charge or open circuit voltage test. If it does not, recharge the battery and test it again.

In the capacity test, a specified load is placed on the battery while the terminal voltage is observed. A good battery should produce current equal to 50% of its cold-cranking rating (or three times its ampere-hour rating) for 15 seconds and still provide 9.6 volts to start the engine.

To do this test using a battery tester with a carbon pile:

1. Charge the battery, if necessary, to at least a specific gravity reading of 1.225 in all cells.

2. Determine the load test specification. This specification is either 50% of the cold-cranking amperage rating, three times the amp-hour rating specified on the battery label, or provided by the vehicle manufacturer.

3. Connect the large load leads across the battery terminals, observing polarity (Figure 8-18).

4. Zero the ammeter.

5. Connect the amps inductive pickup around one of the tester leads.

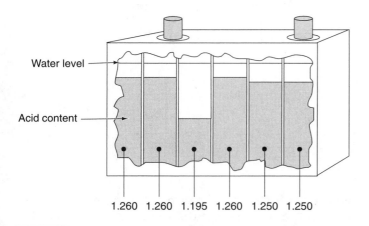

Water level
Acid content

1.260   1.260   1.195   1.260   1.250   1.250

Figure 8-15 A defective cell can be determined by the specific gravity readings.

Open Circuit Voltage	State of Charge
12.6 or greater	100%
12.4 to 12.6	70–100%
12.2 to 12.4	50–75%
12.0 to 12.2	25–50%
11.7 to 12.0	0–25%
11.7 or less	0%

Figure 8-17 The results of the open circuit voltage test indicate the state of charge.

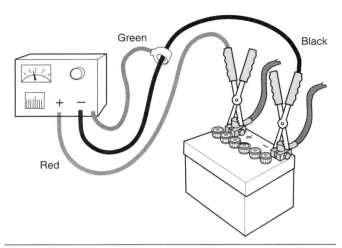

Figure 8-18 Capacity test connections using a battery tester with a carbon pile.

6. Set the test selector to the starting position.

7. Turn the load control knob slowly to apply the amount of load determined in step 2.

8. Read the voltmeter while applying the load for 15 seconds. Do not exceed the 15-second limit. Turn the carbon pile off and record the reading.

9. Check voltage readings against the chart (Figure 8-19).

If voltage level is below the specifications listed in Figure 8-19, observe the battery voltage for the next 10 minutes. If the voltage raises to 12.45 volts or higher, perform the three-minute charge test.

If the voltage does not return to 12.4 volts, recharge the battery until the open circuit test indicates a voltage of 12.66 volts. Repeat the capacity test. If the battery fails again, replace the battery.

If the capacity test readings of a clean and fully charged battery are equal to or above specification, the battery is good. If the battery tests are borderline, perform the three-minute charge test.

Electrolyte Temperature								
F°	70+	60	50	40	30	20	10	0
C°	21+	16	10	4	−1	−7	−12	−18
Minimum Voltage (12 volt Battery)	9.6	9.5	9.4	9.3	9.1	8.9	8.7	8.5

Figure 8-19 Correcting the readings of the capacity test to temperature readings.

Three-minute Charge Test

If the battery fails the load test, it is not always the fault of the battery. It is possible the battery has not been receiving an adequate charge from the charging system. The **three-minute charge test** determines the battery's ability to accept a charge. If the battery is sulfated, the ability of the cells to deliver current and accept a charge are reduced. A battery must have failed the load test to get accurate results from a three-minute charge test.

To conduct the three-minute test:

1. Remove the ground cable, or disconnect the vehicle's computer.

2. Connect a battery charger to the battery, observing polarity.

3. Connect a voltmeter across the battery terminals, observing polarity.

4. Turn on the battery charger to 40 amperes for conventional batteries and 20 to 25 amperes for maintenance-free batteries.

5. Maintain this rate of charge for three minutes.

6. Check the voltage reading at three minutes. If fewer than 15.5 volts, the battery is not sulfated. If the voltmeter reading is above 15.5 volts, the battery is sulfated, or there is a poor internal connection.

7. If the battery passes the three-minute test, slowly recharge the battery and do the load test again.

8. If the battery passes the load test this time, test the charging system.

WARNING *Some battery manufacturers, such as Delco, do NOT recommend the three-minute charge test.*

Figure 8-20 shows a simple but logical troubleshooting sequence for batteries. Whenever a battery's condition is questionable, follow this same logic.

Battery Removal and Cleaning

It is natural for dirt and grease to collect on the top of the battery. If allowed to accumulate, the dirt and grease can form a conductive path between the battery terminals. This may result in a drain on the battery. Also, normal battery gassing will deposit sulfuric acid as the vapors condense. Over a period of time the sulfuric acid will corrode the battery terminals, cable clamps, and hold-down fixtures. As the corrosion builds on the terminals, it adds resistance to the entire electrical system.

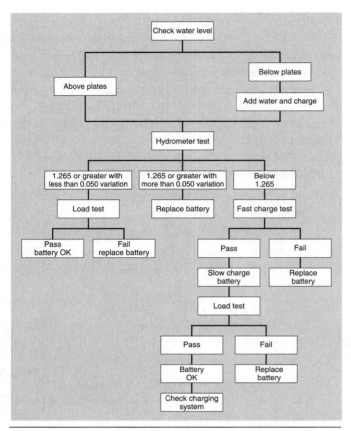

Figure 8-20 A battery troubleshooting flowchart.

Periodic battery cleaning will eliminate these problems. To be able to clean the battery correctly, it is best to remove it from the vehicle. Removing the battery protects the vehicle's finish and other underhood components. Follow Photo Sequence 4 for the procedure for removing the battery from the vehicle. Use special battery terminal pliers to prevent damaging the clamp bolt. Consult the manufacturer's service manual for precautions concerning the vehicle's computer controls.

With the battery removed from the vehicle, wash the entire case (and the hold-down fixtures that were removed from the vehicle) with a solution of one tablespoon of baking soda to one quart of water. Use a cleaning brush to remove heavy corrosion. It may be necessary to use a good detergent to clean the battery of grease and dirt. After the battery case is cleaned, rinse it with clean water followed by drying with paper towels. Dispose of the towels properly. Use the baking soda and water solution to clean the cable clamps. Finally, clean the terminal and cable clamps with the cleaning tool.

WARNING *Do not allow the baking soda solution to enter the battery cells. This will neutralize the acid in the electrolyte and destroy a battery.*

CUSTOMER CARE *Before installing the battery back into the vehicle, it is a good practice to clean the battery tray. First scrape away any heavy corrosion with a putty knife. Next, clean the tray with a baking soda and water solution. Flush with water and allow to dry. After the tray has dried, paint it with rust-resistant paint. After the paint has completely dried, coat the tray with silicone base spray. These extra steps will protect the tray from corrosion.*

When replacing the battery into the vehicle, make sure it is properly seated in the tray. Connect the hold-down fixture and secure, being careful not to overtighten it. Install the positive cable and secure, then install the negative cable. Be sure to observe polarity. Perform a battery terminal test to confirm good connections.

CAUTION *Be careful not to touch the positive terminal with the wrench when tightening the negative cable clamp.*

Spray the cable clamps with a protective coating to prevent corrosion. A little grease or petroleum jelly will also prevent corrosion. Also protective pads are available that go under the clamp and around the terminal (Figure 8-21).

Battery Drain Test

If the customer says the battery is dead every time they attempt to start the vehicle after it has not been used for a short while, the problem may be a current drain from one of the electrical systems. The most common cause for this type of drain is a light that is not turning off, such as glove box, trunk, or engine compartment illumination lights.

Figure 8-21 A protective pad under the battery clamp prevents corrosion of the clamp and terminal.

Removing the Battery

P3-1 Tools needed to remove the battery from the vehicle: rags, baking soda, pan, terminal pliers, cable clamp spreader, terminal puller, assorted wrenches, terminal and clamp cleaning tool, battery lifting strap, safety glasses, heavy rubber gloves, rubber apron, and fender covers.

P3-2 Place the fender covers on the vehicle to protect the finish.

P3-3 Loosen the clamp bolt for the negative cable using terminal pliers and wrench of correct size. Be careful not to put excessive force against the terminal.

P3-4 Use the terminal puller to remove the cable from the terminal. Do not pry the cable off of the terminal.

P3-5 Locate the negative cable away from the battery.

P3-6 Loosen the clamp bolt for the positive cable and use the terminal puller to remove the cable. If the battery has a heat shield, remove it.

Removing the Battery

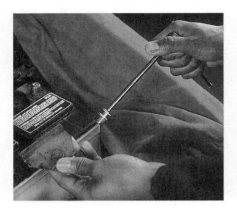

P3-7 Disconnect the hold-down fixture.

P3-8 Using the battery-lifting clamp, remove the battery out of the tray. Keep the battery away from your body. Wear protective clothing to prevent acid spills onto your hands.

P3-9 Transport the battery to the bench. Keep it away from your clothes.

These parasitic drains on the battery can cause various driveability problems. With low battery voltage several problems can result; for example:

1. The computer may go into backup mode or "limp-in" mode of operation.

2. The computer may set false trouble codes.

3. To compensate for the low battery voltage, the computer may raise the engine speed.

The procedure for performing the battery drain test may vary according to the manufacturer. However, battery drain can often be observed by connecting an ammeter in series with the negative battery cable or by placing the inductive ammeter pickup lead around the negative cable. If the meter reads higher than specified amperage, there is excessive drain. Check the service manual for the length of time needed. Most vehicles will time out after a few seconds but in some vehicles, automatic load leveling may require up to 45 minutes. Also, this test is not accurate unless the open circuit voltage reading is 11.5 volts or higher.

If excessive current draw is still present after the time out period, visually check the trunk, glove box, and under-hood lights to see if they are on. If they are, remove the bulb and watch the battery drain. If the drain is now within specifications, find out why the circuit is staying on and repair the problem. If the cause of the drain is not the lights, go to the fuse panel or distribution center and remove one fuse at a time while watching the ammeter. When the drain decreases, the circuit protected by the fuse you removed last is the source of the problem.

Jumping the Battery

WARNING *Before charging a battery that has been in cold weather, check the electrolyte for ice crystals. Do not attempt to charge a frozen battery. Forcing current through a frozen battery may cause it to explode. Allow it to warm at room temperature for a few hours before charging.*

There will be times when you will have to use a boost battery and jumper cables to jump start a vehicle (Figure 8-22). It is important that all safety precautions be followed. Jump starting a dead battery can be dangerous if it is not done correctly. The following steps should be followed to safely jump start most vehicles:

1. Make sure the two vehicles are not touching each other. The excessive current flow through the vehicles' bodies can damage the small ground straps that attach the engine block to the frame. These small wires are designed to carry only 30 amperes. If the vehicles are

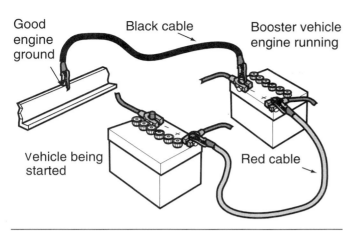

Good engine ground   Black cable   Booster vehicle engine running

Vehicle being started   Red cable

Figure 8-22 Proper jumper cable connections for jump starting a vehicle.

touching, as much as 400 amperes may be carried through them.

2. For each vehicle, engage the parking brake and put the transmission in neutral or park.

3. Turn off the ignition switch and all accessories, on both vehicles.

4. Attach one end of the positive jumper cable to the disabled battery's positive terminal.

5. Connect the other end of the positive jumper cable to the booster battery's positive terminal.

6. Attach one end of the negative jumper cable to the booster battery's negative terminal.

7. Attach the other end of the negative jumper cable to an engine ground on the disabled vehicle.

WARNING *Do not connect this cable end to the battery negative terminal. Doing so may create a spark that will cause the battery to explode.*

8. Attempt to start the disabled vehicle. If the disabled vehicle does not start readily, start the jumper vehicle and run at fast idle to prevent excessive current draw.

9. Once the disabled vehicle starts, disconnect the ground connected negative jumper cable from its engine block.

10. Disconnect the negative jumper cable from the booster battery.

11. Disconnect the positive jumper cable from the booster battery, then from the other battery.

WARNING *Do not use more than 16 volts to jump start a vehicle that is equipped with an engine control module. The excess voltage may damage the electronic components.*

WARNING *A battery that has been rapidly discharged will create hydrogen gas. Do not attach jumper cables to a weak battery if starting the vehicle has been attempted. Wait for at least 10 minutes before connecting the jumper cable and attempting to start the vehicle.*

Summary

❏ Because of its importance, the battery should be checked whenever the vehicle is brought into the shop for service.

❏ Before attempting to do any type of work on or around the battery, the technician must be aware of all precautions associated with battery service.

❏ Before performing any electrical tests, the battery should be inspected, along with the cables and terminals.

❏ Charging the battery is the process of passing an electric current through the battery in an opposite direction than during discharge.

❏ Slow charging recharges the battery at a rate between 3 and 15 amperes for a long period of time.

❏ Fast charging the battery will bring the state of charge up high enough to crank the engine.

❏ The battery terminal test checks for poor electrical connections between the battery cables and terminals.

❏ A battery leakage test is conducted to see if current is flowing across the battery case.

❏ Measuring the state of charge is a check of the battery's electrolyte and plates. It can be determined by testing the specific gravity of the electrolyte using a hydrometer.

❏ The open circuit voltage test is used to determine the battery's state of charge. It is used when a hydrometer is not available or cannot be used.

❏ The capacity test provides a realistic determination of the battery's condition by checking its ability to perform when loaded.

❏ The three-minute charge test determines the battery's ability to accept a charge. If the battery is sulfated, the ability of the cells to deliver current and accept a charge are reduced.

Terms-To-Know

Battery leakage test	Fast charging	Stabilized
Battery terminal test	Multiplying coil	State of charge
Capacity test	Open circuit voltage test	Three-minute charge test
Charging	Slow charging	

ASE Style Review Questions

1. Battery terminal connections are being discussed. Technician A says when disconnecting battery cables, always disconnect the negative cable first. Technician B says when connecting battery cables, always connect the negative cable first. Who is correct?
   a. A only
   b. B only
   c. Both A and B
   d. Neither A nor B

2. A customer's battery is always dead when she attempts to start her car in the morning. After jumping the battery one time in the morning, the car will start throughout the day with no problems. Technician A says that the starter motor is drawing too much current. Technician B says that there may be a glove box light staying on. Who is correct?
   a. A only
   b. B only
   c. Both A and B
   d. Neither A nor B

3. The specific gravity of a battery has been tested. All cells have a corrected reading of about 1.200. Technician A says the battery needs to be recharged before further testing. Technician B says the battery is sulfated and needs to be replaced. Who is correct?
   a. A only
   b. B only
   c. Both A and B
   d. Neither A nor B

4. Battery charging is being discussed. Technician A says to connect a voltmeter across the battery terminals while it is charging and to keep the charge rate so that fewer than 15 volts is read. Technician B says to prevent damage to the alternator and computers, disconnect the negative battery cable before fast charging the battery in the vehicle. Who is correct?
   a. A only
   b. B only
   c. Both A and B
   d. Neither A nor B

5. The battery leakage test is being discussed. Technician A says the battery leakage test is used to determine if the battery can provide current and voltage when loaded. Technician B says a voltmeter reading of 0.05 when performing the battery leakage test is acceptable. Who is correct?
   a. A only
   b. B only
   c. Both A and B
   d. Neither A nor B

6. The open circuit test is being discussed. Technician A says the battery must be stabilized before the open circuit voltage test is performed. Technician B says a test result of 12.4 volts is acceptable. Who is correct?
   a. A only
   b. B only
   c. Both A and B
   d. Neither A nor B

7. A maintenance-free battery has failed the capacity test. Technician A says if the voltage recovers to 12.45 volts, the battery is still good. Technician B says if the voltage level does not return to 12.45 volts, recharge the battery and repeat the capacity test. Who is correct?
   a. A only
   b. B only
   c. Both A and B
   d. Neither A nor B

8. The results of a three-minute charge test are being discussed. Technician A says if the voltmeter indicates fewer than 15.5 volts, the battery must be replaced. Technician B says if the voltmeter reading is above 15.5 volts, the battery is good. Who is correct?
   a. A only
   b. B only
   c. Both A and B
   d. Neither A nor B

9. While jump starting a vehicle, a puff of smoke is observed and the engine ground cable is burned. Technician A says this happened because the two vehicles were touching. Technician B says this was caused by connecting the negative jumper cable to the disabled vehicle's engine ground. Who is correct?
   a. A only
   b. B only
   c. Both A and B
   d. Neither A nor B

10. Ice crystals are found in the electrolyte. Technician A says this indicates the battery is discharged. Technician B says they will melt when the battery is jumped. Who is correct?
   a. A only
   b. B only
   c. Both A and B
   d. Neither A nor B

Direct Current Motors and the Starting System

Objective

Upon completion and review of this chapter, you should be able to:

- ❏ Explain the purpose of the starting system.
- ❏ List and identify the components of the starting system.
- ❏ Explain the principle of operation of the DC motor.
- ❏ Describe the purpose and operation of the armature.
- ❏ Describe the purpose and operation of the field coil.
- ❏ Explain the differences between the types of magnetic switches used.
- ❏ Identify and explain the differences between starter drive mechanisms.
- ❏ Describe the differences between the positive engagement and solenoid shift starter.
- ❏ Describe the operation and features of the permanent magnet starter.

Introduction

The internal combustion engine must be rotated before it will run under its own power. The starting system is a combination of mechanical and electrical parts that work together to start the engine. The starting system is designed to convert electrical energy into mechanical energy. To accomplish this conversion, a starter or cranking motor is used. The starting system includes the following components:

1. Battery
2. Cable and wires
3. Ignition switch
4. Starter solenoid or relay
5. Starter motor
6. Starter drive and flywheel ring gear
7. Starting safety switch

Components in a simplified cranking system circuit are shown (Figure 9-1). This chapter examines both this circuit and the fundamentals of electric motor operation.

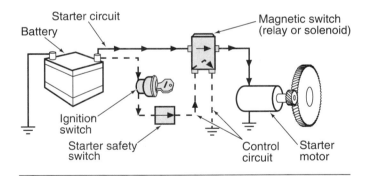

Figure 9-1 Major components of the starting system. The solid line represents the starting circuit. The dashed line indicates the starter control circuit.

A BIT OF HISTORY

In the early days of the automobile, the vehicle did not have a starter motor. The operator had to use a starting crank to turn the engine by hand. Charles F. Kettering invented the first electric "self-starter," which was developed and built by the Delco Electrical Plant. The self-starter first appeared on the 1912 Cadillac and was actually a combination starter and generator.

Motor Principles

DC motors use the interaction of magnetic fields to convert electrical energy into mechanical energy. Magnetic lines of force flow from the north pole to the south pole of a magnet (Figure 9-2). If a current carrying conductor is placed within the magnetic field, two fields will be present. On the left side of the conductor, the lines of force are in the same direction. This will concentrate the flux density of the lines of force on the left side. This will produce a strong magnetic field, because the two fields will reinforce each other. The lines of force oppose each other on the right side of the conductor. This results in a weaker magnetic field. The conductor will tend to move from the strong field to the weak field. This principle is used to convert electrical energy into mechanical energy in a starter motor by electromagnetism.

A simple electromagnet-style starter motor is shown (Figure 9-3). The inside windings are called the **armature**. The armature is the movable component of the motor that consists of a conductor wound around a laminated iron core and is used to create a magnetic field. The armature rotates within the stationary outside windings, called the **field**, which has windings coiled around **pole shoes**. Pole shoes are made of high-magnetic permeability material to help concentrate and direct the lines of force in the field assembly.

When current is applied to the field and the armature, both produce magnetic flux lines. The direction of the windings will place the left pole at a south polarity and the right side at a north polarity. The lines of force move from north to south in the field. In the armature, the flux lines circle in one direction on one side of the loop and in the opposite direction on the other side. Current will now set up a magnetic field around the loop of wire, which will interact with the north and south fields and put a turning force on the loop. This force will cause the loop to turn in the direction of the weaker field (Figure 9-4). However, the armature is limited in how far it is able to turn. When the armature is halfway between the shoe

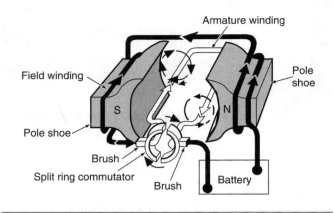

Figure 9-3 Simple electromagnetic motor.

poles, the fields balance one another. The point when the fields balance is called the **static neutral point**.

For the armature to continue rotating, the current flow in the loop must be reversed. To accomplish this, a split-ring **commutator** is in contact with the ends of the armature loops. The commutator is a series of conducting segments located around one end of the armature. Current enters and exits the armature through a set of electrically conductive sliding contacts, usually made of copper and carbon, called **brushes**. The brushes slide over the commutator's sections. As the brushes pass over one section of the commutator to another, the current flow in the armature is reversed. The position of the magnetic fields are the same. However, the direction of current flow through the loop has been reversed. This will continue until the current flow is turned off.

A single loop motor would not produce enough torque to rotate an engine. Power can be increased by the addition of more loops or pole shoes. An armature with its many windings, with each loop attached to corresponding commutator sections, is shown (Figure 9-5). In a typical starter motor (Figure 9-6) there are four brushes that make the electrical connections to the commutator. Two of the brushes are grounded to the starter motor frame, and two are insulated from the frame. Also, the armature is supported by bushings at both ends.

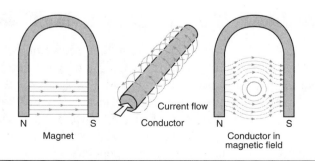

Figure 9-2 Magnetic field interaction.

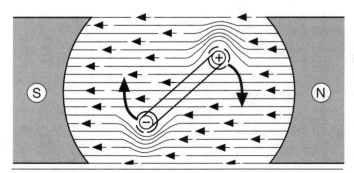

Figure 9-4 Rotation of the conductor is in the direction of the weaker field.

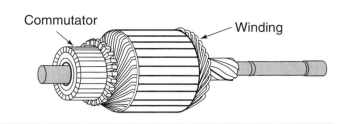

Figure 9-5 Starter armature.

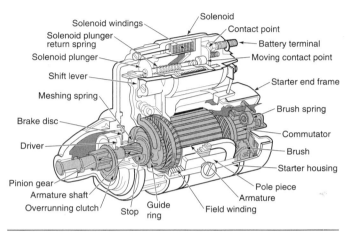

Figure 9-6 Starter and solenoid components. (Reprinted with permission of Robert Bosch Corporation)

WARNING *The high amount of current required to operate the starter motor generates heat very quickly. Continuous operation of the starter motor for longer than 30 seconds causes serious heat damage. The starter motor should not be operated for more than 30 seconds at a time and should have a two-minute wait between cranking attempts.*

Armature

The armature is constructed with a laminated core made of several thin iron stampings that are placed next to each other (Figure 9-7). Laminated construction is used because in a solid iron core the magnetic fields would generate counter voltages in the core called eddy currents. By using laminated construction, **eddy currents** in the core are minimized. Eddy currents cause heat to build up in the core and waste energy.

The slots on the outside diameter of the laminations hold the armature windings. The windings loop around the core and are connected to the commutator. Each commutator segment is insulated from the adjacent segments. A typical armature can have more than 30 commutator segments. A steel shaft is fitted into the center

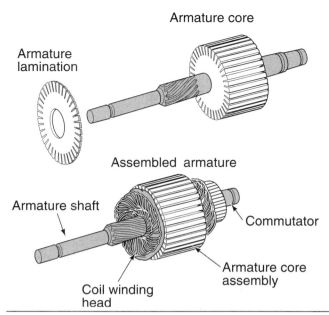

Figure 9-7 Lamination construction of a typical motor armature. (Reprinted with permission of Robert Bosch Corporation)

hole of the core laminations. The commutator is insulated from the shaft.

Two basic winding patterns are used in the armature: lap winding and wave winding. In the lap winding, the two ends of the winding are connected to adjacent commutator segments. In this pattern, the wires passing under a pole field have their current flowing in the same direction.

In the wave winding pattern, each end of the winding connects to commutator segments that are 90 or 180 degrees apart. In this pattern design some windings will have no current flow at certain positions of armature rotation. This occurs because the segment ends of the winding loop are in contact with brushes that have the same polarity. The wave wound pattern is the most common used due to its lower resistance.

Field Coils

The **field coils** are electromagnets constructed of wire ribbons or coils wound around a pole shoe (Figure 9-8). The field coils are attached to the inside of the starter housing (Figure 9-9). The iron pole shoes and the iron starter housing work together to increase and concentrate the field strength of the field coils (Figure 9-10).

When current flows through the field coils, strong stationary electromagnetic fields are created. The fields have a north and south magnetic polarity based on the direction the windings are wound around the pole shoes. The polarity of the field coils alternate to produce opposing magnetic fields.

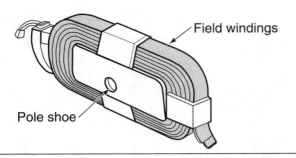

Figure 9-8 Field coil wound around a pole shoe.

In any DC motor, there are three methods of connecting the field coils to the armature: in series, in parallel (shunt), and a compound connection that uses both series and shunt coils. Most starter motors use four field coils.

Series-Wound Motors

Until recently, most starter motors were **series-wound** with current flowing first to the field windings, then to the brushes, through the commutator and the armature winding contacting the brushes at that time, then through the grounded brushes back to the battery source (Figure 9-13). This design permits all of the current that passes through the field coils to also pass through the armature.

A series-wound motor will develop its maximum torque output at the time of initial start. As the motor speed increases, the torque output of the motor will decrease. This decrease of torque output is the result of **counter electromotive force (CEMF)** caused by self-induction. Counter electromotive force is voltage produced in the starter motor itself. This voltage acts against the supply voltage from the battery. CEMF is produced by electromagnetic induction.

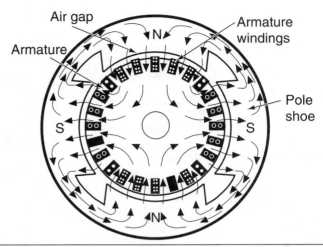

Figure 9-10 Magnetic fields in a four-pole starter motor. (Reprinted with permission of Robert Bosch Corporation)

Shunt-Wound Motors

Electric motors, or **shunt motors**, have the field windings wired in parallel across the armature (Figure 9-12). A shunt-wound field is used to limit the speed that the motor can turn. A shunt motor does not decrease in its torque output as speeds increase. This is because the CEMF produced in the armature does not decrease the field coil strength. Due to a shunt motor's inability to produce high torque, it is not typically used as a starter motor. However, shunt motors may be found as wiper motors, power window motors, power seat motors, and so on.

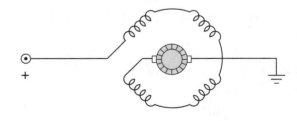

Figure 9-11 Series-wound motor schematic.

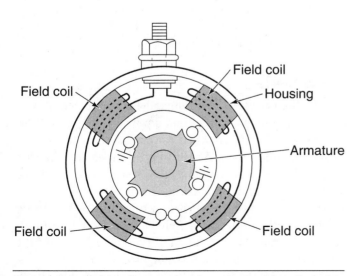

Figure 9-9 Field coils mounted to the inside of starter housing.

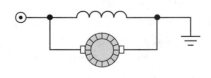

Figure 9-12 Shunt-wound motor schematic.

Compound Motors

A motor that uses the characteristics of a series motor and a shunt motor is called a **compound motor**. In a compound motor some of the field coils are connected to the armature in series, and some field coils are connected in parallel with the battery and the armature (Figure 9-13). This configuration allows the compound motor to develop good starting torque and constant operating speeds. The field coil that is shunt wound is used to limit the speed of the starter motor. Also, on Ford's positive engagement starters, the shunt coil is used to engage the starter drive. This is possible because the shunt coil is energized as soon as battery voltage is sent to the starter.

Permanent Magnet Motors

Some motors use permanent magnets in place of the field coils. These motors are used in many different applications, including starter motors. When a permanent magnet is used instead of coils, there is no field circuit in the motor. By eliminating this circuit, potential electrical problems have also been eliminated, such as field to housing shorts. Another advantage to using permanent magnets is weight savings; the weight of a typical starter motor is reduced by 50%. Most permanent magnet starters are gear reduction type starters.

Multiple permanent magnets are positioned in the housing, around the armature. These permanent magnets are an alloy of boron, neodymium, and iron. The field strength of these magnets is much greater than typical permanent magnets. The operation of these motors is the same as other electric motors, except there is no field circuit or windings.

Starter Drives

The **starter drive** is the part of the starter motor that engages the armature to the engine flywheel ring gear (Figure 9-14). It includes a **pinion gear** set that meshes with the flywheel ring gear on the engine's crankshaft. To prevent damage to the pinion gear or the ring gear, the pinion gear must mesh with the ring gear before the

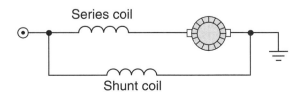

Figure 9-13 Compound motor uses both series and shunt coils.

Figure 9-14 Starter drive pinion gear is used to turn the engine's flywheel.

starter motor rotates. To help assure smooth engagement, the ends of the pinion gear teeth are tapered. Also, the action of the armature must always be from the motor to the engine. The engine must not be allowed to spin the armature. The **ratio** of the number of teeth on the ring gear and the starter drive pinion gear is usually between 15:1 and 20:1. The ratio of the starter drive is determined by dividing the number of teeth on the drive gear (pinion gear) into the number of teeth on the driven gear (flywheel). This means the starter motor is rotating 15 to 20 times faster than the engine. Normal cranking speed for the engine is about 200 rpm. If the starter drive had a ratio of 18:1, the starter would be rotating at a speed of 3,600 rpm. If the engine started and was accelerated to 2,000 rpm, the starter speed would increase to 36,000 rpm. This would destroy the starter motor if it was not disengaged from the engine.

Bendix Inertia Drive

The **bendix drive** depends on **inertia** to provide meshing of the drive pinion with the ring gear (Figure 9-15). Inertia is the tendency of an object that is at rest to stay at rest, and an object that is in motion to stay in motion. The screwshaft threads are a part of the armature, and will turn at armature speed. At the end of the pinion and barrel is the pinion gear that will mesh with the ring gear. The pinion and barrel have internal threads that match those of the screwshaft. When current flows through the starter motor, the armature will begin to spin. Torque from the armature is transmitted via a shock-absorbing drive spring and drive head to the screwshaft. This causes the screwshaft to rotate. However, the barrel does not rotate. The barrel has a weight on one side to increase its inertial effect. The barrel tends to stay at rest, and the screwshaft rotates inside the barrel. As a result, the barrel is threaded down the length of the screwshaft to the end. At the end of the screwshaft, the pinion gear

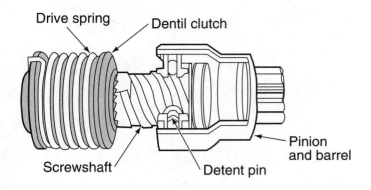

Figure 9-15 Bendix inertia drive. (Courtesy of General Motors Corporation, Service Technology Group)

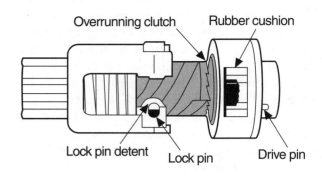

Figure 9-17 Bendix folo-thru starter drive. (Courtesy of General Motors Corporation, Service Technology Group)

engages the ring gear. Here the pinion gear locks to the screwshaft and transfers torque from the armature to the ring gear and engine.

Once the engine starts and is running under its own power, it will rotate faster than the armature. This causes the barrel to screw back down the screwshaft and bring the pinion gear out of engagement with the ring gear.

There are a couple different variations of the inertia drive. One of these variations is the **barrel-type drive** (Figure 9-16). The barrel-type starter drive is similar to the inertia drive except:

1. The pinion is mounted on the end of the barrel.

2. It has a higher gear ratio.

3. It works directly off screw threads at the end of the armature, instead of a screwshaft.

Another variation is the **folo-thru drive**. This drive is much like the barrel drive with the addition of a detent pin and a detent clutch (Figure 9-17).

When the pinion barrel has moved to the end of the screwshaft, the detent pin locks the barrel to the screwshaft. The detent pin operates on centrifugal force. As the shaft is turning rapidly, centrifugal force throws the pin into engagement between the barrel and the screwshaft.

The detent clutch disengages the sections of the screwshaft when the engine is running faster than the armature, and the drive is still engaged. Disengaging the clutch protects the motor from being damaged.

Overrunning Clutch Drive

The most common type of starter drive is the **overrunning clutch**. The overrunning clutch is a roller-type clutch that transmits torque in one direction only and freewheels in the other direction. This allows the starter motor to transmit torque to the ring gear, but prevents the ring gear from transferring torque to the starter motor.

In a typical overrunning-type clutch (Figure 9-18), the clutch housing is internally splined to the starter armature shaft. The drive pinion turns freely on the armature shaft within the clutch housing. When torque is transmitted

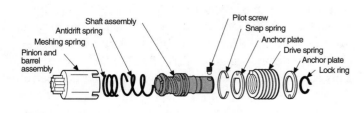

Figure 9-16 Bendix barrel-type starter drive. (Courtesy of General Motors Corporation, Service Technology Group)

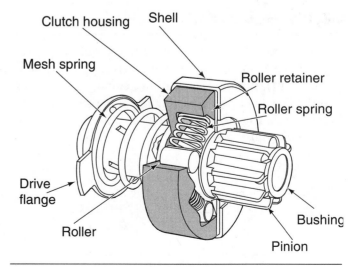

Figure 9-18 Overrunning clutch starter drive. (Reprinted with permission of Ford Motor Company)

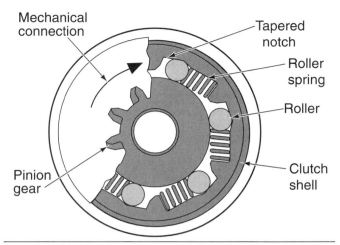

Figure 9-19 When the armature turns, it locks the rollers into the tapered notch. (Reprinted with permission of Robert Bosch Corporation)

through the armature to the clutch housing, the spring-loaded rollers are forced into the small ends of their tapered slots (Figure 9-19). They are then wedged tightly against the pinion barrel. The pinion barrel and clutch housing are now locked together; torque is transferred through the starter motor to the ring gear and engine.

When the engine starts and is running under its own power, the ring gear attempts to drive the pinion gear faster than the starter motor. This unloads the clutch rollers and releases the pinion gear to rotate freely around the armature shaft.

Cranking Motor Circuits

The starting system of the vehicle consists of two circuits: the starter control circuit and the motor feed circuit. These circuits are separate but related. The control circuit consists of the starting portion of the ignition switch, the starting safety switch (if applicable), and the wire conductor to connect these components to the relay or solenoid. The motor feed circuit consists of heavy battery cables from the battery to the relay and the starter or directly to the solenoid if the starter is so equipped.

A BIT OF HISTORY

The integrated key starter switch was introduced in 1949 by Chrysler. Before this, the key turned the system on and the driver pushed a starter button.

Starter Control Circuit Components

Magnetic Switches

The starter motor requires large amounts of current (up to 300 amperes) to generate the torque needed to turn the engine. The conductors used to carry this amount of current (battery cables) must be large enough to handle the current with very little voltage drop. It would be impractical to place a conductor of this size into the wiring harness to the ignition switch. To provide control of the high current, all starting systems contain some type of magnetic switch. There are two basic types of magnetic switches used: the solenoid and the relay.

As discussed in Chapter 3, a solenoid is an electromagnetic device that uses movement of a plunger to exert a pulling or holding force. In the solenoid-actuated starter system, the solenoid is mounted directly on top of the starter motor (Figure 9-20). The solenoid switch on a starter motor performs two functions: It closes the circuit between the battery and the starter motor, then it shifts the starter motor pinion gear into mesh with the ring gear. This is accomplished by a linkage between the solenoid plunger and the shift lever on the starter motor. In the past the most common method of energizing the solenoid was directly from the battery through the ignition switch. However, most of today's vehicles use a starter relay in conjunction with a solenoid. The relay is used to reduce the amount of current flow through the ignition and is usually controlled by the powertrain control module (PCM).

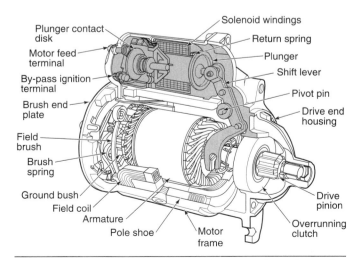

Figure 9-20 Solenoid operated starter has the solenoid mounted directly on top of the motor.

When the circuit is closed and current flows to the solenoid, current from the battery is directed to the **pull-in windings** and the **hold-in windings** (Figure 9-21). Because it may require up to 50 amperes to create a magnetic force large enough to pull the plunger in, both windings are energized to create a combined magnetic field. Once the plunger is moved, the current required to hold the plunger is reduced. This allows the current that was used to pull the plunger in to be used to rotate the starter motor.

When the ignition switch is placed in the START position, voltage is applied to the S terminal of the solenoid (Figure 9-22). The hold-in winding has its own ground to the case of the solenoid. The pull-in winding's ground is through the starter motor. Current will flow through both windings to produce a strong magnetic field. When the plunger is moved into contact with the main battery and motor terminals, the pull-in winding is de-energized. The pull-in winding is not energized because the contact places battery voltage on both sides of the coil (Figure 9-23). The current that was directed through the pull-in winding is now sent to the motor.

Because the contact disc does not close the circuit from the battery to the starter motor until the plunger has moved the shift lever, the pinion gear is in full mesh with the flywheel before the armature starts to rotate.

After the engine is started, releasing the key to the RUN position opens the control circuit. Voltage no longer is supplied to the hold-in windings and the return spring causes the plunger to return to its neutral position.

In the above figures there is an R terminal illustrated. This terminal provides current to the ignition bypass circuit that is used to provide for full battery voltage to the ignition coil while the engine is cranking. This circuit bypasses the ballast resistor. The bypass circuit is not used on most ignition systems today.

A common problem with the control circuit is that low system voltage or an open in the hold-in windings

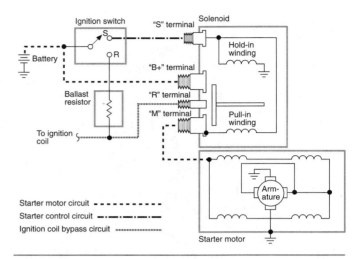

Figure 9-22 Schematic of solenoid-operated starter motor circuit.

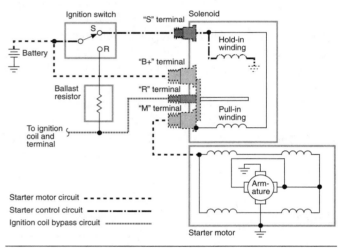

Figure 9-23 Once the contact disc closes the terminals, the hold-in winding is the only one that is energized.

will cause an oscillating action to occur. The combination of the pull-in winding and the hold-in winding is sufficient to move the plunger. However, once the contacts are closed, there is insufficient magnetic force to hold the plunger in place. This condition is recognizable by a series of clicks when the ignition switch is turned to the START position. Before replacing the solenoid, check the battery condition; a low battery charge will cause the same symptom.

Many manufacturers use a **starter relay** (Figure 9-24) instead of, or in addition to, a solenoid. The relay is usually mounted near the battery on the fender well or radiator support. Unlike the solenoid, the relay does not move the pinion gear into mesh with the flywheel ring gear.

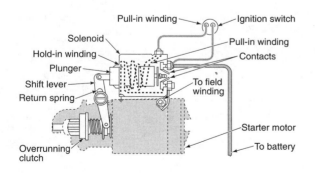

Figure 9-21 The solenoid uses two windings. Both are energized to draw the plunger, then only the hold-in winding is used to hold the plunger in position.

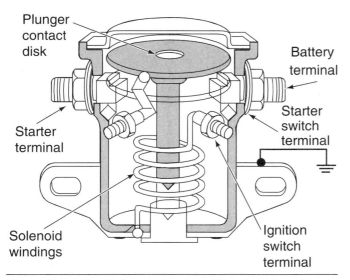

Figure 9-24 Typical starter relay. (Reprinted with permission of Ford Motor Company)

When the ignition switch is turned to the START position, current is supplied through the switch to the relay coil. The coil produces a magnetic field, so it pulls the movable core into contact with the internal contacts of the battery and starter terminals (Figure 9-25). With the contacts closed, full battery current is supplied to the starter motor.

A secondary function of the starter relay is to provide for an alternate path for current to the ignition coil during cranking. This is done by an internal connection that is energized by the relay core when it completes the circuit between the battery and the starter motor.

Ignition Switch

The **ignition switch** is the power distribution point for most of the vehicle's primary electrical systems (Figure 9-26). Most ignition switches have five positions:

1. ACCESSORIES: Supplies current to the vehicle's electrical accessory circuits. It will not supply current to the engine control circuits, starter control circuit, or the ignition system.

2. LOCK: Mechanically locks the steering wheel and transmission gear selector. All electrical contacts in the ignition switch are open. Most ignition switches must be in this position to insert or remove the key from the cylinder.

3. OFF: All circuits controlled by the ignition switch are opened. The steering wheel and transmission gear selector are unlocked. Starting in 1993, Chrysler (and some other manufacturers) began to use ignition switches that would power up the vehicle's computers and instrument panel in the OFF position.

4. ON or RUN: The switch provides current to the ignition, engine controls, and all other circuits controlled by the switch. Some systems will power a chime or light with the key in the ignition switch. Other systems power an anti-theft system when the key is removed and turn it off when the key is inserted.

5. START: The switch provides current to the starter control circuit, ignition system, and engine control circuits.

The ignition switch is spring loaded in the START position. This momentary contact automatically moves the contacts to the RUN position when the driver

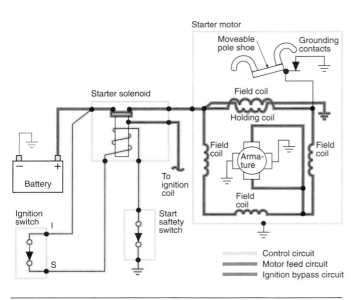

Figure 9-25 Current flow when the starter relay is energized.

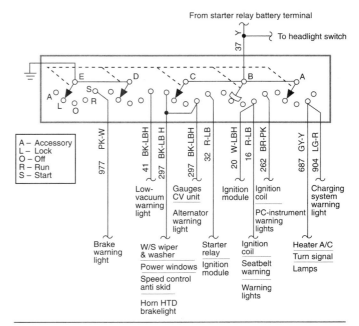

Figure 9-26 Ganged ignition switch. (Reprinted with permission of Ford Motor Company)

releases the key. All other ignition switch positions are detent positions.

Starting Safety Switch

The **neutral safety switch** is used on vehicles that are equipped with automatic transmissions. It opens the starter control circuit when the transmission shift selector is in any position except PARK or NEUTRAL. Actual location of the neutral safety switch depends on the kind of transmission and the location of the shift lever. Some manufacturers place the switch in the transmission (Figure 9-27).

Vehicles equipped with automatic transmissions require a means of preventing the engine from starting while the transmission is in gear. Without this feature the vehicle would lunge forward or backward once it was started, causing personal or property damage. The normally open neutral safety switch is connected in series in the starting system control circuit and is usually operated by the shift lever. When in the PARK or NEUTRAL position, the switch is closed, allowing current to flow to the starter circuit. If the transmission is in a gear position, the switch is opened and current cannot flow to the starter circuit. A transmission range (TR) sensor can be used to inform the PCM what gear the transmission is in to prevent starting in any range other than PARK or NEUTRAL.

Many vehicles equipped with manual transmissions use a similar type of safety switch. The **start/clutch interlock switch** is usually operated by movement of the clutch pedal. When the clutch pedal is pushed downward, the switch closes and current can flow through the starter circuit. If the clutch pedal is left up, the switch is open and current cannot flow.

Some General Motors vehicles use a mechanical linkage that blocks movement of the ignition switch cylinder unless the transmission is in PARK or NEUTRAL (Figure 9-28).

Figure 9-27 The neutral safety switch can be combined with the back-up light switch and installed in the transmission case.

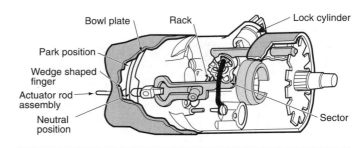

Figure 9-28 Mechanical linkage used to prevent starting the engine while the transmission is in gear. (Courtesy of General Motors Corporation, Service Technology Group)

Most starter safety switches are adjustable. Sometimes a no-start problem can be corrected by checking and adjusting (or replacing) the starter safety switch.

Cranking Motor Designs

The most common type of starter motor used today incorporates the overrunning clutch starter drive instead of the old inertia-engagement bendix drive. There are four basic groups of starter motors:

1. Direct drive
2. Gear reduction
3. Positive-engagement (movable pole)
4. Permanent magnet

Direct Drive Starters

The most common type of starter motor is the solenoid-operated **direct drive** unit (Figure 9-29). Although there are construction differences between applications, the operating principles are the same for all solenoid-shifted starter motors.

When the ignition switch is placed in the START position, the control circuit energizes the pull-in and hold-in windings of the solenoid. The solenoid plunger moves and pivots the shift lever, which in turn locates the drive pinion gear into mesh with the engine flywheel.

When the solenoid plunger is moved all the way, the contact disc closes the circuit from the battery to the starter motor. Current now flows through the field coils and the armature. This develops the magnetic fields that cause the armature to rotate, thus turning the engine.

Gear Reduction Starters

Some manufacturers use a **gear reduction starter** to provide increased torque (Figure 9-30). The gear reduction starter differs from most other designs in that the

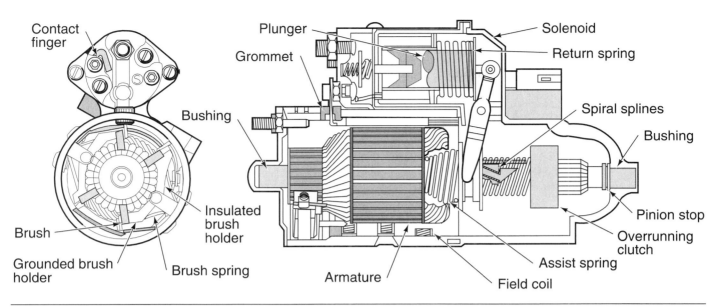

Figure 9-29 Solenoid operated Delco MT series starter motor. (Courtesy of General Motors Corporation, Service Technology Group)

armature does not drive the pinion gear directly. In this design, the armature drives a small gear that is in constant mesh with a larger gear. Depending on the application, the ratio between these two gears is between 2:1 and 3.5:1. The additional reduction allows for a small motor to turn at higher speeds and greater torque with less current draw. Another characteristic of most gear reduction starters is that the commutator and brushes are located in the center of the motor.

The solenoid operation is similar to that of the solenoid-shifted direct drive starter in that the solenoid moves the plunger, which engages the starter drive.

Positive-engagement Starters

One of the most commonly used starters on Ford applications is the **positive-engagement starter** (Figure 9-31). Positive-engagement starters are also called **movable-pole shoe starters**. Positive-engagement starters

use the shunt coil windings of the starter motor to engage the starter drive. The high starting current is controlled by a starter solenoid mounted close to the battery. When the solenoid contacts are closed, current flows through a hollowed field coil used to attract the movable pole shoe. The hollowed field coil is referred to as the **drive coil**. The drive coil creates an electromagnetic field that attracts a movable pole shoe. The movable pole shoe is attached to the starter drive through the plunger lever. When the movable pole shoe moves, the drive gear engages the engine flywheel.

As soon as the starter drive pinion gear contacts the ring gear, a contact arm on the pole shoe opens a set of normally closed grounding contacts (Figure 9-32). With the return to ground circuit opened, all the starter current

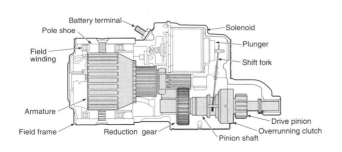

Figure 9-30 Gear reduction starter motor construction. (Reprinted with permission of Ford Motor Company)

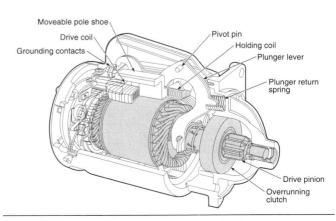

Figure 9-31 Positive engagement starters use a movable pole shoe. (Reprinted with permission of Ford Motor Company)

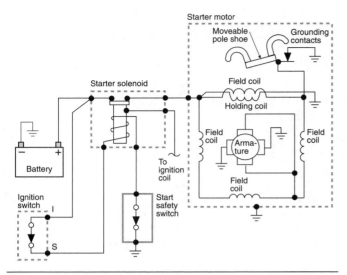

Figure 9-32 Schematic of positive-engagement starter.

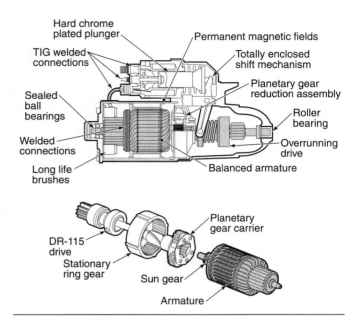

Figure 9-33 The PMGR motor uses a planetary gear set and permanent magnets. (Courtesy of General Motors Corporation, Service Technology Group)

flows through the remaining three field coils and through the brushes to the armature. The starter motor then begins to rotate. To prevent the starter drive from disengaging from the ring gear if battery voltage drops while cranking, the movable pole shoe is held down by a holding coil. The holding coil is a smaller coil inside the main drive coil and is strong enough to hold the starter pinion gear engaged.

Permanent Magnet Starters

The **permanent magnet gear reduction (PMGR)** starter design provides for less weight, simpler construction, and less heat generation as compared to conventional field coil starters (Figure 9-33). The PMGR starter uses four or six permanent magnet field assemblies in place of field coils. Because there are no field coils, current is delivered directly to the armature through the commutator and brushes.

The permanent magnet starter also uses gear reduction through a planetary gear set. The planetary gear train transmits power between the armature and the pin-

ion shaft. This allows the armature to rotate at greater speed and increased torque. The planetary gear assembly consists of a sun gear on the end of the armature, and three planetary carrier gears inside a ring gear. The ring gear is held stationary. When the armature is rotated, the sun gear causes the carrier gears to rotate about the internal teeth of the ring gear. The planetary carrier is attached to the output shaft. The gear reduction provided for by this gear arrangement is 4.5:1. By providing for this additional gear reduction, the demand for high current is lessened.

The electrical operation between the conventional field coil and PMGR starters remains basically the same.

CAUTION *Special care must be taken when handling the PMGR starter. The permanent magnets are very brittle and are easily destroyed if the starter is dropped or struck by another object.*

Summary

❏ The starting system is a combination of mechanical and electrical parts that work together to start the engine.

❏ The starting system components include the battery, cable and wires, the ignition switch, the starter solenoid or relay, the starter motor, the starter drive and flywheel ring gear, and the starting safety switch.

❏ The armature is the movable component of the motor that consists of a conductor wound around a laminated iron core. It is used to create a magnetic field.

❏ Pole shoes are made of high magnetic permeability material to help concentrate and direct the lines of force in the field assembly.

❏ Within an electromagnetic style of starter motor, the inside windings are called the armature. The armature rotates within the stationary outside windings, called the field, which has windings coiled around pole shoes.

❏ The commutator is a series of conducting segments located around one end of the armature.

❏ A split-ring commutator is in contact with the ends of the armature loops. So, as the brushes pass over one section of the commutator to another, the current flow in the armature is reversed.

❏ Two basic winding patterns are used in the armature: lap winding and wave winding.

❏ The field coils are electromagnets constructed of wire coils wound around a pole shoe.

❏ In any DC motor, there are three methods of connecting the field coils to the armature: in series, in parallel (shunt), and a compound connection that uses both series and shunt coils.

❏ A starter drive includes a pinion gear set that meshes with the engine flywheel ring gear on the engine.

❏ The bendix drive depends on inertia to provide meshing of the drive pinion with the ring gear.

❏ The most common type of starter drive is the overrunning clutch. This is a roller-type clutch that transmits torque in one direction only and freewheels in the other direction.

❏ The starting system consists of two circuits called the starter control circuit and the motor feed circuit.

❏ There are four basic groups of starter motors: direct drive, gear reduction, positive engagement (movable pole), and permanent magnet.

Terms-To-Know

Armature	Field coils	Pinion gear
Barrel-type drive	Folo-thru drive	Pole shoes
Bendix drive	Gear reduction starter	Positive-engagement starter
Brushes	Hold-in windings	Pull-in windings
Commutator	Ignition switch	Ratio
Compound motor	Inertia	Series-wound
Counter electromotive force (CEMF)	Movable-pole shoe starters	Shunt motors
Direct drive	Neutral safety switch	Start/clutch interlock switch
Drive coil	Overrunning clutch	Starter drive
Eddy currents	Permanent magnet gear reduction (PMGR)	Starter relay
Field		Static neutral point

Review Questions

Short Answer Essays

1. What is the purpose of the starting system?
2. List and describe the purpose of the major components of the starting system.
3. Explain the principle of operation of the DC motor.
4. Describe the types of magnetic switches used in starting systems.
5. List and describe the operation of the different types of starter drive mechanisms.
6. Describe the differences between the positive-engagement and solenoid shift starter.
7. Explain the operating principles of the permanent magnet starter.
8. Describe the purpose and operation of the armature.
9. Describe the purpose and operation of the field coil.
10. Describe the operation of the two circuits of the starter system.

Fill-in-the-Blanks

1. DC motors use the interaction of magnetic fields to convert the _____ energy into _____ energy.

2. The _____ is the movable component of the motor, which consists of a conductor wound around a _____ iron core and is used to create a _____ field.

3. Pole shoes are made of high magnetic _____ material to help concentrate and direct the _____ _____ _____ in the field assembly.

4. The starter motor electrical connection that permits all of the current that passes through the field coils to also pass through the armature is called the _____ motor.

5. _____ _____ _____ is voltage produced in the starter motor itself. This current acts against the supply voltage from the battery.

6. A starter motor that uses the characteristics of a series motor and a shunt motor is called a _____ motor.

7. The _____ _____ is the part of the starter motor that engages the armature to the engine flywheel ring gear.

8. The _____ _____ is a roller-type clutch that transmits torque in one direction only and freewheels in the other direction.

9. The two circuits of the starting system are called the _____ _____ circuit and the _____ _____ circuit.

10. There are two basic types of magnetic switches used in starter systems: the _____ and the _____.

ASE Style Review Questions

1. The purpose of the starter system is being discussed. Technician A says the starting system is a combination of mechanical and electrical parts that work together to start the engine. Technician B says the starting system is designed to change the mechanical energy into electrical energy. Who is correct?
   a. A only
   b. B only
   c. Both A and B
   d. Neither A nor B

2. The components of the starting system are being discussed. Technician A says the drive belt is part of the starting system. Technician B says the starter drive and flywheel ring gear are components of the starting system. Who is correct?
   a. A only
   b. B only
   c. Both A and B
   d. Neither A nor B

3. The operation of the DC motor is being discussed. Technician A says DC motors use the interaction of magnetic fields. Technician B says DC motors use a mechanical connection to the engine that turns the armature. Who is correct?
   a. A only
   b. B only
   c. Both A and B
   d. Neither A nor B

4. The starter motor armature is being discussed. Technician A says the armature is the stationary component of the motor that consists of a conductor wound around a pole shoe. Technician B says the commutator is a series of conducting segments located around one end of the armature. Who is correct?
   a. A only
   b. B only
   c. Both A and B
   d. Neither A nor B

5. The starter motor field coils are being discussed. Technician A says the field coil is made of wire wound around a non-magnetic pole shoe. Technician B says the field coils are always shunt wound to the armature. Who is correct?
   a. A only
   b. B only
   c. Both A and B
   d. Neither A nor B

6. The starter solenoid is being discussed. Technician A says the solenoid is an electromagnetic device that uses movement of a plunger to exert a pulling or holding force. Technician B says the solenoid makes an electrical connection, which allows voltage to the starter motor. Who is correct?
   a. A only
   b. B only
   c. Both A and B
   d. Neither A nor B

7. Technician A says both windings of the solenoid are energized anytime the ignition switch is in the START position. Technician B says the two windings of the solenoid are called the pull-in and the hold-in windings. Who is correct?
   a. A only
   b. B only
   c. Both A and B
   d. Neither A nor B

8. Permanent magnet starters are being discussed. Technician A says the permanent magnet starter uses four or six permanent magnet field assemblies in place of field coils. Technician B says the permanent magnet starter uses a planetary gear set. Who is correct?
   a. A only
   b. B only
   c. Both A and B
   d. Neither A nor B

9. Technician A says the R terminal of the starter
    solenoid provides current to the ignition bypass
    circuit. Technician B says the ignition switch is a
    component of the starter feed circuit. Who is
    correct?
    a. A only
    b. B only
    c. Both A and B
    d. Neither A nor B

10. A customer's no-start complaint is being
    discussed. Technician A says the problem may be
    that the starter safety switch needs adjusting.
    Technician B says the battery should be tested
    before condemning starter system components.
    Who is correct?
    a. A only
    b. B only
    c. Both A and B
    d. Neither A nor B

10 Starting System Diagnosis and Service

Objective

Upon completion and review of this chapter, you should be able to:

❏ Perform a systematic diagnosis of the starting system.

❏ Determine what can cause slow crank and no-crank conditions.

❏ Perform a quick check test series to determine the problem areas in the starting system.

❏ Perform and accurately interpret the results of a current draw test.

❏ Perform and accurately interpret the results of an insulated circuit resistance test and a ground circuit test.

❏ Perform the solenoid test series and accurately diagnose the solenoid.

❏ Perform the no-crank test and recommend needed repairs as indicated.

❏ Diagnose the starter motor condition by use of the free speed test.

❏ Remove and reinstall a starter motor.

❏ Disassemble, clean, inspect, repair, and reassemble a starter motor.

Introduction

Perhaps one of the most aggravating experiences to a car owner is to have an engine that will not start. However, not all starting problems are caused by the starting system. The ignition and fuel systems must also be in proper condition to perform their functions. In addition, the internal condition of the engine must be such that compression, correct valve timing, and free rotation are all obtained.

The starter motor must be capable of rotating the engine fast enough to start and run under its own power. The starting system is a combination of mechanical and electrical parts that work together to start the engine. In this chapter you will perform the required tests to make a decision concerning the condition and operation of these components. You will also remove, disassemble, reassemble, and reinstall the starter motor.

Starting System Service Cautions

Before beginning any service on the starter system, some precautions must be observed. Along with the precautions outlined in Chapter 2, when servicing the battery, several other precautions should be followed:

1. Refer to the manufacturer's manuals for correct procedures for disconnecting a battery. Some vehicles with onboard computers must be supplied with an auxiliary power source.

2. Disconnect the battery ground cable before disconnecting any of the starter circuit's wires or removing the starter motor.

3. Be sure the vehicle is properly positioned on the hoist or on safety jack stands.

4. Before performing any cranking test, be sure the vehicle is in park or neutral and the parking brakes are applied.

5. Follow manufacturer's directions for disabling the ignition system.

6. Be sure the test leads are clear of any moving engine components.

7. Never clean any electrical components in solvent or gasoline. Clean with low pressure compressed air, denatured alcohol, or wipe with clean rags only.

Starting System Troubleshooting

Customer complaints concerning the starting system generally fall into four categories: no-crank, slow cranking, starter spins but does not turn engine, and excessive noise. As with any electrical system complaint, a systematic approach to diagnosing the starting system will make the task easier. First, the battery must be in good condition and fully charged. Perform a complete battery test series to confirm the battery's condition. Many starting system complaints are actually attributable to battery problems. If the starting system tests are performed with a weak battery, the results can be misleading. The conclusions may be erroneous and costly.

Before performing any tests on the starting system, first begin with a visual inspection of the circuit. Repair or replace any corroded or loose connections, frayed wires, or any other trouble sources. The battery terminals must be clean and the starter motor must be properly grounded.

No-crank means that when the ignition switch is placed in the START position, the starter does not turn the engine. This may be accompanied with a buzzing noise that indicates the starter motor drive has engaged the ring rear, but the engine does not rotate. There may also be no sounds (clicking) from the starter motor or solenoid. If the customer complains of a no-crank situation, attempt to rotate the engine by the crankshaft pulley nut. Rotate the crankshaft two full rotations in a clockwise direction, using a large socket wrench. If the engine does not rotate, it may be seized due to its being operated with no oil, broken engine components, or **hydrostatic lock**. Liquid cannot be compressed. If there is a leak that allows antifreeze from the cooling system to enter the cylinder, the cylinder can fill to a level that the piston is unable to move upward. This condition is referred to as hydrostatic lock.

Slow cranking means that the starter drive engages the ring gear, but the engine turns at too slow of a speed

to start. A slow crank or no-crank complaint can be caused by several potential trouble spots in the circuit (Figure 10-1). Excessive voltage drops in these areas will cause the starter motor to operate slower than required to start the engine. The speed at which the starter motor rotates the engine is important to engine starting. If the speed is too slow, compression is lost and the air/fuel mixture draw is impeded. Most manufacturers require a speed of approximately 250 rpm during engine cranking.

If the starter spins but the engine does not rotate, the most likely cause is a faulty starter drive. If the starter drive is at fault, the starter motor will have to be removed to install a new drive mechanism. Before faulting the starter drive, also check the starter ring gear teeth for wear or breakage, and for incorrect gear mesh of the ring gear and starter motor pinion gear.

Most noises can be traced to the starter drive mechanism. The starter drive can be replaced as a separate component of the starter.

| CUSTOMER CARE | *Always treat the customer's car with respect.* |

Place fender covers over the fenders when performing tasks under the hood. Do not lay tools on the vehicle's finish. Clean your hands before entering the vehicle. Place a seat protector over the seats and paper mats on the floor boards. Give the car back to the customer at least as clean as when you received it.

Testing the Starting System

As with the battery testing series, the tests for the starting system are performed with a starting/charging system tester. The starter performance and battery performance are so closely related that it is important for a full battery test series to be done before trying to test the starter system. If the battery fails the load test and is fully charged, it must be replaced before doing any other tests.

Quick Testing

If the starter does not turn the engine at all, and the engine is in good mechanical condition, the **quick test** can be performed to locate the problem area. The quick test will isolate the problem area to determine if the starter motor, solenoid or control circuit is at fault. To perform this test, make sure the transmission is in neutral and set the parking brake. Turn on the headlights. Next, turn the ignition switch to the START position while observing the headlights.

There are three things that can happen to the headlights during this test:

1. They will go out.

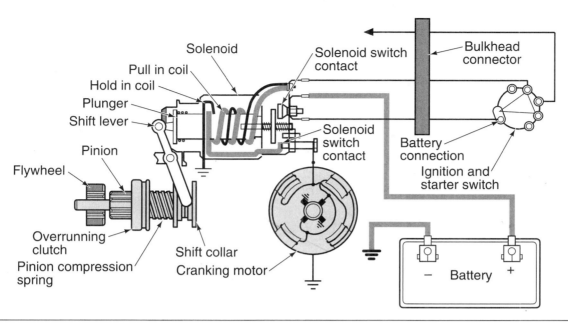

Figure 10-1 Excessive wear, loose electrical connections, or excessive voltage drop in any of these areas can cause a slow crank or no-crank condition.

2. They will dim.

3. They will remain at the same brightness level.

If the lights go out completely, the most likely cause is a poor connection at one of the battery terminals. Check the battery cables for tight and clean connections. It will be necessary to remove the cable from the terminal and clean the cable clamp and battery terminals of all corrosion.

If the headlights dim when the ignition switch is turned to the START position, the battery may be discharged. Check the battery condition. If it is good, then there may be a mechanical condition in the engine that is preventing it from rotating. If the engine rotates when turning it with a socket wrench on the pulley nut, the starter motor may have internal damage. A bent starter armature, worn bearings, thrown armature windings, loose pole shoe screws, or any other worn component in the starter motor that will allow the armature to drag can cause a high current demand.

If the lights stay brightly lit, and the starter makes no sound (listen for a deep clicking noise), there is an open in the circuit. The fault is in either the solenoid or the control circuit. To test the solenoid, bypass the solenoid by bridging the BAT and S terminals on the back of the solenoid (Figure 10-2). The most common fault in the control circuit is a faulty starter safety switch or a burned fusible link.

WARNING *A starter can draw up to 400 amperes. The tool used to jump the terminals must be able to carry this high current and must have an insulated handle.*

If the starter rotates with the solenoid bypassed, the control circuit is at fault. If the starter does not rotate and the lights do not dim, the solenoid is at fault. (Also listen for the starter drive engaging.) If the starter rotates slowly and the headlights dim, there is excessive current draw and the system will have to be tested further.

CUSTOMER CARE *If the starter windings are thrown, this indicates several different problems. The most common is that the driver is keeping the ignition switch in the START position too long after the engine has started. Other causes include the driver opening the throttle plates too wide while starting the engine, which results in excessive armature speeds when the engine does start. Also, the windings can be thrown because of excessive heat build-up in the motor. The motor is designed to operate for very short periods of time. If it is operated for longer than 15 seconds, heat begins to build up at a very fast rate. If the engine does not start after a 15-second crank, the starter motor needs to cool for about two minutes before the next attempt to start the engine.*

Current Draw Test

If the starter motor cranks the engine, the technician should perform the **current draw test**. The current draw test measures the amount of current the starter draws when actuated. It determines the electrical and mechanical condition of the starting system. The following pro-

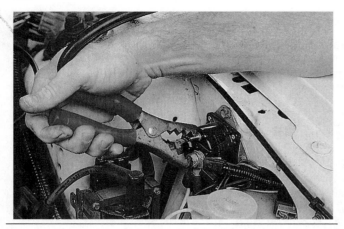

Figure 10-2 Bypassing the solenoid to determine if the solenoid or the control circuit is faulty.

cedure uses a VAT-40 and is similar to the procedure for other starting and charging system testers.

1. Connect the large red and black test leads across the battery, observing polarity.

2. Select INT 18 V.

3. Zero ammeter.

4. Connect the green amps inductive probe around the battery ground cable. If more than one ground cable is used, clamp the probe around all of them (Figure 10-3).

5. Make sure all loads are turned off (lights, radio, etc.).

6. Place the test selector in the STARTING position.

7. Disable the ignition system to prevent the vehicle from starting. This may be done by removing the ignition coil secondary wire from the distributor cap and putting it to ground (Figure 10-4). Remove the battery lead to the distributor on vehicles with high energy ignition (HEI) (Figure 10-5). Do not ground this lead.

8. Crank the engine and note the voltmeter reading.

9. With the ignition switch in the OFF position, turn the load control knob slowly until the voltage reading is the same as obtained in step 8.

10. Read the ammeter scale to determine the amount of current draw.

WARNING *Always refer to the manufacturer's service manual for the correct procedure for disabling the ignition system.*

WARNING *Do not operate the starter motor for longer than 15 seconds. Allow the motor to cool between cranking attempts.*

After recording the readings from the current draw test, compare them with the manufacturer's specifications. The specifications for current draw is the maximum allowable and the specification for cranking

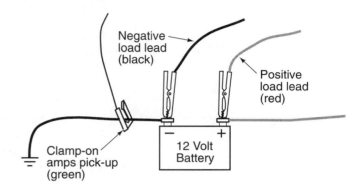

Figure 10-3 Connecting the VAT-40 test leads to perform the starter current draw test. (Courtesy Of Sun Electrical Corporation)

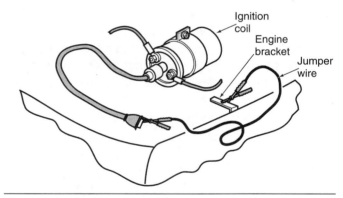

Figure 10-4 Grounding the coil secondary cable to prevent the engine from starting and to protect the coil during testing.

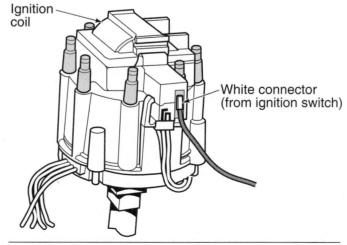

Figure 10-5 To disable the HEI ignition system, disconnect the BAT wire from the coil connector.

voltage is the minimum allowable. If specifications are not available, correctly functioning systems as a rule will crank at 9.6 volts or higher. Current draw is dependent on engine size. Most V-8 engines will have a current draw of about 200 amperes, six-cylinder engines about 150 amperes, and four-cylinder engines about 125 amperes.

If the readings obtained from the current draw test are out of specifications, then additional testing will be required to isolate the problem. If the readings were on the borderline of the specifications or there is an intermittent problem, then detailed testing for bad components will pinpoint potential failures.

Higher than specified current draw test results indicate an impedance to rotation of the starter motor. This includes worn bushings, a mechanical blockage, and excessively advanced ignition timing. Lower than normal current draw test results indicate excessive resistance in the circuit due to poor connections, faulty relay or solenoid, or worn brushes.

Because the readings obtained from the current draw test were taken at the battery, these readings may not be an exact representation of the actual voltage and current delivered to the starter. Voltage losses due to bad cables, connections, and relays (or solenoids) may diminish the amount of voltage to the starter. These should be tested before removing the starter from the vehicle.

Insulated Circuit Resistance Test

The **insulated circuit resistance test** is a voltage drop test that is used to locate high resistance in the starter circuit. An electrical resistance will have a different pressure or voltage on each side of the resistance. Voltage is dropped when current flows through resistance. Most manufacturers design their starting systems to have very little resistance to the flow of current to the starter motor. Most have less than 0.2 volt dropped on each side of the circuit. This means the voltage across the starter input terminal to the starter ground should be within 0.4 volt of battery voltage (Figure 10-6).

Voltage drops are measured by connecting a voltmeter in parallel with the circuit section being tested. In order to obtain a voltage drop reading, a load on the circuit must be applied. The following is the test procedure using a VAT-40 tester:

1. Set the volt selector to the EXT 3 V position.
2. Connect the test leads as shown (Figure 10-7), depending on the type of system being tested.
3. The voltmeter should read off the scale to the right until a load is put on the circuit. If the meter reads zero, reverse the leads.

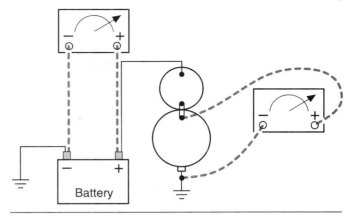

Figure 10-6 Voltage drop testing to determine excessive resistance.

4. Disable the ignition system as discussed in current draw testing.
5. Crank the engine and observe the voltmeter scale.

This tests for voltage drop in the entire circuit, so if voltage drop is excessive the cause of the drop must be located. To locate the cause of the excessive voltage drop, move the voltmeter lead on the starter toward the battery. Check each connection while moving toward the battery. With each move of the test lead, crank the engine while observing the voltmeter reading. Continue to test each connection until a noticeable decrease in voltage drop is detected. The cause of the excessive voltage drop will be located between that point and the preceding point.

SERVICE TIP *As a general rule, allow up to 0.2 volt per cable and 0.1 volt per connection to be dropped. Switches can be as high as 0.3 volt. Use the wiring diagram for the vehicle to determine the number of conductors and connections used in the circuit. This will provide a specification for you if no other specifications are available.*

Ground Circuit Test

A **ground circuit test** is performed to measure the voltage drop in the ground side of the circuit (Figure 10-8). If the starter motor connection to ground is broken or loose, the circuit would be opened. This could cause an intermediate starter system problem, or a starter motor that will not crank the engine. To perform the ground circuit test, connect the voltmeter leads across the ground circuit and read voltage drop while cranking the engine. Follow these directions:

1. Set the volt selection to EXT 3 V.

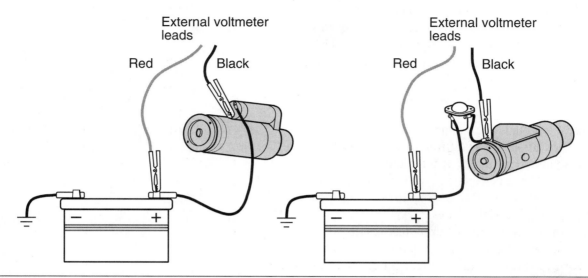

Figure 10-7 (A) Test lead connections for starter-mounted solenoids. (B) Test lead connections for relay-controlled systems. (Courtesy Of Sun Electrical Corporation)

2. Connect the positive volt test lead to the starter motor case and the negative test lead to the ground battery terminal. Make sure any paint is removed from the area where the lead is connected to the case.

3. Crank the engine while observing the voltmeter.

Less than 0.2 volt indicates the ground circuit is good. If more than 0.2 volt is observed, then there is a poor ground circuit connection. A poor ground circuit connection could be the result of loose starter mounting bolts, paint on the starter motor case, or a bad battery ground terminal post connection. Also check the ground cable for high resistance or for being undersized.

Solenoid Circuit Resistance Testing

The **solenoid circuit resistance test** determines the electrical condition of the solenoid and the control circuit. High resistance in the solenoid will reduce the current flow through the solenoid windings and cause the solenoid to function improperly. If the solenoid has high resistance in the windings it may result in the contacts burning and causing excessive resistance to the starter motor.

To perform this test, first disable the ignition system. With the VAT-40 set on EXT 3 V, connect the positive voltmeter lead to the BAT terminal of the solenoid. Connect the negative test lead to the field coil terminal (M terminal). The voltmeter should read off of the scale. If the voltmeter reads zero, reverse the test leads. Crank the engine while observing the voltmeter reading. If the voltmeter reading indicates a voltage drop of greater than 0.1 volt, then the solenoid is defective. If this test proves the starter solenoid is good, then the solenoid switch circuit should be tested. Using a VAT-40, follow these steps:

1. Disable the ignition system.

2. Set the volt selector to the EXT 3 V position.

3. Connect the voltmeter leads to both solenoid switch terminals, observing polarity.

4. Crank the engine and observe the voltmeter reading.

The total voltage drop should be less than 0.5 volt. If the indicated voltage drop is in excess of 0.5 volt, move the voltmeter leads up the circuit and test each component. The voltage drop across each component should be less than 0.1 volt.

Continue to move the voltmeter leads to test for voltage drop through the wires, starter relay, neutral safety switch, and ignition switch.

No-crank Test

The **no-crank test** is performed to locate any opens in the starter or control circuits. In order to perform voltage

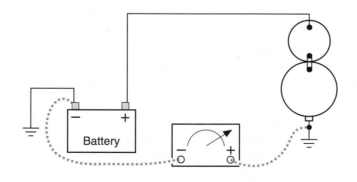

Figure 10-8 Voltage drop testing of the ground side circuit.

drop tests, a load must be placed on the circuit. If there is a no-crank complaint, a voltage drop test cannot be performed. Most no-crank problems are the result of opens in the circuit. However, as with any starter system complaint, check the battery first and replace it if necessary.

The easiest way to diagnose this problem is with the use of a test light. On a system that uses a starter motor-mounted solenoid, the M terminal is the end of the circuit (Figure 10-10). Connecting the positive lead of the test light to the M terminal and the negative lead to a good ground, the light should be on when the ignition switch is located in the START position (Figure 10-10). If the light comes on, then the complete circuit (including the ignition switch, wires, neutral safety switch, solenoid, and all connections) is operating properly. A voltmeter can be used in place of the test light. At the M terminal there should be more than 9.6 volts present when the ignition switch is in the START position.

If the test light comes on very dim, then there is very high resistance in the circuit. By working the test light through the circuit, back to the battery, the reason for the

high resistance should be found. If the test light did not come on, follow the same procedure of backtracking the circuit toward the battery until the open is found. Also check for voltage at the B+ terminal (Figure 10-11). Connecting the test light as shown should light the bulb with the key off. The light should stay on when the ignition switch is turned to the START position. If the light goes out in the START position, repair the cable or end connections.

Once the open has been found, it can be verified by using jumper wires to jump across the defective component or connection. If jumping across the solenoid, for example, and the starter spins, then there is an open in the solenoid. The same procedure can be used to jump across the ignition switch, neutral safety switch, or open wires.

If the test light did not come on when connected to the M terminal, then make a simple test of the ground circuit. This is done by connecting the ground lead of the test light onto the starter body and the positive lead to the M terminal of the solenoid. The light should come on bright with the ignition switch in the START position. If the ground circuit is good, then the starter is suspect and should be bench tested.

Starting systems using a remote-mounted starter relay are tested in the same manner, except there is an additional battery cable to test.

Free Speed Test

Every starter should be **bench tested** before it is installed. The **free speed test** determines the free rotational speed of the armature. This test is also referred to as the **no-load test**. Some manufacturers recommend this test procedure over the current draw test. The starter must be removed from the vehicle, as described in the

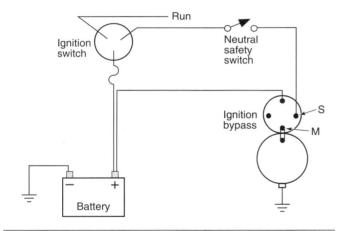

Figure 10-9 Starting system using a solenoid shift.

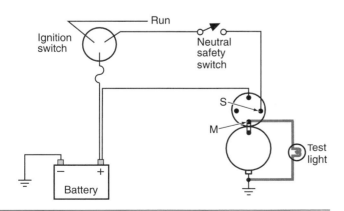

Figure 10-10 Test light connections for testing the solenoid and control circuit.

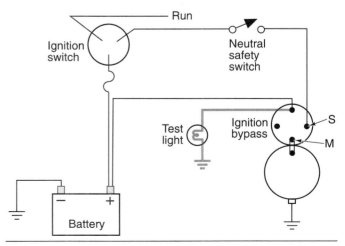

Figure 10-11 Test light connection for checking voltage at the BAT terminal.

next section. With the starter removed from the vehicle, perform the test as follows (Figure 10-12):

WARNING *Do not overtighten the vise against the starter frame assembly. It is possible to crack the frame or the pole shoes.*

1. Place the starter motor into a secure vise.
2. Attach an rpm indicator to the armature shaft at the drive housing end.
3. Connect a remote starter switch between the BAT and S terminals of the solenoid.
4. Connect the jumper cables, as shown in Figure 10-12.
5. Connect the large red and black test leads of the tester across the battery, observing polarity.
6. Select INT 18 V.
7. Zero ammeter.
8. Connect the green amps inductive probe around the jumper cable from the battery negative terminal to starter frame.
9. Place the test selector to the STARTING position.
10. Load the battery by rotating the load control knob until a voltage reading of 10 volts is obtained.
11. Switch to the EXT 18 V position.
12. Close the remote starter switch while reading the ammeter, voltmeter, and tachometer scales.

WARNING *Failure to load the battery to this level can result in the armature windings being thrown. Because there is no load on the starter, the rpms will be excessive if more than 10 volts are used.*

Compare the test results with manufacturer's specifications. General specifications will be about 6,000 to 12,000 rpm with a current draw of 60 to 85 amperes. Voltage should remain at 10 volts. If the test results are within specifications, the starter motor is ready to be reinstalled into the vehicle.

If the current draw was excessive and rpm slower than specifications, there is excessive resistance to rotation. This could be caused by:

1. Worn bushings or bearings
2. Shorted armature
3. Grounded armature
4. Shorted field windings
5. Bent armature

If there was no current draw, and the starter did not rotate, this could be caused by one of the following:

1. Open field windings
2. Open armature coils
3. Broken brush or brush spring

Low armature speed with low current draw indicates excessive resistance. There may be a poor connection between the commutator and the brushes. Also, any connection in the starter and to the starter may be faulty.

If the armature speed and current draw readings are high, check for a shorted field winding.

Starter Motor Removal

If the tests indicate the starter motor must be removed, the first step is to disconnect the battery from the system. It may be necessary to place the vehicle on a lift to gain access to the starter motor. Before lifting the vehicle, disconnect all wires, fasteners, and so forth, that can be reached from the top of the engine compartment.

WARNING *Remove the negative battery cable. It is a good practice to wrap the cable clamp with tape or enclose it in a rubber hose to prevent accidental contact with the battery terminal.*

WARNING *Check for proper pad to frame contact after the vehicle is a few inches above the ground. Shake the vehicle. If there are any unusual noises or movement of the vehicle, lower it down and reset the pads.*

Disconnect the wires leading to the solenoid terminals. To prevent confusion, it is a good practice to use a piece of tape to identify the different wires.

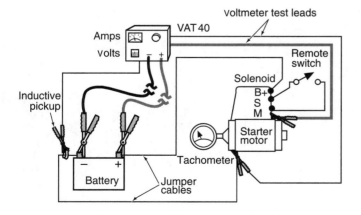

Figure 10-12 Starting/charging system tester connections for free speed test.

On some vehicles it may be necessary to disconnect the exhaust system to be able to remove the starter motor. Spray the exhaust system fasteners with a penetrating oil to assist in removal. Loosen the starter mounting bolts and remove all but one. Support the starter motor; remove the remaining bolt. Then remove the starter motor.

WARNING *The starter motor is heavy; make sure it is secured before removing the last bolt.*

Reverse the procedure to install the starter motor. Be sure all electrical connections are tight. If you are installing a new or remanufactured starter, remove any paint that may prevent a good ground connection. Be careful not to drop the starter. Make sure it is properly supported.

Some General Motors starters use shims between the starter motor and the mounting pad (Figure 10-13). To check this clearance, insert a flat blade screwdriver into the access slot on the side of the drive housing. Pry the drive pinion gear into the engaged position. Use a piece of 0.020" (0.508 mm) diameter wire to check the clearance between the pinion gear and the starter ring gear (Figure 10-14).

If the clearance between the two gears is excessive, the starter will produce a high-pitched whine while the engine is being cranked. If the clearance is too small, the starter will make a high-pitched whine after the engine starts and the ignition switch is returned to the RUN position.

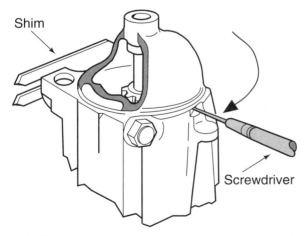

A 0.015" shim will increase the clearance approximately 0.005". More than one shim may be required.

Figure 10-13 Shimming the starter to obtain the correct pinion-to-ring gear clearance. (Courtesy of General Motors Corporation, Service Technology Group)

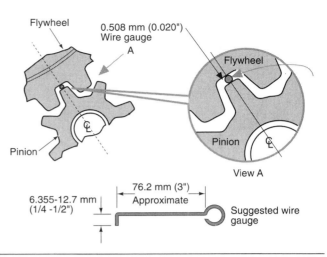

Figure 10-14 Checking the clearance between the pinion gear and ring gear.

SERVICE TIP *The major cause of drive housing breakage is too small of a clearance between the pinion and ring gears. It is always better to have a little more clearance than too small of a clearance.*

Starter Motor Disassembly

If it is determined that the starter is the defective part, it can be disassembled and bench tested. To reduce vehicle down time to a minimum, many repair facilities do not rebuild starters. They replace them instead. However, many shops will replace the starter drive mechanism, which may require several of the following disassembling steps. The decision to rebuild or replace the starter motor is based on several factors:

1. What is best for the customer
2. Shop policies
3. Cost
4. Time
5. Type of starter

If the starter is to be rebuilt, the technician should study the manufacturer's service manual to become familiar with the disassembly procedures for the particular starter. Photo Sequence 5 illustrates a typical procedure for disassembly of a Delco Remy starter. Always refer to the specific manufacturer's service manual for the starter motor you are working on.

WARNING *Do not clean the starter motor components in solvent or gasoline. The residue left can ignite and destroy the starter. Use compressed air that is regulated to 25 psi, wipe with clean rags, or use denatured alcohol to clean the starter components.*

The starter motor can be cleaned and inspected when it is disassembled. Inspect the end frame and drive housing for cracks or broken ends. Check the frame assembly for loose pole shoes and broken or frayed wires. Inspect the drive gear for worn teeth and proper overrunning clutch operation. The commutator should be free of flat spots and should not be excessively burned. Check the brushes for wear. Replace them if worn past manufacturer's specifications.

Starter Motor Component Tests

With the starter motor disassembled and the components cleaned, you are ready to perform tests that will isolate the reason for the failure. The armature and field coils are checked for shorts and opens. In most cases, the whole starter motor assembly is replaced if the armature or field coils are bad.

Field Coil Testing

The field coil and frame assembly should be tested for opens and shorts to ground. In most cases, if one of these conditions is found, the starter is considered unrebuildable in the field and will need to be replaced with a new unit.

Field coils can be wired in a number of different ways. Accurately testing the coils for opens and shorts depends on how they are wired. There are two things to do to determine the best way to check the field coils: refer to a service manual for specific instructions and/or refer to the wiring diagram for the starting circuit. By looking at the wiring diagram you will be able to tell where the coils receive their power and where they ground. Knowing these things is critical to testing the coils. The following procedure is valid for many, but not all, vehicles.

With the ohmmeter on the lowest scale, place one lead on the starter motor input terminal. Connect the other lead to the insulated brushes (Figure 10-15). The ohmmeter should indicate zero resistance. If there is resistance in the field coil, replace the coil and/or the frame assembly.

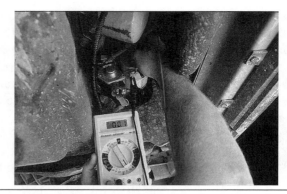
Figure 10-15 Testing the field coils for opens.

With the ohmmeter on the 2K scale, place one lead on the starter motor input terminal. Connect the other lead to the starter frame (Figure 10-16). An infinite reading should be obtained. If the ohmmeter indicates continuity, there is a short to ground in the field coil.

Armature Short Test

A **growler** produces a very strong magnetic field that is capable of inducing a current flow and magnetism in a conductor. It is used to test the armature for shorts and grounds (Figure 10-17).

To test the armature for shorts, place the armature in the growler and hold a thin steel blade parallel to the core (Figure 10-18). Slowly rotate the armature and observe the steel blade. If the blade begins to vibrate or pull toward the core, the armature is shorted and in need of replacement.

Armature Ground Test

With the armature placed in the growler, use a continuity tester to check for continuity between the armature core and any bar of the commutator. If there is continuity, then the armature is grounded and in need of replacement.

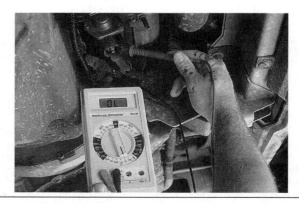
Figure 10-16 Testing the field coils for shorts to ground.

Typical Procedure for Delco Remy Starter Disassembly

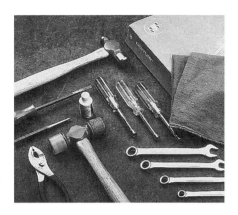

P4-1 Always have a clean and organized work area. Tools required to disassemble the Delco Remy starter: rags, assorted wrenches, snap ring pliers, flat blade screw driver, ball-peen hammer, plastic head hammer, punch, scribe, safety glasses, and arbor press.

P4-2 Clean the case and scribe reference marks at each end of the starter end housings and the frame.

P4-3 Disconnect the field coil connection at the solenoid's M terminal.

P4-4 Remove the two screws that attach the solenoid to the starter drive housing.

P4-5 Rotate the solenoid until the locking flange of the solenoid is free, then remove the solenoid.

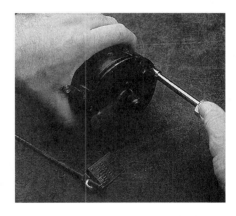

P4-6 Remove the through bolts from the end frame.

Typical Procedure for Delco Remy Starter Disassembly

P4-7 Remove the end frame.

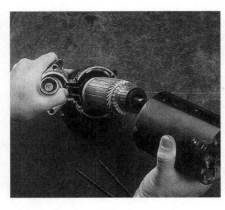

P4-8 Remove the frame.

P4-9 Remove the armature from the drive housing. Note: On some units it may be necessary to remove the shift lever from the drive housing before removing the armature.

P4-10 Remove the thrust washer from the end of the armature shaft Place a 5/8" deep socket over the armature shaft until it contacts the retaining ring of the starter drive. Tap end of socket with a plastic hammer to drive the retainer toward the armature. Move it only far enough to access the snap ring.

P4-11 Remove the snap ring.

P4-12 Remove the retainer from the shaft and remove the clutch and spring from the shaft. Press out the drive housing bushing.

Figure 10-17 A growler is used to test the armature for shorts.

Commutator Tests

If a growler is not available, the armature commutator can be tested for opens and grounds using an ohmmeter. The commutator should be cleaned with crocus cloth. To check for continuity, place the ohmmeter on the lowest scale. Connect the test leads to any two commutator sectors (Figure 10-19). There should be zero ohms of resistance. The armature will have to be replaced if there is resistance.

Place the ohmmeter on the 2K scale and connect one of the test leads to the armature shaft. Connect the other lead to the commutator segments (Figure 10-20). Check each sector. There should be no continuity to ground. The armature will have to be replaced if there is continuity.

Starter Reassembly

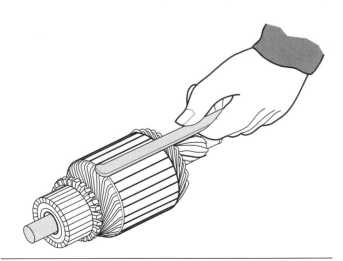

Figure 10-18 The growler generates a magnetic field. If there is a short, the hacksaw blade will vibrate over the area of the short.

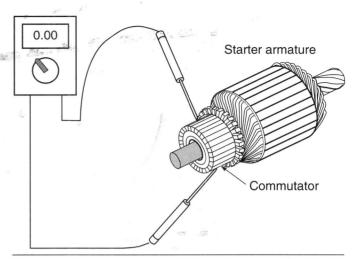

Figure 10-19 Testing the armature for opens. There should be zero resistance between the segments.

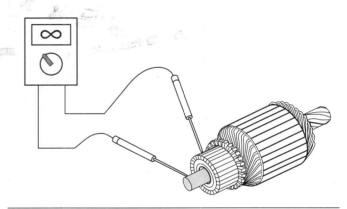

Figure 10-20 Testing the armature for shorts to ground. The meter should read infinite when placed on the shaft and different segments of the commutator.

If the brushes are worn beyond specifications, they must be replaced. Manufacturers use two methods of connecting the brushes: they are either soldered to the coil leads or screwed to terminals.

If the brushes are soldered to the coil leads, cut the old leads (Figure 10-21). Place a piece of heat-shrink tube over the brush connector. Crimp the new brush lead connector to the coil leads. Solder the brush connector to the coil lead with rosin core solder (Figure 10-22). Slide the heat-shrink tube over the soldered connection and use a heating gun to shrink the tube.

SERVICE TIP *There is no provision on most starters to adjust the pinion clearance. However, if the clearance is excessive, it may indicate excessive wear of the solenoid linkage or shift lever.*

Figure 10-21 Removing the worn brushes.

Figure 10-22 Soldering the new brush to the field coil lead.

To reassemble the starter motor, basically reverse the disassembly procedures. Additional steps are listed here:

1. Lubricate the splines on the armature shaft that the drive gear rides on with a high temperature grease.

2. To install the snap ring onto the armature shaft, stand the commutator end of the armature on a block of wood. Position the snap ring onto the shaft and hold in place with a block of wood. Hit the block of wood with a hammer to drive the snap ring onto the shaft (Figure 10-23.

3. Lubricate the bearings with high temperature grease.

4. Apply sealing compound to the solenoid flange before installing the solenoid to the frame (Figure 10-24).

5. Use the scribe marks to locate the correct position of the frame-to-frame end and drive housing.

6. Check the pinion gear clearance. Disconnect the M terminal to the starter motor's field coils. Connect a

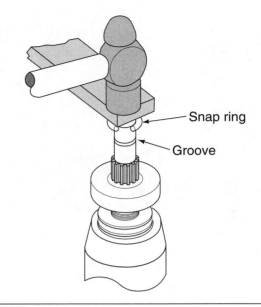

Figure 10-23 Once the snap ring is centered on the shaft, a hammer and block of wood can be used to install the ring onto the shaft.

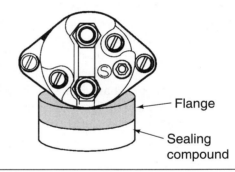

Figure 10-24 Apply sealing compound to the solenoid flange.

jumper cable from the battery positive terminal to the S terminal of the solenoid. Connect the other jumper cable from the battery negative terminal to the starter frame (Figure 10-25). Connect a jumper wire from the M terminal and momentarily touch the other end of the jumper wire to the starter motor frame. This will shift the pinion gear into the cranking position and hold it there until the battery is disconnected. Once the solenoid is energized, push the pinion back toward the armature; this removes any slack. Check the clearance with a feeler gauge (Figure 10-26). Compare clearance with specifications; normally specifications call for a clearance of 0.010 to 0.140 inches (0.25 to 0.35 mm).

7. Perform the free spin test before installing the starter into the vehicle.

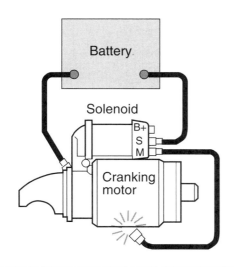
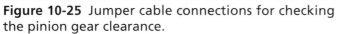

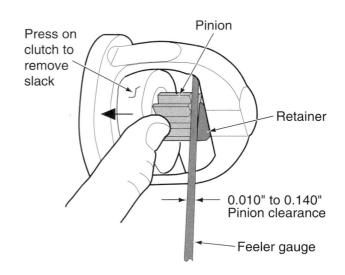

Figure 10-25 Jumper cable connections for checking the pinion gear clearance.

Figure 10-26 Checking the pinion gear to drive housing clearance. (Courtesy of General Motors Corporation, Service Technology Group)

Summary

❑ Before beginning any service on the starter system, the technician must be aware of all precautions.

❑ Customer complaints concerning the starting system generally fall into four categories: no-crank, slow cranking, starter spins but does not turn engine, and excessive noise.

❑ Before performing any tests on the starting system, first begin with a visual inspection of the circuit. Repair or replace any corroded or loose connections, frayed wires, or any other trouble sources. The battery terminals must be clean and the starter motor must be properly grounded.

❑ No-crank means that when the ignition switch is placed in the START position, the starter does not turn the engine.

❑ Slow cranking means that the starter drive engages the ring gear, but the engine turns at too slow of a speed to start.

❑ If the starter spins but the engine does not rotate, the most likely cause is a faulty starter drive.

❑ The starter performance and battery performance are so closely related that it is important for a full battery test series to be done before trying to test the starter system. If the battery fails the load test and is fully charged, it must be replaced before doing any other tests.

❑ If the starter does not turn the engine at all, and the engine is in good mechanical condition, the quick test can be performed to locate the problem area. The quick test will isolate the problem area to determine if the starter motor, solenoid or control circuit is at fault.

❑ If the starter motor cranks the engine, the technician should perform the current draw test to measure the amount of current the starter draws when actuated. This test determines the electrical and mechanical condition of the starting system.

❑ The insulated circuit resistance test is a voltage drop test that is used to locate high resistance in the starter circuit.

❑ A ground circuit test is performed to measure the voltage drop in the ground side of the circuit.

❑ The solenoid circuit resistance test determines the electrical condition of the solenoid and the control circuit.

❑ The no-crank test is performed to locate any opens in the starter or control circuits.

❑ Every starter should be bench tested before it is installed.

❑ The free speed test determines the free rotational speed of the armature.

Terms-To-Know

Bench tested	Hydrostatic lock	Quick test
Current draw test	Insulated circuit resistance test	Slow cranking
Free speed test	No-crank	Solenoid circuit resistance test
Ground circuit test	No-crank test	
Growler	No-load test	

ASE Style Review Questions

1. Brush replacement is being discussed. Technician A says manufacturers use two different methods of connecting the brushes to the field coil leads. Technician B says if new brushes are soldered on, use heat-shrink tubing to insulate the connection. Who is correct?
   a. A only
   b. B only
   c. Both A and B
   d. Neither A nor B

2. Pinion gear to ring gear clearance is being discussed. Technician A says if the clearance is excessive, the starter will produce a high-pitched whine while the engine is being cranked. Technician B says if the clearance is too small, the starter will make a high-pitched whine after the engine starts. Who is correct?
   a. A only
   b. B only
   c. Both A and B
   d. Neither A nor B

3. Voltage drop testing of the starter circuit is being discussed. Technician A says an electrical resistance will have the same voltage on each side of the resistance. Technician B says voltage drops are measured by connecting a voltmeter in series with the circuit that is being tested. Who is correct?
   a. A only
   b. B only
   c. Both A and B
   d. Neither A nor B

4. Technician A says on a no-crank complaint, perform a voltage drop test to find the location of the resistance. Technician B says in order to obtain a voltage drop reading, a load on the circuit must be activated. Who is correct?
   a. A only
   b. B only
   c. Both A and B
   d. Neither A nor B

5. Armature testing is being discussed. Technician A says to test for shorts, place the armature in the growler and hold a thin steel blade parallel to the core and watch for blade vibrations that would indicate a short. Technician B says there should be zero resistance between the commutator sectors and the armature shaft. Who is correct?
   a. A only
   b. B only
   c. Both A and B
   d. Neither A nor B

6. Current draw test results are being discussed. Technician A says a bent starter armature, worn bearings, thrown armature windings, or loose pole shoe screws can cause high current draw. Technician B says lower than normal current draw test results indicate excessive voltage drop in the circuit. Who is correct?
   a. A only
   b. B only
   c. Both A and B
   d. Neither A nor B

7. Technician A says it is important that a full battery test series be done before trying to test the starter system. Technician B says the internal condition of the engine has little effect on the operation of the starting system. Who is correct?
   a. A only
   b. B only
   c. Both A and B
   d. Neither A nor B

8. The starter motor has been rebuilt and is ready to install in the vehicle. Technician A says to perform the free spin test before installing the starter into the vehicle. Technician B says to remove the M terminal connector before installing the starter motor. Who is correct?
   a. A only
   b. B only
   c. Both A and B
   d. Neither A nor B

9. Voltage drop testing of the control circuit is being discussed. Technician A says the maximum amount of voltage drop allowed is 0.9 volt. Technician B says the voltage drop across each wire should be less than 0.1 volt. Who is correct?
   a. A only
   b. B only
   c. Both A and B
   d. Neither A nor B

10. Technician A says most starter noises come from the armature. Technician B says the starter drive cannot be replaced on most starters. Who is correct?
   a. A only
   b. B only
   c. Both A and B
   d. Neither A nor B

11 Charging Systems

Objective

Upon completion and review of this chapter, you should be able to:

❑ Explain the purpose of the charging system.

❑ Identify the major components of the charging system.

❑ Explain the function of the major components of the AC generator.

❑ Describe the two styles of stators.

❑ Describe how AC current is rectified to DC current in the AC generator.

❑ Describe the three principle circuits used in the AC generator.

❑ Explain the relationship between regulator resistance and field current.

❑ Explain the relationship between field current and AC generator output.

❑ Identify the differences between A circuit, B circuit, and isolated circuit.

❑ Explain the operation of charge indicators, including lamps, electronic voltage monitors, ammeters, and voltmeters.

Introduction

The automotive storage battery is not capable of supplying the demands of the electrical system for an extended period of time. Every vehicle must be equipped with a means of replacing the current being drawn from the battery. A **charging system** is used to restore the electrical power to the battery that was used during engine starting. In addition, the charging system must be able to react quickly to high load demands required of the electrical system. It is the vehicle's charging system that generates the current to operate all of the electrical accessories while the engine is running.

Two basic types of charging systems have been used. The first was a DC generator, which was discontinued in the 1960s. Since that time the AC generator has been the predominant charging device. The DC generator and the AC generator both use similar operating principles.

The purpose of the charging system is to convert the mechanical energy of the engine into electrical energy to recharge the battery and run the electrical accessories. When the engine is first started, the battery supplies all the current required by the starting and ignition systems.

As the battery drain continues, and engine speed increases, the charging system is able to produce more voltage than the battery can deliver. When this occurs, the electrons from the charging device are able to flow in a reverse direction through the battery's positive terminal. The charging device is now supplying the electrical system's load requirements; the reserve electrons build up and recharge the battery (Figure 11-1).

If there is an increase in the electrical demand and a drop in the charging system's output equal to the voltage of the battery, the battery and charging system work together to supply the required current.

The entire charging system consists of the following components:

1. Battery

2. AC or DC generator

3. Drive belt

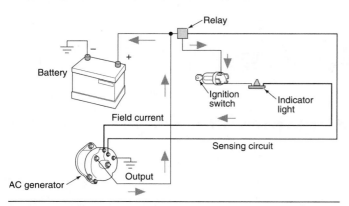

Figure 11-1 Current flow when the charging system is operating.

4. Voltage regulator
5. Charge indicator (lamp or gauge)
6. Ignition switch
7. Cables and wiring harness
8. Starter relay (some systems)
9. Fusible link (some systems)

Principle of Operation

All charging systems use the principle of electromagnetic induction to generate electrical power (Figure 11-2). Electromagnetic principle states that a voltage will be produced if motion between a conductor and a magnetic field occurs. The amount of voltage produced is affected by:

1. The speed at which the conductor passes through the magnetic field.
2. The strength of the magnetic field.
3. The number of conductors passing through the magnetic field.

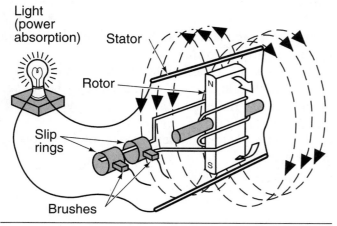

Figure 11-2 Simplified AC generator indicating electromagnetic induction.

To see how electromagnetic induction produces an AC voltage, refer to the illustration (Figure 11-3). When the conductor is parallel with the magnetic field, the conductor is not cut by any flux lines (Figure 11-3A). At this point in the revolution there is zero voltage and current being produced.

As the magnetic field is rotated 90 degrees, the magnetic field is at a right angle to the conductor (Figure 11-3B). At this point in the revolution the maximum number of flux lines cut the conductor at the north pole. With the maximum amount of flux lines cutting the conductor, voltage is at its maximum positive value.

When the magnetic field is rotated an additional 90 degrees, the conductor returns to being parallel with the magnetic field (Figure 11-3C). Once again no flux lines cut the conductor, and voltage drops to zero.

An additional 90-degree revolution of the magnetic field results in the magnetic field being reversed at the top conductor (Figure 11-3D). At this point in the revolution, the maximum number of flux lines cuts the conductor at the south pole. Voltage is now at maximum negative value.

When the magnetic field completes one full revolution, it returns to a parallel position with the magnetic field. Voltage returns to zero. The sine wave is determined by the angle between the magnetic field and the conductor. It is based on the trigonometry sine function of angles. The sine wave shown (Figure 11-4) plots the voltage generated during one revolution. The sine wave that is produced by a single conductor during one revolution is called **single-phase voltage**.

It is the function of the drive belt to turn the magnetic field. Drive belt tension should be checked periodically to assure proper charging system operation. A loose belt can inhibit charging system efficiency, and a belt that is too tight can cause early bearing failure.

DC Generators

The **DC generator** is similar to the DC starter motor used to crank the engine. The housing contains two field coils that create a magnetic field. Output voltage is generated in the wire loops of the armature as it rotates inside the magnetic field. This voltage sends current to the battery through the brushes. A DC generator is actually an AC generator whose current is rectified by a commutator.

The components must be **polarized** whenever a replacement DC generator or voltage regulator is installed. Polarizing will make sure the polarity for both the generator and the regulator is the same. To polarize an externally grounded field circuit (A-type field circuit), use a jumper wire and connect between the BAT terminal

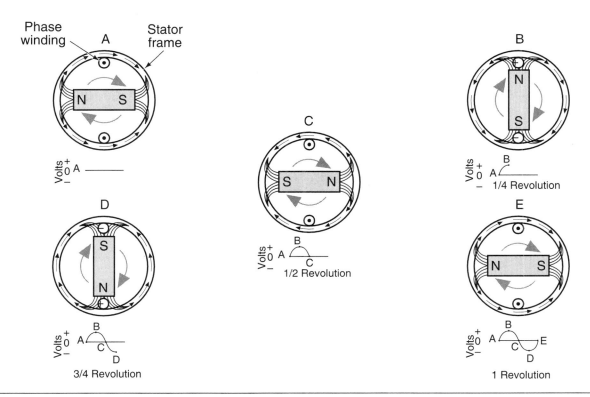

Figure 11-3 Alternating current is produced as the magnetic field is rotated.

and the ARM terminal of the voltage regulator. Make this jumper connection for just an instant. Do not hold the jumper wire on the terminals. For an internally grounded field circuit (B-type), jump the F terminal and the BAT terminal.

AC Generators

The DC generator was unable to produce the sufficient amount of current required when the engine was

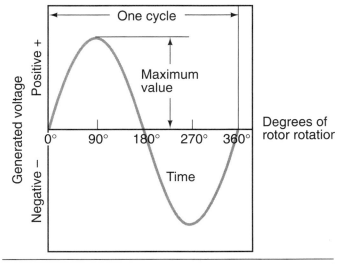

Figure 11-4 Sine wave produced in one revolution of the conductor.

operating at low speeds. With the addition of more electrical accessories and components, the **AC generator**, or alternator, replaced the DC generator. The main components of the AC generator are (Figure 11-5):

1. The rotor
2. Brushes
3. The stator
4. The rectifier bridge
5. The housing
6. Cooling fan

For many years the AC generator was referred to as an **alternator** to depict its production of AC voltage. In recent years there has been an attempt to standardize the industry and the terminology used in literature. Today the alternator is called a generator. To prevent confusion between a generator that produces DC voltage and one that produces AC voltage, this book will signify the type of voltage output.

The **rotor** creates the rotating magnetic field of the AC generator. It is the portion of the AC generator that is rotated by the drive belt. The rotor is constructed of many turns of copper wire around an iron core. There are metal plates bent over the windings at both ends of the rotor windings. The metal plates are called **fingers** or **poles**. The poles do not come into contact with each other, but they are interlaced. When current passes

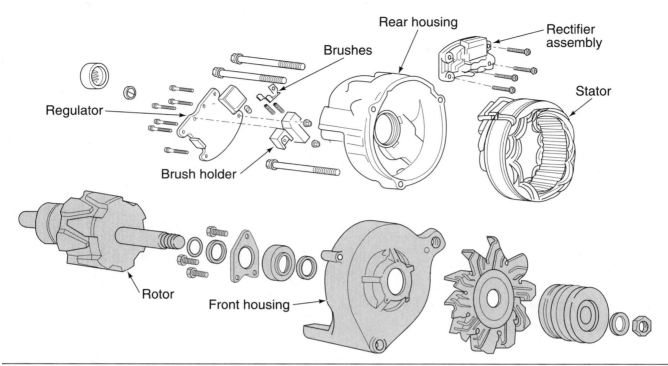

Figure 11-5 Components of an AC generator. (Reprinted with permission of Ford Motor Company)

through the coil (1.5 to 3.0 amperes), a magnetic field is produced. The strength of the magnetic field is dependent on the amount of current flowing through the coil and the number of windings. The current flow through the coil is referred to as **field current**. Most rotors have 12 to 14 poles.

The poles will take on the polarity (north or south) of the side of the coil they touch. The right-hand rule will show whether a north or south pole magnet is created. When the rotor is assembled, the poles alternate north-south around the rotor (Figure 11-6). As a result of this alternating arrangement of poles, the magnetic flux lines will move in opposite directions between adjacent poles

(Figure 11-7). This arrangement provides for several alternating magnetic fields to intersect the stator as the rotor is turning. These individual magnetic fields produce a voltage by induction in the stationary stator windings.

The wires from the rotor coil are attached to two **slip rings** that are insulated from the rotor shaft. The insulated stationary carbon brush passes field current into a slip ring, then through the field coil, and back to the other slip ring. Current then passes through a grounded stationary brush (Figure 11-8) or to a voltage regulator.

The field winding of the rotor receives current through a pair of brushes that ride against the slip rings. The brushes and slip rings provide a means of maintain-

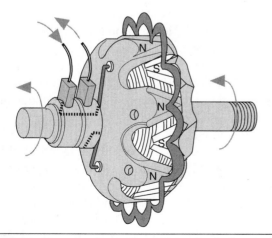

Figure 11-6 The north and south poles of a rotor's field alternate.

Figure 11-7 Magnetic flux lines move in opposite directions between the rotor poles. (Reprinted with permission of Ford Motor Company)

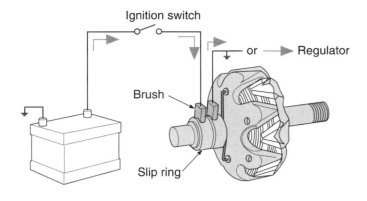

Figure 11-8 The slip rings and brushes provide a current path to the rotor coil.

ing electrical continuity between stationary and rotating components. The brushes ride the surface of the slip rings on the rotor and are held tight against the slip rings by spring tension provided by the brush holders. The brushes conduct only the field current (2 to 5 amperes). The low current that the brushes must carry contributes to their longer life.

Direct current from the battery is supplied to the rotating field through the field terminal and the insulated brush. The second brush may be the ground brush, which is attached to the AC generator housing or to a voltage regulator.

The **stator** is the stationary coil in which electrical voltage is produced. The stator contains three main sets of windings wrapped in slots around a laminated, circular iron frame (Figure 11-9). Each of the three windings has the same number of coils as the rotor has pairs of north and south poles. The coils of each winding are

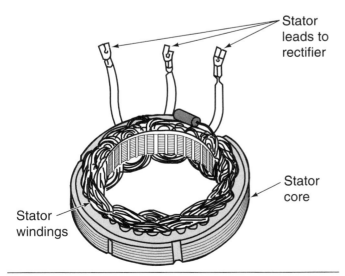

Figure 11-9 Components of a typical stator. (Reprinted with permission of Ford Motor Company)

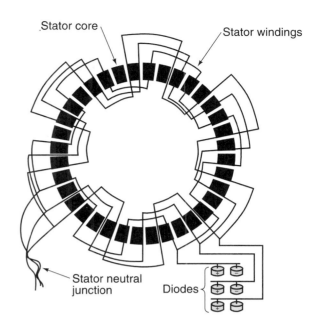

Figure 11-10 Overlapping stator windings produce the required phase angles.

evenly spaced around the core. The three sets of windings alternate and overlap as they pass through the core (Figure 11-10). The overlapping is needed to produce the required phase angles.

The rotor is fitted inside the stator. A small air gap (approximately 0.015 inch) is maintained between the rotor and the stator. This gap allows the rotor's magnetic field to energize all of the windings of the stator at the same time and to maximize the magnetic force.

Each group of windings has two leads. The first lead is for the current entering the winding. The second lead is for current leaving. There are two basic means of connecting the leads. In the past, the most common method was the **wye connection** (Figure 11-11). In the wye connection one lead from each winding is connected to one common junction (**stator neutral junction**). From this junction the other leads branch out in a Y pattern. Today the most common method of connecting the windings is called the **delta connection** (Figure 11-12). The delta connection connects the lead of one end of the winding to the lead at the other end of the next winding.

Each group of windings occupies one third of the stator, or 120 degrees of the circle. As the rotor revolves in the stator, a voltage is produced in each loop of the stator at different phase angles. The resulting overlap of sine waves that is produced is shown (Figure 11-13). Each of the sine waves is at a different phase of its cycle at any given time. As a result, the output from the stator is divided into three phases.

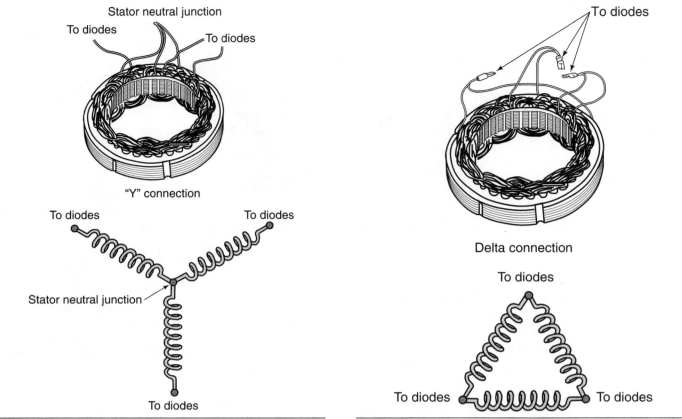

Figure 11-11 Wye-connected stator winding.

Figure 11-12 Delta-connected stator winding.

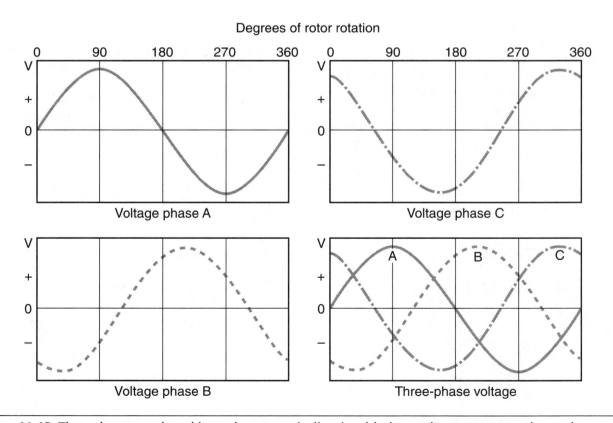

Figure 11-13 The voltage produced in each stator winding is added together to create a three-phase voltage.

The battery and the electrical system cannot accept or store AC voltage. For the vehicle's electrical system to be able to use the voltage and current generated in the AC generator, the AC current needs to be converted to DC current. This conversion is called **rectification**. A split-ring commutator cannot be used to rectify AC current to DC current because the stator is stationary in an AC generator. Instead, a **diode rectifier bridge** is used to change the current in an AC generator. Acting as a one-way check valve, the diodes switch the current flow back and forth so that it flows from the AC generator in only one direction.

When AC current reverses itself, the diode blocks and no current flows. If AC current passes through a positively biased diode, the diode will block off the negative pulse. The result is the scope pattern shown in Figure 11-14. The AC current has been changed to a pulsing DC current. The passing of half of the AC pulses is called **half- wave rectification**.

An AC generator usually uses a pair of diodes for each stator winding, for a total of six diodes (Figure 11-15). Three of the diodes are positive-biased and are mounted in a heat sink. The three remaining diodes are negative-biased and are attached directly to the frame of the AC generator. By using a pair of diodes that are reverse- biased to each other, **full-wave rectification** of the AC sine wave is achieved (Figure 11-16). The negative-biased diodes allow for conducting current from the negative side of the AC sine wave and putting this current into the circuit. Diode rectification changes the negative current into positive output.

With each stator winding connected to a pair of diodes, the resultant waveform of the rectified voltage would be similar to that shown. With six peaks per revolution, the voltage will vary only slightly during each cycle.

The examples used so far have been for single-pole rotors in a three-winding stator. Most AC generators use

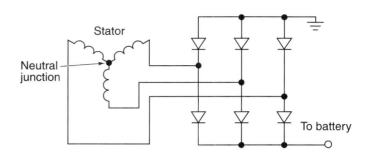

Figure 11-15 A simplified schematic of the AC generator windings connected to the diode rectifier bridge.

either a twelve- or fourteen-pole rotor. Each pair of poles produces one complete sine wave in each winding per revolution. During one revolution a fourteen-pole rotor will produce seven sine waves. The rotor generates three overlapping sine wave voltage cycles in the stator. The total output of a fourteen-pole rotor per revolution would be twenty-one sine wave cycles. With final rectification, the waveform would be similar to the one shown (Figure 11-17).

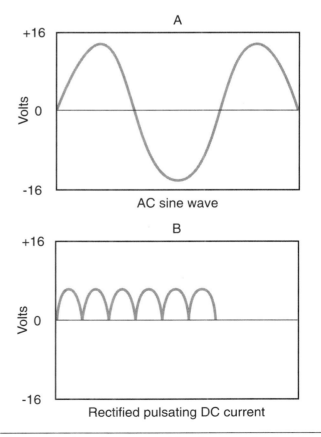

Figure 11-16 Full-wave rectification uses both sides of the AC sine wave to create a pulsating DC current.

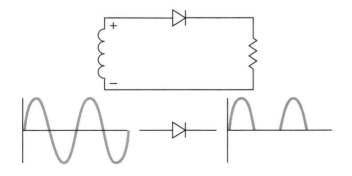

Figure 11-14 AC current rectified to a pulsating DC current after passing through a positive-biased diode. This is called half-wave rectification.

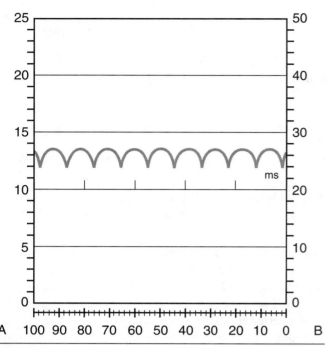

Figure 11-17 The rectified AC output has a ripple that can be shown on an oscilloscope.

Full-wave rectification is desired because using only half-wave rectification wastes the other half of the AC current. Full-wave rectification of the stator output uses the total potential by redirecting the current from the stator windings so that all current is in one direction.

A wye-wound stator with each winding connected to a pair of diodes is shown (Figure 11-18). Each pair of diodes has one negative and one positive diode. During rotor movement, two stator windings will be in series and the third winding will be neutral. As the rotor revolves, it will energize a different set of windings. Also, current flow through the windings is reversed as the rotor passes. Current in any direction through two windings in series will produce DC current.

The action that occurs when the delta-wound stator is used is shown (Figure 11-19). Instead of two windings in series, the three windings of the delta stator are in parallel. This makes more current available because the parallel paths allow more current to flow through the diodes.

Besides rectifying the output voltage, the diodes used in the AC generator also block battery drain from the battery to the generator when the engine is not running.

Most AC generator housings are a two-piece construction, made from cast aluminum. The two end frames provide support of the rotor and the stator. In addition, the end frames contain the diodes, regulator, heat sinks, terminals, and other components of the AC generator. The two end pieces are referred to as:

1. The drive end housing: This housing holds a bearing to support the front of the rotor shaft. The rotor shaft extends through the drive end housing and holds the drive pulley and cooling fan.

2. The slip ring end housing: This housing also holds a rotor shaft that supports a bearing. In addition, it contains the brushes and has all of the electrical terminals. If the AC generator has an integral regulator, it is also contained in this housing.

The cooling fan draws air into the housing through the openings at the rear of the housing. The air leaves through openings behind the cooling fan.

WARNING *Do not pry on the AC generator housing because any excess force can damage the housing.*

General Motors introduced an AC generator called the CS (charging system) series. This generator is smaller than previous designs. Additional features include two cooling fans (one external and one internal) and terminals designed to permit connections to an external computer (Figure 11-20).

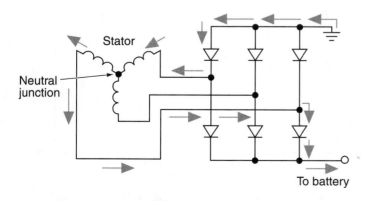

Figure 11-18 Current flow through a wye-wound stator.

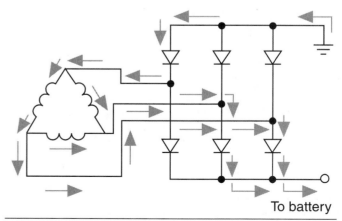

Figure 11-19 Current flow through a delta-wound stator.

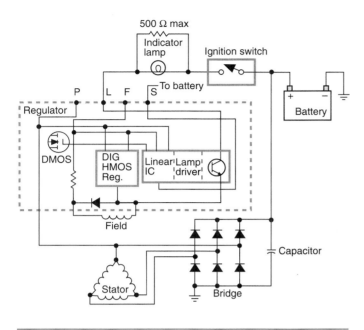

Figure 11-20 CS series AC generator circuit diagram.

AC Generator Circuits

There are three principal circuits used in an AC generator:

1. The charging circuit: Consists of the stator windings and rectifier circuits.

2. The excitation circuit: Consists of the rotor field coil and the electrical connections to the coil.

3. The pre-excitation circuit: Supplies the initial current for the field coil that starts the build-up of the magnetic field.

For the AC generator to produce current, the field coil must develop a magnetic field. The AC generator creates its own field current in addition to its output current.

For excitation of the field to occur, the voltage induced in the stator rises to a point that it overcomes the forward voltage drop of at least two of the rectifier diodes. Before the **diode trio** can supply field current, the anode side of the diode must be at least 0.6 volt more positive than the cathode side. When the ignition switch is turned on, the warning lamp current acts as a small magnetizing current through the field. This current pre-excites the field, reducing the speed required to start its own supply of field current. The diode trio is used by some manufacturers to rectify the stator current so that it can be used to create the magnetic field in the field coil of the rotor. This eliminates extra wiring.

If the battery is completely discharged, the vehicle cannot be push started because there is no excitation of the field coil. Even though the rotor will be rotated by the engine, the generator will not be able to produce any voltage to run the ignition system.

AC Generator Operation Overview

When the engine is running, the drive belt spins the rotor inside the stator windings. This magnetic field inside the rotor generates a voltage in the windings of the stator. Field current flowing through the slip rings to the rotor creates alternating north and south poles on the rotor.

The induced voltage in the stator is an alternating voltage because the magnetic fields are alternating. As the magnetic field begins to induce voltage in the stator's windings, the induced voltage starts to increase. The amount of voltage will peak when the magnetic field is the strongest. As the magnetic field begins to move away from the stator windings, the amount of voltage will start to decrease. Each of the three windings of the stator generates voltage, so the three combine to form a three-phase voltage output.

In the past, the most common type of stator was the wye connection (Figure 11-21). The output terminals (A, B, and C) apply voltage to the rectifier. Because only two stator windings apply voltage (because the third winding is always connected to diodes that are reverse-biased), the voltages come from points A to B, B to C, and C to A.

To determine the amount of voltage produced in the two stator windings, find the difference between the two points. For example, to find the voltage applied from points A and B subtract the voltage at point B from the voltage at point A. If the voltage at point A is 8 volts positive and the voltage at point B is 8 volts negative, the difference is 16 volts. This procedure can be performed for each pair of stator windings at any point in time to get the sine wave patterns (Figure 11-22). The voltages in the windings are designated as VA, VB, and VC. Designations of VAB, VBC, and VCA refer to the voltage difference in the two stator windings. In addition, the numbers refer to the diodes used for the voltages generated in each winding pair.

The current induced in the stator passes through the diode rectifier bridge consisting of three positive and three negative diodes. At this point there are six possible paths for the current to follow. The path that is followed depends on the stator terminal voltages. If the voltage from points A and B is positive (point A is positive in

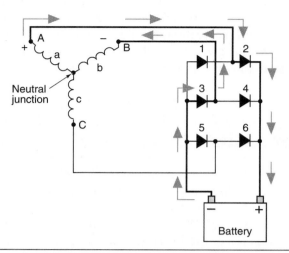

Figure 11-21 Current flow when terminals A and B are positive.

respect to point B), current is supplied to the positive terminal of the battery from terminal A through diode 2 (Figure 11-22). The negative return path is through diode 3 to terminal B.

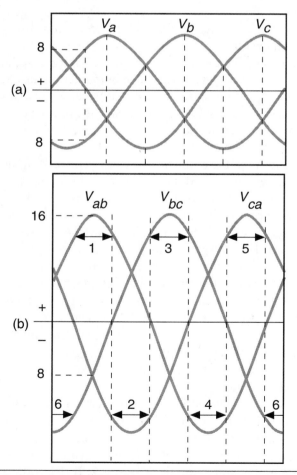

Figure 11-22 (A) Individual stator winding voltages (B) Voltages across the stator terminal A, B, and C.

Both diodes 2 and 3 are forward-biased. The stator winding labeled C does not produce current because it is connected to diodes that are reverse-biased. The stator current is rectified to DC current to be used for charging the battery and supplying current to the vehicle's electrical system.

When the voltage from terminals C and A is negative (point C is negative in respect to point A), current flow to the battery positive terminal is from terminal A through diode 2. The negative return path is through diode 5 to terminal C.

This procedure is repeated through the four other current paths.

Regulation

The battery, and the rest of the electrical system, must be protected from excessive voltages. To prevent early battery and electrical system failure, regulation of the charging system voltage is very important. Also the charging system must supply enough current to run the vehicle's electrical accessories when the engine is running.

AC generators do not require current limiters; because of their design they limit their own current output. Current limit is the result of the constantly changing magnetic field because of the induced AC current. As the magnetic field changes, an opposing current is induced in the stator windings. The **inductive reactance** in the AC generator limits the maximum current that the AC generator can produce. Even though current (amperage) is limited by its operation, voltage is not. The AC generator is capable of producing as high as 250 volts, if it were not controlled.

Voltage regulation is done by varying the amount of field current flowing through the rotor. The higher the field current, the higher the output voltage. By controlling the amount of resistance in series with the field coil, control of the field current and the AC generator output is obtained. To insure a full battery charge, and operation of accessories, most regulators are set for a system voltage between 13.5 and 14.5 volts.

The regulator must have system voltage as an input in order to regulate the output voltage. The input voltage to the AC generator is called **sensing voltage**. If sensing voltage is below the regulator setting, an increase in charging voltage output results by increasing field current. Higher sensing voltage will result in a decrease in field current and voltage output. A vehicle being driven with no accessories on and a fully charged battery will have a high sensing voltage. The regulator will reduce the charging voltage until it is at a level to run the ignition system while trickle charging the battery. If a heavy load is turned on (such as the headlights), the additional draw will cause a drop in the battery voltage. The regulator will sense this low system voltage and will increase current to the rotor. This will allow more current to the field windings. With the increase of field current, the magnetic field is stronger and AC generator voltage output is increased. When the load is turned off, the regulator senses the rise in system voltage and cuts back the amount of field current and ultimately AC generator voltage output.

Since sensing voltage is what determines the amount of system output, the condition and voltage of the battery determines the charge rate. This is why a charging system cannot be tested with a bad battery. System testing must be done with a known good battery that is at least 75% charged. Testing with a defective battery may suggest a faulty AC generator or regulator when these units are good.

Another input that affects regulation is temperature. Because ambient temperatures influence the rate of charge that a battery can accept, regulators are temperature compensated (Figure 11-23). Temperature compensation is required because the battery is more reluctant to accept a charge at lower ambient temperatures. The regulator will increase the system voltage until it is at a higher level so the battery will accept it.

Field Circuits

To properly test and service the charging system, it is important to identify the field circuit being used. Automobile manufacturers use three basic types of field circuits. The first type is called the **A circuit** or **external grounded field circuit**. It has the regulator on the ground side of the field coil. The B+ for the field coil is picked up from inside the AC generator. By placing the regulator on the ground side of the field coil, the regulator will allow the control of field current by varying the current flow to ground. A regulator can be located anywhere in the series circuit and have the same effect.

The second type of field circuit is called the **B circuit** or **internally grounded circuit**. In this case, the voltage regulator controls the power side of the field circuit. Also the field coil is grounded from inside the AC generator.

The third type of field circuit is called the **isolated field**. The isolated field AC generator has two field wires attached to the outside of the case. Both B+ and ground are picked up externally. The voltage regulator can be located on either the ground (A circuit) or on the B+ (B circuit) side (Figure 11-24).

Electromechanical Regulators

There are two basic types of regulators: electromechanical and electronic. Also, on many newer model vehicles regulation is controlled by the computer. Even though the electromechanical regulator is obsolete, a study of its operation will help you to understand the more complex systems.

The external **electromechanical regulator** is a vibrating contact point design. The regulator uses electromagnetics to control the opening and closing of the contact points. Inside the regulator are two coils. The first coil is the **field relay** which applies battery voltage to the AC generator field coil after the engine has started. The second coil is the **voltage limiter** which is connected through the resistor network and determines whether the field will receive high, low, or no voltage. It automatically controls the field voltage for the required amount of charging.

	Volts	
Temperature	Minimum	Maximum
20° F	14.3	15.3
80° F	13.8	14.4
140° F	13.3	14.0
Over 140° F	Less than 13.3	–

Figure 11-23 Chart indicating relationship between temperature and charge rate.

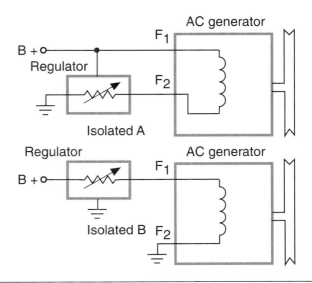

Figure 11-24 In the isolated circuit field AC generator, the regulator can be installed on either side of the field.

The field relay coil and contact with no current flowing through the coil are shown (Figure 11-25). The contact points are open, preventing current flow. An electromagnetic field develops when current flows through the field relay coil. It pulls the contact arm down. Once the contact points close, current will flow.

The voltage regulator coil uses an electromagnetic field to open the contact points. With the points closed, current flows from the battery to the rotor. Current also flows to the regulator coil. As the battery charges, the battery's voltage increases. This increase in voltage strengthens the coil's attraction for the contact points. At a preset voltage level, the coil will overcome the contact point spring tension and open the points. This will stop current from flowing to the rotor. Once the points open, voltage output of the AC generator drops. As the battery voltage decreases, so does the regulator coil's magnetic strength. Spring tension will overcome the magnetic attraction and the points will close again. Once again current flows through the rotor and the AC generator is producing voltage. This action occurs several times per second.

With the ignition switch in RUN, current flow for the field will go through the ignition switch through the resistor and bulb, through the lower contacts of the voltage regulator (closed), and out of the F terminal to the AC generator, through the field coil, and to ground (Figure 11-26). The indicator lamp bulb will light because the bulb is in series with the field coil.

Once the engine is started, the AC generator will begin to produce voltage. At this time the system voltage may be below 13.5 volts, however because the generator's rotor is revolving, there is some production of voltage. This allows voltage out of the R terminal to energize the field relay coil. With the relay coil energized, the contact points close and direct battery voltage flows through terminal 3 and out terminal 4. The bulb will go out with battery voltage on both sides of the bulb. Simultaneously, battery voltage flows through the voltage regulator contact points to the field coil in the AC generator. Because the voltage regulator coil is also connected to the battery (above terminal 4), the lower than 13.5 volts is unable to produce a sufficient electromagnetic field to pull the contact points open. This condition will allow maximum AC generator output. When maximum current is flowing to the rotor it is called **full field current**.

As the battery receives a charge from the AC generator, the battery voltage will increase to over 13.5 volts. The increased voltage will strengthen the electromagnetic field of the voltage regulator coil. The coil will attract the points and cause the lower contacts to open. Current will now flow through the resistor and out terminal F. Because an additional series resistance is added to the field coil circuit of the AC generator, field current is reduced, and AC generator output is also reduced. When current flows through the series resistor to the field coil, it is called **half field current**.

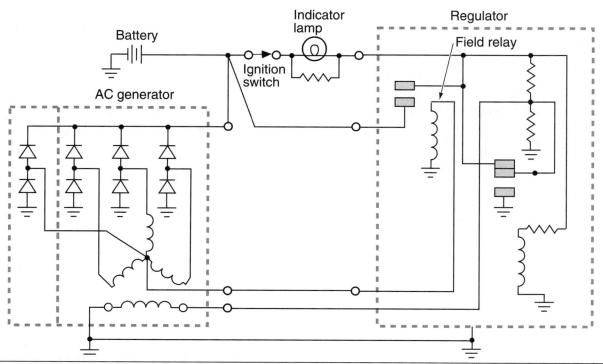

Figure 11-25 Schematic of an electromechanical regulator with no current flow through the coil. (Reprinted with permission of Ford Motor Company)

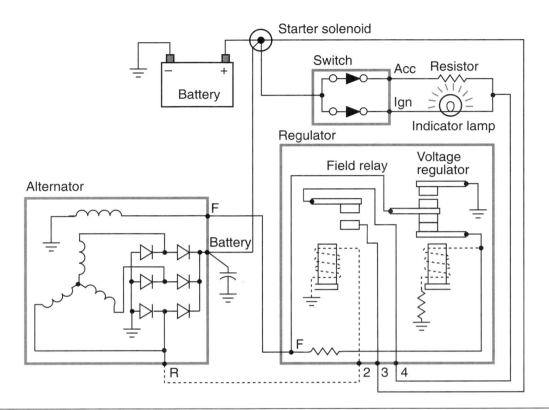

Figure 11-26 Regulator circuit with the ignition switch in the RUN position and the engine not running.

If the AC generator is producing more voltage than the system requires, both battery and system voltage will increase. As this voltage increases to a level above 14.5 volts, the magnetic strength of the coil increases. This increase in magnetic strength will close the top set of contact points and apply a ground to the F terminal and the AC generator's field coil. Both sides of the field coil are grounded, thus no current will flow through it and there is no output.

Once the engine is shut off, there is no current from the R terminal; the field relay coil is de-energized, allowing the points to open. This prevents battery discharge through the charging system.

Electronic Regulators

The second type of regulator is the **electronic regulator**. An electronic regulator uses solid state circuitry to perform the regulatory functions. There are no moving parts, so it can cycle between 10 and 7,000 times per second. This quick cycling provides more accurate control of the field current through the rotor. Electronic regulation control is through the ground side of the field current (A circuit). Electronic regulators can be mounted either externally or internally of the AC generator.

Pulse width modulation controls AC generator output by varying the amount of time the field coil is energized.

The period of time for each cycle does not change, only the amount of on time in each cycle changes. For example, assume that a vehicle is equipped with a 100-ampere generator. If the electrical demand placed on the charging system requires 50 amperes of current, the regulator would energize the field coil for 50% of the time. If the electrical system's demand was increased to 75 amperes, the regulator would energize the field coil 75% of the cycle time.

The electronic regulator uses a zener diode that blocks current flow until a specific voltage is obtained, at which point it allows the current to flow. An electronic regulator is shown (Figure 11-27).

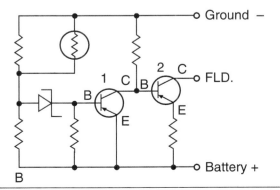

Figure 11-27 A simplified circuit diagram of an electronic regulator utilizing a zener diode.

AC generator current from the stator and diodes first goes through a thermistor. Current then flows to the zener diode. When the upper voltage limit (14.5 volts) is reached, the zener diode will conduct current to flow to the base of transistor 2. This turns transistor 2 on and switches off transistor 1. Transistor 1 controls field current to the rotor. If transistor 1 is off, no current can flow through the field coil and the AC generator will not have any output. When no voltage is applied to the zener diode, current flow stops, transistor 2 is turned off, transistor 1 is turned on, and the field circuit is closed. The magnetic field is restored in the rotor, and the AC generator produces output voltage.

Many manufacturers are installing the voltage regulator internally in the AC generator. This eliminates some of the wiring needed for external regulators. The diode trio rectifies AC current from the stator to DC current that is applied to the field windings (Figure 11-28).

With the engine off and the ignition in RUN position battery voltage is applied to the field through the common point above R1. TR_1 conducts the field current coming from the field coil, producing a weak magnetic field. The indicator lamp lights because TR_1 directs current to ground and completes the lamp circuit.

With the engine running the AC generator starts to produce voltage, the diode trio will conduct and battery voltage is available for the field and terminal 1 at the common connection. Placing voltage on both sides of the lamp gives the same voltage potential at each side; therefore, current does not flow and the lamp goes out.

When voltage output is being regulated the sensing circuit from terminal 2 passes through a thermistor to the zener diode (D2). When the system voltage reaches the upper voltage limit of the zener diode, the zener diode conducts current to TR_2. When TR_2 is biased it opens the field coil circuit and current stops flowing through the field coil. Regulation of this switching on and off is based on the sensing voltage received through terminal 2. With the circuit to the field coil opened, the sensing voltage decreases and the zener diode stops conducting. TR_2 is turned off and the circuit for the field coil is closed.

Computer-Controlled Regulation

On many vehicles after the mid-1980s, the regulator function has been incorporated into the vehicle's engine computer (Figure 11-29). The operation is the same as the internal electronic regulator. Regulation of the field circuit is through the ground (A circuit).

The logic board's decisions concerning voltage regulation are based on output voltages and battery temperature. When the desired output voltage is obtained (based on battery temperature), the logic board duty-cycles a switching transistor. This transistor grounds the AC generator's field to control output voltage.

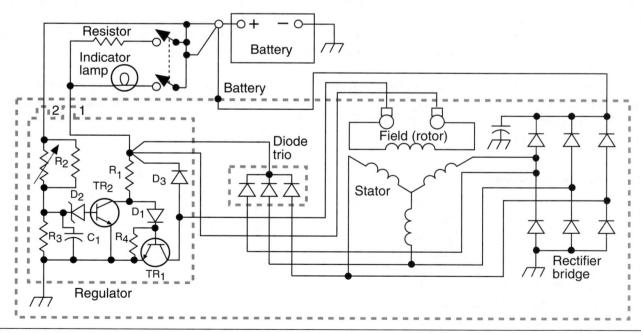

Figure 11-28 AC generator circuit diagram with internal regulator. Uses a diode trio to rectify stator current to be applied to the field coil. The resistor above the indicator lamp is used to ensure current will flow through the terminal 1 if the lamp burns out.

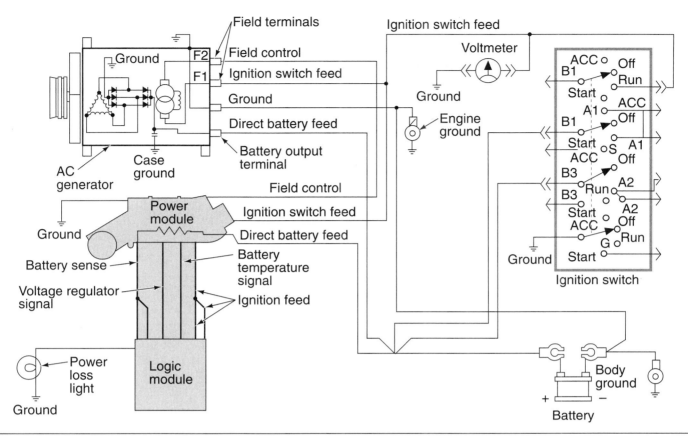

Figure 11-29 Computer-controlled voltage regulator circuit diagram. (Courtesy of Chrysler Corporation)

General Motors' CS series generators may be connected directly to the PCM through terminals L and F at the generator. The voltage regulator portion of the PCM switches the field current on and off at a frequency of about 400 times per second. Varying the on and off time of the field current controls the voltage output of the generator.

Charging Indicators

There are four basic methods of informing the driver of the charging system's condition: indicator lamps, electronic voltage monitor, ammeter, and voltmeter.

Indicator Light Operation

As discussed earlier, most indicator lamps operate on the basis of opposing voltages. If the AC generator output is less than battery voltage, there is an electrical potential difference in the lamp circuit and the lamp will light. In the electromechanical regulator system, if the stator is not producing a sufficient amount of current to close the field relay contact points, the lamp will light (Figure 11-30). If the voltage at the battery is equal to the output voltage, the two equal voltages on both sides of the lamp result in no electrical potential and the lamp goes out.

Electronic regulators that use an indicator lamp operate on the same principle. If there is no stator output through the diode trio, then the lamp circuit is completed to ground through the rotor field and TR_1 (Figure 11-31).

On most systems, the warning lamp will be "proofed" when the ignition switch is in the RUN position before the engine starts. This indicates that the bulb and indicator circuit are operating properly. Proofing the bulb is accomplished because there is no stator output without the rotor turning.

Electronic Voltage Monitor

The electronic voltage monitor module is used to monitor the system voltage. The lamp will remain off if the system voltage is above 11.2 volts. Once system voltage drops below 11.2 volts, a transistor amplifier in the module turns on the indicator lamp. This system can use either a lamp or a light-emitting diode.

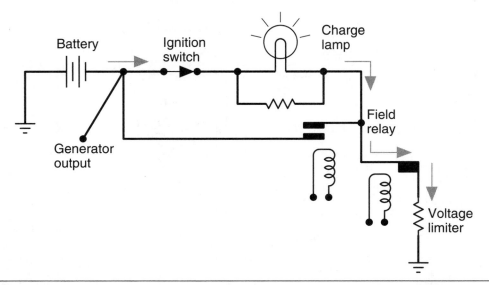

Figure 11-30 Electromechanical regulator with indicator light on with no AC generator output.

Ammeter Operation

In place of the indicator light, some manufacturers install an ammeter. The ammeter is wired in series between the AC generator and the battery (Figure 11-32). Most ammeters work on the principle of D'Arsonval movement.

If the charging system is operating properly, the ammeter needle will remain within the normal range. If the charging system is not generating sufficient current, the needle will swing toward the discharge side of the gauge. When the charging system is recharging the battery, or is called on to supply high amounts of current, the needle deflects toward the charge side of the gauge.

It is normal for the gauge to read a high amount of current after initial engine start up. As the battery is recharged, the needle should move more toward the normal range.

Voltmeter Operation

Because the ammeter is a complicated gauge for most people to understand, many manufacturers use a voltmeter to indicate charging system operation. The voltmeter is usually connected between the battery positive and negative terminals (Figure 11-33).

When the engine is started, it is normal for the voltmeter to indicate a reading between 13.2 and 15.2 volts.

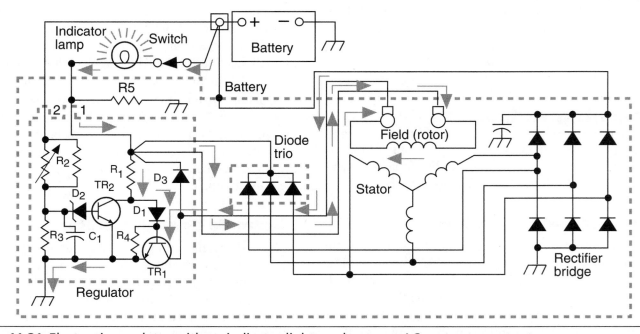

Figure 11-31 Electronic regulator with an indicator light on due to no AC generator output.

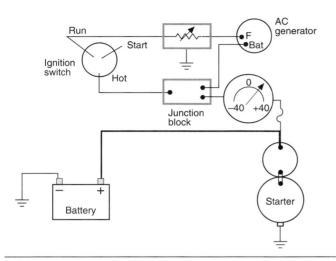

Figure 11-32 Ammeter connected in series to indicate charging system operation.

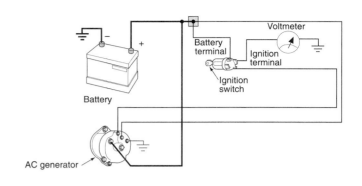

Figure 11-33 Voltmeter connected to the charging circuit to monitor operation.

If the voltmeter indicates a voltage level that is below 13.2, it may mean the battery is discharging. If the volt-

meter indicates a voltage reading that is 15.2 volts, the charging system is overcharging the battery. The battery and electrical circuits can be damaged as a result of higher than normal charging system output.

Summary

❏ The purpose of the charging system is to convert the mechanical energy of the engine into electrical energy to recharge the battery and run the electrical accessories.

❏ All charging systems use the principle of electromagnetic induction to generate the electrical power.

❏ The rotor creates the rotating magnetic field of the AC generator. It is the portion of the AC generator that is rotated by the drive belt.

❏ The stator is the stationary coil in which current is produced. The stator contains three main sets of windings wrapped in slots around a laminated, circular iron frame.

❏ There are two methods of stator connection: the wye connection and the delta connection.

❏ The diode rectifier bridge provides reasonably constant DC voltage to the vehicle's electrical system and battery. The diode rectifier bridge is used to change the current in an AC generator.

❏ The three main circuits used in the AC generator are the charging circuit, which consists of the stator windings and rectifier circuits; the excitation circuit, which consists of the rotor field coil and the electrical connections to the coil; and the pre-excitation circuit, which supplies the initial current for the field coil that starts the build-up of the magnetic field.

❏ The voltage regulator controls the output voltage of the AC generator, based on charging system demands, by controlling field current. The higher the field current, the higher the output voltage.

❏ The A circuit is called an external grounded field circuit, and is always an electronic-type regulator. In the A circuit, the regulator is on the ground side of the field coil.

❏ Usually the B circuit regulator is mounted externally of the AC generator. The B circuit is an internally grounded circuit. In the B circuit, the voltage regulator controls the power side of the field circuit.

❏ Isolated field AC generators pick up battery voltage and ground externally. The AC generator has two field wires attached to the outside of the case. The voltage regulator can be located on either the ground (A circuit) or on the B1 (B circuit) side.

❑ In the electromechanical regulator, the field relay applies voltage to the field coil. The voltage limiter is connected through the resistor network and determines whether the field will receive high, low, or no voltage.

❑ An electronic regulator uses solid-state circuitry to perform the regulatory functions. Electronic regulators can be mounted either externally or internally of the AC generator. Because there are no moving parts, this type of regulator can cycle between 10 and 7,000 times per second.

❑ The electronic regulator uses a zener diode that blocks current flow until a specific voltage is obtained, at which point it allows the current to flow.

❑ On many vehicles after the mid-1980s, the regulator function has been incorporated into the vehicle's engine computer. Regulation of the field circuit is through the ground (A circuit).

❑ There are three basic methods of informing the driver of the charging system's condition: indicator lamps, ammeter, and voltmeter.

Terms-To-Know

A circuit	Field current	Rectification
AC generator	Field relay	Rotor
Alternator	Fingers	Sensing voltage
B circuit	Flat-type rectifier	Single-phase voltage
Charging system	Full field current	Slip rings
DC generator	Full-wave rectification	Stacked-type rectifier
Delta connection	Half field current	Stator
Diode rectifier bridge	Half-wave rectification	Stator neutral junction
Diode trio	Inductive reactance	Voltage limiter
Electromechanical regulator	Internally grounded circuit	Voltage regulation
Electronic regulator	Isolated field	Wye connection
External grounded field circuit	Polarized	
External voltage regulator (EVR)	Poles	

Review Questions

Short Answer Essays

1. List the major components of the charging system.
2. List and explain the function of the major components of the AC generator.
3. How does the regulator control the charging system's output?
4. What is the relationship between field current and AC generator output?
5. Identify the differences between A, B, and isolated circuits.
6. Explain the operation of charge indicator lamps.
7. Describe the two styles of stators.
8. What is the difference between half-wave and full-wave rectification?
9. Describe how AC voltage is rectified to DC voltage in the AC generator.
10. What is the purpose of the charging system?

Fill-in-the-Blanks

1. The charging system converts the _____ energy of the engine into _____ energy to recharge the battery and run the electrical accessories.

2. All charging systems use the principle of _____ _____ to generate the electrical power.

3. The _____ creates the rotating magnetic field of the AC generator.

4. _____ are electrically conductive sliding contacts, usually made of copper and carbon.

5. In the _____ connection stator, one lead from each winding is connected to one common junction.

6. The _____ _____ controls the output voltage of the AC generator, based on charging system demands, by controlling _____ current.

7. In an electronic regulator, _____ _____ _____ controls AC generator output by varying the amount of time the field coil is energized.

8. In most electronic regulators that use an indicator lamp, if there is no _____ _____ , then the lamp circuit is completed to ground.

9. Full-wave rectification in the AC generator requires _____ pair of diodes.

10. The _____ is the stationary coil that current is produced in the AC generator.

ASE Style Review Questions

1. The purpose of the charging system is being discussed. Technician A says the charging system is used to restore the electrical power to the battery that was used during engine starting. Technician B says the charging system must be able to react quickly to high load demands required of the electrical system. Who is correct?
   a. A only
   b. B only
   c. Both A and B
   d. Neither A nor B

2. The operation of the charging system is being discussed. Technician A says that battery condition has no effect on the charging system operation. Technician B says the stator is the rotating component that creates the magnetic field. Who is correct?
   a. A only
   b. B only
   c. Both A and B
   d. Neither A nor B

3. Rectification is being discussed. Technician A says the AC generator uses a segmented commutator to rectify AC current. Technician B says the AC generator uses a set of diodes to rectify AC current. Who is correct?
   a. A only
   b. B only
   c. Both A and B
   d. Neither A nor B

4. Rotor construction is being discussed. Technician A says the poles will take on the polarity of the side of the coil that they touch. Technician B says the magnetic flux lines will move in opposite directions between adjacent poles. Who is correct?
   a. A only
   b. B only
   c. Both A and B
   d. Neither A nor B

5. Stator construction is being discussed. Technician A says the wye connection is the most common. Technician B says the wye connection connects the lead of one end of the winding to the lead at the other end of the next winding. Who is correct?
   a. A only
   b. B only
   c. Both A and B
   d. Neither A nor B

6. AC generator regulation is being discussed. Technician A says AC generators require current limiters. Technician B says the higher the field current, the lower the output voltage. Who is correct?
   a. A only
   b. B only
   c. Both A and B
   d. Neither A nor B

7. Indicator lamp operation is being discussed.
   Technician A says in a system with an electronic
   regulator, if there is no stator output through the
   diode trio the lamp will light. Technician B says
   when there is stator output then the lamp circuit
   has voltage applied to both sides and the lamp will
   not light. Who is correct?
   a. A only
   b. B only
   c. Both A and B
   d. Neither A nor B

8. Technician A says electronic regulators generally
   use pulse width modulation to vary the amount of
   time the field coil is energized. Technician B says
   the electronic regulator will use a LED to maintain
   voltage output control. Who is correct?
   a. A only
   b. B only
   c. Both A and B
   d. Neither A nor B

9. Technician A says only two stator windings apply
   voltage because the third winding is always
   connected to diodes that are reverse-biased.
   Technician B says AC generators that use half-
   wave rectification are the most efficient. Who is
   correct?
   a. A only
   b. B only
   c. Both A and B
   d. Neither A nor B

10. The operation of the electromechanical regulator is
    being discussed. Technician A says the field relay
    applies battery voltage to the AC generator field
    coil after the engine has started. Technician B says
    the voltage limiter is connected through the
    resistor network and determines whether the field
    will receive high, low, or no voltage. Who is
    correct?
    a. A only
    b. B only
    c. Both A and B
    d. Neither A nor B

12 Charging System Testing and Service

Objective

Upon completion and review of this chapter, you should be able to:

❑ Diagnose charging system problems that cause an undercharge or no-charge condition.

❑ Diagnose charging system problems that cause an overcharge condition.

❑ Inspect, adjust, and replace generator drive belts, pulleys, and fans.

❑ Perform charging system output tests and determine needed repairs.

❑ Perform charging system circuit voltage drop tests and determine needed repairs.

❑ Perform voltage regulator tests and determine needed repairs.

❑ Test and replace AC generator diodes and/or rectifier bridge.

❑ Remove and replace the AC generator.

❑ Disassemble, clean, and inspect AC generator components.

❑ Inspect and replace AC generator brushes and brush holders.

❑ Test and diagnose the rotor.

❑ Test and diagnose the stator.

 CAUTION *Always wear safety glasses when performing charging system tests.*

Introduction

Whenever there is a charging system problem, make sure the battery is thoroughly checked first. The battery supplies the electrical power for the charging system. If the battery is bad, the charging system cannot be expected to work its best. In addition, AC generators are designed to maintain the charge of a battery, not to charge a dead battery. The battery's state of charge must be considered before faulting any of the charging system components. If the battery passes the state of charge test, a load test should be performed to determine the capacity. If the battery fails this test, then perform the three-minute charge test. It is important that the battery be in good condition in order to obtain accurate charging system test results. In addition, the battery must be fully charged before proceeding with the diagnosis of the charging system.

There are many different types of testers that can be used to test the charging system and AC generators. Some hand-held multimeters have the ability to conduct many tests; however, the best testers to use are those designed to test the entire system. These testers are commonly referred to as starting/charging system testers. Often in this chapter a reference is made to using a VAT-40. This tester, made by Sun Electric Corporation, is commonly found in service departments. Although a VAT-40 is mentioned in the text, this does not mean that this is the only tester that can be used. Any starting/charging system tester can be used. Always follow the operating procedures for the specific tester being used.

When performing the tests, be sure of the connections you are making. Refer to the service manual for identification of the various terminals for the AC generator and regulator. Connecting a test lead to the wrong terminal can result in AC generator damage, as well as damage to other electrical and electronic components.

It is also important to perform a preliminary inspection. Many problems can be detected during this simple step. Check the following items:

1. Condition of the drive belt (Figure 12-1). If the drive belt is worn or glazed, it will not allow enough rotor rpm to produce sufficient current.

2. Drive belt tension.

3. Electrical connections to the AC generator.

4. Electrical connections to the regulator.

5. Ground connections at the engine and chassis.

6. Battery cables and terminals.

7. Fuses and fusible links.

8. Excessive current drain caused by a light or other electrical component remaining on after the ignition switch is turned off.

9. Check for symptoms of undercharging. These include slow cranking, discharged battery, low instrument panel ammeter or voltmeter readings, and charge indicator lamp on.

10. Check for symptoms of overcharging. These include high ammeter and voltmeter readings, battery boiling, and charge indicator lamp on.

WARNING *Do not over tighten the drive belt. Early bearing failure can occur if the belt is tightened beyond manufacturer's specifications.*

SERVICE TIP *To check the fusible link to the AC generator, use a voltmeter and test for voltage at the BAT terminal. If the battery is good, voltage should be present. If there is no voltage, the fusible link is probably burned out. A better test would be to measure the voltage drop across the link; this will identify any high resistance in the circuit.*

The vehicle manufacturer may have several additional tests to perform. It is important to always follow the procedures outlined by the manufacturer for the vehicle being tested.

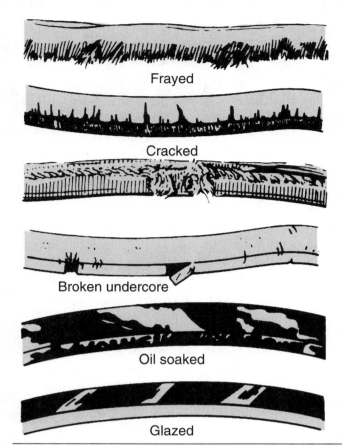

Figure 12-1 If the AC generator drive belt has any of these conditions, it must be replaced.

CAUTION *Many charging system tests require that the vehicle be operated in the shop area. Always place wheel blocks against the drive wheels. Be sure there is proper ventilation of the vehicle's exhaust. Also, be aware of the drive belts and cooling fan. Be sure of where your hands and tools are at all times.*

Charging System Service Cautions

The following are some of the general rules that should be followed when servicing the charging system:

1. Do not run the vehicle with the battery disconnected. The battery acts as a buffer and stabilizes any voltage spikes that may cause damage to the vehicle's electronics.

2. Do not allow output voltage to increase over 16 volts when performing charging system tests.

3. If the battery needs to be recharged, disconnect the cables while charging.

4. Do not attempt to remove electrical components from the vehicle with the battery connected.

5. Before connecting or disconnecting any electrical connections, the ignition switch must be in the OFF position.

6. Avoid contact with the BAT terminal of the AC generator while the battery is connected. Battery voltage is always present at this terminal.

AC Generator Noises

Noises that come from the AC generator can be from three sources. The causes of the noises are identifiable by the type of noise they make. A loose belt will make a squealing noise. Check the belt condition and tension. Replace the belt if necessary.

SERVICE TIP *With the engine off, rub a piece of bar soap on the pulley surface of the drive belts. Do this one belt at a time until the noise stops. This way you will know which belt is the cause of the noise.*

A squealing noise can also be caused by faulty bearings. The bearings are used to support the rotor in the housing halves. To test for bearing noises, use a length of hose, a long screwdriver, or a technician's stethoscope. By placing the end of the probe tool close to the bearings and listening on the other end, any bearing noise will be transmitted so you will be able to hear it. Bearing replacement will require disassembly of the AC generator.

CAUTION *This test is performed with the engine running. Use caution around the drive belts, fan, and other moving components.*

A whining noise can be caused by shorted diodes or stator, or by a dry rotor bearing. A quick way to test for the cause of a whining sound is to disconnect the wiring to the generator. Then start and run the engine. If the noise is not there, the cause of the noise is a magnetic whine due to shorted diodes or stator windings. Use a scope to verify the condition of the diodes and stator. If the noise remains, the cause is mechanical and probably due to worn bearings.

Charging System Troubleshooting

Voltage Output Testing

Once the visual inspection and preliminary checks are completed, the next step is to perform a **voltage output test**. This is used to make a quick determination if the charging system is working properly. If the charging system is operating correctly, then check for battery drain. The following procedure is for performing the test:

1. Connect the voltmeter across the battery terminals, observing polarity.

2. Connect the tachometer, following the manufacturer's procedure.

3. With the engine off, record the base voltage value across the battery.

4. Start the engine. Because most AC generators do not produce maximum voltage output until 1,500 to 2,000 engine rpm, the engine speed needs to be brought up to this level.

5. Observe the voltmeter reading. It should read between 13.5 and 14.5 volts.

If the charging voltage was too high, there may be a problem in the following areas:

1. Defective voltage regulator.

2. Poor voltage regulator ground connection.

3. Defective wiring between the voltage regulator and the AC generator.

If the charging voltage was too low, the fault might be:

1. Loose or glazed drive belt.

2. Defective voltage regulator.

3. Defective AC generator.

4. Discharged battery.

5. Loose or corroded battery cable terminals.

If the voltage reading was correct, perform a load test to check the voltage output under a load condition:

1. With the engine running at idle, turn on the headlights and the heater fan motor to high speed.

2. Increase the engine speed to approximately 2,000 rpm.

3. Check the voltmeter reading. It should increase a minimum of 0.5 volt over the base voltage reading taken previously.

4. If the voltage increases, the charging system is operating properly. If the voltage did not increase, perform the following test series to locate the fault.

SERVICE TIP *If the charging system passes the no-load test but fails the load test, check the condition and tension of the drive belt closely.*

Voltage Drop Testing

The voltage drop of all wires and connections combined should not exceed 3% of the system voltage. Any particular wire or connection should not exceed 0.2 volt; total system drops should be less than 0.7 volt. The ground side voltage drop should be less than 0.2 volt. To perform the voltage drop test using the VAT-40, follow these steps:

1. Connect the large red cable to the battery positive terminal.

2. Connect the large black cable to the battery negative terminal.

3. Select CHARGING.

4. Select EXT 3 V.

5. Zero the ammeter.

6. Clamp the inductive pickup around the AC generator output wire.

7. Using the small red and black test leads, connect at the following locations:

   Insulated circuit: Red lead to AC generator output terminal. Black lead to the battery positive terminal.

   Ground circuit: Red lead to battery negative terminal. Black lead to AC generator housing.

8. Start the engine and hold the engine speed between 1,500 and 2,000 rpm.

9. Using the carbon pile knob, load the system to 9 to 20 amperes. Some manufacturers recommend measuring voltage drop when the generator is putting out its maximum; always follow the recommendations of the manufacturer. Remember if the AC generator is not putting out any current, there will be no voltage drop even if the circuit is very corroded.

 **SERVICE TIP** *Turning on the headlights may be substituted for the carbon pile.*

10. Read voltmeter.

If a higher voltage drop is observed anywhere in the circuit, work up the circuit to find the fault. Check every wire and connection.

If a VAT-40 is not available, the voltage drop test can be performed with just a voltmeter. Simply follow steps 7 through 10 in the previous procedure.

Field Current Draw Test

Because field current is required to create a magnetic field, it is necessary to determine if current is flowing to the field coil. The **field current draw test** determines if there is current available to the field windings. To perform the field current draw test using a starting/charging system tester, follow these steps:

1. Connect the large red and black cables across the battery, observing polarity.

2. Select CHARGING.

3. Select INT 18 V.

4. Zero the ammeter.

5. Disconnect the field wire from either the AC generator or the regulator.

6. Connect the multiplying coil to the field terminal. Make the connection toward the AC generator.

7. Connect the field lead of the tester to the multiplying coil.

8. Clamp the inductive pickup around the loop of the multiplying coil.

9. Move the toggle switch to the proper field-type position (A or B).

10. Read ammeter while toggle switch is depressed.

11. Compare results with manufacturer's specifications.

A multiplying coil is made of 10 wraps of wire. This multiplies the ammeter reading so that a starting/charging system tester's scale can be used to read lower current. For example, if the needle is pointing to 25 amperes, when using the multiplying coil the actual reading is 2.50 amperes.

For GM A circuit systems, steps 1 through 4 are the same. Remove the field plug from the AC generator and connect a Y-type connector between terminals 1 and 2 (Figure 12-2). Connect the multiplying coil to the Y connector. Connect the field lead of the tester to the multi-

Figure 12-2 Connecting the multiplying coil and amp pickup to a generator.

plying coil. The inductive pickup is clamped around the loop of the multiplying coil. Press the toggle switch to B and read the field current draw on the ammeter.

If the readings are within the specification limits, then the field circuit is good. If the readings are over specifications, a shorted field circuit or bad regulator may be the problem. If the readings were too low, then there is high electrical resistance that may be caused by worn brushes.

To test Ford's integral alternator/regulator (IAR) system, use a voltmeter as follows:

1. Connect the negative voltmeter lead to the housing.

2. With the ignition switch in the OFF position, connect the positive voltmeter lead to the F terminal screw of the regulator (Figure 12-3).

3. Check the voltmeter reading. It should indicate battery voltage. If it reads battery voltage, the field circuit is normal.

4. If the voltmeter reading is less than battery voltage, disconnect the wiring plug from the regulator.

5. Connect the positive voltmeter lead to the I terminal of the plug (Figure 12-4).

6. Check the voltmeter reading. It should indicate zero volts. If there is voltage present, repair the I lead from the ignition switch. The I lead is receiving voltage from another source.

7. If there was no voltage present in step 6, connect the positive voltmeter lead to the S terminal of the regulator wiring plug.

8. Check the voltmeter reading. If there are zero volts, replace or service the regulator.

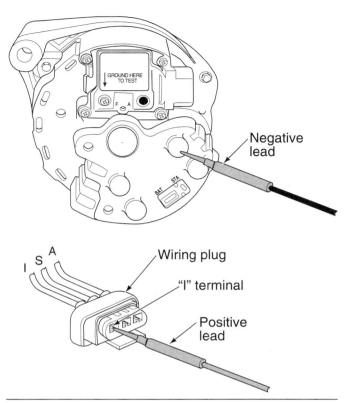

Figure 12-4 Wiring plug (connector) terminal identification. (Reprinted with permission of Ford Motor Company)

9. If voltage is indicated, disconnect the wiring plug from the AC generator.

10. Check for voltage to the regulator wiring plug S terminal.

11. If voltage is still present, repair the S terminal wire lead to the AC generator. The S terminal wire is receiving voltage from another source.

12. Replace the rectifier bridge if no voltage is present.

Current Output Testing

Current output testing will determine the maximum output of the AC generator. The system must be loaded in order to obtain AC generator current output. By connecting a carbon pile to maintain system voltage at 12 volts, the signal voltage to the regulator will be reduced. When this occurs, the regulator attempts to recharge the battery by full fielding. This will produce the maximum current output to the battery. Follow these steps to perform the output test:

1. Connect the large red and black cables across the battery, observing polarity.

2. Select CHARGING.

3. Zero the ammeter.

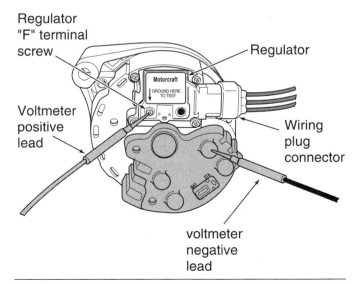

Figure 12-3 Testing the IAR generator for field current draw. (Reprinted with permission of Ford Motor Company)

4. Connect the inductive pickup around all battery ground cables (Figure 12-5).

5. With the ignition switch in the RUN position, engine not running, observe the ammeter reading. This reading indicates how much current is required to operate any full-time accessories.

6. Start the engine and hold between 1,500 and 2,000 rpm.

7. Turn the load knob for the carbon pile slowly, until the highest ammeter reading possible is obtained.

8. Return the load control knob to the OFF position.

WARNING *Do not reduce battery voltage below 12 volts.*

9. The highest reading indicates maximum current output.

10. Add the maximum output reading to the reading obtained in step 4.

11. This total should be within 10% of the rated output of the AC generator.

If the ammeter reading indicates that output is 2 to 8 amperes below the specification, then an open diode or slipping belt may be the problem. If the output reading indicates 10 to 15 amperes below specifications, this may be caused by a shorted diode or slipping belt. If the output is below specifications, perform the full field test.

To test General Motors CS-130 and 144 AC generator, first use a voltmeter to test for voltage at terminals L and I as follows:

1. With the ignition switch in the OFF position, disconnect the electrical connector from the AC generator.

2. Connect a voltmeter between terminal L and ground.

3. Turn the ignition switch to the RUN position with the engine not running. Record the voltmeter reading.

4. Move the test lead from the L terminal to the I terminal and measure the voltage. Turn off the ignition switch. If the voltmeter reading was zero at either terminal, there is an open in the circuit.

5. Reconnect the electrical connector to the AC generator.

6. Connect the large red and black cables across the battery, observing polarity.

7. Select CHARGING.

8. Select INT 18 V.

9. Zero the ammeter.

10. Clamp the inductive pickup around the negative battery cables.

11. Start the engine. Let it idle with no electrical load. If the voltmeter reads more than 16 volts, replace or service the AC generator.

12. Turn the electrical accessories on.

13. Load the battery by rotating the load control knob until maximum amperage is obtained.

14. While maintaining 13 volts or more, note the current reading.

15. Return the load control knob to the OFF position.

If the current reading is not within 15 amperes of the rated AC generator output, replace or service the AC generator.

To test Bosch and Nippondenso AC generators used with Chrysler's SMEC and SBEC computer controllers, begin by connecting the tester in the normal fashion. Select EXT 18 V and connect the external voltmeter positive test lead to the AC generator B+ terminal and the negative lead to a good ground. Remove the air hose between the SMEC and air cleaner. Connect a jumper wire to ground and the dash side of the R3 terminal of the black eight-way connector (Figure 12-6). Start the engine and allow it to idle. Rotate the load control knob. Increase engine rpm in increments until a speed of 1,250 rpm and a voltmeter reading of 15 volts are obtained. Note the ammeter reading. If the current output is less than specified, perform the voltage drop tests. If the circuit passes the voltage drop test, replace or service the AC generator. If the current output is within specifications, go to the diagnostic fault codes procedure listed in the service manual.

WARNING *Do not allow system voltage to increase over 16 volts or damage to the electrical system will result.*

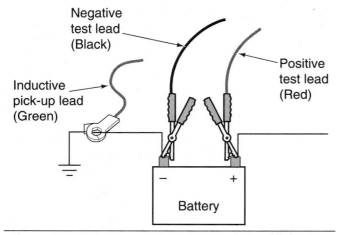

Figure 12-5 Common test connections for performing most charging system tests.

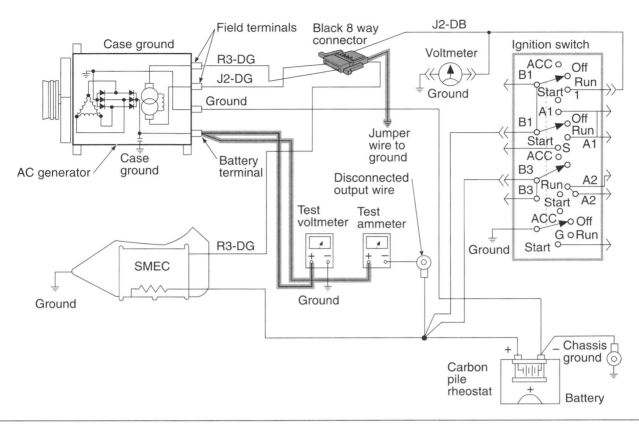

Figure 12-6 Wiring diagram of Chrysler's SMEC-controlled charging system. (Courtesy of Chrysler Corporation)

Full Field Test

Full fielding the AC generator means that the field windings are constantly energized with full battery current. Full fielding will produce maximum AC generator output. The **full field test** will isolate if the detected problem lies in the AC generator or the regulator. The full field test needs to be performed only if the charging system failed the output test. This test is performed by full fielding the AC generator with the regulator bypassed. If this test still produces lower than specified output, the AC generator is the cause of the problem. If the output is within specifications with the regulator bypassed, then the problem is within the regulator.

> **WARNING** The full field test is to be performed on AC generator systems only. Do not use this test on vehicles equipped with a DC generator system.

When full fielding the system, the battery should be loaded to protect vehicle electronics and computers. With the voltage regulator bypassed there is no control of voltage output. The AC generator is capable of producing well over 30 volts. This increased voltage will damage the circuits not designed to handle that high of voltage. Never allow the voltage to increase over 16 volts.

> **WARNING** Not all AC generators can be full fielded. Check the manufacturer's procedures before attempting to full field an AC generator.

General Motors Full Field Testing

To full field a GM A circuit SI-type AC generator with an internal voltage regulator, insert a screwdriver into the D-shaped test hole (Figure 12-7). This test hole lines up with a small tab that is attached to the negative brush. By inserting a screwdriver into the D hole about 1/2 in. (12.7 mm) and grounding it to the housing, the regulator is bypassed. Perform the output test with the regulator bypassed. If the output is within specifications, the regulator is at fault.

> **WARNING** Do not force the screwdriver into the D hole more than 1 in. (25.4 mm). Damage to several electrical systems can result. Do not full field for longer than 10 seconds.

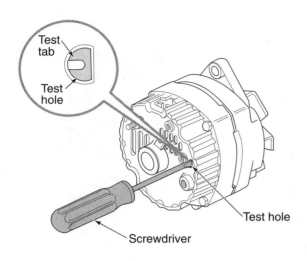

Figure 12-7 Full fielding the GM 10SI AC generator by grounding the tab in the "D" test hole.

A variation of the test calls for shorting the negative brush in the D hole while the ignition switch is in the RUN position. If the brushes and rotor are good, then the rear bearing should be magnetized and attract a metal screwdriver.

On GM charging systems equipped with an external voltage regulator, full fielding is done by unplugging the regulator and using a jumper wire to bypass the regulator. Connect the jumper wire between the F connector lead and connector lead 3 (Figure 12-8). With the engine running, output of the AC generator should be within specifications.

WARNING *Before attempting to remove the connector to the regulator, be sure the ignition switch is in the OFF position. This will prevent damage to any electrical systems.*

If the AC generator output is within 10% of specifications with the regulator bypassed, service or replace the regulator. If the output is below specifications with the regulator bypassed, service or replace the AC generator.

Ford Full Field Testing

Ford Motor Company has utilized different designs of the integral regulator. The early design had one terminal, called the exciter, which was connected to the outside of the regulator (Figure 12-9). The wiring schematic for this type of design is illustrated (Figure 12-10). By removing the protective cover from the field terminal (closest to the rear bearing), the field circuit can be grounded and the regulator bypassed.

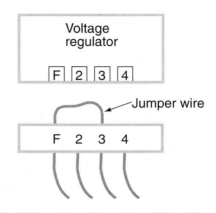

Figure 12-8 Jumper wire connection to bypass the regulator and full field the generator.

Before full fielding the IAR AC generator, check the rotor and field circuit resistance:

1. Disconnect the wiring plug to the regulator.
2. Connect an ohmmeter between the regulator A and F terminals.
3. Read the ohmmeter. The resistance should not be below 2.4 ohms.

If the resistance is less than 2.4 ohms, there is a short to ground somewhere in the circuit. Check for the following:

1. A failed regulator.
2. A shorted rotor circuit.
3. A shorted field circuit.

WARNING *Do not replace the regulator without first repairing any shorts in the rotor or field circuits. To do so may damage the new regulator.*

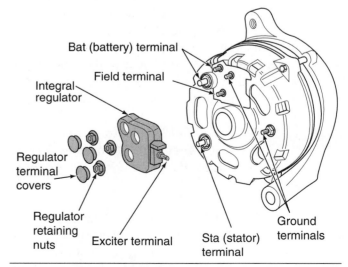

Figure 12-9 Integral regulator with exciter terminal. (Reprinted with permission of Ford Motor Company)

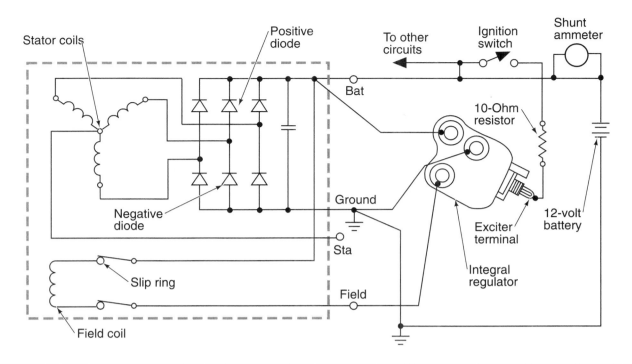

Figure 12-10 Wiring schematic of integral regulator with exciter terminal. (Reprinted with permission of Ford Motor Company)

The illustration (Figure 12-11) shows the wiring of the IAR system. To full field this system, disconnect the wiring connector to the AC generator and install a twelve-gauge wire jumper between the B+ terminal blades (Figure 12-12). Connect another jumper wire from the regulator F terminal screw to ground. Connect a voltmeter with the positive lead connected to one of the B+ jumper wire terminals and the negative test lead to a good ground. Start the engine and perform the load output test. The regulator is faulty if the voltage rises to specifications. If the voltage does not rise to specifications, the AC generator needs to be serviced or replaced.

Ford charging systems that use external voltage regulators are full fielded in the following manner:

1. Disconnect the wiring plug from the regulator.

2. Use an ohmmeter to measure the resistance between the F terminal of the wiring plug and ground (Figure 12-13).

3. The reading should be higher than 2.4 ohms. If the reading is lower than 2.4 ohms, the field circuit is grounded and needs to be repaired before continuing. The ground can be in the wiring harness or the AC generator field circuit.

WARNING *Before attempting to remove the connector to the regulator, be sure the ignition switch is in the OFF position. This will prevent damage to electrical system components.*

4. If the resistance is higher than 2.4 ohms, connect a jumper wire between the F and A terminals of the regulator wiring plug (Figure 12-14).

5. Run the engine between 1,500 and 2,000 rpm.

WARNING *Do not allow output voltage to increase over 16 volts. Load the battery by turning on the headlights to prevent excessive voltage output.*

6. Compare the test results with manufacturer's specifications.

Chrysler Full Field Testing

Most late-model Chryslers use an A-type field with the voltage regulator inside the power train control module. Early Chrysler vehicles used an isolated field with two field leads. To full field these systems, disconnect the green field wire from the AC generator. Then connect a jumper wire from the AC generator field terminal to ground (Figure 12-15). Start the engine and check the output.

CAUTION *When disconnecting the field wire from the AC generator, be sure the ignition switch is in the OFF position.*

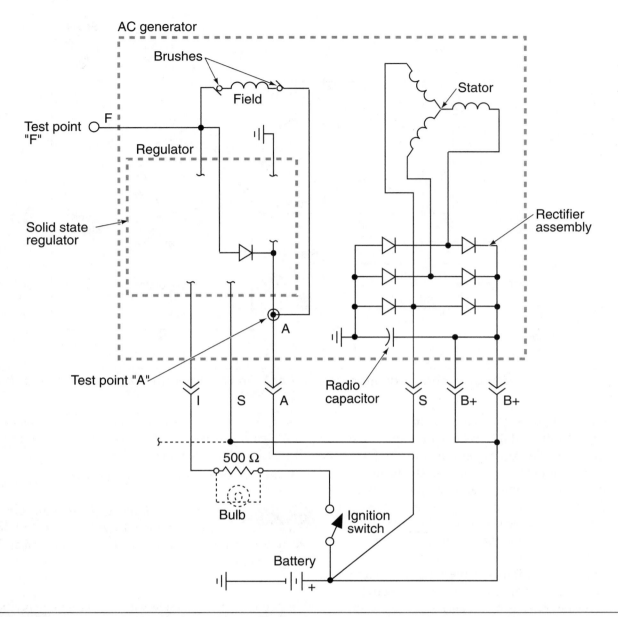

Figure 12-11 Wiring diagram of Ford's IAR charging system. (Reprinted with permission of Ford Motor Company)

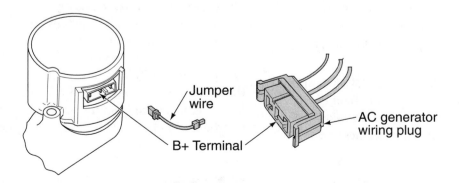

Figure 12-12 Jumper wire connections between B+ terminals. (Reprinted with permission of Ford Motor Company)

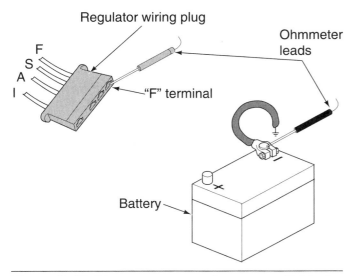

Figure 12-13 Ohmmeter connections to test field wire lead resistance. (Reprinted with permission of Ford Motor Company)

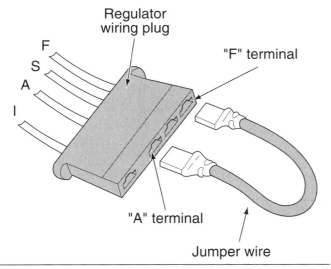

Figure 12-14 Jumper wire connections to bypass regulator and full field the generator. (Reprinted with permission of Ford Motor Company)

Special Full Field Testing

Some AC generators that use internal regulators, or computer-controlled regulators, do not provide for a means of full fielding. This can be determined by looking at the wiring diagram for the charging system. In fact, by studying the circuit in the wiring diagram, you should be able to full field any AC generator that can be full fielded. The following procedure uses a starting/charging system tester to full field an AC generator while observing the output of the system. If the AC generator fails this test, it should be repaired or replaced.

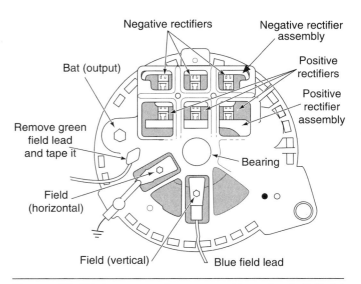

Figure 12-15 Jumper wire connections to full field Chrysler's isolated field generator. (Courtesy of Chrysler Corporation)

WARNING *Check the manufacturer's service manual to see if there are instructions that prohibit full fielding the AC generator in this manner. If the AC generator cannot be full fielded, it will have to be disassembled and bench tested.*

When using the VAT-40, the full field test is performed as follows:

1. Connect the large red and black leads across the battery, observing polarity.
2. Disconnect the field wire from either the AC generator or the regulator.
3. Connect the blue field test lead from the VAT-40 to the field terminal. Make the connection toward the AC generator.

WARNING *To prevent AC generator damage, do not use the blue test lead while the regulator is still connected to the system.*

4. Select CHARGING.
5. Select INT 18 V.
6. Zero the ammeter.
7. Clamp the inductive pickup around the AC generator output wire.
8. Turn the ignition switch to the RUN position. Do not start the engine. Record the ammeter reading.
9. Start the engine and hold the speed between 1,500 and 2,000 rpm.
10. Turn the load control knob until the voltmeter reads the voltage specified for maximum output.

11. Read the current output on the 100 amp scale and compare to specifications. If the reading is within specifications, the regulator and AC generator are fine. If the reading is below specifications, continue testing.

12. Release the load control knob and allow the engine to run at 1,500 to 2,000 rpm.

13. Turn the load control knob until the voltmeter indicates 2 volts less than system voltage.

14. Full field the AC generator using the field switch. Use the load control knob to prevent system voltage from exceeding 14 volts.

15. Adjust the load control knob to the voltage reading that is required in the manufacturer's specifications.

16. Read the output on the 100 ampere scale while depressing the toggle switch.

17. Release the field selector switch and return the load knob to the OFF position.

18. Add the reading obtained in step 8 to the reading in step 16. Compare to manufacturer's specifications.

WARNING *Always refer to the manufacturer's service manual for the correct location of the AC generator terminals. Incorrect test connections can damage the AC generator and the vehicle's electrical system components.*

Regulator Test

The **regulator test** is used to determine if the regulator is maintaining the correct voltage output under different load demands. To perform the regulator voltage test using a VAT-40, follow these steps:

1. Connect the large red and black cables across the battery, observing polarity.

2. Select REGULATOR.

3. Select INT 18 V.

4. Zero the ammeter.

5. Clamp the inductive pickup around the AC generator output wire.

6. Start the engine and hold between 1,500 and 2,000 rpm.

7. Allow the engine to run until the ammeter reads 10 amperes or less. This indicates the battery is fully charged.

8. Voltage should read regulated voltage (13.5–14.5 volts).

9. Load the system to between 10 and 20 amperes.

10. Voltmeter should still read regulated voltage.

Diode/Stator Test

An AC generator may have an open diode, yet test close to manufacturer's specifications. If there is an open diode that is not determined in testing, a newly installed regulator may fail. In addition, an open diode can lead to the failure of other diodes. The **diode/stator test** is performed to determine the condition of the diodes. This test is not valid for AC generators that failed the full field test. This test is performed in the following manner:

1. Connect the large red and black cables across the battery, observing polarity.

2. Select the CHARGING position.

3. Select INT 18 V.

4. Zero the ammeter.

5. Clamp the inductive pickup around all of the negative battery cables.

6. Start the engine and hold between 1,500 and 2,000 rpm.

7. Adjust the load control knob to obtain an indicated charge rate of 15 amperes.

8. Set the selector to the DIODE/STATOR position while observing the red and blue DIODE/STATOR scale.

9. Return the load control knob to the off position.

If the meter was in the blue section of the scale, the diodes and stator are good. If the meter was in the red section of the scale, the diodes or the stator is bad. The AC generator will need to be disassembled to perform bench testing of these units.

Charging System Requirement Test

It is possible to have a charging system that is working properly, yet not meet the requirements of the vehicle's electrical system. If an AC generator is installed on the vehicle without sufficient output to meet the demands of the vehicle, the customer may have complaints that are identical to those of a charging system that is not functioning at all. The actual AC generator output should be at least 10% to 20% greater than the load demand. The **charging system requirement test** is used to determine the total electrical demand of the vehicle's electrical system.

To determine the vehicle's electrical requirement:

1. Connect the large red and black leads across the battery, observing polarity.

2. Select the CHARGING position.

3. Select INT 18 V.

4. Zero the ammeter.

5. Clamp the inductive pickup around all of the negative battery cables.

6. Turn the ignition switch to the RUN position. Do not start the engine.

7. Turn on all accessories to their highest position.

8. Read the ammeter. The indicated amperage is the total load demand of the vehicle.

AC Generator Removal and Replacement

AC generator removal varies according to the engine size, engine placement, and vehicle accessories (such as power steering and air conditioning). The following is the typical procedure for removal and replacement of the AC generator:

1. Place fender covers over the fenders.

2. Disconnect the battery ground cable.

CAUTION *Never attempt to remove the AC generator or disconnect any wires to the generator without first disconnecting the battery negative cable. Always wear safety glasses when working around the battery.*

3. Disconnect the wiring harness connections to the AC generator.

4. Loosen the AC generator pivot bolt.

5. Remove the bolt that attaches the AC generator to the adjustment arm.

6. Rotate the AC generator to loosen the drive belt and disengage the belt from the pulley.

7. Remove the AC generator pivot bolt while supporting the AC generator.

8. Remove the AC generator from the vehicle.

Reverse the removal procedure to install the AC generator. To adjust the belt tension, leave the pivot and adjusting arm bolts loose. Look up the correct belt tension specification for the vehicle you are working on. Install a belt tension gauge on the belt (Figure 12-16). Apply pressure to the AC generator front housing only when adjusting the belt tension. Once the correct tension reading is obtained, tighten the bolts to specified torque value.

Bench Testing the AC Generator

After the AC generator is removed from the vehicle, there are some tests that should be performed before dis-

Figure 12-16 Using a belt tension gauge is the only way to be sure of proper belt tension.

assembly. The AC generator can be bench tested for rectifier and stator grounds, and for field opens or shorts.

Rectifier and Stator Ground Test

Some DMMs have a feature for testing diodes and stator windings. If this type of meter is available, it should be used. These meters are designed to force current through a diode, thereby allowing more accurate testing. If this type of tester is not available, use an analog ohmmeter. A rectifier and stator ground test is conducted with the ohmmeter set on the 1X scale. Connect one test lead from the ohmmeter to the B+ terminal and the other lead to the stator (STA) terminal (Figure 12-17). Record

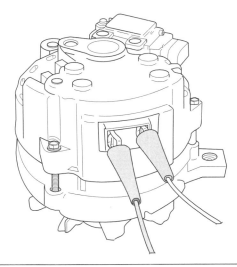

Figure 12-17 Ohmmeter test lead connections to test the rectifier bridge. (Reprinted with permission of Ford Motor Company)

the ohmmeter reading. Then, reverse the test connections and record the ohmmeter reading.

There should be no needle movement in one direction, and a reading of about 6.5 ohms in the other direction. A reading in both directions indicates a bad positive diode. If there is no needle movement in either direction (or higher than 6.5 ohms) there is a bad connection in the rectifier assembly. Repeat the same test from the STA terminal to the AC generator housing (Figure 12-18).

A reading in both directions indicates either a bad negative diode or a grounded stator winding. If there is no needle movement in either direction (or higher than 6.5 ohms) there is a bad connection in the rectifier assembly.

Field Open or Short Circuit Test

The field open or short test procedure will vary depending on whether the AC generator uses an external voltage regulator or an integral unit. Use an analog ohmmeter on the 1X scale to test an AC generator with an external regulator:

1. Connect one of the ohmmeter test leads to the field terminal.

2. Connect the other ohmmeter lead to the ground terminal.

3. Spin the pulley while observing the ohmmeter.

4. Compare results with manufacturer's specifications. The readings should fluctuate while the rotor is spinning.

For most AC generators the indicated reading on the ohmmeter should be between 2.5 and 100 ohms. If there is an infinite reading, this indicates an open brush lead,

worn or stuck brushes, or a bad rotor assembly. If the reading is less than 2.5 ohms, there is a grounded brush assembly, grounded field terminal, or a bad rotor.

To test AC generators with an integral regulator:

1. Connect one lead from the ohmmeter to the regulator A terminal and the other lead to the regulator F terminal (Figure 12-19).

2. Spin the AC generator while observing the ohmmeter. Record the test results.

3. Reverse the ohmmeter leads and repeat step 2. In one direction there should be an ohmmeter reading between 2.2 and 100 ohms. In the opposite direction there should be a reading between 2.2 and approximately 9 ohms. An infinite reading in one direction and approximately 9 ohms in the other indicates an open brush lead, worn or stuck brushes, defective rotor, or a loose regulator to brush attaching screw. If the ohmmeter reading is less than 2.2 ohms in both directions, the regulator is defective and needs to be replaced. If the ohmmeter reading was greater than 9 ohms in both directions, check the F terminal screw for looseness. If it is okay, replace the regulator.

4. Connect one of the ohmmeter leads to the AC generator housing. Connect the other to the regulator F terminal. Record the test results.

5. Reverse the ohmmeter leads and record the results. The ohmmeter reading should be infinite in one direction and about 9 ohms in the other direction. If the ohmmeter reading is less than infinite, there is a grounded brush lead, grounded rotor, or a defective regulator.

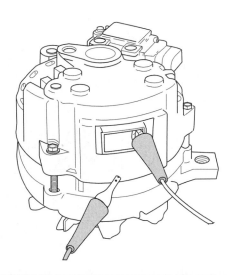

Figure 12-18 Testing for a grounded stator winding. (Reprinted with permission of Ford Motor Company)

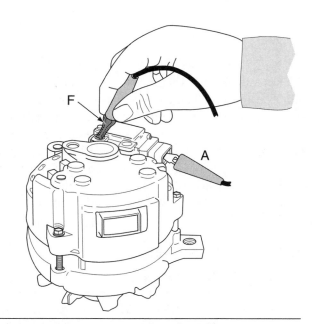

Figure 12-19 Field open or short test connections. (Reprinted with permission of Ford Motor Company)

AC Generator Disassembly

If the AC generator fails the previous tests, the technician must decide whether to rebuild or replace the AC generator. This decision is based on several factors:

1. What is best for the customer

2. Shop policies

3. Cost

4. Time

5. Type of AC generator

Once the decision is made to disassemble the AC generator, the technician should study the manufacturer's service manual and become familiar with the procedure for the particular AC generator being rebuilt. Photo Sequence 6 shows the procedure for disassembling the Ford IAR AC generator. A disassembled view of this AC generator is shown (Figure 12-20).

Once the AC generator is disassembled, the components must be cleaned and inspected. Using a clean cloth, wipe the stator, rotor, and front bearing. Inspect the front and rear bearings by rotating them on the rotor shaft. Check for noises, looseness, or roughness. Replace the defective bearing if any of these conditions are present.

Check the rotor shaft rear bearing surface. If the surface is not smooth, the rotor will have to be replaced. Visual inspection of the rotor includes checking the slip rings for smoothness and roundness. If the rings are discolored, dirty, scratched, nicked, or have burrs, they may be cleaned with fine grit emery cloth. Caution must be observed to prevent creating flat spots while polishing the slip rings. The minimum diameter of the slip rings is 1.22 in. (31 mm).

Inspect the terminals and wire leads of the rotor and stator. Also check both the rotor and stator for signs of burnt insulation of the windings. If there is damage to the insulation, replace the component.

Inspect the housing halves for cracks. Also check the fan and pulley for looseness on the rotor shaft and for cracks. Replace any part that does not pass inspection. Remove the heat transfer grease that is in the rectifier mounting area with a clean cloth.

The AC generator's brushes should be inspected and tested anytime the unit is disassembled. Brushes should be replaced whenever they are worn shorter than 1/4 in. (6.35 mm) in length. Also the brush springs must be checked for sufficient strength to keep constant contact of brushes with slip rings. Brush continuity may be checked using an ohmmeter. There should be zero resistance through the brush path. Replace the brushes if there is any resistance indicated.

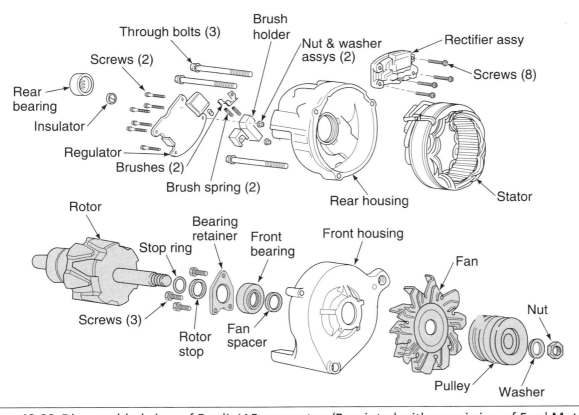

Figure 12-20 Disassembled view of Ford's IAR generator. (Reprinted with permission of Ford Motor Company)

AC Generator Component Testing

Once the AC generator is disassembled and cleaned, the individual components can be tested. The chart (Figure 12-21) illustrates the test connections and results for the major components.

Testing the Rotor

The most important test of a rotor is a complete visual inspection. Carefully check the rotor windings for signs of discoloration or overheating. If these signs are present, the rotor is no good. Also carefully inspect the slip rings; they should be flat, smooth, and free of damage. If the rotor passes the visual inspection, proceed to test it with an ohmmeter.

An ohmmeter can be used to measure the resistance between the slip rings (Figure 12-22). Always check the manufacturer's service manual for the correct specification for the unit you are working on. If specifications are not available, the following are some typical values:

GM	2.4 to 3.5V
Ford	3.0 to 5.5V
Chrysler	3.0 to 6.0V

If the resistance reading is below specifications, then the rotor is shorted. If the resistance is high, the rotor connections are badly corroded or open. The rotor must

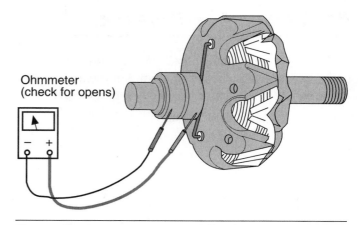

Ohmmeter (check for opens)

Figure 12-22 Test connections for checking for opens in the rotor windings.

be replaced if any of these conditions are found.

Connecting the ohmmeter from each of the slip rings to the rotor shaft should show infinite resistance (Figure 12-23). If the reading was very low, the field coil is grounded and the rotor will have to be replaced.

SERVICE TIP *In many instances it is less expensive to replace the AC generator than to replace the rotor only.*

Testing the Stator

When testing a stator, a visual inspection is the most productive test. Look for discoloration or other damage to the windings. Often the assembly will look fine but

COMPONENT	TEST CONNECTION	NORMAL READING	IF READING WAS:	TROUBLE IS:
Rotor	Ohmmeter from slip ring to rotor shaft	Infinite resistance	Very low	Grounded
	Test lamp from slip ring to shaft	No light	Lamp lights	Grounded
	Test lamp across slip rings	Lamp lights	No light	Open
Stator	Ohmmeter from any stator lead to frame	Infinite resistance	Very low	Grounded
	Test lamp from lead to frame	No light	Lamp lights	Grounded
	Ohmmeter across any pair of leads	Less than $\frac{1}{2}\,\Omega$	Any very high reading	Open
Diodes	Ohmmeter across diode, then reverse leads	Low reading one way; high reading other way	Both readings low / Both readings high	Shorted / Open
	12-V test lamp across diode, then reverse leads	Lamp lights one way, but not other way	No light either way / Lamp lights both ways	Open / Shorted

Figure 12-21 Guidelines for bench testing a generator.

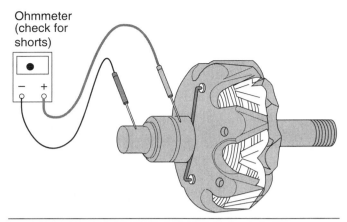

Figure 12-23 Testing the rotor for shorts to ground.

will actually be damaged due to excessive heat. One quick way of checking for this is to take the blade of a knife and scrape the windings. If the coating or varnish flakes off, the windings overheated and the varnish is baked. Pay special attention to the connectors. Any signs of damage or breakage indicate the stator should not be reused. If the stator passes the visual inspection, it should be checked for opens and shorts to ground with an ohmmeter. To test for an open, connect the ohmmeter test leads to any pair of stator leads (Figure 12-24). Continue to test the stator until all combinations of pair connections are completed. On all of these connections, the resistance should be less than 0.5V. If the ohmmeter reads infinity between any two leads, the stator has an open and it must be replaced.

To test the stator for a short to ground, connect the ohmmeter to the stator leads and stator frame (Figure 12-25). The ohmmeter should read infinity on all three stator leads. If the reading is less than infinity, the stator is shorted to ground and it must be replaced.

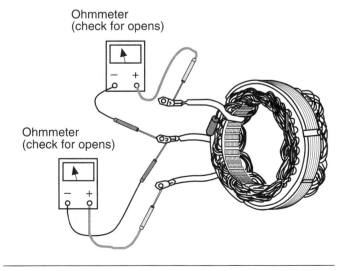

Figure 12-24 Testing the stator for opens.

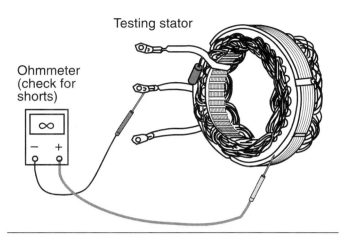

Figure 12-25 Testing the stator for shorts to ground.

SERVICE TIP *A shorted stator is difficult to test for because the resistance is very low for a normal stator. If all other components test okay, but output was low, a shorted stator is the probable cause.*

Testing the Diodes (Rectifier Bridge)

Because a diode should allow current to flow in only one direction, it must be tested for continuity in both directions. Using the ohmmeter, connect the test leads to the diode lead and case (Figure 12-26). Read the ohmmeter scale. If the diode is good it will show high resistance in one direction and low resistance in the opposite direction. If both readings are low, the diode is shorted. If there is high resistance in both directions, the diode is open. Test all six diodes and replace any that are defective.

Testing the Diode Trio

If the AC generator is equipped with a diode trio, it must be tested for opens and shorts. The procedure is much the same as with the rectifier bridge test. Connect one of the ohmmeter test leads to the signal connector. Connect the other test lead to one of the three connectors (Figure 12-27). Test each of the three connectors. Record your readings.

Reverse the ohmmeter leads and record the readings obtained on each of the three connectors. The ohmmeter should read less than 300V in one direction and above 300V in the other direction. These results should be obtained on all three connections. If not, replace the diode trio.

Testing GM Internal Regulator

The GM internal regulator can be tested on the vehicle using the full fielding test. If the AC generator is

PHOTO SEQUENCE 5

Typical Procedure for IAR Alternator Disassembly

P5-1 Always have a clean and organized work area. Tools required to disassemble the Ford IAR AC generator: rags, T20 TORX wrench, plastic hammer, arbor press, 100-watt soldering iron, soft jaw vise, safety glasses, and assorted nut drivers.

P5-2 Using a T20 TORX, remove the four attaching screws that hold the regulator to the AC generator rear housing and remove the regulator and brush assembly as one unit.

P5-3 Using a T20 TORX, remove the two screws that attach the regulator to the brush holder. Separate the regulator from the brush holder. Remove the A terminal insulator from the regulator.

P5-4 Scribe or mark the two end housings and the stator core for reference during assembly, then remove the three through bolts that attach the two housings. Separate the front housing from the rear housing. The rotor will come out with the front housing, while the stator will stay with the rear housing.

P5-5 Separate the three stator lead terminals from the rectifier bridge.

P5-6 Remove the stator coil from the housing.

Typical Procedure for IAR Alternator Disassembly

P5-7 Using a T20 TORX, remove the four attaching bolts that hold the rectifier bridge and remove the rectifier bridge from the housing.

P5-8 Use a socket to tap out the bearing from the housing.

P5-9 Clamp the rotor in a soft jaw vise and remove the pulley attaching nut, flatwasher, drive pulley, fan, and fan spacer from the rotor shaft.

P5-10 Separate the front housing from the rotor. If the stop ring is damaged remove it from the rotor. If not, leave it on the rotor shaft.

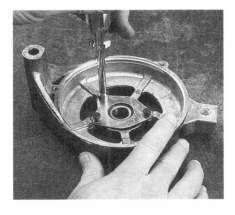

P5-11 Remove the three screws that hold the bearing retainer to the front housing and remove the retainer.

P5-12 Remove the front bearing from the housing. NOTE: It may be necessary to use an arbor press to remove the bearing if it does not slide out easily.

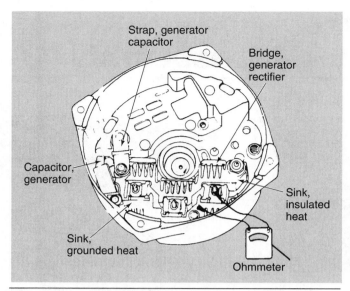

Figure 12-26 Testing the rectifier bridge. (Courtesy of General Motors Corporation, Service Technology Group)

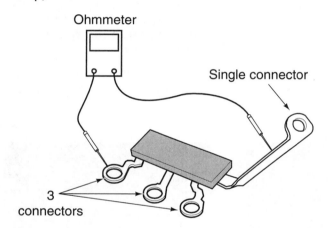

Figure 12-27 Diode trio test connections. (Courtesy of General Motors Corporation, Service Technology Group)

removed from the vehicle it can be tested once it is disassembled (Figure 12-28).

 *When the test connections are made, the charger must be turned off.*

Make the connections, as shown, with the charger turned off. The test light should light because it is receiving only battery voltage as a signal. Turn on the charger and increase the charging rate until the test light goes out. The voltmeter should read between 13.5 and 16 volts. If the test light did not light when connected, the regulator has an open. If the test light fails to go out, the regulator is always in a full field mode. In either case the regulator must be replaced.

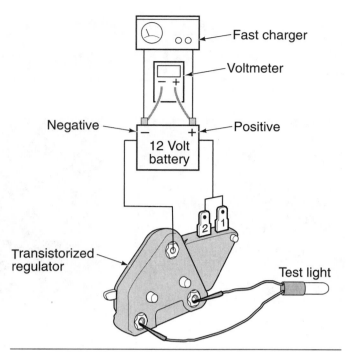

Figure 12-28 Bench testing the GM internal voltage regulator.

When reinstalling the voltage regulator into the housing, the current insulated screws must be installed in the proper location (Figure 12-29). If the insulated screws are installed into the wrong locations, then the regulator will either produce maximum output or zero output.

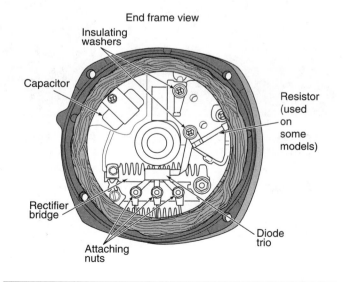

Figure 12-29 Correct location of the isolators is important to proper AC generator operation. (Courtesy of General Motors Corporation, Service Technology Group)

AC Generator Reassembly

The reassembly procedure is basically the reverse order of the disassembly. The following are suggestions to assist in the assembly process:

1. Always check the manufacturer's specifications for the proper torque values of the attaching screws.

2. Use high temperature grease on the bearings.

3. Check for free rotor rotation after installing and torquing the pulley to the rotor shaft.

4. Apply heat sink grease across the rectifier bridge base plate.

5. Protect the diodes from excess heat while soldering the connections. Use a pair of needle nose pliers as a heat sink (Figure 12-30).

When assembling a generator, always follow the recommendations of the manufacturer. It is critical that all screws and bolts be installed with the insulating washers that were present before disassembly. These insulators maintain proper circuit polarity. If a washer is left out, a short circuit will exist.

When installing the rotor into the brushes, most generators are equipped with a hole that allows a pin or paper clip to be inserted into the brush holder to keep them back and allow the rotor to fit into the brushe. After the rotor is in place, the pin can be removed and the brushes will snap into place on the slip rings. Before installing the pin to hold the brushes, make sure the brush springs are properly positioned behind the brushes.

Diode Pattern Testing

CUSTOMER CARE *It is good practice to check the diode pattern of the AC generator anytime an electronic component fails. Because the electronics of the vehicle cannot accept AC current, the damage to the replaced component could have been the result of a bad diode. By performing this check it is possible to find the cause of the problem.*

To test the diodes using an oscilloscope, set the oscilloscope on the lowest scale available. Connect the pri-

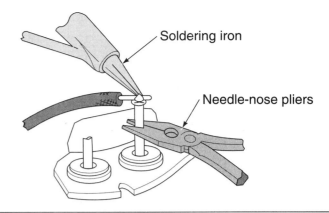

Figure 12-30 Protect the diodes from excessive heat by providing a heat sink.

mary test leads on the AC generator output terminal and ground. Start the engine and place a moderate load on the charging system (15 to 20 amperes). Different patterns may appear. What is considered normal depends on the load placed on the system.

The diode pattern (Figure 12-31) illustrates a good pattern. However, the second pattern shown (Figure 12-34) is also a good pattern if the AC generator is under a full load. The third pattern shown (Figure 12-33) is a good pattern for some AC generators.

Patterns that have high resistance, open, and shorted diodes are illustrated (Figures 13-34 through 13-37). Remember to check the waveforms for noise. If the diodes don't rectify all of the AC, some will ride on the DC output.

SERVICE TIP *Instead of using a carbon pile, it is possible to place a moderate load on the charging system by turning on the headlights and a few other electrical accessories.*

An alternate way to check the action of the diodes is to check for AC voltage at the battery. Do this with a DMM set to a low AC voltage scale. Connect the meter across the battery; load the charging system by turning on the headlights. Ideally there should be zero volts AC at the battery. A voltage reading of more than 0.5 VAC is excessive and indicates the diodes are not rectifying the AC output of the AC generator.

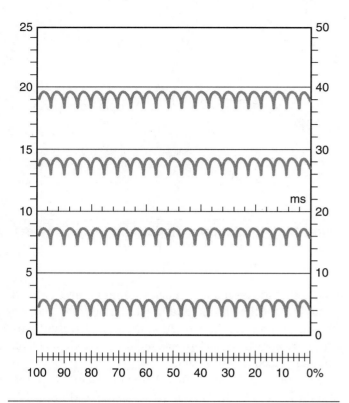

Figure 12-31 Good diode test pattern. (Courtesy of Sun Electrical Corporation).

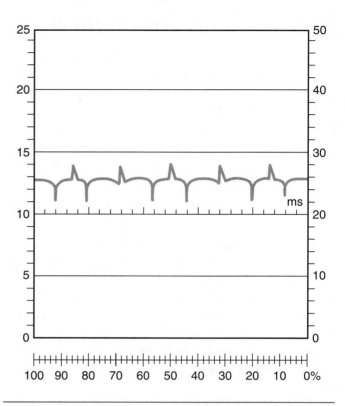

Figure 12-33 Good diode test pattern when AC generator has no load demands. (Courtesy of Sun Electrical Corporation)

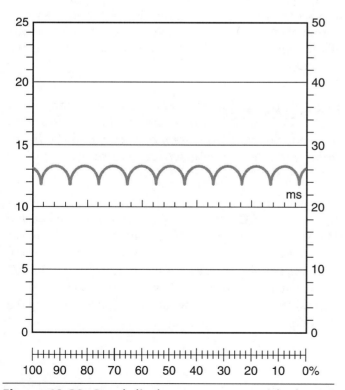

Figure 12-32 Good diode test pattern with the AC generator under full load. (Courtesy of Sun Electrical Corporation)

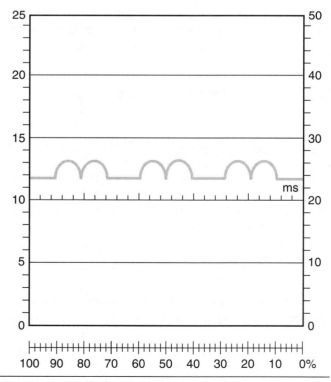

Figure 12-34 Shorted diodes or shorted stator winding when the AC generator is placed under full load. (Courtesy of Sun Electrical Corporation)

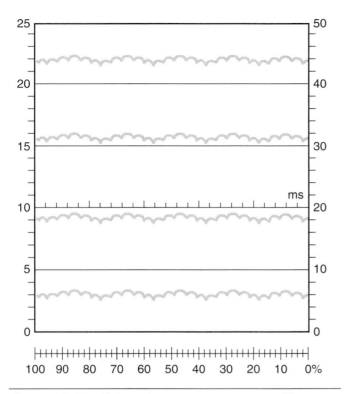

Figure 12-35 High resistance test pattern. (Courtesy of Sun Electrical Corporation)

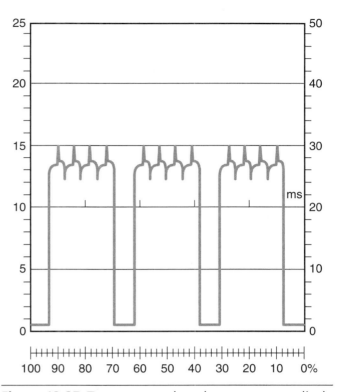

Figure 12-37 Test pattern that shows an open diode in the diode trio. (Courtesy of Sun Electrical Corporation)

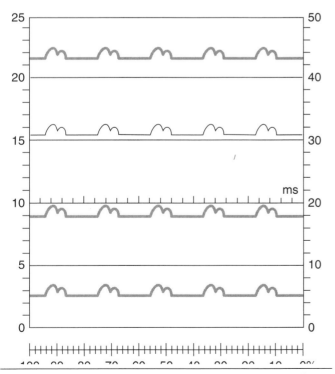

Figure 12-36 Test pattern indicates one open and one shorted diode. (Courtesy of Sun Electrical Corporation)

Summary

❑ Whenever there is a charging system problem, make sure the battery is thoroughly checked first.

❑ It is also important to perform a preliminary inspection. Many problems can be detected during this simple step.

❑ There are several general precautions the technician must be aware of when servicing the charging system.

❑ The causes of the noises coming from the generator are identifiable by the type of noise they make.

❑ The voltage output test is used to make a quick determination if the charging system is working properly.

❑ The voltage drop of all wires and connections combined should not exceed 3% of the system voltage. Any particular wire or connection should not exceed 0.2 volt; total system drops should be less than 0.7 volt. The ground side voltage drop should be less than 0.2 volt.

❑ The field current draw test determines if there is current available to the field windings.

❑ Current output testing will determine the maximum output of the AC generator.

❑ Full fielding the AC generator means that the field windings are constantly energized with full battery current. Full fielding will produce maximum AC generator output.

❑ The full field test will isolate if the detected problem lies in the AC generator or the regulator.

❑ The regulator test is used to determine if the regulator is maintaining the correct voltage output under different load demands.

❑ The diode/stator test is performed to determine the condition of the diodes.

❑ The charging system requirement test is used to determine the total electrical demand of the vehicle's electrical system.

Terms to Know

Charging system requirement test	Field current draw test	Regulator test
Current output testing	Full fielding	Voltage output test
Diode/stator test	Full field test	

ASE Style Review Questions

1. Charging system testing is being discussed: Technician A says before attempting to test the charging system, the battery must be checked first. Technician B says the state of charge of the battery is not a concern to charging system testing. Who is correct?
   a. A only
   b. B only
   c. Both A and B
   d. Neither A nor B

2. AC generator noise complaints are being discussed: Technician A says a loose belt will make a grumbling noise. Technician B says a whining noise can be caused by a shorted diode. Who is correct?
   a. A only
   b. B only
   c. Both A and B
   d. Neither A nor B

3. Technician A says the no-load/load voltage output test is used to make a quick determination concerning whether or not the charging system is working properly. Technician B says the first step is to perform a visual inspection and preliminary checks of the charging system. Who is correct?
   a. A only
   b. B only
   c. Both A and B
   d. Neither A nor B

4. Test results of the voltage output test are being discussed: Technician A says if the charging voltage is too high, there may be a loose or glazed drive belt. Technician B says if the charging voltage was too low, the fault might be a grounded field wire from the regulator (full fielding). Who is correct?
   a. A only
   b. B only
   c. Both A and B
   d. Neither A nor B

5. Voltage drop testing is being discussed: Technician A says the total system drops should be less than 0.7 volt. Technician B says the ground side voltage drop should be less than 0.2 volt. Who is correct?
   a. A only
   b. B only
   c. Both A and B
   d. Neither A nor B

6. Technician A says the field current draw test determines if there is current available to the field windings. Technician B says a slipping belt can cause a low reading when performing the field current draw test. Who is correct?
   a. A only
   b. B only
   c. Both A and B
   d. Neither A nor B

7. Full field test procedures are being discussed: Technician A says to full field a GM A circuit AC generator, insert a screw driver into the D-shaped test hole and ground the tab. Technician B says check the rotor and field circuit resistance before full fielding the IAR AC generator. Who is correct?
   a. A only
   b. B only
   c. Both A and B
   d. Neither A nor B

8. Technician A says full fielding means the field windings are constantly energized with full battery voltage. Technician B says the full fielding test should be performed only if the charging system passes the output test. Who is correct?
   a. A only
   b. B only
   c. Both A and B
   d. Neither A nor B

9. Technician A says if full fielding with the regulator bypassed produces lower than specified output, the regulator is the cause of the problem. Technician B says the full field test will isolate whether the detected problem lies in the AC generator or the regulator. Who is correct?
   a. A only
   b. B only
   c. Both A and B
   d. Neither A nor B

10. Technician A says all AC generators have a means of full fielding and bypassing the regulator. Technician B says many import and domestic AC generators must be disassembled to determine the cause of the charging system failure. Who is correct?
    a. A only
    b. B only
    c. Both A and B
    d. Neither A nor B

13 Lighting Circuits

Objective

Upon completion and review of this chapter, you should be able to:

❏ Describe the operation and construction of automotive lamps.

❏ Describe the differences between conventional sealed-beam, halogen, and composite headlight lamps.

❏ Describe the operation and controlled circuits of the headlight switch.

❏ Describe the operation of the dimmer switch.

❏ Explain the operation of the most common styles of concealed headlight systems.

❏ Describe the operation of the various exterior light systems, including parking, tail, brake, turn, side, clearance, and hazard warning lights.

❏ Explain the operating principles of the turn signal and hazard light flashers.

❏ Describe the operation of the various interior light systems, including courtesy and instrument panel lights.

Introduction

Today's technician is required to understand the operation and purpose of the various lighting circuits on the vehicle. The addition of computers and their many sensors and actuators (some that interlink to the lighting circuits) make it impossible for technicians to just bypass part of the circuit and rewire the system to their own standards. If a lighting circuit is not operating properly the safety of the driver, the passengers, people in other vehicles, and pedestrians are in jeopardy. When today's technician performs repairs on the lighting systems, the repairs must meet at least two requirements: they must assure vehicle safety and meet all applicable laws.

The lighting circuits of today's vehicles can consist of more than 50 light bulbs and hundreds of feet of wiring.

Incorporated within these circuits are circuit protectors, relays, switches, lamps, and connectors. In addition, more sophisticated lighting systems use computers and sensors. The lighting circuits consist of an array of interior and exterior lights, including headlights, taillights, parking lights, stop lights, marker lights, dash instrument lights, courtesy lights, and so on.

The lighting circuits are largely regulated by federal laws, so the systems are similar between the various manufacturers. However, there are variations. Before attempting to do any repairs on an unfamiliar circuit, the technician should always refer to the manufacturer's service manuals. This chapter provides information about the types of lamps used, describes the headlight circuit, discusses different types of concealed headlight systems, and explores the various exterior and interior light circuits individually.

Lamps

A **lamp** is a device that produces light as a result of current flow through a **filament**. The filament is generally made from tungsten and is enclosed within a glass envelope (Figure 13-1). A lamp generates heat as current flows through the filament. This causes it to get very hot. The changing of electrical energy to heat energy in the resistive wire filament is so intense that the filament starts to glow and emits light. The process of changing energy forms to produce light is called **incandescence**. The lamp must have a vacuum surrounding the filament to prevent it from burning so hot that the filament burns through. The glass envelope that encloses the filament maintains the presence of vacuum. When the lamp is manufactured, all the air is removed and sealed outside of the lamp by the glass envelope. If air is allowed to enter the lamp, the oxygen would cause the filament to oxidize and burn up.

Many lamps are designed to execute more than one function. A **double filament lamp** (Figure 13-2) can be used in the stoplight, taillight, and the turn signal circuits combined.

It is important that any burned-out lamp be replaced with the correct lamp. The technician can determine what lamp to use by checking the lamp's standard trade number.

Headlights

There are four basic types of headlights used on automobiles today: (1) standard sealed beam, (2) halogen sealed beam, (3) composite, and (4) high intensity discharge (HID).

Sealed-Beam Headlights

From 1939 to about 1975 the headlights used on vehicles remained virtually unchanged. During this time the headlight was a round lamp. The introduction of the rectangle headlight in 1975 enabled the vehicle manufacturers to lower the hood line of their vehicles. Both the round and rectangle headlights were **sealed-beam** construction (Figure 13-3). The sealed-beam headlight is a self contained glass unit made up of a filament, an inner reflector, and an outer glass lens. To prevent the filament from becoming oxidized, all oxygen must be removed. In the standard sealed-beam headlight, this is done by filling the inside of the lamp with argon gas.

The standard sealed-beam headlight does not surround the filament with its own glass envelope (bulb). The glass lens is fused to the **parabolic reflector** that is sprayed with vaporized aluminum. The vaporized aluminum gives a reflecting surface that is comparable to silver. The reflector intensifies the light produced by the filament, and the **lens** directs the light to form the required light beam pattern.

WARNING *Because of the construction and placement of prisms in the lens, it is important that the technician install the headlight in its proper position. The lens is usually marked "TOP" to indicate the proper installation position*

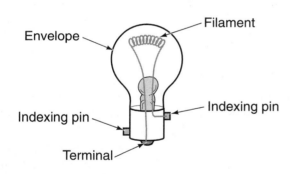

Figure 13-1 A single filament bulb.

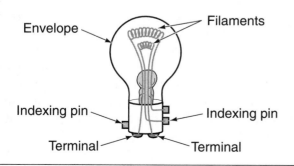

Figure 13-2 A double filament lamp.

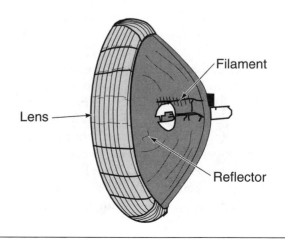

Figure 13-3 Sealed-beam headlight construction.

(A) Top view

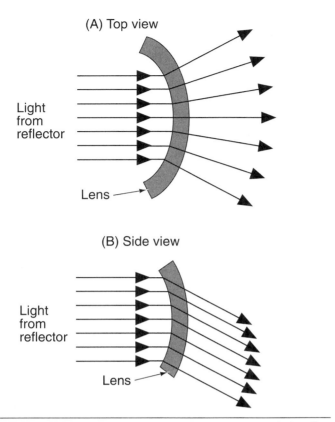

Light from reflector

Lens

(B) Side view

Light from reflector

Lens

Figure 13-4 The prism directs the beam into (A) a flat horizontal pattern and (B) downward.

The lens is designed to produce a broad flat beam. The light from the reflector is passed through concave prisms in the glass lens. The prisms control the horizontal spreading and the vertical control of the light beam to prevent upward glaring (Figure 13-4).

By placing the filament in different locations on the reflector, the direction of the light beam is controlled (Figure 13-5). In a dual filament lamp, the lower filament is used for the high beam and the upper filament is used for the low beam.

Halogen Headlights

The **halogen lamp** most commonly used in automotive applications consists of a small bulb filled with iodine vapor. Halogen is the term used to identify a group of chemically related non-metallic elements. These elements include chlorine, fluorine, and iodine. The bulb is made of a high temperature resistant glass or plastic surrounding a tungsten filament. This inner bulb is installed in a sealed glass housing (Figure 13-6). With the halogen added to the bulb, the tungsten filament is capable of withstanding higher temperatures than that of conventional sealed-beam lamps. The higher temperatures enable the halogen bulb to burn brighter.

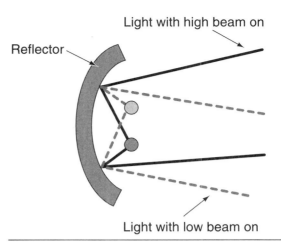

Reflector

Light with high beam on

Light with low beam on

Figure 13-5 Filament placement controls the light beam projection.

In a conventional sealed-beam headlight, the heating of the filament causes atoms of tungsten to be released from the surface of the filament. These released atoms deposit on the glass envelope and create black spots that affect the light output of the lamp. In a halogen lamp, the iodine vapor causes the released tungsten atoms to be redeposited back onto the filament. This virtually eliminates any black spots. It also allows for increased high beam output of 25% over conventional lamps.

An additional advantage of the halogen bulb is cracking or breaking of the lens does not prevent halogen headlight operation since the filament is enclosed in its own bulb. As long as the filament envelope has not been broken the filament will continue to operate, however, a broken lens will result in poor light quality and should be replaced.

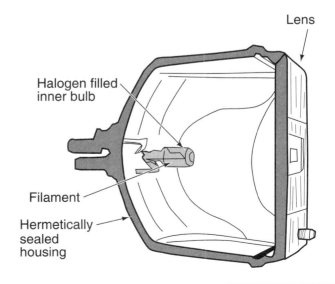

Lens

Halogen filled inner bulb

Filament

Hermetically sealed housing

Figure 13-6 Halogen-sealed beam headlight with iodine vapor bulb.

WARNING *It is not recommended that halogen sealed-beam and standard sealed-beam headlights be mixed on the vehicle. Also, if the vehicle was originally equipped with halogen headlights, do not replace these with standard sealed beams. Doing so may result in poor light quality.*

Composite Headlights

Many of today's vehicles have a halogen headlight system using a replaceable bulb. This system is called **composite headlights**. By using the composite headlight system, vehicle manufacturers are able to produce any style of headlight lens they desire (Figure 13-7). This improves the aerodynamics, fuel economy, and styling of the vehicle.

Many manufacturers vent the composite headlight housing because of the increased amount of heat developed by these bulbs. Because the housings are vented, condensation may develop inside of the lens assembly. This condensation is not harmful to the bulb and does not affect headlight operation. When the headlights are turned on, the heat generated from the halogen bulbs will dissipate the condensation quickly. On systems using nonvented composite headlights, condensation is not considered normal. The assembly should be replaced.

WARNING *Whenever technicians replace a composite lamp, care must be taken not to touch the envelope with the fingers. Staining the bulb with normal skin oil can substantially shorten the life of the bulb. Handle the lamp only by its base (Figure 13-8). Also dispose of the lamp properly.*

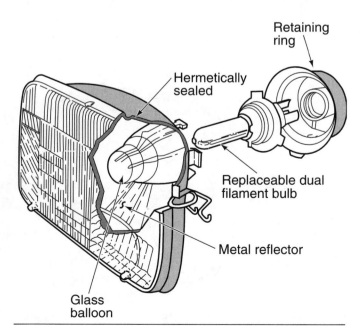

Figure 13-7 A composite headlight system with a replaceable halogen bulb.

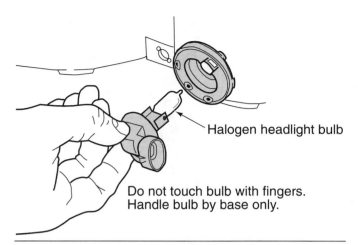

Do not touch bulb with fingers.
Handle bulb by base only.

Figure 13-8 The correct method of handling the composite bulb during replacement. (Reprinted with permission of Ford Motor Company)

HID Headlamps

High intensity discharge (HID) headlamps are the latest headlight development. These headlamps put out three times more light and twice the light spread on the road than conventional halogen headlamps. They also use about two-thirds less power to operate and will last two to three times longer. HID lamps produce light in both ultraviolet and visible wavelengths. This advantage allows highway signs and other reflective materials to glow. This type lamp first appeared on select models from BMW in 1993, Ford in 1995, and Porsche in 1996.

These lamps do not rely on a glowing filament for light. Rather light is provided as a high voltage bridges an air gap between two electrodes. The presence of an inert gas amplifies the light given off by the arcing. More than 15,000 volts are used to jump the gap between the electrodes. To provide this voltage, a voltage booster and controller is required. Once the gap is bridged by the high voltage, only about 80 volts is required to keep current flow across the gap.

The great light output of these lamps allows the headlamp assembly to be smaller and lighter. These advantages allow designers more flexibility in body designs as they attempt to make their vehicles more aerodynamic and efficient.

Headlight Switches

The **headlight switch** controls most of the vehicle's lighting systems. The headlight switch may be located either on the dash by the instrument panel or on the steering column (Figure 13-9). The most common style of headlight switch is the three-position type with OFF,

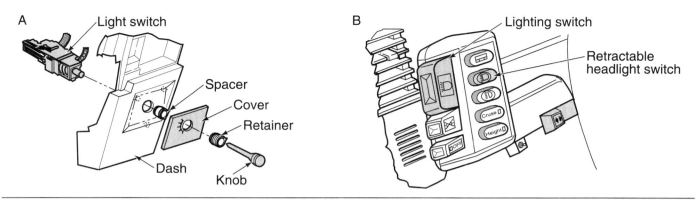

Figure 13-9 (A) Instrument panel mounted headlight switch. (Reprinted with permission of Ford Motor Company) (B) Steering column mounted headlight switch. (Courtesy of Toyota Motor Corp.)

PARK, and HEADLIGHT positions. The headlight switch will generally receive direct battery voltage to two terminals of the switch. This allows the light circuits to be operated without having the ignition switch in the RUN or ACC (accessory) position.

When the headlight switch is in the OFF position, the open contacts prevent current from flowing to the lamps (Figure 13-10). When the switch is in the PARK position, current will flow from terminal 5 through the closed contacts to the side marker, taillight, license plate, and instrument cluster lights. This circuit is usually protected by a 15- to 20-ampere fuse that is separate from the headlight circuit.

When the switch is located in the HEADLIGHT position, current will flow from terminal 1 through the circuit breaker and the closed contacts to light the headlights. Current from terminal 5 continues to light the lights that were on in the PARK position. The circuit breaker is used to prevent temporary overloads in the system from totally disabling the headlights.

The rheostat is a variable resistor that the driver uses to control the instrument cluster illumination lamp brightness. As the driver turns the light switch knob, the

resistance in the rheostat is changed. The greater the resistance, the dimmer the instrument panel illumination lights glow.

Dimmer Switches

The **dimmer switch** provides the means for the driver to select either high beam or low beam operation, and to switch between the two. The dimmer switch is connected in series within the headlight circuit and controls the current path for high and low beams.

In the past, the most common location of the dimmer switch was on the floor board next to the left kick panel. This switch is operated by the driver pressing on it with a foot. Positioning the switch on the floor board made the switch subject to damage because of rust, dirt, and so forth. Most newer vehicles locate the dimmer switch on the steering column (usually operated by pulling on the turn signal stock) to prevent early failure and to increase driver accessibility.

A BIT OF HISTORY

Foot operated dimmer switches became standard equipment in 1923.

Headlight Circuits

The complete headlight circuit consists of the headlight switch, dimmer switch, high beam indicator, and the headlights. When the headlight switch is pulled to the HEADLIGHT position, current flows to the dimmer

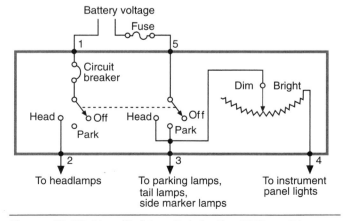

Figure 13-10 Headlight switch in the OFF position.

switch through the closed contacts (Figure 13-11). If the dimmer switch is in the LOW position, current flows through the low beam filament of the headlights. When the dimmer switch is placed on the HIGH position, current flows through the high beam filaments of the headlights.

The headlight circuits just discussed are designed with insulated side switches and grounded bulbs. In this system, battery voltage is present to the headlight switch. The switch must be closed in order for current to flow through the filaments and to ground. The circuit is complete because the headlights are grounded to the vehicle body or chassis. Many import manufacturers use insulated bulbs and ground side switches. In this system, when the headlight switch is located in the HEADLIGHT position the contacts are closed to complete the circuit path to ground. The headlight switch is located after the headlight lamps in the circuit. In addition, many circuits use relays.

No matter if the headlights use insulated side switches or ground side switches, each system is wired in parallel to each other. This prevents total headlight failure if one filament burns out.

Concealed Headlights

A vehicle equipped with a **concealed headlight system** hides the lamps behind doors when the headlights are turned off. This is done for improved fuel economy and styling. When the headlight switch is turned to the HEADLIGHT position, the headlight doors open. The headlight doors can be controlled by either electric motors or by vacuum.

Vacuum-Controlled Systems

Systems that utilize vacuum to operate the headlight doors use a headlight switch with a **vacuum distribution valve** (Figure 13-12). The vacuum distribution valve controls the direction of vacuum to the vacuum motors at the headlight doors or to vent.

With the headlight switch in the OFF position, engine vacuum is supplied to the vacuum doors to keep them closed. When the headlight switch is pulled out to turn on the headlights, the distribution valve vents the vacuum trapped in the vacuum motors. This allows the springs to open the doors. Using vacuum to close the doors and springs to open the doors assures the doors will be open if vacuum to the system is lost. The vacuum reservoir stores vacuum to keep the doors closed when the engine is shut off or when low vacuum is supplied from the engine (such as during wide open throttle). Also, a bypass valve is installed into the system to provide a means of manually opening the doors in the event the system fails (Figure 13-13).

Figure 13-12 A vacuum distribution valve located at the back of the headlight switch.

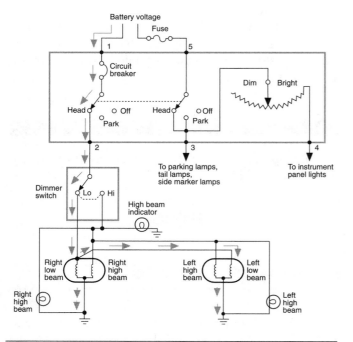

Figure 13-11 Headlight circuit indicating current flow with the dimmer switch in the LOW beam position.

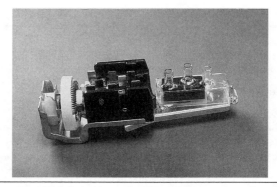

Figure 13-13 A bypass valve is used to open the doors if the system fails. (Courtesy of Chrysler Corporation)

Electrically Controlled Systems

Electrically controlled systems can use either a torsion bar and a single motor to open both headlight doors, or a separate motor for each headlight door. Most systems will use **limit switches** to stop current flow when the doors are full up or full down. These switches generally operate from a cam on the reaction motor (Figure 13-14). Only one limit switch can be closed at a time. When the door is full up, the opening limit switch opens and the closing limit switch closes. When the door is full down, the closing limit switch is open and the opening limit switch closes. This prepares the reaction motor for the next time that the system is activated or deactivated.

The electrically operated concealed headlight system must provide a provision for manually opening the doors in the event of a system failure (Figure 13-15).

In a system that incorporates an integrated chip, each motor has its own relay and limiting switches. When the limit switches are in the A-B position, the doors are fully open. When the switches are in the A-C position, the doors are fully closed.

When the headlights are turned on, terminal 14 is grounded through the headlight switch. This signals the IC to bias the leg of transistor TR2 for 10 seconds. When TR2 is biased it closes the circuits for relays 1 and 2, through position A-C of the limit switches and terminal 18. With the relay coils energized, battery voltage is supplied to the motors and they operate until the limit switches close to A-B or the 10-second timer turns off. TR3 and TR4 are also biased to complete the circuit for the headlight and taillight relays. When the relays are energized, the circuit is completed to the headlights and taillights, and they illuminate.

The motors used in this system provide for 360-degree rotation. The first 180 degrees of rotation opens the doors and an additional 180 degrees closes them. The

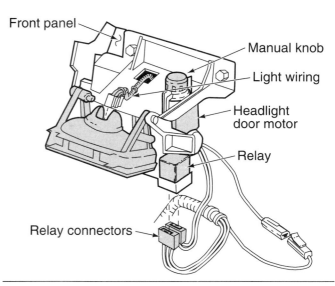

Figure 13-15 Electrically controlled concealed headlight system with a manual control valve. (Courtesy of General Motors Corporation, Service Technology Group)

timer circuit is used to protect the motors. The diodes in the motor relays are called clamping diodes. These protect the IC from induced voltage when the relay coils collapse.

When the headlights are turned off, ground through terminal 14 is opened by the headlight switch. This signals the IC to bias transistor TR1 for 10 seconds. TR1 closes the circuit to ground for the relay circuit through the A-B positions of the limit switches and terminal 18. With the relays energized, battery voltage is sent to the motors. Also the IC ceases to bias TR3 and TR4. As a result, the relays are de-energized and the lamps turn off.

Figure 13-16 illustrates one of the methods used by the Chrysler Corporation to operate the electric motors for their concealed headlight system. When the ignition switch is in the RUN position but the headlight switch is off, current flows through the ignition switch to the relay. The relay is energized because the coil is grounded through the headlight filaments. With the coil energized the relay points close. However, the door closing limit switch is open. This results in a de-energized door closing field winding.

When the headlight switch is turned to the HEADLIGHT position, current continues to flow to the relay coil through the ignition switch. However, current is also sent to the other side of the relay coil from the headlight switch. Voltage is equal on both sides of the relay coil, so there is no voltage potential and the coil is de-energized. The relay contact points close to the door opening field winding. With the door opening limit switch closed, the motor operates until the limit switch is opened.

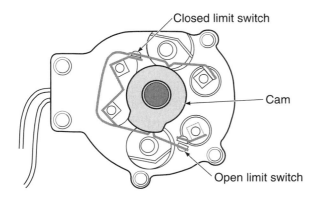

Figure 13-14 Most limit switches operate off of a cam on the motor. (Courtesy of Chrysler Corporation)

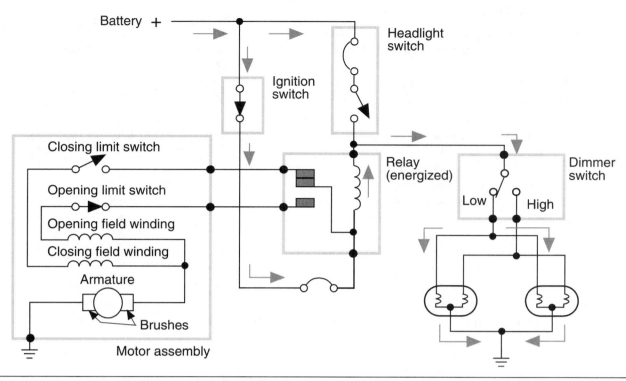

Figure 13-16 Current flow with the headlight switch OFF and the headlight doors closed.

When the headlights are turned off again, voltage is applied to only one side of the coil through the ignition switch. Ground is provided through the headlight filaments. This energizes the relay, closing the contact point to the door closing winding. Since the door closing limit switch is closed, the motor operates until the limit switch opens again.

Flash to Pass

Many steering column-mounted dimmer switches have an additional feature called **"flash to pass."** This circuit illuminates the high beam headlights even with the headlight switch in the OFF or PARK position. When the driver activates the flash to pass feature, the contacts in the dimmer switch complete the circuit to the high beam filaments.

Exterior Lights

When the headlight switch is pulled to the PARK or HEADLIGHT position, the front parking lights, taillights, side marker lights, and rear license plate light are all turned on. The front parking lights usually use dual filament bulbs. The other filament is used for the turn signals and hazard lights.

Most taillight assemblies include the brake, parking, rear turn signal, and rear hazard lights. The center high mounted stop lamp (CHMSL), back-up lights, and license plate lights can be included as part of the taillight circuit design. Depending on the manufacturer, the taillight assembly can be wired to use single filament or dual filament bulbs. When single filament bulbs are used, the taillight assembly is wired as a three-bulb circuit. A three-bulb circuit uses one bulb each for the tail, brake, and turn signal lights on each side of the vehicle. When dual filament bulbs are used, the system is wired as a two-bulb circuit. Each bulb can perform more than one function.

Taillight Assemblies

Parking and taillights are controlled by the headlight switch. They can be turned on without having to turn on the headlights. Usually the first detent on the headlight switch is provided for this function. Figure 13-17 illustrates a parking and taillight circuit controlled by the headlight switch. Thus the lights can be operated with the ignition switch in the OFF position.

A BIT OF HISTORY

Taillights on both sides of the car did not appear until 1929.

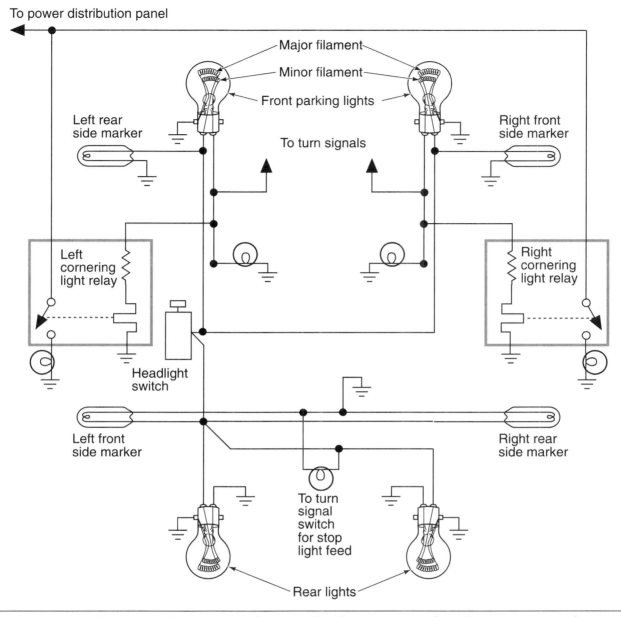

Figure 13-17 A parking and taillight circuit. (Reprinted with permission of Ford Motor Company)

In a three-bulb taillight system, the brake lights are controlled directly by the brake light switch (Figure 13-20). In many applications, the brake light switch is attached to the brake pedal, with a small clearance between the pedal arm pin and the eye of the pushrod (Figure 13-21). When the brakes are applied, the pedal moves down and the switch plunger moves away from the pushrod. This closes the contact points and lights the brake lights. On some vehicles the brake light switch may be a pressure sensitive switch located in the brake master cylinder. When the brakes are applied, the pressure developed in the master cylinder closes the switch to light the lamps.

The brake light switch receives direct battery voltage through a fuse, which allows the brake lights to operate when the ignition switch is in the OFF position. Once the switch is closed, current flows to the brake lights. The brake lights on both sides of the vehicle are wired in parallel. The bulb is grounded to complete the circuit.

Most brake light systems use dual filament bulbs that perform multifunctions. Usually the filament of the dual filament bulb, which is also used by the turn signal and hazard lights, is used for the brake lights (the high intensity filament). In this type of circuit, the brake lights are wired through the turn signal and hazard switches (Figure 13-20). If neither turn signal is on, the current is sent

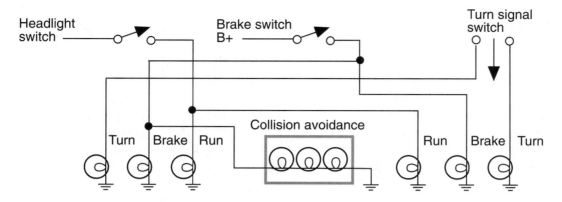

Figure 13-18 The three-bulb taillight circuit. One bulb is used for each function.

to both of the brake lights. If the left turn signal is on, current for the right brake light is sent to the lamp through the turn signal switch and wire designated as 18BR/RD. The left brake light does not receive any voltage from the brake switch because the turn signal switch opens that circuit.

Since the turn signal switches used in a two-bulb system also controls a portion of the operation of the brake lights, they have a complex system of contact points. The technician must remember that many brake light problems are caused by worn contact points in the turn signal switch.

All brake lights must be red and, starting in 1986, the vehicle must have a **center high mounted stop lamp (CHMSL)**, it is also referred to as a **collision avoidance light**. This lamp must be located on the center line of the vehicle and no lower than three inches below the bottom of the rear window (six inches on convertibles). In a three-bulb system, wiring for the CHMSL is in parallel to the brake lights.

In a two-bulb circuit the CHMSL can be wired in one of two common methods. The first method is to connect to the brake light circuit between the brake light switch

and the turn signal switch (Figure 13-22). This method is simple to perform. However, it increases the number of conductors needed in the harness.

The most common method used by manufacturers is to install diodes in the conductors that are connected between the left and right side bulbs. If the brakes are applied with the turn signal switch in the neutral position, the diodes will allow voltage to flow to the CHMSL. If the turn signal switch is placed in the left turn position, the left light must receive a pulsating voltage from the flasher. However, the steady voltage being applied to the right brake light would cause the left light to burn steady if the diode was not used. Diode 1 will block the voltage from the right lamp, preventing it from reaching the left light. Diode 2 will allow this voltage from the right brake light circuit to be applied to the CHMSL.

A BIT OF HISTORY

In 1921, turn signals were made standard equipment by Leland Lincoln. This marque later joined Ford Motor

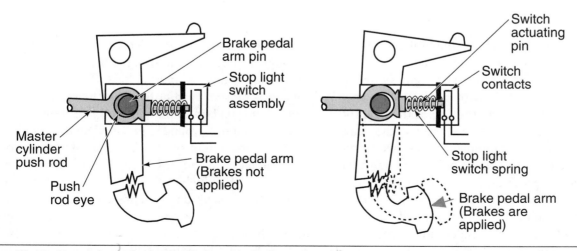

Figure 13-19 Operation of a brake light switch. (Reprinted with permission of Ford Motor Company)

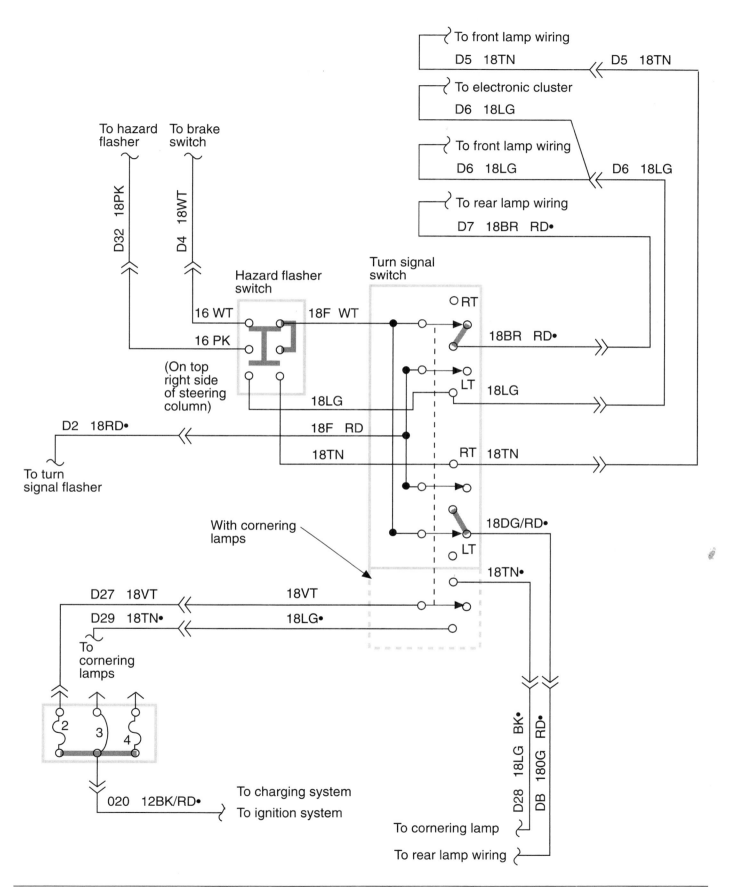

Figure 13-20 Turn signal switch used in a two- bulb taillight circuit. (Courtesy of Chrysler Corporation)

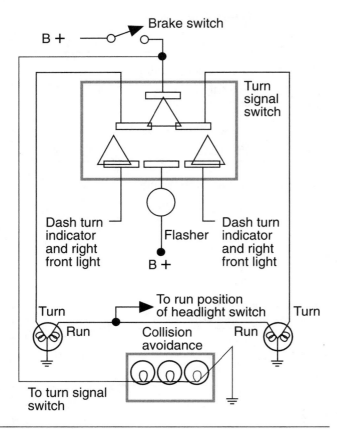

Figure 13-21 Wiring the CHMSL into the two-bulb circuit between the brake light switch and the turn signal.

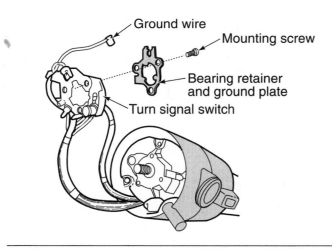

Figure 13-22 Typical turn signal switch location. (Courtesy of Chrysler Corporation)

Company. Leland Lincolns were built by Henry Leland who was the originator of Cadillac. Early turn signals were not like those used today; many were steel arms with reflective material on them. These arms pivoted out on the side of the car as it was turning. This style contin-

ued for many years until Buick introduced electric turn signals to the public in 1939.

Turn signals are used to indicate the driver's intention to change direction or lanes. Most turn signal switches receive their voltage from the ignition switch when it is in the RUN position only. This prevents the turn signals from operating while the ignition switch is in the OFF position. The turn signal switch is located in the steering column behind the lock and horn plates (Figure 13-22).

Figure 13-23 illustrates the operation of the turn signal circuit. In the neutral position, the contacts are opened, preventing current flow (Figure 13-24). When the driver moves the turn signal lever to indicate a left turn, the turn signal switch closes the contacts to direct voltage to the front and rear lights on the left side of the vehicle.

When the turn signal switch is moved to indicate a right turn, the contacts are moved to direct voltage to the front and rear turn signal lights on the right side of the vehicle.

SERVICE TIP *If the turn signals operate properly in one direction but do not flash in the other, the problem is not in the flasher unit. A burned-out lamp filament will not cause enough current to flow to sufficiently heat the bimetallic strip to cause it to open. Thus, the lights do not flash. Locate the faulty bulb and replace it.*

With the contacts closed, power flows from the **flasher** through the turn signal switch to the lamps. The flasher (Figure 13-24) consists of a set of normally closed contacts, a bimetallic strip, and a coil heating element (Figure 13-25). These three components are wired in series. As current flows through the heater element, it increases in temperature, which heats the bimetallic strip. The strip then bends and opens the contact points. Once the points are open, current flow stops. The bimetallic strip cools and the contacts close again. With current flowing again the process is repeated. Because the flasher is in series with the turn signal switch, this action causes the turn signal lights to turn on and off.

WARNING *The incorrect selection of the turn signal flasher may result in fast, slow, or no turn signal operation.*

The **hazard warning system** is part of the turn signal system (Figure 13-26). It has been included on all vehi-

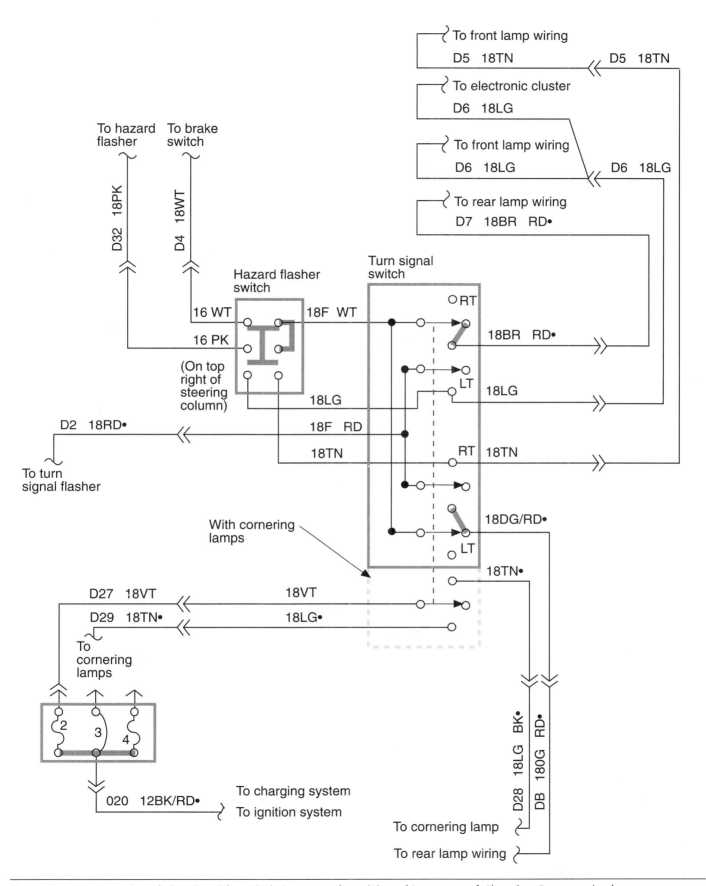

Figure 13-23 Turn signal circuit with switch in neutral position. (Courtesy of Chrysler Corporation)

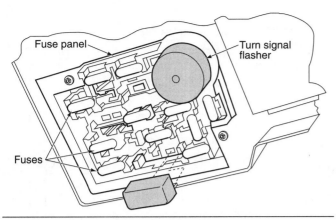

Figure 13-24 A turn signal flasher located into the fuse box. (Reprinted with permission of Ford Motor Company)

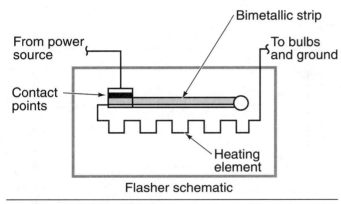

Flasher schematic

Figure 13-25 The flasher uses a bimetallic strip and a heating coil to flash the turn signal lights. (Reprinted with permission of Ford Motor Company)

cles sold in North America since 1967. All four turn signal lamps flash when the hazard warning switch is turned on. Depending on the manufacturer, a separate flasher can be used for the hazard lights than the one used for the turn signal lights. The operation of the hazard flasher is identical to that of the turn signal. The only difference is that the hazard flasher is capable of carrying the additional current drawn by all four turn signals. And, it receives its power source directly from the battery.

Neon Third Brake Lights

In 1995 Ford Motor Company began equipping select models with rear high-mount brake lights that use **neon lights**. These lights are more energy efficient and turn on more quickly than the regular lights. Behind the third brake light lens is a single neon bulb. Since neon bulbs have no filament, the neon bulb should last much longer than a regular bulb.

The neon bulbs turn on within 3 milliseconds after being activated. Halogen lamps require 300 milliseconds. The importance of this quickness is that it gives the drivers behind the vehicle an earlier warning to stop. This early warning can give the approaching driver 19 more feet for stopping when driving at 60 miles per hour.

Cornering Lights

Cornering lights are lamps that illuminate when the turn signals are activated. They burn steady when the turn signal switch is in a turn position to provide additional illumination of the road in the direction of the turn.

Vehicles equipped with cornering lights have an additional set of contacts in the turn signal switch. These con-

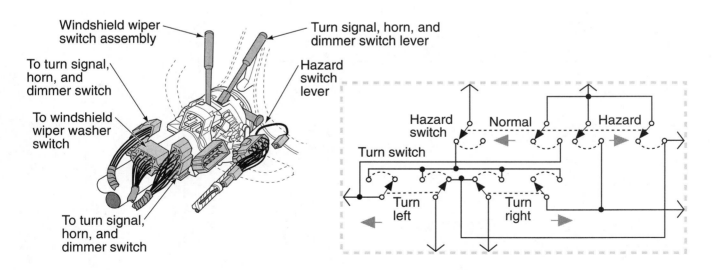

Figure 13-26 The hazard warning system is incorporated into the turn signal system. (Reprinted with permission of Ford Motor Company)

tacts operate the cornering light circuit only. The contacts can receive voltage from either the ignition switch or the headlight switch. If the ignition switch provides the power, the cornering lights will be activated any time the turn signals are used. If the contacts receive the voltage from the headlight switch, the cornering lights do not operate unless the headlight switch is in the PARK or HEADLIGHT position (Figure 13-27).

Backup Lights

All vehicles sold in North America after 1971 are required to have **backup lights**. Backup lights illuminate the road behind the vehicle and warn other drivers and pedestrians of the driver's intention to back up. Figure 13-28 illustrates a backup light circuit. Power is supplied through the ignition switch when it is in the RUN position. When the driver shifts the transmission into reverse, the backup light switch contacts are closed and the circuit is completed.

Many automatic transmission equipped vehicles incorporate the backup light switch into the neutral safety switch. Most manual transmissions are equipped with a separate switch. Either style of switch can be located on the steering column, on the floor console, or on the transmission. Depending on the type of switch used, there may be a means of adjusting the switch to assure that the lights are not on when the vehicle is in a forward gear selection.

A BIT OF HISTORY

The 1921 Wills-St. Claire was the first car to display a backup lamp.

Side Marker and Clearance Lights

Side marker lights are installed on all vehicles sold in North America since 1969. These lamps permit the vehicle to be seen when entering a roadway from the side.

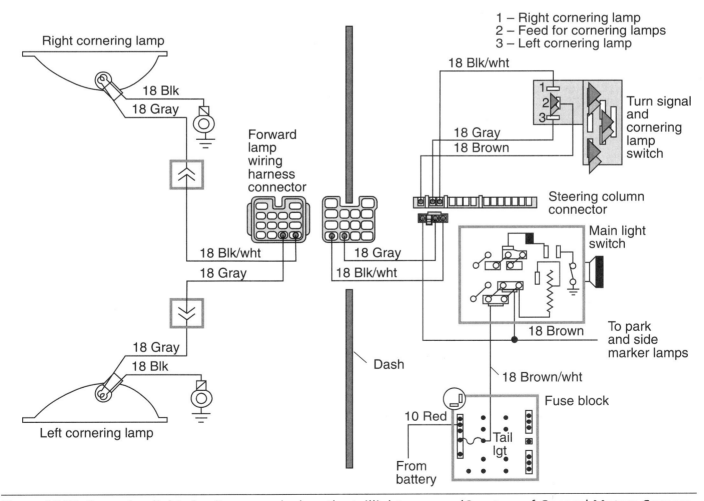

Figure 13-27 Cornering light circuit powered when the taillights are on. (Courtesy of General Motors Corporation, Service Technology Group)

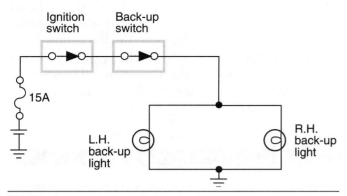

Figure 13-28 Backup light circuit.

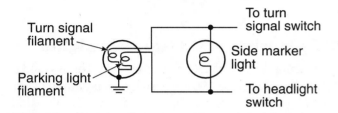

Figure 13-29 Side marker light wired across two circuits.

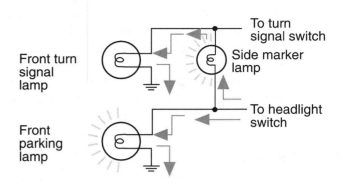

Figure 13-30 Current flow to the side marker light with the parking light on.

This also provides a means for other drivers to determine vehicle length. The front side marker light lens must be amber and the rear lens must be red. Vehicles that use wrap-around headlight and taillight assemblies also use this lens for the side marker lights. Vehicles that surpass certain length and height limits are also required to have clearance lights that face both to the front and rear of the vehicle.

The common method of wiring the side marker lights is in parallel with the parking lights. Wired in this manner, the side marker lights would only illuminate when the headlight switch is in the PARK or HEADLIGHT position.

Many vehicle manufacturers use a method of wiring in the side marker lights so that they flash when the turn signals are activated. The side marker light is wired across the parking light and turn signal light (Figure 13-29). If the parking lights are on, voltage is applied to the side marker light from the parking light circuit. Ground for the side marker light is provided through the turn signal filament. Because of the large voltage drop across the side marker lamp, the turn signal bulb will barely illuminate. In this condition the side marker light stays on constantly (Figure 13-30).

If the parking lights are off and the turn signal is activated, the side marker light receives its voltage source from the turn signal circuit. Ground for the side marker light is provided through the parking light filament. The

voltage drop over the side marker light is so high that the parking light will not illuminate. The side marker light will flash with the turn signal light.

If the turn signal is activated while the parking lights are illuminated, the side marker light will flash alternately with the turn signal light. When both the turn signal light and the parking light are on, there is equal voltage applied to both sides of the side marker light. There is no voltage potential across the bulb, so the light does not illuminate. The turn signal light turns off as a result of the flasher opening. Then the turn signal light filament provides a ground path and the side marker light comes on. The side marker light will stay on until the flasher contacts close to turn on the turn signal light again.

Interior Lights

Interior lighting includes courtesy lights, map lights, and instrument panel lights.

Courtesy Lights

Courtesy lights illuminate the vehicle interior when the doors are open. Courtesy lights operate from the headlight and door switches and receive their power source directly from a fused battery connection. The switches can be either ground switch circuit (Figure 13-31) or insulated switch circuit design (Figure 13-32). In the insulated switch circuit, the switch is used as the power relay to the lights. In the grounded switch circuit, the switch controls the grounding portion of the circuit for the lights. The courtesy lights may also be activated by the headlight switch. When the headlight switch knob is turned to the extreme counterclockwise position, the contacts in the switch close and complete the circuit.

The 1913 Spaulding touring car had such luxuries as four seats with folding backs, air mattresses, and electric reading lamps.

Instrument Cluster and Panel Lights

Consider the following three types of lighting circuits within the instrument cluster:

1. **Warning lights** alert the driver to potentially dangerous conditions such as brake failure or low oil pressure.

2. **Indicator lights** include turn signal indicators.

3. **Illumination lights** provide indirect lighting to illuminate the instrument gauges, speedometer, heater controls, clock, ashtray, radio, and other controls.

The power source for the instrument panel lights is provided through the headlight switch. The contacts are closed when the headlight switch is located in the PARK or HEADLIGHT position. The current must flow through a variable resistor (rheostat) that is either a part of the

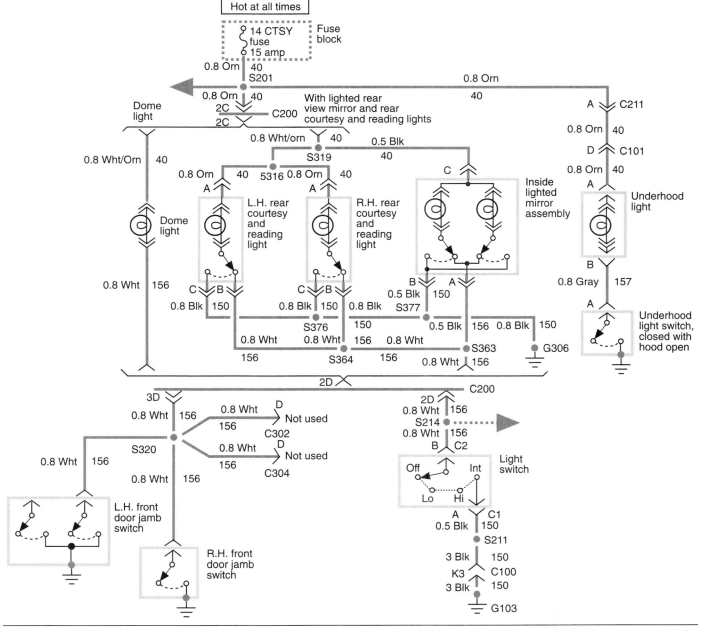

Figure 13-31 Courtesy lights using ground side switches. (Courtesy of General Motors Corporation, Service Technology Group)

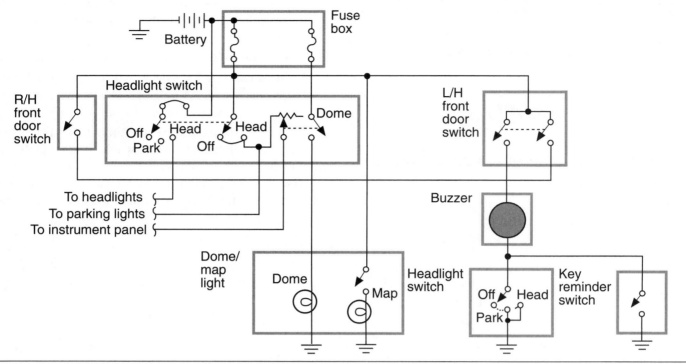

Figure 13-32 Courtesy lights using insulated side switches.

headlight switch or a separate dial on the dash. The resistance of the rheostat is varied by turning the knob. By varying the resistance, changes in the current flow to the lamps control the brightness of the lights (Figure 13-33).

Lighting System Complexity

Today's vehicles have a sophisticated lighting system and electrical interconnections. It is possible to have problems with lights and accessories that cause them to operate when they are not supposed to. This is through a

condition called **feedback**. Feedback is a condition that can occur when electricity seeks a path of lower resistance, but the alternate path operates another component than that intended. Feedback can be classified as a short. For example, if there is an open in the circuit, electricity will seek another path to follow. This may cause any lights or accessories in that path to turn on.

An example of feedback and how it may cause undesired operations are illustrated in Figure 13-34. The illustration shows a system that has the dome light, taillight, and brake light circuits on one fuse; the cigar lighter circuit has its own fuse. The two fuses are located in the main fuse block and share a common bus bar on the power side.

If the dome light fuse blew, and the headlight switch was in the PARK or HEADLIGHT position, the courtesy lights, dome light, taillight, parking lights, and instrument lights would all be very dimly lit. Current would flow through the cigar lighter fuse to the courtesy light and on to the door light switch. Current would then continue through the dome light to the headlight switch. Because the headlight switch is now closed, the instrument panel lights are also in the circuit. The lights are dim because all the bulbs are now connected in series.

If the dome light fuse is blown and the headlight switch is in the OFF position, all lights will turn off. However, if the door is opened, the courtesy lights will come on but the dome light will not.

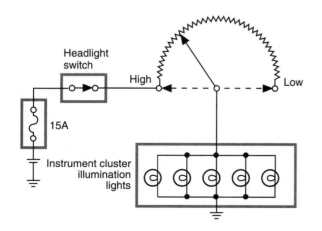

Figure 13-33 A rheostat controls the brightness of the instrument panel lights.

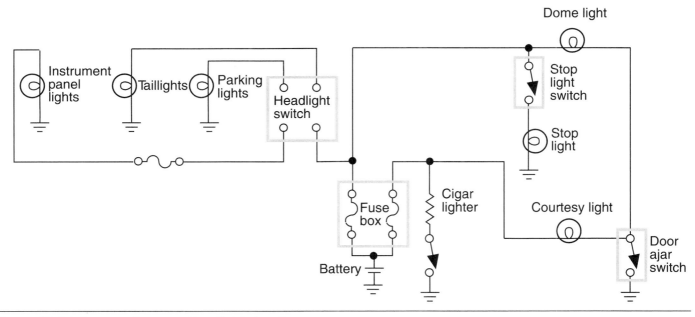

Figure 13-34 A normally operating light circuit.

With the same blown fuse and the brake light switch closed, the dome light, courtesy light, and brake light will all illuminate dimly because the loads are in series.

In this example, if the dome light and courtesy lights come on dimly when the cigar lighter is pushed in, the problem can be caused by a blown cigar lighter fuse. With the cigar lighter pushed in, a path to ground is completed. The lights and cigar lighter are now in series, thus the lights are dim and there is not enough current to heat and release the cigar lighter. If the cigar lighter was left in this position the battery would eventually drain down.

A blown cigar lighter fuse will also cause the dome light to get brighter when the doors are open, and the courtesy lights will go out. Also, if the lighter is pushed in and the brake light switch is closed, the dome and courtesy lights will go out.

Feedback can also be the result of a conductive corrosion that is developed at a connection. If the corrosion allows for current flow from one conductor to an adjacent conductor in the connection, the other circuit will also be activated. The illustration (Figure 13-35) shows how corrosion in a common connector can cause the dome light to illuminate when the wiper motor is turned on. Because the wiper motor has a greater resistance than the light bulb, more current will flow through the bulb than through the motor. The bulb will light brightly, but the motor will turn very slowly or not at all. The same effect will result if the courtesy light switch is turned on with the motor switch off.

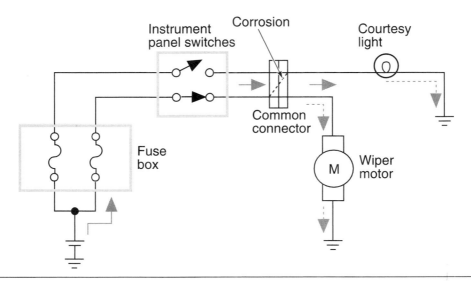

Figure 13-35 ■ A corroded common connection can cause feedback.

Summary

❏ The most commonly thought of light circuit is the headlights. But there are many lighting systems in the vehicle.

❏ Different types of lamps are used to provide illumination for the systems. The lamp may be either a single filament bulb that performs a single function, or a double filament that performs several functions.

❏ The headlight lamps can be one of four designs: standard sealed beam, halogen sealed beam, composite, or high intensity discharge (HID).

❏ The headlight filament is located on a reflector that intensifies the light, which is then directed through the lens. The lens is designed to change the circular light pattern into a broad, flat light beam. Placement of the filament in the reflector provides for low and high beam light patterns.

❏ Some manufacturers use concealed headlights to improve the aerodynamics of the vehicle. The concealed headlight doors can operate from vacuum or by electrically controlled motors. Some systems incorporate the use of IC chips into the concealed headlight door control.

❏ In addition to the headlight system, the lighting systems include:

Stop lights	Turn signals	Hazard lights	Parking lights
Taillights	Backup lights	Side marker lights	Courtesy lights

Instrument panel lights

❏ The headlight switch can be used as the control of many of these lighting systems. Most headlight switches have a circuit breaker that is an integral part of the switch. The circuit breaker provides protection of the headlight system without totally disabling the headlight operation if a circuit overload is present.

❏ A rheostat is used in conjunction with the headlight switch to control the brightness of the instrument panel illumination lights.

Terms-To-Know

Backup lights	Feedback	Indicator lights
Center high mounted stop light (CHMSL)	Filament	Lamp
	Flash to pass	Lens
Collision avoidance light	Flasher	Limit switches
Composite headlights	Halogen lamp	Neon lights
Concealed headlight system	Hazard warning system	Parabolic reflector
Cornering lights	Headlight switch	Sealed-beam
Courtesy lights	High intensity discharge (HID)	Turn signals
Dimmer switch	Illumination lights	Vacuum distribution valve
Double filament lamp	Incandescence	Warning lights

Review Questions

Short Answer Essays

1. What lighting systems are controlled by the headlight switch?

2. How is the brightness of the instrument cluster lamps controlled?

3. What is CHMSL?

4. What is the purpose of the lamp filament?

5. What three lighting circuits are incorporated within the instrument cluster?

6. List and describe the three types of headlight lamps used.

7. Describe the influence that the turn signal switch has on the operation of the brake lights in a two-bulb taillight assembly.

8. What is the purpose of the diodes on some CHMSL circuits?

9. How are most cornering light circuits wired to allow the cornering light to be on steady when the turn signal switch is activated?

10. Describe what the term feedback means and how it can affect the operation of the electrical system.

Fill-in-the-Blanks

1. When today's technician performs repairs on the lighting systems, the repairs must meet at least two requirements: They must assure vehicle _____ and meet all _____ _____.

2. A _____ is a device that produces light as a result of current flow through a _____ .

3. _____ _____ redirect the light beam and create a broad, flat beam.

4. The _____ _____ controls most of the vehicle's lighting systems.

5. The _____ _____ provides the means for the driver to select either high or low beam operation.

6. The complete headlight circuit consists of the _____ _____ , _____ _____ , _____ _____ _____ , and the _____ .

7. The headlight doors of a concealed system can be controlled by either _____ _____ or by _____ .

8. On vehicles equipped with cornering lights, the turn signal switch has an additional set of _____ that operate the cornering light circuit only.

9. Most limit switches operate off of a _____ on the motor.

10. In most automatic transmission equipped vehicles, the backup light switch is part of the _____ _____ _____ . Most manual transmissions are equipped with a _____ switch.

ASE Style Review Questions

1. Two-bulb taillight assemblies are being discussed. Technician A says the brake lights use the high intensity filament of the taillight bulb. Technician B says the current from the brake lights flows through the turn signal switch before going to the brake lights. Who is correct?
   a. A only
   b. B only
   c. Both A and B
   d. Neither A nor B

2. Composite headlights are being discussed. Technician A says composite headlights have replaceable bulbs. Technician B says a cracked or broken lens will prevent the operation of the composite headlight. Who is correct?
   a. A only
   b. B only
   c. Both A and B
   d. Neither A nor B

3. The brake light circuit is being discussed. Technician A says the brake light switch receives current from the ignition switch. Technician B says the brake light switch receives current from the headlight switch. Who is correct?
   a. A only
   b. B only
   c. Both A and B
   d. Neither A nor B

4. The turn signal circuit is being discussed. Technician A says the dimmer switch is a part of the circuit. Technician B says some flashers use a bimetallic strip to open and close the circuit. Who is correct?
   a. A only
   b. B only
   c. Both A and B
   d. Neither A nor B

5. The turn signals work properly in one direction, but in the other direction the indicator light stays on steadily. Technician A says a burned-out light bulb may be the fault. Technician B says the flasher is at fault. Who is correct?
   a. A only
   b. B only
   c. Both A and B
   d. Neither A nor B

6. The concealed headlight system is being discussed. Technician A says the system can use vacuum to operate the doors. Technician B says the system can use electric motors to operate the doors. Who is correct?
   a. A only
   b. B only
   c. Both A and B
   d. Neither A nor B

7. The CHMSL circuit is being discussed. Technician A says the diodes are used to assure proper turn signal operation. Technician B says the diodes are used to prevent radio static when the brake light is activated. Who is correct?
   a. A only
   b. B only
   c. Both A and B
   d. Neither A nor B

8. The exterior lights of a vehicle are being discussed. Technician A says the cornering lights use an additional set of contacts in the turn signal switch. Technician B says the cornering light circuit can receive its voltage from the ignition switch. Who is correct?
   a. A only
   b. B only
   c. Both A and B
   d. Neither A nor B

9. Technician A says the side marker lights can be wired to flash with the turn signals. Technician B says wrap-around lenses can be used for side marker lights. Who is correct?
   a. A only
   b. B only
   c. Both A and B
   d. Neither A nor B

10. Technician A says feedback is normal during the operation of the electrical system. Technician B says feedback is a form of a short. Who is correct?
    a. A only
    b. B only
    c. Both A and B
    d. Neither A nor B

14 Lighting Circuits Repair and Diagnosis

Objective

Upon completion and review of this chapter, you should be able to:

❏ Correctly replace sealed-beam and composite headlights.

❏ Correctly aim sealed-beam and composite headlights.

❏ Diagnose the cause of brighter than normal lights.

❏ Diagnose the cause of dimmer than normal lights.

❏ Diagnose lighting systems that do not operate.

❏ Determine causes for incorrect concealed headlight operation.

❏ Remove and replace dash- and steering column-mounted headlight switches.

❏ Replace multifunction switches.

❏ Test and determine needed repairs of the dimmer switch and related circuits.

❏ Replace the dimmer switch.

❏ Diagnose incorrect taillight assembly operation.

❏ Diagnose the turn signal system for improper operation.

❏ Replace the turn signal switch.

❏ Diagnose the interior lights, including courtesy, instrument, and panel lights.

Introduction

The lighting system of the vehicle is becoming very complex. There may be over 50 light bulbs and hundreds of feet of wiring in the lighting circuits (Figure 14-1). The circuits include circuit protectors, switches, lamps, and connectors. Any failure requires a systematic approach to diagnose, locate, and correct the fault in the minimum amount of time.

The importance of the lighting system cannot be overemphasized. The lighting system should be checked whenever the vehicle is brought into the shop for repairs. Often a customer may not be aware of a light failure. If a lighting circuit is not operating properly there is a potential danger to the driver and other people. When today's technician performs repairs on the lighting systems, the repairs must assure vehicle safety and meet all applicable laws. Be sure to use the correct lamp type and size for the application.

Before performing any lighting system tests, check the battery for state of charge. Also be sure all cable connections are clean and tight. Visually check the wires for damaged insulation, loose connections, and improper routing.

When troubleshooting the lighting system, if only one bulb is not operating it is usually faster to replace it with a known good unit first. Check the connector for signs of corrosion. When testing the circuit with a voltmeter, ohmmeter, or test light, check those components that can be easily accessed first.

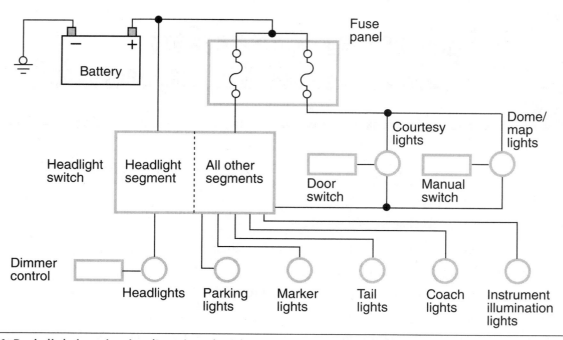

Figure 14-1 Basic lighting circuits. (Reprinted with permission of Ford Motor Company)

Headlights

The most common service performed on the headlights is lamp replacement and aiming. These procedures vary depending on the type of headlights used.

Headlight Replacement

One of the most common lighting system repairs is replacing the headlight. After a period of time the filament may burn through or the lens may be broken. Before the headlight is replaced, however, a voltmeter or test light should be used to confirm that voltage is present. Next check the ground for proper connections. If these tests are positive, the headlight is probably faulty and needs to be replaced. If there is no voltage present at the connector, work back toward the switch and battery until the fault is located.

SERVICE TIP *If it is necessary to replace the headlight lamp often, check the charging system. A too-high voltage output will cause the filament to burn hotter than designed and will shorten the life of the lamp.*

The procedure for replacing the headlight differs depending on the type of bulb used. Most conventional and halogen sealed-beam headlights are replaced in the same manner. Composites require different procedures. Always refer to the service manual for the vehicle you are working on.

WARNING *Because of the construction and placement of the prisms in the lens, it is important that the headlight is installed in its proper position. The lens is usually marked "TOP" to indicate the proper installed position.*

Sealed Beam

To replace a sealed-beam headlight:

1. Place fender covers around the work area.
2. This type of replacement usually requires the removal of the **bezel** (Figure 14-2). The bezel is the retaining trim around a component.

Figure 14-2 Remove the bezel to gain access to the headlight.

3. Remove the retaining ring screws and the retaining trim (Figure 14-3). Do not turn the two headlight aiming adjustment screws (Figure 14-4).

4. Remove the headlight from the shell assembly.

5. Disconnect the wire connector from the back of the lamp.

6. Check the wire connector for corrosion or other foreign material. Clean as needed.

7. Coat the connector terminals and the prongs of the new headlight with dielectric grease to prevent corrosion.

8. Install the wire connector onto the headlight prongs and place the headlight into the shell assembly. When positioning the headlight, be sure the embossed number is at the top.

9. Install the retainer trim and fasteners.

10. Check operation of the headlight.

11. Check headlight aiming as described in the next section.

12. Install the headlight bezel.

WARNING *It is not recommended that halogen sealed-beam and standard sealed-beam headlights be mixed on the vehicle. Also, if the vehicle was equipped with halogen headlights as original equipment, do not replace the headlights with standard sealed beams. Doing so may result in poor light quality.*

CUSTOMER CARE *Because the filament of the halogen lamp is contained in its own bulb, cracking or breaking of the lens does not prevent headlight operation. The filament will continue to operate as long as the filament envelope has not been broken. However, a broken lens will result in poor light quality and should be replaced for the safety of the customer.*

Composite Headlights

To replace a composite headlight:

1. Place fender covers around the work area.

2. Remove the wire connector from the bulb.

3. Unlock the bulb retaining ring by rotating it 1/8 of a turn (Figure 14-5).

4. Slide off the retaining ring from the base.

5. Gently pull the bulb straight back out of the socket. To prevent breaking the bulb and locating tabs, do not rotate it while pulling.

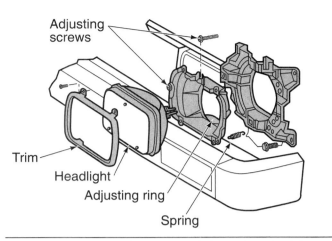

Figure 14-3 Headlight removal. (Reprinted with permission of Ford Motor Company)

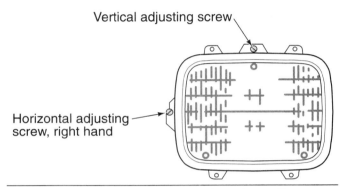

Figure 14-4 Typical locations for the adjustment screws.

6. Check the wire connector for corrosion or other foreign material. Clean as needed.

7. Coat the connector terminals and the prongs of the new headlight with dielectric grease to prevent corrosion.

WARNING *Do not get any of the dielectric grease onto the bulb. The bulb's life will be shortened.*

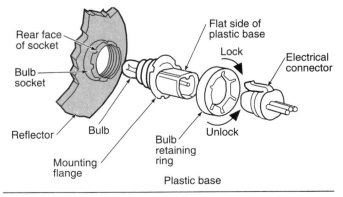

Figure 14-5 Composite headlight bulb replacement. (Reprinted with permission of Ford Motor Company)

8. Place the new bulb in the socket. The flat part of the base should face up. It may be necessary to turn the bulb slightly to align the locating tabs.

WARNING *Take care not to touch the bulb envelope with the fingers. Staining the bulb with normal skin oil can substantially shorten the life of the bulb. Handle the lamp only by its base. Also dispose of the old lamp properly.*

9. The mounting flange on the bulb base should contact the rear of the socket.
10. Install the retaining ring over the base and lock the ring into the socket.
11. Reconnect the wire connector to the bulb. It will snap and lock into position when properly installed.
12. Check headlight operation.

WARNING *To prevent early failure, do not energize the bulb unless it is installed into the socket.*

13. Check and adjust headlight aiming as needed.

Headlight Aiming

The headlights should be checked for proper aiming whenever the lamps are replaced. Proper aiming is important for good light projection onto the road and to prevent discomfort and dangerous conditions for oncoming drivers.

Correct headlight beam position is so critical that government regulations control limits for headlight aiming. For example, a headlight that is mis-aimed by one degree downward will reduce the vision distance by 156 feet. The following are maximum allowable limits that have been established by all states:

1. Low beam: In the horizontal plane, the left edge of the headlight high intensity area should be within 4 inches (102 mm) to the right or left of the vertical centerline of the lamp. In the vertical plane, the top edges of the headlight high intensity area should be within 4 inches (102 mm) above or below the horizontal centerline of the lamp (Figure 14-6).

2. High beam: In the horizontal plane, the center of the headlight high intensity area should be within 4 inches (102 mm) to the right or left of the vertical centerline of the lamp. In the vertical plane, the center of the headlight high intensity area should be within 4 inches (102 mm) above or below the horizontal centerline of the lamp (Figure 14-7).

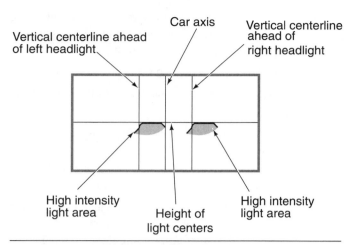

Figure 14-6 Low beam headlight aiming adjustment pattern.

Before the headlights are aimed, the vehicle must be checked for proper **curb height**. Curb height is the height of the vehicle when it has no passengers or loads, and normal fluid levels and tire pressure. Curb height checks include checking the springs, tire inflation, removing any additional load in the vehicle, a half-filled fuel tank, and the removal of dirt, ice, snow, and so on, from the vehicle.

Most shops use portable mechanical aiming units (Figure 14-8). These are secured to the headlight lens by suction cups (Figure 14-9). The aiming unit should have a variety of adapters to attach to the various styles of headlights. Before using the aiming equipment, be sure to follow the manufacturer's procedure for calibration.

To aim sealed-beam headlights, begin by placing the vehicle on a level floor area. Place fender covers around the work area. It may be necessary to remove the trim and bezel from around the headlight. Using the correct adapter, connect the calibrated aimer units to the head-

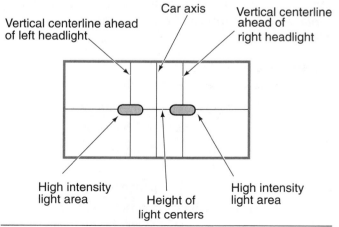

Figure 14-7 High beam headlight aiming adjustment pattern.

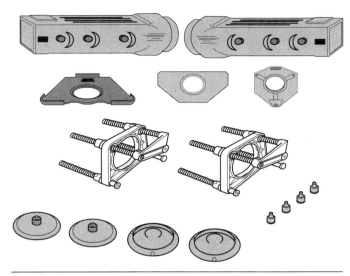

Figure 14-8 Typical portable mechanical headlight aiming equipment and adapters. (Courtesy of General Motors Corporation, Service Technology Group)

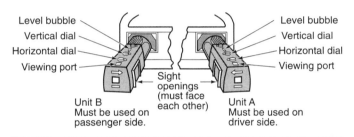

Figure 14-9 The aiming units attach to the headlight lens with suction cups. (Courtesy of Chrysler Corporation)

lights. Be sure the adapters fit the headlight aiming pads on the lens (Figure 14-10). Zero the horizontal adjustment dial. Confirm that the split image target lines are visible in the view port (Figure 14-11). If the target lines are not seen, rotate the aimer unit. Turn the headlight horizontal adjusting screw until the split image target lines are aligned. Repeat for the headlight on the other side.

To set the vertical aim of the headlight, turn the vertical adjustment dial on the aiming unit to zero. Turn the

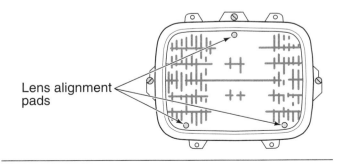

Figure 14-10 Headlight aiming pads.

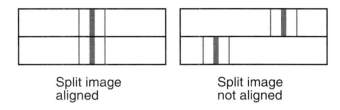

Figure 14-11 Split image target. (Courtesy of Chrysler Corporation)

vertical adjustment screw until the spirit level bubble is centered (Figure 14-12). Recheck the horizontal aiming on each headlight. The vertical adjustment may have altered the original adjustments.

If the vehicle is equipped with a four-headlight system, repeat the procedures for the other pair of lamps.

To adjust many composite and HID headlight designs, special adapters are required. Also, the lens must have headlight aiming pads to be able to use a mechanical aiming unit. The headlight assembly will have a number molded on it. The adjustment rod setting must be set to that number and locked in place. The aiming unit is attached to the headlight lens in the same manner as with sealed-beam headlights (Figure 14-13).

The adjustment procedure is identical to that of the sealed-beam headlights. The illustration (Figure 14-14) shows the typical location of the headlight adjusting screws.

Some composite headlight and HID designs do not have alignment pads on the lens. Usually these systems do not adjust the beam location by moving the lens. Rather the reflector position is changed. Since the lens does not move, conventional headlight aimers are not used. Usually these systems are aimed by locating the vehicle 25 feet away from a blank wall, with the vehicle on a level surface. The wall is marked based on the centerline of the vehicle then the location of the beam on the wall is adjusted to meet manufacturer's specifications. Some manufacturers may require that the headlights be adjusted with the high beams.

Many late-model vehicles have spirit levels built into the headlamp assembly (Figure 14-15). These are not always to be used for initial headlamp adjustment. Most

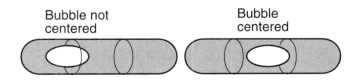

Figure 14-12 Center the spirit level by turning the vertical aiming screw. (Courtesy of Chrysler Corporation)

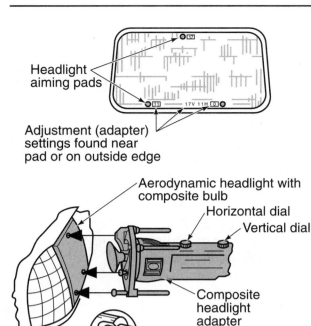

Figure 14-13 Connect the aiming equipment to the headlight lens. The lens must have aiming pads. (Courtesy of General Motors Corporation, Service Technology Group)

of these are supplied for the driver to be able to adjust their headlights based on the load in the vehicle. For example, if the trunk is loaded the front of the vehicle is lifted and the light beam is too high. By turning the adjuster wheels until the bubble is in the middle of the level, the beam is returned to its original position. After the load is removed from the trunk, the headlights are adjusted until the bubble is returned to the middle again. Whenever the headlights are adjusted, the technician should adjust the spirit level also.

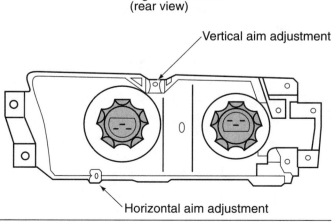

Figure 14-14 Composite headlight aiming screw locations. (Courtesy of General Motors Corporation, Service Technology Group)

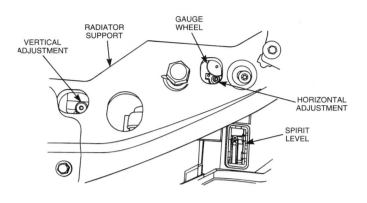

Figure 14-15 Some headlight assemblies are equipped with spirit levels to provide easy adjustments to offset vehicle loads. (Courtesy of Chrysler Corporation)

Dimmer or Brighter than Normal Lights

The complete headlight circuit consists of the headlight switch, dimmer switch, high beam indicator, and the headlights (Figure 14-16). Excessive resistance in these units, or at their connections, can result in lower illumination levels of the headlights.

The extra resistance can be on the insulated side or the ground side of the circuit. To locate the excessive resistance, perform a voltage drop test (Figure 14-17). Consult the wiring diagram to determine the number of connectors and switches. This will provide you with the specification for maximum voltage drop. Start at the light and work toward the battery.

All headlight systems are wired in parallel to each other. If both headlights are dim, then the excessive resistance is in the common portions of the circuit. Dim headlights can also be the result of low generator output.

Other causes of dim lights can be the use of the wrong lamps, improper circuit routing, the addition of extra electrical loads to the circuit, and the wrong size conductors.

SERVICE TIP *Headlights do not wear out and get dimmer with age. If one of the headlights is dimmer than the other, there is excessive voltage drop in that circuit. If a new headlight is installed, the breaking and making of the socket connection may clean the terminals enough to make a good contact. Once the new headlight is installed, it may operate properly. Do not be fooled. It was not the headlight that was at fault. It was the connection.*

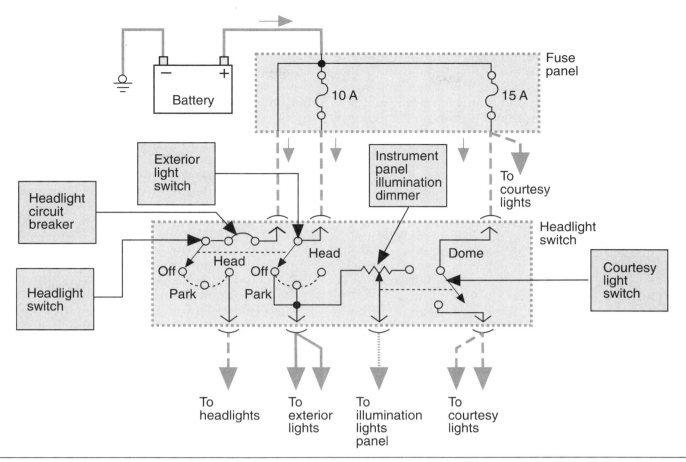

Figure 14-16 Headlight circuit components. (Reprinted with permission of Ford Motor Company)

Brighter than normal lights can be the result of higher than specified generator output or improper lamp application. It is also possible that the dimmer switch contacts are "welded" into the high beam position.

Concealed Headlights

A vehicle equipped with a concealed headlight system can use either electric or vacuum motors to operate the headlight doors. Electrically controlled systems use either a torsion bar to open both headlight doors from a single motor or a separate motor for each headlight door.

If the electrically operated doors fail to operate, check the fuse first. If the fuse is good, check for voltage at the connection to the motor. If voltage is not present, then trace the circuit back to the switch and battery.

If there was voltage at the connector, check the ground circuit. Also check the operation of the limit switches before condemning the motor.

Vacuum-controlled doors are tested for the presence of vacuum through the distribution valve to the vacuum motors. Use a vacuum gauge to check the amount of vacuum being applied to the motor. If vacuum is present to

the motor, use a vacuum pump to check the operation of the motor.

CAUTION *Be careful not to get hands and fingers caught in the door if it should suddenly open.*

All concealed headlight systems have a means of manually opening the doors. Check the service manual for the correct procedure for opening the doors. Most electrical doors have a knob that is rotated to open the doors.

CAUTION *Be sure the headlight switch is in the OFF position before manually opening the doors. The doors may snap open, catching fingers between the door and the vehicle body.*

SERVICE TIP *Most vacuum-controlled doors use the vacuum to close the door. If the headlight doors open while the vehicle is sitting overnight, check for a leak and test the one-way check valve.*

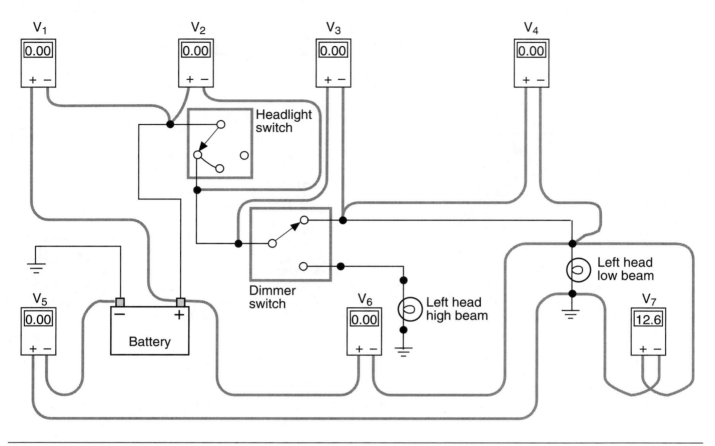

Figure 14-17 Voltage drop testing the headlight system.

Headlight Switch Testing and Replacement

The headlight switch controls most of the vehicle's lighting systems. The headlight switch will generally receive direct battery voltage to two of its terminals. Disconnect the battery before removing the headlight switch.

In the headlight switch circuit a rheostat is used to control the instrument cluster illumination lamp brightness. Most dash-mounted headlight switches incorporate the rheostat into the switch assembly. Steering column-mounted switches may have the rheostat located on the dash.

Many customer complaints concerning the lighting systems can be the result of a faulty headlight switch. For example, dim or no instrument panel lights, dim or no headlights, dim or no parking lights, and improperly operating dome lights can all be caused by the headlight switch.

SERVICE TIP *Headlights that flash on and off as the vehicle goes over road irregularities indicate a loose connection. Headlights that flash on and off at a constant rate indicate that the circuit breaker is being tripped. There is an overload in the circuit that must be traced and repaired.*

Dash-Mounted Switches

There are many methods used to retain the headlight switch to the dash. Consult the service manual of the vehicle you are working on. The following is one of the most common methods of removing the headlight switch:

1. Place fender covers on the fenders.
2. Disconnect the battery ground cable.
3. Remove any vent control cables, ducts, trim panels, or other components that may interfere with switch removal.
4. Pull the headlight switch knob all the way out to the ON position.

5. Reach under the dash to the backside of the switch and depress the knob release button (Figure 14-18). Keep the button depressed and pull the knob out of the switch.

6. Remove the headlight switch retaining bezel.

7. Lower the switch far enough to disconnect the connector plug and remove the switch.

With the switch removed it can be tested for continuity and the connector plug will serve as a test point for the lighting circuits. First test at the connector.

In this procedure the terminals are identified as follows:

- Terminal B= battery
- Terminal A=fuse
- Terminal H=headlights
- Terminal R=rear park and side marker lights
- Terminal I=instrument panel lights
- Terminal D1=dome light feed
- Terminal D2=dome light

Consult the service manual for terminal identification for the vehicle you are working on. If this is not listed in a separate chart, you should be able to identify the circuits from the wiring diagrams.

1. Connect the 12-volt test light across terminal B and ground (Figure 14-19). The test light should light. If not, there is an open in the circuit back to the battery.

2. Connect the test light across terminal A and ground. The test light should come on. If not, repair the circuit back to the fuse panel.

3. Connect a jumper wire between terminals B and H. The headlights should come on. If the headlights fail to turn on, trace the H circuit to the headlights. Also check the ground circuit side from the headlights.

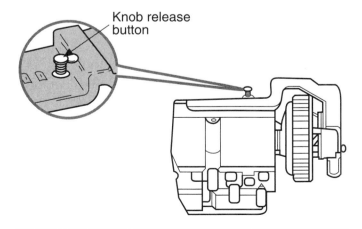

Figure 14-18 Pushing the release button will allow the knob to be removed. (Reprinted with permission of Ford Motor Company)

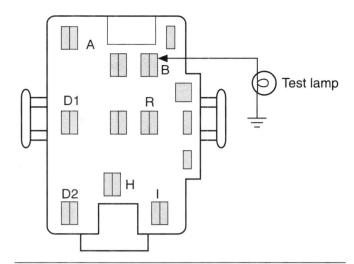

Figure 14-19 Testing for battery voltage to the connector. (Reprinted with permission of Ford Motor Company)

4. Connect a jumper wire between terminals A and R. The rear lamps should illuminate. If not, trace the circuit to the rear lights. Also check the ground return path.

5. Connect a jumper wire between terminals A and I. The instrument panel lights should come on. If not, trace the circuit to the panel lights.

The chart (Figure 14-20) indicates the test results that should be obtained when testing the headlight switch for continuity. Use an ohmmeter or a self-powered test light to test the switch.

Steering Column-Mounted Switches

On some vehicles it is possible to test the steering column-mounted switch without removing it. The test is conducted at the connector at the base of the column (Figure 14-21). However, on some models it is necessary to remove the column cover and/or the steering wheel to gain access. A common procedure for replacing the multifunction switch is shown in Photo Sequence 7.

Dimmer Switch Testing and Replacement

The dimmer switch is connected in series within the headlight circuit and controls the current path for high and low beams. The dimmer switch is located either on the floor board next to the left kick panel, or on the steering column. Testing of the switch is done by using a set of jumper wires to bypass the switch (Figure 14-22). If the headlights operate with the switch bypassed, the

Note:
A self-powered test lamp or ohmmeter will be required.
Terminal identification used on test procedure corresponds
with actual identification on headlamp switch.

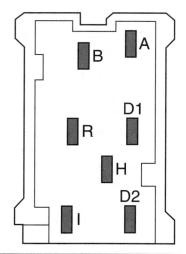

Switch terminals	Switch positions		
	Off	Park	Headlamp
B to H	No continuity	No continuity	Continuity
B to R	No continuity	No continuity	No continuity
B to A	No continuity	No continuity	No continuity
R to H	No continuity	No continuity	No continuity
R to A	No continuity	Continuity	Continuity
H to A	No continuity	No continuity	No continuity
D to D	Continuity should only exist with rheostat in full counterclockwise position		
I to R	Continuity should be measured with rheostat in full counterclockwise position. Then slowly rotate rheostat clockwise and test lamp should slowly dim.		

Figure 14-20 Continuity test chart for the headlight switch. (Reprinted with permission of Ford Motor Company)

switch is faulty. This is a common problem with older vehicles that have the dimmer switch on the floor board because the switch is subjected to damage due to rust, dirt, and so on.

 CAUTION *Remove the battery negative cable before replacing the dimmer switch.*

Floor-Mounted Switches

Removal and replacement of the floor-mounted dimmer switch is done by first pulling back on the floor mat to expose the switch. Disconnect the wire plug, and remove the hold-down fasteners. Install the new dimmer switch and relocate the mat so it does not interfere with switch operation.

Steering Column-Mounted Switches

The steering column-mounted dimmer switch can be operated by an actuator control rod from the lever to a remotely mounted switch (Figure 14-23). Another style incorporates the dimmer switch into the multifunction switch.

To remove the remote switch, first place fender covers on the fenders of the vehicle and disconnect the battery negative cable. Disconnect the wire connector at the switch. Remove the two switch mounting screws and disengage the switch from the actuator rod.

When installing the switch, make sure the actuator rod is firmly seated into the switch. During the installa-

tion, adjust the position of the switch so that all actuator rod slack is taken up. If the switch has alignment holes, compress the switch until two appropriately sized dowels can be inserted into the alignment holes. While applying a slight rearward pressure, install and tighten the mounting bolts.

When the switch is adjusted properly it will click when the lever is lifted and again when it is returned to its downward position. The second click should occur just before the stop.

If the dimmer switch is a part of the multifunction switch, follow the general procedure shown in Photo Sequence 7.

To reinstall the multiswitch, reverse the procedure. Torque the steering column attaching nuts to the amount specified in the service manual.

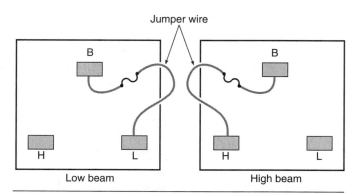

Figure 14-22 Jumper wire connections to bypass the dimmer switch.

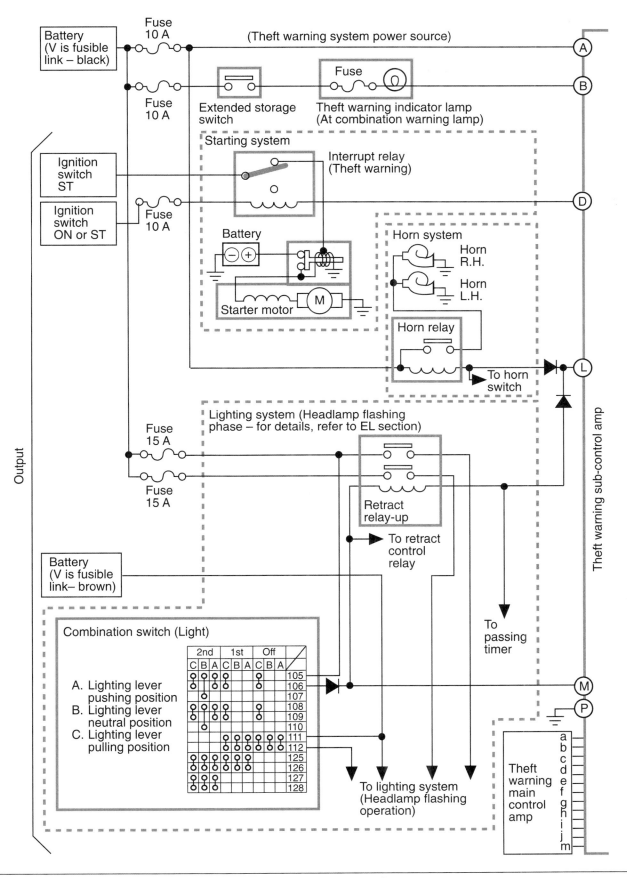

Figure 14-21 Using the connector to test the multifunction switch.

PHOTO SEQUENCE

6 # Removal of the Multifunction Switch

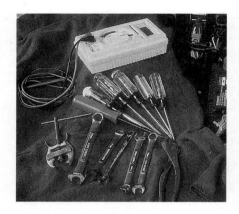

P6-1 Tools required to test and remove the multifunction switch: fender covers, battery terminal pliers, terminal pullers assorted combination wrenches, torx drivers, and ohmmeter.

P6-2 Place the fender covers around the battery work area. Loosen the negative battery clamp bolt and remove the clamp using terminal pullers. Place the battery cable where it cannot contact the battery.

P6-3 Remove the shroud retaining screws and remove the lower shroud from the column.

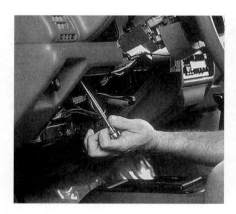

P6-4 Loosen the steering column attaching nuts. Do not remove the nuts.

P6-5 Lower the steering column enough to remove the upper shroud.

P6-6 Remove the turn signal lever by slightly rotating the outer end of the lever then pulling straight out on the lever.

Removal of the Multifunction Switch

P6-7 Peel back the foam shield from the turn signal switch then disconnect the turn signal switch electrical connectors.

P6-8 Remove the screws attaching the turn signal switch to the lock cylinder assembly and disengage the switch from the lock assembly.

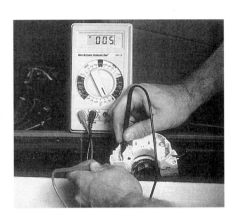

P6-9 Use an ohmmeter to test the switch.

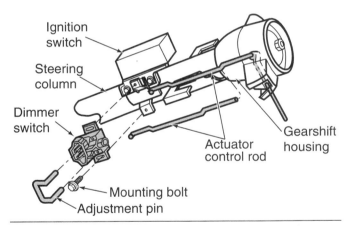

Figure 14-23 Steering column-mounted dimmer switch. (Reprinted with permission of Ford Motor Company)

Taillight Assemblies

In a three-bulb taillight system, the brake lights are controlled directly by the brake light switch (Figure 14-24). The brake lights on both sides of the vehicle are wired in parallel. Most brake light systems use dual filament bulbs that perform multifunctions. In this type of circuit, the brake lights are wired through the turn signal and hazard switches (Figure 14-25).

If all of the taillights do not operate, check the condition of the fuse. If it is good, use a voltmeter to test the circuit. With the headlight switch in the PARK position (first detent), check for voltage at the last common connection between the switch and the lamps. If battery voltage is present, then the problem is in the individual circuits from that connector to the lamps. If no battery voltage is present, test for voltage from the switch terminal. If no voltage is present at this terminal, yet the headlights operate when in the ON position, replace the switch. If battery voltage is present, the problem is between the switch and the last common connection. If there was no voltage present at the switch terminal, check for battery voltage into the switch.

Most taillight bulbs can be replaced without removal of the lens assembly. The bulb and socket are removed by twisting the socket slightly and pulling it out of the lens assembly (Figure 14-26). To remove the bulb from

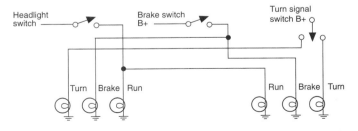

Figure 14-24 Three-bulb taillight circuit has individual control for each bulb.

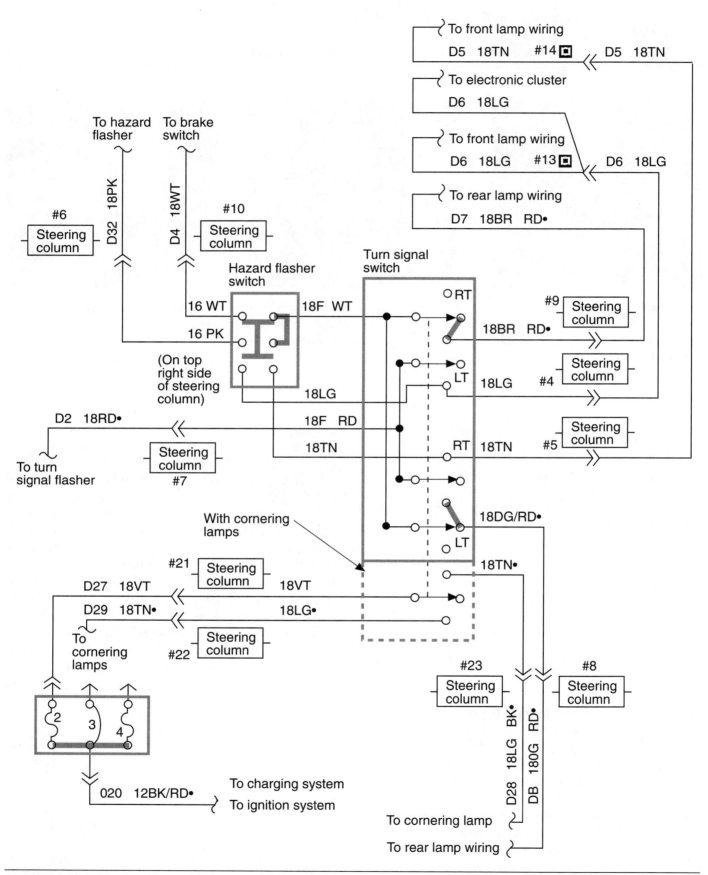

Figure 14-25 Turn signal switch used in a two-bulb taillight circuit (Courtesy of Chrysler Corporation)

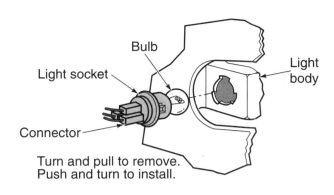

Figure 14-26 Bulb and socket removal from the tail-light lens assembly. (Courtesy of Nissan Corporation)

the socket, push in on the bulb slightly while turning it. When the lugs align with the channels of the socket, pull the bulb from the socket (Figure 14-27).

The illustration shows how the lens assembly is fastened to the vehicle body. Remove the attachment nuts from the back of the assembly to remove it.

WARNING *Using the wrong type of lamp for the socket and application can result in "crazy lights." This is a result of feedback caused by the incorrect bulb.*

Turn Signals and Brake Lights

To test the turn signal switches, use a 12-volt test light to probe for voltage into and out of the switch. The ignition switch must be in the RUN position for the circuit to operate. If voltage is present on the input side of the switch but not on the output side, the switch is faulty.

Check brake light operation through the turn signal switch in the same manner. Also, check the brake light switch for proper adjustment and operations.

Many vehicles use a turn signal switch that is separate from the multifunction switch. The steering wheel will have to be removed to gain access to the turn signal switch on these vehicles. The following procedure is a common method of turn signal switch replacement:

1. Place the fender covers on the fenders.

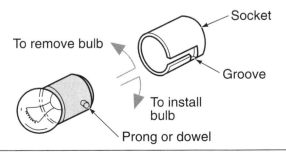

Figure 14-27 Removing the bulb from the socket. (Courtesy of Nissan Corporation)

2. Disconnect the battery negative cable.
3. Remove the steering column trim.
4. Remove the horn pad from the steering wheel (Figure 14-28).
5. Remove the steering shaft nut and horn collar (if equipped).
6. Use a suitable puller to remove the steering wheel.
7. With a suitable compressor, compress the preload spring to the lock plate (Figure 14-29). Compress the spring only enough to remove the snap ring.
8. Use a pick and a small flat blade screwdriver to remove the snap ring.

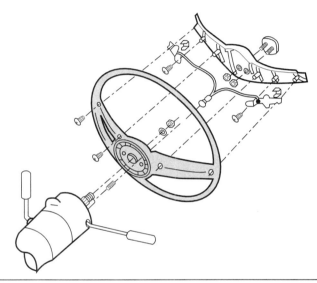

Figure 14-28 Steering wheel attachment. (Courtesy of General Motors Corporation, Service Technology Group)

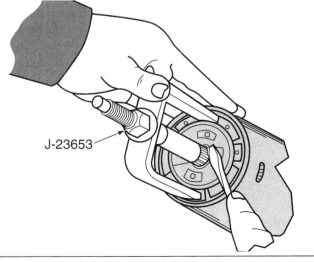

Figure 14-29 To remove the snap ring, use the compressing tool to relieve the pressure against the snap ring. (Courtesy of General Motors Corporation, Service Technology Group)

9. Remove the lock plate, horn contact carrier, and spring.

10. Remove the bolts at the upper steering column support and the upper mounting bracket from the column.

11. Disconnect the turn signal wiring connector.

12. Wrap tape around the wire and connector.

13. Remove the hazard warning knob from the column.

14. Remove the switch retaining screws and remove the switch (Figure 14-30).

On vehicles equipped with cornering lights, the turn signal switch has an additional set of contacts that operate the cornering light circuit only. The contacts can receive voltage from either the ignition switch or the headlight switch. If the ignition switch provides the power, the cornering lights will be activated any time the turn signals are used. If the contacts receive the voltage from the headlight switch, the cornering lights do not operate unless the headlight switch is in the PARK or HEADLIGHT position.

The most common complaints attributable to the flasher are too fast or too slow of a flashing rate. If the flasher is of the wrong type and rating, the amount of time required to heat the coil will differ from what the manufacturer designed into the circuit. Also, newer flashers that use electronic circuits will flash at an increased speed if one of the turn signal bulbs is burned out or the circuit is defective. If the flasher is rated higher than required, the flashing rate is reduced because it takes longer for the current to heat the coil.

Check the size and type of light bulbs in the circuit. Use only the lamp size recommended by the manufacturer. If these checks do not correct the problem, test the

generator output. Higher or lower voltage output than specified may cause the flasher rate to be incorrect. If the charging system output is within specifications, check for excessive resistance in the turn signal circuit. Check both sides of the circuit.

If none of the turn signals operate, check the fuse. Next check the flasher. Remove the flasher from the fuse box. Connect a jumper wire across the fuse box terminals (Figure 14-31). If the turn signal lamp operates with the lever in either indicator position, the flasher is faulty. If the lights still do not illuminate, test the turn signal switch.

SERVICE TIP *If the turn signals operate properly in one direction but do not flash in the other, the problem is not in the flasher unit. A burned-out lamp filament will not cause enough current to flow to heat the bimetallic strip sufficiently to cause it to open. Thus the lights do not flash. Locate the faulty bulb and replace it.*

Interior Lights

Interior lighting includes courtesy lights, map lights, and instrument panel lights.

Courtesy Lights

Figure 14-32 provides a systematic approach to troubleshooting courtesy lights. Follow the steps in order to locate the fault.

If all the lights of the circuit do not light, begin by checking the fuses. If the fuse is good, then use a voltmeter to check for battery voltage at the last common connection. If voltage is present at the fuse box but not at the common connection, the problem is between these two points. If battery voltage is present at the common connection, trace the individual circuits until the cause(s) for the open is located.

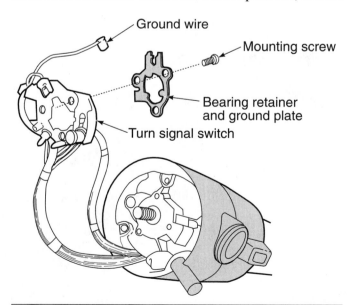

Figure 14-30 Remove the turn signal switch from the column. (Courtesy of Chrysler Corporation)

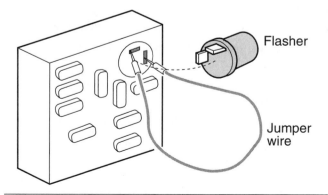

Figure 14-31 Connecting jumper wires across the terminals to bypass the flasher unit.

COURTESY LAMPS DO NOT TURN ON WHEN ONE DOOR
IS OPENED—OK WHEN OTHER DOORS ARE OPEN

	TEST STEP	RESULT ►	ACTION TO TAKE
A0	VERIFY THE CONDITION		
		►	GO TO **A1**.
A1	CHECK POWER		
	• Check for power at door switch.	► ~~OK~~	SERVICE the power circuit back to the fuse.
		► OK	GO to **A2**.
A2	CHECK THE DOOR SWITCH		
	• Check the door switch for proper operation.	► ~~OK~~	REPLACE the switch.
		► OK	SERVICE the circuit from the switch to the lamp(s).

CK5998-A

COURTESY LAMP(S) DOES NOT COME ON WHEN
HEADLAMP SWITCH IS TURNED COUNTERCLOCKWISE TO STOP

	TEST STEP	RESULT ►	ACTION TO TAKE
B0	VERIFY CONDITION		
		►	GO TO **B1**.
B1	CHECK OPERATION OF DOOR SWITCHES		
	• Check to see if courtesy lamps operate from door switches.	► ~~OK~~	GO to Diagnostic Chart **C**—Courtesy Lamp(s) does not come on when all doors are open
		► OK	GO to **B2**.
B2	CHECK FOR POWER		
	• Check for power at headlamp switch.	► ~~OK~~	SERVICE circuits back to fuse panel.
		► OK	GO to **B3**.
B3	CHECK FOR CONTINUITY		
	• Check continuity of headlamp switch.	► ~~OK~~	REPLACE headlamp switch.
		► OK	SERVICE circuit from switch to lamps.

CK5999-C

Figure 14-32 Courtesy light diagnostic procedures. (Reprinted with permission of Ford Motor Company)

Instrument Cluster and Panel Lights

The power source for the instrument panel lights is provided through the headlight switch. The contacts are closed when the headlight switch is located in the PARK or HEADLIGHT position. The current must flow through a variable resistor (rheostat) that is either a part of the headlight switch or a separate dial on the dash. The resistance of the rheostat is varied by turning the knob. By varying the resistance, changes in the current flow to the lamps control the brightness of the lights.

Test for voltage output from the headlight switch to determine if the switch is operating properly. Vary the amount of resistance in the rheostat while observing the voltmeter. The voltage reading from the rheostat should vary as the knob is turned. If voltage is present to the printed circuit, check the ground.

If all connections are good, remove the dash and test the printed circuit board. Use an ohmmeter to check for opens and shorts in the printed circuit board from the connector plug to the lamp sockets. If the printed circuit is bad, it must be replaced. There are no repairs to the board.

WARNING *Be careful when testing the printed circuit. Do not touch the circuit paths with your fingers. Do not scratch the lamination with the test probes. Doing so may destroy a good circuit board.*

Summary

❑ When troubleshooting the lighting system, if only one bulb is not operating it is usually faster to replace it with a known good unit first.

❑ The most common service performed on the headlights is lamp replacement and aiming.

❑ Before the headlight is replaced a voltmeter or test light should be used to confirm that voltage is present. Next check the ground for proper connections.

❑ The headlights should be checked for proper aiming whenever the lamps are replaced.

❑ Correct headlight beam position is so critical that government regulations control limits for headlight aiming.

❑ Some composite headlight and HID designs do not have alignment pads on the lens. Usually these systems do not adjust the beam location by moving the lens. Rather the reflector position is changed.

❑ Excessive resistance in any component of the headlight circuit, or at their connections, can result in lower illumination levels of the headlights.

❑ All headlight systems are wired in parallel to each other. If both headlights are dim, then the excessive resistance is in the common portions of the circuit.

❑ Dim headlights can also be the result of low generator output.

❑ Brighter than normal lights can be the result of higher than specified generator output or improper lamp application. It is also possible that the dimmer switch contacts are "welded" into the high beam position.

❑ Many customer complaints concerning the lighting systems can be the result of a faulty headlight switch.

❑ On some vehicles it is possible to test the steering column-mounted switch without removing it.

❑ Most taillight bulbs can be replaced without removal of the lens assembly. The bulb and socket are removed by twisting the socket slightly and pulling it out of the lens assembly. To remove the bulb from the socket, push in on the bulb slightly while turning it. When the lugs align with the channels of the socket, pull the bulb from the socket.

❑ To test the turn signal switches, use a 12-volt test light to probe for voltage into and out of the switch. The ignition switch must be in the RUN position for the circuit to operate. If voltage is present on the input side of the switch but not on the output side, the switch is faulty.

❑ The most common complaints attributable to the flasher are too fast or too slow of a flashing rate.

❑ The power source for the instrument panel lights is provided through the headlight switch. Test for voltage output from the headlight switch to determine if the switch is operating properly. In addition, vary the amount of resistance in the rheostat while observing the voltmeter.

Terms-To-Know

Bezel   Curb height

ASE Style Review Questions

1. The dimmer switch is being discussed. Technician A says that all dimmer switches are mounted on the floor board. Technician B says the dimmer switch can be incorporated into the multifunction switch. Who is correct?
   a. A only
   b. B only
   c. Both A and B
   d. Neither A nor B

2. The courtesy lights of a vehicle equipped with insulated door plunger switches are remaining "on" after all of the doors are closed. Technician A says one of the switches may be shorted to ground. Technician B says one of the switches is electrically open. Who is correct?
   a. A only
   b. B only
   c. Both A and B
   d. Neither A nor B

3. A customer is concerned that the headlights are brighter than normal and that she is having to replace the lamps regularly. Technician A says this can be caused by too high charging system output. Technician B says this can be caused by excessive resistance in the circuit. Who is correct?
   a. A only
   b. B only
   c. Both A and B
   d. Neither A nor B

4. The left turn signals of a vehicle flash very slowly; the right turn signals operate properly. Technician A says this may be caused by excessive circuit resistance in the left turn signal circuit. Technician B says this may be caused by someone installing higher than specified wattage rated bulbs into the left turn signal circuit. Who is correct?
   a. A only
   b. B only
   c. Both A and B
   d. Neither A nor B

5. The taillight assembly is being discussed. Technician A says all bulbs used in taillight assemblies are signal filament. Technician B says if all of the taillights do not operate, probe for voltage at a common connection. Who is correct?
   a. A only
   b. B only
   c. Both A and B
   d. Neither A nor B

6. The turn signals of a vehicle operate in the left direction only. Technician A says the flasher is bad. Technician B says the fuse is blown. Who is correct?
   a. A only
   b. B only
   c. Both A and B
   d. Neither A nor B

7. Diagnosis of the instrument panel lights is being discussed. Technician A says the power source for the instrument panel lights is provided through the headlight switch and rheostat. Technician B says the printed circuit board is repairable. Who is correct?
   a. A only
   b. B only
   c. Both A and B
   d. Neither A nor B

8. Composite headlight bulb replacement is being discussed. Technician A says care must be taken not to touch the envelope with your fingers. Technician B says not to energize the bulb unless it is installed into the socket. Who is correct?
   a. A only
   b. B only
   c. Both A and B
   d. Neither A nor B

9. Technician A says that the dimmer switch is connected in parallel to the headlight circuit. Technician B says that concealed headlight doors can not be manually opened. Who is correct?
   a. A only
   b. B only
   c. Both A and B
   d. Neither A nor B

10. Technician A says if only one lamp in the circuit is not operating, the fastest check is to replace the bulb with a known good one. Technician B says to start the diagnostic tests at the easiest location to access. Who is correct?
    a. A only
    b. B only
    c. Both A and B
    d. Neither A nor B

15 Conventional Analog Instrumentation, Indicator Lights, and Warning Devices

Objective

Upon completion and review of this chapter, you should be able to:

❏ Describe the operation of mechanical speedometers and odometers.

❏ Describe the function and operation of the tachometer.

❏ Explain the purpose of an instrument voltage regulator.

❏ Describe the operation of bimetallic gauges.

❏ Describe the operation of electromagnetic gauges, including D'Arsonval, three-coil, two-coil, and air core.

❏ Explain the function and operation of the various gauge sending units, including thermistors, piezoresistive, and mechanical variable resistors.

❏ Explain the operation of various warning lamp circuits.

❏ Explain the operation of various audible warning systems.

Introduction

Analog instrument gauges and indicator lights monitor the various vehicle operating systems. They provide information to the driver about the current operation of the monitored system (Figure 15-1). Warning devices also provide information to the driver, however, these are usually associated with an audible signal. Some vehicles use a voice module to alert the driver to certain conditions. This chapter details the operation of analog electrical gauges, indicator lights, and warning devices.

Speedometers

The **speedometer** indicates the speed of the vehicle. A conventional speedometer (Figure 15-2) uses a cable that is connected to the output shaft of the transmission (or transfer case if four-wheel drive). The rotation of the output shaft causes the speedometer cable to rotate within its housing (Figure 15-3). The cable then transfers rotation to the speedometer assembly. In the speedometer

assembly, the cable is connected to a permanent magnet surrounded by an aluminum drum (Figure 15-4). The speedometer needle is attached to the drum. There is no solid or mechanical connection between the speedometer needle and the cable.

The rotating permanent magnet produces a rotating magnetic field around the drum which generates circulating currents in the drum. The aluminum drum is not attracted to the magnetic field. However, it is a conductor of electric current. These eddy currents produce a small magnetic field that interacts with the field of the rotating magnet. This interaction of the two magnetic fields pulls the drum and needle around with the rotating magnet. A fine hairspring holds back the drum to prevent the needle from bouncing. As the drum rotates, the needle moves across the speedometer scale.

One of the early styles of speedometers used a regulated amount of air pressure to turn a speed dial. The air

Indicator/Warning Light Key

1. Turn signals
2. Charging system (AMP)
3. Door ajar
4. Malfunction of anti-lock brake system
5. High beam
6. Brake
7. Engine temp. gauge
8. Fuel gauge
9. Seat belt
10. Check engine
11. Speedometer
12. Tachometer
13. Boost gauge
14. Low fuel *
15. Low oil pressure *
16. Trac-control active

* Low oil pressure/hi coolant temp/low fuel (below 1/8) redundant with gauges

Figure 15-1 Instrument panel. (Courtesy of Chrysler Corporation)

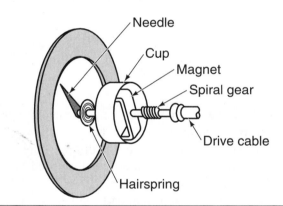

Figure 15-2 The main elements of a conventional-type speedometer.

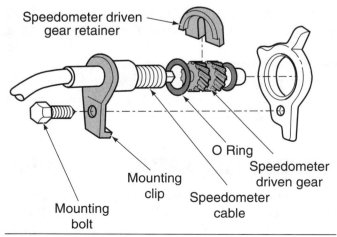

Figure 15-3 Speedometer cable connection to the transmission.

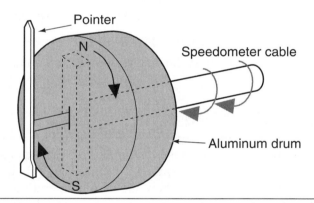

Figure 15-4 Eddy currents cause the needle to move across the scale.

pressure was generated in a chamber containing two intermeshing gears. The gears were driven by a flexible shaft that was connected to a front wheel or the driveshaft. The air was applied against a vane inside of the speed dial. The amount of air applied was proportional to the speed of the vehicle.

Odometers

The **odometer** is a mechanical counter in the speedometer unit that indicates total miles accumulated on the vehicle. Many vehicles also have a second odometer that can be reset to zero, this is referred to as a **trip odometer**. The odometer is driven by the speedometer cable through a spiral cut gear on a shaft called a **worm gear** (Figure 15-5). The odometer usually has six or seven wheels; each wheel is numbered 0 through 9. The wheels are geared together so as the right wheel makes one full revolution, the next wheel to the left moves one position. This action is repeated for all of the wheels.

WARNING *Federal and state laws prohibit tampering with the correct mileage as indicated on the odometer. If the odometer must be replaced, it must be set to the reading of the original odometer. Or, a door sticker must be installed indicating the reading of the odometer when it was replaced.*

Tachometers

A **tachometer** is an instrument used to measure the speed of the engine in revolutions per minute (RPM). An electrical tachometer receives voltage pulses from the ignition system; usually the coil (Figure 15-6). Each of

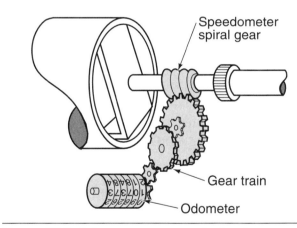

Figure 15-5 Odometer gears and wheels.

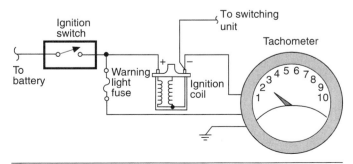

Figure 15-6 Electrical tachometer wired into the ignition system.

the voltage pulses represents the generation of one spark at the spark plug. The rate of spark plug firing is in direct relationship to the speed of the engine. A circuit within the tachometer converts the ignition pulse signal into a varying voltage. The voltage is applied to a voltmeter that serves as the engine speed indicator.

Gauges

Most instrument panels will include at least one **gauge**. A gauge is a device that displays the measurement of a monitored system by the use of a needle or pointer that moves along a calibrated scale. The **electromechanical gauge** acts as an ammeter because the gauge reading changes with variations in the **sender unit** resistance. The gauge is called an electromechanical device because it is operated electrically, but its movement is mechanical. The sender unit is the sensor for the gauge. It is a variable resistor that changes resistance values with changing monitored conditions.

There are two basic types of electromechanical gauges: the bimetallic gauge and the electromagnetic gauge. The basic principles of operation of electrical gauges are very simple. Once these principles are mastered, the technician should have no difficulty in locating and repairing a faulty gauge circuit.

WARNING *These gauges are called analog because they use needle movement to indicate current levels. However, many newer instrument panels use computer driven analog gauges that operate under different principles. It is important for the technician to follow the manufacturer's procedures for testing the gauges or gauge damage will result.*

A BIT OF HISTORY

In 1914, Studebaker introduced a dash-mounted fuel gauge. Before that gas levels were measured with a wooden dipstick.

Instrument Voltage Regulator

Some gauge systems require the use of an **instrument voltage regulator (IVR)** to provide a constant voltage to the gauge so it will read accurately under all charging conditions. The IVR consists of a set of normally closed contacts and a heating coil wrapped around a bimetallic strip consisting of two different types of metals (Figure 15-7). Heating one of the metals more rapidly than the other causes the strip to flex in proportion to the amount of heat.

When voltage is applied to the IVR, current flows through the contacts and the heating coil. Because the heating coil is connected to a ground, the flowing current heats the coil and the bimetallic strip bends. When the strip bends, the contacts open and current flows through the IVR and the heater coil stops. With no current flowing through the coil, the bimetallic strip cools and straightens. This closes the contact points and the process is repeated. This constant opening and closing of the contact points causes a pulsating DC current with a varying potential equivalent in heating effects to approximately 5 volts. The equivalent voltage is constant, and is applied to the gauge. The IVR output is approximately 5 volts, regardless of the generator output.

Opening and closing of the contact points creates induced voltage (like that in an ignition coil). This voltage is directed through a **radio choke** that absorbs it and prevents static in the vehicle's radio.

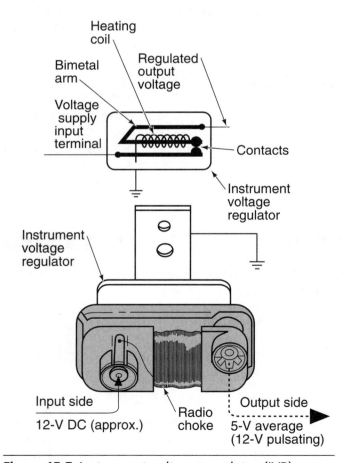

Figure 15-7 Instrument voltage regulator (IVR).

Bimetallic Gauges

Bimetallic gauges (or thermoelectric gauges) are simple dial and needle indicators that transform the heating effect of electricity into mechanical movement. The movement of the needle is slow and smooth, so the needle avoids responding to sudden changes that could cause a flickering needle. This prevents the gauges from constantly moving as a result of vehicle movement (as would be the case with a fuel gauge for example).

The construction of the bimetallic gauge features an indicating needle that is linked to the free arm of a U-shaped bimetallic strip (Figure 15-8). The free arm has a heater coil that is connected to the gauge terminal posts. When current flows through the heater coil, it heats the bimetallic arm. This causes the arm to bend and moves the needle across the gauge dial.

The amount that the bimetallic strip bends is proportional to the heat produced in the heater coil. When the current flow through the heater coil is small, less heat is created and needle movement is slight. If the current through the heater coil is increased, more heat is created and the needle will move more. The variable resistance used to control the current flow through the heater coil is in the sending unit (Figure 15-9).

The bimetallic gauge uses heat to move the needle. Thus it is influenced by changes as a result of high or low ambient temperatures. Change in ambient temperatures is compensated for by the U-shape of the bimetallic element. When ambient temperature bends the free arm, it also affects the fixed arm. The effect on the fixed arm is equal to the free arm but in the opposite direction. This action then cancels the effects of outside temperatures.

Electromagnetic Gauges

Electromagnetic gauges produce needle movement by magnetic forces instead of heat. There are four types of electromagnetic gauges: the D'Arsonval, the three-coil, the two-coil, and the air core.

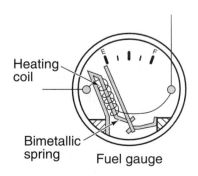

Figure 15-8 Bimetallic gauge construction. (Reprinted with permission of Ford Motor Company)

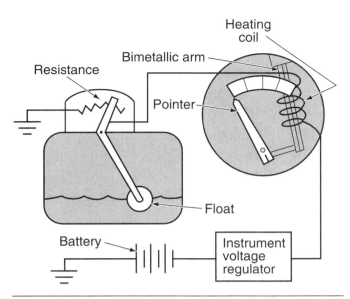

Figure 15-9 Typical bimetallic fuel gauge circuit. (Courtesy of Chrysler Corporation)

The **D'Arsonval gauge** uses the interaction of a permanent magnet and a electromagnet, and the total field effect to cause needle movement. The D'Arsonval gauge consists of a permanent horseshoe-type magnet that surrounds a movable electromagnet (armature) that is attached to a needle (Figure 15-10). When current flows through the armature it becomes an electromagnet and is repelled by the permanent magnet. When current flow through the armature is low, the strength of the electromagnet is weak and needle movement is small. When the current flow is increased, the magnetic field created in the armature is increased and needle movement is greater. The armature has a small spring attached to it to return the needle to zero when current is not applied to the armature. An IVR is used because voltage to the armature must remain constant to give accurate readings.

The **three-coil** gauge consists of an electromagnet or a permanent magnet with a needle attached to it. The three-coil gauge uses the interaction of three electromagnets and the total field effect upon a permanent magnet to cause needle movement. The three-coil gauge is also known as a magnetic bobbin gauge.

The permanent magnet is surrounded by three electromagnets. There may also be a quantity of silicone dampening fluid to restrict needle movement due to vehicle movement (Figure 15-11). The voltage drop at the variable resistor-type sending unit determines the magnetic strength of the coils.

The three coils of fine wire are wound on a square plastic frame. The needle shaft is supported by a bearing sleeve extending from the frame. The needle shaft connects the pointer.

The three coils are the low-reading coil, a bucking coil, and a high-reading coil (Figure 15-12). The **bucking coil** produces a magnetic field that bucks or opposes the low-reading coil. The **low-reading coil** and the bucking coil are wound together, but in opposite directions. The **high-reading coil** is positioned at a 90° angle to the low-reading and bucking coils. To compensate for production tolerances in the coils, a selective **shunt resistor** is attached to the back of the gauge housing. This selective resistor bypasses a certain amount of current past the coils.

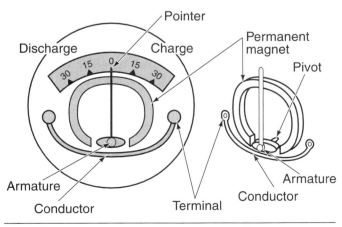

Figure 15-10 D'Arsonval gauge needle movement. (Reprinted with permission of Ford Motor Company)

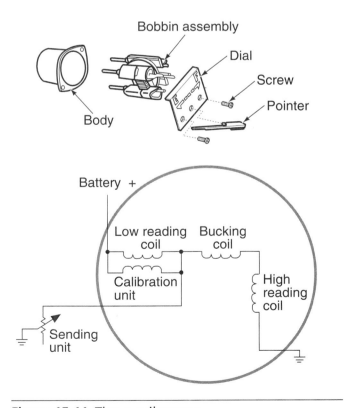

Figure 15-11 Three-coil gauge.

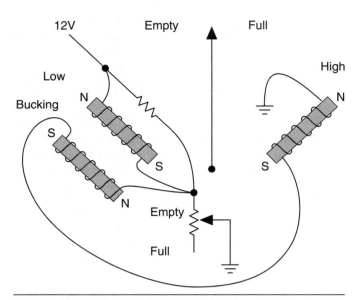

Figure 15-12 Three-coil gauge circuit.

When voltage is applied to the gauge, there are two paths for current flow. One path is through the low-reading coil and through the sending unit to ground (Figure 15-13). The second path is through the low-reading coil to the bucking coil and the high-reading coil to ground (Figure 15-14). The amount of resistance that the sending unit has determines the path. If there is less resistance to ground through the sending unit than through the bucking and high-reading coils, the current will take the path through the sending unit. If the resistance through the sending unit is greater than the resistance through the coils, the current will not flow through the sending unit.

When sending unit resistance is low, more current flows through the low-reading coil than through the bucking and high-reading coils. This causes the needle to be attracted to the left and the gauge reads toward zero. When sending unit resistance is high, very little current will flow through the sending unit. The current now

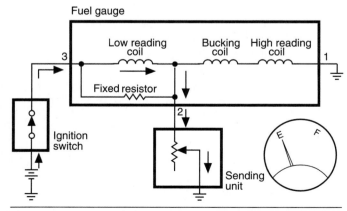

Figure 15-13 With sending unit resistance low, the needle is attracted to the low-reading coil.

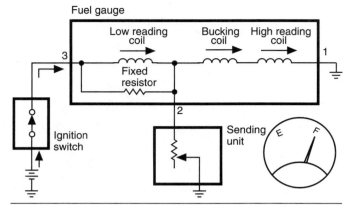

Figure 15-14 With sending unit resistance high, the needle is attracted to the high-reading coil.

flows through the three coils. The magnetic field created by the bucking coil cancels the magnetic field of the low-reading coil. The high-reading coil's magnetic field then attracts the needle and it swings toward maximum.

At an intermediate sending unit resistance value, the current can flow through both paths. If resistance was equal in the two paths, the needle would point at the mid-range. As the magnetic field(s) changes as the result of resistance change in the sending unit, the needle will swing toward the more powerful magnetic field.

Variation in the charging system output will not affect the gauge readings. This is because of the current flowing through both the low-reading and high-reading coils. Any variation in charging system output affects the strength of both coils. Also, it is normal for the three-coil gauge needle to remain in its last position when the ignition switch is turned off.

The **two-coil gauge** uses the interaction of two electromagnets and the total field effect upon an armature to cause needle movement. There are different designs of the two-coil gauge, depending on the gauge application. For example, in a coolant temperature gauge (Figure 15-15), both coils receive battery voltage. One of the coils is grounded directly, and the other is grounded through the sending unit. As the resistance in the sending unit varies as a result of temperature changes, so too does the cur-

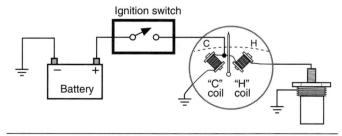

Figure 15-15 A two-coil temperature gauge.

rent flow through that coil. The two magnetic fields have different strengths depending on the amount of current flow through the sending unit. Since the majority of current flows the path with the least resistance, the sending unit resistance dictates the needle movement. The strength of the "H" coil's magnetic field changes with the amount of resistance in the sending unit. The stronger of the two magnetic fields causes the armature, with the needle attached to it, to move in proportion to the engine coolant temperature.

With the engine temperature cold, the sending unit resistance is high. More current will flow through the "C" coil than the "H" coil. This will cause the needle to be attracted to the "C" coil's magnetic field and indicate a cold temperature on the gauge face. As the resistance in the sending unit decreases, more current will flow through the "H" coil. This will result in needle movement toward the hot temperature range on the gauge face.

A two-coil gauge that is constructed to be used as a fuel gauge will use a different method (Figure 15-16). The "E" coil receives battery voltage. At the end of the coil, the voltage is divided. One path is through the "F" coil to ground; the other is through the sending unit to ground. With the fuel level low, there is less resistance in the sending unit resistor than in coil "F." Current flow will be through coil "E," to the sending unit, through the variable resistor, and to ground. Because most of the current is flowing through coil "E," it will produce a stronger magnetic field and attract the needle toward it. Higher sending unit resistance causes more current to flow through the "F" coil. As more current bypasses the sending unit and flows through the "F" coil, the magnetic field becomes stronger and the needle moves toward the full position.

Two-coil gauges do not require the use of an IVR. Also, it is normal for the gauge to remain at its last position when the ignition is turned off.

The **air core gauge** works on the same principle as the two coil. However, the pointer is connected to a permanent magnet (Figure 15-17). The air core gauge uses the interaction of two electromagnets and the total field effect upon a permanent magnet to cause needle movement. Two windings are placed at different angles, one wound around the other. There is no core inside of the windings. Instead, the permanent magnet is placed inside of the windings. The magnet aligns itself to a resultant field, in the field windings, according to the resistance of the sending unit. The sending unit resistance varies the strength of the field winding, which opposes the strength of the reference winding. The strength of the electromagnetic field depends on the resistance in the sending unit.

The air core gauge does not require an IVR. Also, it is normal for the gauge to remain at its last position when the ignition is turned off.

Gauge Sending Units

There are three types of sending units associated with the gauges just described: (1) a thermistor, (2) a piezoresistive sensor, and (3) a mechanical variable resistor.

In the coolant temperature sensing circuit, current is sent from the gauge unit into the top terminal of the sending unit, through the variable resistor (thermistor), and to the engine block (ground). The resistance value of the **thermistor** changes in proportion to coolant temperature (Figure 15-18). As the temperature rises, the resistance decreases and the current flow through the gauge increases. As the coolant temperature lowers, the resistance value increases and the current flow decreases.

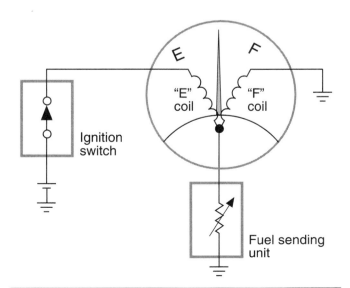

Figure 15-16 A two-coil fuel gauge.

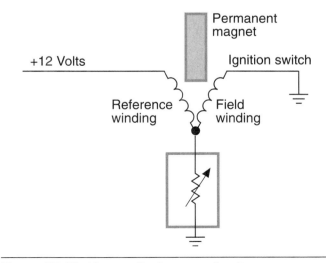

Figure 15-17 Air core fuel gauge circuit.

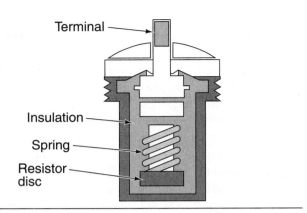

Figure 15-18 A thermistor used to sense engine temperature.

Gauges used to measure changes in pressure usually use a **piezoresistive sensor**. The piezoresistive sensor sending unit is threaded into the oil delivery passage of the engine. The pressure exerted by the oil causes the flexible diaphragm to move (Figure 15-19). The diaphragm movement is transferred to a contact arm that slides along the resistor. The position of the sliding contacts on the arm in relation to the resistance coil determines the resistance value, and the amount of current flow through the gauge to ground.

A fuel gauge sending unit is an example of a mechanical variable resistor (Figure 15-20). The sending unit is located in the fuel tank and has a float that is connected to the wiper of a variable resistor. The floating arm rises and falls with the difference in fluid level. This movement of the float is transferred to the sliding contacts. The position of the sliding contacts on the resistor determines the resistor value.

A sending unit used with bimetallic gauges will have a low resistance value when the fuel level is high. The low resistance value causes a high current to flow

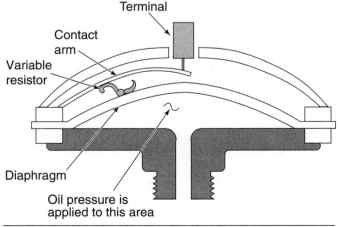

Figure 15-19 Piezoresistive sensor used for measuring engine oil pressure.

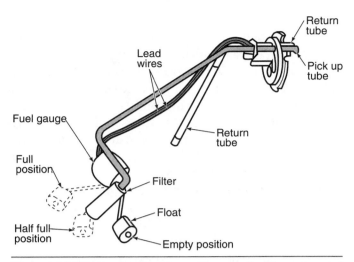

Figure 15-20 Fuel gauge sending unit. (Courtesy of Chrysler Corporation)

through the gauge, thus heating the bimetallic strip and causing the needle to move toward FULL. A sending unit used with an electromagnetic gauge has high resistance when the float level is high.

WARNING *It is important to know that the sending unit operation is different for electromagnetic gauges than bimetallic gauges. When resistance is low in most bimetallic gauges the gauge reads high; with many electromagnetic gauges the gauge reads low.*

Warning Lamps

A **warning light** may be used to warn of low oil pressure, high coolant temperature, defective charging system, or a brake failure. A warning light can be operated by two methods: a sending unit circuit or controlling the voltage being applied to the lamp.

Sending Unit Controlled Lights

Unlike gauge sending units, the sending unit for a warning light is nothing more than a simple switch. The style of switch can be either normally open or normally closed, depending on the monitored system.

Most oil pressure warning circuits use a normally closed switch (Figure 15-21). A diaphragm in the sending unit is exposed to the oil pressure. The switch contacts are controlled by the movement of the diaphragm. When the ignition switch is turned to the RUN position with the engine not running, the oil warning light turns on. Because there is no pressure to the diaphragm, the contacts remain closed and the circuit is complete to

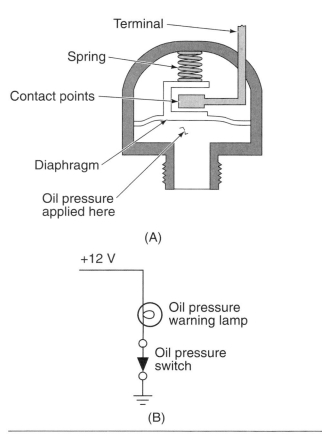

Figure 15-21 (A) Oil pressure light sending unit. (B) Oil pressure warning lamp circuit.

ground. When the engine is started, oil pressure builds and the diaphragm moves the contacts apart. This opens the circuit and the warning light goes off. The amount of oil pressure required to move the diaphragm is about 3 psi. If the oil warning light comes on while the engine is running, it indicates that the oil pressure has dropped below the 3 psi limit.

Most coolant temperature warning light circuits use a normally open switch (Figure 15-22). The temperature sending unit consists of a fixed contact and a contact on

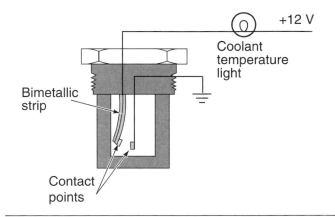

Figure 15-22 Temperature indicator light circuit.

a bimetallic strip. As the coolant temperature increases, the bimetallic strip bends. As the strip bends, the contacts move closer to each other. Once a predetermined temperature level has been exceeded, the contacts are closed and the circuit to ground is closed. When this happens the warning light is turned on.

With normally open-type switches, the contacts are not closed when the ignition switch is turned to ON. In order to perform a bulb check on normally open switches a **prove-out circuit** is included (Figure 15-23). A prove-out circuit completes the warning light circuit to ground through the ignition switch when it is in the START position. The warning light will be on during engine cranking to indicate to the driver that the bulb is working properly.

It is possible to have more than one sending unit connected to a single bulb. The illustration (Figure 15-24) shows a wiring circuit of a dual purpose warning light. The light will come on whenever oil pressure is low or coolant temperature is too high.

Another system that is monitored with a warning light is the braking system. The illustration (Figure 15-25) shows a brake system combination valve. The center portion of the valve senses differences in the hydraulic

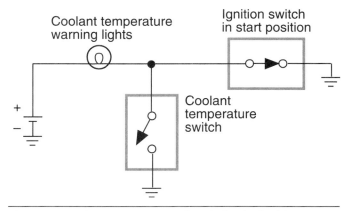

Figure 15-23 A prove-out circuit included in a normally open (NO) coolant temperature light system.

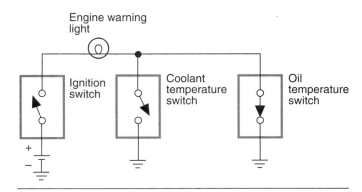

Figure 15-24 One warning lamp used with two sensors.

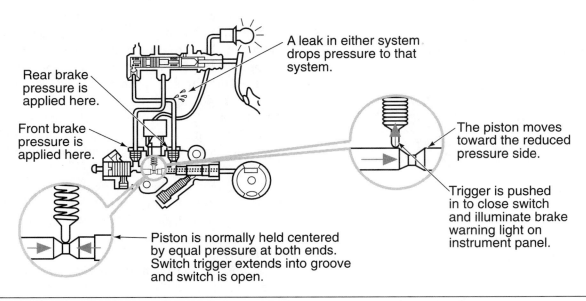

Rear brake pressure is applied here.

Front brake pressure is applied here.

A leak in either system drops pressure to that system.

The piston moves toward the reduced pressure side.

Trigger is pushed in to close switch and illuminate brake warning light on instrument panel.

Piston is normally held centered by equal pressure at both ends. Switch trigger extends into groove and switch is open.

Figure 15-25 Brake warning light switch as part of the combination valve.

pressures on both sides of the valve. With the differential valve centered, the plunger on the warning light switch is in the recessed area of the valve. If the pressure drops in either side of the brake system, the differential valve will be forced to move by hydraulic pressure. When the differential valve moves, the switch plunger is pushed up and the switch contacts close.

Opposing Voltage-Controlled Lights

A lamp lights when there is a voltage drop or potential across its terminals. If equal voltage is applied to

both sides of the lamp, the lamp remains off because current cannot flow. This is the basic principle of most precomputerized instrumentation charging system indicator lights (Figure 15-26).

When the ignition switch is turned to the ON position and the engine is off, the light is "proofed." The current flows through the ignition switch and warning light to the input terminal 1. The current continues to the common connector at R1 and through the field coil. Transistor TR1 is biased to conduct through R1 and the circuit is completed to ground. When the engine is started, the generator begins to produce voltage. The voltage pro-

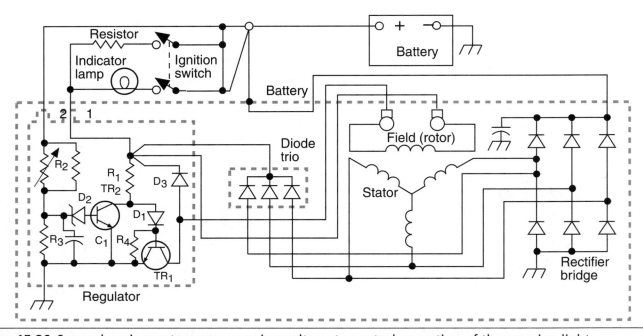

Figure 15-26 Some charging system use opposing voltage to control operation of the warning light.

duced in the stator is sent through the diode trio. This places output voltage at the common connection at R1, putting equal voltage to both sides of the indicator light. Because there is no potential difference, current cannot flow and the light goes out. The resistor is wired in parallel to the light to assure that field current will flow if the lamp burns out. If the indicator light comes on while the engine is running, it indicates that charging system output voltage has dropped. With a potential at both light terminals, the lamp lights.

Audible Warning Systems

Many warning systems also include an audible signal to alert the driver of potentially dangerous conditions. This could include seatbelt warning systems, low oil pressure, key in ignition, headlights on, and so forth.

Seatbelt Warning Systems

When the ignition switch is turned on and the seatbelt is not fastened, a buzzer will sound and a warning light will come on for approximately 8 seconds. The seatbelt switch is a normally closed switch to ground (Figure 15-27). When the seatbelt is fastened, the switch is opened. With the key in the RUN position, battery voltage is supplied to the timer. If the seatbelt is not buckled, the cir-

cuit is complete to ground, and the buzzer and light operate. When the driver's seatbelt is not fastened, the bimetallic strip heats up and opens the circuit. Because battery current flows through the heater as long as the ignition switch is in the RUN position, the bimetallic strip stays hot enough to remain open until the ignition switch is turned to the OFF position.

Headlight and Key In Ignition Warning Systems

If drivers fail to turn off the headlights and they open the door, a buzzer or chime will sound. The driver door switch is a normally closed switch to ground. The switch opens when the door is closed. If the headlight switch is in the HEADLIGHT or PARK position, battery voltage is applied through the buzzer and the headlight switch to the driver door switch (Figure 15-27). If the driver's door is opened, the circuit is completed and the buzzer sounds until either the headlight switch is turned off or the driver door is shut.

The "key in ignition" warning system operates in the same manner. When the ignition key is inserted into the ignition switch tumblers, the key warning switch is closed. If the driver's door is opened while the key is in the ignition switch, the circuit is completed and the buzzer will sound.

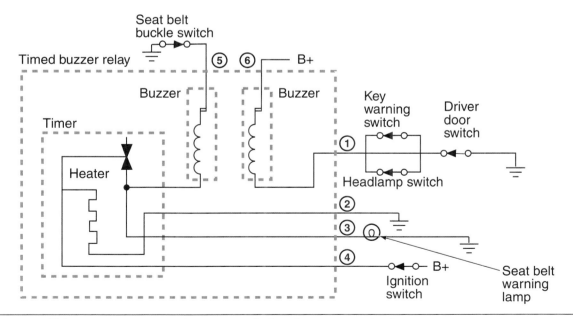

Figure 15-27 Audible warning system. (Courtesy of Chrysler Corporation)

Summary

❑ Through the use of gauges and indicator lights, the driver is capable of monitoring several engine and vehicle operating systems.

❑ The gauges include speedometer, odometer, tachometer, oil pressure, charging indicator, fuel level, and coolant temperature.

❑ Some gauge types require the use of an IVR to assure constant voltage to the gauge coils.

❑ The most common types of gauges are the bimetallic, D'Arsonval, three-coil, two-coil, and air core.

❑ All gauges require the use of a variable resistance sending unit. Styles of sending units include thermistors, piezoresistive sensors, and mechanical variable resistors.

❑ In the absence of gauges, important engine and vehicle functions are monitored by warning lamps. These circuits generally use an On-Off switch-type sensor. The exception would be the use of voltage-controlled warning lights that use the principle of voltage drop.

Terms-To-Know

Air core gauge	Low-reading coil	Tachometer
Bucking coil	Odometer	Thermistor
D'Arsonval gauge	Piezoresistive sensor	Three-coil gauge
Electromagnetic gauges	Prove-out circuit	Trip odometer
Electromechanical gauge	Radio choke	Two-coil gauge
Gauge	Sender unit	Warning light
High-reading coil	Shunt resistor	Worm gear
Instrument voltage regulator (IVR)	Speedometer	

Review Questions

Short Answer Essays

1. What are the most common types of electromagnetic gauges?

2. What type of electromagnetic gauge uses an IVR?

3. Describe the operation of the IVR.

4. Describe the operation of the piezoresistive sensor.

5. Describe the operation of the conventional speedometer.

6. Where do most tachometers receive the signal from?

7. Describe the principle of operation of the bimetallic gauge.

8. Describe what occurs in the warning light circuit of the charging system when the light is being proofed.

9. What is a thermistor used for?

10. What is meant by electromechanical?

Fill-In-The-Blanks

1. The purpose of the tachometer is to indicate _____ _____ .

2. A piezoresistive sensor is used to monitor _____ changes.

3. The most common style of fuel level sending unit is _____ variable resistor.

4. If equal voltage is applied to both sides of a lamp, the lamp is _____.

5. The brake warning light is activated by _____ pressure in the brake hydraulic system.

6. The IVR sends a constant _____ to the gauges, regardless of _____ output.

7. _____ gauges transform the heating effect of electricity into mechanical movement.

8. In a three-coil gauge, the _____ produces a magnetic field that bucks or opposes the low-reading coil. The _____ coil and the bucking coil are wound together, but in opposite directions. The _____ _____ coil is positioned at a 90-degree angle to the low-reading and bucking coils.

9. The _____ gauge consists of a permanent horseshoe-type magnet that surrounds a movable electromagnet (armature), which is attached to a needle.

10. A _____ _____ circuit completes the warning light circuit to ground through the ignition switch when it is in the START position.

ASE Style Review Questions

1. Odometer replacement is being discussed. Technician A says that it is permissible to turn back the reading on a odometer as long as the customer is notified. Technician B says that if an odometer is replaced it must be set to the same reading as the original odometer. Who is correct?
   a. A only
   b. B only
   c. Both A and B
   d. Neither A nor B

2. The IVR is being discussed. Technician A says that an IVR is used to assure a constant 12 volts is applied to the gauge coils. Technician B says an IVR is used on bimetallic gauges. Who is correct?
   a. A only
   b. B only
   c. Both A and B
   d. Neither A nor B

3. The operation of the speedometer is being discussed. Technician A says there is a mechanical link between the speedometer cable and the needle. Technician B says the system operates on magnetism. Who is correct?
   a. A only
   b. B only
   c. Both A and B
   d. Neither A nor B

4. Electromagnetic gauges are being discussed. Technician A says the D'Arsonval gauge uses the interaction of a permanent magnet and an electromagnet, and the total field effect to cause needle movement. Technician B says the three-coil gauge uses the interaction of three electromagnets and the total field effect upon another magnet to cause needle movement. Who is correct?
   a. A only
   b. B only
   c. Both A and B
   d. Neither A nor B

5. Electric tachometers are being discussed. Technician A says the tachometer receives its signal from the 1 side of the ignition coil. Technician B says the tachometer is used to measure drive shaft speed. Who is correct?
   a. A only
   b. B only
   c. Both A and B
   d. Neither A nor B

6. Technician A says the three-coil gauge uses magnetic field strength for operation. Technician B says the three coils used in some gauges are called the low-reading coil, a bucking coil, and a high-reading coil. Who is correct?
   a. A only
   b. B only
   c. Both A and B
   d. Neither A nor B

7. Sending unit operation is being discussed. Technician A says in most bimetallic gauges if sending unit resistance is low, the gauge reads high. Technician B says with most electromagnetic gauges if sending unit resistance is low, the gauge reads low. Who is correct?
   a. A only
   b. B only
   c. Both A and B
   d. Neither A nor B

8. Warning light circuits are being discussed. Technician A says most oil pressure warning circuits use a normally closed switch. Technician B says most coolant temperature warning light circuits use a normally open switch. Who is correct?
   a. A only
   b. B only
   c. Both A and B
   d. Neither A nor B

9. The brake failure warning system is being discussed. Technician A says if the pressure drops in either side of the brake system, the switch plunger is pushed up and the switch contacts close. Technician B says the warning light comes on if the pressure is low on both sides of the brake system. Who is correct?
   a. A only
   b. B only
   c. Both A and B
   d. Neither A nor B

10. Technician A says the trip odometer is able to be reset to zero. Technician B says the odometer is driven by the speedometer cable through a worm gear. Who is correct?
    a. A only
    b. B only
    c. Both A and B
    d. Neither A nor B

16 Conventional Instrument Cluster Diagnosis and Repair

Objective

Upon completion and review of this chapter, you should be able to:

❑ Remove and replace the instrument cluster.

❑ Remove and replace the printed circuit.

❑ Diagnose and repair causes of noisy, erratic, and inaccurate speedometer readings.

❑ Diagnose and repair causes of tachometer malfunctions.

❑ Diagnose and repair faulty gauge circuits.

❑ Diagnose and repair the cause of multiple gauge failure.

❑ Diagnose sender units, including thermistors, piezoresistive, and mechanical variable styles.

❑ Diagnose and repair warning light circuits.

❑ Diagnose and repair the cause of multiple warning light failures.

❑ Diagnose and repair audible warning systems.

Introduction

The instrument panel gauges and warning lamps monitor the various vehicle operating systems and provide information about their operation to the driver. Most problems in the gauges or warning lamps are usually caused by an open circuit in the wiring or the printed circuit; improper gauge calibration; loose connections; excessive resistance; or defective bulbs, gauges, or sending units.

In this chapter you will learn how to diagnose the gauge, lamp, and sending unit of the various styles of conventional instrument panels. You will also learn how to remove the instrument panel to replace the printed circuit, and how to repair the speedometer cable core.

Instrument Panel and Printed Circuit Removal

Many times it may be necessary to remove the instrument panel to replace defective gauges, lamps, or printed circuits. Before removing the instrument panel, always disconnect the battery negative cable. Consult the service manual for the procedure for the vehicle you are working on. The following is a common method of removing the instrument cluster and printed circuit:

1. Place fender covers on the fenders and disconnect the battery negative cable.

2. Remove the retaining screws to the steering column opening cover. Then remove the cover (Figure 16-1).

3. Remove the finish panel retaining screws. On some models it may be necessary to remove the radio knobs.

4. Remove the finish panel.

5. Remove the retaining bolts that hold the cluster to the dash.

6. Reach behind the instrument panel and disconnect the speedometer cable.

7. Gently pull the cluster away from the dash.

8. Disconnect the cluster feed plug from the printed circuit receptacle. Be careful not to damage the printed circuit.

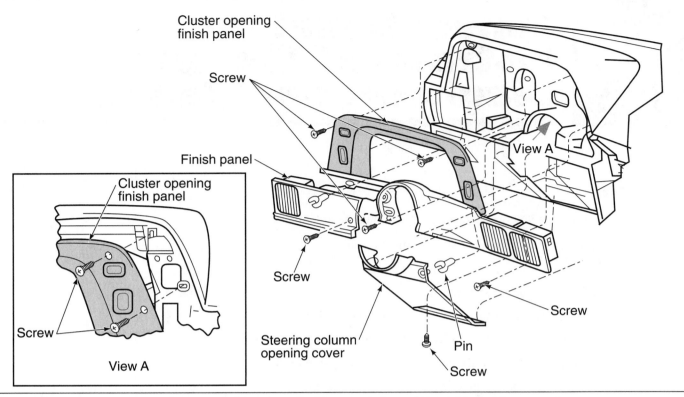

Figure 16-1 Instrument panel removal. (Reprinted with permission of Ford Motor Company)

9. Remove the IVR and all illumination and indicator lamp sockets (Figure 16-2).

10. Remove the charging system warning lamp resistor if applicable.

11. Remove all printed circuit attaching nuts and remove the printed circuit.

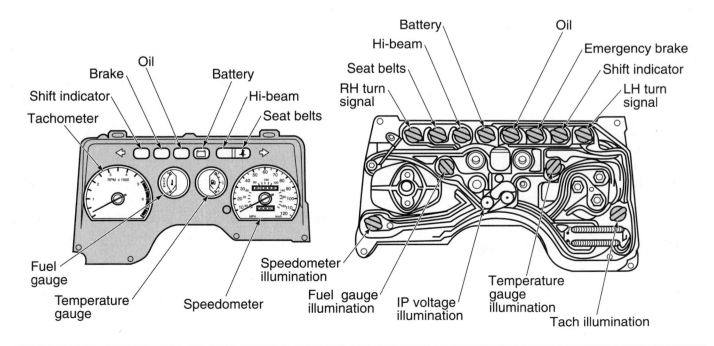

Figure 16-2 Instrument panel printed circuit board. (Reprinted with permission of Ford Motor Company)

Speedometers

Speedometer complaints range from chattering noises when cold, to inaccurate readings, to not operating at all. Sometimes the problem can be corrected by lubricating the cable with an approved lubricant. In other cases, the speedometer head may be faulty and needs to be replaced. The illustration (Figure 16-3) shows a diagnostic chart for the speedometer.

NOISY, ERRATIC, INOPERATIVE, OR INACCURATE

	TEST STEP	RESULT ▶	ACTION TO TAKE
A0	VERIFY CONDITION		
	• Make sure quick connect is properly attached at speedometer head. Make sure cable is connected at the speed sensor, if applicable.	Noisy ▶	GO to **A1**.
		Erratic or pointer waver ▶	GO to **A6**.
		Inoperative speed indication ▶	GO to **A11**.
		Inoperative odometer ▶	GO to **A12**.
		Inaccurate speed indication ▶	GO to **A18**.
A1	CHECK FOR NOISE		
	• With engine running in neutral, check for noise.	▶ OK	GO to **A2**.
		▶ O̸K̸	CHECK for other causes of vehicle noise.
A2	CHECK SENSOR		
	• Check sensor for erratic or noisy operation.	▶ OK	GO to **A3**.
		▶ O̸K̸	REPLACE sensor.
A3	CHECK CABLE		
	• Check cable for kinks or bends.	▶ OK	GO to **A4**.
		▶ O̸K̸	If kinks are severe, REPLACE cable. For minor bends, ADJUST cable routing to obtain generous curves and RECHECK for condition resolution.
A4	CHECK CABLE (CONTINUED)		
	• Disconnect cable and check core for kinks, burns or bent tips.	▶ OK	GO to **A5**.
		▶ O̸K̸	REPLACE core.
A5	CHECK DRIVEN GEAR		
	• Check for damaged driven gear.	▶ OK	REPLACE speedometer head.
		▶ O̸K̸	REPLACE gear.

CK5966-B

Figure 16-3 Speedometer diagnostic chart. (Reprinted with permission of Ford Motor Company)

NOISY, ERRATIC, INOPERATIVE, OR INACCURATE (Continued)

TEST STEP	RESULT ▶	ACTION TO TAKE
A6 CHECK SPEED SENSOR • Check speed sensor for erratic rotation or binding.	▶ OK ▶ ⊘OK	GO TO **A7.** REPLACE sensor.
A7 CHECK CABLE • Check cable for kinks or bends in routing.	▶ OK ▶ ⊘OK	GO to **A3.** If kinks or bends are severe, REPLACE cable. For minor bends, ADJUST cable routing to obtain generous curves and RECHECK for condition resolution.
A8 CHECK DRIVEN GEAR • Check for damaged driven gear.	▶ OK ▶ ⊘OK	GO to **A9.** REPLACE gear.
A9 CHECK CORE • Disconnect cable and check core for kinks, burrs, or bent tips.	▶ OK ▶ ⊘OK	GO to **A10.** REPLACE core.
A10 RECHECK CORE • Reinstall core and turn by hand to feel for rough or irregular motion.	▶ OK ▶ ⊘OK	REPLACE speedometer head. REPLACE cable.
A11 CHECK ODOMETER • Check to see that odometer is operating.	▶ OK ▶ ⊘OK	REPLACE speedometer head. GO to **A13.**
A13 CHECK POINTER OPERATION • Check to see that pointer operates.	▶ ⊘OK ▶ OK	GO to **A13.** REPLACE speedometer head.

CK6052-B

Figure 16-3 (continued)

NOISY, ERRATIC, INOPERATIVE, OR INACCURATE (Continued)

TEST STEP	RESULT ▶	ACTION TO TAKE
A13 CHECK MAGNET SHAFT • Disconnect cable and check that magnet shaft in speedometer head turns freely.	▶ OK ▶ OK̸	GO TO **A14**. REPLACE speedometer head.
A14 CHECK DRIVE AND DRIVEN GEAR • Check drive and driven gear for damage or wear.	▶ OK ▶ OK̸	GO to **A15**. REPLACE damaged gear.
A15 CHECK CABLE • Check speedometer cable for kinks or improper routing.	▶ OK̸ ▶ OK̸	GO to **A16**. REPLACE cable.
A16 CHECK SENSOR SHAFT • Disconnect cable from speed sensor and check that shaft in sensor turns freely.	▶ OK̸ ▶ OK	GO to **A17**. REPLACE sensor.
A17 CHECK CORE • Check for broken core.	▶ OK ▶ OK̸	If core is seized and will not turn, REPLACE cable. REPLACE core.
A18 CHECK ODOMETER • Check accuracy or odometer over a measured distance.	▶ OK̸ ▶ OK	GO to **A19**. REPLACE speedometer head.
A19 CHECK DRIVEN GEAR • Check for proper driven gear.	▶ OK ▶ OK̸	GO to **A20**. REPLACE gear.
A20 CHECK DRIVE GEAR, AXLE AND TIRES • Check for proper drive gear, axle and tires.	▶ OK ▶ OK̸	REPLACE speedometer head. REPLACE incorrect component or driven gear.

CK5968-B

Figure 16-3 (continued)

If the customer says the speedometer is noisy, it is possible that the cable is "dry" and in need of lubricant. However, the cause of the noise must be isolated because just applying lubricant may only stop the noise temporarily. It is a good practice to remove the cable core to clean and inspect it before adding lubricant.

The speedometer uses eddy currents instead of a direct mechanical connection from the cable to the head. If the cause of the noise is not in the cable, the bushings in the head may be worn. This would allow the cup and magnet to come in contact, which would result in noisy operation and inaccurate readings. The speedometer head will have to be replaced.

WARNING *It is possible to short out wires while reaching behind the instrument panel. Disconnect the battery negative cable before removing the speedometer cable assembly.*

To remove the core, disconnect the speedometer cable assembly from the back of the speedometer head. For most vehicles this is done by reaching under the instrument panel and pressing down on the flat surface of the plastic quick connect. On some vehicles it may be necessary to remove the instrument panel to gain access to the speedometer cable. With the cable assembly disconnected, visually inspect the cable for kinks or other damage. Raise the drive wheels from the ground and start the engine. Place the transmission in gear and allow the drive wheels to rotate at engine idle. Observe the cable rotation inside the housing. It should be smooth and constant. Shut off the engine. Be sure to apply the brakes to stop transmission movement before attempting to return the shift lever to the PARK position.

WARNING *Do not allow the drive wheels to rotate faster than 50 mph. Because only one wheel will rotate, one of the differential side gears is remaining stationary as the pinion gears "walk" around it. Excessive speed may result in differential damage.*

It may be possible to remove the core by pulling it out of the housing. If the core cannot be removed in this way, disconnect the speedometer retainer from the transaxle (Figure 16-4). Pull the core out of the speed sensor.

WARNING *If the cable attaches to a speed control sensor, do not attempt to remove the spring retainer clip with the speedometer cable in the sensor.*

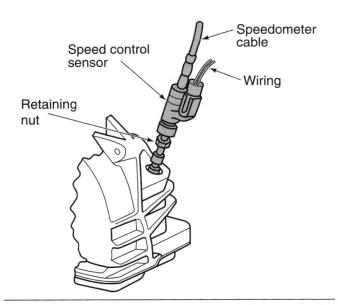

Figure 16-4 Speedometer connection at the transaxle. (Reprinted with permission of Ford Motor Company)

With the cable core removed from the housing, clean it with solvent and wipe it dry. Place the core on a flat surface and stretch it out straight. Roll it back and forth while looking for signs of kinks or other damage.

If the core is damaged, it must be replaced. Determine the exact length of the old core and cut the new core to that size. Do not cut from the squared end of the core. Install the tip onto the core and crimp it in place using a crimping die and hammer (Figure 16-5). Apply a light film of approved lubricant to the cable core and install the new core into the housing.

Check the drive and driven gears for damage. Whenever a drive gear is replaced, also replace the driven gear regardless of its apparent condition. To assure proper

Figure 16-5 Crimping the tip to the speedometer cable core.

gear mesh during installation, rotate the drive shaft while inserting the cable assembly into the transaxle.

When installing the cable to the speedometer head, apply a small amount of approved lubricant to the drive hole. Check that the speedometer cable takes virtually no change of direction for at least 5 inches (127 mm) from the speedometer head.

WARNING *Changing tire size and differential gear ratios from original equipment specifications will result in speedometer inaccuracy. In many states, it is illegal to calibrate the speedometer unless it is performed by a shop that is certified to perform this task.*

CUSTOMER CARE *The speedometer cable should be lubricated every 10,000 miles. This practice will reduce speedometer cable problems that will cause noisy, erratic, or inaccurate readings.*

If the speedometer assembly must be replaced, usually a new odometer is included with the assembly. Be sure to follow the manufacturer's procedures for setting the odometer to the correct reading.

WARNING *Federal and state laws prohibit tampering with the correct mileage as indicated on the odometer. If the odometer must be replaced, it must be set to the reading of the original odometer. Or, a door sticker must be installed indicating the reading of the odometer when it was replaced (Figure 16-6).*

Tachometers

Most electrically operated tachometers receive their reference pulses from the ignition system. If the

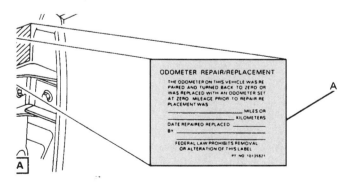

Figure 16-6 An odometer repair label. (Courtesy of General Motors Corporation, Service Technology Group)

tachometer is faulty, it must be replaced. There is no servicing the meter itself.

Gauges

WARNING *These gauges are called analog because they use needle movement to indicate current levels. However, many modern instrument panels use computer driven analog gauges that operate under different principles. It is important that the technician follow the manufacturer's procedures for testing the gauges or gauge damage will result.*

With the exception of the voltmeter and ammeter, all electromechanical gauges (whether bimetallic or electromagnetic) use a variable resistance sending unit. The types of tests performed will depend on the nature of the problem and if the system uses an IVR.

Single Gauge Failure

If the gauge system does not use an IVR, check the gauge for proper operation as follows:

1. Check the fuse panel for any blown fuses. The gauge that is not operating may share a fuse with some other circuit that is separate from the other gauges.
2. Disconnect the wire connector from the sending unit of the malfunctioning gauge.
3. Check the terminal connectors for signs of corrosion or damage.
4. Use a test light to confirm that voltage is present to the connector with the ignition switch in the RUN position. If the test light does not illuminate, check the circuit back to the gauge and battery.
5. Connect the 10-ohm resistor to the lead. Connect the lead to ground (Figure 16-7).
6. With the ignition switch in the RUN position, watch the gauge. Depending on the gauge design, the needle should indicate either high or low on the scale. Check the service manual for the correct results.
7. Connect a 73-ohm resistor between the sensor lead and ground. Repeat step 6.
8. If the test results are in the acceptable range, the sending unit is faulty.
9. If the gauge did not operate properly in step 5, check the wiring to the gauge. If the wiring is good, replace the gauge.

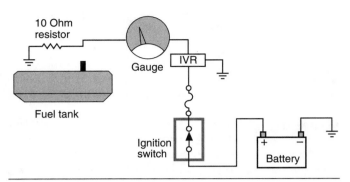

Figure 16-7 Testing the gauge operation. The resistor is used to protect the circuit.

WARNING *It is important to know that the sending unit operation is different for electromagnetic gauges than bimetallic gauges. In most bimetallic gauges the gauge reads high if resistance is low. With many electromagnetic gauges the gauge reads low if sending unit resistance is low.*

WARNING *Grounding the sender terminal lead directly may damage the gauge. Use a resistor to protect the circuit.*

If the gauge circuits use an IVR, follow steps 1 through 4 as described. The test light should flicker on and off. If it did not illuminate, reconnect the sending unit lead and check for voltage at the sender unit side of the gauge (Figure 16-8). If there is voltage at this point, repair the circuit between the gauge and the sending unit. If voltage is not shown, test for voltage at the battery side of the gauge. If voltage is present at this point, the gauge is defective and must be replaced. If no voltage is present

at this terminal, continue to check the circuit between the battery and the gauge.

If the IVR was working properly and voltage was present to the sender unit, follow steps 5 through 9.

SERVICE TIP *It is common for the fuel gauge sender unit to get corrosion on the ground wire connection. Before replacing the sending unit, clean the ground connections and test for proper operation.*

Multiple Gauge Failure

If all gauges fail to operate properly, begin by checking the circuit fuse. Next, test for voltage at the last common circuit point (Figure 16-9). If voltage is not present at this point, work toward the battery to find the fault.

If the system uses an IVR, use a voltmeter to test for regulated voltage at a common point to the gauges (Figure 16-10). If the voltage is out of specifications, check the ground circuit of the IVR. If that is good, replace the IVR. If there is no voltage present at the common point, check for voltage on the battery side of the IVR. If voltage is present at this point, then replace the IVR.

If regulated voltage is within specifications, test the printed circuit from the IVR to the gauges. If there is an open in the printed circuit, replace the board.

SERVICE TIP *It is unlikely that all of the gauges would fail at the same time. If the diagnostic tests indicate that the gauges are defective, bench test the gauges before replacing them. Use an ohmmeter to check the resistance. Most electromagnetic gauges should read between 10 and 14 ohms. On systems that do not use an IVR and all of the gauges are defective, check the charging system for excessive output.*

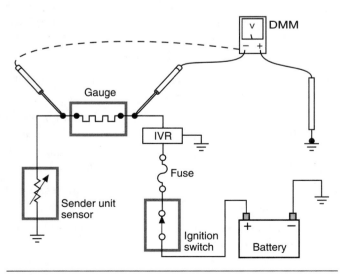

Figure 16-8 Checking for regulated voltage on the sender unit side of the gauge.

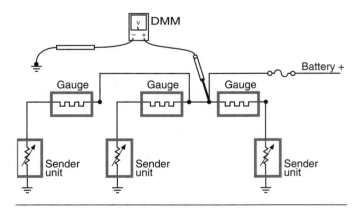

Figure 16-9 Check the last common connection in the circuit for voltage.

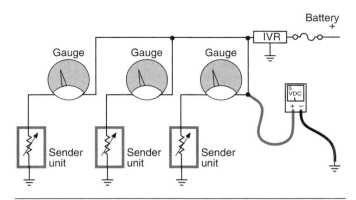

Figure 16-10 Testing for correct IVR operation.

Gauge Sending Units

There are three types of sending units associated with electromechanical gauges: a thermistor, a piezoresistive sensor, and a mechanical variable resistor. Most of these can be tested before replacement to confirm the fault.

The fuel level sending unit can be tested in or out of the tank. If it is tested in the tank, add and remove fuel to change the level. The easiest method is to bench test the unit. Photo Sequence 8 illustrates how to bench test a mechanical variable resistor sending unit.

To test the coolant temperature sensing unit, use an ohmmeter to measure the resistance between the terminal lead and ground (Figure 16-11). The resistance value of the variable resistor should change in proportion to coolant temperature. Check the test results with manufacturer specifications.

To test a piezoresistive sensor sending unit used for oil pressure gauges, connect the ohmmeter to the sending unit terminal and ground (Figure 16-12). Check the resistance with the engine off and compare to specifications. Start the engine and allow it to idle. Check the resistance value and compare to specifications. Before replacing the

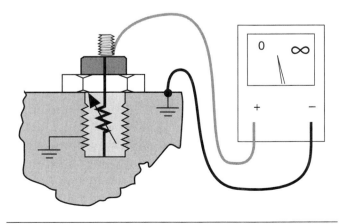

Figure 16-11 Testing a thermistor with an ohmmeter.

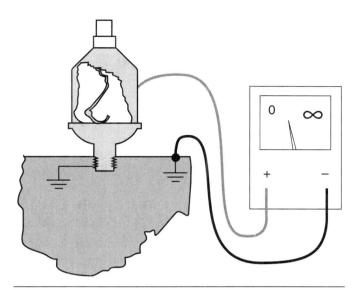

Figure 16-12 Using an ohmmeter to test a piezoresistive sensor.

sending unit, connect a shop oil pressure gauge to confirm that the engine is producing adequate oil pressure.

WARNING *If the engine is making knocking or other noises, do not test the sending unit first. Immediately connect a shop oil pressure gauge and check for oil pressure. Do not increase the engine speed over idle unless instructed to do so.*

Warning Lamps

A warning light may be used to warn of low oil pressure, high coolant temperature, defective charging system, or a brake failure. Unlike gauge sending units, the sending units for a warning light are nothing more than simple switches. The style of switch can be either normally open or normally closed, depending on the monitored system.

SERVICE TIP *With the bulbs used for most warning light circuits it is hard to determine whether or not the filament is good. When a test procedure requires that a bulb be checked, it is usually easier to replace the bulb with a known good one.*

It is not likely that all of the warning lights would fail at the same time. Check the fuse if all of the lights are not operating properly. Next, check for voltage at the last common connection. If voltage is not present, then trace the circuit back toward the battery. If voltage is present at the common connection, test each circuit branch in the same manner as described here for individual lamps.

Bench Testing the Fuel Level Sender Unit

P7-1 Tools required to perform this task: DVOM and jumper wires service manual.

P7-2 With the DOM set on the ohmmeter function, connect the negative test lead to the ground terminal and the positive test lead to the variable resistor terminal of the sender unit.

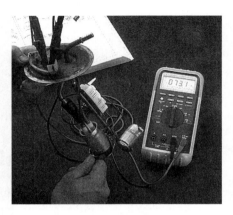

P7-3 Holding the sender unit in its normal position, place the float rod against the empty stop. Read the ohmmeter and check the results with specifications.

P7-4 Slowly move the float toward the full stop while observing the ohmmeter. The resistance change should be smooth and consistent.

P7-5 Check the resistance value while holding the float against the full stop. Check the results with specifications.

P7-6 Check the float that it is not filled with fuel, distorted, or loose.

To test a faulty warning lamp, turn the ignition switch to the START position. The proving circuit should light the warning lamp. If the light does not come on during the prove-out, disconnect the sender switch lead. Use a jumper wire to connect the sender switch lead to ground. With the ignition switch in the RUN position, the warning lamp should light. If the lamp is illuminated, replace the sending switch. If the light does not come on, either the bulb is burned out or the wiring is damaged. Use a test light to confirm voltage is present at the sensor terminal connector. If there is voltage, the bulb is probably bad.

WARNING *Check the manufacturer's service manual to confirm the location of the sensor switch you are testing. For example, there may be a coolant temperature switch for the warning light and a coolant temperature sensor for the engine computer. Grounding the computer terminal lead may result in damage to the computer. Usually a warning light sensor switch will have one lead. The computer sensor will have two to four leads and is contained in a weather-pack connector.*

If the customer says the warning light stays on, test the system in the following manner. Disconnect the lead to the sender switch. The light should go out with the ignition switch in the RUN position. If it does not, there is a short to ground in the wiring from the sender switch to the lamp. If the light goes out, replace the sender switch. However, if the oil pressure warning light is always on use a shop oil pressure gauge to confirm adequate oil pressure. If oil pressure is less than specifications, shut the engine off immediately. If oil pressure is good at idle check it again at 2,000 rpm. If oil pressure is good, the fault is in the warning light circuit.

Testing Charging Indicator Lamp Circuits

The basic principle of most charging system indicator lights is that a lamp only lights when there is a voltage drop or potential across its terminals. If equal voltage is applied to both sides of the lamp, the lamp remains off because current cannot flow.

Before checking the charging warning lamp circuit, check that the AC generator drive belt is in good condition and properly tensioned. Perform a battery test series, and test the AC generator. If all of these test results are good, then check the lamp circuit.

Most charging warning lamps are proofed when the ignition switch is placed in the RUN position with the engine not running. If the lamp does not come on, or if it remains on after the engine is started, perform the following tests based on the type of AC generator.

Delco-Remy SI System

If the charging warning light fails to come on during the prove-out, check for a blown fuse, damaged printed circuit, defective bulb, or an open in the circuit from the ignition switch to terminal 1 (Field) of the AC generator.

If the lamp remains on after the engine is started, check the fuses. If the fuse is good, there may be a defective diode in the lamp circuit. General Motors has used a

diode in this circuit. If it is defective, the light will glow brighter as engine speed is increased.

SERVICE TIP *If the charging warning light does not go out when the ignition key is in the OFF position, the most likely cause is a faulty diode bridge*

Delco-Remy CS System

If the light fails to come on when proofed, disconnect the wiring connector at the AC generator. Using a jumper wire, ground the L terminal lead (Figure 16-13). If the light comes on, the AC generator is faulty. If the light remains off, either the bulb is blown or there is an open in the wire between the ignition switch and the L terminal.

If the light does come on during the prove-out test, start the engine while observing the lamp. The lamp should light momentarily when the engine is started and then go out. If it remains on, disconnect the wire connector from the AC generator. If the warning lamp does go out, the AC generator is faulty. If the light remains on, there is a short to ground in the L terminal wire circuit.

Ford Electronic Voltage Regulator (EVR) System

If the warning light bulb does not come on during the prove-out test, turn off the ignition switch and disconnect the regulator wiring connector. Use a jumper wire to ground the I terminal lead (Figure 16-14). Turn the ignition switch to the RUN position while observing the warning lamp. If the lamp lights, check the ground for the regulator. If good, test the regulator lamp circuit. If the lamp still does not illuminate, the problem is one of the following:

1. Faulty bulb.
2. Open in the resistor circuit that is connected in parallel to the bulb (Figure 16-15).

Figure 16-13 Testing the lamp circuit of a CS alternator system.

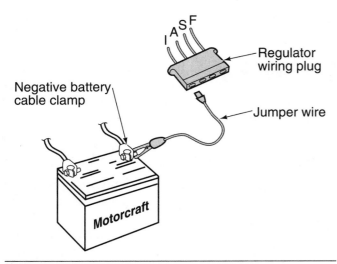

Figure 16-14 Testing the EVR lamp indicator circuit. (Reprinted with permission of Ford Motor Company)

3. An open in the circuit between the ignition switch and bulb.

Ford Integral Alternator Regulator (IAR) System

If the light does not illuminate when it is proofed, turn off the ignition switch. Disconnect the wiring con-

nector from the regulator. Use a jumper wire to ground the I terminal of the connector (Figure 16-16).

Turn the ignition switch to the RUN position while observing the warning lamp. If the lamp comes on, the regulator will need to be tested. If the lamp does not come on, check the bulb. If the bulb is good, there is an

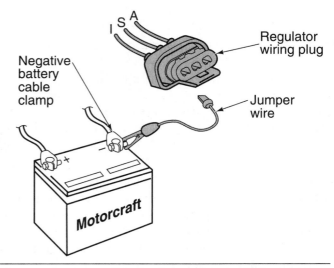

Figure 16-16 Testing the lamp circuit of the IAR system. (Reprinted with permission of Ford Motor Company)

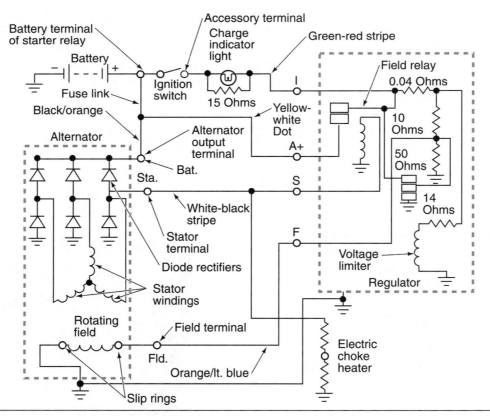

Figure 16-15 Wiring diagram of the light circuit for an EVR system. The resistor is used to bypass the bulb in the event the bulb burns out. If the resistor was not used, the charging system would not produce voltage if the bulb burned out.

open in the circuit between the ignition switch and the regulator or in the resistor across the lamp.

Chrysler LED Warning Indicator

The light-emitting diode (LED) is an integral part of the ammeter gauge. The LED will glow if system voltage drops below a predetermined value. To test the circuit, place the ignition switch to the RUN position. Turn on the headlights, windshield wipers, and activate the brake lights. If the LED does not come on within one minute, the gauge must be replaced. If the LED comes on within the one-minute time span, start the engine and gradually increase the engine speed to 2,000 rpm. If the LED does not turn off (and the charging system output is within specifications), replace the gauge.

Even though the LED operates separate of the gauge, if it is defective the entire gauge unit must be replaced.

Audible Warning Systems

Many warning systems also include an audible signal to alert the driver of potentially dangerous conditions. This could include seatbelt warning systems, low oil pressure, key in ignition, headlights on, and so on. Most customers complain because systems continue to sound after the normal amount of time has passed.

The buzzer may have a timer circuit to control how long it is activated. If the buzzer does not shut off, replace the timer with a known good unit. If the buzzer

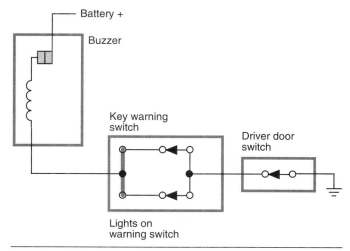

Figure 16-17 Typical buzzer warning system that is not timer controlled.

does not have a timer circuit, and remains active after the indicated problem is corrected (headlights on with the door open, for example), check for a short to ground on the switch side of the buzzer and for a faulty switch (Figure 16-17). If the buzzer does not sound, remove it and apply 12 volts across the terminals. If the buzzer sounds, it is good and the problem is in the circuit.

A chime module or tone generator can be used to alert the driver of several different conditions. This reduces the number of buzzers and timer circuits that would be required to operate each system separately.

Chrysler uses a chime module or timed buzzer relay depending on the year of manufacture (Figure 16-18). To test these units:

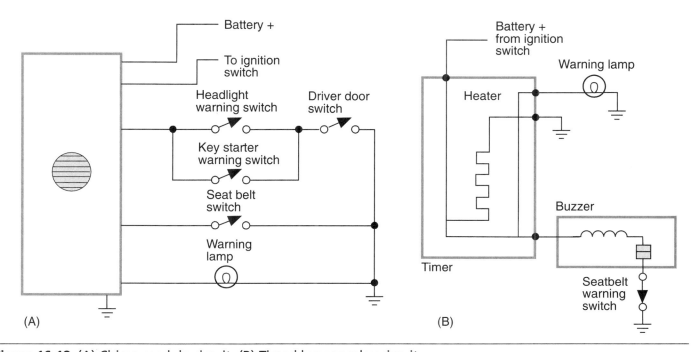

Figure 16-18 (A) Chime module circuit. (B) Timed buzzer relay circuit.

1. Remove the module from the vehicle.

2. Use a test light made with a 194 bulb to connect between terminal 3 of the module and ground.

3. Use jumper wires to ground terminals 2 and 5 of the module.

4. Connect a jumper wire that is fitted with an in-line 20-amp fuse to the battery positive post and module terminal 4.

5. The test light and audible signal should come on for approximately eight seconds. If not, the module is bad.

A schematic of Ford's audible warning system is shown (Figure 16-19). The schematic can be used to test the tone generator in the same manner as described earlier.

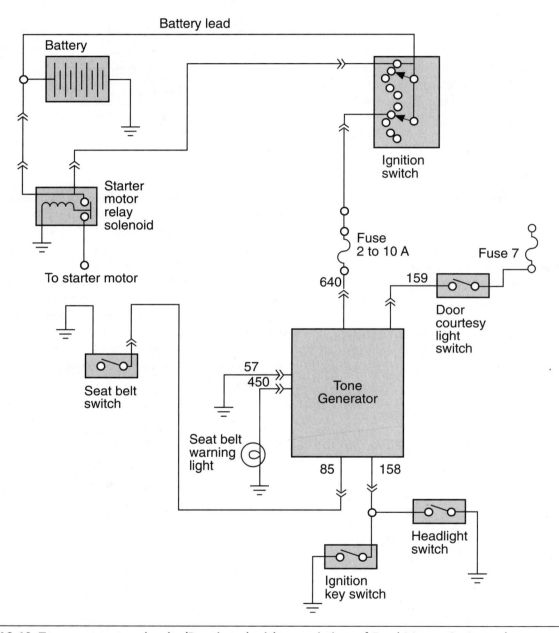

Figure 16-19 Tone generator circuit. (Reprinted with permission of Ford Motor Company)

Summary

❑ Before removing the instrument panel, always disconnect the battery negative cable.

❑ Speedometer complaints range from chattering noises when cold, to inaccurate readings, to not operating at all.

❑ If the speedometer assembly must be replaced, usually a new odometer is included with the assembly. Be sure to follow the manufacturer's procedures for setting the odometer to the correct reading.

❑ With the exception of the voltmeter and ammeter, all electromechanical gauges (whether bimetallic or electromagnetic) use a variable resistance sending unit. The types of tests performed will depend on the nature of the problem and if the system uses an IVR.

❑ To test the coolant temperature sensing unit, use an ohmmeter to measure the resistance between the terminal lead and ground. The resistance value of the variable resistor should change in proportion to coolant temperature.

❑ To test a piezoresistive sensor sending unit used for oil pressure gauges, connect the ohmmeter to the sending unit terminal and ground. Check the resistance with the engine off and compare to specifications. Start the engine and allow it to idle. Check the resistance value and compare to specifications.

❑ It is not likely that all of the warning lights would fail at the same time. Check the fuse if all of the lights are not operating properly.

❑ The basic principle of most charging system indicator lights is that a lamp only lights when there is a voltage drop or potential across its terminals.

❑ Before checking the charging warning lamp circuit, check that the AC generator drive belt is in good condition and properly tensioned. Perform a battery test series, and test the AC generator. If all of these test results are good, then check the lamp circuit.

❑ Many warning systems also include an audible signal to alert the driver of potentially dangerous conditions.

❑ A chime module or tone generator can be used to alert the driver of several different conditions. This reduces the number of buzzers and timer circuits that would be required to operate each system separately.

ASE Style Review Questions

1. Gauge malfunction is being discussed. Technician A says that if the IVR is defective, only one gauge will be affected. Technician B says that if all gauges are not operating, check the fuse. Who is correct?
   a. A only
   b. B only
   c. Both A and B
   d. Neither A nor B

2. A customer complains that their speedometer makes noise constantly. Technician A says that the problem can be a kinked cable core. Technician B says that the problem can be worn drive gear. Who is correct?
   a. A only
   b. B only
   c. Both A and B
   d. Neither A nor B

3. Technician A says that using a lubricant on the speedometer cable will cause the speedometer to be inaccurate. Technician B says that if the drive gear is replaced the driven gear should also be replaced. Who is correct?
   a. A only
   b. B only
   c. Both A and B
   d. Neither A nor B

4. Technician A says warning lamp circuits use variable resistor sensors. Technician B says gauges use switches as sensors. Who is correct?
   a. A only
   b. B only
   c. Both A and B
   d. Neither A nor B

speed brush (Figure 17-13). The wiper will operate until the park switch swings back to the PARK position.

The delay between wiper sweeps is determined by the amount of resistance the driver puts into the potentiometer control. By rotating the intermittent control knob, the resistance value is altered. The module contains a capacitor that is charged through the potentiometer. Once the capacitor is saturated, the electronic switch is "triggered" to send current to the wiper motor. The capacitor discharge is long enough to start the wiper operation and the park switch is returned to the RUN position. The wiper will continue to run until one sweep is completed and the park switch opens. The amount of time between sweeps is based on the length of time required to saturate the capacitor. As more resistance is added to the potentiometer, it takes longer to saturate the capacitor.

Many manufacturers incorporate the intermittent wiper function into the body computer. Also, some man-

ufacturers are equipping their vehicles with speed sensitive wiper systems. The delay between wiper sweeps is determined by the speed of the vehicle. This feature is discussed in a later chapter.

Depressed Wiper Systems

Wiper systems with a **depressed-park** feature forces the blades to drop down below the lower windshield molding to hide them. These systems use a second set of contacts with the park switch. These contacts are used to reverse the rotation of the motor for about 15 degrees after the wipers have reached the normal park position. The circuitry of the depressed circuit is different from standard wiper motors.

The operation of a depressed-park wiper system in the LOW SPEED position is shown (Figure 17-14). Current flows through the number 3 wiper to the common brush. Ground is provided through the low speed brush and switch wiper 2.

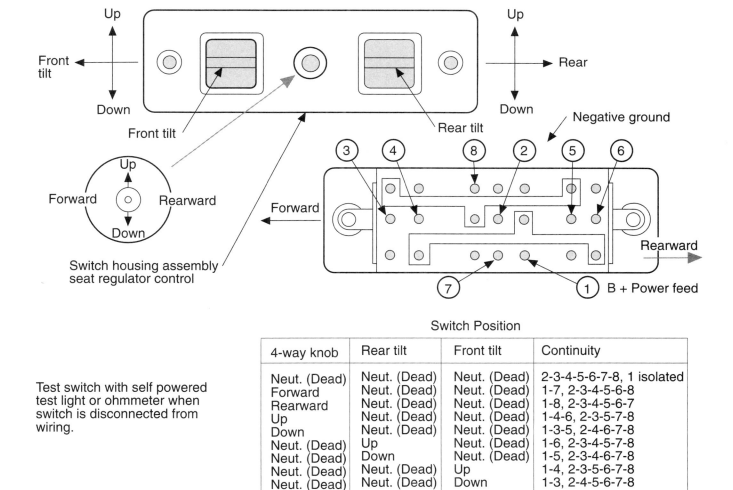

Test switch with self powered test light or ohmmeter when switch is disconnected from wiring.

Switch Position

4-way knob	Rear tilt	Front tilt	Continuity
Neut. (Dead)	Neut. (Dead)	Neut. (Dead)	2-3-4-5-6-7-8, 1 isolated
Forward	Neut. (Dead)	Neut. (Dead)	1-7, 2-3-4-5-6-8
Rearward	Neut. (Dead)	Neut. (Dead)	1-8, 2-3-4-5-6-7
Up	Neut. (Dead)	Neut. (Dead)	1-4-6, 2-3-5-7-8
Down	Neut. (Dead)	Neut. (Dead)	1-3-5, 2-4-6-7-8
Neut. (Dead)	Up	Neut. (Dead)	1-6, 2-3-4-5-7-8
Neut. (Dead)	Down	Neut. (Dead)	1-5, 2-3-4-6-7-8
Neut. (Dead)	Neut. (Dead)	Up	1-4, 2-3-5-6-7-8
Neut. (Dead)	Neut. (Dead)	Down	1-3, 2-4-5-6-7-8

Figure 17-13 Current flow when intermittent wiper mode is initiated. (Reprinted with permission of Ford Motor Company)

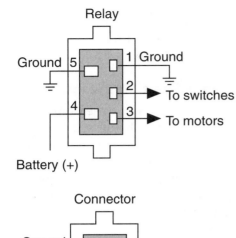

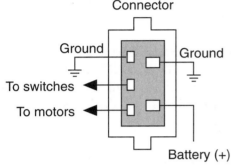

Figure 17-14 Depressed-park wiper system in LOW-SPEED position.

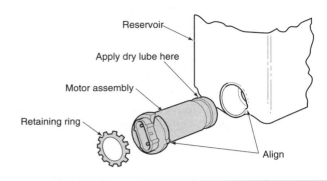

Figure 17-15 Washer motor installed into the reservoir. (Reprinted with permission of Ford Motor Company)

Figure 17-16 General Motors' pulse-type washer system incorporates the washer motor into the wiper motor.

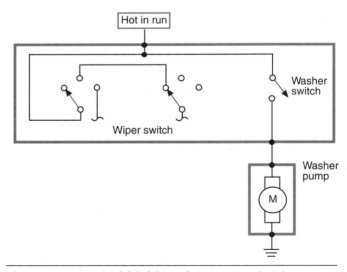

Figure 17-17 Windshield washer motor circuit.

When the switch is placed in the OFF position, current is supplied through the park switch wiper B and switch wiper 3. Ground is supplied through the low speed brush and switch wiper 1, then to park switch wiper A.

When the wipers reach their park position, the park switch swings to the PARKING position. Current flow is through the park switch wiper A, to switch wiper 1. Wiper 1 directs the current to the low speed brush. The ground path is through the common brush, switch wiper 3, and park switch wiper B. This reversed current flow is continued until the wipers reach the depressed-park position, when park switch wiper A swings to the PARKED position.

Washer Pumps

Windshield washers spray a washer fluid solution onto the windshield and work in conjunction with the wiper blades to clean the windshield of dirt. Some vehicles equipped with composite headlights incorporate a headlight washing system along with the windshield washer. Most systems have the washer pump motor installed into the reservoir (Figure 17-15). General Motors uses a pulse-type washer pump that operates off the wiper motor (Figure 17-16).

The system is activated by holding the washer switch (Figure 17-17). If the wiper/washer system also has an intermittent control module, a signal is sent to the module when the washer switch is activated. An override circuit in the module operates the wipers on low speed for a programmed length of time. The wipers will either return to the parked position or will operate in intermittent mode, depending on system design.

Blower Motor Circuit

The **blower motor** is used to move air inside the vehicle for air conditioning, heating, defrost, and ventilation. The motor is usually a permanent magnet, single-speed motor and is located in the heater housing assembly (Figure 17-18). A blower motor switch mounted on the dash controls the fan speed (Figure 17-19). The switch position directs current flow to a **resistor block** that is wired in series between the switch and the motor. The resistor block consists of two or three helically wound wire resistors wired in series (Figure 17-20).

The blower motor circuit includes the control assembly, blower switch, resistor block, and the blower motor (Figure 17-21). This system uses an insulated side switch and a grounded motor. Battery voltage is applied to the control head when the ignition switch is in the RUN or ACC positions. The current can flow from the control head to the blower switch and resistor block in any control head position except OFF.

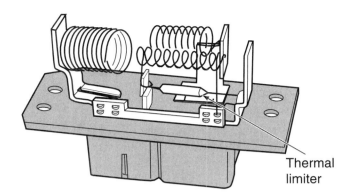

Figure 17-20 Fan motor resistor block. (Reprinted with permission of Ford Motor Company)

When the blower switch is in the LOW position, the blower switch wiper opens the circuit. Current can only flow to the resistor block directly through the control head. The current must pass through all the resistors before reaching the motor. With the voltage dropped over the resistors, the motor speed is slowed.

When the blower switch is placed in the MED 1, MED 2, or HIGH position, the current flows through the blower switch to the resistor block. Depending on the speed selection, the current must pass through one, two, or none of the resistors. With more applied voltage to the motor, the fan speed is increased as the amount of resistance decreases.

Current through the circuit will remain constant; varying the amount of resistance changes the voltage applied to the motor. Because the motor is a single-speed motor, it obtains its fastest rotational speed with full battery voltage. The resistors drop the amount of voltage to the motor, resulting in slower speeds.

Some manufacturers use ground side switching with an insulated motor (Figure 17-22). The switch completes the circuit to ground. Depending on wiper position, current flow is directed through the resistor block. The operating principles are identical to those of the insulated switch already discussed.

Many of today's vehicles are using the body computer to control fan speed by pulse width modulation. This principle is discussed in a later chapter.

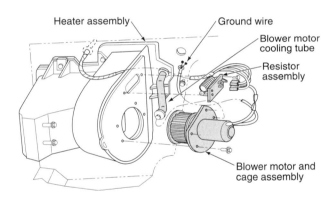

Figure 17-18 The blower motor is usually installed into the heater assembly. Mode doors control if vent, heater, or A/C cooled air is blown by the motor cage.

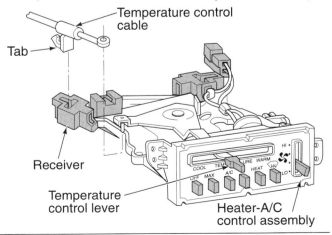

Figure 17-19 Comfort control assembly. (Courtesy of Chrysler Corporation)

A BIT OF HISTORY

Automotive electric heaters were introduced at the 1917 National Auto Show. Hot water in-car heaters were introduced in 1926.

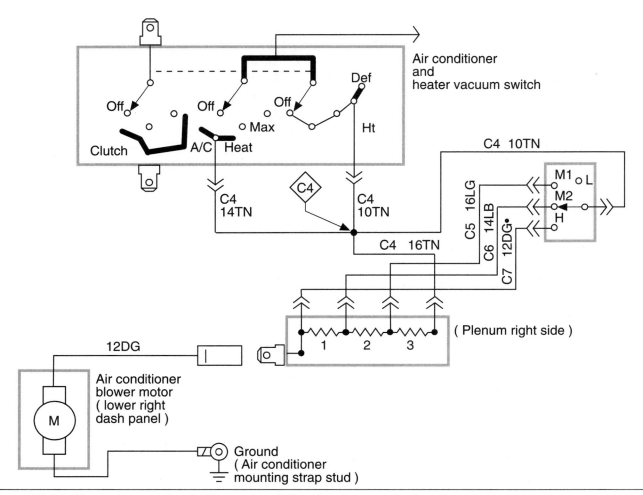

Figure 17-21 Blower motor circuit. (Courtesy of Chrysler Corporation)

Electric Defoggers

When electrons are forced to flow through a resistance, heat is generated. Rear window **electric defoggers** use this principle of controlled resistance to heat the glass to remove ice and/or condensation. This system is also referred to as a **heated backlight**. The resistance is through an **electric grid** which is a series of horizontal ceramic silver-compounded lines baked into the surface of the window (Figure 17-23). Terminals are soldered to the vertical bus bars. One terminal supplies the current from the switch; the other provides the ground (Figure 17-24).

The system may incorporate a timer circuit to control the system relay (Figure 17-25). The timer is used due to the high amount of current required to operate the system (approximately 30 amperes). If this drain were allowed to continue for extended periods of time, battery and charging system failure could result. Because of the high current draw, most vehicles equipped with a rear window defogger use a high output AC generator.

The control switch is a three-position, spring-loaded switch that returns to the center position after making momentary contact to the ON or OFF terminals. Activation of the switch energizes the electronic timing circuit, which energizes the relay coil. With the relay contacts closed, current flows to the heater grid. At the same time, voltage is applied to the ON indicator. The timer is activated for 10 minutes. At the completion of the timed cycle, the relay is de-energized and the circuit to the grid and indicator light is broken. If the switch is activated again, the timer will energize the relay for five minutes.

The timer sequence can be aborted by moving the switch to the OFF position or by turning off the ignition switch. If the ignition switch is turned off while the timer circuit is activated, the rear window defogger switch will have to be returned to the ON position to activate the system again.

Ambient temperatures have an effect on electrical resistance, thus the amount of current flow through the grid depends on the temperature of the grid. As the ambient temperature decreases, the resistance value of the grid also decreases. A decrease in resistance increases

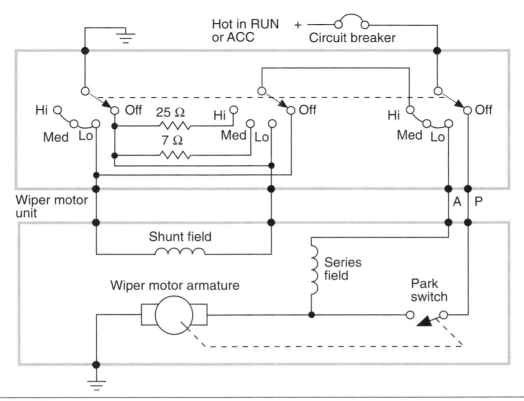

Figure 17-22 Ground side switch to control the blower motor system. (Reprinted with permission of Ford Motor Company)

the current flow and results in quick warming of the window. The defogger system tends to be self-regulated to match the requirements for defogging.

Many manufacturers also have heater grids in the outside mirrors. The principle of operation is the same as the rear window heater grid. The mirror heaters are usually wired into the rear window circuit so they both operate at the same time whenever the rear window defogger switch is activated.

Power Mirrors

The electrically controlled **power mirror** allows the driver to position the outside mirrors by use of a switch

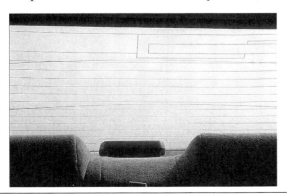

Figure 17-23 Rear window defogger grid.

located inside the vehicle. The mirror assembly will use built-in dual drive, reversible permanent magnet (PM) motors (Figure 17-26).

A single switch for controlling both the left and right side mirrors may be used. On many systems, selection of the mirror to be adjusted is by rotating the knob counterclockwise for the left mirror and clockwise for the right mirror. After the mirror is selected, movement of the joystick (up, down, left, or right) moves the mirror in the corresponding direction. The illustration (Figure 17-27) shows a logic table for the mirror switch and motors.

Automatic Rear View Mirror

Some manufacturers have developed interior rear view mirrors that automatically tilt when the intensity of light striking the mirror is sufficient enough to cause discomfort to the driver.

The system has two photocells mounted in the mirror housing. One of the photocells is used to measure the intensity of light inside the vehicle. The second is used to measure the intensity of light being received by the mirror. When the intensity of the light striking the mirror is greater than ambient light by a predetermined amount, a solenoid is activated that tilts the mirror.

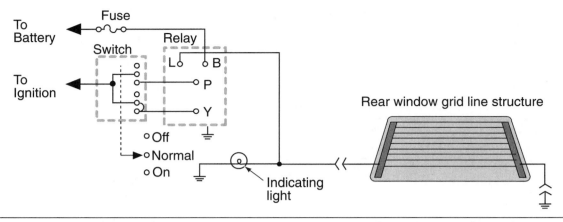

Figure 17-24 Rear window defogger circuit schematic. (Courtesy of Chrysler Corporation)

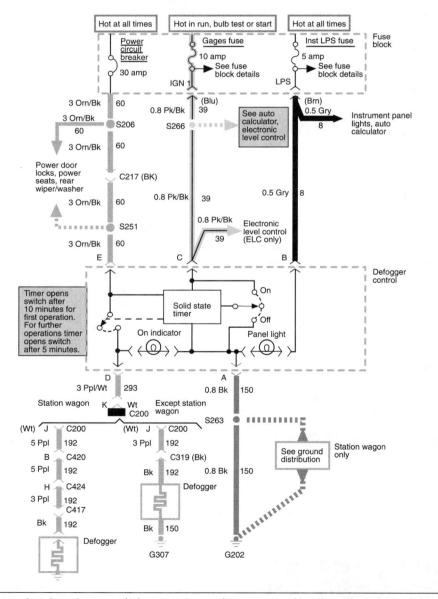

Figure 17-25 Defogger circuit using a solid-state timer. (Courtesy of General Motors Corporation, Service Technology Group)

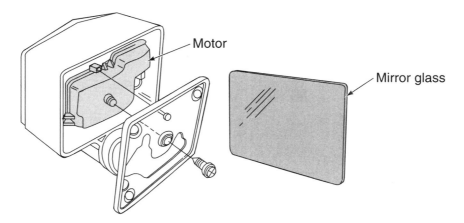

Figure 17-26 Power mirror motor. (Courtesy of General Motors Corporation, Service Technology Group)

Electrochromic Mirrors

Electrochromic mirrors automatically adjust the amount of reflectance based on the intensity of glare (Figure 17- 29). The process is similar to that used by photo-gray or photochromic sunglasses. If the glare is heavy, the mirror darkens to about 6% reflectivity. The electrochromic mirror has the advantage of providing a comfort zone where the mirror will provide 20% to 30% reflectivity. When no glare is present, the mirror changes to the daytime reflectivity rating of up to 85%. The reduction of the glare by darkening of the mirror does not impair visibility.

Electrochromic mirrors can be installed as the outside mirrors and/or inside mirrors. Usually if both inside and outside electrochromic mirrors are used, the outside mirror is controlled by the inside mirror. The mirror is constructed of a thin layer of electrochromic material that is placed between two plates of conductive glass. There are two photocell sensors that measure light intensity in front and in back of the mirror. During night driving, the headlight beam striking the mirror causes the mirror to gradually become darker as the light intensity increases. The darker mirror absorbs the glare. On some systems the sensitivity of the mirror can be adjusted by the driver through a three-position switch.

When the ignition switch is placed in the RUN position, battery voltage is applied to the three-position switch. If the switch is in the MIN position battery voltage is applied to the solid-state unit, and sets the sensitivity to a low level. The MAX setting causes the mirror to darken more at a lower glare level. When the transmission is placed in reverse, the reset circuit is activated. This returns the mirror to daytime setting for clearer viewing to back up.

Power Windows

Many vehicles replace the conventional window crank with electric motors that operate the side windows. In addition, most station wagon models are equipped with electric rear tailgate windows. The motor used in the **power window system** is a reversible PM or two-field winding motor.

The power window system usually consists of the following components:

1. Master control switch
2. Individual control switches
3. Individual window drive motors
4. Lock-out or disable switch

Another design is rack-and-pinion gears. The rack is a flexible strip of gear teeth with one end attached to the window (Figure 17-29).

A BIT OF HISTORY

Power windows were introduced in 1939.

The motor operates the **window regulator** either through a cable or directly. The window regulator converts the rotary motion of the motor into the vertical movement of the window. On direct drive motors, the motor pinion gear meshes with the **sector gear** on the regulator (Figure 17-30). As the window is lowered, the spiral spring is wound. The spring unwinds as the window is raised to assist in raising the window. The spring reduces the amount of current required to raise the window by the motor itself.

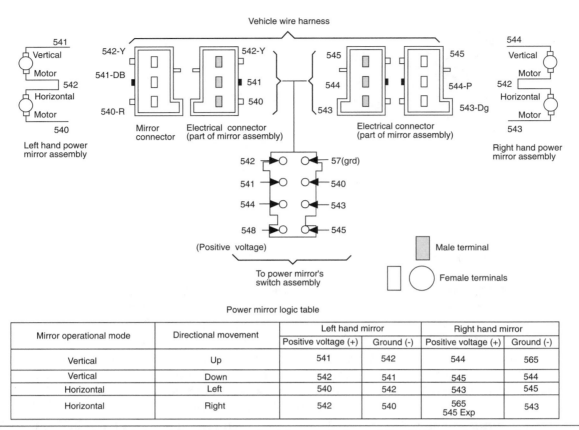

Figure 17-27 Power mirror logic table. (Reprinted with permission of Ford Motor Company)

		Left hand mirror		Right hand mirror	
Mirror operational mode	Directional movement	Positive voltage (+)	Ground (-)	Positive voltage (+)	Ground (-)
Vertical	Up	541	542	544	565
Vertical	Down	542	541	545	544
Horizontal	Left	540	542	543	545
Horizontal	Right	542	540	565 545 Exp	543

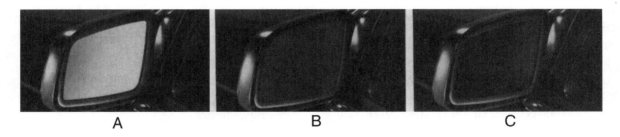

Figure 17-28 Electrochromic mirror operation: (A) day time; (B) mild glare; and (C) high glare. (Courtesy of Gentex Corporation).

The master control switch provides the overall control of the system. Power to the individual switches is provided through the master switch. The master switch may also have a safety lock switch to prevent operation of the windows by the individual switches. When the safety switch is activated, it opens the circuit to the other switches and control is only by the master switch. As an additional safety feature, some systems prevent operation of the individual switches unless the ignition switch is in the RUN or ACC position (Figure 17-31).

Wiring circuits depend on motor design. Most PM-type motors are insulated, with ground provided through the master switch (Figure 17-32). When the master control switch is placed in the UP position, current flow is

from the battery, through the master switch wiper, through the individual switch wiper, to the top brush of the motor. Ground is through the bottom brush and circuit breaker to the individual switch wiper, to the master switch wiper and ground.

When the window is raised from the individual switch, battery voltage is supplied directly to the switch and wiper from the ignition switch. The ground path is through the master control switch.

When the window is lowered from the master control switch, the current path is reversed. The current flows through the individual switch to lower the window.

Some manufacturers use a two-field coil motor that is grounded with insulated side switches. The two field

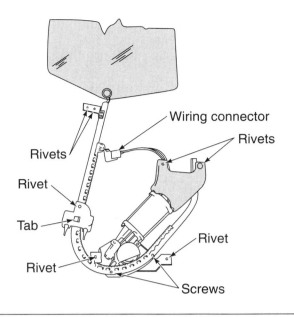

Figure 17-29 Rack and pinion style power window motor and regulator. (Courtesy of Chrysler Corporation)

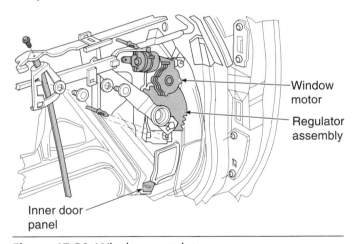

Figure 17-30 Window regulator.

coils are wired in opposite directions, and only one coil is energized at a time. Direction of the motor is determined by which coil is activated.

Power Seats

The power seat system is classified by the number of ways in which the seat is moved. The most common classifications are:

1. Two-way: Moves the seat forward and backward.
2. Four-way: Moves the seat forward, backward, up, and down.
3. Six-way: Moves the seat forward, backward, up, down, front tilt, and rear tilt.

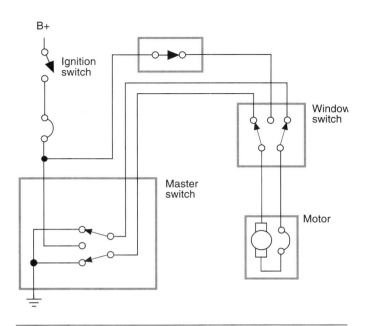

Figure 17-31 Power window wiring diagram.

All modern six-way power seats use a reversible, permanent magnet three-armature motor called **trimotor** (Figure 17-33). The motor may transfer rotation to a rack-and-pinion or to a worm gear drive transmission. A typical control switch consists of a four-position knob and a set of two-position switches (Figure 17-34). The four-position knob controls the forward, rearward, up, and down movement of the seat. The separate two-position switches are used to control the front tilt and rear tilt of the seat.

Current direction through the motor determines the rotation direction of the motor. The switch wipers control the direction of current flow (Figure 17-35). If the driver pushes the four-way switch into the down position, the entire seat lowers. Switch wipers 3 and 4 are swung to the left and battery voltage is sent through wiper 4 to wipers 6 and 8. These wipers direct the current to the front and rear height motors. The ground circuit is provided through wipers 5 and 7, to wiper 3 and ground.

Some manufacturers equip their seats with adjustable support mats that shape the seat to fit the driver. The lumbar support mat provides the driver with additional comfort by supporting the back curvature.

Power Door Locks

Some manufacturers use a **child safety latch** in the door lock system to prevent the door from being opened from the inside, regardless of the position of the door

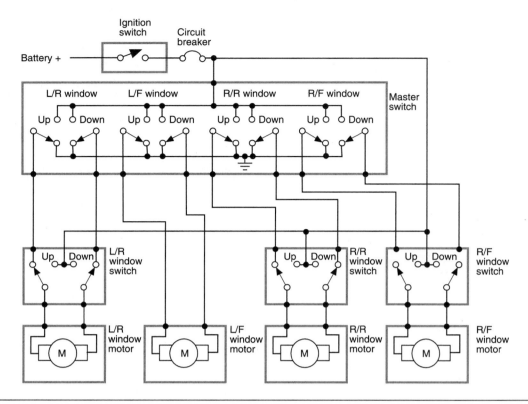

Figure 17-32 Typical power window circuit using PM motors.

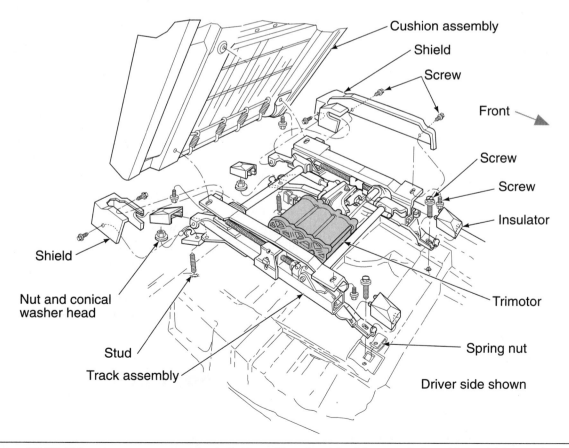

Figure 17-33 Trimotor power seat installation. (Reprinted with permission of Ford Motor Company)

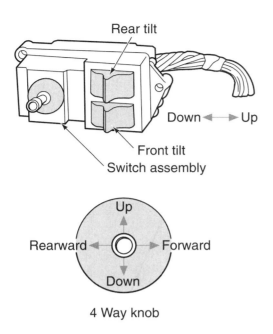

Figure 17-34 Power seat control switch. (Reprinted with permission of Ford Motor Company)

lock knob. The child safety latch is activated by a switch designed into the latch bellcrank.

Electric power locks use either a solenoid or a permanent magnet reversible motor. Due to the high current demands of solenoids, most modern vehicles use PM motors (Figure 17-36). Depending on circuit design, the system may incorporate a relay (Figure 17-37). The relay has two coils and two sets of contacts to control current direction. In this system the door lock switch energizes one of the door lock relay coils to send battery voltage to the motor.

In the system that does not use relays the switch provides control of current flow in the same manner as power seats or windows.

Many vehicles are equipped with automatic door locks that are activated when the gear shift lever is placed in the DRIVE position. The doors unlock when the selector is returned to the PARK position.

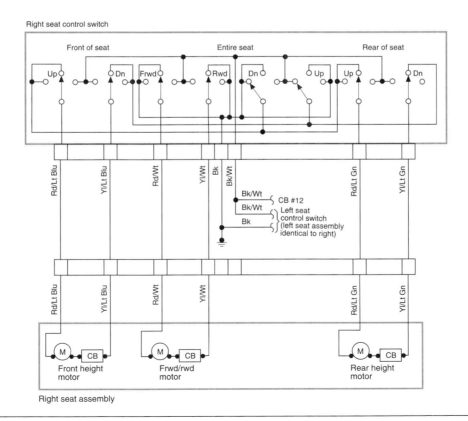

Figure 17-35 Six-way power seat circuit diagram.

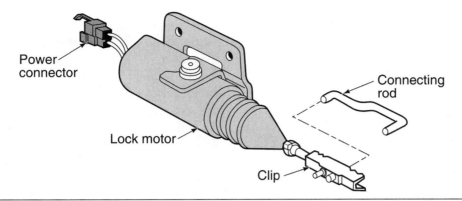

Figure 17-36 PM power door lock motor. (Courtesy of Chrysler Corporation)

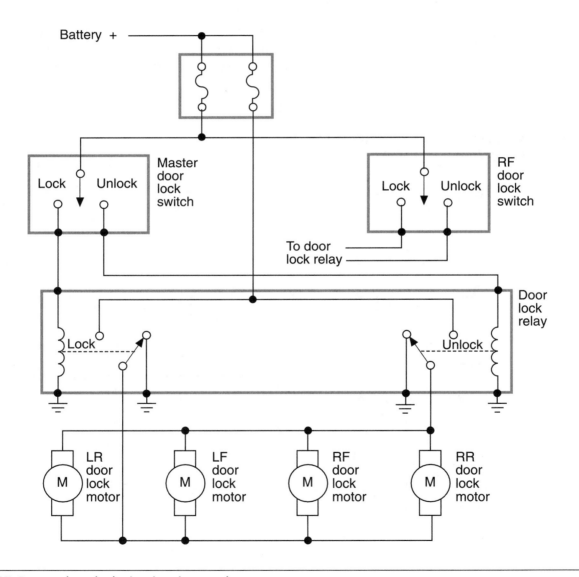

Figure 17-37 Power door lock circuit using a relay.

Summary

❏ Automotive electrical horns operate on an electromagnetic principle that vibrates a diaphragm to produce a warning signal.

❏ Horn switches are either installed in the steering wheel or as a part of the multifunction switch. Most horn switches are normally open switches.

❏ Horn switches that are mounted on the steering wheel require the use of sliding contacts. The contacts provide continuity for the horn control in all steering wheel positions.

❏ The most common type of horn circuit control is a relay.

❏ Most two-speed windshield wiper motors use permanent magnet fields whereby the motor speed is controlled by the placement of the brushes on the commutator.

❏ Some two-speed, and all three-speed wiper motors use two electromagnetic field windings: series field and shunt field. The two field coils are wound in opposite directions so that their magnetic fields will oppose each other. The strength of the total magnetic field will determine at what speed the motor will operate.

❏ Park contacts are located inside the wiper motor assembly, and supply current to the motor after the switch has been turned to the PARK position. This allows the motor to continue operating until the wipers have reached the PARK position.

❏ Intermittent wiper mode provides a variable interval between wiper sweeps and is controlled by a solid-state module.

❏ Systems that have a depressed-park feature use a second set of contacts with the park switch, which are used to reverse the rotation of the motor for about 15 degrees after the wipers have reached the normal park position.

❏ Blower fan motors use a resistor block that consists of two or three helically wound wire resistors that are connected in series to control fan speed.

❏ The blower motor circuit includes the control assembly, blower switch, resistor block, and the blower motor.

❏ Electric defoggers heat the rear window by means of a resistor grid.

❏ Electric defoggers may incorporate a timer circuit to prevent the high current required to operate the system from damaging the battery or charging system.

❏ The electrically controlled mirror allows the driver to position the outside mirrors by use of a switch that controls dual drive, reversible PM motors.

❏ Power windows, seats, and door locks usually use reversible PM motors, whereby motor rotational direction is determined by the direction of current flow through the switch wipers.

Terms-To-Know

Blower motor	Heated backlight	Power window
Child safety latch	High-note horn	Resistor block
Depressed-park	Horn	Sector gear
Diaphragm	Intermittent wiper	Trimotor
Electric defoggers	Low-note horn	Window regulator
Electric grid	Park contacts	Windshield wipers
Electrochromic mirrors	Power mirror	

Review Questions

Short Answer Essays

1. Describe the operation of a relay-controlled horn circuit.
2. Explain how brush placement determines the speed of a two-speed, permanent magnet motor.
3. How do wiper motor systems that use a three-speed, electromagnetic motor control wiper speed?
4. What is the purpose of the capacitor in some intermittent wiper systems?
5. Describe the method used to control blower fan motor speeds.
6. Describe the operation of electric defoggers.
7. Briefly explain the principles of operation for power windows.
8. Describe the operation of a six-way, trimotor power seat.
9. Explain how depressed-park wipers operate.
10. What is the advantage of PM motors over solenoids in the power lock system?

Fill-in-the-Blanks

1. Electrical accessories provide for additional _____ and _____ .
2. The _____ is a thin, flexible, circular plate that is held around its outer edge by the horn housing, allowing the middle to flex.
3. Horn switches that are mounted on the steering wheel require the use of _____ _____ to provide continuity in all steering wheel positions.
4. In systems equipped with _____ _____ , the blades drop down below the lower windshield molding to hide them.
5. PM windshield wiper motors use _____ brushes.
6. The operational speed of electromagnetic field winding motors used in wiper systems depends on the strength of the _____ _____ _____ .
7. Most blower motor fan speeds are controlled through a _____ _____ that is wired in series.
8. Electric defoggers operate on the principle that when electrons are forced to flow through a _____ , heat is generated.
9. A window _____ converts the rotary motion of the motor into the vertical movement of the window.
10. The three-armature motor is called a _____ .

ASE Style Review Questions

1. The sound generation from the horn is being discussed. Technician A says electrical horns operate by the generation of heat to vibrate a diaphragm. Technician B says the vibration of the column of air in the horn produces the sound. Who is correct?
   a. A only
   b. B only
   c. Both A and B
   d. Neither A nor B

2. The horn circuit is being discussed. Technician A says if the circuit does not use a relay, the horn switch carries the total current requirements of the horns. Technician B says that most systems that use a relay have battery voltage present to the lower contact plate of the horn switch and that the switch closes the path for the relay coil. Who is correct?
   a. A only
   b. B only
   c. Both A and B
   d. Neither A nor B

3. Permanent magnet wiper motors are being discussed. Technician A says that motor speed is controlled by the placement of the brushes on the commutator. Technician B says that the motor uses two electromagnetic fields. Who is correct?
   a. A only
   b. B only
   c. Both A and B
   d. Neither A nor B

4. Technician A says if there more armature windings connected between the common and high speed brushes there is less magnetism in the armature and a lower counterelectromotive force, thus the motor turns faster. Technician B says the less CEMF in the armature, the greater the armature current. Who is correct?
   a. A only
   b. B only
   c. Both A and B
   d. Neither A nor B

5. Electromagnetic field wiper motors are being discussed. Technician A says the two field coils are wound in opposite directions so that their magnetic fields will oppose each other. Technician B says the strength of the total magnetic field will determine the speed of the motor. Who is correct?
   a. A only
   b. B only
   c. Both A and B
   d. Neither A nor B

6. Intermittent wiper systems are being discussed. Technician A says the system uses a solid-state module or is incorporated into the body computer. Technician B says that the capacitor operates the wiper motor directly until it is fully discharged. Who is correct?
   a. A only
   b. B only
   c. Both A and B
   d. Neither A nor B

7. The blower motor circuit is being discussed. Technician A says the motor is usually a permanent magnet, single speed motor. Technician B says the switch position directs current flow to one of the three different brushes in the motor. Who is correct?
   a. A only
   b. B only
   c. Both A and B
   d. Neither A nor B

8. Technician A says that a resistor block that is wired in series between the switch and the motor controls fan speed. Technician B says that the thermal limiter acts as a circuit breaker to protect the circuit if the resistor block gets too hot. Who is correct?
   a. A only
   b. B only
   c. Both A and B
   d. Neither A nor B

9. The electric rear window defogger is being discussed. Technician A says the grid is a series of controlled voltage amplifiers. Technician B says the system may incorporate a timer circuit to protect the vehicle electrical systems. Who is correct?
   a. A only
   b. B only
   c. Both A and B
   d. Neither A nor B

10. Technician A says that the master control switch for the power windows provides the overall control of the system. Technician B says that current direction through the power seat motors determines the rotation direction of the motor. Who is correct?
    a. A only
    b. B only
    c. Both A and B
    d. Neither A nor B

18 Electrical Accessories Diagnosis and Repair

Objective

Upon completion and review of this chapter, you should be able to:

❏ Identify the causes of no operation, intermittent operation, or constant horn operation.

❏ Diagnose the cause of poor sound quality from the horn system.

❏ Perform diagnosis and repair of no windshield wiper operation in one speed only or in all speeds.

❏ Identify causes for slower than normal wiper operation.

❏ Determine the cause for improper park operation.

❏ Identify causes for continuous wiper operation.

❏ Diagnose faulty intermittent wiper system operation.

❏ Remove and install wiper motors and wiper switches.

❏ Determine the causes for improper operation of the windshield washer system and be able to replace the pump if required.

❏ Perform diagnosis of problems associated with blower motor circuits.

❏ Diagnose and repair electric rear window defoggers.

❏ Diagnose the power window system.

❏ Diagnose common problems associated with the power seat system.

❏ Perform diagnosis of the power door lock system.

Introduction

The electrical accessories included in this chapter represent the most often performed electrical repairs. Most of the systems discussed do not provide for rebuilding of components. The technician must be capable of diagnosing the fault and then replacing the defective part. As with any electrical system, always use a systematic diagnostic approach to finding the cause. Refer to the service manual to obtain information concerning correct system operation. The fault is easier to locate once you understand how the system is supposed to operate.

Included in this chapter are diagnostic and repair procedures for horn systems, windshield wipers and washer systems, blower motors, electric defogger systems, power seat and window systems, and power door locks. Although the procedures here are typical, always refer to the service manual for the vehicle you are working on.

Horn Diagnosis

Customer concerns associated with the horn system can include no operation, intermittent operation, continuous operation, or poor sound quality. Testing of the horn system varies between systems that do and do not use a relay.

No Horn Operation

When a customer complains of no horn operation, first confirm the complaint by depressing the horn button. If it is mounted in the steering wheel, rotate the steering wheel from stop to stop while depressing the horn button. If the horn sounds intermittently while the steering wheel is turned, the problem is probably in the sliding contact ring in the steering column, or the tension spring is worn or broken. If the horn does not sound during this test, continue to check the system as outlined below.

If the system uses a relay, follow this typical procedure:

1. Check the fuse or fusible link. If defective, replace as needed. After replacement of the fuse, operate all other circuits protected by it. It is possible that another circuit is faulty but the customer has not noticed.

2. Connect a jumper wire from the battery positive terminal to the horn terminal. If the horn sounds, continue testing; if the horn does not sound, check the ground connection. Replace the horn if the ground is good.

3. In a multiple horn system, test for voltage at the last common connection between the horn relay and the horns (Figure 18-1). On a single horn system, test for voltage at the horn terminal. Voltage should be present at this connection only when the horn button is depressed. Continue testing if there is no voltage at this connection. If voltage is present, check the individual circuits from the connection to the horns; repair as needed.

4. Check for voltage at the power feed terminal of the relay (Figure 18-2). If there is no voltage at this point, trace the wiring from the relay to the battery to locate the problem. Continue testing if voltage is present at the power feed terminal.

5. Check for voltage at the switch control terminal of the relay. If voltage is not present at this terminal, the relay is defective and must be replaced. Continue testing if voltage is present.

6. Use a jumper wire to ground the switch control terminal of the relay. If the horn does not sound, replace the relay. Continue to test if the horn sounds.

7. Check for voltage on the battery side of the horn switch. If there is no voltage at this location, the fault is between the relay and the switch. Continue testing if voltage is present.

8. Check for continuity through the switch. If good, check the ground connection for the switch, then recheck operation. Replace the horn switch if there is no continuity when the button is depressed.

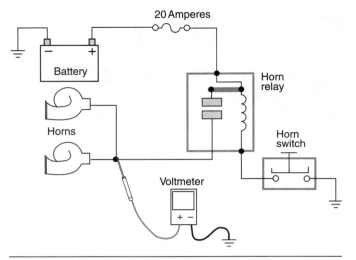

Figure 18-1 Testing for voltage at the last common connection.

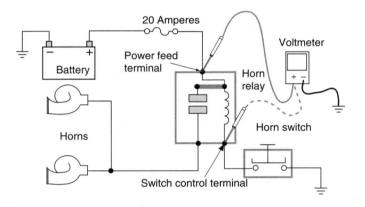

Figure 18-2 Testing for power into the relay and the relay coil.

If the system does not use a relay, perform the following tests to locate the fault:

1. Check the fuse or fusible link. If defective, replace as needed. After replacing the fuse, operate all other circuits protected by it. It is possible that another circuit is faulty but the customer has not noticed.

2. Connect a jumper wire from the battery positive terminal to the horn terminal. If the horn sounds, continue testing; if the horn does not sound, check the ground connection. Replace the horn if the ground is good.

3. In a multiple horn system, test for voltage at the last common connection between the horn switch and the horns. On a single horn system, test for voltage at the horn terminal. Voltage should be present at this connection only when the horn button is depressed. Continue testing if there is no voltage at this connection. If voltage is present, check the individual circuits from the connection to the horns; repair as needed.

4. Check for voltage at the horn side of the switch when the button is depressed. If voltage is present, the problem is in the circuit from the switch to the horn(s). Continue testing if there is no voltage at this connection.

5. Check for voltage at the battery side of the switch. If voltage is present, the switch is faulty and must be replaced. If there is no voltage at this terminal, the problem is in the circuit from the battery to the switch.

Poor Sound Quality

Poor sound quality can be the result of several factors. In a multiple horn system, if one of the horns is not operating, the horn sound quality may suffer. Other reasons include damaged diaphragms, excessive circuit resistance, poor ground connections, or improperly adjusted horns.

If one horn of a multiple horn system is not operating, use a jumper wire from the battery positive terminal to the horn terminal to determine whether the fault is in the horn or in the circuit. If the horn sounds, the problem is in the circuit between the last common connection and the affected horn.

If one or all horns are producing poor quality sound, use a voltmeter to measure the voltage at the horn terminal when the horn switch is closed. The voltage should be within 0.5 volt of battery voltage. If the voltage measured is less than battery voltage, there is excessive voltage drop in the circuit. Work back through the circuit measuring voltage drop across connectors, relays, and switches to find the source of the high resistance.

If the applied voltage to the horn terminal is good, connect a jumper wire between the battery positive terminal and the horn terminal to activate the horn. With the horn sounding, turn the adjusting screw counterclockwise one quarter to three eighths of a turn. Replace the horn if the sound quality cannot be improved.

Horn Sounds Continuously

A horn that sounds continuously is usually caused by a sticking horn switch or sticking contact points in the relay. To find the fault, disconnect the horn relay from the circuit. Use an ohmmeter to check for continuity from the battery feed terminal of the relay to the horn circuit terminal (Figure 18-3). If there is continuity, the relay is defective. If there is no continuity through the relay, test the switch for continuity; there should be no continuity through the switch when the horn button is in the rest position.

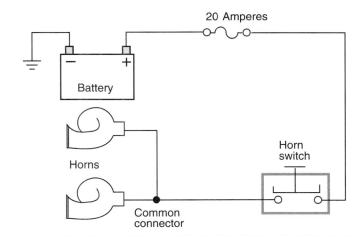

Figure 18-3 Ohmeter testing to find the cause of continuous horn operation.

Wiper System Service

Customer complaints concerning windshield wiper operation can include no operation, intermittent operation, continuous operation, and wipers will not park. Other complaints have to do with blade adjustment (such as blades slapping the molding or one blade parks lower than the other).

When a customer brings the vehicle into the shop because of faulty windshield wiper operation, the technician needs to determine if it is an electrical or mechanical problem. To do this, simply disconnect the arms to the wiper blades from the motor (Figure 18-4). Turn on the wiper system. Observe operation of the motor. The problem is mechanical if the motor is operating properly.

No Operation in One Speed Only

Problems that cause the system to not operate in only one switch position are usually electrical. Use the service

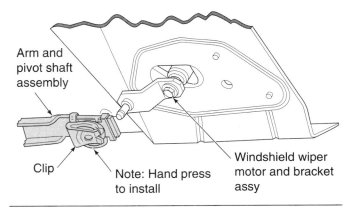

Figure 18-4 Disconnecting the mechanical arms from the motor. (Reprinted with the permission of Ford Motor Company)

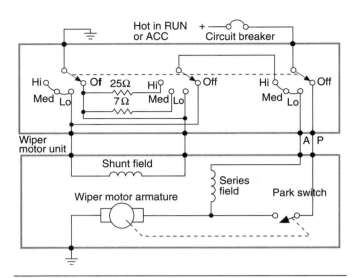

Figure 18-5 Three-speed windshield wiper system schematic.

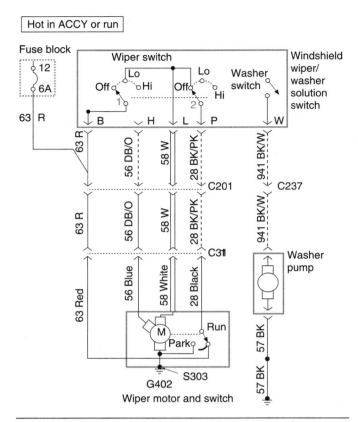

Figure 18-6 Two-speed wiper circuit. (Reprinted with the permission of Ford Motor Company)

manual's wiring schematic to determine proper operation. For example, use the three-speed wiper system schematic illustrated (Figure 18-5) to determine the cause of a motor that does not operate in the MEDIUM speed position only. The problem is that the 7-ohm resistor is open. The problem could not be the shunt field in the motor because LOW and HIGH speeds operate; nor could it be in the wiring to the motor because this is shared by all speeds.

An opened resistor can be verified by using a voltmeter to measure voltage at the terminal leading to the shunt field. If it drops to zero volts in the MEDIUM position, the switch must be replaced. By proper use of the wiring schematic, and by understanding the correct operation of the system, you are able to diagnose this problem without having to use any test equipment. The voltmeter is used only to verify your conclusions.

In two-speed systems, the motor operating in only one speed position can be caused by several different faults. It will require the use of wiring schematics and test equipment to locate. Use Figure 18-6 to step through a common test sequence to locate the reason why the motor does not operate in the HIGH position. This instance is used as an example of troubleshooting the wiper system. Be sure to use the correct schematic for the vehicle that you are working on.

WARNING *Do not leave the ignition switch in the RUN position for extended periods of time without the engine running. This may result in damage to the ignition system components.*

Turn the ignition switch to the ACC position if the wipers will operate in this position. If not, place the ignition switch in the RUN position. Place the wiper switch in the HIGH position. Use a voltmeter to test for voltage at circuit connector 56 of the motor. If voltage is present at this point, the high speed brush is worn or the wire from the terminal to the brush is open. Most shops do not rebuild the wiper motor; replacement is usually the preferred service. If there is no voltage present at circuit connector 56, check for voltage at the connector for the switch at circuit 56. If voltage is present at this point, the fault is in the circuit from the switch to the motor. If there is no voltage at this point, replace the switch.

To test for no LOW speed operation only, use the same procedure to test circuit 58.

No Wiper Operation

If the wiper motor does not operate in any speed position, check the fuse. If the fuse is blown, replace it and test the operation of the motor. Also check for binding in the mechanical portion of the system. This can cause an overload and blow the fuse.

If the fuse is good, check the motor ground by using a jumper wire from the motor body to a good chassis ground. If the motor operates when the ignition switch is in the RUN position and the wiper switch is placed in all speed positions, repair the ground connection. Continue testing if the motor does not operate.

Use a voltmeter to check for voltage at the low speed terminal of the motor with the ignition switch in the RUN or ACC position and the wiper switch in LOW position. If there is no voltage at this point, test for voltage on the LOW speed terminal of the wiper switch. If the voltmeter indicates battery voltage at this terminal, the fault is in the circuit between the switch and motor. Look for indications of burned insulation or other damage that would affect both the HIGH and LOW speed circuit.

No voltage at the LOW speed terminal of the wiper switch indicates the fault may be in the switch or the power feed circuit. Test for battery voltage at the battery supply terminal of the switch. If there is voltage at this point, the switch is faulty and needs to be replaced. If no voltage is at the supply terminal, trace the circuit back to the battery to locate the fault.

If battery voltage was present at the LOW speed terminal of the motor, check for voltage at the HIGH speed terminal. Voltage at both of these terminals indicates the motor is faulty and needs to be replaced. If there is no voltage at the HIGH speed terminal, use the procedure just described to trace the HIGH speed circuit.

SERVICE TIP *No voltage at either terminal means the problem is probably at a shared location. In most systems this would be the power supply portion of the wiper switch and the switch itself. If there is no voltage at either terminal, go directly to the power supply terminal of the switch. If the switch is good, and power is through the switch, the problem is in the wiring loom. Check all connectors. If good, remove the harness protector and inspect the wires for burned insulation or breaks.*

Slower than Normal Wiper Speeds

Slower than normal wiper speeds can be caused by electrical or mechanical faults. An ammeter can be used to determine the current draw of the motor with and without the mechanical portion connected. If the draw changes substantially when the mechanical portion is removed, the fault is in the arms and/or wiper blades.

Most electrical circuit faults that result in slow wiper operation are caused by excessive resistance. If the complaint is that all speeds are slow, use the voltage drop test procedure to check for resistance in the power feed supply circuit to the wiper switch. If the power supply circuit is good, then check the switch for excessive resistance.

If the insulated side voltage tests fail to locate the problem, check the voltage drop on the ground side of the wiper motor. Connect the voltmeter positive lead to the ground terminal of the motor (or motor body) and the negative lead to the vehicle chassis. The voltage drop should be no more than 0.1 volt. If excessive, repair the ground circuit connections. Check voltage drop through the motor by connecting the positive voltmeter lead to the LOW speed terminal and the negative lead to the vehicle chassis. Operate the motor on LOW speed and observe the voltmeter. Check the results against specifications (usually about 0.4 volt max.). If the voltage drop is excessive during this test, the motor is faulty and should be replaced.

Wipers Will Not Park

The most common complaint associated with a faulty park switch is that the wipers stop in the position they are in when the switch is turned off. This may not be the direct fault of the park switch, however. The operation of the park switch can usually be observed by removing the motor cover (Figure 18-7). Operate the wipers through three or four cycles while observing the latch arm. When the wiper switch is placed in the OFF position, the park switch latch must be in position to catch the drive pawl. Check to make sure the drive pawl is not bent. If good, replace the park switch.

A faulty wiper switch can also cause the park feature to not operate. Using the illustration (Figure 18-8), if wiper 2 is bent or broken so it does not make an electrical connection with the contacts, the wipers will not park even with the park switch in the PARK position, as shown. To test the switch, check for voltage at circuit 58 when the switch is moved from the LOW to the OFF position. If the switch is operating properly, there should be voltage present for a few seconds after the switch is in

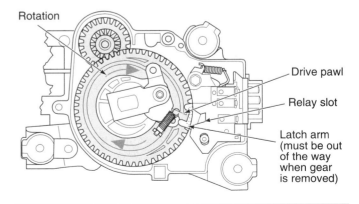

Figure 18-7 Checking the operation of the park switch while the motor is operating. (Courtesy of General Motors Corporation)

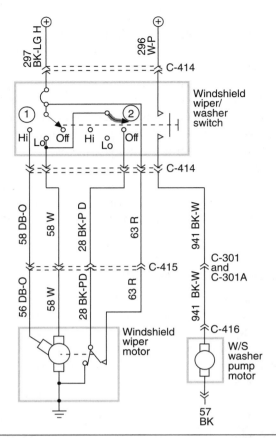

Figure 18-8 A faulty wiper in the switch can cause the park feature to not operate. (Reprinted with the permission of the Ford Motor Company)

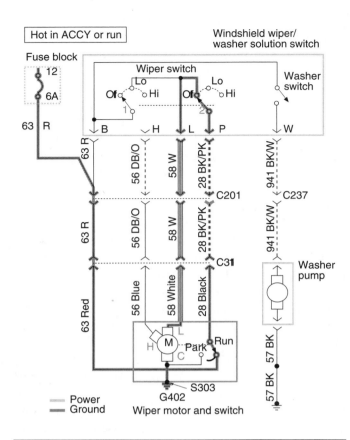

Figure 18-9 Sticking contacts in the park switch can cause the wipers to operate even after the switch is turned off. (Reprinted with the permission of Ford Motor Company)

the OFF position. No voltage at this circuit when the wiper switch is turned off indicates that the problem may be in the switch.

The park switch operation can also be checked by using a test light to probe for voltage at circuit 28 when the wiper switch is turned off. Probing for voltage at this circuit should produce a pulsating light when the motor is running.

If the wiper blades continue to operate with the wiper switch in the OFF position, the most probable cause is "welded" contacts in the park switch (Figure 18-9). If the park switch does not open, current will continue to flow to the wiper motor. The only way to turn off the wipers is to turn off the ignition switch or to physically remove the wires to the motor.

Some motors provide for replacement of the park switch, however, most shops replace the motor.

Intermittent Wiper System Diagnosis

The illustration shown (Figure 18-10) is a schematic of the interval wiper system used by Ford. If the interval function is the only portion of the system that fails to

operate properly, begin by checking the ground connection (circuit 57). If the ground is good, perform a continuity test of the switch using an ohmmeter (Figure 18-11). If the switch is good and all wires and connections are good, then replace the module.

WARNING *Some manufacturers have incorporated the intermittent wiper system into the body computer. These systems may incorporate self-diagnosis through the body-control module (BCM). Do not measure resistance through the module; damage to the circuits may result.*

Wiper Motor Removal and Installation

Removal procedures differ between manufacturers. Some motors are situated in areas that may require the removal of several engine compartment components. Always refer to the correct service manual for the recommended procedure.

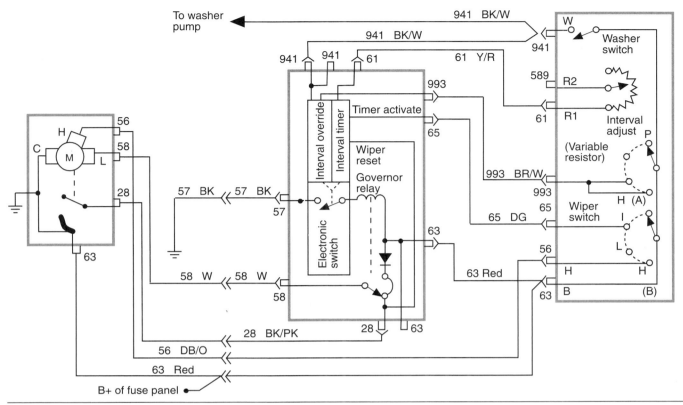

Figure 18-10 Interval windshield wiper system schematic. (Reprinted with the permission of Ford Motor Company)

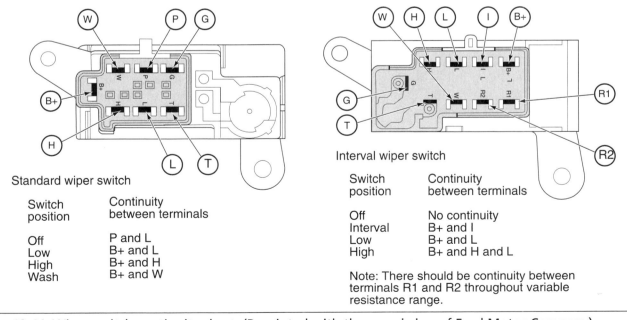

Standard wiper switch

Switch position	Continuity between terminals
Off	P and L
Low	B+ and L
High	B+ and H
Wash	B+ and W

Interval wiper switch

Switch position	Continuity between terminals
Off	No continuity
Interval	B+ and I
Low	B+ and L
High	B+ and H and L

Note: There should be continuity between terminals R1 and R2 throughout variable resistance range.

Figure 18-11 Wiper switch continuity chart. (Reprinted with the permission of Ford Motor Company)

WARNING *The internal permanent magnets of the motor are constructed of ceramic material. Use care in handling the motor to avoid damaging the magnets.*

Place the fender covers over the vehicle's fenders and disconnect the battery negative cable. To gain access to the motor, remove the shield cover from the cowl. Remove the wire connector to the motor. To remove the linkage, lift the locking tab and pull the clip away from the pin (Figure 18-12). Remove the attaching bolts from the motor assembly. Remove the motor (Figure 18-13). On some motors it may be necessary to first remove the operating arm from the motor.

Installation is basically the reverse of the removal procedure, but be sure to attach the ground wire to one of the mounting bolts during installation.

Wiper Switch Removal and Installation

The wiper switch removal procedure differs between manufacturers and depends on switch location. The procedures presented here are common, however, always refer to the manufacturer's service manual for correct procedures. Always protect the customer's investment by using fender covers while disconnecting the battery negative cable.

The following is a typical procedure for removing dash-mounted switches. Depending on the location of the switch control, it may be necessary to remove the finish panel. Usually the finish panel is held in place by a combination of fasteners and clips.

Remove the switch housing retaining screws. Then remove the housing and pull off the wiper switch knob. Disconnect the wire connectors from the switch. Remove the switch from the dash.

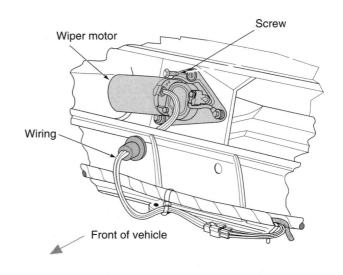

Figure 18-13 Once the retaining screws are removed, the motor can be removed from the vehicle. (Reprinted with the permission of Ford Motor Company)

Reverse the procedure to install the switch.

To remove steering column-mounted switches begin by removing the upper and lower steering column shrouds to expose the switch. Disconnect the wire connectors to the switch. It may be necessary to peel back the foam to gain access to the retaining screws. Remove the screws and the switch (Figure 18-14).

Windshield Washer System Service

Many windshield washer problems are due to restrictions in the delivery system. To check for restrictions, remove the hose from the pump and operate the system.

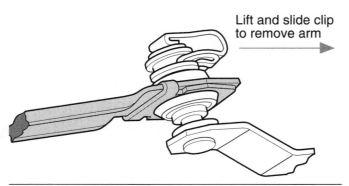

Figure 18-12 To remove the clip, lift up the locking tab and pull the clip. (Reprinted with the permission of Ford Motor Company)

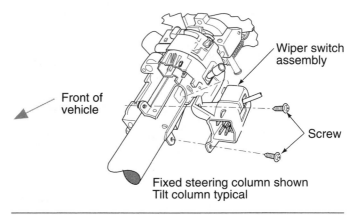

Figure 18-14 Steering column-mounted wiper switch removal. (Reprinted with the permission of Ford Motor Company)

If the pump ejects a stream of fluid, then the fault is in the delivery system. If the pump does not deliver a spray of fluid, continue testing using the following procedure:

1. Make checks of obvious conditions such as low fluid level, blown fuses, or disconnected wires.

2. Activate the washer switch while observing the motor. If the motor operates but does not squirt fluid, check for blockage at the pump. Remove any foreign material. If there is no blockage, then replace the motor.

3. If the motor does not operate, use a voltmeter or test light to check for voltage at the washer pump motor with the switch closed. If there is voltage, then check the ground circuit with an ohmmeter. If the ground connection is good, then replace the pump motor.

4. If there is no voltage to the pump motor in step 3, trace the circuit back to the switch. Test the switch for proper operation. If there is power into the switch but not out of it to the motor, replace the switch.

If the motor is in need of replacement, follow this procedure for pumps installed in the reservoir. Disconnect the wire connector and hoses from the pump. Remove the reservoir assembly from the vehicle. Use a small blade screwdriver to pry out the retainer ring. Use a pair of pliers to grip one of the walls that surrounds the terminals (Figure 18-15). Pull out the motor, seal, and impeller.

CAUTION *Wear safety glasses to prevent the ring from striking your eyes. Also be careful to position the palm of your hand so that if the screwdriver slips, it will not puncture your skin.*

Before installing the pump assembly, lubricate the seal with a dry lubricant. The lubricant is used to prevent the seal from sticking to the wall of the reservoir. Align the small projection on the motor with the slot in the reservoir and assemble. Make sure the seal sits against the bottom of the motor cavity. Use a 12-point, 1-inch socket to hand press the retaining ring into place.

Replace the reservoir assembly in the vehicle. Reconnect the hose and wires. When refilling the reservoir, do so slowly to prevent air from being trapped in the reservoir. Check system operation while checking for leaks.

WARNING *Do not operate the washer pump without fluid. Doing so may damage the new pump motor.*

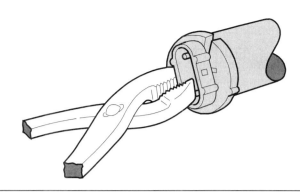

Figure 18-15 Use a pair of pliers to pull the motor out of the reservoir. (Reprinted with the permision of Ford Motor Company)

Blower Motor Service

Conventional blower motor speed is controlled by sending current through a resistor block. The higher the resistance value, the slower the fan speed. The position of the switch determines which resistor will be added to the circuit. The switch can be either ground side or insulated side switches.

If the customer states the fan operates only in a couple of speed positions, the most likely cause is an open resistor in the resistor block. Using the illustration (Figure 18-16), if resistor 1 is open, the motor will not operate in any position except HIGH speed. If resistor 2 was

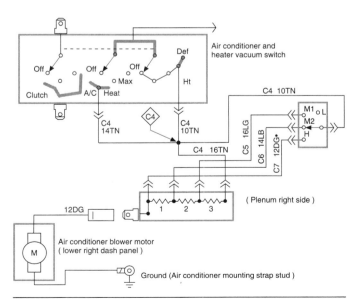

Figure 18-16 Typical wiring for a four-speed fan motor circuit. (Courtesy of Chrysler Corporation)

open, the motor would operate in HIGH and M2 speeds only. If resistor 3 was open, the motor would operate in all speeds except LOW.

If the motor operates in any one of the speed select positions, the fault is not in the motor. If the motor fails to operate at all, begin by inspecting the fuse. If the fuse is good, use the correct wiring schematic to determine whether ground or insulated side switches are used. The diagnostic procedure used depends on the circuit design.

Inoperable Motor with Insulated Switches

Use a jumper wire to bypass the switch and resistor block to check motor operation. Connect the jumper wire from a battery positive supply to the motor terminal. If the motor does not operate, connect a second jumper wire from the motor body to a good ground. Replace the motor if it still does not operate.

If the motor operated when the switch and resistor block were bypassed, trace up the circuit to the switch. Use a voltmeter or test light to check for voltage in and out of the blower speed control switch. The switch is faulty if voltage is at the input terminal but not at any of the output terminals. No voltage at the input terminal indicates an open in the circuit between the battery and the switch.

SERVICE TIP *Because the high speed circuit bypasses the resistor block, it is doubtful that no motor operation would be the fault of open resistors. However, an open in the wire from the block to the motor can cause the problem. Most likely the switch is bad and in need of replacement. Always confirm your diagnosis by doing a continuity test on the switch.*

Inoperable Motor with Ground Side Switches

Using the illustration (Figure 18-17) as an example of negative side switch blower motor circuit, to test the motor, connect the jumper wire from the motor negative terminal to a good ground. This bypasses the switch and resistor block. If the motor does not operate, use a voltmeter or test light to check for voltage at the battery terminal of the motor. If voltage is present, then the motor is defective.

If there is no voltage at the battery terminal of the motor, the problem is in the circuit from the battery to the motor. Be sure to check the circuit breaker.

Operation of the motor when the jumper wire is connected to the ground terminal indicates the problem is in

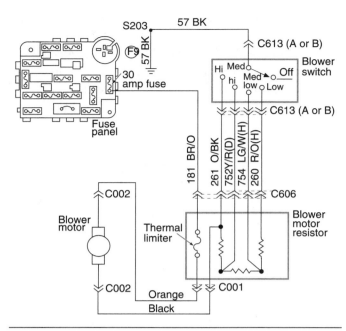

Figure 18-17 Blower motor circuit using negative side switch. (Reprinted with the permission of Ford Motor Company)

the switch side of the circuit. Use an ohmmeter to check the ground connection for the switch. A jumper wire or test light can also be used to test this connection.

If the ground connection is good, use a voltmeter or test light to probe for voltage at any of the circuits from the resistor block to the switch. Replace the switch if there is power to this point. Replace the resistor block if there is no power at these points.

Constantly Operating Blower Fans

General Motors, and other manufacturers, designed many of their blower fans to constantly operate. Do not confuse this normal operation with a circuit defect. However, systems that use ground side switches can develop a circuit defect resulting in the fan motor not shutting off. A short to ground at any point on the ground side of the circuit will cause the motor to run all the time. Other areas to check include the switch and the circuit between the switch and the resistor block.

In insulated side switch circuits, check for copper-to-copper shorts in the power side of the system. If the motor is receiving power from another circuit due to a copper-to-copper short, the motor will continue to run whenever current is flowing through that circuit (Figure 18-18). Some systems may incorporate a relay, and if the contact points fuse together, the motor will continue to operate.

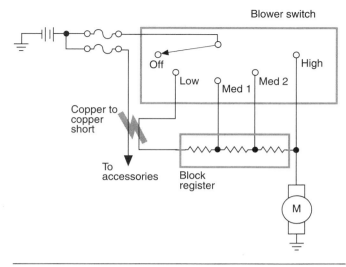

Figure 18-18 A copper-to-copper short can cause the fan motor to run when the switch is turned off.

Electric Defogger Diagnosis and Service

If the rear window defogger fails to operate when the switch is activated, use the test light to test the grids. Under normal conditions the test light should be bright on one side of the grid and off on the other side. If the test light has full brilliance on both sides of the grid, then the ground connection for the grid is broken.

If the test light does not illuminate at any position on the grid, use normal test procedures to check the switch and relay circuits. There may be several fuses involved in the system. Use the correct wiring diagram to determine the fuse identification.

Most rear window defogger complaints will be associated with broken grids. These will generally be complaints that only a portion of the window is cleared while the rest remains foggy. Some grid wire breaks are easily detected by visual inspection. However, many are too small to see. To test the grid lines, start the engine and activate the system. (Remember the system is controlled by a timer.) Use a test light to check each grid wire to locate the breaks. Test each grid in at least two places one on each side of the center line. The test results that should be obtained on each grid are illustrated in Figure 18-19.

WARNING *Be careful not to tear the grid with the test light. Only a light touch on the grid should be required.*

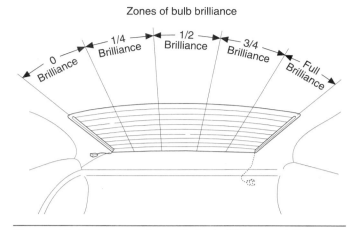

Figure 18-19 Zones of test light brilliance while probing a rear window defogger. (Courtesy of General Motors Corporation)

If the test light does not indicate normal operation on a specific grid line, place the test light probe on the grid at the left bus bar and work toward the right until the light goes out. The point where the light goes out is the location of a break (Figure 18-20). Mark the location of the break with a grease pencil on the outside of the glass.

The rear window defogger should turn off about 10 minutes after activation. If the circuit fails to turn off, check the ground for the control module. If the ground is good, replace the module.

Grid Wire Repair

If the grid wire is broken, it is possible to repair the grid with a special repair kit.

Bus Bar Lead Repair

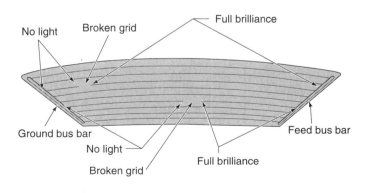

Figure 18-20 Test light brilliance when it is passed over an opening in the grid. (Courtesy of General Motors Corporation)

The bus bar lead wire can be resoldered using a solder containing 3% silver and a rosin flux paste. Clean the repair area using a steel wool pad. Apply the rosin flux paste in small quantities to the wire lead and bus bar. Tin the solder iron tip with the solder. Finish the repair by soldering the wires to the bus bar. Be careful not to overheat the wire.

Power Window Diagnosis

Use the illustration (Figure 18-21) as a guide to diagnosing the power window circuit. The circuit used in this example is typical, however, use the service manual for the vehicle that you are working on to get the correct wiring schematic.

If the window does not operate, begin by testing the circuit breaker. Use a test light or voltmeter to test for voltage at both sides of the circuit breaker. If voltage is present at both sides, then the circuit breaker is good. If there is voltage into the circuit breaker but not out of it, the circuit breaker is bad. If there is no voltage into the circuit breaker, then there is an open in the feed from the battery.

SERVICE TIP *It does not matter if the motor or the switches are tested next. Test the unit that is easiest to get to first.*

If the circuit breaker is good, use jumper wires to test the motor. The motor is a reversible motor, so connections to the motor terminals are not polarity sensitive. Disconnect the wire connectors to the motor. Connect battery positive to one of the terminals and ground the other. If the motor does not operate, reverse the jumper wire connections. The motor should reverse directions when the polarity is reversed. If the motor does not operate in one or both directions, it is defective and needs to be replaced.

CAUTION *Do not place your hands into the window's operating area. Make the final test connections outside of the door where there is no danger of getting caught in the window track.*

If the motor operates when the switches are bypassed, the problem is in the control circuit. To test the master switch, connect the test light between terminals 1 and 2 (Figure 18-22). When the master switch is in the rest (OFF) position, the test light should illuminate. If the light does not glow, there is an open in the circuit to the

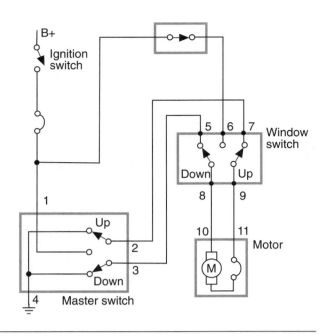

Figure 18-21 Simplified power window circuit.

window switch or from the window switch to ground at terminal 4. Check the ground at terminal 4 for good connections. If good, continue testing.

If the test light illuminates when connected across terminals 1 and 2, place the switch in the UP position. The test light will go out if the switch is good. Repeat the test between terminals 1 and 3. Place the switch in the DOWN position.

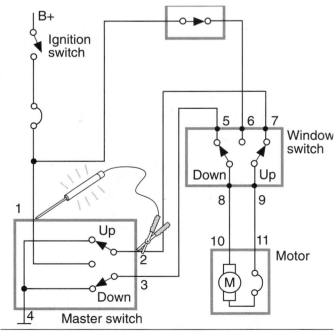

Figure 18-22 Using a test light to check the operation of a power window master switch.

If the master switch is good, test the window switch. Battery voltage should be present at terminal 5. If not, check to see if the lockout switch is closed. Check the circuit from terminal 5 to the circuit breaker. Connect the test light across terminals 5 and 6 (Figure 18-23). The light should remain on until the switch is placed into the UP position. Test the DOWN position by connecting the test light across terminals 6 and 7.

Slower than normal operating speeds are an indication of excessive resistance or of binding in the mechanical linkage. Use the voltage drop test method to locate the cause of excessive resistance. Excessive resistance can be in the switch circuits, the ground circuit, or in the motor. If the problem is mechanical, lubricate the track and check for binding or bent linkage.

CAUTION *Follow the manufacturer's recommended procedure when removing the power window motor. The springs used in window regulators can cause serious injury if removed improperly.*

Power Seat Diagnosis

The power seat system is usually very simple to troubleshoot. Test for voltage to the input of the switch control. If voltage is available to the switch, remove it from the seat or arm rest. Using a continuity chart from the service manual (Figure 18-24), test the switch for proper

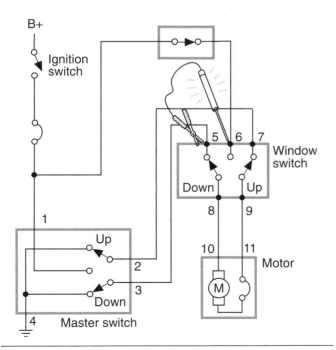

Figure 18-23 Test light connections for testing a window switch.

operation. If the switch is operating properly, it may be necessary to remove the seat to test the motors and circuits to the motors.

The power seat motors are tested in the same manner as the power window motor. Be sure to test each armature of the trimotor. If any of the armatures fail to operate, the trimotor must be replaced as a unit.

CAUTION *Be careful when making the jumper wire connection to test the motor. Do not place your hands in locations where they can become pinched or trapped when the seat moves.*

CAUTION *If the trimotor needs to be replaced, follow the manufacturer's service procedures closely. Improper removal of the springs may result in personal injury.*

Noisy operation of the seat can generate from the motor, transmission, or cable. If the motor or transmission is the cause of the noise, it must be replaced. A noisy cable can usually be cured with a dry lubricant.

Power Door Lock Diagnosis

To test the door motor, apply 12 volts directly to the motor terminals. The actuator rod should complete its travel in less than one second. Reverse polarity to test operation in both directions.

The switch is checked for continuity using an ohmmeter. There should be no continuity between any terminals when the switch is in its neutral position. Use the circuit schematic to determine when there should be continuity between terminals.

If the system uses a relay, use the schematic to determine relay circuit operation (Figure 18-25). In this example, battery voltage should be present at terminal 4 of the connector. Using an ohmmeter, check the ground connections of terminals 1 and 5 of the connector. To test the relay, connect a test light across terminal 3 and ground. Ground terminal 1 and apply power to terminals 2 and 4. The test light will light if the relay is good.

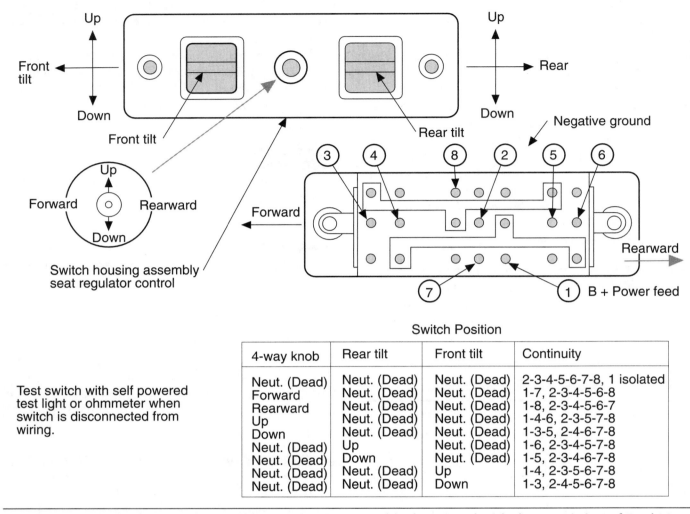

Test switch with self powered test light or ohmmeter when switch is disconnected from wiring.

Switch Position

4-way knob	Rear tilt	Front tilt	Continuity
Neut. (Dead)	Neut. (Dead)	Neut. (Dead)	2-3-4-5-6-7-8, 1 isolated
Forward	Neut. (Dead)	Neut. (Dead)	1-7, 2-3-4-5-6-8
Rearward	Neut. (Dead)	Neut. (Dead)	1-8, 2-3-4-5-6-7
Up	Neut. (Dead)	Neut. (Dead)	1-4-6, 2-3-5-7-8
Down	Neut. (Dead)	Neut. (Dead)	1-3-5, 2-4-6-7-8
Neut. (Dead)	Up	Neut. (Dead)	1-6, 2-3-4-5-7-8
Neut. (Dead)	Down	Neut. (Dead)	1-5, 2-3-4-6-7-8
Neut. (Dead)	Neut. (Dead)	Up	1-4, 2-3-5-6-7-8
Neut. (Dead)	Neut. (Dead)	Down	1-3, 2-4-5-6-7-8

Figure 18-24 Continuity chart for a six-way power seat assembly. (Reprinted with the permission of Ford Motor Company)

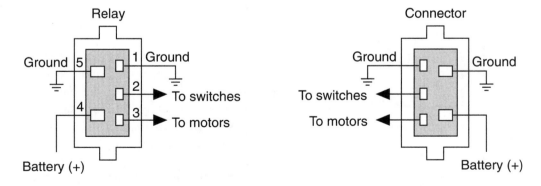

Figure 18-25 Testing a door lock relay and connector. (Reprinted with the permission of Ford Motor Company)

Summary

❏ The purpose of the charging system is to convert the mechanical energy of the engine into electrical energy to recharge the battery and run the electrical accessories.

❏ Horn system diagnosis varies based on whether the system is equipped with a relay or not.

❏ Poor sound quality can be the result of several factors. In a multiple horn system, if one of the horns is not operating, the horn sound quality may suffer. Other reasons include damaged diaphragms, excessive circuit resistance, poor ground connections, or improperly adjusted horns.

❏ A horn that sounds continuously is usually caused by a sticking horn switch or sticking contact points in the relay.

❏ Customer complaints concerning windshield wiper operation can include no operation, intermittent operation, continuous operation, and wipers will not park. Other complaints have to do with blade adjustment (such as blades slapping the molding or one blade parks lower than the other).

❏ When a customer brings the vehicle into the shop because of faulty windshield wiper operation, the technician needs to determine if it is an electrical or mechanical problem. To do this, simply disconnect the arms to the wiper blades from the motor.

❏ Slower than normal wiper speeds can be caused by electrical or mechanical faults. An ammeter can be used to determine the current draw of the motor with and without the mechanical portion connected. If the draw changes substantially when the mechanical portion is removed, the fault is in the arms and/or wiper blades.

❏ The most common complaint associated with a faulty park switch is that the wipers stop in the position they are in when the switch is turned off.

❏ If the customer states the blower fan operates only in a couple of speed positions, the most likely cause is an open resistor in the resistor block.

❏ Blower motor operation can be checked using a jumper wire to bypass the switch and resistor block.

❏ If the rear window defogger fails to operate when the switch is activated, use the test light to test the grids. Under normal conditions the test light should be bright on one side of the grid and off on the other side. If the test light has full brilliance on both sides of the grid, then the ground connection for the grid is broken.

❏ If the grid wire is broken, it is possible to repair the grid with a special repair kit.

❏ Power window problems can be caused by faulty motors, master switches, or individual switches.

❏ The power seat system is usually very simple to troubleshoot. Test for voltage to the input of the switch control. If voltage is available to the switch, remove it from the seat or arm rest. Using a continuity chart from the service manual, test the switch for proper operation.

ASE Style Review Questions

1. A customer says the horn will not turn off. Technician A says that the cause could be welded diaphragm contacts inside of the horn. Technician B says that the relay may be faulty. Who is correct?
   a. A only
   b. B only
   c. Both A and B
   d. Neither A nor B

2. The two-speed windshield wiper operates in high position only. Technician A says the low speed brush may be worn. Technician B says that the motor has a faulty ground connection. Who is correct?
   a. A only
   b. B only
   c. Both A and B
   d. Neither A nor B

3. The reasons for slower than normal wiper operation are being discussed. Technician A says the problem may be in the mechanical linkage. Technician B says the problem may be excessive electrical resistance. Who is correct?
   a. A only
   b. B only
   c. Both A and B
   d. Neither A nor B

4. The wiper motor does not park when the wiper switch is placed in the OFF position. Technician A says the park switch is faulty. Technician B says the wiper switch is faulty. Who is correct?
   a. A only
   b. B only
   c. Both A and B
   d. Neither A nor B

5. The intermittent wiper function is not operating. Technician A says to check the resistance of the module in systems using the body computer. Technician B says to check the ground connection of the interval module. Who is correct?
   a. A only
   b. B only
   c. Both A and B
   d. Neither A nor B

6. The heater fan motor does not operate in HIGH speed position. Technician A says the cause is a faulty resistor block. Technician B says the motor is defective. Who is correct?
   a. A only
   b. B only
   c. Both A and B
   d. Neither A nor B

7. The grid of a rear window defogger is only removing some areas of fog from the window. Technician A says the timer circuit is faulty. Technician B says the grid is damaged. Who is correct?
   a. A only
   b. B only
   c. Both A and B
   d. Neither A nor B

8. The passenger side power window does not operate in either direction, despite which switch is used (master or window). Technician A says the problem is a faulty master switch. Technician B says the problem is a worn motor. Who is correct?
   a. A only
   b. B only
   c. Both A and B
   d. Neither A nor B

9. The six-way power seat is not operating in any switch position. Technician A says to check the circuit breaker. Technician B says to use a continuity chart to test the switch. Who is correct?
   a. A only
   b. B only
   c. Both A and B
   d. Neither A nor B

10. The power door locks will lock the door, but they do not unlock it. Technician A says the motor is faulty. Technician B says the unlock relay is faulty. Who is correct?
    a. A only
    b. B only
    c. Both A and B
    d. Neither A nor B

19 Basic Distributor and Electronic Ignition Systems

Objective

Upon completion and review of this chapter, you should be able to:

❏ Describe the three major functions of an ignition system.

❏ Name the operating conditions of an engine that affect ignition timing.

❏ Name the two major electrical circuits used in ignition systems and their common components.

❏ Describe the operation of ignition coils, spark plugs, and ignition cables.

❏ Describe the various types of spark timing systems, including electronic switching systems and their related engine position sensors.

❏ Describe the operation of distributor-based ignition systems.

Introduction

One of the requirements for an efficient running engine is the correct amount of heat delivered into the cylinders at the right time. This requirement is the responsibility of the ignition system. The ignition system supplies properly timed, high-voltage surges to the spark plugs. These voltage surges cause an arc across the electrodes of a spark plug, and this heat begins the combustion process inside the cylinder. For each cylinder in an engine, the ignition system has three main jobs. First, it must generate an electrical spark with enough heat to ignite the air/fuel mixture in the combustion chamber. Second, it must maintain the spark long enough to allow for the combustion of all the air and fuel in the cylinders. Third, it must deliver the spark to each cylinder so combustion can begin at the right time during the compression stroke of each cylinder.

When the combustion process is completed, a very high pressure is exerted against the top of the piston. This pressure pushes the piston down on its power stroke. This pressure is the force that gives the engine power. In order for an engine to produce maximum power, the maximum pressure from combustion should be present when the piston is at 10° to 23° **after top dead center (ATDC)**. **Top dead center (TDC)** is the uppermost position of the piston in its cylinder.

Because combustion of the air/fuel mixture within a cylinder takes a short period of time, usually measured in thousandths of a second (milliseconds), the combustion process must begin before the piston is on its power stroke. Therefore, the delivery of the spark must be timed to arrive at some point before the piston reaches top dead center.

Determining how much before TDC the spark should begin gets complicated by the fact that the speed of the piston increases as it moves from its compression stroke to its power stroke, while the time needed for combustion stays about the same. This means the spark should be delivered earlier as the engine's speed increases (Figure 19-1). However, as the engine has to provide more power to do more work, the load on the crankshaft tends to slow down the acceleration of the piston and the spark needs to be somewhat delayed.

Calculating when the spark should begin gets more complicated with the fact that the rate of combustion varies according to certain factors. Higher compression pressures tend to accelerate the combustion process. Higher octane gasolines ignite less easily and require more burning time. Increased vaporization and turbu-

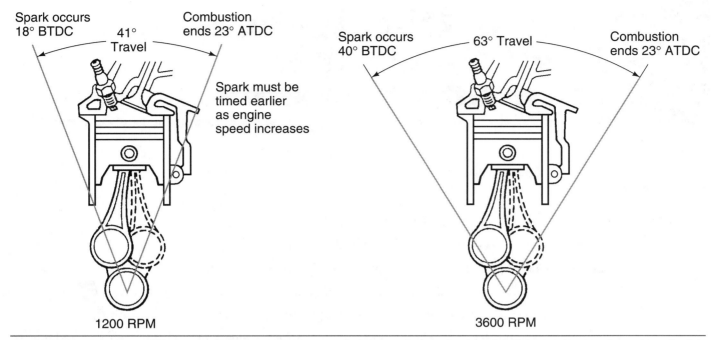

Figure 19-1 With an increase in speed, ignition must begin earlier to end by 23 degrees ATDC. (Reprinted with the permission of Ford Motor Company)

lence tend to decrease combustion times. Other factors, including intake air temperature, humidity, and barometric pressure, also affect combustion. Because of all of these complications, delivering the spark at the right time is a difficult task.

Basic Circuitry

All ignition systems consist of two interconnected electrical circuits: a **primary** (low voltage) circuit and a **secondary** (high voltage) circuit (Figure 19-2).

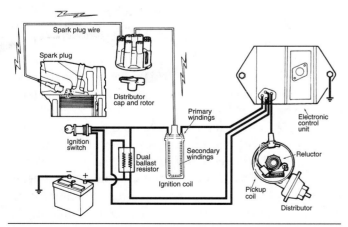

Figure 19-2 Typical primary and secondary ignition circuits. (Courtesy of Chrysler Corporation)

Depending on the exact type of ignition system, components in the primary circuit include the following:
- battery
- ignition switch
- ballast resistor or resistance wire (some systems)
- starting bypass (some systems)
- ignition coil primary winding
- triggering device
- switching device or control module
- ground

The secondary circuit includes these components:
- ignition coil secondary winding
- distributor cap and rotor (some systems)
- ignition (spark plug) cables
- spark plugs

Primary Circuit Operation

When the ignition switch is in the RUN position, current from the battery flows through the ignition switch and primary circuit resistor to the primary winding of the ignition coil. From there it passes through some type of switching device and back to ground. The switching device can be mechanically or electronically controlled by a triggering device. The current flow in the ignition coil's primary winding creates a magnetic field. The switching device or control module interrupts this current flow at predetermined times. When it does, the magnetic field in the primary winding collapses. This collapse generates a high-voltage surge in the secondary winding of the ignition coil. The secondary circuit of the system begins at this point.

Secondary Circuit Operation

The secondary circuit carries high voltage to the spark plugs. The exact manner in which the secondary circuit delivers these high-voltage surges depends on the system design. Until 1984 all ignition systems used some type of distributor to accomplish this job. However, in an effort to reduce emissions, improve fuel economy, and boost component reliability, most auto manufacturers are now using distributorless or **electronic ignition (EI)** systems.

In a **distributor ignition (DI)** system, high voltage from the secondary winding passes through an ignition cable running from the coil to the distributor. The distributor then distributes the high voltage to the individual spark plugs through a set of ignition cables. The cables are arranged in the distributor cap according to the firing order of the engine. A rotor, which is driven by the distributor shaft, rotates and completes the electrical path from the secondary winding of the coil to the individual spark plugs. The distributor delivers the spark to match the compression stroke of the piston. The **distributor** assembly may also have the capability of advancing or retarding ignition timing.

The **distributor cap** is mounted on top of the distributor assembly, and an alignment notch in the cap fits over a matching lug on the housing. The cap can only be installed in one position, thereby assuring the correct firing sequence.

The rotor is positioned on top of the distributor shaft, and a projection inside the rotor fits into a slot in the shaft. This allows the rotor to only be installed in one position. A metal strip on the top of the rotor makes contact with the center distributor cap terminal, and the outer end of the strip rotates past the cap terminals as it rotates. This action completes the circuit between the ignition coil and the individual spark plugs according to the firing order.

Electronic ignition (EI) systems have no distributor; spark distribution is controlled by an electronic control unit and/or the vehicle's computer. Instead of a single ignition coil for all cylinders, each cylinder may have its own ignition coil, or two cylinders may share one coil. The coils are wired directly to the spark plug they control. An ignition control module, tied into the vehicle's computer control system, controls the firing order and the spark timing and advance. This module is typically located under the coil assembly.

Ignition Components

All ignition systems share a number of common components. Some, such as the battery and ignition switch, perform simple functions. The battery supplies battery voltage to the ignition primary circuit. Current flows when the ignition switch is in the START or the RUN position.

Ignition Coils

The ignition coil (Figure 19-3) is the heart of the ignition system. It works in the manner of a **pulse transformer** to build up the low battery voltage of 12.6 volts to a voltage high enough to **ionize** (electrically charge) the spark plug gap and ignite the air/fuel mixture. The coil is capable of producing approximately 30,000 to 60,000 volts. However, the amount of voltage produced is dependent on many factors. The coil will produce only enough voltage required to overcome these factors: plug gap, air/fuel ratio, plug wire resistance, engine speed, compression ratio, and so forth. The margin of voltage which can be produced above that which is required to fire the spark plug represents the **electrical reserve** built into the ignition system. As plugs wear and other resistances in the system increase, the ignition system is capable of compensating for this through the electrical reserve.

The center of the ignition coil contains a core of laminated soft iron or steel. Adding a core to a coil increases the magnetic strength of the coil. The core is surrounded by approximately 22,000 turns of very fine wire. This winding is called the **secondary coil windings**.

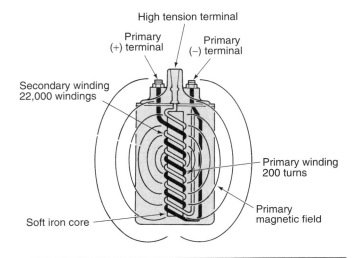

Figure 19-3 Cutaway of a typical ignition coil. (Courtesy of Chrysler Corporation)

Surrounding the secondary coil windings is the **primary coil windings**. The primary windings are made of approximately 200 turns of 20-gauge wire. Because of the build-up of heat through the windings due to resistance to current flow, some coils contain oil to help cool them. If oil is not used then an air-cooled, epoxy-sealed E coil is used. It is called an **E coil** because of the shape of the core is in the figure of an E.

The primary windings extend through the case of the coil and are marked. The markings differ from manufacturer to manufacturer. Some of the most common markings are: BAT and DIST, or positive (battery) and negative (distributor), or + and -. The naming of the terminals as positive and negative refers to the most positive and the least positive terminals. These labels assure polarity through the coil is correct. The plug polarity is **center pole negative**. Negative plug polarity requires less voltage to ionize the plug than does positive polarity. This is because of the relative temperatures between the center electrode (hot) and the ground electrode (cold). Electricity will more easily jump from hot to cold than from cold to hot. If the primary wires are incorrectly connected to the coil, the coil will have reverse polarity and the voltage required to fire the plug will be increased by 40%. This may cause poor engine performance and a high speed miss.

The negative terminal is attached to a switching device (either mechanical or electronic) which opens and closes the primary circuit. With the ignition switch in the "run" position, voltage should be present at both sides of the primary windings. Whenever current starts to flow into a coil, an opposing current is created in the windings of the coil. This opposing current is the result of self-induction and is called inductive reactance. As explained earlier, inductive reactance is similar to resistance since it resists any increase in current flow in a coil. This results in a very fast build-up to maximum magnetic field strength. At this point there can be no more build-up of magnetic strength and is referred to as saturation.

Since there are two windings within the ignition coil, whenever there is a change in the magnetic field of one winding it affects the other winding. If the current to the primary winding is stopped, then the magnetic field collapses and cuts across the secondary winding. This creates a high voltage in the secondary winding. Also there is a build-up of about 250 volts within the primary winding. This process of creating high voltage in the secondary windings by changing the magnetic field in the primary windings is called **mutual induction**.

As discussed earlier, the positive side of the coil is connected to the battery through the ignition switch and the negative side of the coil is connected to a switching device. This switching device controls the current flow through the coil by turning on or off the return path to the battery source. Figure 19-4 illustrates a basic ignition system that uses contact points for the means of turning on or off the current flow. When the switch is closed current flowing through the primary windings of the coil generates a high magnetic field inside of the coil. When the switch opens the primary windings return path, the magnetic field collapses and induces a high voltage current into the secondary windings. This high voltage is then directed to the correct spark plug to ignite the air/fuel mixture in that cylinder. For each spark plug, the coil must be charged then discharged. All of the components regulating the current in the coil primary windings are referred to as the **primary ignition circuit**. This would include the ignition switch, primary coil windings, the switching device, and all conductors that connect these components. All of the components that create and distribute the high voltage produced in the coil are called the **secondary ignition circuit**. This would include the coil secondary windings, coil high tension lead, distributor cap, rotor, plug wires, and the spark plugs.

A BIT OF HISTORY

A turning point in the automotive industry came in 1902 when the Humber, an automobile from England, used a magneto electric ignition system.

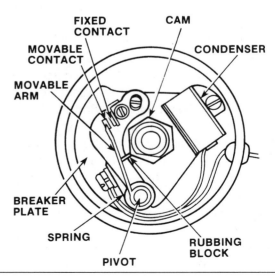

Figure 19-4 Older ignition systems used breaker points as the switching device in the primary circuit.

Secondary Voltage

The typical amount of secondary coil voltage required to jump the spark plug gap is 10,000 volts. Most coils have a maximum secondary voltage of 25,000 volts. The difference between the required voltage and the maximum available voltage is referred to as **secondary reserve voltage**. This reserve voltage is necessary to compensate for high cylinder pressures and increased secondary resistances as the spark plug gap increases through use. The maximum available voltage must always exceed the required firing voltage or ignition misfire will occur. If there is an insufficient amount of voltage available to ionize the spark plug gap, the spark plug will not fire.

Since DI and EI systems are both firing spark plugs with approximately the same air gaps, the amount of voltage required to fire the spark is nearly the same in both systems. However, EI systems have higher voltage reserves. If the additional voltage in an EI system is not used to create the spark across the plug's gap, it is used to maintain the spark for a longer period of time. The average length of time an EI system can maintain the spark is 1.5 milliseconds. A typical DI system maintains the spark for 1 millisecond.

The number of ignition coils used in an ignition system varies with the type of ignition system. In most distributor ignition systems, only one ignition coil is used. The high voltage of the secondary winding is directed by the distributor to the various spark plugs in the system. Therefore, there is one secondary circuit with a continually changing path.

While distributor systems have a single secondary circuit with a continually changing path, distributorless systems have several secondary circuits, each with an unchanging path.

Spark Plugs

The **spark plug** is designed to transfer the high-voltage current to the combustion chamber, where the electrical spark is the motivating force behind establishing combustion of the air/fuel mixture. In order to produce this spark, a gap (between .020 and .080 in.) is established by the spark plug that must be bridged by the high-voltage current (Figure 19-5).

Because of it's environment, the spark plug is subject to more stress than most other components in the engine. The spark plug must deliver high voltage, with split second accuracy, thousands of times per second, in temperatures that can exceed 2,500 degrees.

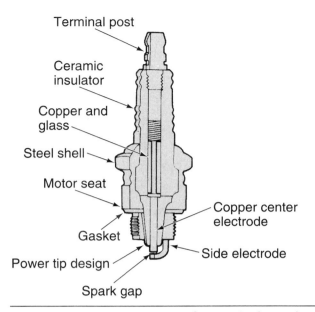

Figure 19-5 Construction of a typical spark plug. (Courtesy of Chrysler Corporation)

Also the spark plug must be designed to seal the cylinder from compression losses. To perform this task there are three important aspects to the spark plug:

1. Thread diameter. The spark plug is manufactured with one of the following thread sizes: 10, 12, 14, or 18mm diameters. The most common sizes for automotive use are 14mm and 18mm.

2. Reach. It is important proper reach be used when replacing a plug (Figure 19-6). Reach determines the location of the plug gap in the combustion chamber. If the reach is too short, the engine could ping (pre-ignition) as a result of carbon build-up on the exposed cylinder head threads. Also the shrouded plug gap may cause a misfire. If the reach is too long, the piston may contact the spark plug and cause severe damage. In addition the engine may ping due to overheating of the exposed threads.

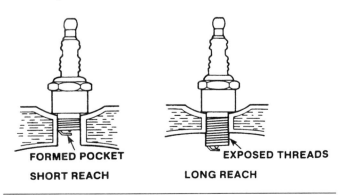

Figure 19-6 Spark plug reach: long versus short.

3. Method of sealing. There are two methods used to seal the spark plug in the cylinder head (Figure 19-7). The first is the use of a compressible gasket against a flat surface. The second uses a tapered seat which provides for a wedge seal between the head and the spark plug.

Heat Range

The **heat range** of the spark plug refers to its heat dissipation properties (Figure 19-8). The spark plug must be able to retain some of the heat developed during combustion to burn away carbon and oil deposits that form on the electrode. If the heat range was too low, these deposits would accumulate and foul the plug and cause the engine to misfire. If the heat range is too high, the insulator tip will become overheated to the point of igniting the air/fuel mixture prematurely.

The heat range of a spark plug is determined by how far the center insulator goes up into the shell around the center electrode. The greater the distance, the longer it will take to dissipate heat and the hotter the plug will get. When replacing a spark plug it is important to use the heat range designated by the manufacturer.

There are many operational factors that affect the temperature at which the spark plug must work. Figure 19-9 charts many of these factors. Insulator tip temperature is shown on the left axis of each chart while the different factors are listed on the bottom axis.

Resistor Plugs

Resistor plugs are spark plugs with a resistor built into the electrode core (Figure 19-10). The purpose of this resistor is to reduce the electromagnetic radiation created within the ignition system. If this radiation is not reduced by use of resistor plugs or plug wires, then radio and television interference would result.

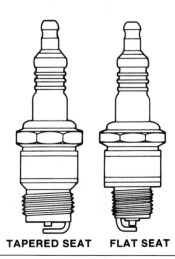

Figure 19-7 Spark plug seats: tapered versus flat.

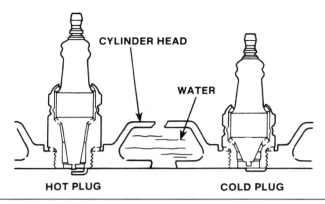

Figure 19-8 Spark plug heat range: hot versus cold.

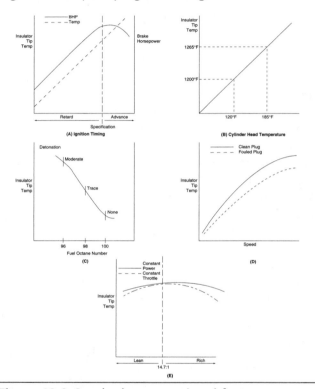

Figure 19-9 Spark plug operational factors.

A BIT OF HISTORY

Louis S. Clark of the Auto Car Company designed porcelain spark plug insulation in 1902.

Distributor Caps and Rotors

The distributor cap and rotor distribute the high voltage created in the secondary windings of the coil to the correct spark plug (Figure 19-11). These units are manufactured from bakelite, alkyd plastic, phenol resin, or fiberglass reinforced polyester resin plastic. Inside the cap are ribs between the side tower inserts. These ribs are

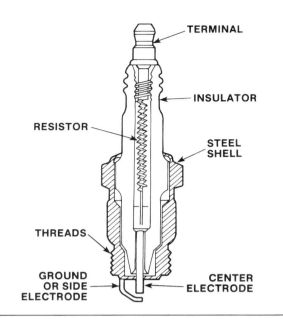

Figure 19-10 Components of a typical spark plug.

called **"antiflashover ribs"** and prevent sparks from arcing to the wrong tower insert.

WARNING *Always replace spark plugs with the same design specified by the manufacturer.*

The high voltage induced in the coil when the switching device opens is sent to the center tower of the distributor cap through the secondary cable. The high voltage is then sent to inside the cap and to the rotor. The rotor is being turned by the distributor shaft. The high voltage is then directed to each cylinder according to the firing order.

An air gap of a few thousandths of an inch exists between the tip of the rotor electrode and the spark plug electrode inside the cap. This gap is necessary in order to prevent the two electrodes from making contact. If they did make contact, both would wear out rapidly. This gap cannot be measured when the distributor is assembled; therefore, the gap is usually described in terms of the voltage needed to create an arc between the electrodes.

Ignition Cables

Ignition cables carry the high voltage from the distributor or the multiple coils (EI systems) to the spark plugs. The cables are not solid wire; rather, they contain carbon-impregnated fiber cores which act as resistors (Figure 19-12). The resistance reduces radio and television interference, increases firing voltages, and reduces spark plug wear by decreasing current. Metal terminals on each end of the spark plug wires contact the spark plug and the distributor cap terminals. Insulated boots on the ends of the cables strengthen the connections as well as prevent dust and water infiltration and voltage loss.

Ignition Timing

Ignition timing refers to the precise time a spark is sent to the cylinder relative to the piston position. This spark is sent the instant the coil primary circuit is opened. As the piston approaches top dead center (TDC) on the compression stroke, the air/fuel mixture in the cylinder is compressed and ready to be ignited. The precise timing for the ignition of the compressed air/fuel mixture is critical for maximum power, increased fuel economy, reduced emissions, and reduced engine wear.

Ignition timing is specified by referring to the position of the number one piston in relationship to crankshaft rotation. Ignition timing reference marks can be located on a pulley or flywheel to indicate the position of

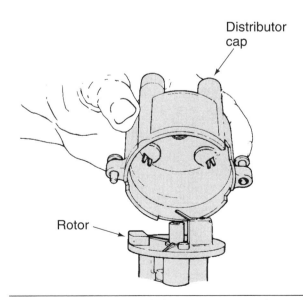

Figure 19-11 A typical distributor cap. (Courtesy of Chrysler Corporation)

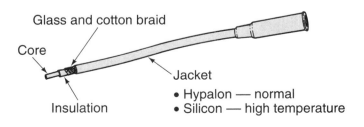

Figure 19-12 Construction of a typical ignition cable. (Courtesy of Chrysler Corporation)

the number one piston (Figure 19-13). Vehicle manufacturers specify initial, or basic ignition timing.

If the spark is sent at the time that the piston is at TDC, it is called **zero degrees advance**. If the spark is sent to the cylinder after the piston has passed TDC, it is called firing after top dead center (ATDC). Most engines are designed to receive the spark as the piston approaches TDC. This allows the air/fuel mixture to burn as the piston starts its downward stroke (power stroke). When the spark is sent as the piston approaches TDC it is called **before top dead center (BTDC)**.

If the spark is sent after the point in time that it is supposed to be sent, it is **retarded timing**. For example, if the specification is 12° BTDC but the spark is sent at 8° BTDC, the spark is slow to get to the cylinder in reference to the piston position. If the spark is sent earlier than it is required, the timing is **advanced**. For example, if the spark is sent at 10° BTDC when the specification requires 6°, the spark is too early into the cylinder in reference to the piston position.

If the spark timing is too far advanced, a condition called **ping (detonation)** may be experienced. Ping is caused by two explosions occurring in the cylinder. One explosion is from the spark plug while the second is caused by **heat of compression** (Figure 19-14). The second explosion is called an uncontrolled explosion. As the air/fuel mixture is compressed, the molecules are packed close together. This is required to receive enough power from the ignited air/fuel mixture to drive the piston downward, however the temperature can get so hot that the mixture will ignite from the heat alone. When these two flame fronts collide in the middle of the cylinder, a hard knock is heard. This can cause severe damage to the piston. Ping can be caused by other influences such as low octane fuel, carbon build-up, high engine temperatures, or faulty EGR system; however, check and adjust

the ignition timing before condemning any of the other possibilities. The timing that is set when the engine is at curb idle is called **initial timing** or **base timing**.

If optimum engine performance is to be maintained, the ignition timing of the engine must change as the operating conditions of the engine change. All of the different operating conditions affect the speed of the engine and the load on the engine. All ignition timing changes are made in response to these primary factors.

Engine Speed

At higher engine speeds, the crankshaft turns through more degrees in a given period of time. The time required to burn the air/fuel mixture in the cylinder is about 2 to 3 milliseconds. This requires that the air/fuel mixture burn be completed by 23° ATDC on the power stroke of the piston. As engine speed increases, the timing of the spark must be advanced in order to complete the burn by 23° ATDC.

However, air/fuel mixture turbulence increases with rpm. This causes the mixture inside the cylinder to burn faster. Increased turbulence requires that ignition must occur slightly later or be slightly retarded.

These two factors must be balanced for best engine performance. Therefore, while the ignition timing must be advanced as engine speed increases, the amount of advance must be decreased to compensate for the increased turbulence.

Engine Load

The load on an engine is related to the work it must do. Driving up hills or pulling extra weight increases engine load. Under load, the pistons accelerate more slowly and the engine runs less efficiently. A good indi-

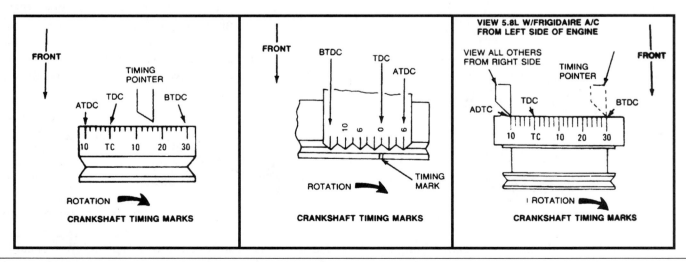

Figure 19-13 Various crankshaft pulleys with timing marks. (Reprinted with the permission of Ford Motor Company)

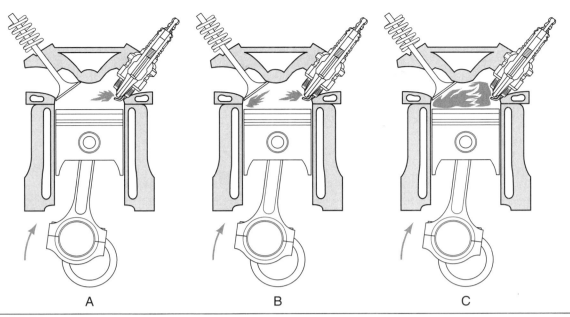

Figure 19-14 Detonation results from an uncontrolled flame front that is started after the spark plug is fired. (A) Combustion begins. (B) Detonation or postspark begins a second flame front. (C) The two flame fronts collide to create a knocking sound.

cation of engine load is the amount of vacuum formed during the intake stroke.

Under light loads and with the throttle plate(s) partially opened, a high vacuum exists in the intake manifold. The amount of air/fuel mixture drawn into the manifold and cylinders is small. This means the air and fuel molecules are relatively far apart. On compression, this thin mixture produces less combustion pressure and combustion time is slow. To complete combustion by 23 degrees ATDC, ignition timing must be advanced to allow for a longer burning time.

Under heavy loads, when the throttle is opened fully, a larger mass of the air/fuel mixture can be drawn in, and the vacuum in the manifold is low. Combustion is fast because the air and fuel molecules are close together. High combustion pressure and rapid burning results. In such a case, the ignition timing must be advanced less or retarded to prevent complete burning from occurring before 23 degrees ATDC.

Firing Order

Up to this point, the primary focus of discussion has been ignition timing as it relates to any one cylinder. However, the function of the ignition system extends beyond timing the arrival of a spark to a single cylinder. It must perform this task for each cylinder of the engine in a specific sequence.

Each cylinder of an engine produces power once every 720 degrees of crankshaft rotation. Each cylinder must have a power stroke at its own appropriate time during the rotation. To make this possible, the pistons and rods are arranged in a precise fashion. This is called the engine's **firing order**. The firing order is arranged to reduce rocking and imbalance problems. Because the potential for this rocking is determined by the design and construction of the engine, the firing order varies from engine to engine. Vehicle manufacturers simplify identifying each cylinder by numbering them (Figure 19-15). Regardless of the particular firing order used, the number one cylinder always starts the firing order, with the rest of the cylinders following in a fixed sequence.

The ignition system must be able to monitor the rotation of the crankshaft and the relative position of each piston in order to determine which piston is on its compression stroke. It must also be able to deliver a high-voltage surge to each cylinder at the proper time during its compression stroke. How the ignition system does these things depends on the design of the system.

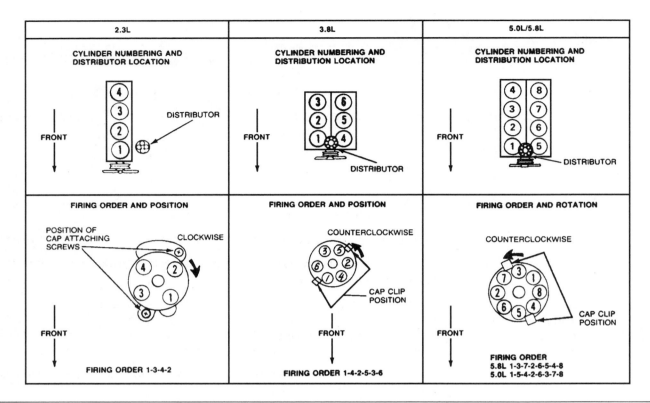

Figure 19-15 Cylinder numbering and firing orders. (Reprinted with the permission of Ford Motor Company)

The man credited with the development of the ignition system, as we know it today, was Charles Kettering. He is also credited with the development of electric starters for Cadillac in 1911, quick drying paint, and ethyl gasoline.

Spark Timing Systems

To help understand the basis of spark timing control, a study in distributors and early ignition systems is presented.

The distributor is responsible for distributing the secondary spark to the correct combustion camber. It consists of a housing, shaft, drive connection, sub plate, advance plate, centrifugal advance unit, vacuum advance unit, rotor, and a cap (Figure 19-16). The distributor shaft is rotated at camshaft speed, or half of the crankshaft speed. In many engines the distributor drive connection also drives the engine oil pump.

Early ignition systems used mechanical points to operate the primary circuit. The point-type ignition system worked well for many years, but was replaced with a more reliable electronic system in the mid 1970s. Even though the point-type system is now obsolete, understanding the principles of this system will help in understanding the operation of more complex systems.

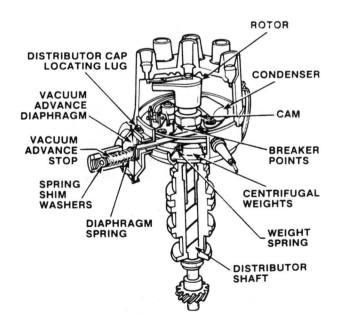

Figure 19-16 Typical nonelectronic distributor with mechanical timing advance units.

Located under the distributor cap and rotor is the mechanical switch that is commonly referred to as **breaker point** (Figure 19-17). The breaker point assembly, which was mounted on the **breaker plate** inside the distributor, consisted of a fixed contact, movable contact, movable arm, rubbing block, pivot, and spring. The fixed contact was grounded through the distributor housing, and the movable contact was connected to the negative terminal of the coil's primary winding. As the cam was turned by the camshaft, the movable arm opened and closed, which opened and closed the primary circuit in the coil. When the points were closed, primary current flow attempted to saturate the coil. When the points opened, primary current stopped. This caused the magnetic field to collapse across the secondary and the primary windings. When it collapsed across the primary windings, it induced a voltage into the windings. A condenser was used to absorb this voltage spike and prevent it from arcing across the points. The collapsing across the secondary windings caused high voltage to be induced. The firing of the plug was the result of opening the points.

A primary, or ballast resistor, was located in series between the battery and the primary coil winding and was responsible for keeping the primary voltage at the desired level (about 9 or 10 volts). This reduced the chances of the contact points burning due to high voltage. The ballast resistor could be either a separate unit or a specially made wire. During starting, the ballast resistor was bypassed to provide maximum current flow to the primary circuit.

The distributor also mechanically adjusted the time the spark arrived at the cylinder through the use of two mechanisms: the centrifugal advance and the vacuum advance units. This improved engine performance, fuel efficiency, and emission levels.

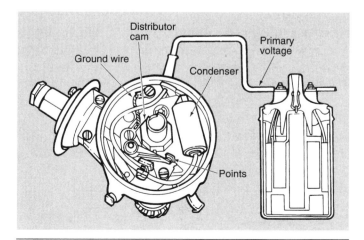

Figure 19-17 A breaker point ignition system. (Courtesy of Chrysler Corporation)

As its name implies, the distributor mechanically distributed the spark so that it arrived at the right time during the compression stroke of each cylinder. The distributor's shaft, rotor, and cap performed this function.

Rise Time

According to SAE, **rise time** is the amount of time (measured in microseconds) for the output of the coil to rise from 10% to 90% of its maximum output. The faster the rise time the easier it is to fire a fouled plug. An ignition spark that reaches its maximum voltage quickly has no time to flow through foul spark plug deposits to ground. Typical rise times for various ignition systems are:

1. Point type: 200 microseconds (s)
2. Electronic ignition: 20 to 50 s
3. Capacitor discharge: 1 to 3 s.

Timing Advance Systems

A BIT OF HISTORY

The first use of automatic spark advance was revealed in 1900. In spite of this development, many cars, for many years, had driver operated timing advance controls.

As stated earlier, engine speed and load affects when the ignition spark must be introduced in the combustion chamber. Early ignition systems changed the timing mechanically using centrifugal and vacuum advance mechanisms. At idle, the firing of the spark plug usually occurs just before the piston reaches top dead center. At higher engine speeds however, the spark must be delivered to the cylinder much earlier in the cycle to achieve maximum power from the air/fuel mixture since the engine is moving through the cycle more quickly. To change the timing of the spark in relation to rpm, the **centrifugal advance** mechanism is used (Figure 19-18).

This mechanism consists of a set of weights and springs connected to the distributor shaft and a distributor armature assembly. During idle speeds, the springs keep the weights in place and the armature and distributor shaft rotate as one assembly. When speed increases, centrifugal force causes the weights to slowly move out against the tension of the springs. This allows the armature assembly to move ahead in relation to the distributor shaft rotation.

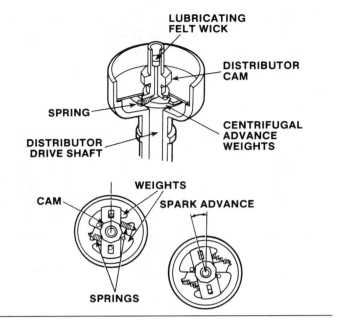

Figure 19-18 Typical centrifugal advance mechanism.

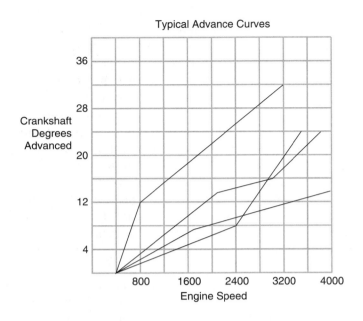

Figure 19-19 Typical advance curves.

The ignition's triggering device is mounted to the armature assembly. Therefore, as the assembly moves ahead, ignition timing becomes more advanced.

The amount of centrifugal advance is controlled by the size and shape of the centrifugal weights and the amount of spring tension. The advance curve can be altered by the technician by changing these units.

Most centrifugal advance systems provide between 7 and 15 distributor degrees (14 and 30 crankshaft degrees) of timing advance. Timing advance does not occur suddenly, but is increased gradually over a specific engine speed. This results in a **timing curve** (Figure 19-19).

Most factory shop manuals give the specifications for the centrifugal advance in distributor degrees and rpm. Since the distributor is driven off the camshaft, it rotates at half the crankshaft speed. To convert the distributor degrees and rpm to crankshaft degrees and rpm, multiply by 2. For example, 8 distributor degrees and 1000 distributor rpm equals 16 crankshaft degrees and 2000 crankshaft rpm.

During part-throttle engine operation, high vacuum is present in the intake manifold. To get the most power and the best fuel economy from the engine, the plugs must fire even earlier during the compression stroke than is provided by a centrifugal advance mechanism.

The heart of the **vacuum advance** mechanism is the spring-loaded diaphragm, which fits inside a metal housing and connects to the breaker plate (Figure 19-20). Vacuum is applied to one side of the diaphragm in the housing chamber while the other side of the diaphragm is open to the atmosphere. Any increase in vacuum allows

atmospheric pressure to push the diaphragm. In turn, this causes the breaker plate to rotate in the opposite direction of distributor rotation to advance the timing. The more vacuum present on one side of the diaphragm, the more atmospheric pressure is able to cause a change in timing. The rotation of the breaker plate moves the location of the breaker points so they open earlier. These units are also equipped with a spring that allows the timing to return to only centrifugal timing advance as vacuum decreases.

Most vacuum advance units use **ported vacuum** to sense engine load. Ported vacuum is manifold vacuum that is controlled by the throttle plates of the carburetor or throttle body fuel injection plates. Ported vacuum is proportional to engine load and throttle position. Since ported vacuum is taken from above the throttle plates, there is no vacuum present at idle. When the throttle plates open, vacuum is present and continues to be unless the plates are opened suddenly or the engine is under a heavy load causing the vacuum to drop. The full vacuum advance of about 30 crankshaft degrees is available with the throttle plates only partially open, such as when the engine speed and load is in the cruise range.

Total Advance

The centrifugal advance and vacuum advance units operate together to provide the optimum timing for the engine under all speeds and loads (Figure 19-21). **Total**

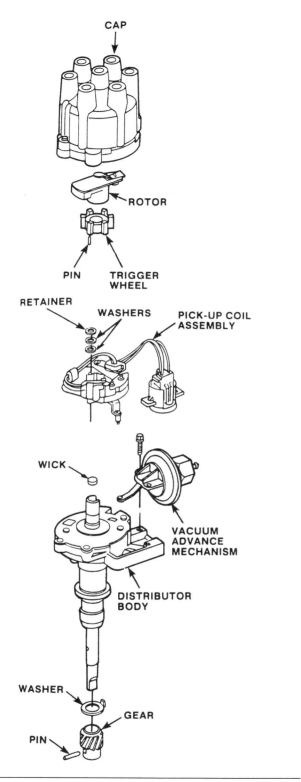

Figure 19-20 Typical vacuum advance mechanism.

advance is the total of base timing plus centrifugal and vacuum advance. For example, if base timing is set at 8 degrees BTDC with a combined centrifugal and vacuum advance of 22 degrees, total advance is 30 degrees. Total

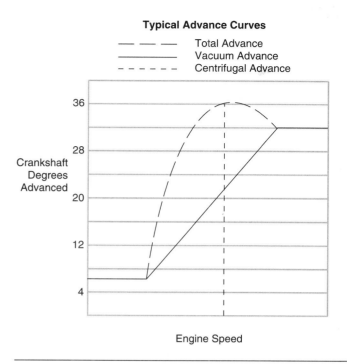

Figure 19-21 Total advance is the sum of base timing plus centrifugal and vacuum advance.

advance is affected by base timing, thus base timing must be correct for the total timing advance curve to be correct (Figure 19-22).

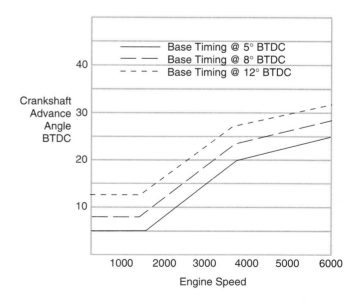

Figure 19-22 The amount of spark advance at any engine speed is affected by the base timing.

Ignition Timing Controls

When timing is at maximum advance, peak combustion temperatures increase. High temperatures and pressures increase the formation of **oxides of nitrogen (NOx)**. NOx is formed when combustion temperatures exceed about 2,500 degrees F (1.370° C) and oxygen and nitrogen combine in large quantities. NOx is a contributor to photochemical smog.

Maximum advance also contributes to the increase of **hydrocarbon (HC)** emissions. Hydrocarbons are unburned fuel. Even though the combustion temperatures are higher with advanced timing, the exhaust gases are cooler. This is because peak temperatures are reached earlier in the engine cycle. Hydrocarbons are emitted as a result of the cooler exhaust gases not heating the exhaust ports and manifolds sufficiently enough to oxidize the unburned fuel. Exhaust gas temperature must be high to cause the unburned fuel (HC) to combine with any remaining oxygen.

Retarded timing results in lower NOx and HC emissions because combustion chamber temperatures are lower and exhaust gas temperature is higher. However, retarded timing makes the combustion less efficient. With retarded timing, performance and economy will be reduced.

To help decrease the amount of HC and NOx emissions, some manufacturers use a **spark-delay valve (SDV)**. The spark-delay valve will delay the timing advance for about 30 seconds. By holding the timing in a less advanced condition at low and intermediate speeds, combustion temperatures are reduced and exhaust gas temperatures are increased.

The spark-delay valve is constructed of a plastic housing with a sintered metal restrictor and a check valve (Figure 19-23). One end of the valve is connected to the ported vacuum source, and the other end to the vacuum advance unit. Vacuum that is applied to the metal restrictor and passage is delayed as the vacuum must filter through the sintered restrictor. The amount of time required to filter through the sintered restrictor delays vacuum advance. Most SDVs are color coded to indicate the amount of delay provided.

The check valve opens when vacuum drops. This allows the vacuum in the vacuum advance unit to be released. When the check valve releases the vacuum, the timing will retard. This will occur when vacuum is reduced or dropped as a result of heavy engine load, deceleration, or when the engine is running at idle.

Chrysler used a similar system to control spark delay. Their system was called **orifice spark advance control (OSAC)**. The OSAC system used an orifice in place of the sintered restrictor used in the SDV. In later years they

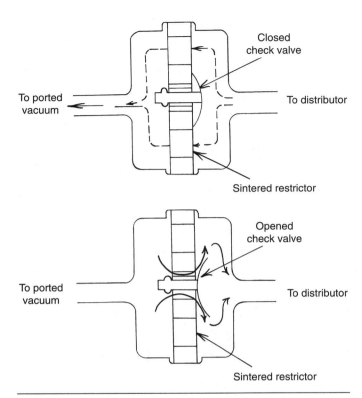

Figure 19-23 The spark delay valve slows the vacuum to the distributor to delay advance.

utilized a **thermostatic vacuum valve**. This valve senses engine coolant temperature and opens when coolant temperature is high. When open, the valve allows manifold vacuum to the vacuum advance unit. With full advance, the exhaust temperatures are cooler and engine temperature is lowered.

Ford, and some other manufacturers, used a valve bypass system that would allow for vacuum advance when the engine was cold. A solenoid vacuum valve was installed in a by-pass line around the SDV. An ambient temperature switch would energize the solenoid vacuum valve to open the by-pass line when ambient temperatures were low. When open, the solenoid would allow for full timing advance to aid in engine warm-up and cold engine driveability.

Some later applications incorporated a **high temperature bypass** system in place of the solenoid vacuum valve. This system operated similar to Chrysler's and allowed for full advance when engine temperatures were too high. The high temperature by-pass system used a thermostatic vacuum valve that sensed engine coolant temperature. The valve would open to allow ported vacuum to the vacuum advance unit. The difference between this system and Chrysler's is that Chrysler used manifold vacuum.

Many manufacturers used a system of sensing vehicle speed or transmission gear selection to reduce vacuum advance at low and cruise speeds. These systems used a **solenoid vacuum valve** that would prohibit vacuum advance at low and cruise speeds. They also used either a transmission switch or speed sensor to control the solenoid. The solenoid could be either normally energized or normally de-energized and the switch could be either normally open or normally closed.

Speed controlled spark systems use a speed sensor in the speedometer cable (Figure 19-24). These units use a rotating magnet that is connected to the speedometer. The magnet rotates inside a stationary winding. As the magnet rotates in the windings, an induced low voltage is produced. The induced voltage is proportional to the speed of the vehicle. This low voltage signal is sent to the electronic amplifier that amplifies the low voltage into higher voltage. The low voltage signal is amplified proportionally to the input signal. If the amplified voltage is high enough, it will energize the solenoid.

Electronic or Solid-State Ignition

From the fully mechanical breaker point system, ignition technology progressed to basic electronic or solid-state ignitions (Figure 19-25). Breaker points were replaced with electronic triggering and switching devices. The electronic switching components are normally inside a separate housing known as an electronic control unit (ECU) or control module (Figure 19-26). The original (solid-state) electronic ignitions still relied on mechanical and vacuum advance mechanisms in the distributor.

As technology advanced, many manufacturers expanded the ability of the ignition control modules. For example, by tying a manifold vacuum sensor into the ignition module circuitry, the module could now detect when the engine was under heavy load and retard the timing automatically. Similar add-on sensors and circuits

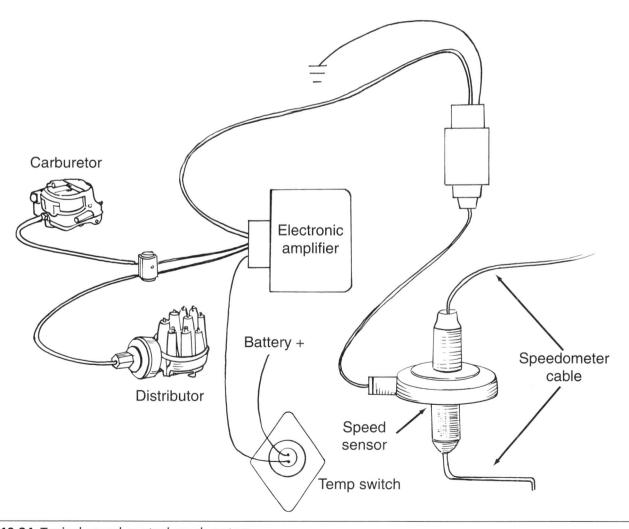

Figure 19-24 Typical speed control spark system.

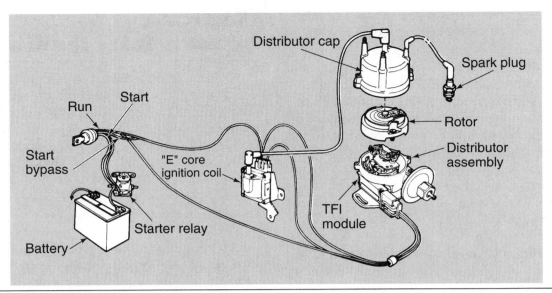

Figure 19-25 One design of an electronic distributor-type ignition system. (Reprinted with the permission of Ford Motor Company)

were designed to control spark knock, start-up emissions, and altitude compensation.

Electronic ignition systems control the primary circuit using an NPN transistor instead of breaker contact points. The transistor's emitter is connected to ground and takes the place of the fixed contact point. The collector is connected to the negative (-) terminal of the coil, taking the place of the movable contact point. When the triggering device supplies a small amount of current to the base of the switching transistor, the collector and emitter act as if they are closed contact points (a conductor), allowing current to build up in the coil primary circuit. When the current to the base is interrupted by the switching device, the collector and emitter act as an open contact (an insulator), interrupting the coil primary current. An example of how this works is shown in Figure 19-27, which is a simplified diagram of an electronic ignition system.

In a breaker-point ignition system, the air gap between the two contacts determined the dwell. As the rubbing block of the points wore down, point gap changed and so did dwell. Since dwell is the length of time current flows through the primary windings of the coil, maintaining a proper dwell is important to having sufficient secondary coil voltage output. Electronic switching devices have no rubbing block, and dwell tends to be maintained over long periods of time.

Electronic DI System Operation

The primary circuit of a DI system is controlled electronically by input sensors and an electronic control unit (module) that contains some type of switching device. Figure 19-28 shows a basic electronic ignition system. The system consists of a distributor with a **magnetic pulse pickup** unit and reluctor, an electronic control module, and a ballast resistor.

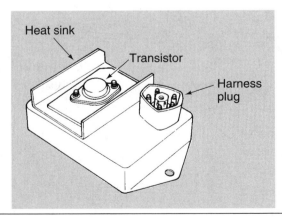

Figure 19-26 A typical electronic control module. (Courtesy of Chrysler Corporation)

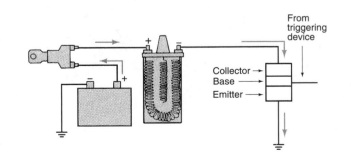

Figure 19-27 The switching device in an electronic ignition system is a transistor.

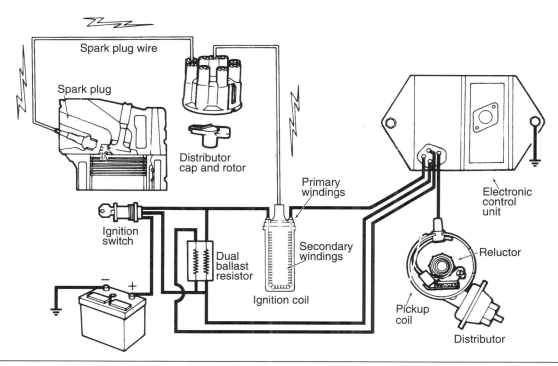

Figure 19-28 A basic electronic DI system. (Courtesy of Chrysler Corporation)

Basically, a magnetic pulse generator consists of two parts: a timing disc and a pickup coil. The **pickup** coil consists of a length of wire wound around a weak permanent magnet. Depending on the type of ignition system used, the **timing disc** may be mounted on the distributor shaft (Figure 19-29), at the rear of the crankshaft (Figure 19-30), or on the crankshaft vibration damper (Figure 19-31).

The magnetic pulse or PM generator operates on basic electromagnetic principles. Remember that a voltage can only be induced when a magnetic field is moved across a conductor. The magnetic field is provided by the pickup

Figure 19-30 A flywheel crank timing sensor. (Courtesy of Chrysler Corporation)

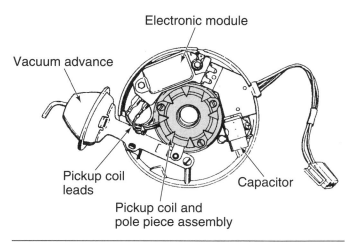

Figure 19-29 A magnetic pulse generator mounted in a distributor. (Courtesy of Oldsmobile Division—GMC)

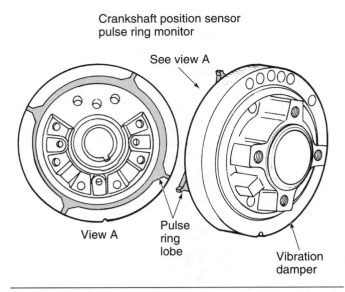

Crankshaft position sensor
pulse ring monitor

See view A

View A

Pulse
ring
lobe

Vibration
damper

Figure 19-31 The pulse ring on a crankshaft pulley.

unit, and the rotating timing disc provides the movement of the magnetic field needed to induce voltage.

As the disc teeth approach the pickup coil, they repel the magnetic field, forcing it to concentrate around the pickup coil (Figure 19-32A). Once a tooth passes by the pickup coil, the magnetic field is free to expand or dissipate (Figure 19-32B), until the next tooth on the disc approaches. Approaching teeth concentrate the magnetic lines of force, while passing teeth allow them to expand. This pulsation of the magnetic field causes the lines of magnetic force to cut across the winding in the pickup coil, inducing a small amount of AC voltage that is sent to the switching device in the primary circuit.

When a disc tooth is directly in line with the pickup coil, the magnetic field is not expanding or contracting. Since there is no movement or change in the field, voltage at this precise moment drops to zero. At this point,

the switching device inside the ignition module reacts to the zero voltage signal by turning the ignition's primary circuit current off. As explained earlier, this forces the magnetic field in the primary coil to collapse, discharging a secondary voltage to the distributor or directly to the spark plug.

As soon as the tooth rotates past the pickup coil, the magnetic field collapses and a voltage signal is induced. The only difference is that the polarity of the charge is reversed. Negative becomes positive or positive becomes negative. Upon sensing this change in voltage, the switching device turns the primary circuit back on and the process begins all over.

The slotted disc is mounted on the crankshaft, vibration damper, or distributor shaft in a very precise manner. The disc teeth align with the pickup coil at the exact time certain pistons are nearing TDC. This means the zero voltage signal needed to trigger the secondary circuit occurs at precisely the correct time.

The pickup coil might have only one pole as shown in Figure 19-33. Other magnetic pulse generators have pickup coils with two or more poles. The one shown in Figure 19-34 has as many poles as it has trigger teeth.

The pulse signals the transistor to open the primary circuit, firing the plug. Once the plug stops firing, the transistor closes the primary coil circuit. The length of time the transistor allows current flow in the primary ignition circuit is determined by the electronic circuitry in the control module. Some systems used a dual ballast resistor. The ceramic ballast resistor assembly is mounted on the firewall and has a ballast resistor for primary current flow and an auxiliary resistor for the control module. The ballast resistor has a 0.5-ohm resistance that maintains a constant primary current. The auxiliary ballast resistor uses a 5-ohm resistance to limit voltage to the electronic control module.

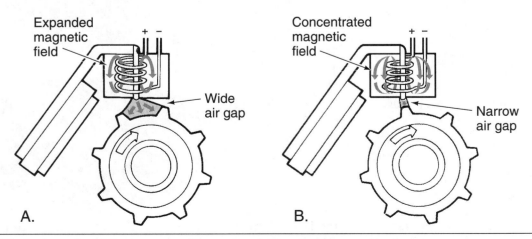

Expanded magnetic field

Wide air gap

A.

Concentrated magnetic field

Narrow air gap

B.

Figure 19-32 Action of a PM generator. (Courtesy of Chrysler Corporation)

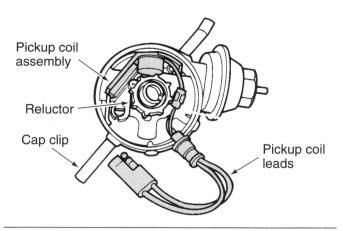

Figure 19-33 A single pole pickup coil. (Courtesy of Chrysler Corporation)

There are some DI systems that do not require a ballast resistor. For instance, some control units directly regulate the current flow through the primary of the coil.

Computer-Controlled Electronic Ignition

Computer-controlled ignition systems offer continuous spark timing control through a network of engine sensors and a central microprocessor. Based on the inputs it receives, the central microprocessor or computer makes decisions regarding spark timing and sends signals to the ignition module to fire the spark plugs according to those inputs and according to the programs in its memory.

Computer-controlled ignition systems may or may not use a distributor to distribute secondary voltage to the spark plugs. As mentioned earlier, distributorless systems use multiple coils and modules to provide and distribute high secondary voltages directly from the coil to the plug. These systems will be discussed in greater detail in a later chapter.

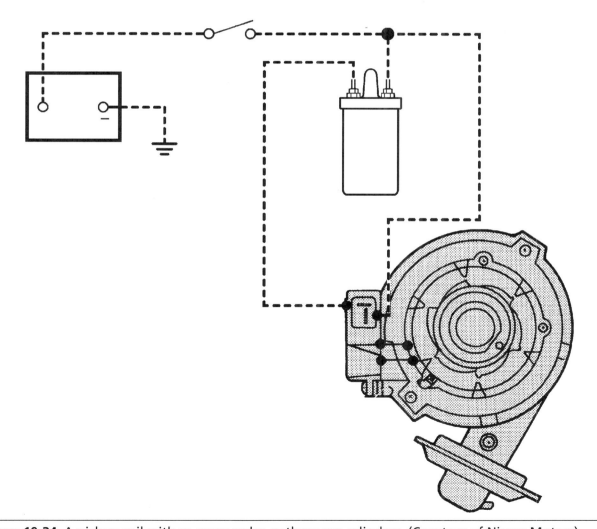

Figure 19-34 A pickup coil with as many poles as there are cylinders. (Courtesy of Nissan Motors)

Summary

❏ The ignition system supplies high voltage to the spark plugs to ignite the air/fuel mixture in the combustion chambers.

❏ The arrival of the spark is timed to coincide with the compression stroke of the piston. This basic timing can be advanced or retarded under certain conditions, such as high engine rpm or extremely light or heavy engine loads.

❏ The ignition system has two interconnected electrical circuits: a primary circuit and a secondary circuit.

❏ The primary circuit supplies low voltage to the primary winding of the ignition coil. This creates a magnetic field in the coil.

❏ A switching device interrupts primary current flow, collapsing the magnetic field, and creating a high-voltage surge in the ignition coil secondary winding.

❏ The switching device used in electronic systems is an NPN transistor. Old ignitions use mechanical breaker point switching.

❏ The secondary circuit carries high-voltage surges to the spark plugs. On some systems, the circuit runs from the ignition coil through a distributor to the spark plugs.

❏ The distributor may house the switching device plus centrifugal or vacuum timing advance mechanisms. Some systems locate the switching device outside the distributor housing.

Terms-To-Know

Advanced	High temperature bypass	Rise time
After top dead center (ATDC)	Hydrocarbon (HC)	Run pickup
Antiflashover ribs	Ignition cables	Secondary circuit
Base timing	Ignition timing	Secondary coil windings
Before top dead center (BTDC)	Initial timing	Secondary ignition circuit
Breaker plate	Ionize	Secondary reserve voltage
Breaker points	Magnetic pulse pickup	Solenoid vacuum valve
Center pole negative	Mutual induction	Spark delay valve (SDV)
Centrifugal advance	Orifice spark advance control	Spark plug
Detonation	(OSAC)	Speed controlled spark
Distributor	Oxides of nitrogen (NOX)	Starting pickup
Distributor cap	Pickup coil	Thermostatic vacuum valve
Distributor ignition (DI)	Ping	Thick film integrated (TFI)
Dwell	Ported vacuum	Timing curve
E coil	Primary circuit	Timing disc
Electrical reserve	Primary coil windings	Top dead center (TDC)
Electronic ignition (EI)	Primary ignition circuit	Total advance
Firing order	Pulse transformer	Vacuum advance
Heat of compression	Reach	Zero degrees advance
Heat range	Resistor plugs	
High energy ignition (HEI)	Retarded timing	

Review Questions

Short Answer Essay

1. Under what condition is the ballast resistor in an ignition system's primary circuit bypassed?

2. Under light loads, what must be done to complete air/fuel combustion in the combustion chamber by the time the piston reaches 23 degrees ATDC?

3. At high engine rpm, what must be done to complete air/fuel combustion in the combustion chamber by the time the piston reaches 10 degrees ATDC?

4. Describe the three major functions of an ignition system.

5. Explain the components and operation of a magnetic pulse generator.

6. Name the engine operating conditions that affect ignition timing requirements.

7. Briefly describe the operation of the ignition coil.

8. Describe the operation of spark plugs and ignition cables.

9. What is the basic difference between the primary and secondary ignition circuits?

10. What primary role does a rotor have in the ignition system?

Fill-in-the-Blanks

1. Modern ignition cables contain fiber cores that act as a _____ in the secondary circuit to cut down on radio and television interference and reduce spark plug wear.

2. The magnetic field surrounding the pickup coil in a magnetic pulse generator moves when the _____ .

3. The arrival of the spark is timed to coincide with the _____ stroke of the piston.

4. Basic ignition timing is typically _____ with increases in engine speed and _____ with increases of engine load.

5. The ignition system has two interconnected electrical circuits: a _____ circuit and a _____ circuit.

6. A _____ device interrupts primary current flow, collapsing the magnetic field, and creating a high-voltage surge in the ignition coil secondary winding.

7. The switching device used in electronic systems is a(n) _____ .

8. The PM generator produces a(n) _____ voltage.

9. The difference between the required voltage and the maximum available voltage is referred to as _____ _____ voltage.

10. The _____ _____ of the spark plug refers to its heat dissipation properties

ASE Style Review Questions

1. While discussing what happens when the low-voltage current flow in the coil primary winding is interrupted by the switching device, Technician A says the magnetic field collapses. Technician B says a high-voltage surge is induced in the coil secondary winding. Who is correct?
   a. A only
   b. B only
   c. Both A and B
   d. Neither A nor B

2. Technician A says an ignition system must generate sufficient voltage to force a spark across the spark plug gap. Technician B says the ignition system must time the arrival of the spark to coincide with the movement of the engine's pistons and vary it according to the operating conditions of the engine. Who is correct?
   a. A only
   b. B only
   c. Both A and B
   d. Neither A nor B

3. While discussing electronic ignition systems: Technician A says a transistor actually controls primary current flow through the coil. Technician B says a reluctor controls the primary coil current. Who is correct?
   a. A only
   b. B only
   c. Both A and B
   d. Neither A nor B

4. Technician A says that the pulse generator signals the electronic control unit. Technician B say that the pulse generator replaces the electronic control unit. Who is correct?
   a. A only
   b. B only
   c. Both A and B
   d. Neither A nor B

5. Technician A says that the Ford TFI module is located on the fender well. Technician B say the TFI module is located on the distributor. Who is correct?
   a. A only
   b. B only
   c. Both A and B
   d. Neither A nor B

6. Technician A says the coil on GM's HEI system is located on top of the distributor cap. Technician B says that the coil is external to the cap. Who is correct?
   a. A only
   b. B only
   c. Both A and B
   d. Neither A nor B

7. Technician A says the high voltage produced in the secondary winding of the ignition coil is a result of magentic reluctance. Technician B says the elctrical reserve of the ignition coil is the margin of voltage that can be produced above that required to fire the spark plugs. Who is correct?
   a. A only
   b. B only
   c. Both A and B
   d. Neither A nor B

8. Technicain A says the heat range of the spark plug determines the amount of voltage required to jump the gap. Technician B says ignition timing refers to the delivery of the ignition spark relative to the delivery of the air/fuel mixture. Who is correct?
   a. A only
   b. B only
   c. Both A and B
   d. Neither A nor B

9. Technician A says that the pickup coils used in all HEI systems are the same. Technician B says that the pickup coils are wound in opposite directions for different applications. Who is correct?
   a. A only
   b. B only
   c. Both A and B
   d. Neither A nor B

10. Technician A says that the voltage produced in the pickup coil when the reluctor aligns with the coil is used to signal engine speed. Technician B says that this voltage is used to impress the base leg of the transistor. Who is correct?
    a. A only
    b. B only
    c. Both A and B
    d. Neither A nor B

20 Basic Ignition System Diagnosis and Service

Objective

Upon completion and review of this chapter, you should be able to:

- ❏ Use symptoms to identify probable problem areas.
- ❏ Perform a visual inspection of ignition system components, primary wiring, and secondary wiring to locate obvious trouble areas.
- ❏ Identify and describe the major sections of primary circuit and secondary circuit trace patterns, including the firing line, spark line, intermediate area, and dwell zone.
- ❏ Perform cranking output, spark duration, coil polarity, spark plug firing voltage, rotor, secondary resistance, and spark plug load tests using the oscilloscope.
- ❏ Test the components of the primary and secondary ignition circuits.
- ❏ Test individual ignition components using test equipment such as a voltmeter, ohmmeter, and test light.
- ❏ Service and install spark plugs.
- ❏ Test and set (when possible) ignition timing.
- ❏ Perform a no-start diagnosis, and determine the cause of the no-start condition.
- ❏ Remove, inspect, and service distributors.
- ❏ Inspect, service, and test ignition modules.
- ❏ Check ignition spark advance.
- ❏ Perform pickup tests on distributor ignition (DI) systems.

Introduction

This chapter concentrates on testing ignition systems and their individual components. It must be stressed, however, that there are many variations in the ignition systems used by automotive manufacturers. The tests covered in this chapter are those generally used as basic troubleshooting procedures. Exact test procedures and the ideal troubleshooting sequence will vary between vehicle makers and individual models. Always consult the vehicle's service manual when performing ignition system service. There are two important precautions that should be taken during all ignition system tests:

1. Turn the ignition switch off before disconnecting any system wiring.
2. Do not touch any exposed connections while the engine is cranking or running.

Logical Troubleshooting

The importance of logical troubleshooting cannot be overemphasized. The ability to diagnose a problem (to find its cause and its solution) is what separates an automotive technician from a parts changer. No matter if you are servicing the electronic controls of a late-model vehicle or an older vehicle with very basic systems, you need to approach your troubleshooting efforts in an orderly and sensible way.

The first step is to gather as much information about the problem as possible from the owner. What happens? When does it happen? Always? Sometimes? What is the weather like when it happens? These are just a few of the many questions you should ask. After you have gathered all this information, try to duplicate the problem. Sometimes this is difficult, especially when the problem is intermittent. Wiggling wires and vacuum hoses will sometimes help to locate an intermittent problem.

The next step is simply a visual inspection and road test. Check all wires, connections, and vacuum hoses. Sometimes an owner does not notice something or does not describe it well. This is an important step; gather as much information as you possibly can.

Based on the symptoms you know of so far, look through the manufacturer's service bulletins to see if this problem is listed for the vehicle you are working on. If there is a match, follow the repair instructions. Then road test the vehicle to see if the problem is fixed. If it is not, continue with your diagnostics.

Look over the ignition and related systems on the vehicle. Think about what each component does and when it does it. Then, based on the information you have gathered, make a list of those parts or systems that could be causing the problem. By doing this, you are making a list of those items that should be checked or tested, and eliminating the rest. Do not waste your time checking something that cannot cause the problem.

Begin your tests with the most likely and the easiest to get to. The purpose of your testing should be to eliminate possible causes. Looking for the cause may prejudice your thinking. Assume nothing; let the test results take you to the cause of the problem. As you test, you are eliminating components and circuits that are good. Take those items off your list of possible causes. If something fails a test, don't assume that it is the cause of the problem unless you have no other items on your list. Often a part fails or cannot work properly because another part is not working as it should. If you replace or attempt to repair the first item that does not seem to work right, you may be wasting time and money. Eventually you will have one or a few items on your list. These are most likely the causes of the problem. Repair or replace as necessary; then check to make sure the problem is corrected.

The best automotive technicians use this same logical process to diagnose engine problems. When faced with a driveability problem, they compare clues (such as meter readings, oscilloscope readings, and visible problems) with their knowledge of proper conditions and discover a logical reason for the way the engine is performing. Logical diagnosis means following a simple basic procedure. Start with the most likely cause and work to the most unlikely. In other words, check out the easiest, most obvious solutions first before proceeding to the less likely, and more difficult, solutions. Do not guess at the problem or jump to a conclusion before considering all of the factors.

This logical approach has a special application to troubleshooting electronic engine controls. Always check all non-electronic engine control possibilities before attempting to diagnose the electronic engine control itself. For example, low battery voltage might result in faulty sensor readings. The distributor could also be sending faulty signals to the computer, resulting in poor ignition timing.

Visual Inspection

As stated earlier, all diagnoses of ignition systems should include a visual inspection. The system should be checked for obvious problems, such as disconnected, loose, or damaged secondary cables; disconnected, loose, or dirty primary wiring; a loose or damaged distributor cap or rotor; a worn or damaged primary system switching mechanism; and an improperly mounted electronic control unit.

Inspection of Secondary and Primary Wiring

Spark plug and ignition coil wires should be firmly inserted into the distributor cap and coil and onto spark plugs. Inspect all secondary wires for cracks and worn insulation, which can cause high-voltage leaks. Inspect all of the boots on the ends of the secondary wires for cracks and hard, brittle conditions. Replace the wires and boots if they show evidence of these conditions. When checking the coil wire, check the ignition coil. The coil should be inspected for cracks or any evidence of arcing or burning at the coil tower. The coil container should be inspected for oil leaks. If oil is leaking from the coil, air space is present in the coil, which allows condensation to form internally. Condensation in an ignition coil causes high voltage leaks and engine misfiring. Ignition coils with the windings exposed should be carefully inspected for signs of burning.

White or grayish powdery deposits on secondary wires at the point where they cross or near metal parts indicate that the cable's insulation is faulty. The deposits occur because the high voltage in the wire has burned the dust collected on the cable. Such faulty insulation may produce a spark that sometimes can be heard and seen in the dark. An occasional glow around the spark plug wires, known as a **corona effect,** is not harmful but indicates that the wire should be replaced.

Make sure the spark plug wires are arranged according to the firing order of the engine. Spark plug wires from consecutively firing cylinders should cross rather than run parallel to one another (Figure 20-1). When they are parallel to each other, **crossfiring** can occur. Crossfire is the electromagnetic induction spark that can be transmitted in another close wire. When induction crossfire occurs, no spark is jumped from one wire to the other, but the spark is the result of induction from another field. Crossfire induction is most common in two parallel wires which fire one after the other in the firing order. To prevent crossfire the plug wires must be installed in the proper separator, and any two parallel wires next to each other in the firing order should be positioned as far away from each other as possible.

Primary ignition system wiring should be checked for tight connections, especially on vehicles with electronic or computer-controlled ignitions. Electronic circuits operate on very low voltage. Voltage drops caused by corrosion or dirt will cause running problems. Missing tab locks on wire connectors are often the cause of intermittent ignition problems due to vibration or thermal related failure. Also, many late model vehicles require the use of a special dielectric grease on connector terminals. This grease seals the connections from outside contaminants.

Test the integrity of a suspect connection by tapping, tugging, and wiggling the wires while the engine is running. Be gentle. The object is to recreate an ignition interruption, not to cause permanent circuit damage. With the engine off, separate the suspect connectors and check them for dirt and corrosion. Clean the connectors according to the manufacturer's recommendations.

Do not overlook the ignition switch as a source of intermittent ignition problems. A loose mounting rivet or poor connection can result in erratic spark output. To check the switch, gently wiggle the ignition key and connecting wires with the engine running. If the ignition cuts out or dies, the problem is located. Also check the battery connection to the starter solenoid. Some vehicles use this connection as a voltage source for the coil. A bad connection can result in ignition interruption.

Ground Circuits

Keep in mind that automakers use body panels, frame members, and the engine block as the ground for the electrical system. Ground straps are often neglected, or worse, left disconnected after routine service. With the increased use of plastics in today's vehicles, ground straps may mistakenly be reconnected to a nonmetallic surface. The result of any of these problems is a poor or no ground. Seeking a ground, current may attempt to feedback through another circuit. This may cause the circuit to operate erratically or fail all together. The current may also be forced through other components, such as wheel bearings or shift and clutch cables that are not meant to handle current flow, causing them to wear prematurely or become seized in their housing.

Examples of bad ground circuit induced ignition failures include burnt ignition modules resulting from missing or loose coil ground straps, erratic performance caused by a poor distributor-to-engine block ground, and intermittent ignition operation resulting from a poor ground at the control module.

Distributor Cap and Rotor

A defective distributor cap or rotor can cause the engine to have a no-start condition. In addition it may also cause hard starting during high-moisture conditions, missing on acceleration, reduced fuel economy, and reduced driveability. An improperly installed rotor or distributor cap can cause any of the above conditions plus a potentially damaging engine backfire that may start an engine fire or personal injury.

When inspecting the distributor rotor, check the following items:

1. Rotor locating tab and hold-downs. All rotors have a means of assuring they are installed in the proper direction in reference to the distributor shaft. This tab also

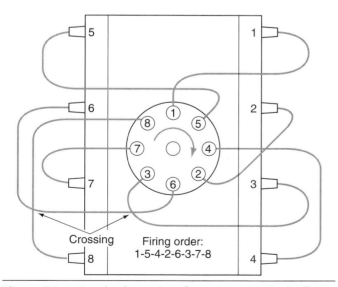

Figure 20-1 Spark plug wires from consecutively firing cylinders should cross each other rather than run parallel to one another.

assures that the rotor turns with the shaft. If the tab is broken then the rotor may not spin with the shaft or not be located on the correct cylinder.

2. Eroded rotor tip.

3. Rotor spring arm. When the high voltage enters the distributor cap it then jumps to the spring arm of the rotor. The spring arm contacts the center insert of the cap, then directs the voltage to the rotor tip. If the spring arm is eroded away because of electrical arcing, it will increase the voltage required to fire the spark plug. If available voltage is not sufficient, the spark plug may misfire or not fire at all.

4. Punch through. Inspect the top and bottom of the rotor carefully for grayish, whitish, or rainbow- hued spots. Such discoloration indicates that the rotor has lost its insulating qualities. High voltage is being conducted to ground through the plastic. Rotor **punch through** is caused when an electrical current has burned a hole to the metal distributor shaft (ground) thus bypassing the spark plugs.

5. Cracks in the rotor (Figure 20-2). A cracked rotor must be replaced.

Check the distributor cap for the following (Figure 20-3):

1. Distributor cap locating tab. This tab assures proper location of the cap onto the distributor housing.

2. Distributor cap hold-down (Figure 20-4). The distributor cap must be securely attached to the distributor housing to prevent contamination and to assure its location.

3. Dirt and water in the cap. These contaminants will require additional voltage in order to fire the spark plug. If available voltage is not enough to overcome this resistance then no spark will occur at the spark plug.

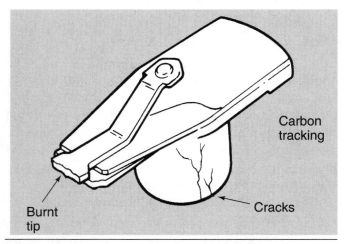

Figure 20-2 Inspect the rotor for cracks and evidence of high voltage leaks. (Courtesy of Chrysler Corporation)

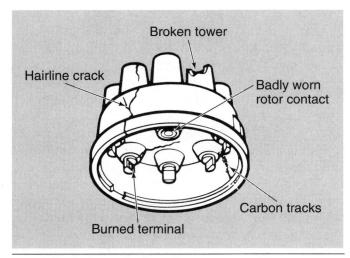

Figure 20-3 Types of distributor cap problems.

4. Center carbon insert. All of the high voltage coming through the coil on its way to the spark plug must pass through this insert. Any additional resistance caused by wear or contamination can cause the engine not to start or to misfire.

5. Tower inserts. The inserts are made from aluminum or brass so it is possible for them to oxidize. Aluminum oxide is an abrasive that can cause excessive damage to the distributor. If the insert is made from brass then the oxidation will be green in color.

6. Cracks. If any cracks are present on the distributor cap then replacement of the part is required.

7. Carbon tracking. **Carbon tracking** indicates that high-voltage electricity has found a low-resistance conductive path over or through the plastic. The result is a cylinder that fires at the wrong time, or a misfire.

If the distributor cap or rotor has a mild build-up of dirt or corrosion, it should be cleaned. If it cannot be cleaned, it should be replaced. Small round brushes are available to clean cap terminals. Wipe the cap and rotor with a clean shop towel, but avoid cleaning these components in solvent or blowing them off with compressed air, which may contain moisture. Cleaning these components with solvent or compressed air may result in high-voltage leaks.

Timing Advance Mechanisms

The hoses to the vacuum advance unit should be securely attached. Disconnected hoses can cause poor fuel economy, poor performance, poor idle, and stalling.

Centrifugal advance mechanisms should be checked for free motion. Move the rotor on the distributor shaft clockwise and counterclockwise. The rotor should rotate in one direction approximately 1/4 inch, then spring back. If not, the centrifugal advance mechanism might be binding or rusty. A lubricant might be necessary on

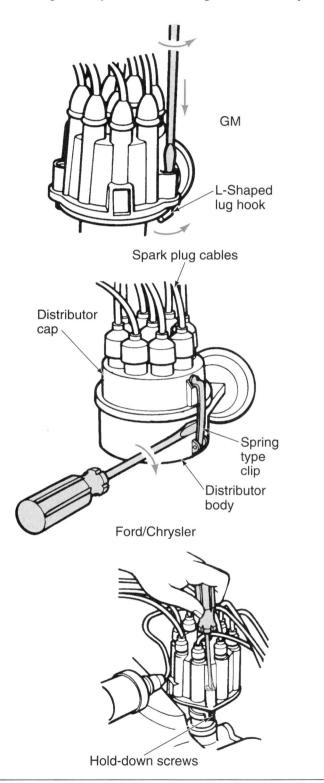

Figure 20-4 Distributor cap hold-down devices. (Courtesy of Chrysler Corporation)

advance mechanism pivots and rubbing surfaces after the mechanism is cleaned.

Electronically controlled advance mechanisms eliminate centrifugal and vacuum advance/retard mechanisms within the distributor. However, all wiring connections to and from the distributor and ignition module should be thoroughly inspected for damage.

Primary Circuit Switches and Sensors

Electronic and computer-controlled ignitions use transistors as switches. These transistors are contained inside a control module housing that can be mounted to or in the distributor or remotely mounted in the vehicle's engine compartment. Control modules should be tightly mounted to clean surfaces. A loose mounting can cause a heat build-up that can damage and destroy transistors and other electronic components contained in the module. Some manufacturers recommend the use of a special heat-conductive silicone grease between the control unit and its mounting (Figure 20-5). This helps conduct heat away from the module, reducing the chance of heat-related failure. During the visual inspection, check all electrical connections to the module. They must be clean and tight.

The transistor in the control module is controlled by a voltage pulse from a crankshaft position sensor. In most ignition systems, this sensor is either a magnetic pulse generator or Hall-effect sensor. These sensors are mounted either on the distributor shaft or the crankshaft. Magnetic pulse generators are relatively trouble free. The reluctor (pole piece) is replaced only if it is broken or cracked. The pickup coil wire leads can become

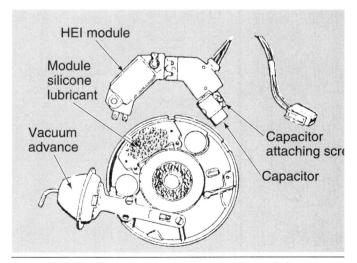

Figure 20-5 When replacing control modules, some manufacturers specify applying a dielectric grease to the module's mounting surface. (Courtesy of Oldsmobile Division—GMC)

grounded if their insulation wears off as the breaker plate moves with the vacuum advance unit. Inspect these leads carefully (Figure 20-6). Position these wires so they do not rub the breaker plate as it moves.

Under unusual circumstances, the nonmagnetic reluctor can become magnetized and upset the pickup coil's voltage signal to the control module. Use a steel feeler gauge to check for signs of magnetic attraction and replace the reluctor if the test is positive. On some systems, the gap between the pickup and the reluctor must be checked and adjusted to manufacturer's specifications.

Pickup Gap Adjustment

When the pickup coil is bolted to the pickup plate, such as on Chrysler distributors, the pickup air gap may be measured with a nonmagnetic feeler gauge positioned between the reluctor high points and the pickup coil (Figure 20-7).

If a pickup gap adjustment is required, loosen the pickup mounting bolts and move the pickup coil until the manufacturer's specified air gap is obtained. Re-tighten

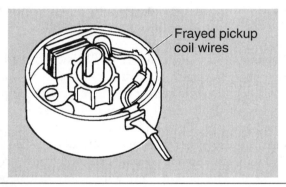

Figure 20-6 Inspect the pickup coil's wiring for damage and worn insulation.

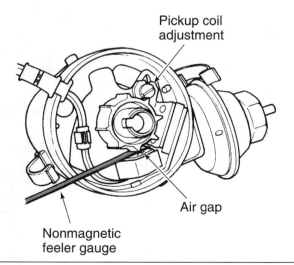

Figure 20-7 Pickup coil air gap adjustment. (Courtesy of Chrysler Corporation)

the pickup coil retaining bolts to the specified torque. Some pickup coils are riveted to the pickup plate, and a pickup gap adjustment is not required.

When checking the pickup coil, always check the distributor bushing for horizontal movement, which changes the pickup gap and may cause engine misfiring.

Ignition System Diagnosis

As discussed, ignition system problems may result in a no-start condition, poor fuel economy, and poor driveability. Diagnosis of the ignition system depends on the type of problem the customer is experiencing.

No-Start Diagnosis

The following ignition defects may cause a no-start condition or hard starting:

1. Defective coil
2. Defective cap and rotor
3. Defective pickup coil
4. Open secondary coil wire
5. Low or zero primary voltage at the coil
6. Fouled spark plugs

No-Start Ignition Diagnosis, Primary Ignition Circuit

The same no-start diagnosis may be performed on most electronic ignition systems. Follow these steps for the no-start diagnosis:

1. Connect a 12-V test lamp from the coil tachometer (tach) terminal to ground, and turn on the ignition switch (Figure 20-8). On General Motors DI systems, the test light should be on because the module primary circuit is open. If the test light is off, there is an open circuit in the coil primary winding or in the circuit from the ignition switch to the coil battery terminal. On some other systems the test light should be off because the module primary circuit is closed, and since there is primary current flow, most of the voltage is dropped across the primary coil winding. This action results in very low voltage at the tach terminal, which does not illuminate the test light. On these systems, if the test light is illuminated, there is an open circuit in the module or in the wire between the coil and the module.

2. Crank the engine and observe the test light. If the test light flutters while the engine is cranked, the pickup coil signal and the module are satisfactory. When the test lamp does not flutter, one of these components is defective. The pickup coil may be tested with an ohm-

meter. If the pickup coil is satisfactory, the module is defective.

No-Start Ignition Diagnosis, Secondary Ignition Circuit

1. If the test light flutters in the primary circuit no-start diagnosis, connect a test spark plug to the coil secondary wire, and ground the spark plug case (Figure 20-9). The test spark plug must have the correct voltage requirement for the ignition system being tested. For example, test spark plugs for General Motors DI systems have a 25,000-V requirement compared to a 20,000-V requirement for many other test spark plugs. A short piece of vacuum hose may be used to connect the test spark plug to the center distributor cap terminal on General Motors DI systems with an integral coil in the distributor cap.

2. Crank the engine and observe the spark plug. If the test spark plug fires, the ignition coil is satisfactory. If the test spark plug does not fire, the coil is probably defec-

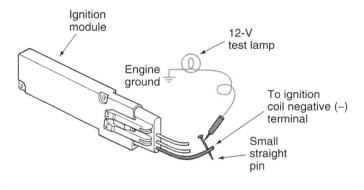

Figure 20-8 Test light connected to the negative primary coil terminal and ground. (Reprinted with the permission of Ford Motor Company)

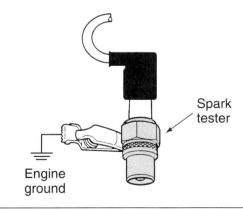

Figure 20-9 Test spark plug connected to coil secondary wire. (Reprinted with the permission of Ford Motor Company)

tive because the primary circuit no-start test proved the primary circuit is triggering on and off.

3. Connect the test spark plug to several spark plug wires and crank the engine while observing the spark plug. If the test spark plug fired in step 2 but does not fire at some of the spark plugs, the secondary voltage and current are leaking through a defective distributor cap, rotor, or spark plug wires, or the plug wire is open. If the test spark plug fires at all the spark plugs, the ignition system is satisfactory.

Engine Misfiring Diagnosis

If engine misfiring occurs, check the following items:

1. Engine compression
2. Intake manifold vacuum leaks
3. High resistance in spark plug wires, coil secondary wire, or cap terminals
4. Electrical leakage in the distributor cap, rotor, plug wires, coil secondary wire, or coil tower
5. Defective coil
6. Defective spark plugs
7. Low primary voltage and current
8. Improperly routed spark plug wires
9. Worn distributor bushings

Power Loss

Check the following items to diagnosis a power loss condition:

1. Engine compression
2. Restricted exhaust or air intake
3. Late ignition timing
4. Insufficient spark advance
5. Cylinder misfiring

Engine Detonation, Spark Knock

If the engine detonates, check the following items:

1. Engine compression higher than specified
2. Ignition timing too far advanced
3. Excessive spark advance
4. Spark plug heat range too hot
5. Improperly routed spark plug wires
6. Defective knock sensor

Reduced Fuel Mileage

When the fuel consumption is excessive, check the following components:

1. Engine compression
2. Late ignition timing
3. Lack of spark advance
4. Cylinder misfiring

Oscilloscope Testing

No discussion of ignition troubleshooting would be complete without a comprehensive discussion of oscilloscope use. The job of the oscilloscope is to convert the electrical signals of the ignition system into a visual image showing voltage changes over a given period of time. This information is displayed on a cathode ray tube (CRT) screen in the form of a continuous voltage line called a pattern or trace. By studying the pattern, a technician can determine what is happening inside the ignition system.

The information on the design and use of oscilloscopes given in this text is general in nature and is based on a typical "tune-up" scope, not a lab scope. Although some components in this chapter will be tested with a lab scope, most will be done on a tune-up scope. Always follow the oscilloscope manufacturer's specific instructions when connecting test leads or conducting test procedures.

Scales

On the face of the CRT screen is a voltage versus time graph (Figure 20-10). The screen has four scales: two vertical scales (one on the left and one on the right), a horizontal dwell scale, and a horizontal millisecond scale. The proper scale must be selected for the test being conducted.

The left and right vertical scales measure voltage. The vertical scale on the left side of the graph is normally divided into increments of 1 kilovolt (1,000 volts) and ranges from 0 to 25 kilovolts (kV). This scale is useful for testing secondary voltage. It can also be used to measure primary voltage by interpreting the scale in volts rather than kilovolts.

Normally the vertical scale on the right side of the graph is divided into increments of 2 kilovolts and has a range of 0 to 50kV. This scale is also used for testing secondary voltage and to measure primary voltage in the 0 to 500 volt range.

The horizontal percent of dwell scale is located at the bottom of the scope screen. This scale is used for checking the dwell angle in both the primary and secondary

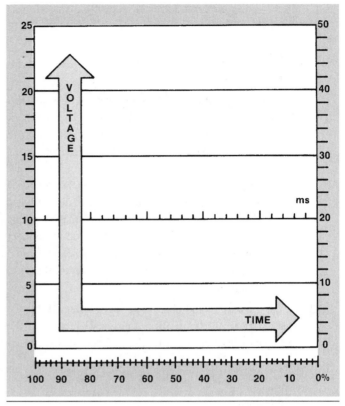

Figure 20-10 A scope displays voltage over time.

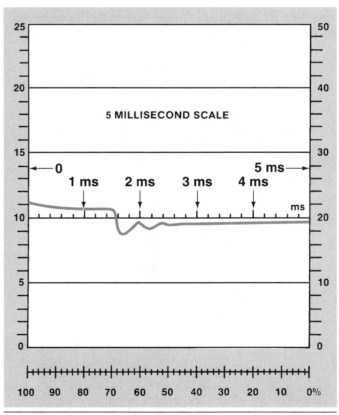

Figure 20-11 A 5-millisecond scale showing spark duration.

circuits. The dwell scale is divided into increments of 2 percentage points and ranges from 0 to 100% or in degrees. The degree graduations reflect the number of camshaft degrees allowed for the ignition cycle of each cylinder. For example, a four-cylinder engine scale goes from 0 to 90 degrees (360 degrees divided by 4).

The fourth scale (the millisecond scale) is a horizontal line that runs along the center of the voltage versus time graph. Depending on the test mode selected, the millisecond scale shows a range of 0 to 5 milliseconds (ms) or 0 to 25 milliseconds. The 5 ms scale is often used to measure the duration of the spark (Figure 20-11). In the 25-millisecond mode, the complete firing pattern can normally be displayed.

An oscilloscope or scope displays the trace from left to right, similar to reading a book. Oscilloscopes normally have four leads (Figure 20-12): a primary pickup that connects to the primary circuit, or negative terminal of the ignition coil; a ground lead that connects to a good engine ground; a secondary pickup, which clamps around the coil's high tension wire; and a trigger pickup, which clamps around the spark plug wire of the number one cylinder.

Trace Interpretation

Depending on the function selected, the scope can display a pattern for either the secondary or primary circuit. A typical secondary pattern is shown in Figure 20-13. A secondary pattern displays the following types of information: firing voltage, spark duration, coil and condenser oscillations, primary circuit on/off switching, dwell and cylinder timing accuracy, and secondary circuit accuracy.

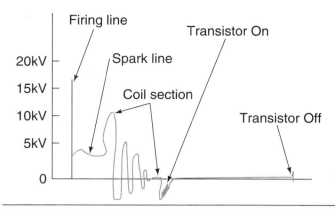

Figure 20-13 Typical secondary pattern.

A primary circuit pattern is normally used when something in the secondary pattern indicates a need to take a closer look at the primary circuit. It is also useful for observing the condition and action of the primary circuit switching devices and for providing a means to observe cylinder timing problems.

Pattern Display Modes

The oscilloscope has several ways to display the voltage patterns of the primary and secondary circuits. When the **display pattern** is selected, the oscilloscope displays the patterns of all the cylinders in a row from left to right. Each cylinder's ignition cycle is displayed according to the engine's firing order. In Figure 20-14, the firing order is 1, 8, 4, 3, 6, 5, 7, 2. The pattern begins with the spark line of the number one cylinder and ends with the firing line for the number one cylinder. This display pattern allows the technician to compare the voltage peaks for each cylinder.

A second choice of patterns is the **raster pattern** (Figure 20-15). A raster pattern stacks the voltage patterns of the cylinders one above the other. The number one pattern is displayed at the bottom of the screen and the rest of the cylinder's firing patterns are arranged above it in the engine's firing order. The pattern begins with the spark line and ends with the firing line for the next cylinder. This allows for a much closer inspection of the voltage and time trends than is possible with the display pattern.

The patterns for the cylinders can also be displayed in a **superimposed pattern**. A superimposed pattern displays all the patterns one on top of the other. Like the raster pattern, the patterns are displayed the full width of the screen, beginning with the spark line and ending with the firing line. A superimposed pattern allows a technician to compare one cylinder's pattern to the others. A superimposed pattern is shown in Figure 20-16.

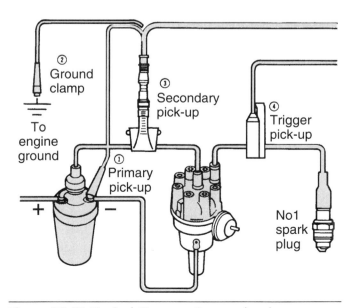

Figure 20-12 Typical "tune-up" scope hook-up.

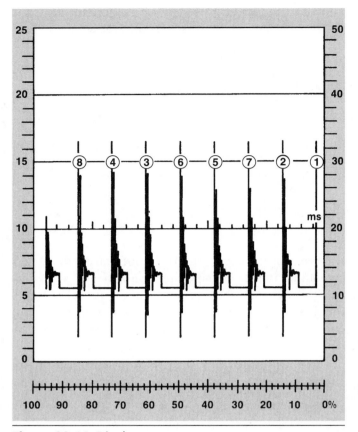

Figure 20-14 Display pattern.

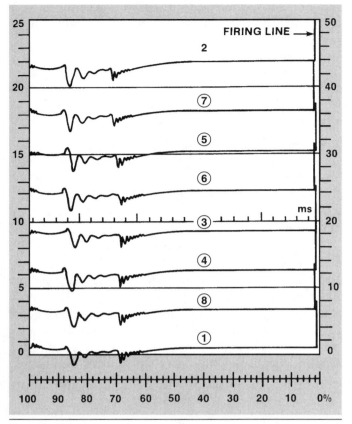

Figure 20-15 Raster pattern.

The **firing line** of a pattern is an upward line that signifies voltage. The firing line indicates the voltage needed to overcome the resistance in the secondary. Typically, around 10,000 volts is needed to overcome this resistance and initiate a spark. Keep in mind that cylinder conditions have an effect on this resistance. Leaner air/fuel mixtures, high compression, and increases in heat will increase resistance and the required firing voltages.

Once secondary resistance is overcome, the spark jumps the plug gap, establishing current flow and igniting the air/fuel mixture in the cylinder. This process is known as **gap bridging**. The length of time the spark actually lasts is represented by the **spark line** of the pattern. The spark line begins at the firing line and continues to the right until the coil's voltage drops below the level needed to keep current flowing across the gap.

After the plug fires, the next major section of the pattern begins. This section is called the **intermediate section** or coil-condenser zone. It shows the remaining coil voltage as it dissipates or drops to zero. Remember that once the spark has ended there is quite a bit of voltage stored in the ignition coil. The voltage remaining in the coil then oscillates or alternates back and forth within the primary circuit until it drops to zero. Notice that the lines

representing the coil-condenser section steadily drop in height until the coil's voltage is zero.

The next section of the trace pattern begins with the primary circuit current ON signal. It appears as a slight downward turn followed by several small oscillations. The slight downward curve occurs just as current begins to flow through the coil's primary winding. The oscillations that follow indicate the beginning of the magnetic field build-up in the coil. This curve marks the beginning of a period known as the dwell section. The end of the dwell section occurs when the primary current is turned off by the switching device (transistor or breaker points). The trace turns sharply upward at the end of the dwell section; this is the firing line for the next cylinder. The length of the dwell section represents the amount of time that current is flowing through the primary.

In general, most scope patterns look more or less like the one just described. The patterns produced by some systems have fewer oscillations in the intermediate section. Patterns may also vary slightly in the dwell section. The length of this section depends on when the control module turns the transistor on and off. Several variables may affect this timing.

Older breaker point systems and some electronic ignition systems use a fixed dwell period. In a fixed dwell

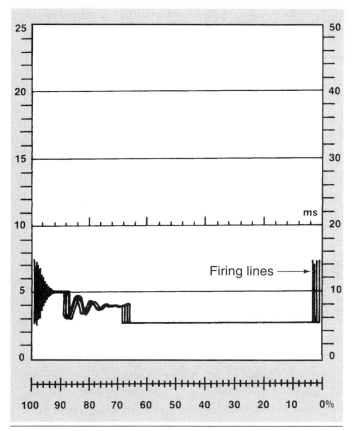

Figure 20-16 Superimposed pattern.

Most electronic ignition systems have a variable dwell function built into their control modules. In these systems, dwell changes significantly with engine speed. At idle and low rpm speeds, a short dwell provides enough time for complete ignition coil saturation (Figure 20-18A). The current on and current off signals appear very close to each other, usually less than 20 degrees.

As engine speed increases, the control module lengthens the dwell time (Figure 20-18B). This, of course, increases the available time for coil saturation. Although not common, it is possible for the variable dwell function of the control module to fail and still allow the engine to run. If testing indicates a lack of variable dwell on a system equipped with it, the control module must be replaced.

Many modern electronic ignitions feature **current limiting**. These systems saturate the ignition coil quickly by passing high current through the primary winding for a fraction of a second. Once the coil is saturated, the need for high current is eliminated, and only a small amount of current is used to keep the coil saturated. This type of system extends coil life.

The point at which the control module cuts back from high to low current appears as a small blip or oscillation during the dwell section of the pattern (Figure 20-19). At

system, the number of dwell degrees remain the same during all engine speeds. One exception to this is early Chrysler electronic ignition systems. These systems are characterized by a very long, distinctive dwell trace. The dwell section begins just as the spark line ends. At higher engine speeds, the spark line increases slightly, forcing dwell to begin later (Figure 20-17). The change in dwell time is slight, usually less than 5 degrees. This change in dwell has no effect on ignition timing because timing is controlled by switching off primary current.

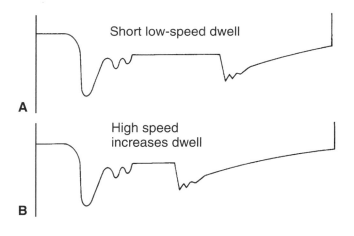

Figure 20-18 Example of a variable dwell ignition system.

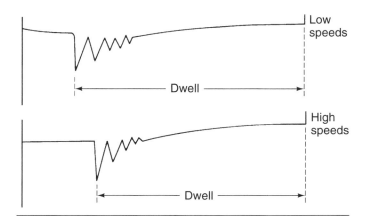

Figure 20-17 Speed reduces dwell slightly in this long dwell electronic ignition system.

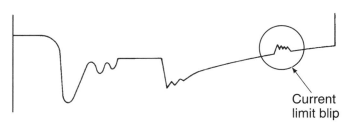

Figure 20-19 Pattern showing a current limiting "blip" during dwell.

very high engine speeds, this telltale blip may be missing, because the module keeps high current flow going to keep the coil continually saturated for fast firing.

If the primary winding of the coil has developed excessive resistance or the coil is otherwise faulty, the cutback blip may never occur. This is because the primary winding never becomes fully saturated. Further testing of the coil would be needed to pinpoint the cause of the missing blip. As with variable dwell, it is possible for the vehicle to run with the current limit function of the control module burned out or faulty.

Open Circuit Precautions

Precautions must be taken to prevent open circuits in an electronic ignition system during electrical test procedures. This point is brought up because a common test performed on breaker point ignition systems involved seeing how much available voltage the ignition coil could produce. The coil wire was removed and the engine cranked or a plug wire was pulled with the engine running. Both procedures forced the ignition coil to produce its maximum output to try to overcome the great amount of resistance created in the circuit. Typically, the output would exceed 20,000 volts, and the technician would have a good idea of what the coil was capable of producing under adverse conditions.

Open circuiting a modern ignition system, however, can damage the system. Features such as variable dwell and current limiting circuitry bypass all primary resistance if a spark does not occur. This causes tremendous amounts of current to flow through the primary circuit, producing an extremely high-voltage spark that must search out a path to ground. A frequent path to ground chosen by the spark is through the side of the coil. This results in an insulation breakdown at the coil and a site where arcing to ground is likely to occur in the future.

To prevent these tremendous voltages from occurring during coil output testing, a special test plug is used when checking the coil voltage output. The test plug is connected to the coil wiring running to the distributor or to an ignition cable. A grounding clip on the test plug is then connected to a good ground source.

The test plug looks like a spark plug, but it has a very large electrode gap. When the engine is cranked to check for voltage output, the ignition coil is forced to produce a higher voltage to overcome the added resistance created by the wide gap. Typically, about 35,000 volts are needed to fire the test plug. This is enough to stress the system without damaging it.

To safely perform a coil output test, proceed as follows:

1. Install the test plug in the coil wire or a plug wire if there is not a coil wire.

2. Set the oscilloscope on display and a voltage range of 50kV.

3. Crank the engine over and note the height of the firing line.

The firing line should exceed 35kV and be consistent. Lower than specified firing line voltages may indicate that the test plug did not fire. This could be the result of lower than normal available voltage to the primary circuit. The control module may have developed high internal resistance. The coil may also be faulty. Further testing is needed to help pinpoint the problem.

Spark Duration Testing

Spark duration is measured in milliseconds, using the millisecond sweep of the oscilloscope. Most vehicles have a spark duration of approximately 1.5 milliseconds. A spark duration of approximately 0.8 millisecond is too short to provide complete combustion. A short spark also increases pollution and causes power loss. If the spark duration is too long (over approximately 2.0 milliseconds), the spark plug electrodes might wear prematurely. When the oscilloscope shows a long spark duration, it normally follows a short firing line, which may indicate a fouled spark plug, low compression, or a rich air/fuel mixture. Spark duration is normally measured two time: during engine cranking and during engine running.

Some oscilloscopes do not have a millisecond sweep. Instead, they are equipped with a percent of dwell scale. In that case, the percent of dwell must be converted to milliseconds. A conversion chart is given in Table 21-1. The table lists percent of dwell at various rpm and the corresponding spark duration in milliseconds. On oscilloscopes without a millisecond sweep, the pattern selector should be set to superimpose or raster when performing the spark duration cranking and running tests.

Spark Plug Firing Voltage

The coil must generate sufficient voltage to overcome the total resistance in the secondary circuit and to establish a spark across the spark plug electrodes. On the oscilloscope, this spark plug **firing voltage** is seen as the highest spike in the pattern. The firing voltage might be affected by the condition of the spark plugs or the secondary circuit, engine temperature, fuel mixture, and compression pressures. To test the spark plug firing voltage on an oscilloscope, observe the firing line of all cylinders for height and uniformity. The normal height of the firing voltages should be between 7 and 13kV with no more than a 3kV variation between cylinders.

	ms	600 rpm	1,200 rpm	2,400 rpm	Spark Duration
8 cylinder	0.5	2%	4%	8%	too short
	1.0	4%	8%	17%	minimum
	1.5	6%	13%	25%	average
	2.0	8%	17%	33%	too long
6 cylinder	0.5	1.5%	3%	6%	too short
	1.0	3%	6%	12%	minimum
	1.5	4.5%	9%	18%	average
	2.0	6%	12%	24%	too long
4 cylinder	0.5	1%	2%	4%	too short
	1.0	2%	4%	8%	minimum
	1.5	3%	6%	13%	average
	2.0	4%	8%	17%	too long

Table 20-1 Converting Percent of Dwell to Milliseconds.

When high firing voltages are present in one or more of the cylinders, perform a rotor air gap voltage drop test. The purpose of this test is to determine the amount of secondary voltage that is required to bridge the rotor gap.

To perform a rotor air gap voltage drop test:

1. Set the pattern selector to the display position, the function selector to secondary, and the pattern height control to the 0 to 25kV scale.

2. Start the engine and adjust the speed to 1000 rpm.

3. Observe the height of the firing lines. Record the height and firing order number of any abnormal cylinder.

4. Shut the engine off.

5. Remove the spark plug wire of an abnormal cylinder from the distributor cap. Connect one end of the jumper lead to ground and the other end to the large portion of a grounding probe. Place the other end of the grounding probe in the distributor cap tower terminal.

6. Start the engine and adjust the speed to 1000 rpm.

7. Observe the firing line of the abnormal cylinder previously recorded. There should be a drop in the firing voltage when using the grounding probe.

The observed voltage represents voltage needed to overcome the resistance in the coil wire and of the rotor air gap. If during the test the firing line remains high, the rotor or distributor cap might be defective. Visually inspect both and replace as necessary. A bent distributor shaft or worn shaft bushings will cause excessive rotor air gaps on about half the cylinders.

WARNING *Do not remove the spark plug wire from the distributor cap tower terminal while the engine is running. This causes an open circuit and might damage the ignition system components.*

Secondary Circuit Resistance

Analysis of the spark line of a secondary pattern reveals the condition of the secondary circuit. The amount of resistance in the secondary circuit is indicated by the slope of the spark line. Excessive resistance in the secondary circuit causes the spark line to have a steep slope with a shorter firing duration.

A good spark line should be relatively flat and measure 2 to 4kV in height. High resistance in the secondary circuit produces a firing line and spark line that are higher in voltage with shorter firing durations.

To pinpoint the cause of high resistance, use a grounding probe and jumper wire to bypass each component of the secondary circuit on all abnormal cylinders. Connect one end of the jumper lead to ground and the other end to the large portion of a grounding probe. Start the engine and adjust the speed to 1000 rpm. Touch each secondary connector with the point of the grounding probe and observe the spark lines.

If, after grounding, the abnormal spark lines appear normal, the part just bypassed is the cause of the problem.

Spark Plugs Under Load

The voltage required to fire the spark plugs increases when the engine is under load. The voltage increase is moderate and uniform if the spark plugs are in good condition and properly gapped. However, if any unusual characteristics are displayed on the scope patterns when load is applied to the engine, the spark plugs are probably faulty. This condition is most evident in the firing voltages displayed on the scope. To test spark plug operation under load, note the height of the firing lines at idle speed. Then, quickly open and release the throttle (snap accelerate), and note the rise in the firing lines while checking the voltages for uniformity. A normal rise would be between 3 and 4kV upon snap acceleration. A problem with the secondary ignition circuit or the air/fuel mixture is noted if one or more of the cylinders show a voltage rise of over 4kV, or if one or more cylinders show a voltage rise less than 3kV or no voltage rise at all.

Coil Condition

The energy remaining in the coil after the spark plugs fire gradually diminishes in a series of oscillations. These oscillations are observable in the intermediate sections of both the primary and secondary patterns. If the scope pattern shows an absence of normal oscillations in the intermediate section, check for a possible short in the coil by testing the resistance of the primary and secondary windings.

Individual Component Testing

Tables 21-2 and 21-3 outline a procedure for quickly isolating an ignition-related problem when an oscilloscope is not available. The first troubleshooting tree determines whether a spark is generated in the secondary system. If no spark is available, the second troubleshooting tree leads step by step through individual component tests until the problem is located.

The following sections briefly outline common test procedures for individual system components. For accurate testing, always refer to the wiring diagrams and testing instructions given in the appropriate service manual.

Ignition Switch

A faulty ignition switch or faulty wiring may not supply adequate power to the ignition control module or ignition coil. The ignition system shown in Figure 20-20 has two wires connected to the run terminal of the ignition switch. One is connected to the module. The other is connected to the primary resistor and coil. The start terminal of the switch is also wired to the module.

SERVICE TIP *The secret to component testing is to use good troubleshooting practices. Work systematically through a circuit, testing each wire, connector, and component. Do not jump around back and forth between components. The component inadvertently overlooked is probably the one causing the trouble. Checks must be made for available voltage, voltage output, resistance of wires and connectors, and available ground. Always compare the readings with specifications given in the manufacturer's service manual.*

You can check for voltage using either a 12-volt test light or a digital voltmeter. To use a test light, turn the ignition key off and disconnect the wire connector at the module. Also, disconnect the S terminal of the starter solenoid to prevent the engine from cranking. Turn the key to the run position and probe the red wire connection to check for voltage. Also check for voltage at the BAT terminal of the ignition coil. Next, turn the key to the start position and check for voltage at the white wire connector at the module and the BAT terminal of the ignition coil.

To do the same test using a digital voltmeter, turn the ignition switch to the off position and install a small straight pin into the appropriate module wire. Connect the digital voltmeter's positive lead to the straight pin and ground the negative lead to the distributor base. Turn the ignition to the run or start position as needed and measure voltage. Do not allow the straight pin to contact the engine or ground. Reading should be within 90% of battery voltage.

Primary Resistor

Measure the resistance of the primary resistor using an ohmmeter. Remember that the key must be off when this test is performed or power in the circuit will damage

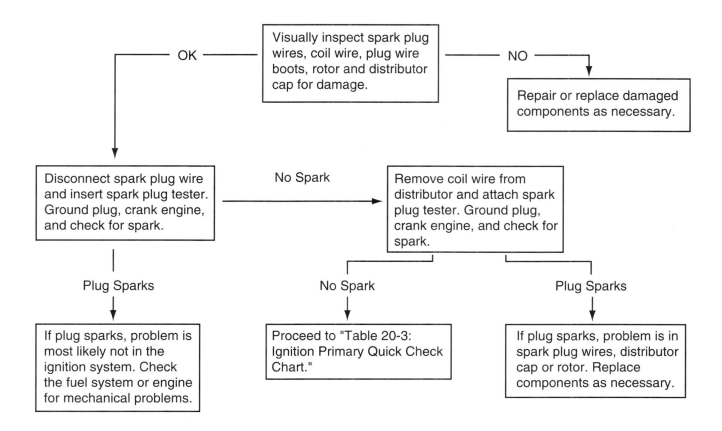

Table 20-2 Ignition Secondary Quick Check Chart.

the meter. The ohmmeter leads should be connected at the BAT terminal of the coil and the wiring harness connector wire that joins the red wire in the ignition module connector.

Coil Input Voltage

The purpose of this test is to determine if the primary circuit to the coil has excessive resistance. Follow the these steps to perform this test:

1. Connect the voltmeter positive lead to the positive terminal of the ignition coil. Connect the negative lead to engine ground.

2. Set the voltmeter to a scale that will read 12 volts accurately.

3. Disconnect and ground the coil high tension lead from the distributor to disable the engine from starting.

4. Turn the ignition switch to the "start" position while observing the voltmeter reading. Check manufacturer's

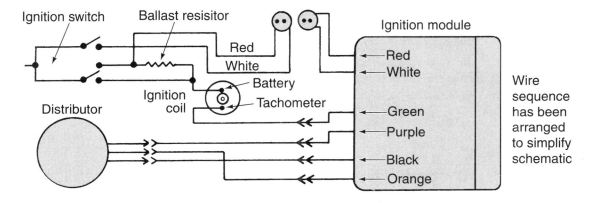

Figure 20-20 Simple ignition switch circuit. (Reprinted with the permission of Ford Motor Company)

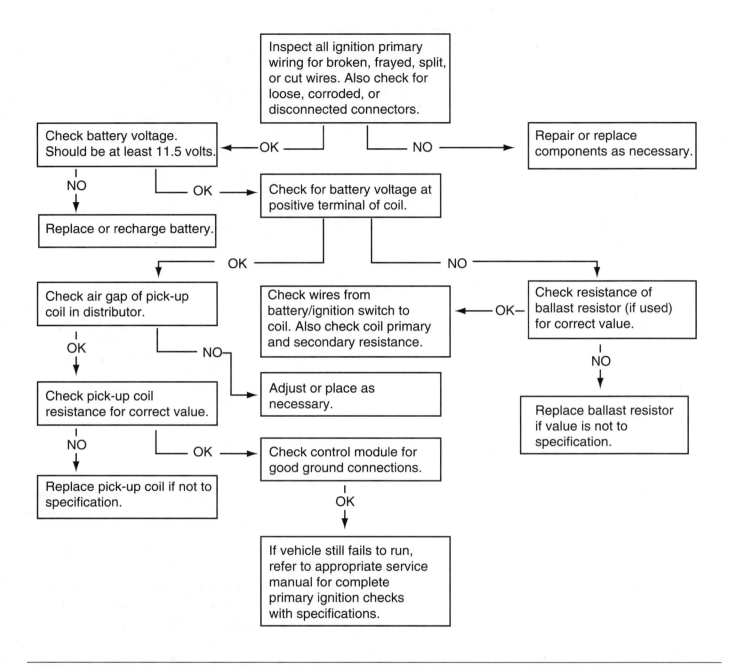

Table 20-3 Ignition Primary Quick Check Chart.

specifications for proper voltage reading (usually about 9.5 volts).

5. If the voltage is too low then there is a resistance in the supply circuit to the coil. This could be caused by any of the following:

   (a) Battery condition.

   (b) Excessive starter draw.

   (c) Excessive resistance in the ballast bypass circuit.

   (d) Excessive resistance in the ignition switch start circuit.

   (e) Excessive resistance in the starter relay to the bypass circuit.

6. With the voltmeter still connected as in step one, turn the ignition switch to the "run" position. The voltmeter reading should now be about 7.5 volts (or factor specifications).

7. If the reading is below specifications it could be due to:

   (a) Battery condition.

   (b) Ignition switch resistance.

   (c) Excessive resistance in the ballast resistor circuit.

If the results of this test show that the coil supply circuit is operating properly then the coil should be tested.

Ignition Coil Resistance

With the key off and the battery lead to the ignition coil disconnected, use an ohmmeter to measure the resistance of the primary and secondary windings of the ignition coil. Calibrate the ohmmeter on the X1 scale and connect the meter leads across the BAT (+) and tach (-) terminals of the ignition coil to check the primary winding for an open, short, or high resistance condition (Figure 20-21). Primary resistance usually ranges from 0.5 to 2 ohms. Check the factory specifications listed in the service manual.

To measure secondary coil resistance, calibrate the ohmmeter on the X1,000 (X1K) scale and connect it across the BAT (+) terminal and the ignition coil's center (high voltage) tower (Figure 20-22). Secondary resistance usually ranges from 8,000 to 20,000 ohms. Again, check the service manual specifications.

The results from the resistance checks on the primary and secondary windings give a general view of the coil's condition. These tests do not guarantee that the coil will work correctly. Some defects will only show up on scope tests, such as defective insulation around the coil windings, which causes high-voltage leaks.

Pickup Coil

The pickup coil of a magnetic pulse generator or metal detection sensor is also checked for proper resistance using an ohmmeter. Connect the ohmmeter to the pickup coil terminals to test the pickup coil for an open or a shorted condition. While the ohmmeter leads are connected, pull on the pickup leads and watch for a change in the readings. Any change would indicate an intermittent open in the pickup leads. Most pickup coils

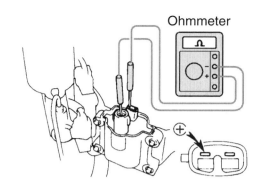

Figure 20-22 Ohmmeter connected across secondary windings of an ignition coil. (Courtesy of Toyota Motors)

have 150 to 900 ohms resistance, but always refer to the manufacturer's specifications. If the pickup coil is open, the ohmmeter will display an infinite reading. When the pickup coil is shorted, the ohmmeter will display a reading lower than specifications.

Connect the ohmmeter from one of the pickup leads to ground to test the pickup for a short to ground. If there is no short to ground, the ohmmeter will give an infinite reading. Refer to Figure 20-23; ohmmeter #1 is testing for a short to ground, while ohmmeter #2 is testing the resistance of the pickup coil.

A lab scope can be used to check pickup coil operation. Connect the scope's leads across the pickup coil and set the scope to a low AC voltage scale. When the distributor shaft is spun, an AC waveform should appear on the screen (Figure 20-24). The trace is not a true sine wave, but it should have both a positive and negative pulse.

Another method of measuring the pickup coil's AC signal is with a voltmeter set on its low AC voltage scale. The meter registers AC voltage during cranking. Measure this voltage as close to the control module as possi-

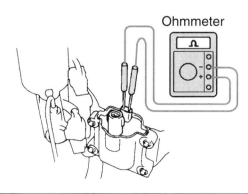

Figure 20-21 Ohmmeter connected across primary windings of an ignition coil. (Courtesy of Toyota Motors)

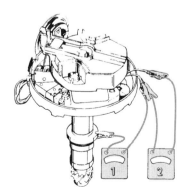

Figure 20-23 Ohmmeter tests on a pickup coil. Testing for opens and shorts to ground. (Courtesy of Oldsmobile Division—GMC)

ble to account for any resistance in the connecting wire from the pickup coil to the module.

Ignition Module

Use an ohmmeter to ensure that the ignition module connection to ground is good. One lead of the meter should be connected to the ground terminal at the module and the other to a good ground. Zero resistance indicates good continuity in the ground circuit. Any resistance reading during this test is unacceptable.

The most effective method of testing for a defective ignition module is to use an ignition module tester, if one is available for that module. This electronic (Figure 20-25) tester evaluates and determines if the module is operating within a given set of design parameters. It does so by simulating normal operating conditions while looking for faults in key components.

Some ignition module testers are able to perform an ignition coil spark test (actually firing the coil) and a distributor pickup test. Test selection is made by pushing the appropriate button. The module tester usually responds to these tests with a pass or fail indication.

Unfortunately, many module testers are designed to troubleshoot specific makes and models of ignitions. Many shops find it impractical to have testers for every type of system they service.

Keep in mind that ignition modules are very reliable. They are also one of the most expensive ignition system components. So, if a module tester is not available, check out all other system components before condemning the ignition module.

Ignition Module Removal and Replacement

The ignition module removal and replacement procedure varies depending on the ignition system. Always follow the procedure in the vehicle manufacturer's service manual. Follow these steps for module removal and replacement on an HEI distributor:

1. Remove the battery wire from the coil battery terminal, and remove the inner wiring connector on the primary coil terminals. Remove the spark plug wires from the cap.
2. Rotate the distributor latches one-half turn and lift the cap from the distributor.
3. Remove the two rotor retaining bolts and the rotor.
4. Remove the primary leads and the pickup leads from the module.
5. Remove the two module mounting screws, and remove the module from the distributor housing.

WARNING *Lack of silicone grease on the module mounting surface may cause module overheating and damage.*

6. Wipe the module mounting surface clean, and place a light coating of silicone heat-dissipating grease on the module mounting surface.
7. Install the module and tighten the module mounting screws to the specified torque.
8. Install the primary leads and pickup leads on the module.

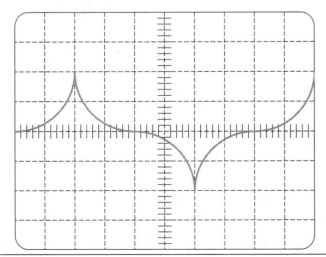

Figure 20-24 An AC waveform from a PM generator pickup unit.

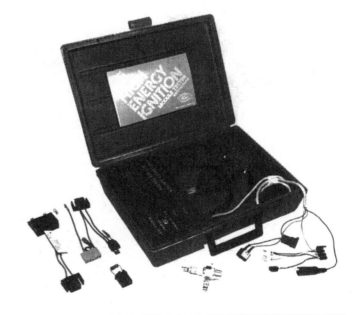

Figure 20-25 Ignition module tester. (Courtesy of Automotive Group, Kent-Moore Division, SPX Corporation)

9. Be sure the lug on the centrifugal advance mechanism fits into the rotor notch while installing the rotor, and tighten the rotor mounting screws to the specified torque.

10. Install the distributor cap, and be sure the projection in the cap fits in the housing notch.

11. Push down on the cap latches with a screwdriver, and rotate the latches until the lower part of the latch is hooked under the distributor housing.

12. Install the coil primary leads and battery wire on the coil terminals. Be sure the notch on the primary leads fits onto the cap projection. Install the spark plug wires.

HEI Ignition Modules

Whenever replacing General Motors' HEI ignition modules, be sure to match the module with the one being replaced. Besides a rare three-pin unit there are:

1. Four-pin either white or black color modules. These modules are interchangeable with each other. The scope patterns between these two models are slightly different.

2. Five-pin module. This module has an extra circuit that is used to retard ignition timing when the engine is cold.

3. Seven-pin module. This module is used on computer command control systems and includes three extra pins that are used for electric spark control. There are several seven-pin modules, and they are not interchangeable. Be certain that the correct replacement module is being used.

Also, whenever replacing the module be sure to install the special heat-dissipating silicone grease under the module to conduct heat away from the module.

Spark Plug Wires

The spark plug wires may be left in the distributor cap for test purposes, so the cap terminal connections are tested with the spark plug wires. Calibrate an ohmmeter on the X1,000 scale, and connect the ohmmeter leads from the end of a spark plug wire to the distributor cap terminal inside the cap to which the plug wire is connected (Figure 20-26).

If the ohmmeter reading is more than specified by the vehicle manufacturer, remove the wire from the cap and check the wire alone. If the wire has more resistance than specified, replace the wire. When the spark plug wire resistance is satisfactory, check the cap terminal for corrosion. Repeat the ohmmeter tests on each spark plug wire and the coil secondary wire.

Distributor Tests and Service

All service procedures for a distributor depend on the manufacturer and design of the distributor. Always follow the recommended procedures given in the service manual. In most cases, removal of the distributor begins any repairs to the distributor.

Before removing the distributor, check the condition of the distributor's bushing. Do this by grasping the distributor shaft and moving it toward the outside of the distributor. If any movement is detected, remove the distributor and check the bushing on the bench. Many manufacturers recommend complete distributor replacement rather than bushing or shaft replacement.

To remove a typical distributor, begin by disconnecting the distributor wiring connector and vacuum advance hoses. Then, remove the distributor cap and note the position of the rotor. Remove the distributor hold-down bolt and clamp. Note the position of the vacuum advance unit in relation to the engine. Then pull the distributor from the engine. After the distributor has been removed, install a shop towel in the distributor opening to keep foreign material out of the engine.

Distributor pickups can be tested in the way discussed earlier. It is important that all wires and connectors between the distributor and module and module to the engine control computer be visually checked as well as checked for excessive resistance with an ohmmeter.

If the resistance checks are within specifications, the circuit should be checked with a digital voltmeter. Turn the ignition ON and connect the voltmeter across the voltage input wire and ground. Compare the reading to specifications. The voltmeter should also be used to check the resistance across the ground circuit of the distributor. Do this by measuring the voltage.

Distributor Disassembly

With the distributor removed, it can be disassembled for cleaning and inspection. Follow these steps for a typical distributor disassembly procedure:

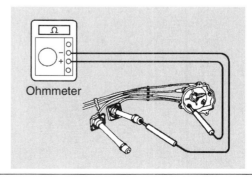

Figure 20-26 Ohmmeter connected to the spark plug wire and the distributor cap terminal to test the plug wire. (Courtesy of Toyota Motor Corporation)

1. Mark the gear in relation to the distributor shaft so the gear may be installed in the original position.

2. Support the distributor housing on top of a vise, and drive the roll pin from the gear and shaft with a pin punch and hammer (Figure 20-27).

3. Pull the gear from the distributor shaft, and remove any spacers between the gear and the housing. Note the position of these spacers so they may be installed in their original position.

4. Wipe the lower end of the shaft with a shop towel and inspect this area of the shaft for metal burrs. Remove any burrs with fine emery paper.

5. Pull the distributor shaft from the housing.

6. Remove the pickup coil leads from the module and the pickup retaining clip. Lift the pickup coil from the top of the distributor bushing.

7. Remove the two vacuum advance mounting screws, and remove this advance assembly from the housing.

With the distributor disassembled, inspect it in the following manner:

1. Inspect all lead wires for worn insulation, and loose terminals. Replace these wires as necessary.

2. Inspect the centrifugal advance mechanism for wear, particularly check the weights for wear on the pivot holes. Replace the weights, or complete shaft assembly, if necessary.

3. Inspect the pickup plate for wear and rotation. If this plate is loose or seized, replacement is required.

4. Connect a vacuum hand pump to the vacuum advance outlet and apply 20 inches of vacuum. The advance diaphragm should hold this vacuum without leaking.

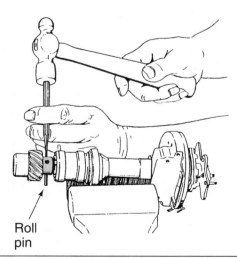

Figure 20-27 Driving the roll pin from the distributor gear (Courtesy of Oldsmobile Division, General Motors Corporation)

WARNING *Distributor electrical components, and the vacuum advance may be damaged by washing them in solvent. The housing may be washed in solvent, but do not wash electrical components or the vacuum advance.*

5. Check the distributor gear for worn or chipped teeth.

6. Inspect the reluctor for damage. If the high points are damaged, the distributor bushing is likely worn, allowing the high points to hit the pickup coil.

Follow these steps for a typical distributor assembly procedure:

1. Install the vacuum advance and tighten the mounting screws to the specified torque.

2. Install the pickup coil and the retaining clip. Connect the pickup leads to the module.

3. Install the module and mounting screws as discussed previously.

4. Place some bushing lubricant on the shaft and install the shaft in the distributor.

5. Install the spacers between the housing and gear in their original position.

6. Install the gear in its original position, and be sure the hole in the gear is aligned with the hole in the shaft.

7. Support the housing on top of a vise and drive the roll pin into the gear and shaft.

8. Install a new O-ring or gasket on the distributor housing.

See the section on ignition timing to properly set base timing after the distributor is installed.

Stress Testing Components

Often an intermittent ignition problem only occurs under certain conditions such as vibrations, extremes in heat or cold, or during rainy or humid weather. Careful questioning of the customer should lead to determining if the problem is stress condition related. Does the problem occur on cold mornings? Does it occur when the engine is fully warmed up? Is it a rainy day problem? If the answer to any of these questions is positive, you can reproduce the same conditions in the shop during stress testing.

If the problem seems to be related to cold weather, begin by setting the scope for a raster pattern, then cool major ignition components such as the control module, pickup coil, and major connections one at a time using a liquid cool-down agent.

After cooling a component, watch the pattern for any signs of malfunction, particularly in the dwell section. If there is no sign of malfunction, cool down the next component after the first has warmed to normal operating temperature. Cooling (or heating) more than one component at a time provides inconclusive results.

CAUTION *When using cool-down sprays, wear eye protection and avoid spraying your skin or clothing. Use extreme caution.*

If the problem is heat related, use a heat gun or hair dryer to direct hot air onto or into the component. Heat guns intended for stripping paint and other household jobs can become extremely hot and melt plastic, wire insulation, and other materials. Use a moderate setting and proceed cautiously. Look for changes in the dwell section of the trace, particularly in the variable dwell or current limiting areas. If connections appear to be the problem, disconnect them, clean the terminals, and coat them with dielectric compound to seal out dirt and moisture.

When the problem seems to be moisture related, lightly spray the components, coil and ignition cables, and connections with water. Do not flood the area; a light mist does the job. A scope set on raster or display helps pinpoint problems, but it is often possible to hear and feel the miss or stutter without the use of a scope. As with heat and cold testing, do not spray down more than one area at a time or results could be misleading. If you suspect a poor connection, clean and seal it; then retest it.

Spark Plug Service

Manufacturers recommend that spark plugs be replaced after 20,000 to 100,000 miles of use. This service interval depends on a number of factors, including the type of ignition system, engine design, spark plug design, operating conditions, type of fuel used, and types of emission control devices used.

A spark plug socket is essential for plug removal and installation. Spark plug sockets are available in two sizes: 13/16-inch (for 14-millimeter gasketed and 18-millimeter tapered-seat plugs) and 5/8-inch (for 14-millimeter tapered-seat plugs). They can be either 3/8- or 1/2-inch drive, and many feature an external hex so that they can be turned using an open end or box wrench.

To remove spark plugs:

1. Remove the cables from each plug, being careful not to pull on the cables. Instead, grasp the boot and twist it off gently.
2. Using a spark plug socket and ratchet, loosen each plug a couple of turns.

3. Use compressed air to blow any dirt away from the base of the plugs.
4. Remove the plugs, making sure to remove the gasket as well (if applicable).

Once the spark plugs have been removed, it is important to "read" them (Figure 20-28). In other words, inspect them closely, noting in particular any deposits on the plugs and the degree of electrode erosion. A plug in good working condition can still have minimal deposits on it. They are usually light tan or gray in color. However, there should be no evidence of electrode burning, and the increase of the air gap should be no more than 0.001 inch for every 10,000 miles of engine operation. A plug that exceeds this wear should be replaced and the cause of excessive wear repaired.

SERVICE TIP *To save time and avoid confusion later, use masking tape to mark each of the cables with the number of the plug it attaches to.*

It is possible to diagnose a variety of engine conditions by examining the firing end of the spark plugs. If an engine is in good shape, they should all look alike. Whenever plugs from different cylinders look different, a problem exists somewhere in the engine or its systems. Following are examples of plug problems and how they should be dealt with.

If the insulator is dark or black on all spark plugs, possible causes may be:

1. Excessively rich air/fuel mixture resulting from an improperly adjusted carburetor or fuel injection system.
2. Stuck choke system or fuel enrichment system.
3. Defective valve seals.
4. Defective piston rings.
5. Wrong heat range for the driving conditions.

A **cold fouling** condition is the result of an excessively rich air/fuel mixture. It is characterized by a layer of dry, fluffy black carbon deposits on the tip of the plug. Cold fouling is caused by a rich air/fuel mixture or an ignition fault causing the spark plug not to fire. If only one or two of the plugs show evidence of cold fouling, sticking valves or leaking injectors are the likely causes. Correct the cause of the problem before reinstalling or replacing the plugs. When the tip of the plug is almost drowned in excess oil, this condition is known as **wet fouling**. In an engine, the oil may be entering the combustion chamber past worn valve guides or valve guide

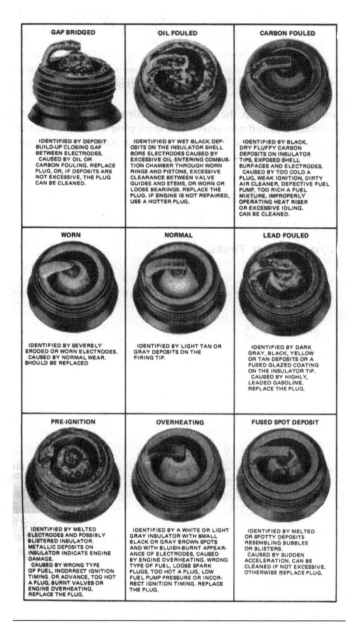

GAP BRIDGED	OIL FOULED	CARBON FOULED
IDENTIFIED BY DEPOSIT BUILD-UP CLOSING GAP BETWEEN ELECTRODES. CAUSED BY OIL OR CARBON FOULING. REPLACE PLUG, OR, IF DEPOSITS ARE NOT EXCESSIVE, THE PLUG CAN BE CLEANED.	IDENTIFIED BY WET BLACK DEPOSITS ON THE INSULATOR SHELL BORE ELECTRODES CAUSED BY EXCESSIVE OIL ENTERING COMBUSTION CHAMBER THROUGH WORN RINGS AND PISTONS, EXCESSIVE CLEARANCE BETWEEN VALVE GUIDES AND STEMS, OR WORN OR LOOSE BEARINGS. REPLACE THE PLUG. IF ENGINE IS NOT REPAIRED, USE A HOTTER PLUG.	IDENTIFIED BY BLACK, DRY FLUFFY CARBON DEPOSITS ON INSULATOR SURFACES AND ELECTRODES. CAUSED BY TOO COLD A PLUG, WEAK IGNITION, DIRTY AIR CLEANER, DEFECTIVE FUEL PUMP, TOO RICH A FUEL MIXTURE, IMPROPERLY OPERATING HEAT RISER OR EXCESSIVE IDLING. CAN BE CLEANED.

WORN	NORMAL	LEAD FOULED
IDENTIFIED BY SEVERELY ERODED OR WORN ELECTRODES. CAUSED BY NORMAL WEAR. SHOULD BE REPLACED	IDENTIFIED BY LIGHT TAN OR GRAY DEPOSITS ON THE FIRING TIP.	IDENTIFIED BY DARK GRAY, BLACK, YELLOW OR TAN DEPOSITS OR A FUSED GLAZED COATING ON THE INSULATOR TIP. CAUSED BY HIGHLY LEADED GASOLINE. REPLACE THE PLUG.

PRE-IGNITION	OVERHEATING	FUSED SPOT DEPOSIT
IDENTIFIED BY MELTED ELECTRODES AND POSSIBLY BLISTERED INSULATOR. METALLIC DEPOSITS ON INSULATOR INDICATE ENGINE DAMAGE. CAUSED BY WRONG TYPE OF FUEL, INCORRECT IGNITION TIMING OR ADVANCE, TOO HOT A PLUG, BURNT VALVES OR ENGINE OVERHEATING. REPLACE THE PLUG.	IDENTIFIED BY A WHITE OR LIGHT GRAY INSULATOR WITH SMALL BLACK OR GRAY BROWN SPOTS AND WITH BLUISH-BURNT APPEARANCE OF ELECTRODES, CAUSED BY ENGINE OVERHEATING. WRONG TYPE OF FUEL, LOOSE SPARK PLUGS, TOO HOT A PLUG, LOW FUEL PUMP PRESSURE OR INCORRECT IGNITION TIMING. REPLACE THE PLUG.	IDENTIFIED BY MELTED OR SPOTTY DEPOSITS RESEMBLING BUBBLES OR BLISTERS. CAUSED BY SUDDEN ACCELERATION. CAN BE CLEANED IF NOT EXCESSIVE. OTHERWISE REPLACE PLUG.

Figure 20-28 Normal and abnormal spark plug conditions. (Reprinted with the permission of Ford Motor Company)

seals. If the vehicle has an automatic transmission, a likely cause of wet-fouled plugs is a defective vacuum modulator that is allowing transmission fluid into the chamber. On high-mileage engines, check for worn rings or excessive cylinder wear. The best solution is to correct the problem and replace the plugs with the specified type.

Splash fouling occurs immediately following an overdue tune-up. Deposits in the combustion chamber, accumulated over a period of time due to misfiring, suddenly loosen when the temperature in the chamber returns to normal. During high-speed driving, these

deposits can stick to the hot insulator and electrode surfaces of the plug. These deposits can actually bridge across the gap, stopping the plug from sparking.

Under high engine speed conditions, the combustion chamber deposits can form a shiny, yellow glaze over the insulator. When it gets hot enough, the glaze acts as an electrical conductor causing the current to follow the deposits and short out the plug. **Glazing** can be prevented by avoiding sudden wide-open throttle acceleration after sustained periods of low-speed or idle operation. Because it is virtually impossible to remove glazed deposits, glazed plugs should be replaced.

Plug overheating is characterized by white or light gray blistering of the insulator. There may also be considerable electrode gap wear. Overheating can result from using too hot a plug, over-advanced ignition timing, detonation, a malfunction in the cooling system, an overly lean air/fuel mixture, using too-low octane fuel, an improperly installed plug, or a heat-riser valve that is stuck closed. Overheated plugs must be replaced.

Preignition damage is caused by excessive engine temperatures. Preignition damage is characterized by melting of the electrodes, or chipping of the electrode tips. When this problem occurs, look for the general causes of engine overheating, including over-advanced ignition timing, a burned head gasket, and using too-low octane fuel. Other possibilities include loose plugs or using plugs of the improper heat range. Do not attempt to reuse plugs with preignition damage.

Regapping Spark Plugs

Both new and used spark plugs should have their air gaps set to the engine manufacturer's specifications. Round wire-type feeler gauges work best for measuring the air gap. Adjustments are made by bending the side electrode of the plug. Keep the side electrode in line with the center electrode while bending it to adjust the gap. Never assume the gap is correct just because the plug is new. Do not try to reduce a plug's air gap by tapping the side electrode on a bench. Never attempt to set a wide gap, electronic ignition-type plug to a small gap specification. Likewise, never attempt to set a small gap, breaker point ignition-type plug to the wide gap necessary for electronic ignitions. In either case, damage to the electrodes results. Never try to bend the center electrode to adjust the air gap. This cracks the insulation.

To install spark plugs:

1. Wipe all dirt and grease from the plug seats with a clean cloth.

2. Be sure the gaskets on gasketed plugs are in good condition and properly placed on the plugs. If reusing a

plug, install a new gasket on it. Be sure that there is only one gasket on each plug.

3. Adjust the air gap as needed.

4. Install the plugs and finger tighten. If the plugs cannot be installed easily by hand, the threads in the cylinder head may require cleaning with a thread-chasing tap. Be especially careful not to cross-thread the plugs when working with aluminum heads.

5. Tighten the plugs with a torque wrench, following the vehicle manufacturer's specifications, or the values listed in Table 21-4.

When spark plug wires are being installed, make sure they are routed as indicated in the service manual. When removing the spark plug wires from a spark plug, grasp the spark plug boot tightly and twist while pulling the cable from the end of the plug. When installing a spark plug wire, make sure the boot is firmly seated around the top of the plug; then squeeze the boot to expel any air that is trapped inside.

Setting Ignition Timing

Pre-computerized ignition distributors (and some computerized units) require the timing to be adjusted periodically. The correct ignition timing procedure varies depending on make and year of the vehicle and the type of ignition system. Ignition timing specifications and instructions are included on the underhood emission decal, and more detailed instructions are provided in the vehicle's service manual; these must be followed. Many late-model vehicles have computer-controlled ignition timing. On these systems, ignition timing is not adjustable and an ignition timing problem indicates a problem in the computer circuit.

Spark plug gap, idle speed, and dwell must be correct before setting ignition timing. Also, the engine must be at operating temperature. The following are some general points common to many systems:

- Base timing is normally checked and adjusted at idle speed or low engine speeds, such as 650 rpm. At this low speed, any mechanical advance mechanism is not activated and does not affect the base reading.
- Many systems require checking timing with the automatic transmission placed in drive. This is often abbreviated as DR on the instruction sticker.
- The vacuum advance mechanism must also be disabled so it does not affect the base reading. This is done by removing and plugging the vacuum advance line or hose from the manifold.
- In computer-controlled systems, the microprocessor controls ignition timing advance. To set correct base timing, the computer must be eliminated from the timing control circuit. This is normally done by disconnecting the appropriate connector at the distributor. For example, Ford calls this connector the spout (spark out) connector. GM and other manufacturers have similar methods of disabling computer timing control. Setting base timing is very important in computer-controlled systems. If base timing is incorrect, the computer makes further adjustments based on incorrect data.

Installing the Distributor

When installing the distributor, remember the distributor gear easily goes into mesh with the camshaft gear, but many distributors also drive the oil pump with drive in the lower end of the distributor gear or shaft. It may be necessary to hold down on the distributor housing and crank the engine to get the distributor shaft into mesh with the oil pump drive. When this action is required, repeat steps 2 and 3 and be sure the rotor is under the number 1 spark plug wire terminal in the distributor cap with the timing marks aligned.

This procedure indexes the distributor so the engine can start. This process is called **still timing** since the engine is not running. It is necessary to start the engine and check base timing after the distributor is installed.

Setting Base Timing

If possible, use a timing light to check the timing. Follow these steps for ignition timing adjustment:

1. Connect the timing light, and start the engine.

Plug Type	Cast-Iron Head	Aluminum Head
14–mm Gasketed	25 to 30	15 to 22
14–mm Tapered Seat	7 to 15	7 to 15
18–mm Tapered Seat	15 to 20	15 to 20

Table 20-4 Spark Plug Installation Torque Values

2. The engine must be idling at the manufacturer's recommended rpm and all other timing procedures must be followed.

3. Aim the timing light marks at the timing indicator, and observe the timing marks (Figure 20-29). Compare this reading to the manufacturer's specifications. For example, if the specification reads 10 degrees before top dead center (Figure 20-30A) and the reading found is 3 degrees before top dead center (Figure 20-30B), the timing is retarded or off by 7 degrees.

4. If the timing mark is not at the specified location, rotate the distributor until the mark is at the specified location (Figure 20-31).

5. Tighten the distributor hold-down bolt to the specified torque, and recheck the timing mark position.

6. Connect the vacuum advance hose and any other connectors, hoses, or components that were disconnected for the timing procedure.

Many later model vehicles have a magnetic timing probe receptacle near the timing indicator (Figure 20-32). The probe holder is generally located close to TDC, however it is offset for TDC by a set number of degrees. This is called **magnetic offset angle**. Some equipment manufacturers supply a magnetic **timing mete**r with a pickup that fits in the probe hole. The meter pickup must be connected to the number 1 spark plug wire, and the power supply leads connected to the battery terminals (Figure 20-33). Many timing meters have two scales, timing degrees and engine rpm.

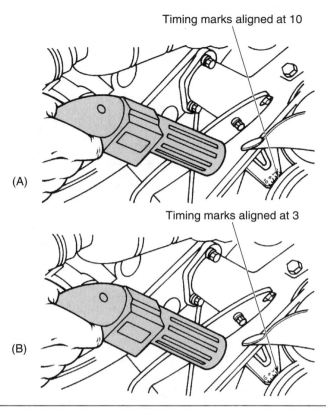

Timing marks aligned at 10

(A)

Timing marks aligned at 3

(B)

Figure 20-30 (A) Timing marks illuminated by a timing light and showing 10 degrees BTDC. (B) Timing marks at 3 degrees BTDC.

A magnetic timing **offset** knob on the meter must be adjusted to the vehicle manufacturer's specifications to compensate for the position of the probe receptacle. The magnetic offset angle varies between the manufacturers. Listed are typical offset angles:

GM	9.5 degrees
AMC	10 degrees
Chrysler	10 degrees
Mercedes	15 degrees
BMW	20 degrees
VW	20 degrees
Volvo	20 degrees
Ford	Ford does not recommend magnetic timing since offset angles vary between assembly plants and years of manufacture.

Once the offset is adjusted, the engine may be started and the timing is indicated on the meter scale. The timing adjustment procedure is the same with the timing meter or light.

On pre-computerized distributor systems, it may be necessary to compensate for higher altitudes when adjusting base timing. Timing specifications are given for sea level, since air is thinner at higher elevations the spark needs to be advanced to receive the same quality of

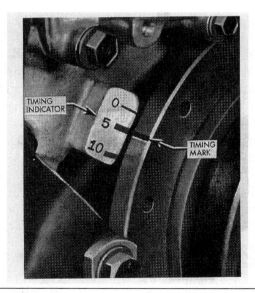

Figure 20-29 When the timing light flashes, the timing mark on the crankshaft pulley must appear at the specified location on the timing indicator above the pulley. (Courtesy of Buick Motor Division, General Motors Corporation)

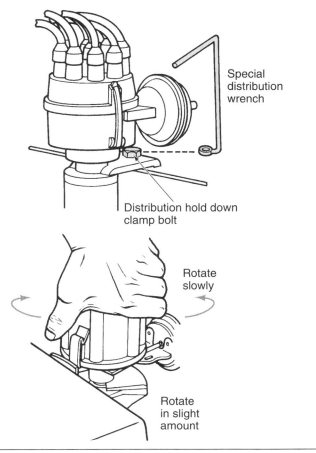

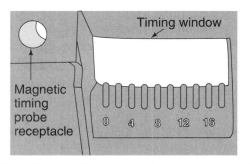

Figure 20-32 Magnetic timing probe receptacle near the timing indicator. (Courtesy of Chrysler Corporation)

burn as at sea level. A general rule of thumb is to increase (advance) base timing half a degree for every 1,000 feet of altitude. For example if the specifications require 8° BTDC @ 900 rpm and the vehicle is being driven at an elevation of 5,000 feet, the required setting would be 10 1/2° @ 900 rpm. On computerized ignition systems, do not advance timing over the manufacturer's specified base timing.

A distributor rotor can rotate in either a clockwise or counterclockwise rotation. To determine which direction the rotor is turning on a particular engine, grasp the distributor housing with your hand with the vacuum advance unit in your palm. The direction your fingers are pointing is the direction of rotation.

Figure 20-31 Set the timing by rotating the distributor. (Courtesy of Chrysler Corporation)

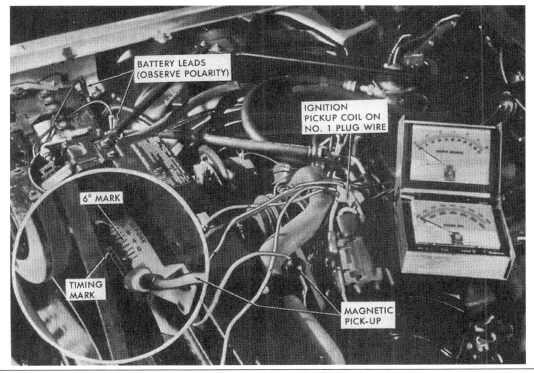

Figure 20-33 Magnetic timing meter and related connections. (Courtesy of Cadillac Motor Car Division, General Motors Corporation)

Checking Mechanical Advance Units

Mechanical or centrifugal units advance ignition timing in response to increases in engine speed. To check the operation of this unit, loosen the distributor hold-down bolt or clamp. Disable any vacuum advance system. Raise engine speed to 2000 rpm or the speed specified in the service manual. Then shine the light on the vibration damper timing marks, and turn the timing knob on the advance timing light until the base timing reading of 10 degrees is again registering (Figure 20-34). Observe the reading on the timing meter of the timing light. It tells how many degrees the flash of light has been delayed. This meter reading should match the specified advance at that rpm level.

If the reading does not match, the mechanical advance mechanism is faulty. The vacuum advance can be checked in the same manner. Use a hand-held vacuum pump to draw on the diaphragm until the vacuum specification is reached, and adjust the advance timing light until correct base timing is again registering. Once again, the amount of flash delay matches the amount of vacuum advance present. The normal test procedure is to compare vacuum and mechanical advance to manufacturer specifications at several different engine speeds.

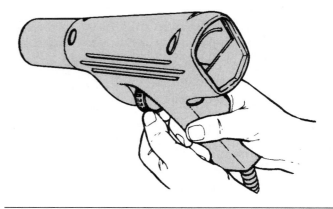

Figure 20-34 Timing light with advance control for checking timing advance. (Courtesy of Chrysler Corporation)

Symptoms of overly advanced timing include pinging or engine knock. Insufficient advance or retarded timing at higher engine rpms could cause hesitation and poor fuel economy.

Summary

❏ The ability to diagnose a problem (to find its cause and its solution) is what separates an automotive technician from a parts changer.

❏ The first step in diagnostics is to gather as much information about the problem as possible from the owner.

❏ All diagnoses of ignition systems should include a visual inspection.

❏ The coil should be inspected for cracks or any evidence of arcing or burning at the coil tower.

❏ Make sure the spark plug wires are arranged according to the firing order of the engine. Spark plug wires from consecutively firing cylinders should cross rather than run parallel to one another.

❏ Primary ignition system wiring should be checked for tight connections, especially on vehicles with electronic or computer-controlled ignitions.

❏ Examples of bad ground circuit induced ignition failures include burnt ignition modules resulting from missing or loose coil ground straps, erratic performance caused by a poor distributor-to-engine block ground, and intermittent ignition operation resulting from a poor ground at the control module.

❏ The distributor cap and rotor must be carefully inspected for signs of wear and damage.

❏ Centrifugal advance mechanisms should be checked for free motion.

❏ The job of the oscilloscope is to convert the electrical signals of the ignition system into a visual image showing voltage changes over a given period of time.

❏ A secondary pattern displays the following types of information: firing voltage, spark duration, coil and condenser oscillations, primary circuit on/off switching, dwell and cylinder timing accuracy, and secondary circuit accuracy.

❏ A primary circuit pattern is normally used when something in the secondary pattern indicates a need to take a closer look at the primary circuit.

❏ When the display pattern is selected, the oscilloscope displays the patterns of all the cylinders in a row from left to right.

❏ A raster pattern stacks the voltage patterns of the cylinders one above the other.

❏ A superimposed pattern displays all the patterns one on top of the other.

❏ The firing line of a pattern is an upward line that signifies voltage. The firing line indicates the voltage needed to overcome the resistance in the secondary.

❏ The length of time the spark actually lasts is represented by the spark line of the pattern.

❏ The intermediate section shows the remaining coil voltage as it dissipates or drops to zero.

❏ The length of the dwell section represents the amount of time that current is flowing through the primary.

❏ Spark duration is measured in milliseconds, using the millisecond sweep of the oscilloscope.

❏ On the oscilloscope, the spark plug firing voltage is seen as the highest spike in the pattern.

❏ The ignition coil should be checked for primary winding resistance, secondary winding resistance, and continuity between the primary and secondary windings.

❏ The pickup coil of a magnetic pulse generator or metal detection sensor is checked for proper resistance using an ohmmeter.

❏ The most effective method of testing for a defective ignition module is to use an ignition module tester.

❏ The spark plug wires are checked for excessive resistance using an ohmmeter.

❏ Often an intermittent ignition problem only occurs under certain conditions such as vibrations, extremes in heat or cold, or during rainy or humid weather. Often it is possible to reproduce the same conditions in the shop during stress testing.

❏ It is possible to diagnose a variety of engine conditions by examining the firing end of the spark plugs.

❏ Both new and used spark plugs should have their air gaps set to the engine manufacturer's specifications.

❏ Pre-computerized ignition distributors (and some computerized units) require the timing to be adjusted periodically. Spark plug gap, idle speed, and dwell must be correct before setting ignition timing. Also, the engine must be at operating temperature.

Terms-To-Know

Carbon tracking	Gap bridging	Spark line
Cold fouling	Glazing	Splash fouling
Corona effect	Intermediate section	Still timing
Crossfiring	Magnetic offset angle	Superimposed pattern
Current limiting	Offset	Timing meter
Display pattern	Punch through	Wet fouling
Firing line	Raster pattern	
Firing voltage	Spark duration	

ASE Style Review Questions

1. Upon inspection, a spark plug reveals a layer of dry, fluffy black carbon deposits on its tip. Technician A says the plug cannot be used again. Technician B says the plug can be cleaned and serviced, then reused. Who is correct?
   a. A only
   b. B only
   c. Both A and B
   d. Neither A nor B

2. The firing line on an oscilloscope pattern extends below the waveform. Technician A says the polarity of the coil is reversed. Technician B says the problem is fouled spark plugs. Who is correct?
   a. A only
   b. B only
   c. Both A and B
   d. Neither A nor B

3. The firing lines on an oscilloscope pattern are all abnormally low. Technician A says the problem is probably low coil output. Technician B says the problem could be an overly rich air/fuel mixture. Who is correct?
   a. A only
   b. B only
   c. Both A and B
   d. Neither A nor B

4. Technician A says that dielectric grease is used to seal dirt and moisture out of electrical terminals. Technician B says it is used to dissipate heat away from sensitive electrical components. Who is correct?
   a. A only
   b. B only
   c. Both A and B
   d. Neither A nor B

5. While discussing timing of the distributor to the engine with the number 1 piston at TDC compression and the timing marks aligned: Technician A says the distributor must be installed with the rotor under the number 1 spark plug terminal in the distributor cap and one of the reluctor high points aligned with the pickup coil. Technician B says the distributor must be installed with the rotor under the number 1 spark plug terminal in the distributor cap and the reluctor high points out of alignment with the pickup coil. Who is correct?
   a. A only
   b. B only
   c. Both A and B
   d. Neither A nor B

6. Ignition coil ohmmeter tests are being discussed. Technician A says the ohmmeter should be placed on the X1,000 scale to test the secondary winding. Technician B says the ohmmeter tests on the coil check the condition of the winding insulation. Who is correct?
   a. A only
   b. B only
   c. Both A and B
   d. Neither A nor B

7. Basic ignition timing adjustment on vehicles with computer-controlled distributor ignition (DI) is being discussed. Technician A says on some DI systems, a timing connector must be disconnected. Technician B says the distributor must be rotated until the timing mark appears at the specified location on the timing indicator. Who is correct?
   a. A only
   b. B only
   c. Both A and B
   d. Neither A nor B

8. No-start diagnosis using a test spark plug is being discussed. Technician A says if the test light flutters at the coil tach terminal, but the test spark plug does not fire when connected from the coil secondary wire to ground with the engine cranking, the ignition coil is defective. Technician B says if the test spark plug fires when connected from the coil secondary wire to ground with the engine cranking, but the test spark plug does not fire when connected from the spark plug wires to ground, the cap or rotor is defective. Who is correct?
   a. A only
   b. B only
   c. Both A and B
   d. Neither A nor B

9. Possible causes for a higher than normal firing line on a scope is being discussed. Technician A says lean air/fuel mixtures can be the cause. Technician B says a shorted plug wire can be the cause. Who is correct?
   a. A only
   b. B only
   c. Both A and B
   d. Neither A nor B

10. Ohmmeter testing of the pickup coil is being discussed. Technician A says an ohmmeter reading below the specified resistance indicates the pickup coil is grounded. Technician B says an ohmmeter reading below the specified resistance indicates the pickup coil is open. Who is correct?
    a. A only
    b. B only
    c. Both A and B
    d. Neither A nor B

21 Introduction to the Computer

Objective

Upon completion and review of this chapter, you should be able to:

❏ Describe the principle of analog and digital voltage signals.

❏ Explain the principle of computer communications.

❏ Describe the basics of logic gate operation.

❏ Describe the basic function of the central processing unit (CPU).

❏ Explain the basic method by which the CPU is able to make determinations.

❏ List and describe the differences in memory types.

❏ Describe the function of adaptive strategy.

❏ List and describe the functions of the various sensors used by the computer.

❏ List and describe the operation of output actuators.

❏ Explain the principle of multiplexing.

Introduction

This chapter introduces the basic theory and operation of the digital **computer** used to control many of the vehicle's accessories. A computer is an electronic device that stores and processes data then controls other devices based on the data. The aura of mystery surrounding automotive computers is so great that some technicians are intimidated by them. Knowledge is the key to dispelling this myth. Although it is not necessary to understand all of the concepts of computer operation to service the systems they control, knowledge of the digital computer will help you feel more comfortable when working on these systems. It is important to understand that things do not just happen by magic, there is a purpose and method for all actions performed by the computer.

The use of computers on automobiles has expanded to include control and operation of several functions including fuel delivery, ignition systems, emission systems, climate control, lighting circuits, cruise control,

antilock braking, electronic suspension systems, and electronic shift transmissions. Some of these functions are the responsibility of the powertrain control module (PCM) while others are functions of what is known as a **body control module (BCM)**.

What each module does is determined by the manufacturer. Some are quite basic while others include several enhancements. For example, some body computer-controlled systems include direction lights, rear window defoggers, illuminated entry, intermittent wipers, and other systems once thought of as basic.

A computer processes the physical conditions that represent information (data). The operation of the computer is divided into four basic functions:

1. **Input:** A voltage signal sent from an input device. This device can be a sensor or a switch activated by the driver or technician.

2. **Processing:** The computer uses the input information and compares it to programmed instructions. The logic circuits process the input signals into output demands.

3. **Storage:** The program instructions are stored in an electronic memory. Some of the input signals are also stored for later processing.

4. **Output:** After the computer has processed the sensor input and checked its programmed instructions, it will put out control commands to various output devices. These output devices may be the instrument panel display or a system actuator. The output of one computer can also be used as an input to another computer.

Understanding these four functions will help today's technician organize the troubleshooting process. When a system is tested, the technician will be attempting to isolate the problem to one of these functions.

Analog and Digital Principles

Remember, voltage does not flow through a conductor; current flows while voltage is the pressure that "pushes" the current. However, voltage can be used as a signal, for example, difference in voltage levels, frequency of change, or switching from positive to negative values can be used as a signal.

The computer is capable of reading only voltage signals. The programs used by the computer are "burned" into IC chips using a series of numbers. These numbers represent various combinations of voltages that the computer can understand. The voltage signals to the computer can be either analog or digital. Most of the inputs from the sensors are **analog** variables. An analog voltage signal is continuously variable within a defined range. For example, ambient temperature sensors do not change abruptly. The temperature varies in infinite steps from low to high. The same is true for several other inputs such as engine speed, vehicle speed, fuel flow, and so on.

Compared to an analog voltage representation, **digital** voltage patterns are square-shaped because the transition from one voltage level to another is very abrupt (Figure 21-1). A digital signal is produced by an on/off or high/low voltage. The simplest generator of a digital signal is a switch (Figure 21-2). In this illustration, if 5 volts is applied to the circuit, the voltage sensor will read close to 5 volts (a high voltage value) when the switch is open. Closing the switch will result in the voltage sensor reading close to zero volts. This measuring of voltage drops sends a digital signal to the computer. The voltage values are represented by a series of digits, which create a **binary code.**

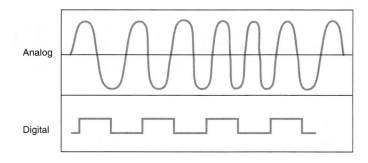

Figure 21-1 Analog voltage signals are constantly variable. Digital voltage patterns are either on or off, high or low; digital signals are referred to as square waves.

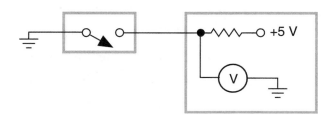

Figure 21-2 Simplified voltage sensing circuit that indicates whether the switch is opened or closed.

Binary Numbers

A transistor that operates as a relay is the basis of the digital computer. As the input signal switches from off to on, the transistor output switches from cutoff to saturation. The on and off output signals represent the binary digits 1 and 0.

The computer converts the digital signal into binary code by translating voltages above a given value to 1, and voltages below a given value to 0. As shown (Figure 21-3), when the switch is open and 5 volts is sensed, the voltage value is translated into a 1 (high voltage). When the switch is closed, lower voltage is sensed and the voltage value is translated into a 0. Each 1 or 0 represents one bit of information. Note, high voltage being translated to a 1 and low voltage to a 0 is given for explanation purposes. In some systems a low voltage is represented by a binary 1 while a high voltage is equal to 0.

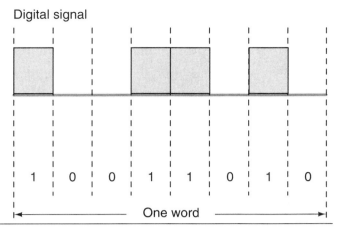

Digital signal

One word

Figure 21-3 Each binary 1 and 0 is one bit of information. Eight bits equal one byte. (Courtesy of General Motors Corporation)

Decimal number	Binary number code 8 4 2 1	Binary to decimal conversion
0	0000	= 0 + 0 = 0
1	0001	= 0 + 1 = 1
2	0010	= 2 + 0 = 2
3	0011	= 2 + 1 = 3
4	0100	= 4 + 0 = 4
5	0101	= 4 + 1 = 5
6	0110	= 4 + 2 = 6
7	0111	= 4 + 2 + 1 = 7
8	1000	= 8 + 0 = 8

Figure 21-4 Binary number code to base ten numbers.

In the binary system, whole numbers are grouped from right to left. Because the system uses only two digits, the first portion must equal a 1 or a 0. To write the value of 2, the second position must be used. In binary, the value of 2 would be represented by 10 (one two and zero ones). To continue, a 3 would be represented by 11 (one two and one one). Figure 21-4 illustrates the conversion of binary numbers to digital base ten numbers. For example, if a thermistor is sensing 150 degrees, the binary code would be 10010110. If the temperature increases to 151 degrees, the binary code changes to 10010111.

The computer contains a crystal oscillator or **clock** that delivers a constant time pulse. The clock is a crystal that electrically vibrates when subjected to current at certain voltage levels. As a result, the chip produces very regular series of voltage pulses. The clock maintains an orderly flow of information through the computer circuits by transmitting one bit of binary code for each pulse (Figure 21-5). In this manner the computer is capable of distinguishing between the binary codes such as 101 and 1001.

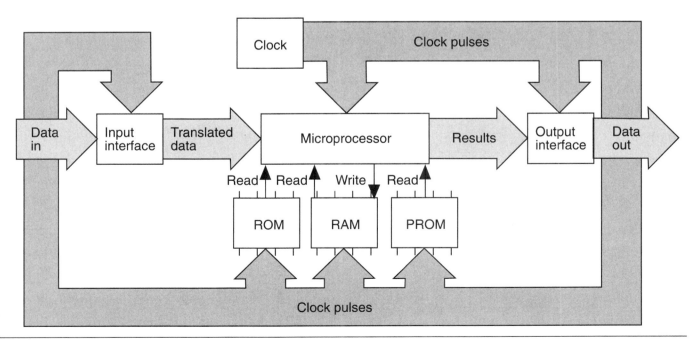

Figure 21-5 Interaction of the main components of the computer. All of the components monitor clock pulses. (Courtesy of General Motors Corporation)

Signal Conditioning

The input and/or output signals may require conditioning in order to be used. This conditioning may include amplification and/or signal conversion.

Some input sensors, such as the oxygen (O_2) sensor, produce a very low voltage signal of less than 1V. This signal also has an extremely low current flow. Therefore, this type of signal must be amplified, or increased, before it is sent to the microprocessor. This amplification is accomplished by the **amplification circuit** in the input

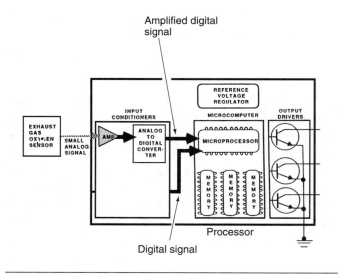

Figure 21-6 Amplification circuit in the computer input conditioning chip. (Reprinted with the permission of Ford Motor Company)

conditioning chip inside the computer (Figure 21-6).

For the computer to receive information from the sensor, and to give commands to actuators, it requires an **interface**. The interface has two purposes: protect the computer from excessive voltage levels, and to translate input and output signals. The computer will have two interface circuits: input and output. The digital computer cannot accept analog signals from the sensors and requires an input interface to convert the analog signal to digital. The **analog to digital (A/D) converter** continually scans the analog input signals at regular intervals. For example, if the A/D converter scans the TPS signal and finds this signal at 5V, the A/D converter assigns a numeric value to this specific voltage. The A/D converter then changes this numeric value to a binary code (Figure 21-7).

Also, some of the controlled actuators may require an analog signal. In this instance, an output **digital to analog (D/A) converter** is used.

Central Processing Unit

The **central processing unit (CPU)** is the brain of the computer. The CPU is also referred to as the **microprocessor**. The CPU is constructed of thousands of transistors placed on a small chip (Figure 21-8). The CPU brings information into and out of the computer's memory. The input information is processed in the CPU and checked against the program in memory. The CPU also checks memory for any other information regarding programmed parameters. The information obtained by the

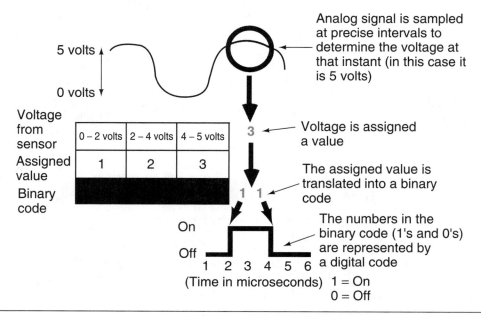

Figure 21-7 The A/D converter in the computer assigns a numeric value to input voltages and changes this numeric value to a binary code. (Reprinted with the permission of Ford Motor Company)

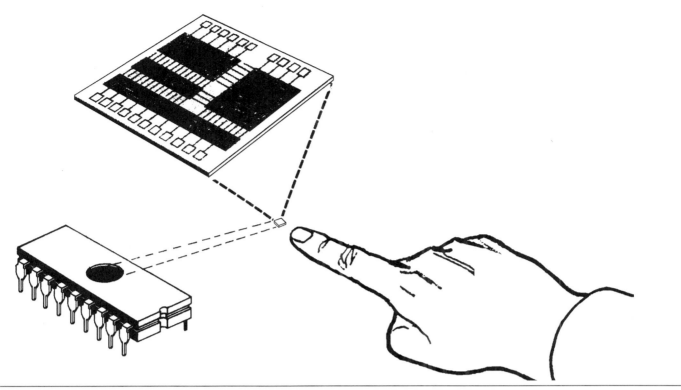

Figure 21-8 Microprocessor components are etched on an integrated circuit chip that is small enough to fit on a fingertip. (Reprinted with the permission of Ford Motor Company)

CPU can be altered according to the program instructions. The program may have the CPU apply logic decisions to the information. Once all calculations are made, the CPU will deliver commands to make the required corrections or adjustments to the operation of the controlled system.

A **program** is a group of instructions that are followed by the microprocessor. The program guides the microprocessor in decision making. For example, the program may inform the microprocessor when sensor information should be retrieved and then tell the microprocessor how to process this information. Finally, the program guides the microprocessor regarding the activation of output control devices such as relays and solenoids. The various memories contain the programs and other vehicle data the microprocessor refers to as it performs calculations. As the microprocessor performs calculations and makes decisions, it works with the memories by either reading information from the memories or writing new information into the memories.

The CPU has several main components (Figure 21-9). The registers used include the accumulator, the data counter, the program counter, and the instruction register. The control unit implements the instructions located in the instruction register. The arithmetic logic unit (ALU) performs the arithmetic and logic functions.

Computer Memory

The computer requires a means of storing both permanent and temporary **memory**. The memories contain many different locations. These locations may be compared to file folders in a filing cabinet, with each location containing one piece of information. An **address** is assigned to each memory location. This address may be compared to the lettering or numbering arrangement on file folders. Each address is written in a binary code, and these codes are numbered sequentially beginning with 0.

While the engine is running, the engine computer receives a large quantity of information from a number of sensors. The computer may not be able to process all this information immediately. In some instances, the computer may receive sensor inputs required to make several different decisions. In these cases, the microprocessor writes information into memory by specifying a memory address and sending information to this address.

When stored information is required, the microprocessor specifies the stored information address and requests the information. When stored information is requested from a specific address, the memory sends a copy of this information to the microprocessor. However, the original stored information is still retained in the memory address.

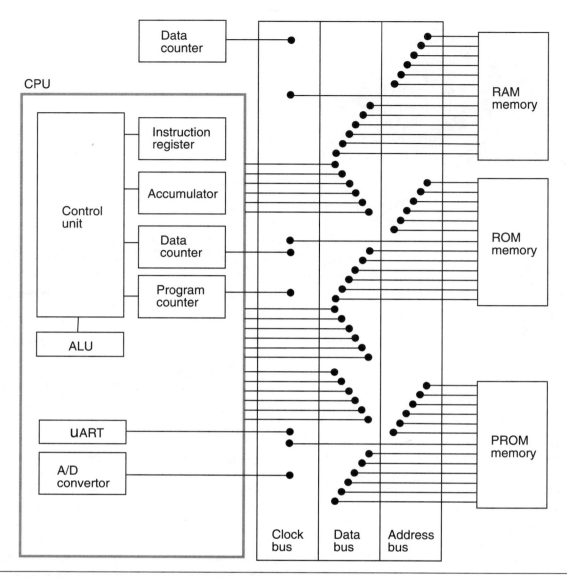

Figure 21-9 Main components of the computer and the CPU.

The memories store information regarding the ideal air/fuel ratios for various operating conditions. The sensors inform the computer about the engine and vehicle operating conditions. The microprocessor reads the ideal air/fuel ratio information from memory and compares this information with the sensor inputs. After this comparison, the microprocessor makes the necessary decision and operates the injectors to provide the exact air/fuel ratio required by the engine.

Several types of memory chips may be used in the computer:

1. **Read only memory (ROM)** contains a fixed pattern of 1s and 0s that represent permanent stored information. ROM contains the basic operating perameters for the vehicle. This information is used to instruct the computer on what to do in response to input data. The CPU reads the information contained in ROM, but it cannot write to it or change it. ROM memory is not lost when power to the computer is lost.

2. **Random access memory (RAM)** is constructed from flip-flop circuits formed into the chip. The RAM will store temporary information that can be read from or written to by the CPU. RAM stores information that is waiting to be acted upon, and it stores output signals that are waiting to be sent to an output device. RAM can be designed as volatile or nonvolatile. In **volatile RAM**, the data will be retained only if current flows through the memory. RAM that is connected to the battery through the ignition switch will lose its data when the switch is turned off. **Nonvolatile RAM (NVRAM)** will retain its information if current is removed. See number 7 below for a more detailed explanation of NVRAM.

3. **Keep alive memory (KAM)** is a variation of RAM. KAM is connected directly to the battery through circuit protection devices. For example, the microprocessor can read and write information to and from the KAM, and erase KAM information. However, the KAM retains information when the ignition switch is turned off. KAM will be lost when the battery is disconnected, if the battery drains too low, or if the circuit opens.

4. **Programmable read only memory (PROM)** contains specific data that pertains to the exact vehicle in which the computer is installed. This information may be used to inform the CPU of the accessories that are equipped on the vehicle. The information stored in the PROM is the basis for all computer logic. The information in PROM is used to define or adjust the operating perimeters held in ROM. In many instances, the computer is interchangeable between models of the same manufacturer; however, the PROM is not. Consequently, the PROM may be replaceable and plug into the computer.

WARNING *Installing a PROM chip backward will immediately destroy the chip. In addition, electrostatic discharge (ESD) will destroy the chip. Static straps should be used when working on a computer to prevent ESD while you are working on the unit.*

5. **Erasable programmable read only memory (EPROM)** is similar to PROM except its contents can be erased to allow new data to be installed. A piece of Mylar tape covers a window. If the tape is removed and the microcircuit is exposed to ultraviolet light, its memory is erased.

6. **Electrically erasable programmable read only memory (EEPROM)** allows changing the information electrically one bit at a time. The flash EEPROM is an integrated circuit (IC) chip inside the computer. This IC contains the program used by the computer to provide control. It is possible to erase and reprogram the EEPROM without removing this chip from the computer. For example, when a modification to the PCM operating strategy is required, it is no longer necessary to replace the PCM. The flash EEPROM may be reprogrammed through the data link connector (DLC) using the manufacturer's specified diagnostic equipment. This process of reprogramming the computer through the EEPROM is called **flashing**.

7. **Nonvolatile RAM (NVRAM)** is a combination of RAM and EEPROM into the same chip. During normal operation, data is written to and read from the RAM portion of the chip. If the power is removed from the chip, or at programmed timed intervals, the data is transferred from RAM to the EEPROM portion of the chip. When the power is restored to the chip, the EEPROM will write the data back to the RAM.

Adaptive Strategy

If a computer has **adaptive strategy** capabilities, the computer can actually learn from past experience. For example, the normal voltage input range from a throttle position sensor (TPS) may be 0.6V to 4.5V. If a worn TPS sends a 0.4-V signal to the computer, the microprocessor interprets this signal as an indication of component wear, and the microprocessor stores this altered calibration in the KAM. The microprocessor now refers to this new calibration during calculations, and thus normal engine performance is maintained. If a sensor output is erratic or considerably out of range, the computer may ignore this input. When a computer has adaptive strategy, a short learning period is necessary under the following conditions:

1. After the battery has been disconnected

2. When a computer system component has been replaced or disconnected

3. On a new vehicle

During this learning period, the engine may surge, idle fast, or have a loss of power. The average learning period lasts for five miles of driving.

Information Processing

The air charge temperature (ACT) sensor input will be used as an example of how the computer processes information. If the air temperature is low, the air is more dense and contains more oxygen per cubic foot. Warmer air is less dense and therefore contains less oxygen per cubic foot. The cold, dense air requires more fuel to mix with it compared to the warmer air that is less dense. The microprocessor must supply the correct amount of fuel in relation to air temperature and density.

An ACT sensor is positioned in the intake manifold where it senses air temperature. This sensor contains a resistive element that has an increased resistance when the sensor is cold. Conversely, the ACT sensor resistance decreases as the sensor temperature increases. When the ACT sensor is cold, it sends a high analog voltage signal to the computer, and the A/D converter changes this signal to a digital signal.

When the microprocessor receives this ACT signal, it addresses the tables in the ROM. The look-up tables list air density for every air temperature. When the ACT sensor voltage signal is very high, the look-up table indicates very dense air. This dense air information is relayed to the microprocessor, and the microprocessor operates the output drivers and injectors to supply the exact amount of fuel required by the engine (Figure 21-10).

Logic Gates

Logic gates are the thousands of field effect transistors (FET) incorporated into the computer circuitry. These circuits are called logic gates because they act as gates to output voltage signals depending on different combinations of input signals. The FETs use the incoming voltage patterns to determine the pattern of pulses leaving the gate. The following are some of the most common logic gates and their operation. The symbols represent functions and not electronic construction:

1. **NOT gate:** A NOT gate simply reverses binary 1s to 0s and vice versa (Figure 21-11). A high input results in a low output and a low input results in a high output. The NOT gate is also called an **inverter**.

2. **AND gate:** The AND gate will have at least two inputs and one output. The operation of the AND gate is similar to two switches in series to a load (Figure 21-12). The only way the light will turn on is if switches A and B are closed. The output of the gate will be high only if both inputs are high.

3. **OR gate:** The OR gate operates similarly to two switches that are wired in parallel to a light (Figure 21-13). If switch A or B is closed, the light will turn on. A high signal to either input will result in a high output.

4. **NAND gate** and **NOR gate:** A NOT gate placed behind an OR or AND gate inverts the output signal (Figure 21-14).

5. **Exclusive-OR (XOR) gate:** A combination of gates that will produce a high output signal only if the inputs are different (Figure 21-15).

These different gates are combined to perform the processing function. The following are some of the most common combinations:

1. **Decoder circuit:** A combination of AND gates used to provide a certain output based on a given combination of inputs. When the correct bit pattern is received by the decoder, it will produce the high voltage signal to activate the relay coil.

2. **Multiplexer (MUX):** The basic computer is not capable of looking at all of the inputs at the same time. A multiplexer is used to examine one of many inputs depending on a programmed priority rating.

3. **Demultiplexer (DEMUX):** Operates similar to the MUX except that it controls the order of the outputs. The process the MUX and DEMUX operate on is called **sequential sampling**. The computer will deal with all of the sensors and actuators one at a time, or sequentially.

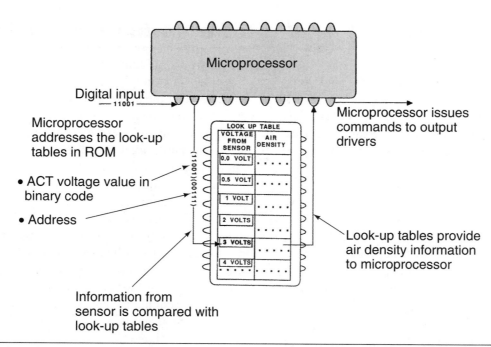

Figure 21-10 The microprocessor addresses the look-up tables in the ROM, retrieves air density information, and issues commands to the output drivers. (Reprinted with the permission of Ford Motor Company)

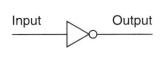

Truth table	
Input	Output
0	1
1	0

Figure 21-11 The NOT gate symbol and a truth table. The NOT gate inverts the input signal.

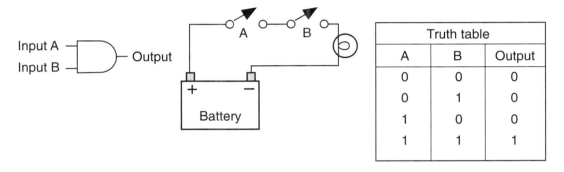

Truth table		
A	B	Output
0	0	0
0	1	0
1	0	0
1	1	1

Figure 21-12 The AND gate symbol and a truth table. The AND gate operates similarly to switches in series.

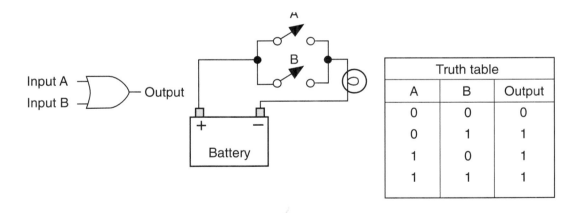

Truth table		
A	B	Output
0	0	0
0	1	1
1	0	1
1	1	1

Figure 21-13 OR gate symbol and a truth table. The OR gate is similar to parallel switches.

4. **RS circuits and clocked RS flip-flop circuits:** Logic circuits that remember previous inputs and do not change their outputs until they receive new input signals. The clocked flip-flop circuit has an inverted clock signal as an input so that circuit operations occur in the proper order. Flip-flop circuits are called **sequential logic circuits** because the output is determined by the sequence of inputs. A given input affects the output produced by the next input.

5. **Driver circuits:** A driver is a term used to describe a transistor device that controls the large current in the output circuit. Drivers are controlled by a computer to operate such things as fuel injectors, ignition coils, and many other high current circuits. The high currents handled by a driver are not really that high; they are just more than what is typically handled by a transistor. Several types of driver circuits are used on automobiles, such as Quad, Discrete, Peak and Hold, and Saturated Switch driver circuits.

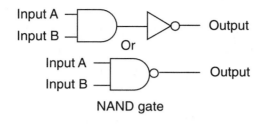

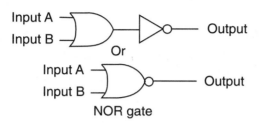

Figure 21-14 Symbols and truth tables for NAND and NOR gates. The small circle represents an inverted output on any logic gate symbol.

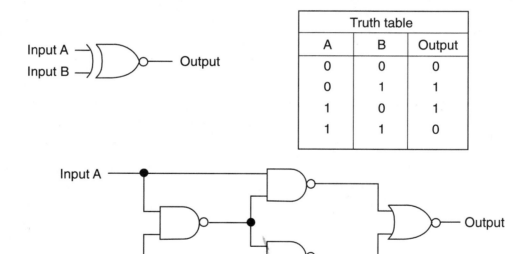

Figure 21-15 XOR gate symbol and truth table. An XOR gate is a combination of NAND and NOR gates.

6. **Registers:** Used in the computer to temporarily store information. A register is a combination of flip-flops that transfer bits from one to another every time a clock pulse occurs.

7. **Accumulators:** Registers designed to store the results of logic operations that can become inputs to other modules.

Inputs

As discussed earlier, the CPU receives inputs that it checks with programmed values. Depending on the input, the computer will control the **actuator(s)** until the programmed results are obtained (Figure 21-16). The inputs can come from other computers, the driver, the technician, or through a variety of sensors.

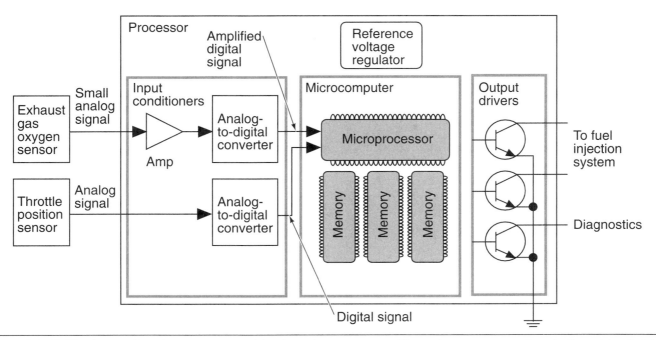

Figure 21-16 The input signals are processed in the microprocessor. The microprocessor directs the output drivers to activate actuators as instructed by the program.

Driver input signals are usually provided by momentarily applying a ground through a switch. The computer receives this signal and performs the desired function. For example, if the driver wishes to reset the trip odometer on a digital instrument panel, the driver would push the reset switch. This switch will provide a momentary ground that the computer receives as an input and sets the trip odometer to zero.

Switches can be used as an input for any operation that only requires a yes-no, or on-off, condition. Other inputs include those supplied by means of a sensor and those signals returned to the computer in the form of feedback.

Sensors

Sensors convert some measurement of vehicle operation into an electrical signal. There are many different designs of sensors. Some are nothing more than a switch that completes the circuit. Others are complex chemical reaction devices that generate their own voltage under different conditions. Repeatability, accuracy, operating range, and linearity are all requirements of a sensor. **Linearity** refers to the sensor signal being as constantly proportional to the measured value as possible. It is an expression of the sensor's accuracy.

Thermistors

A **thermistor** is a solid-state variable resistor made from a semiconductor material that changes resistance in relation to temperature changes. A thermistor is used to

sense engine coolant or ambient temperatures. By monitoring the thermistor's resistance value, the computer is capable of observing very small changes in temperature. The computer sends a reference voltage to the thermistor (usually 5 volts) through a fixed resistor. As the current flows through the thermistor resistance to ground, a voltage sensing circuit measures the voltage drop over the fixed resistor (Figure 21-17). The voltage dropped over the fixed resistor will change as the resistance of the thermistor changes. Using its programmed values, the computer is able to translate the voltage drop into a temperature value.

There are two types of thermistors. **Negative temperature coefficient (NTC)** thermistors reduce their resistance as the temperature increases. **Positive temperature coeffi-**

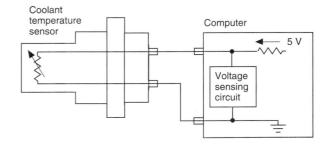

Figure 21-17 A thermistor is used to measure temperature. The sensing unit measures the resistance change and translates the date into temperature values.

cient (PTC) thermistors increase their resistance as the temperature increases. The NTC is the most commonly used.

Wheatstone Bridges

The **Wheatstone bridge** is a series-parallel arrangement of resistors between an input terminal and ground. Wheatstone bridges are used to measure temperature, pressure (**piezoresistive**) and mechanical strain. A Wheatstone bridge is nothing more than two simple series circuits connected in parallel across a power supply (Figure 21-18). Construction design varies between manufacturers but usually three of the resistors are kept at exactly the same value and the fourth is the sensing resistor. When all four resistors have the same value, the bridge is balanced and the voltage sensor will indicate a value of zero volts. The output from the amplifier acts as a voltmeter. Remember since a voltmeter measures electrical pressure between two points, it will display this value. For example, if the reference voltage is 5 volts, and the resistors have the same value, then the voltage drop over each resistor is 2.5 volts. Since the voltmeter is measuring the potential on the line between R_S and R_1, and between R_2 and R_3, it will read 0 volts because both of these lines have 2.5 volts on them. If there is a change in the resistance value of the sensor resistor, a change will occur in the circuit's balance. The sensing circuit will receive a voltage reading that is proportional to the amount of resistance change. A common use of a Wheatstone bridge is the hot wire sensor in a **mass air flow (MAF) sensor**. The sensor consists of a hot wire circuit, a cold wire circuit, and an electronic signal processing area. The hot and cold wire circuits form the Wheatstone bridge. The cold wire circuit is made of a fixed resistor and a thermistor. The amount of voltage dropped across the two resistors is determined by the temperature of the thermistor.

The hot wire circuit is made up of a fixed resistor and a variable resistance heat element (hot wire). The heat element generates heat in proportion to the amount of current flowing through it. This heat, in turn, changes its resistance. As air flows past the hot wire, it moves a small amount of heat from the element. This cooling of the element causes the voltage drop across it to change. The voltage drop across the hot wire is compared to the voltage drop across the fixed resistor in the cold wire circuit and air flow is determined.

Piezoelectric Devices

Piezoelectric devices are used to measure fluid and air pressures by generating their own voltages. The most commonly found piezoelectric device is the engine **knock sensor**. The knock sensor measures engine knock, or vibration, and converts the vibration into a voltage signal. The knock sensor is also referred to as a **detonation sensor**.

The sensor is a voltage generator and has a resistor connected in series with it. The resistor protects the sensor from excessive current flow in case the circuit becomes shorted. The voltage generator is a thin ceramic disc attached to a metal diaphragm. When engine knock occurs, the vibration of the noise puts a pressure on the diaphragm. This puts pressure on the piezoelectric crystals in the ceramic disc. The disc generates a voltage that is proportional to the amount of pressure. The voltage generated ranges from zero to one or more volts. Each time the engine knocks, a voltage spike is generated by the sensor.

Potentiometers

A **potentiometer** is a variable resistor that provides accurate voltage drop readings to the computer. The potentiometer usually consists of a wire wound resistor with a movable center wiper (Figure 21-19). A constant voltage value (usually 5 volts) is applied to terminal A. If the wiper (which is connected to the shaft or movable component of the unit that is being monitored) is located

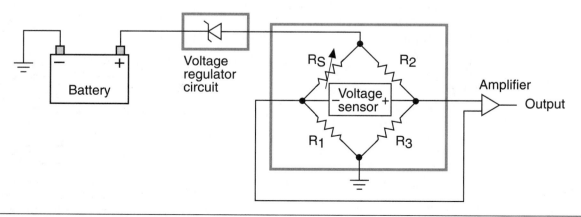

Figure 21-18 Wheatstone bridge.

close to this terminal, there will be low voltage drop represented by a high voltage signal back to the computer through terminal B. As the wiper is moved toward the C terminal, the sensor signal voltage to terminal B decreases. The computer interprets the different voltage values into different shaft positions. The potentiometer can measure linear or rotary movement. As the wiper is moved across the resistor, the position of the unit can be tracked by the computer.

Since applied voltage must flow through the entire resistance, temperature and other factors do not create false or inaccurate sensor signals to the computer. A rheostat is not as accurate and its use is limited in computer systems.

Magnetic Pulse Generators

Magnetic pulse generators (or **permanent magnet generators**) are commonly used to send data to the computer about the speed of the monitored component. For example, this sensor can be used to provide information about vehicle speed and individual wheel speed. The signals from the speed sensors are used for computer-driven instrumentation, cruise control, antilock braking, speed sensitive steering, and automatic ride control systems.

In some applications a magnetic pulse generator is also used to inform the computer of the position of a monitored component. This is common in engine controls where the computer needs to know the position of the crankshaft in relation to rotational degrees.

The components of the pulse generator are:

1. A **timing disc** that is attached to the rotating shaft or cable. The number of teeth on the timing disc is determined by the manufacturer and depends on application. The teeth will cause a voltage generation that is constant per revolution of the shaft. For example, a vehicle speed sensor may be designed to deliver 4,000 pulses per mile. The number of pulses per mile remains constant regardless of speed. The computer calculates

how fast the vehicle is going based on the frequency of the signal.

2. A **pickup coil** consists of a permanent magnet that is wound around by fine wire.

An air gap is maintained between the timing disc and the pickup coil. As the timing disc rotates in front of the pickup coil, the generator sends a pulse signal (Figure 21-20). As a tooth on the timing disc aligns with the core of the pick-up coil, it repels the magnetic field. The magnetic field is forced to flow through the coil and pick-up core (Figure 21-21). When the tooth passes the core, the magnetic field is able to expand. This action is repeated every time a tooth passes the core. The moving lines of magnetic force cut across the coil windings and induce a voltage signal.

When a tooth approaches the core, a positive current is produced as the magnetic field begins to concentrate around the coil. When the tooth and core align, there is

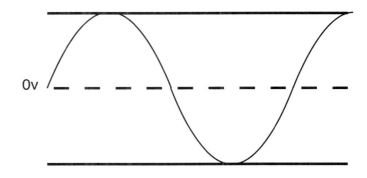

Figure 21-20 Pulse signal sine wave.

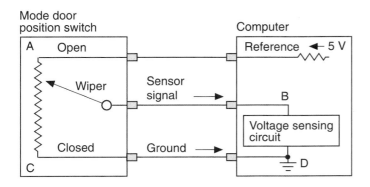

Figure 21-19 A potentiometer sensor circuit measures the amount of voltage drop to determine position.

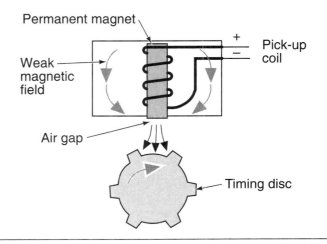

Figure 21-21 A strong magnetic field is produced in the pick-up coil as the teeth align with the core.

no more expansion or contraction of the magnetic field (thus no movement) and the voltage drops to zero. When the tooth passes the core, the magnetic field expands and a negative current is produced. The resulting pulse signal is amplified, digitalized, and sent to the microprocessor.

Hall-Effect Switches

The **Hall-effect switch** performs the same functions as the magnetic pulse generator, however its operation is different. Hall-effect switches operate on the principle that if a current is allowed to flow through thin conducting material that is exposed to a magnetic field, another voltage is produced (Figure 21-22).

The Hall-effect switch contains a permanent magnet and a thin semiconductor layer made of gallium arsenate crystal (Hall layer), and a shutter wheel (Figure 21-23). The Hall layer has a negative and a positive terminal connected to it. Two additional terminals located on either side of the Hall layer are used for the output circuit.

The permanent magnet is located directly across from the Hall layer so that its lines of flux will bisect at right angles to the current flow. The permanent magnet is mounted so that a small air gap is between it and the Hall layer.

A steady current is applied to the crystal of the Hall layer. This produces a signal voltage that is perpendicular to the direction of current flow and magnetic flux. The signal voltage produced is a result of the effect the magnetic field has on the electrons. When the magnetic field bisects the supply current flow, the electrons are deflected toward the Hall layer negative terminal. This results in a weak voltage potential being produced in the Hall switch.

A **shutter wheel** is attached to a rotational component. The shutter wheel consists of a series of alternating windows and vanes. It creates a magnetic shunt that changes the strength of the magnetic field from the permanent magnet. As the wheel rotates, the shutters (vanes) will pass in this air gap. When a shutter vane enters the gap, it intercepts the magnetic field and shields the Hall layer from its lines of force. The electrons in the supply current are no longer disrupted and return to a normal state. This results in low voltage potential in the signal circuit of the Hall switch.

The signal voltage leaves the Hall layer as a weak analog signal. To be used by the computer, the signal must be conditioned. It is first amplified because it is too weak to produce a desirable result. The signal is also inverted so a low input signal is converted into a high output signal. It is then sent through a **Schmitt trigger** where it is digitized and conditioned into a clean square wave signal. The signal is finally sent to a switching transistor. The computer senses the turning on and off of the switching transistor to determine the frequency of the signals and calculates speed.

Feedback Signals

If the computer sends a command signal to open a blend door in an automatic climate control system, a **feedback signal** may be sent back from the actuator to inform the computer the task was performed. Feedback means that data concerning the effects of the computers commands are fed back to the computer as an input signal. The feedback signal will confirm both the door position and actuator operation (Figure 21-24). Another form of feedback is for the computer to monitor voltage as a switch, relay, or other actuator is activated. Changing states of the actuator will result in a predictable change in the computer's voltage sensing circuit. The computer may set a diagnostic code if it does not receive the correct feedback signal.

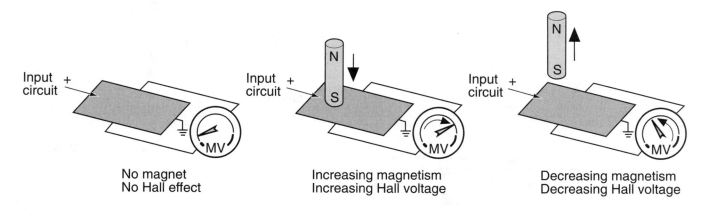

Figure 21-22 Hall-effect principles of voltage induction.

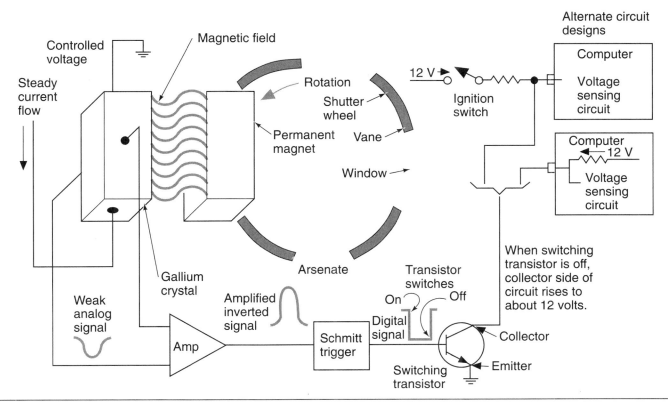

Figure 21-23 Typical circuit of a Hall-effect switch.

Oxygen (O₂) Sensors

One of the most commonly used feedback sensors is the **oxygen (O₂) sensor.** The O_2 sensor is mounted in the exhaust gas stream and provides the PCM with a measurement of the oxygen in the engine's exhaust. The sensor is constructed of a zirconium dioxide ceramic thimble covered with a thin layer of platinum (Figure 21-25).

When the thimble is filled with oxygen-rich outside air and the outer surface of the thimble is exposed to oxygen-depleted exhaust gases, a chemical reaction in the sensor produces a voltage. The generation of voltage is similar to the activity that takes place in a battery, except at much lower voltages. The voltage output varies

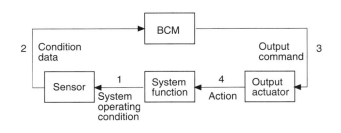

Figure 21-24 Principle of feedback signals.

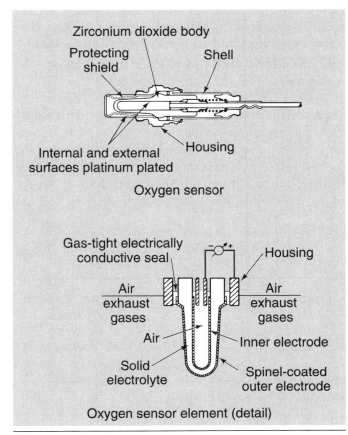

Figure 21-25 Oxygen (O₂) sensor design. (Courtesy of Chrysler Corporation)

with the level of oxygen present in the exhaust. As oxygen in the exhaust decreases, the voltage output increases. Likewise, as the oxygen level in the exhaust increases, the output voltage decreases (Figure 21-26).

Some O_2 sensors have a single wire connected from the oxygen-sensing element to the computer. This wire is the signal wire used to supply O_2 data to the computer. If an O_2 sensor has two wires, the second wire is a ground wire.

Most O_2 sensors today use an internal heater element. Since the O_2 sensor does not produce a satisfactory signal until it reaches about 600°F (315°C), the internal heater provides faster sensor warm-up time and helps to keep the sensor hot during prolonged idle operation. The internal O_2 sensor heater maintains higher sensor temperatures, which helps to burn deposits off the sensor. When the O_2 sensor has an internal heater, the sensor can be placed farther away from the engine in the exhaust stream, thus giving engineers more flexibility in sensor location. These O_2 sensors are identified by having three or four wires. The third wire is connected to an electric heating element in the sensor. Voltage is supplied from the ignition switch (or a relay is energized when the engine is running) to this heater. In a four-wire sensor there is an individual circuit for the signal wire, the heater wire, and the two ground wires. In these four-wire sensors, the heater and the sensing element have individual ground wires. A replacement O_2 sensor must have the same number of wires as the original sensor.

A lean air/fuel ratio provides excess quantities of oxygen in the exhaust stream, because the mixture entering the cylinders has an excessive amount of air in relation to the amount of fuel. Therefore, air containing oxygen is left over after the combustion process. When the exhaust stream has high oxygen content, oxygen from the atmosphere is also present inside the O_2 sensor element. When oxygen is present on both sides of the sensor element, the sensor produces a low voltage.

A rich air/fuel ratio contains excessive fuel in relation to the amount of air entering the cylinders. A rich air/fuel mixture produces very little oxygen in the exhaust stream because the oxygen in the air is all mixed with fuel, and excess fuel is left over after combustion is completed. When the exhaust stream with very low oxygen content strikes the O_2 sensor, there is high oxygen content from the atmosphere inside the sensor element. With different oxygen levels on the inside and outside of the sensor element, the sensor produces up to 1V. As the air/fuel ratio cycles from lean to rich, the O_2 sensor voltage changes in a few milliseconds.

In a gasoline fuel system, the **stoichiometric** (ideal) air/fuel mixture is 14.7:1. This indicates that for every 14.7 pounds of air entering the air intake, the carburetor or fuel injectors supply 1 pound of fuel. At the stoichiometric air/fuel ratio, combustion is most efficient, and nearly all the oxygen in the air is mixed with fuel and burned in the combustion chambers. Computer-controlled carburetor or fuel injection systems maintain the air/fuel ratio at stoichiometric under most operating conditions. As the air/fuel ratio cycles slightly rich and lean from the stoichiometric ratio, the O_2 sensor voltage cycles from high to low. Once the engine is at normal operating temperature, the O_2 sensor voltage should vary between 0.3V and 0.8V if the air/fuel ratio is at, or near, stoichiometric.

WARNING *The use of some RTV sealants will contaminate the O_2 sensor.*

WARNING *The use of leaded gasoline or the presence of coolant leaks into the combustion chamber will contaminate the O_2 sensor.*

Some vehicles are now equipped with a titania-type O_2 sensor. This sensor design is similar to the zirconia-type sensor, but the sensing element is made from titania rather than zirconia.

The titania-type sensor modifies voltage, whereas the zirconia-type sensor generates voltage. The computer supplies battery voltage to the titania-type sensor, and this voltage is lowered by a resistor in the circuit. The resistance of the titania varies as the air/fuel ratio cycles from rich to lean. When the air/fuel ratio is rich, the titania resistance is low, and this provides a higher voltage signal to the computer. If the air/fuel ratio is lean, the

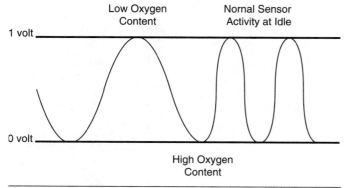

Figure 21-26 A high voltage value from the sensor indicates a low oxygen content in the exhaust. A low voltage indicates high oxygen content.

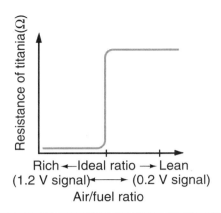

Figure 21-27 Titania-type O_2 sensor resistance and voltage signal. (Reprinted with the permission of Ford Motor Company)

titania resistance is high, and a lower voltage signal is sent to the computer (Figure 21-27).

The titania-type O_2 sensor provides a satisfactory signal almost immediately after a cold engine is started. This action provides improved air/fuel ratio control during engine warm-up.

Outputs

Once the computer's programming instructs that a correction or adjustment must be made in the controlled system, an output signal is sent to an actuator. This involves translating the electronic signals into mechanical motion.

An **output driver** is used within the computer to control the actuators. The circuit driver usually applies the ground circuit of the actuator. The ground can be applied steadily if the actuator must be activated for a selected amount of time. For example, if the BCM inputs indicate that the automatic door locks are to be activated, the actuator is energized steadily until the locks are latched. Then the ground is removed.

Other systems require the actuator to either be turned on and off very rapidly or for a set amount of cycles per second. It is **duty cycled** if it is turned on and off a set amount of cycles per second. The duty cycle is the percentage of on-time to total cycle time. Most duty cycled actuators cycle ten times per second. To complete a cycle it must go from off to on to off again. If the cycle rate is ten times per second, one actuator cycle is completed in one tenth of a second. If the actuator is turned on for 30% of each tenth of a second and off for 70%, it is referred to as a 30% duty cycle (Figure 21-28).

If the actuator is cycled on and off very rapidly, the pulse width can be varied to provide the programmed results. Pulse width is the length of time in milliseconds that an actuator is energized. For example, the computer program will select an illumination level of the digital instrument panel based on the intensity of the ambient light in the vehicle. The illumination level is achieved through pulse width modulation of the lights. If the lights need to be bright, the pulse width is increased, which increases the length of on-time. As the light intensity needs to be reduced, the pulse width is decreased (Figure 21-29).

Actuators

Actuators are devices that perform the actual work commanded by the computer. Most computer-controlled actuators are electromechanical devices that convert the output commands from the computer into mechanical action. These actuators are used to open and close switches, control vacuum flow to other components, and

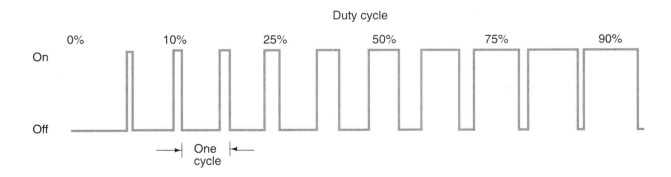

Figure 21-28 Duty cycle is the percentage of on-time per cycle. Duty cycle can be changed; however, the total cycle time remains constant.

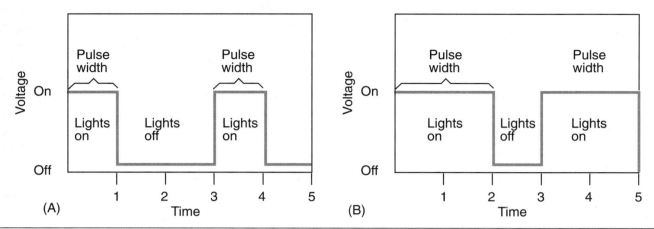

Figure 21-29 Pulse width is the duration of on-time: (A) pulse width modulation to achieve dimmer dash lights, (B) pulse width modulation to achieve brighter dash lights.

operate doors or valves depending on the requirements of the system.

Although it does not fall into the strict definition of an actuator, the BCM can also control lights, gauges, and driver circuits.

A common actuator controlled by the computer is the relay. A relay allows control of a high current draw circuit by a very low current draw circuit. The computer usually controls the relay by providing the ground for the relay coil (Figure 21-30). The use of relays protects the computer by keeping the high current from passing through it. For example, the motors used for power door locks require a high current draw to operate them. Instead of having the computer operate the motor directly, it will energize the relay. With the relay energized, a direct circuit from the battery to the motor is completed.

Computer control of the **solenoid** is usually provided by applying the ground through the output driver. A solenoid is commonly used as an actuator because it operates

well under duty cycling conditions.

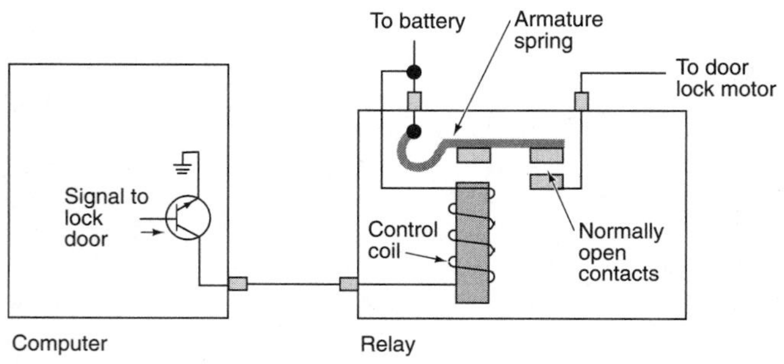

One of the most common uses of the solenoid is the fuel injector. Another common use is to control vacuum to other components. For example, many automatic climate control systems use vacuum motors to move the blend doors. The computer can control the operation of the doors by controlling the solenoid.

Many computer-controlled systems use a **stepper motor** to move the controlled device to whatever location is desired. A stepper motor contains a permanent magnet armature with two, four, or more field coils (Figure 21-31). By applying voltage pulses to selected coils of the motor, the armature will turn a specific number of degrees. When the same voltage pulses are applied to the opposite coils, the armature will rotate the same number of degrees in the opposite direction.

Some applications require the use of a permanent magnet field motor (Figure 21-32). The polarity of the voltage applied to the armature windings determines the direction the motor rotates. The computer can apply a continuous voltage to the armature until the desired result is obtained.

Multiplexing

Multiplexing is a system that enables the transmission of several messages over a single channel or circuit at the same time. Multiplexing provides the ability to use a single circuit to distribute and share data between several control modules throughout the vehicle. Because the data is transmitted through a single circuit, bulky wiring harnesses are eliminated.

Each control module on a MUX wiring system is connected by **bus data links**. The term **bus** refers to the transporting of data. Each module can transmit and

Figure 21-30 The computer's output driver applies the ground for the relay coil.

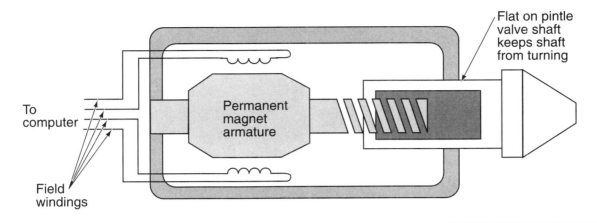

Figure 21-31 Typical stepper motor.

receive digital codes over the bus data links (Figure 21-33). This allows the modules to share their information. The signal sent from a sensor can go to any of the modules and can be used by the other modules. In the past, if information from the same sensing device was needed by several controllers, a wire from each controller needed to be connected in parallel to that sensor. If the sensor signal was analog, the controllers needed an analog to digital (A/D) convertor to be able to "read" the sensor information. By using multiplexing, the need for separate conductors from the sensor to each module is eliminated, and the number of drivers in the controllers is reduced.

Digital signals are used by controllers to communicate messages, both internally and with other controllers. A chip is used to prevent the digital codes from overlapping by allowing only one code to be transmitted at a time. Each digital message is preceded by an identifica-

tion code that establishes its priority. If two modules attempt to send a message at the same time, the message with the higher priority code is transmitted first.

The major difference between a multiplexed system and a nonmultiplexed system is the way data is gathered and processed. In nonmultiplexed systems, the signal from a sensor is sent as an analog signal through a dedicated wire to the computer or computers. At the

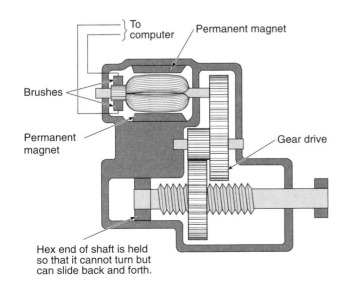

Figure 21-32 Reversible permanent magnet motor.

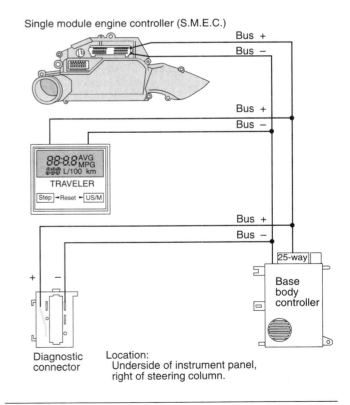

Figure 21-33 Computers use multiplexing to reduce the number of conductors that would be required. (Courtesy of Chrysler Corporation)

computer, the signal is changed from an analog to a digital signal. Because each sensor requires its own dedicated signal wire, the number of wires required to feed data from all of the sensors and transmit control signals to all of the output devices is great.

In a MUX system, the sensors process the information and send a digital signal to the computer. Since the computer or control module of any system can only process one input at a time, it calls for input signals as it needs them. By timing the transmission of data from the sensors to the control module, a single data wire can be used. Between each transmission of data to the control module, the sensor is electronically disconnected from the control module.

Summary

❏ A computer is an electronic device that stores and processes data and is capable of operating other devices.

❏ The operation of the computer is divided into four basic functions: input, processing, storage, and output.

❏ Binary numbers are represented by the numbers 1 and 0.

❏ Logic gates are the thousands of field effect transistors incorporated into the computer circuitry. The FETs use the incoming voltage patterns to determine the pattern of pulses that leave the gate. The most common logic gates are NOT, AND, OR, NAND, NOR, and XOR gates.

❏ There are several types of memory chips used in the body computer; ROM, RAM, and PROM are the most common types.

❏ ROM (read only memory) contains a fixed pattern of 1s and 0s representing permanent stored information used to instruct the computer on what to do in response to input data.

❏ RAM (random access memory) will store temporary information that can be read from or written to by the CPU.

❏ PROM (programmable read only memory) contains specific data that pertains to the exact vehicle in which the computer is installed.

❏ EPROM (Erasable PROM) is similar to PROM except its contents can be erased to allow new data to be installed.

❏ EEPROM (Electrically Erasable PROM) allows changing the information electrically one bit at a time.

❏ Inputs provide the computer with system operation information or driver requests.

❏ Switches can be used as an input for any operation that only requires a yes-no, or on-off, condition.

❏ Sensors convert some measurement of vehicle operation into an electrical signal.

❏ A thermistor is a solid-state variable resistor made from a semiconductor material that changes resistance in relation to temperature changes. Negative temperature coefficient (NTC) thermistors reduce their resistance as the temperature increases. Positive temperature coefficient (PTC) thermistors increase their resistance as the temperature increases.

❏ The Wheatstone bridge is a series-parallel arrangement of resistors between an input terminal and ground. Usually this sensor is used to measure temperature, pressure, and stress.

❏ A potentiometer is a variable resistor that usually consists of a wire-wound resistor with a movable center wiper.

❏ Magnetic pulse generators use the principle of magnetic induction to produce a voltage signal and are commonly used to send data concerning the speed of the monitored component to the computer.

❏ Hall-effect switches operate on the principle that if a current is allowed to flow through thin conducting material that is exposed to a magnetic field, another voltage is produced.

❏ Actuators are devices that perform the actual work commanded by the computer. They can be in the form of a motor, relay, switch, or solenoid.

❏ A servomotor produces rotation of less than a full turn. A feedback mechanism is used to position itself to the exact degree of rotation required.

❏ A stepper motor contains a permanent magnet armature with two, four, or more field coils. It is used to move the controlled device to whatever location is desired by applying voltage pulses to selected coils of the motor.

❏ Multiplexing is a system in which electrical signals are transmitted by a peripheral serial bus instead of conventional wires. This allows several devices to share signals on a common conductor.

Terms-To-Know

Accumulators
Actuators
Adaptive strategy
Address
Amplification circuit
Analog
Analog to digital (A/D) converter
AND gate
Baud rate
Biased
Binary code
Body control module (BCM)
Bus
Bus data links
Central processing unit (CPU)
Clock
Clocked RS flip-flop circuits
Computer
Decoder circuit
Demultiplexer (DEMUX)
Detonation sensor
Digital
Digital to analog (D/A) converter
Driver circuits
Duty cycled
Electrically erasable programmable read only memory (EEPROM)
Erasable programmable read only memory (EPROM)
Exclusive-OR (XOR) gate
Feedback signal

Flashing
Hall-effect switch
Hex code
Input
Interface
Inverter
Jeep/Truck engine controller (JTEC)
Keep alive memory (KAM)
Knock sensor
Linearity
Logic gates
Logic module
Magnetic pulse generators
Mass air flow (MAF) sensor
Memory
Microprocessor
Multiplexer (MUX)
Multiplexing
NAND gate
Negative temperature coefficient (NTC)
Nonvolatile RAM (NVRAM)
NOR gate
NOT gate
OR gate
Output
Output driver
Oxygen (O_2) sensor
Permanent magnet generator
Pickup coil
Piezoelectric devices

Piezoresistive
Positive temperature coefficient (PTC)
Potentiometer
Power module
Processing
Program
Programmable read only memory (PROM)
Random access memory (RAM)
Read only memory (ROM)
Registers
RS circuits
Schmitt trigger
Sensors
Sequential logic circuits
Sequential sampling
Shutter wheel
Single board engine controller (SBEC)
Single module engine controller (SMEC)
Solenoid
Stepper motor
Stoichiometric
Storage
Thermistors
Timing disc
Vehicle control module (VCM)
Volatile RAM
Wheatstone bridge

Review Questions

Short Answer Essays

1. What is binary code?

2. Describe the basics of NOT, AND, and OR logic gate operation.

3. List and describe the four basic functions of the computer.

4. What is the difference between ROM, RAM, and PROM?

5. Explain the principle of multiplexing.

6. How does the Hall-effect switch generate a voltage signal?

7. Describe the basic function of a stepper motor.

8. What is meant by feedback as it relates to computer control?

9. What is the difference between duty cycle and pulse width?

10. What are the purposes of the interface?

Fill-in-the-Blanks

1. In binary code, the number 4 is represented by _____ .

2. The _____ is a crystal that electrically vibrates when subjected to current at certain voltage levels.

3. _____ are registers designed to store the results of logic operations.

4. The _____ _____ _____ is the brain of the computer.

5. _____ contains specific data that pertains to the exact vehicle in which the computer is installed.

6. _____ convert some measurement of vehicle operation into an electrical signal.

7. Negative temperature coefficient (NTC) thermistors _____ their resistance as the temperature increases.

8. _____ switches operate on the principle that if a current is allowed to flow through thin conducting material exposed to a magnetic field, another voltage is produced.

9. Magnetic pulse generators use the principle of _____ _____ to produce a voltage signal.

10. _____ means that data concerning the effects of the computer's commands are fed back to the computer as an input signal.

ASE Style Review Questions

1. Technician A says during the processing function the computer uses input information and compares it to programmed instructions. Technician B says during the output function the computer will put out control commands to various output devices. Who is correct?
   a. A only
   b. B only
   c. Both A and B
   d. Neither A nor B

2. Technician A says analog means the voltage signal is either on-off, yes-no, or high-low. Technician B says digital means the voltage signal is infinitely variable within a given range. Who is correct?
   a. A only
   b. B only
   c. Both A and B
   d. Neither A nor B

3. Logic gates are being discussed. Technician A says NOT gate operation is similar to that of two switches in series to a load. Technician B says an AND gate simply reverses binary 1s to 0s and vice versa. Who is correct?
   a. A only
   b. B only
   c. Both A and B
   d. Neither A nor B

4. Computer memory is being discussed. Technician A says ROM can be written to by the CPU. Technician B says RAM will store temporary information that can be read from or written to by the CPU. Who is correct?
   a. A only
   b. B only
   c. Both A and B
   d. Neither A nor B

5. Technician A says volatile RAM is erased when it is disconnected from its power source. Technician B says nonvolatile RAM will retain its memory if removed from its power source. Who is correct?
   a. A only
   b. B only
   c. Both A and B
   d. Neither A nor B

6. Technician A says EPROM memory is erased if the tape is removed and the microcircuit is exposed to ultraviolet light. Technician B says electrostatic discharge will destroy the memory chip. Who is correct?
   a. A only
   b. B only
   c. Both A and B
   d. Neither A nor B

7. Technician A says negative temperature coefficient thermistors reduce their resistance as the temperature decreases. Technician B says positive temperature coefficient thermistors increase their resistance as the temperature increases. Who is correct?
   a. A only
   b. B only
   c. Both A and B
   d. Neither A nor B

8. Technician A says magnetic pulse generators are commonly used to send data to the computer concerning the speed of the monitored component. Technician B says an on-off switch sends a digital signal to the computer. Who is correct?
   a. A only
   b. B only
   c. Both A and B
   d. Neither A nor B

9. Speed sensors are being discussed. Technician A says the timing disc is stationary and the pickup coil rotates in front of it. Technician B says the number of pulses produced per mile increases as rotational speed increases. Who is correct?
   a. A only
   b. B only
   c. Both A and B
   d. Neither A nor B

10. Technician A says a Hall-effect switch uses a steady supply current to generate a signal. Technician B says a Hall-effect switch consists of a permanent magnet wound with a wire coil. Who is correct?
    a. A only
    b. B only
    c. Both A and B
    d. Neither A nor B

22 Body Computer System Diagnosis

Objective

Upon completion and review of this chapter, you should be able to:

❏ Describe the service precautions associated with servicing the BCM.

❏ Distinguish between hard and intermittent codes.

❏ Perform a complete visual inspection of the problem system.

❏ Enter BCM diagnostics by use of a scan tool.

❏ Enter BCM diagnostics through the ECC panel.

❏ Perform basic actuator tests.

❏ Perform basic sensor tests.

❏ Properly replace computer PROM chips.

Introduction

This chapter focuses on diagnostic procedures for the body control module (BCM). Powertrain control module (PCM) diagnostics will be discussed in later chapters as the different systems are introduced.

Because the body computer controls many of the functions of the vehicle's electrical systems, it is important for today's technician to be able to properly diagnose problems with this system. The use of the body computer has expanded to include the functions of climate control, lighting circuits, cruise control, antilock braking, electronic suspension systems, electronic shift transmissions, and alternator regulation. In some systems the direction light, the rear window defogger, the illuminated entry, and the intermittent wiper systems are included in the body controller function (Figure 22-1).

As discussed in Chapter 21, a computer processes the physical conditions that represent information (data). The operation of the computer is divided into four basic functions: input, processing, storage, and output. Understanding these four computer functions will help you organize the troubleshooting process. When a system is tested, you are attempting to isolate a problem with one of these functions.

In the process of controlling the various electrical systems, the BCM continuously monitors operating conditions for possible system malfunctions. The computer compares system conditions against programmed parameters. If the conditions fall outside of these limits, the computer detects a malfunction. A **trouble code** is set to indicate the portion of the system at fault. The technician can access this code for aid in troubleshooting.

If a malfunction results in improper system operation, the computer may minimize the effects by using **failsoft** action. Failsoft means the computer will substitute a fixed input value if a sensor fails. This provides for system operation, but at a limited function. For example, if the automatic temperature control system has a malfunction from the ambient temperature sensor,

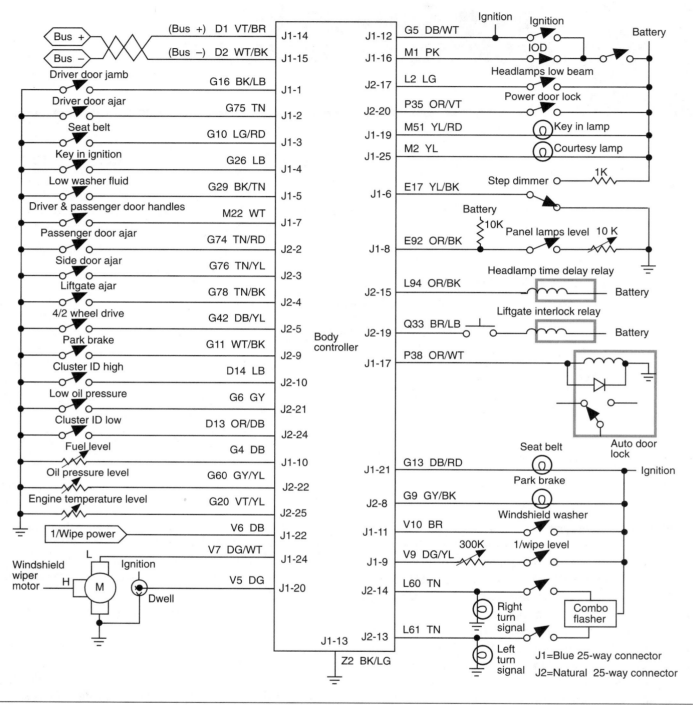

Figure 22-1 The body computer controls many of the vehicle's electrical systems. (Courtesy of Chrysler Corporation)

instead of shutting down the whole system, the computer will provide a fixed value as its own input. This fixed value can be either a programmed value in the computer's memory, or it can be the last received signal from the sensor prior to failure. This allows the system to operate on a limited basis instead of shutting down completely. Some other faults may result in the auto-

matic temperature control system switching to high fan speed, full heat, or defrost mode.

There are several things you need to know before you learn how to access the computer's memory to gain information concerning system operation. You need to become familiar with what you're looking at, and you must follow proper precautions when servicing these systems.

Electronic Service Precautions

The technician must take some precautions before servicing any computer or any of its controlled systems. Computers are designed to withstand normal current draws associated with normal operation. However, overloading any of the system circuits will result in damage to the computer. Follow these service precautions to prevent computer and circuit damage:

1. Do not ground or apply voltage to any controlled circuits unless the service manual instructs you to do so.

2. Use only a high impedance multimeter (10 megohm or greater) to test the circuits. Never use a test light unless specifically instructed to do so in the service manual.

3. Make sure the ignition switch is turned off before making or breaking electrical connections to the computer.

4. Unless instructed otherwise in the service manual, turn off the ignition switch before making or breaking any electrical connections to sensors or actuators.

5. Turn the ignition switch off whenever disconnecting or connecting the battery terminals. Also turn it off when pulling and replacing the fuse.

6. Do not connect any other electrical accessories to the insulated or ground circuits of the computer-controlled systems.

7. Use only manufacturer's specific test and replacement procedures for the year and model of vehicle being serviced.

By following these precautions, plus those listed in the service manual, you can avoid having to replace expensive components.

Electrostatic Discharge

Some manufacturers mark certain components and circuits with a code or symbol to warn technicians that the units are sensitive to **electrostatic discharge** (Figure 22-2). Since static electricity can be as high as 25,000 volts, it can destroy an electronic component or render it useless.

When handling any electronic part, especially those that are static sensitive, follow the guidelines below to reduce the possibility of electrostatic build-up on your body and the inadvertent discharge to the electronic part. If you are not sure if a part is sensitive to static, treat it as if it is.

Figure 22-2 General Motors' Electrostatic Discharge (ESD) symbol that warns the technician that the component or circuit is sensitive to static. (Courtesy of Chevrolet Motor Division—GMC)

1. Always touch a known good ground before handling the part. This should be repeated while handling the part and more frequently after sliding across a seat, sitting down from a standing position, or walking a distance.

2. Avoid touching the electrical terminals of the part unless you are instructed to do so in the written service procedures. It is good practice to keep your fingers off all electrical terminals, as the oil from your skin can cause corrosion.

3. When you are using a voltmeter, always connect the negative meter lead first.

4. Do not remove a part from its protective package until it is time to install the part.

5. Before removing the part from its package, ground yourself and the package to a known good ground on the vehicle.

6. When replacing a PROM, ground your body by putting a metal wire around your wrist and connect the wire to a good ground.

Trouble Codes

Trouble codes are two or three digital characters that are displayed in the diagnostic display if the testing and failure requirements are both met. Most computers are capable of displaying the stored faults in memory. The method used to retrieve the codes varies greatly; the technician must refer to the correct service manual for the procedure. Depending on system design, the computer may store codes for long periods of time; some lose the code when the ignition switch is turned off.

Systems that do not retain the code when the ignition is turned off require the technician to test drive the vehicle and attempt to duplicate the fault. Once the fault is detected by the computer, the code must be retrieved before the ignition switch is turned off again.

The trouble code does not necessarily indicate the faulty component; it only indicates that circuit of the system that is not operating properly (Figure 22-3). For example, the code displayed may be F11, indicating an A/C high side temperature sensor problem. This does not mean the sensor is bad; the fault is in that circuit, which includes the wiring, connections, sensor, and BCM. To locate the problem, follow the diagnostic procedure in the service manual for the code received.

Two types of codes can be displayed: intermittent codes and hard fault codes.

Hard Codes vs Intermittent Codes

Some computers will store trouble codes in their memory until they are erased by the technician or until a set amount of engine starts have passed. Some computers will display two sets of fault codes. Usually, the first set of codes displayed represent all codes stored in memory, including both hard and intermittent codes. The second set of codes displayed are only hard codes. The codes displayed in the first set but not in the second set are intermittent codes.

Hard code failures are those that are detected every time the computer tests the system. An **intermittent code** is one in which a fault had occurred in the past, but is not present during the last test of the circuit. This type of fault is also called a **soft code**.

Most diagnostic charts cannot be used to locate intermittent faults. This is because the testing at various points of the chart depends on the fault being present to locate the problem. If the fault is not present, the technician may be instructed erroneously to replace the BCM module, even though it is not defective.

Many intermittent problems are the result of poor electrical connections. Diagnosis should start with a good visual inspection of the connectors involved with the code. Even on hard codes, visually inspect the circuit before conducting any other tests.

Visual Inspection

Perhaps the most important check to be made before diagnosing a BCM-controlled system is a complete visual inspection. The visual inspection can identify faults that could cause the technician to spend wasted time in diagnostics. In addition, the problem can be pinpointed without any further steps.

Inspect the following:

1. All sensors and actuators for physical damage.

2. Electrical connections into sensors, actuators, and control modules.

3. All ground connections.

4. Wiring for signs of burned or chaffed spots, pinched wires, or contact with sharp edges or hot exhaust manifolds.

5. All vacuum hoses for pinches, cuts, or disconnects.

The time spent performing a visual inspection is worthwhile. Put forth the effort to check wires and hoses that are hidden under other components.

Entering Diagnostics

There are as many methods of entering BCM diagnostics as there are vehicle manufacturers. Most require a scan tool. As discussed in an earlier chapter, a scan tool is a microprocessor designed to communicate with the vehicle's computers. It will access trouble codes, run system operation, actuator, and sensor tests. The scan tool is plugged into the diagnostic connector for the system being tested. Some manufacturers provide a single diag-

BCM diagnostic codes	
Codes	Circuit affected
▼ F10	Outside temperature sensor circuit
▼ F11	A/C high side temperature sensor circuit
▼ F12	A/C low side temperature sensor circuit
▼ F13	In-car temperature sensor circuit
▼ F30	CCP to BCM data circuit
▼ F31	FDC to BCM data circuit
▼ F32	ECM-BCM data circuits
▼ F40	Air mix door problem
▼ F43	Heated windshield problem
☑ F46	Low refrigerant warning
☑ F47	Low refrigerant problem
☑ F48	Low refrigerant pressure
▼ F49	High temperature clutch disengage
▼ F51	BCM prom error
☑ Turn on "service air cond" light	
▼ Does not turn on any light	
Comments	
F11	Turns on cooling fans when A/C clutch is engaged
F12	Disengages A/C clutch
F32	Turns on cooling fans
F47 & 48	Switches from "auto" to "econ"

Figure 22-3 The trouble codes direct the technician to the affected circuits. (Courtesy of General Motors Corporation)

nostic connector and the technician chooses the system to be tested through the scan tool. Always refer to the correct service manual for the vehicle being serviced. Use only the methods identified in the service manual for retrieving trouble codes. Once the trouble codes are retrieved, consult the appropriate diagnostic chart for instructions on isolating the fault. It is also important to check the codes in the order required by the manufacturer.

Using Chrysler's DRBIII as an Example of Scanner Use

WARNING *The following is given as a guide and is intended to complement the service manual. Improper methods of code retrieval may result in damage to the computer.*

Chrysler uses several modules that share information with the body controller through a multiplex system (Figure 22-4). Connecting the DRBIII into the diagnostic connector will give the technician access to information concerning the operation of most vehicle systems. The DRBIII is a diagnostic read-out box (third generation) designed by Chrysler to gain access to the diagnostic function of the controllers (Figure 22-5). Although Chrysler recommends the use of the DRBIII scanner, other scan equipment with the correct adapters and cartridges can be used. The following is a typical method used to enter body controller diagnostics:

1. Locate the diagnostic connection using the component locator (Figure 22-6).
2. Connect the DRBIII to the vehicle by plugging its connector into the vehicle's diagnostic connector. Be sure to use the correct cable for the vehicle.
3. Turn the ignition switch to the RUN position. After the power-up sequence is completed, the copyright date and diagnostic program version should be displayed.
4. The display will change to a menu of selections.
5. Select option 1 (STAND ALONE DRB III)
6. Under the SELECT SYSTEM menu select the BODY option.
7. Enter body system diagnostics by selecting (BODY COMPUTER). The display will change to indicate the BUS test is being performed. If the message is different than that shown in the figure, there is a problem in the CCD bus that must be corrected. No further testing is possible until this problem is corrected.

8. After a few seconds the display will change and will provide information concerning the BCM. The display will give the name of the module selected, along with the version number of the module. After a few seconds the display will display BODY COMP MENU.
9. Use the down arrow key to scroll the menu selection if needed. Press 2 (READ DTCs).

The DRBIII will either display that no faults were detected or provide you with the fault codes. The first screen will tell you the number of fault codes found, the code for the first one, and a description of the code. Use the down arrow key to scroll through the entire list of codes retrieved.

Cadillac BCM Trouble Code Retrieval

WARNING *The procedures may change between models, years, and the type of instrument cluster installed. Refer to the correct service manual for the vehicle being diagnosed.*

Many Cadillac vehicles allow access to trouble codes, and other system operation information, through the electronic climate control panel (ECC). The BCM and engine control module (ECM) share information with each other. Thus, both ECM and BCM codes are retrieved through the ECC. To enter diagnostics, begin by placing the ignition switch in the RUN position. Depress the OFF and WARMER buttons on the ECC panel simultaneously (Figure 22-7). Hold the buttons until all display segments are illuminated.

Cadillac uses the onboard ECC panel to display trouble codes, whereas other GM vehicles use a Tech 1 scan tool. Starting in 1996, the Tech 2 scan tool was used to retrieve codes on certain models. In 1996, Cadillac switched to the use of the Tech 1 scan tool to retrieve class 2 data. When diagnosing GM systems, make sure to follow the procedures designated by GM.

WARNING *Diagnosis should not be attempted if all segments do not illuminate. It is possible to misdiagnose a problem as a result of not receiving the correct code. For example, if two segments of the display fail to illuminate, a code 24 could look like code 21.*

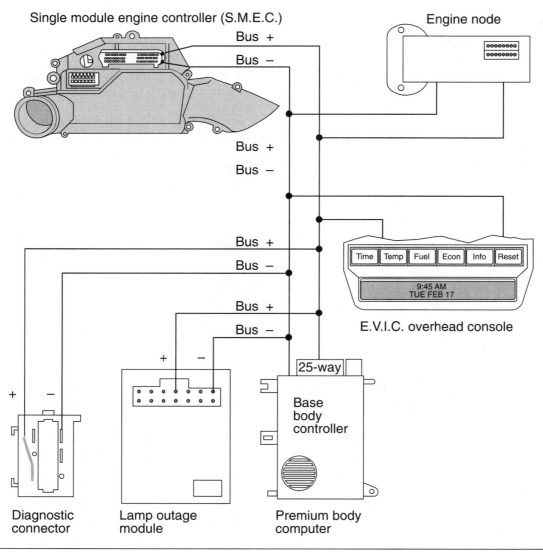

Figure 22-4 Multiplex systems used to interface several different modules. (Courtesy of Chrysler Corporation)

When the segment check is completed, the computer will display any trouble codes in its memory. An "8.8.8" will be displayed for about 1 second. Then an "..E" will appear. This signals the beginning of engine controller trouble codes. The display will show all engine controller trouble codes beginning with the lowest number and progressing through the higher numbers. All codes associated with the engine controller will be prefixed with an "E." If there are no codes, the "..E" will not be displayed. The "E" codes will be displayed twice. The first set are all codes in memory, both hard and intermittent. The second set will be only the hard codes. An ".E.E" is displayed to separate the two sets of codes.

Once all "E" codes are displayed, the computer will display BCM codes. The BCM codes are usually pre-

fixed by an "F." An "..F" will proceed the first set of codes displayed. The first set will be all codes stored in memory since the last 100 engine starts. An ".F.F" will appear to signal the separation of the first pass and the second pass. The second set of codes will be all hard codes.

When all codes are displayed, or there are no trouble codes in memory, ".7.O" will be displayed. This indicates the system is ready for the next diagnostic feature to be selected. To erase the BCM trouble codes, press the OFF and LOW buttons together until "F.O.O" appears. Release the buttons and ".7.O" will reappear. Turn off the ignition switch and wait at least 10 seconds before reentering the diagnostic mode.

Figure 22-5 DRB III

When in the diagnostic mode, it is possible to exit the system without erasing the trouble codes. Press AUTO on the ECC panel and the temperature will reappear in the display panel. This exits the diagnostic mode.

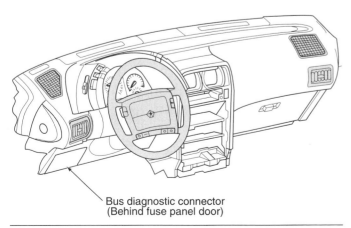

Bus diagnostic connector
(Behind fuse panel door)

Figure 22-6 Disconnect connector location. (Courtesy of Chrysler Corporation)

Testing Actuators

Testing of actuators is included here to orient you to the basic procedures. Specific procedures will be presented throughout this manual for individual systems and types of actuators.

Most computer-controlled actuators are electromechanical devices that convert the output commands from the computer into mechanical action. These actuators are used to open and close switches, control vacuum flow to other components, and operate doors or valves depending on the requirements of the system.

Most systems allow for testing of the actuator through the scan tool or ECC panel while in the correct mode. Actuators that are duty cycled by the computer are more accurately diagnosed through this method. In the example given earlier of retrieving trouble codes from the Chrysler system, at step 8 select ACTUATOR TESTS. This will allow the technician to activate selected actuators to test their operation.

In the General Motors system, the technician can access selected actuator operation. At the .7.0 display (before erasing codes), press the OUTSIDE TEMP button. The display should switch to F.8.O. Use the HIGH and LOW buttons to scroll through the parameters. Use the parameter chart from the service manual. For example, if parameter P.2.3 is selected, the actual air blend door position is displayed in percentage.

If the actuator must be tested by other means than the scanner, follow the manufacturer's procedures very carefully. Because many of the actuators used by the BCM operate with 5 to 7 volts, do not connect a jumper from a 25-volt source unless directed to do so by the service manual. Some actuators are easily tested with a voltmeter by testing for input voltage to the actuator. If there is input voltage of the correct level, check for a good ground connection. If both of these are good, then the actuator is faulty. If an ohmmeter needs to be used to measure the resistance of an actuator, disconnect it from the circuit first.

Testing Actuators with a Lab Scope

Since most actuators are electromechanical devices, when they fail it is because they are electrically faulty or mechanically faulty. By observing the action of an actuator on a lab scope, you will be able to watch its electrical activity. Normally if there is a mechanical fault, it will affect its electrical activity as well. Therefore, you get a good sense of the actuator's condition by watching it on a lab scope.

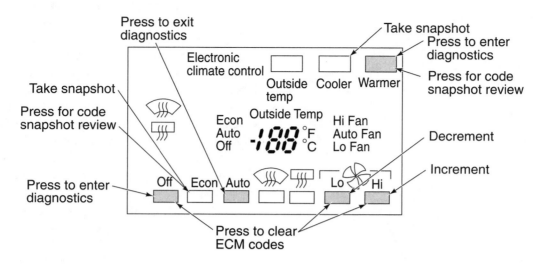

Figure 22-7 The buttons on the ECC panel allow the technician to access information from the computer when it is in the diagnostic mode. (Courtesy of General Motors Corporation)

To test an actuator, you need to know what it is. Most actuators are solenoids. The computer controls the action of the solenoid by controlling the pulse width of the control signal. By watching the control signal, you can see the turning on and off of the solenoid (Figure 22-8). The voltage spikes are caused by the discharge of the coil in the solenoid.

Some actuators are controlled pulse-width modulated signals (Figure 22-9). These signals show a changing pulse width. These devices are controlled by varying the pulse width, signal frequency, and voltage levels.

Both waveforms should be checked for amplitude, time, and shape. You should also observe changes to the pulse width as operating conditions change. A bad wave-

form will have noise, glitches, or rounded corners. You should be able to see evidence that the actuator immediately turns on and off according to the commands of the computer.

Testing Sensors

Testing of sensors is included here to orient you to the basic procedures. Specific procedures will be presented throughout this manual for individual systems and types of sensors.

There are many different designs of sensors, depending on the operation of the system they monitor. Some

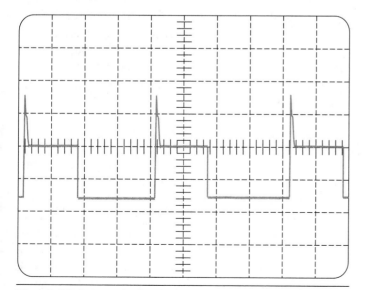

Figure 22-8 A typical solenoid control signal.

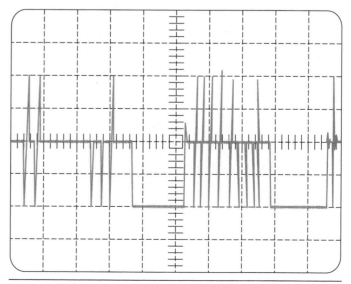

Figure 22-9 A typical pulse-width modulated solenoid control signal.

sensors are nothing more than a switch that completes the circuit. Others are complex chemical reaction devices that generate their own voltage under different conditions.

Thermistor and Potentiometer Testing

Thermistors and potentiometers are tested by measuring the input voltage to the sensor and the feedback voltage to the computer. The feedback voltage to the computer should change smoothly as the resistance value of the sensor changes. To test these voltage signals, a series of jumper wires may be required (Figure 22-10). The jumper wires provide a method of gaining access to the terminals of weather-pack connectors without breaking the wire insulation.

> **WARNING** *Do not use a test light to test for voltage. This may damage the system. Also, do not probe for voltage by sticking the wire insulation.*

An ohmmeter can be used to measure the changes in resistor values. Disconnect the sensor from the system. Connect the ohmmeter leads to the reference and ground terminals (Figure 22-11). Check the results against specifications. If good, connect the leads between the reference terminal and the feedback terminal (Figure 22-12). The resistance should change smoothly and consistently as the wiper position is changed.

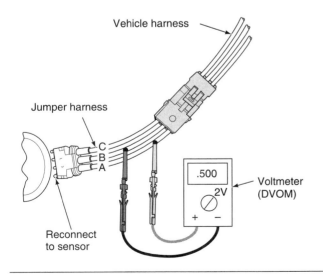

Figure 22-10 Jumper wires connected between the sensor and the harness allow the technician to probe for voltage or to test resistance without damaging the wiring. (Courtesy of General Motors Corporation)

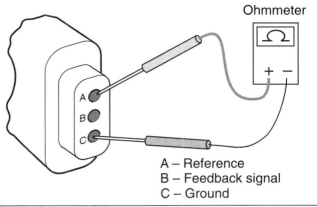

A – Reference
B – Feedback signal
C – Ground

Figure 22-11 Connecting an ohmmeter to test a potentiometer.

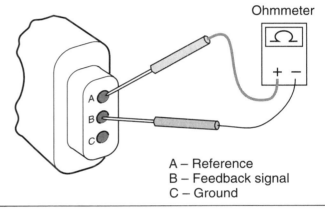

A – Reference
B – Feedback signal
C – Ground

Figure 22-12 Ohmmeter connection to test the movement of the wiper in a potentiometer.

Potentiometer and thermistor testing with a lab scope is a good way to watch the sweep of the resistor. The waveform (Figure 22-13) is a DC signal that moves up

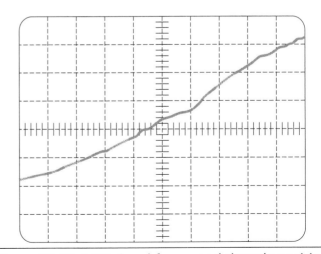

Figure 22-13 A DC signal for a good throttle position sensor, measured from closed throttle to wide-open throttle.

as the voltage increases. Most potentiometers in computer systems are fed a reference voltage of 5 volts. Therefore, the voltage output of these sensors will range from zero to 5 volts. The change in voltage should be smooth. Look for glitches in the signal. These can be caused by changes in resistance or an open.

Testing Magnetic Pulse Generators

To test the magnetic pulse generator, disconnect it from the system. Use an ohmmeter to test the resistance value of the coil and compare the results with specifications. The voltage generation of the sensor can be tested by connecting a voltmeter across the sensor terminals. The voltmeter must be in the AC position and on the lowest scale. Rotate the shaft while observing the voltage signal. It should increase and decrease with changes in shaft speed.

Magnetic pulse generators can also be tested with a lab scope. Instead of connecting a voltmeter across the sensor's terminals, connect the lab scope leads. The expected pattern is an AC signal that should be a perfect sine wave when the speed is constant. When the speed is changing, the AC signal should change in amplitude and frequency as shown in Figure 22-14.

Testing Hall-effect Sensors

To test Hall-effect type sensors, begin by disconnecting the wire harness. Connect a voltage source of correct voltage level across the positive and negative terminals of the Hall layer. Connect a voltmeter across the negative and signal voltage terminals.

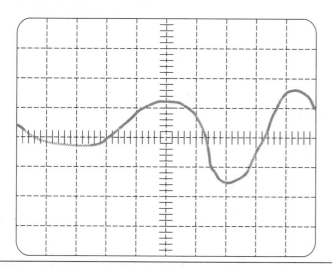

Figure 22-14 An AC signal from an ABS wheel speed sensor. The signal is changing over time because the speed of the wheel is accelerating quickly.

SERVICE TIP *A battery of correct voltage can be constructed from size D flashlight batteries connected together. Also, a 12-volt voltage source can be lowered to the correct level by wiring a rheostat into the jumper wires. Use the rheostat to reduce the voltage to the required level.*

Insert a metal feeler gauge between the Hall layer and the magnet. Make sure the feeler gauge is touching the Hall element. If the sensor is operating properly, the battery will read close to battery voltage. When the feeler gauge blade is removed, the voltage should decrease. On some units, the voltage will drop to near zero. Check the service manual to see what voltage you should observe when installing and removing the feeler gauge.

SERVICE TIP *Testing of the magnetic pulse generator and Hall-effect sensors can also be done with an oscilloscope. The signal from the pulse generator may not be a true sine wave, but it should have both positive and negative pulses. The Hall-effect switch should show a square wave pattern.*

WARNING *Some voltage generating sensors (such as the oxygen sensor used by the engine controller) cannot be tested with an ohmmeter. Refer to the service manual before testing these sensors.*

Sensor Testing with a Scan Tool

Most scan tools will display the voltage values or switch position of many sensors. Access to this information differs depending on the scan tool used. The DRBIII is used as an example of retrieving this information.

While the DRBIII is displaying the BODY COMP MENU, select SENSOR DISPLAY. In this section the value of selected sensors can be viewed. If the technician requires switch position, select INPUTS/OUTPUTS and the display will indicate the various positions of various switches used as inputs to the computer.

Breakout Boxes

The vehicle's wiring harness creates special problems when it comes to diagnosing circuits. Some manufacturers use a **breakout box** connected between the module and the wiring harness that will allow the technician to "see" the exact information the computer is receiving and sending. Breakout boxes (Figure 22-15) allow the technician to test circuits, sensors, and actuators by providing test points. The breakout box is connected in series with the computer so the voltages will be the same as those received by the computer.

Once the breakout box is connected into the system, a DMM can be used to measure the voltage signals and resistance values of the circuit. The diagnostic manual provided with the breakout box will direct the technician through a series of test procedures. Comparing test results with specifications will lead the technician to the problem area.

The breakout box has the advantage that it taps directly into the sensor or actuator circuit. This provides the technician with the exact voltage signal being sent or received. The scan tool provides the technician with an interpretation of these values only. In addition, hard-to-get-to components can be tested without having to disconnect them from the circuit.

PROM Replacement

Some manufacturers provide for PROM replacement in their body computers. To replace the PROM (if the diagnostic chart leads you to this step), remove the BCM from the vehicle. Follow all service precautions while servicing the BCM.

Follow Photo Sequence 8 for the correct method of replacing the PROM.

Figure 22-15 Using the breakout box to test a circuit.

WARNING *Installing a PROM chip backward will immediately destroy the chip. In addition, electrostatic discharge (ESD) will destroy the chip. There are static straps available that will prevent ESD while you are working on the unit.*

Summary

❑ In the process of controlling the various electrical systems, the BCM continuously monitors operating conditions for possible system malfunctions.

❑ A trouble code is set to indicate the portion of the system at fault.

❑ If a malfunction results in improper system operation, the computer may minimize the effects by using failsoft action. Failsoft means the computer will substitute a fixed input value if a sensor fails.

❑ When handling any electronic part, especially those that are static sensitive, follow the suggested guidelines to reduce the possibility of electrostatic build-up on your body and the inadvertent discharge to the electronic part.

❑ Systems that do not retain the code when the ignition is turned off require the technician to test drive the vehicle and attempt to duplicate the fault. Once the fault is detected by the computer, the code must be retrieved before the ignition switch is turned off again.

❑ The trouble code does not necessarily indicate the faulty component; it only indicates that circuit of the system that is not operating properly.

❑ Hard code failures are those that are detected every time the computer tests the system.

❑ An intermittent code is one in which a fault had occurred in the past, but is not present during the last test of the circuit.

❑ Perhaps the most important check to be made before diagnosing a BCM-controlled system is a complete visual inspection. The visual inspection can identify faults that could cause the technician to spend wasted time in diagnostics.

Typical Procedure for Replacing the PROM

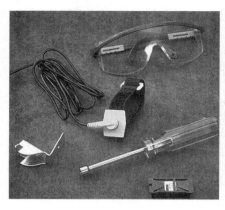

P8-1 Tools required to remove and replace the PROM; rocker-type PROM removal tool, ESD strap, safety glasses, and replacement PROM.

P8-2 Place the BCM onto the work bench with the PROM access cover facing up. Be careful not to touch the electrical connectors with your fingers.

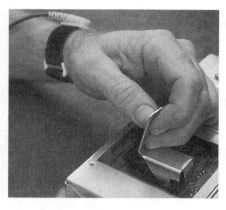

P8-3 Using the rocker-type PROM removal tool, engage one end of the PROM *carrier* with hook end of tool. Grasp the PROM carrier with the tool only at the narrow ends of the carrier.

P8-4 Press on the vertical bar end of the tool. Rock the end of the PROM carrier up as far as possible..

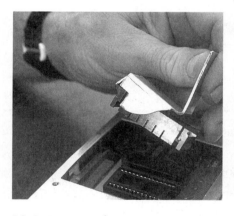

P8-5 Repeat the process on the other end of the carrier until the PROM carrier is removed from the socket.

P8-6 Check for proper orientation of the PROM in the carrier. The notch in the PROM should be referenced to the smaller notch in the carrier. If the replacement PROM does not come in its own carrier, it will be necessary to remove the old PROM and install the replacement PROM into the carrier. Be careful not to bend the pins.

Typical Procedure for Replacing the PROM (cont'd)

P8-7 Align the PROM carrier with the socket. The small notch of the carrier must be aligned with the small notch in the socket.

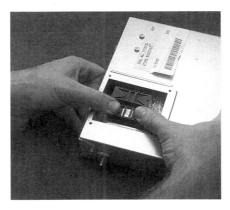

P8-8 Press the PROM carrier until it is firmly seated into the socket. Do not press on the PROM.

❏ A scan tool is a microprocessor designed to communicate with the vehicle's computers. It will access trouble codes, operate actuators, and and prefrom tests on the system's sensors.

❏ Many Cadillac vehicles allow access to trouble codes and other system operation information through the electronic climate control panel (ECC).

❏ Cadillac uses the onboard ECC panel to display trouble codes, whereas other GM vehicles use a Tech 1 scan tool. Starting in 1996, the Tech 2 scan tool began to be used to retrieve codes on certain models. In 1996, Cadillac switched to the use of the Tech 1 scan tool to retrieve class 2 data.

❏ All codes associated with the engine controller will be prefixed with an "E." The BCM codes are usually prefixed by an "F."

❏ Most systems allow for testing of the actuator through the scan tool or ECC panel while in the correct mode.

❏ Thermistors and potentiometers are tested by measuring the input voltage to the sensor and the feedback voltage to the computer. The feedback voltage to the computer should change smoothly as the resistance value of the sensor changes.

❏ Potentiometer and thermistor testing with a lab scope is a good way to watch the sweep of the resistor. The waveform is a DC signal that moves up as the voltage increases.

❏ To test the magnetic pulse generator, disconnect it from the system. Use an ohmmeter to test the resistance value of the coil and compare the results with specifications.

❏ Magnetic pulse generators can also be tested with a lab scope. The expected pattern is an AC signal that should be a perfect sine wave when the speed is constant.

❏ Most scan tools will display the voltage values or switch position of many sensors.

❏ Some manufacturers use a breakout box connected between the module and the wiring harness that will allow the technician to "see" the exact information the computer is receiving and sending.

Terms-To-Know

Breakout box	Failsoft	Soft code
DRBIII	Hard code	Trouble code
Electrostatic discharge	Intermittent code	

ASE Style Review Questions

1. The service precautions associated with servicing the BCM are being discussed. Technician A says to use a test light to check for voltage. Technician B says to make sure the ignition switch is turned off before making or breaking electrical connections to the BCM. Who is correct?
   a. A only
   b. B only
   c. Both A and B
   d. Neither A nor B

2. Trouble codes are being discussed. Technician A says hard code failures are those that have occurred in the past, but were not present during the last BCM test of the circuit. Technician B says intermittent codes are those that were detected the last time the BCM tested the circuit. Who is correct?
   a. A only
   b. B only
   c. Both A and B
   d. Neither A nor B

3. Technician A says that trouble codes will indicate the exact failure in the circuit. Technician B says that trouble codes will direct the technician to the circuit that has a fault in it. Who is correct?
   a. A only
   b. B only
   c. Both A and B
   d. Neither A nor B

4. Retrieving trouble codes is being discussed. Technician A says the scan tool is the only method of pulling the codes. Technician B says to apply 12 volts to the diagnostic connector. Who is correct?
   a. A only
   b. B only
   c. Both A and B
   d. Neither A nor B

5. Accessing trouble codes through the ECC on General Motors vehicles is being discussed. Technician A says to depress the OFF and WARMER buttons simultaneously. Technician B says the body codes are prefixed with an "E". Who is correct?
   a. A only
   b. B only
   c. Both A and B
   d. Neither A nor B

6. Technician A says to erase the BCM trouble codes, press the OFF and LO buttons together. Technician B says to exit the system without erasing the trouble codes, press the AUTO button. Who is correct?
   a. A only
   b. B only
   c. Both A and B
   d. Neither A nor B

7. Testing of sensors is being discussed. Technician A says sensors can only be tested with a scan tool. Technician B says a breakout box can be used to test sensor signals. Who is correct?
   a. A only
   b. B only
   c. Both A and B
   d. Neither A nor B

8. Testing of the magnetic pulse generator is being discussed. Technician A says to use an ohmmeter to test the resistance value of the coil. Technician B says the voltage generation of the sensor can be tested by connecting a voltmeter across the sensor terminals. Who is correct?
   a. A only
   b. B only
   c. Both A and B
   d. Neither A nor B

9. Replacement of the PROM is being discussed. Technician A says the notch in the PROM should be referenced to the smaller notch in the carrier. Technician B says the small notch of the carrier must be aligned with the small notch in the socket. Who is correct?
   a. A only
   b. B only
   c. Both A and B
   d. Neither A nor B

10. Technician A say installing a PROM chip in backwards will immediately destroy the chip. Technician B says electrostatic discharge will destroy the chip. Who is correct?
    a. A only
    b. B only
    c. Both A and B
    d. Neither A nor B

23 Computer-Controlled Ignition Systems

Objective

Upon completion and review of this chapter, you should be able to:

❏ Describe the function of common sensors used in computer-controlled DI and EI systems.

❏ Explain spark control in computerized DI systems.

❏ Describe the difference between DI and EI systems.

❏ Explain spark control in common EI systems.

❏ Describe the advantages of electronic ignition (EI) systems.

❏ Describe the operation of common EI systems.

❏ Describe the coil secondary-to-spark plug wiring connections on an EI system, including an explanation of how the spark plugs fire.

❏ Explain the purpose of the camshaft sensor signal in an EI system.

❏ Describe the design and location of the reluctor ring and magnetic sensor on an EI system.

❏ Explain the coil firing sequencing on a common EI system during different modes of operation.

Introduction

The operation of the ignition system is extremely important to obtain proper engine performance and economy. The purpose of the ignition system is to create a spark, or current flow, across each pair of spark plug electrodes at the proper instant under all engine operating conditions. This purpose sounds relatively simple, but when considering the number of spark plug firings required and the extreme variation in engine operating conditions, the function of the ignition system is very complex.

For example, if a V-8 engine is rotating at 3,000 revolutions per minute (rpm), and the ignition system must fire 4 spark plugs per revolution, the ignition system must supply 12,000 sparks per minute. These plug firings must also occur at the proper instant, without misfiring. If the ignition system misfires or does not fire the spark plugs at the proper time, fuel economy, engine performance, and emission levels are adversely affected.

A distributor is not required in electronic ignition (EI) systems. In many of these systems, a crank sensor located at the front of the crankshaft is used to trigger the ignition system. When a distributor is used in the ignition system, the distributor drive gear, shaft, and bushings are subject to wear. Worn distributor components cause erratic ignition timing and spark advance, which result in reduced economy and performance plus increased exhaust emissions. Since the distributor is eliminated in EI systems, ignition timing remains more stable over the life of the engine, which means improved economy and performance with reduced emissions.

A specific amount of energy is available in a secondary ignition circuit. Electrical energy may be measured in watts, which are calculated by multiplying amperes and volts. In a secondary ignition circuit, the energy is normally produced in the form of voltage required to start firing the spark plug, and then a certain amount of current flow across the spark plug electrodes. Electronic ignition systems are capable of producing much higher energy than point-type or distributor ignition systems.

Since distributor ignition and electronic ignition systems are both firing spark plugs with approximately the same gaps, the voltage required to start firing the spark plugs in both systems is similar. If the additional energy in the EI systems is not produced in the form of voltage, it will be produced in the form of current flow for a longer time across the spark plug electrodes. The average current flow across the spark plug electrodes in an EI system is 1.5 milliseconds compared to approximately 1 millisecond in a DI system.

This extra current flow duration of 0.5 millisecond on an EI system may seem insignificant, but it is very important on today's engines with stricter fuel economy and emission regulations. Today's emission standards demand leaner air/fuel ratios. This additional spark duration on EI systems helps to prevent cylinder misfiring with leaner air/fuel ratios. For these reasons, many car manufacturers have equipped their engines with EI systems.

Common Sensors

Computer-controlled DI and EI ignition systems use a series of sensors to determine the correct timing of the spark. Although the sensors on different systems are designed to provide the same basic information, the type of sensor used can vary between manufacturers and even between different engine applications.

Engine Crankshaft Position Sensors

The time when the ignition primary circuit must be opened and closed is related to the position of the pistons and the crankshaft. Therefore, the position of the crankshaft is used to control the flow of current to the base of the switching transistors in the ignition coil drivers.

A number of different types of sensors are used to monitor the position of the crankshaft. These engine position sensors and generators serve as triggering devices and include magnetic pulse generators, metal detection sensors, Hall-effect sensors, and photoelectric sensors.

The mounting location of the **crankshaft position sensor (CKP)** depends on the design of the ignition system. All four types of sensors can be mounted in the distributor, which is turned by the camshaft.

Magnetic pulse generators and Hall-effect sensors can also be located on the crankshaft (Figures 23-1 and Figure 23-2). Usually the crankshaft position sensor supplies the PCM with information concerning the position of two pistons as they approach TDC. However, the PCM does not know which piston is approaching TDC compression stroke. The function of the **camshaft position sensor (CMP)** is to identify the piston approaching TDC compression stroke. Both Hall-effect sensors and magnetic pulse generators can also be used as camshaft reference sensors to identify which cylinder is the next one to fire.

Metal Detection Sensors

Metal detection sensors are found on many early electronic ignition systems. They work much like a magnetic pulse generator with one major difference. A trigger wheel is pressed over the distributor shaft, and a pickup coil detects the passing of the trigger teeth as the distributor shaft rotates. However, unlike a magnetic pulse generator, the pickup coil of a metal detection sensor does not have a permanent magnet. Instead, the pickup coil is an electromagnet. A low level of current is supplied to the coil by an electronic control unit, inducing a weak magnetic field around the coil. As the reluctor on the distributor shaft rotates, the trigger teeth pass very close to the coil (Figure 23-3). As the teeth pass in and out of the coil's magnetic field, the magnetic field builds

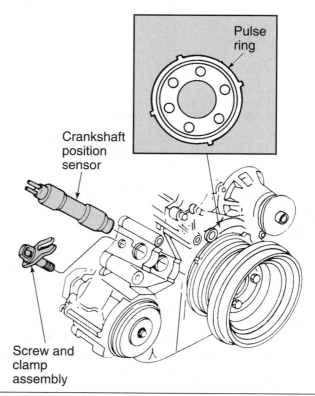

Figure 23-1 A typical crankshaft speed sensor. (Reprinted with the permission of Ford Motor Company)

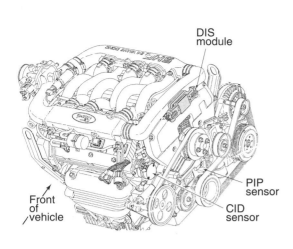

Figure 23-2 An engine with both a camshaft and a crankshaft position sensor. (Reprinted with the permission of Ford Motor Company)

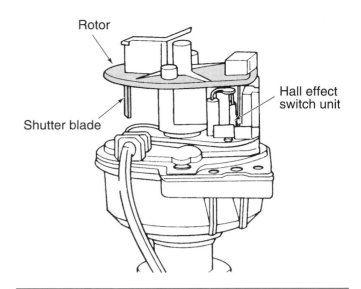

Figure 23-4 A distributor rotor with a shutter blade for the Hall-effect switch. (Courtesy of Chrysler Corporation)

and collapses, producing a corresponding change in the coil's voltage. The voltage changes are monitored by the control unit to determine crankshaft position.

Hall-effect Sensor

Introduced in early 1982, the Hall-effect sensor or switch is now the most commonly used engine crankshaft position sensor. There are several good reasons for this. Unlike a magnetic pulse generator, the Hall-effect sensor produces an accurate digital voltage signal throughout the entire rpm range of the engine, especially at low engine speeds such as during cranking. Furthermore, a Hall-effect switch produces a square wave signal that is more compatible with the digital signals required by onboard computers.

The Hall-effect sensor operation was described in Chapter 22. The shutter wheel is the last major component of some Hall switches. The shutter wheel consists of a series of alternating windows and vanes that pass between the Hall layer and magnet. The shutter wheel may be part of the distributor rotor (Figure 23-4) or be separate from the rotor.

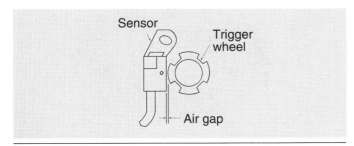

Figure 23-3 A metal detector sensor.

The shutter wheel performs the same function as the timing disc on magnetic pulse generators. The only difference is with a Hall-effect switch there is no electromagnetic induction. Instead, the shutter wheel creates a magnetic shunt that changes the magnetic field strength on the Hall-effect element. When a vane of the shutter wheel is positioned between the magnet and Hall-effect element, the metallic vane blocks the magnetic field and keeps it from penetrating the Hall-effect element. As a result, only a few residual electrons are deflected in the layer, and Hall output voltage is low (Figure 23-5). Conversely, when a window rotates into the air gap, the magnetic field is able to penetrate the Hall-effect layer, which in turn pushes the Hall voltage to its maximum range.

The points where the shutter vane begins to enter and begins to leave the air gap are directly related to primary circuit control. As the leading edge of a vane enters the air gap, the magnetic field is deflected away from the Hall-effect layer and Hall voltage decreases. When that happens, the modified Hall output signal increases abruptly and turns the switching transistor on. Once the transistor is turned on, the primary circuit closes and the coil's energy storage cycle begins.

Primary current continues to flow as long as the vane is in the air gap. As the vane starts to leave the gap, however, the changing Hall voltage signal prompts a parallel decline in the modified output signal. When the output signal goes low, the bias of the transistor changes. Primary current flow stops.

In summary, the ignition module supplies current to the coil's primary winding as long as the shutter wheel's vane is in the air gap. As soon as the shutter wheel

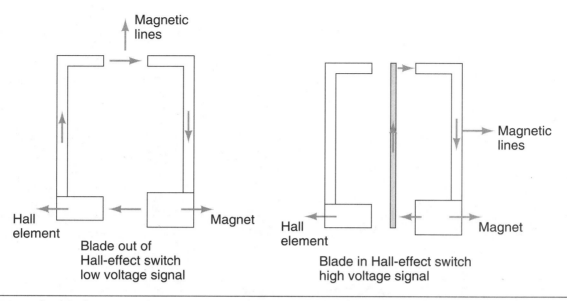

Figure 23-5 As the blade moves through the Hall-effect switch, a high voltage signal is produced.

moves away and the Hall voltage is produced, the control unit stops primary circuit current, high secondary voltage is induced, and ignition occurs.

In addition to ignition control, a Hall-effect switch can also be used to generate precise rpm signals (by determining the frequency at which the voltage rises and falls) and provide the sync pulse for sequential fuel ignition operation.

Another method is to use a slotted flex plate (Figure 23-6). The slots are spaced at manufacturer-set intervals. In the illustration there are three groups of slots. Each group is 120° apart from each other. Each group has four slots spaced 20° apart from each other. As the slots pass below a Hall-effect crankshaft position sensor, a switched voltage between high and low is produced (Figure 23-7).

Photoelectric Sensor

A fourth type of crankshaft position sensor is the **photoelectric sensor**. The parts of this sensor include a light emitting diode, a light sensitive phototransistor, and a slotted disc called a light beam interrupter (Figure 23-8).

The slotted disc is attached to the distributor shaft. The LED and the photo cell are situated over and under the disc opposite each other. As the slotted disc rotates between the LED and photo cell, light from the LED shines through the slots. The intermittent flashes of light are translated into voltage pulses by the photo cell. When the voltage signal occurs, the control unit turns the primary system on. When the disc interrupts the light and the voltage signal ceases, the control unit turns the pri-

mary system off, causing the magnetic field in the coil to collapse and sending a surge of voltage to a spark plug.

The photoelectric sensor sends a very reliable signal to the control unit, especially at low engine speeds. These units have been primarily used on Chrysler and Mitsubishi engines. Some Nissan and General Motors products use them as well.

Figure 23-6 A slotted flex plate used with the crank timimg sensor. (Courtesy of Chrysler Corporation)

Figure 23-7 The switched voltage informs the PCM of pistons approaching TDC. Frequency of signals determines the engine speed.

Knock Sensor

The knock sensor is threaded into the engine block, intake manifold, or cylinder head. The knock sensor contains a piezoelectric sensing element, which changes a vibration to a voltage signal. An internal resistor is connected in parallel with the piezoelectric sensing element.

When the engine detonates, a vibration is present in the engine block and cylinder head castings. The knock sensor changes this vibration to a voltage signal and sends the signal to the computer. When this signal is received, the computer reduces spark advance to eliminate the detonation. A typical knock sensor signal would be 300 millivolts (mV) to 500 mV depending on the severity of detonation. On some General Motors cars and light-duty trucks, the knock sensor signal is sent to an electronic spark control (ESC) module which changes the analog sensor signal to a digital signal.

Computer-Controlled Ignition System Operation

Computer-controlled ignition systems (Figure 23-9) control the primary circuit and distribute the firing voltages in the same manner as other types of electronic igni-

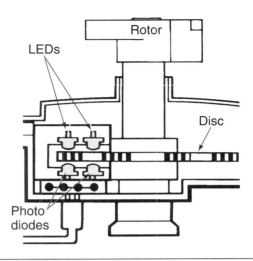

Figure 23-8 A photoelectric-type crankshaft position sensor. (Courtesy of Chrysler Corporation)

tion systems. The main difference between the systems is the elimination of any mechanical or vacuum advance devices from the distributor in the computer-controlled systems. In these systems, the distributor's sole purpose is to generate the primary circuit's switching signal and distribute the secondary voltage to the spark plugs. Timing advance is controlled by a microprocessor, or computer. In fact, some of these systems have even removed the primary switching function from the distributor by using a crankshaft position sensor. In this case, the function of the distributor is to distribute secondary voltage to the spark plugs.

Spark timing on these systems is controlled by a computer that continuously varies ignition timing to obtain optimum air/fuel combustion. The computer mon-

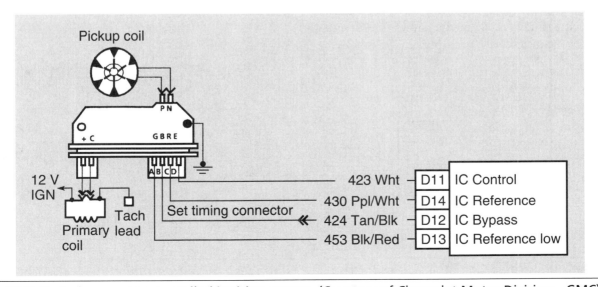

Figure 23-9 Typical computer-controlled ignition system. (Courtesy of Chevrolet Motor Division—GMC)

itors the engine operating parameters with sensors. Based on this input, the computer signals an ignition module to collapse the primary circuit, allowing the secondary circuit to fire the spark plugs (Figure 23-10).

Timing control is selected by the computer's program. During engine starting, computer control is bypassed and the mechanical setting of the distributor controls spark timing. Once the engine is started and running, spark timing is controlled by the computer. This scheme or strategy allows the engine to start regardless of whether the electronic control system is functioning properly or not.

The goal of computerized spark timing is to produce maximum engine power, top fuel efficiency, and minimum emissions levels during all types of operating conditions. The computer does this by continuously adjusting ignition timing. The computer determines the best spark timing based on certain engine operating conditions such as crankshaft position, engine speed, throttle position, engine coolant temperature, and initial and operating manifold or barometric pressure. Once the computer receives input from these and other sensors, it compares the existing operating conditions to information permanently stored or programmed into its memory. The computer matches the existing conditions to a set of conditions stored in its memory, determines proper timing setting, and sends a signal to the ignition module to fire the plugs.

The computer continuously monitors existing conditions, adjusting timing to match what its memory tells it is the ideal setting for those conditions. It can do this very quickly, making thousands of decisions in a single second. The control computer typically has the following types of information permanently programmed into it:

- Speed-related spark advance. As engine speed increases to a particular point, there is a need for more advanced timing. As the engine slows, the timing should be retarded or have less advance. The computer bases speed-related spark advance decisions on engine speed and signals from the TP sensor.

- Load-related spark advance. This is used to improve power and fuel economy during acceleration and heavy load conditions. The computer defines the load and the ideal spark advance by processing information from the TP sensor, MAP, and engine speed sensors. Typically, the more load on an engine, the less spark advance is needed.

- Warm-up spark advance. This is used when the engine is cold, since a greater amount of advance is required while the engine warms up.

- Special spark advance. This is used to improve fuel economy during steady driving conditions. During constant speed and load conditions, the engine will be more efficient with much advance timing.

- Spark advance due to barometric pressure. This is used when barometric pressure exceeds a preset calibrated value.

- Spark advance is used to control engine idle speed.

All of this information is looked at by the computer to determine the ideal spark timing for all conditions. The calibrated or programmed information in the computer is contained in what is called **software look-up tables** (Figure 23-11). In this three-dimensional map, arrows point in the directions of increased speed, load, and spark timing advance. The line intersections represent the spark advance for all combinations of load and speed. These values are stored digitally as a look-up table in the controller's memory.

Ignition timing also works in conjunction with the electronic fuel control system to provide emission control, optimum fuel economy, and improved driveability. They are all dependent on spark advance. An example of this type of system is used by Chrysler.

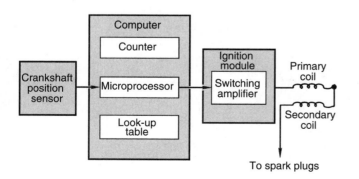

Figure 23-10 Simple diagram of a computer-controlled ignition system.

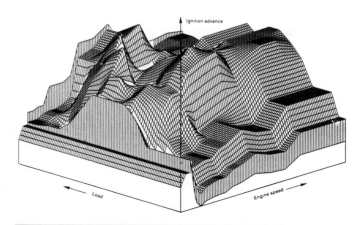

Figure 23-11 Ignition timing map. (Reprinted with permission from Robert Bosch Corporation)

Some Chrysler systems have two Hall-effect switches in the distributor when the engine is equipped with port fuel injection. In some units, the pickup unit used for ignition triggering is located above the pickup plate in the distributor and is referred to as the **reference pickup**. The second pickup unit is positioned below the plate. A ring with two notches is attached to the distributor shaft and rotates through the lower pickup unit. This lower pickup is called the **synchronizer (SYNC) pickup**.

In other designs, the two pickup units are mounted below the pickup plate, and one set of blades rotates through both Hall-effect unit . The shutter blade representing the number one cylinder has a large opening in the center of the blade. When this blade rotates through the SYNC pickup, a different signal is produced compared to the other blades. This number one blade signal informs the powertrain control module (PCM) when to activate the injectors.

Electronic Ignition System Operation

Electronic ignition (EI) systems (Figure 23-12) electronically perform the functions of a distributor. They control spark timing and advance in the same manner as the computer-controlled ignition systems. Yet the EI is a step beyond the computer-controlled system because it also distributes spark electronically instead of mechanically. The distributor is completely eliminated from these systems. When the system is working properly, there is no base timing to adjust and there are no moving parts to wear.

The development and spreading popularity of EI is the result of reduced emissions, improved fuel economy, and increased component reliability brought about by these systems. EI offers advantages in production costs and maintenance considerations. By removing the distributor, the manufacturers realize a substantial savings in ignition parts and related machining costs. By eliminating the distributor, they also do away with cracked caps, eroded carbon buttons, burned-through rotors, moisture misfiring, base timing adjustments, and the like.

The computer, ignition module, and position sensors combine to control spark timing and advance. The computer collects and processes information to determine the ideal amount of spark advance for the operating conditions. The ignition module uses crank/cam sensor data to control the timing of the primary circuit in the coils. Remember that there is more than one coil in a distributorless ignition system. The ignition module synchronizes the coils' firing sequence in relation to crankshaft position and firing order of the engine. Therefore, the ignition module takes the place of the distributor. This function is called **coil synchronizing**.

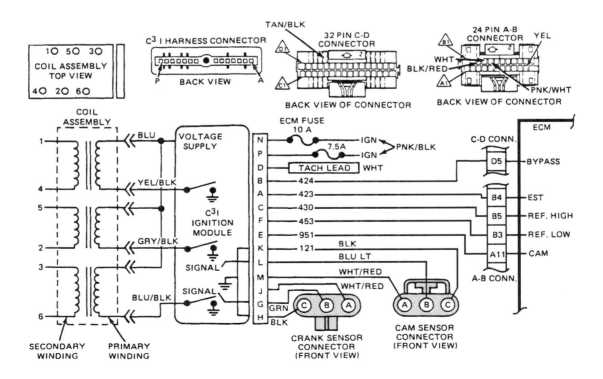

Figure 23-12 A typical electronic ignition (EI) system. (Courtesy of Buick Motor Division—GMC)

The ignition module also adjusts spark timing below 400 rpm (for starting) and when the vehicle's control computer bypass circuit becomes open or grounded. Depending on the exact EI system, the ignition coils can be serviced as a complete unit or separately. The coil assembly is typically called a **coil pack** and consists of two or more individual coils.

On those EI systems that use one coil per spark plug, the electronic ignition module determines when each spark plug should fire and controls the on/off time of each plug's coil. The systems with a coil for every two spark plugs also use an electronic ignition module, but they use the **waste spark** method of spark distribution. Each end of the coil's secondary winding is attached to a spark plug. Each coil is connected to a pair of spark plugs in cylinders whose pistons rise and fall together. When the field collapses in the coil, voltage is sent to both spark plugs that are attached to the coil. In all V-6s, the paired cylinders are 1 and 4, 2 and 5, and 3 and 6 (or 4 and 1 and 3 and 2 on 4-cylinder engines). With this arrangement, one cylinder of each pair is on its compression stroke while the other is on the exhaust stroke. Both cylinders get spark simultaneously, but only one spark generates power while the other is wasted out the exhaust. During the next revolution, the roles are reversed.

Due to the way the secondary coils are wired, when the induced voltage cuts across the primary and secondary windings of the coil, one plug fires in the normal direction (positive center electrode to negative side electrode) and the other plug fires just the reverse, side to center electrode (Figure 23-13). As shown in Figure 23-14, both plugs fire simultaneously, completing the series

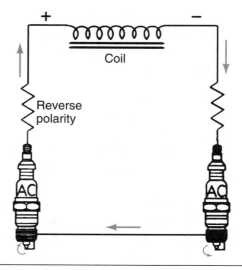

Figure 23-14 Complete circuit for spark plug firing in an EI ignition system.

circuit. Each plug always fires the same way on both the exhaust and compression strokes.

The coil is able to overcome the increased voltage requirements caused by reversed polarity and still fire two plugs simultaneously because each coil is capable of producing up to 100,000 volts. There is very little resistance across the plug gap on exhaust, so the plug requires very little voltage to fire, thereby providing its mate (the plug that is on compression) with plenty of available voltage.

Figure 23-15 shows a waste spark system in which the coils are mounted directly over the spark plugs so that no wiring between the coils and plugs is necessary. This type system operates in the same way as other EI systems. On other systems, the coil packs are mounted remote from the spark plugs. High-tension secondary wires carry high-voltage current from the coils to the plugs (Figure 24-16).

A few EI systems have one coil per cylinder with two spark plugs per cylinder. During starting, only one plug is fired. Once the engine is running, the other plug also fires. One spark plug is located on the intake side of the combustion chamber, while the other is located on the exhaust side. Two coil packs are used, one for the plugs on the intake side and the other for the plugs on the exhaust side. These systems are called **dual plug systems**. During dual plug operation, the two coil packs are synchronized so that each cylinder's two plugs fire at the same time. The coils fire two spark plugs at the same time. Therefore on a four-cylinder engine, four spark plugs are fired at a time, two during the compression stroke of the cylinder and two during the exhaust stroke of another cylinder.

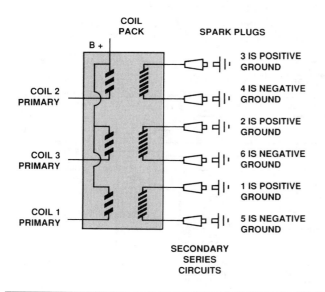

Figure 23-13 Spark plug firing for a six-cylinder engine with EI. (Reprinted with the permission of Ford Motor Company)

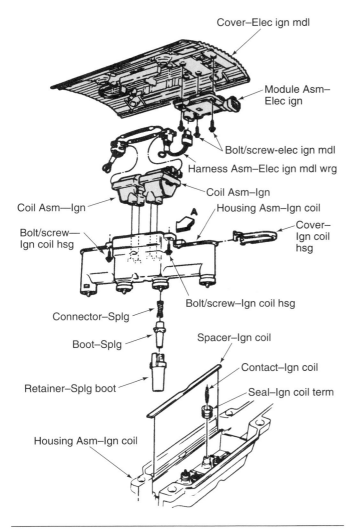

Figure 23-15 GM's Quad 4 with the ignition coils mounted directly over the spark plugs. (Courtesy of Oldsmobile Division—GMC)

Timing References

From a general operating standpoint, most distributorless ignition systems are similar. However, there are variations in the way different distributorless systems obtain a timing reference in regard to crankshaft and camshaft position.

Some engines use separate Hall-effect sensors to monitor crankshaft and camshaft position for the control of ignition and fuel injection firing orders. The crankshaft pulley has interrupter rings that are equal in number to half of the cylinders of the engine (Figure 23-17). The resultant signal informs the PCM as to when to fire the plugs. The camshaft sensor inputs determine when the number one piston is at TDC on the compression stroke.

Other systems use a dual Hall-effect sensor. One sensor generates three signals per crankshaft rotation, at 120° intervals. The other sensor generates one signal per

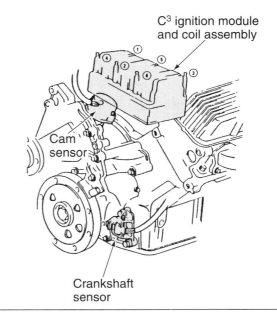

Figure 23-16 A coil pack. (Courtesy of Oldsmobile Division—GMC)

revolution, which tells the computer when the number one cylinder is on TDC. From these signals the computer can calculate the position of the camshaft, as well as know the position of the crankshaft.

Defining the different types of EI systems used by manufacturers focuses on the location and type of sensors used. There are other differences, such as the construction of the coil pack, wherein some are a sealed assembly and others have individually mounted ignition coils. Some EI systems have a camshaft sensor mounted in the opening where the distributor was mounted. The camshaft sensor ring has one notch and produces a leading edge and trailing edge signal once per camshaft revolution. These systems also use a crankshaft sensor. Both the camshaft and crankshaft sensors are Hall-effect sensors.

Some systems have the camshaft sensor mounted in the front of the timing chain cover. A magnet on the

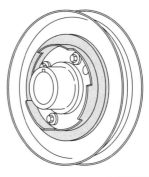

Figure 23-17 Crankshaft pulley with interrupter ring (Courtesy of Buick Motor Division, General Motors Corporation)

camshaft gear rotates past the inner end of the camshaft sensor and produces a signal for each camshaft revolution.

Other systems use a dual crankshaft sensor located behind the crankshaft pulley. When this type of sensor is used, there are two interrupter rings on the back of the pulley that rotate through the Hall-effect switches at the dual crankshaft sensor (Figure 23-18). The inner ring with three equally spaced blades rotates through the inner Hall-effect switch, whereas the outer ring with one opening rotates through the outer Hall-effect (Figure 23-19).

In this dual sensor, the inner sensor provides three leading edge signals, and the outer sensor produces one leading edge during one complete revolution of the crankshaft (Figure 23-20). The outer sensor is the SYNC sensor.

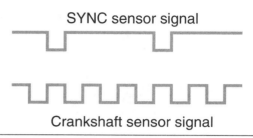

SYNC sensor signal

Crankshaft sensor signal

Figure 23-20 The signals generated by a dual crankshaft position sensor. (Courtesy of Oldsmobile Division—GMC)

Examples of EI Systems

Chrysler V-6 EI Systems

The crankshaft position sensor is mounted in an opening in the transaxle bell housing. The inner end of this sensor is positioned near a series of notches and slots that are integral with the transaxle drive plate.

A group of four slots is located on the transaxle drive plate for each pair of engine cylinders, and thus a total of 12 slots are positioned around the drive plate. The slots in each group are positioned 20° apart. When the slots on the transaxle drive plate rotate past the crankshaft timing sensor, the voltage signal from the sensor changes from 0V to 5V. This varying voltage signal informs the PCM regarding crankshaft position and speed, and the PCM calculates spark advance from this signal. The slots are spaced so the first rise from 0 volt to 5 volts is 69° BTDC. The next is 49° BTDC, then 29° BTDC and the final slot is 9° BTDC. The PCM also uses the crankshaft position sensor signal along with other inputs to determine air/fuel ratio. Base timing is determined by the signal from the 9° slot in each group of slots, and base timing adjustment is not possible.

The camshaft reference sensor is mounted in the top of the timing gear cover. A notched ring on the camshaft gear rotates past the end of the camshaft reference sensor. This ring contains two single slots, two double slots, a triple slot, and an area with no slots.

When a camshaft gear notch rotates past the camshaft reference sensor, the signal from the sensor changes from 0V to 5V. The single, double, and triple notches provide different voltage signals from the camshaft reference sensor as they rotate past the sensor, and these signals are sent to the PCM. Since the ignition module is part of the PCM board, an external module is not used on these systems. The PCM determines the exact camshaft and crankshaft position from the camshaft reference sensor signals, and the PCM uses these signals to sequence the coil primary windings and each pair of injectors at the correct instant.

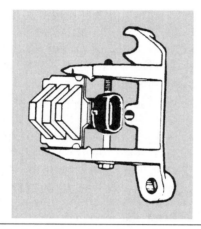

Figure 23-18 A dual crankshaft position sensor. (Courtesy of Oldsmobile Division—GMC)

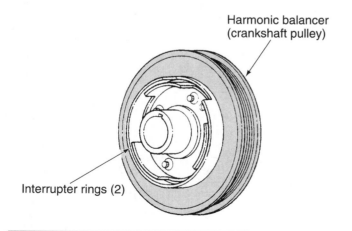

Harmonic balancer (crankshaft pulley)

Interrupter rings (2)

Figure 23-19 A crankshaft pulley fitted with two sets of interrupter rings for a dual sensor. (Courtesy of Oldsmobile Division—GMC)

The PCM is able to make these calculations within one engine revolution during engine starting. While observing both the camshaft and crankshaft position sensors, the PCM determines which cylinder is approaching TDC compression stroke. When the PCM sees one group of slots from the crankshaft position sensor, followed by three slots (for example) from the camshaft position sensor, it knows cylinder number 4 is approaching TDC. If the PCM sees one group of slots from the crankshaft position sensor, followed by no signal from the camshaft position sensor (for example), it knows cylinder number 1 is approaching TDC (Figure 23-21). The PCM is in synchronization once it recognizes number 1 or number 4 piston.

The PCM supplies 8.0 to 9.0 volts from terminal 7 through an orange wire to both the crankshaft timing sensor and the camshaft reference sensor. A black ground wire is connected from both of these sensors to PCM terminal 4. The camshaft reference sensor signal is connected to PCM terminal 44, whereas the crankshaft timing sensor signal is sent to terminal 24 on the PCM.

The coil assembly contains three ignition coils. Two spark plug wires are connected from the spark plugs to the secondary terminals on each coil. In each coil, the ends of the secondary winding are connected to the two secondary terminals on that coil. The PCM supplies 12 volts to one end of the primary windings in each coil when the ignition switch is turned on. The other end of each primary winding is connected to the PCM.

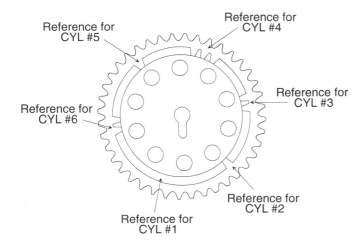

Reference for
CYL #5

Reference for
CYL #4

Reference for
CYL #3

Reference for
CYL #6

Reference for
CYL #2

Reference for
CYL #1

Figure 23-21 Camshaft and crankshaft position sensors signal pattern allows the PCM to synchronize. (Courtesy of Chrysler Corporation)

System Operation

When the engine starts cranking, the spark plugs fire and the injectors discharge fuel within one crankshaft revolution. The PCM determines when to sequence the coils and injectors from the camshaft reference sensor signals. If the camshaft reference sensor or the crankshaft timing sensor are defective, the engine will not start. Each coil fires two spark plugs at the same instant, and the current flows down through one spark plug and up through the other spark plug. One of the cylinders in which a spark plug is firing is on the compression stroke, while the other cylinder is on the exhaust stroke. When the engine is cranking, all spark plug firings are at 9° BTDC on the compression stroke. The PCM also fires all injectors one time. Once the PCM synchronizes the CMP and CKP sensors, it begins firing the appropriate coils.

The spark plug wires from coil number 1 are connected to cylinders 1 and 4, whereas the spark plug wires from coil number 2 go to cylinders 2 and 5, and the spark plug wires on coil number 3 are attached to cylinders 3 and 6. The cylinder firing order for the 3.3-L V-6 engine is 1-2-3-4-5-6.

Once the engine is started, the PCM knows the exact crankshaft position and speed from the camshaft reference sensor and crankshaft timing sensor signals. The leading edges of the slots in the transaxle drive plate rotate past the crankshaft timing sensor at 69, 49, 29, and 9 degrees BTDC. Since the PCM synchronized the CMP and CKP sensors it is now able to recognize the order from the CMP sensor. The CMP sends a series of signals of one high, two high, three high, one high, two high, none. When these signals are received, the PCM is programmed to fire the appropriate coil (Figure 23-22).

With the engine running, the PCM determines the precise spark advance required when the next cylinder fires. This precise spark advance is provided when the PCM opens the primary circuit on the appropriate coil at the right instant.

Since Chrysler engines are now equipped with sequential fuel injection (SFI), the PCM grounds each injector individually, but the proper injector sequencing is determined from the camshaft reference sensor signals.

CAUTION *Since EI systems have considerably higher maximum secondary voltage compared to point-type or electronic ignition systems, greater electrical shocks are obtained from EI systems. Although such shocks may not be directly harmful to the human body, they may cause you to jump or react suddenly, which could result in personal injury.*

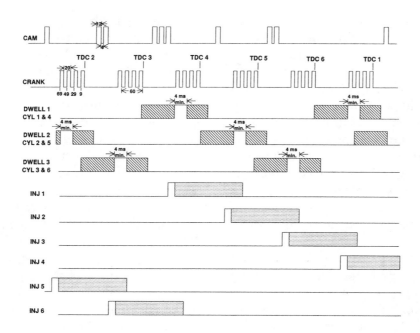

Figure 23-22 Coil firing in response to camshaft and crankshaft position sensor input. (Courtesy Chrysler Corporation)

Summary

❑ Ignition timing is directly related to the position of the crankshaft. Magnetic pulse generators and Hall-effect sensors are the most widely used engine position sensors. They generate an electrical signal at certain times during crankshaft rotation. This signal triggers the electronic switching device to control ignition timing.

❑ EI systems provide longer spark duration at the spark plug electrodes than conventional electronic ignition systems. This helps to fire leaner air/fuel ratios in today's engines.

❑ Compared to electronic ignition systems with distributors, EI systems provide more stable control of ignition timing and spark advance, which reduces emissions and improves fuel economy.

❑ Direct ignition systems eliminate the distributor. Each spark plug, or in some cases, pair of spark plugs, has its own ignition coil. Primary circuit switching and timing control are done using a special ignition module tied into the vehicle control computer.

❑ Computer-controlled ignition eliminates centrifugal and vacuum timing mechanisms. The computer receives input from numerous sensors. Based on this data, the computer determines the optimum firing time and signals an ignition module to activate the secondary circuit at the precise time needed.

Terms-To-Know

3X signal	Electronic engine control (EEC)	Slow-start
18X signal	Fast-start	Software look-up tables
Camshaft position sensor (CMP)	Ignition control module (ICM)	Spark output (SPOUT)
Coil pack	Ignition diagnostic monitor	Synchronizer (SYNC) pickup
Coil peak current cutoff	(IDM)	Thick-film integrated IV
Coil synchronizing	Metal detection sensors	(TFI-IV)
Crankshaft position sensor (CKP)	Photoelectric sensor	Three-coil ignition (C3I)
DIS module	Profile ignition pickup (PIP)	Variable reluctance sensor (VRS)
Dual plug systems	Reference pickup	Waste spark

Review Questions

Short Answer Essay

1. Describe the coil secondary-to-spark plug wiring connections on an EI system.

2. Explain the purpose of the camshaft position sensor in an EI system.

3. List the different types of crankshaft sensors.

4. Explain the purpose of the SYNC pickup in an electronic ignition system with computer-controlled spark advance and two distributor pickups.

5. What is the purpose of the knock sensor.

6. Describe the advantages of an EI system.

7. Describe how each pair of spark plugs is fired in an EI system.

8. Explain how the PCM fires the spark plugs in the proper firing order in a Chrysler EI system.

9. How do EI ignition systems differ from conventional electronic ignition systems?

10. Explain why a distributorless ignition system has more than one ignition coil.

Fill-in-the-Blanks

1. In an EI system coil, two spark plug wires are connected to the ends of each _____ winding.

2. In an EI system coil, 12V are supplied to each primary coil winding, and the other end of each primary winding is connected to the _____ .

3. EI systems help to prevent cylinder misfiring because these systems have a longer _____ _____ .

4. On EI systems, if the crankshaft sensor signal is defective, the engine will not _____ .

5. Computer-controlled ignition systems rely on the inputs from various _____ to control ignition timing.

6. The calibrated or programmed information in the computer concerning spark advance is contained in what is called _____ _____ _____.

7. The knock sensor changes sensed vibration to a _____ signal.

8. On EI systems that use a coil for everyt two spark plugs use the _____ _____ method of spark distribution.

9. A few EI systems have one coil per cylinder with two spark plugs per cylinder. Two coil packs are used, one for the plugs on the intake side and the other for the plugs on the exhaust side. These systems are called

   _____ _____ _____ .

10. The type of crankshaft position sensor that uses LED's is called a _____ .

ASE Style Review Questions

1. Technician A says EI systems typically use one coil for two spark plugs. Technician B says some EI systems rely on a waste spark system to fire the spark plug. Who is correct?
   a. A only
   b. B only
   c. Both A and
   d. Neither A nor B

2. Technician A says metal detection sensors use a premanent magnet system. Technician B says the Hall-effect switch cannot be used on EI systems. Who is correct?
   a. A only
   b. B only
   c. Both A and B
   d. Neither A nor B

3. Technician A says waste spark refers to a system that fires two plugs at the same time. Technician B says waste spark systems have a spark plug attached to each end of the secondary coil. Who is correct?
   a. A only
   b. B only
   c. Both A and B
   d. Neither A nor B

4. In EI systems using one ignition coil for every two cylinders. Technician A says two plugs fire at the same time, with one wasting the spark on the exhaust stroke. Technician B says one plug fires in the normal direction (center to side electrode) and the other in reversed polarity (side to center electrode). Who is correct?
   a. A only
   b. B only
   c. Both A and B
   d. Neither A nor B

5. Technician A says a magnetic pulse generator uses a permanent magnet. Technician B says a Hall-effect switch is equipped with an electromagnet. Who is correct?
   a. A only
   b. B only
   c. Both A and B
   d. Neither A nor B

6. PM generators are being discussed. Technician A says the pickup coil does not produce a voltage signal when a reluctor tooth approaches the coil. Technician B says the pickup coil does not produce a voltage signal when a reluctor tooth moves away from the coil. Who is correct?
   a. A only
   b. B only
   c. Both A and B
   d. Neither A nor B

7. Electronic ignition (EI) systems are being discussed. Technician A says when a pair of spark plugs is firing, one of the cylinders is on the exhaust stroke and the other cylinder is on the power stroke. Technician B says each pair of spark plugs is fired in series. Who is correct?
   a. A only
   b. B only
   c. Both A and B
   d. Neither A nor B

8. EI systems are being discussed. Technician A says the basic ignition timing is not adjustable on most EI systems. Technician B says the crankshaft sensor may be moved to adjust basic ignition timing on an EI system. Who is correct?
   a. A only
   b. B only
   c. Both A and B
   d. Neither A nor B

9. The Chrysler V-6 EI system is being discussed. Technician A says when the engine is cranking, the spark plugs are fired at 9 degrees before TDC. Technician B says it takes four engine revolutions before the camshaft and crankshaft position sensors come into sync. Who is correct?
   a. A only
   b. B only
   c. Both A and B
   d. Neither A nor B

10. Technician A says computer-controlled ignition systems use computer controls to change spark advance according to engine speed, but use a vacuum system on the distributor to measure engine load. Technician B says computer-controlled ignition systems use a series of sensors and look-up tables to determine ideal spark timing. Who is correct?
    a. A only
    b. B only
    c. Both A and B
    d. Neither A nor B

24 Computer-Controlled Ignition System Diagnosis

Objectives

Upon completion and review of this chapter, you should be able to:

❏ Use symptoms to identify and locate probable ignition system problem areas.

❏ Test individual DI and EI system components using a lab scope or DMM.

❏ Perform a no-start diagnosis, and determine the cause of the no-start condition.

❏ Perform pickup tests on distributor ignition (DI) systems.

❏ Perform tests on optical-type pickups.

❏ Perform no-start ignition tests on the cam and crankshaft sensors on electronic ignition (EI) systems.

❏ Perform no-start ignition tests on the coil and powertrain control module (PCM) on EI systems.

❏ Replace cam and crankshaft sensors on EI systems.

❏ Perform coil tests on EI systems.

❏ Adjust crankshaft sensors on EI systems.

❏ Install and time the cam sensor on an EI system, 3.8-L turbocharged engine.

❏ Perform magnetic sensor tests on EI systems.

❏ Perform no-start tests on EI systems with magnetic sensors.

❏ Diagnose engine misfiring on EI-equipped engines.

Introduction

When you are working on a vehicle with electronic engine controls, you need to think about how the computer processes information. Most often the computer thinks the way you should when diagnosing problems. In order to control an engine system, the computer makes a series of decisions. Decisions are made in a step-by-step fashion until a conclusion is reached. Generally, the first decision is to determine the engine mode. For example, to control spark timing the computer first determines whether the engine is cranking, idling, cruising, or accelerating. Then, the computer can choose the best system strategy for the present engine mode. In a typical example, sensor inputs indicate the engine temperature, rpm, manifold absolute pressure, and the throttle plate opening. The computer determines the engine load then it determines the goal to be reached. In a final series of decisions, the computer determines how the goal can be achieved.

If a sensor input is missing or is corrupted, the decisions made will reflect the lack of information. Testing procedures of the input sensors are typically the same between the different manufacturers. However, it is always recommended that you use the correct service manual to determine the expected values. If all of the input sensors test good, then move to the outputs. Finally, if all inputs and outputs are good the module is suspect. However, never replace a module until connects, power, and grounds have been tested and confirmed good.

Isolating Computerized Engine Control Problems

Regardless of the type of computerized ignition system used (DI or EI), determining the defective part or area of the system requires a thorough knowledge of how the system works and following a logical troubleshooting process.

There are many variations in the operation of electronic engine controls. Therefore, always check the appropriate factory service manual before beginning any diagnosis and repair. Here are some guidelines to follow:

1. Check all mechanical and non-computer controlled systems before progressing into a computer system check. Quite often sensors are working properly, but their signal is being affected by a non-computer system component.

2. Electronic engine control problems are usually caused by defective sensors and, to a lesser extent, output devices. The logical procedure in most cases is, therefore, to check the input sensors and wiring first, then the output devices and their wiring, and, finally the computer.

3. Most late-model computerized engine controls have self-diagnostic capabilities. A malfunction in any sensor, output device, or in the computer itself is stored in the computer's memory as a trouble code. Stored codes can be retrieved and the indicated problem areas are then checked.

4. Five methods are used to check individual components: visual, ohmmeter, voltmeter, lab scope, and scan tool checks.

5. Most sensors and output devices can be checked with an ohmmeter. For example, Figure 24-1 shows an ohmmeter used to check a temperature sensor. Normally the ohmmeter reading is low on a cold engine and high or infinity on a hot engine if the sensor is a PTC. If the sensor is an NTC, the opposite readings would be expected. Output devices such as solenoids or motors can also be checked with an ohmmeter.

6. Many sensors, output devices, and their wiring can be diagnosed by checking the voltage to them, and in some cases, from them. Even some oxygen sensors can be checked in this manner. All oxygen sensors, as well as other sensors, can be checked with a lab scope. Watching voltage over time will give you a clear view of how they are working. A scan tool allows you to watch sensor and output device activity.

7. In some cases, a final check on the computer can be made only by substitution. Also some vacuum, MAP,

and barometric pressure sensors can only be checked by substitution. Substitution is not an allowable diagnostic method under the mandates of OBD-II. Nor is it the most desirable way to diagnose problems. To substitute, however, replace the suspected part with a known good unit and recheck the system. If the system now operates normally, the original part was defective.

Self-Diagnostic Systems

Most DI and EI ignition systems have **onboard diagnostics**. These systems test the input sensors and actuators for continuity and in some case for functionality. By entering into a self-test mode, the computer is able to evaluate the condition of the entire electronic engine control system, including itself. If problems are found, they are identified as either hard faults (on-demand) or intermittent failures. Each type of fault or failure is assigned a trouble code that is stored in computer memory. Figure 24-2 shows an example of trouble codes from pre-OBD-II equipped vehicles. Onboard diagnostics second generation (OBD-II) vehicles use a new set of trouble codes. OBD-II is discussed in greater detail in a later chapter.

A hard fault is also referred to as a current fault and means a problem has been found somewhere in the system at the time of the self-test. An intermittent or history fault, on the other hand, indicates a malfunction occurred (for example, a poor connection causing an intermittent open or short), but is not present at the time of the self-test. Nonvolatile RAM allows intermittent faults to be stored for up to a specific number of ignition key on/off cycles. If the trouble does not reappear during that period, it is erased from the computer's memory.

There are various methods of assessing the trouble codes generated by the computer. Most manufacturers

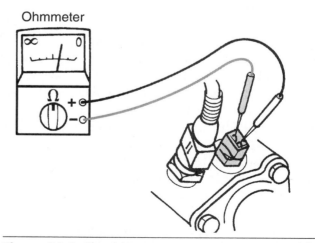

Ohmmeter

Figure 24-1 Checking a temperature sensor with an ohmmeter. (Courtesy of Toyota Motors)

SENSOR TROUBLE CODES

Sensor	Trouble Codes		
	GM	Ford	Chrysler
Oxygen (EGO)	13,44, 45,55	43,91, 92,93	21,51,52
Throttle Position (TPS)	21,22	23,53, 63,73	24
Engine Vacuum (MAP)	31,33,34	22,72	13,14
Barometric Pressure (BARO)	32	—	37
Coolant (ECT)	14,15	21,51,61	17,22
Knock	42,43	25	17 (some only)
Vehicle Speed (VSS)	24	—	15
Air Temperature (MAT, VAT, ACT)	23,25	24,54,64	23
Airflow (VAF, MAF)	33,34, 44,45	26,56, 66,76	—
EGR Valve (EGR, EVP)	—	31,32,33, 34,83,84	31

Figure 24-2 Different manufacturers use different numerical codes to indicate problems in input and output circuits. (Courtesy of *Counterman* Magazine)

have diagnostic equipment designed to monitor and test the electronic components of their vehicles. Aftermarket companies also manufacture scan tools that have the capability to read and record the input and output signals passing to and from the computer. Some vehicles also flash trouble codes on dash lights or display them on CRT screens.

For each individual make and model, it is important to check the appropriate service manual to find out how trouble codes should be accessed and how to interpret the codes. The following sections give several examples of retrieving trouble codes, each using a different method.

Before reading self-diagnostic or trouble codes, do a visual check to make sure the problem is not a result of wear, loose connections, or vacuum hoses. Inspect the air cleaner, throttle body, or injection system. Do not forget the PCV system and vacuum hoses. Make sure the vapor canister is not flooded. Inspect all wiring harnesses and connectors. Today's electronic circuits cannot tolerate the increased resistance caused by corrosion in connectors or wires. Also, make sure the charging system is working properly since low system voltage will cause erroneous fault codes.

General Motors' Systems

The main components of General Motors' onboard diagnostic system include the PCM, a MIL, and a data link connector (DLC). The MIL serves two purposes (Figure 24-3). First, it is a signal to the driver that a detectable system failure has occurred. Second, it can also be used by a technician as an aid to identify any problems in the system.

Each time the ignition key is turned to the on position, the system does a self-check. The self-check makes sure that all of the bulbs, fuses, and electronic modules are working. If the self-test finds a problem, it might store a code for later servicing. It may also instruct the computer to turn on the MIL to show that service is needed.

To find the problem, you must perform a diagnostic circuit check to be sure the diagnostic system is working. Turn the ignition key on, but do not start the engine. If the check engine light does not come on, follow the diagnostic chart in the service manual. If the light comes on, ground the diagnostic test terminal. This terminal is located on the DLC connector (Figure 24-4). Connect pin A (A ground) to pin B (test terminal). Watch the MIL to see if it displays code 12. Code 12 is one flash, a short pause and two flashes (Figure 24-5). If the computer's diagnostic program is working properly, it flashes code 12 three times. If code 12 does not appear, follow the diagnostic chart in the manual. The system then displays any fault codes in the same manner.

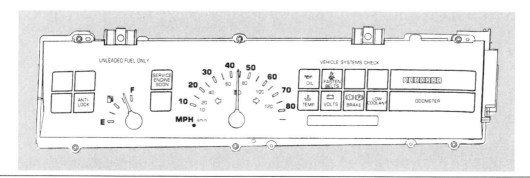

Figure 24-3 General Motors' CCC warning light system.

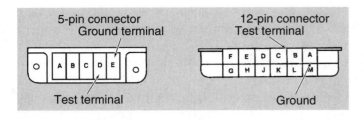

Figure 24-4 To initiate self-diagnostics, connect the ground and test terminals on the DLC connector.

Codes are identified through the flashing MIL as follows. The number of times the light flashes on before it pauses tells the number to which it refers. In other words, if the MIL flashes like this: "on-off-on-off-on-off, pause," that means number 3 because it was on three times before it paused. If it flashed "on-off-on-off, pause," that means number 2. If it flashes "on-off, pause," that means number 1.

Because each code has two numbers to it, such as 12, 23, or 32, the flashes are read as a set. This means you always see two series of flashes separated by a pause: first number, pause, second number. If the PCM has stored more than one trouble code, the codes flash starting with the lowest number code first and then work their way to the highest number code. For example, if the codes were 13 and 24, then 13 would flash first, then 24 would flash.

If the PCM is good, the first code that flashes when the connector is grounded and the key turned on is going to be 12. This is the tachometer code. It is a normal condition when the key is on but the engine is not running. If a code 12 is not received, then there is something wrong in the DLC, or related circuits.

Once the codes have been identified, check the service manual (Figure 24-6). Each code directs you to a specific troubleshooting tree. If you have a code, follow

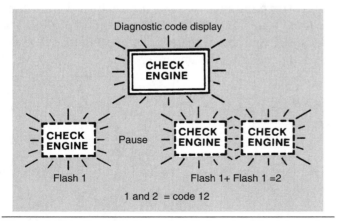

Figure 24-5 The check engine light signals the codes. A code 12 signals that the computer's diagnostic program is working properly. (Courtesy of General Motors Corporation)

the tree exactly. Never try to skip a step, or you'll be certain to miss the problem.

A hand-held scan tool can be used by itself to draw codes or in conjunction with MIL. Most scan tools will display more information than just codes; this is their advantage. When using a scan tool, follow the tool manufacturer's directions for all testing.

After correcting all faults or problems, clear the PCM memory of any current codes by pulling the PCM fuse at the fuse panel for 10 seconds. Then, to make sure the codes have cleared, remove the test terminal ground, and set the parking brake. Put the transmission in park, and run the engine for 2 to 5 minutes. Watch the check engine light. If the light comes on again, ground the test lead and note the flashing trouble code. If no light comes on, check the codes recorded earlier, if any. It is a good idea to recheck the problem on a road test. Some parts such as the vehicle speed sensor do not show any problems with the engine just idling.

Chrysler's System

At the heart of early Chrysler computer-controlled systems is a digital microprocessor known as the logic module (LM). From its location behind the right front kick pad (in the passenger compartment), the LM issues commands affecting fuel delivery, ignition timing, idle speed, and the operation of various emission control devices.

The logic module operates in conjunction with a subordinate control unit called the power module (PM). The power module, located inside the engine compartment, has the responsibility of controlling the injector and ignition coil ground circuit (based on the LM's commands). It is also in charge of supplying the ground to the **automatic shutdown (ASD)** relay. The ASD relay controls the voltage supply to the fuel pump, logic module, injectors, and coil drive circuits and is energized through the PM when the ignition switch is turned on.

Current Chrysler products use a single unit that consists of the logic and power modules. This unit is called the Single Board Electronic Controller (SBEC). On these systems, the LM and PM function as a single unit. However, each of these still has the same primary purpose as it did when they were in separate units. During normal operating conditions, the LM bases its decisions on information from several input devices (Figure 24-7).

Chrysler has a simple method for checking trouble codes. Without starting the engine, turn the ignition key from OFF to RUN three times within 5 seconds, ending with it in the RUN position. The MIL, check engine light, or power loss light glows a short time to test the bulb, then starts flashing. Count the first set of flashes as tens. There is a half second pause before the light starts

DIAGNOSTIC CODE IDENTIFICATION

The "Service Engine Soon" light will only be "ON" if the malfunction exists under the conditions listed below. If the malfunction clears, the light will go out and the code will be stored in the ECM/PCM. Any codes stored will be erased if no problem reoccurs within 50 engine starts.

CODE AND CIRCUIT	PROBABLE CAUSE	CODE AND CIRCUIT	PROBABLE CAUSE
Code 13 - Oxygen O_2 Sensor Circuit (Open Circuit)	Indicates that the oxygen sensor circuit or sensor was open for one minute while off idle.	Code 33 - Manifold Absolute Pressure (MAP) Sensor Circuit (Signal Voltage High- Low Vacuum)	MAP sensor output to high for 5 seconds or an open signal circuit.
Code 14 - Coolant Temperature Sensor (CTS) Circuit (High Temperature Indicated)	Sets if the sensor or signal line becomes grounded for 3 seconds.	Code 34 - Manifold Absolute Pressure (MAP) Sensor Circuit (Signal Voltage Low-High Vacuum)	Low or no output from sensor with engine running.
Code 15 - Coolant Temperature Sensor (CTS) Circuit (Low Temperature Indicated)	Sets if the sensor, connections, or wires open for 3 seconds.	Code 35 - Idle Air Control (IAC) System	IAC error
Code 21 - Throttle Position Sensor (TPS) Circuit (Signal Voltage High)	TPS voltage greater than 2.5 volts for 3 seconds with less than 1200 RPM.	Code 42 - Electronic Spark Timing (EST)	ECM/PCM has seen an open or grounded EST or bypass circuit.
Code 22 - Throttle Position Sensor (TPS) Circuit (Signal Voltage Low)	A shorted to ground or open signal circuit will set code in 3 seconds.	Code 43 - Electronic Spark Control (ESC) Circuit	Signal to the ECM/PCM has remained low for too long or the system has failed a functional check.
Code 23 - Intake Air Temperature (IAT) Sensor Circuit (Low Temperature Indicated)	Sets if the sensor, connections, or wires open for 3 seconds.	Code 44 - Oxygen (O_2) Sensor Circuit (Lean Exhaust Indicated)	Sets if oxygen sensor voltage remains below .2 volt for about 20 seconds.
Code 24 - Vehicle Speed Sensor (VSS)	No vehicle speed present during a road load decel.	Code 45 - Oxygen (O_2) Sensor Circuit (Rich Exhaust Indicated)	Sets if oxygen sensor voltage remains above .7 volt for about 1 minute.
Code 25 - Intake Air Temperature (IAT) Sensor Circuit (High Temperature Indicated)	Sets if the sensor or signal line becomes grounded for 3 seconds.	Code 51 - Faulty MEM-CAL or PROM Problem	Faulty MEM-CAL, PROM, or ECM/PCM.
Code 32 - Exhaust Gas Recirculation (EGR) System	Vacuum switch shorted to ground on start up OR Switch not closed after the ECM/PCM has commanded EGR for a specified period of time. OR EGR solenoid circuit open for a specified period of time.	Code 52 - Fuel CALPAK Missing	Fuel CAL-PAK missing or faulty.
		Code 53 - System Over Voltage	System overvoltage. Indicates a basic generator problem.
		Code 54 - Fuel Pump Circuit (Low Voltage)	Sets when the fuel pump voltage is less than 2 volts when reference pulses are being received.
		Code 55 - Faulty ECM/PCM	Faulty ECM/PCM

Figure 24-6 These are typical GM trouble codes listed with their probable causes. (Courtesy of Chevrolet Motor Division—GMC)

flashing again. This time count by ones. Add the two sets of flashes together to obtain the trouble code. For example, 3 flashes, a half second pause, followed by 5 flashes would be read as code 35. Watch carefully, because each trouble code is displayed only once. Look the code up in the service manual. This tells you which circuit to check. Once the trouble codes are flashed, the computer will flash a code 55. If the light does not flash at all, there are no trouble codes stored.

The same diagnostic or trouble code tests can be done using a scan tool. Connect the tool to the diagnostic con-

nectors on the left fender apron. Follow the same sequence with the ignition key. The trouble codes are displayed on the scan tool readout. You can also check circuits, switches, and relays with the scan tool.

Ford's System

The main computer of Ford's EEC-IV system is called an electronic control assembly (ECA) or PCM. Like the other computer-controlled systems, the PCM has self-diagnostic capabilities. By entering a mode known as self-test (Figure 24-8), the computer is able to

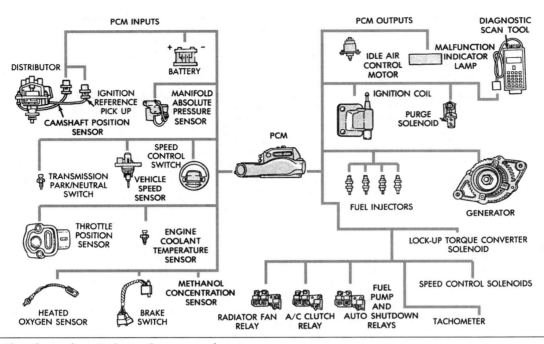

Figure 24-7 Chrysler's electronic engine control system.

evaluate the condition of the entire electronic system, including itself. If problems are found, a code will be displayed as either hard faults or intermittent failures.

On 1988 and newer model vehicles, the diagnostic or trouble codes are displayed by the MIL on the instrument panel. Since 1994, Ford has equipped some models with the EEC-V system, which is an OBD-II compliant system. This system does not display codes on the MIL.

The Ford self-test diagnostic procedure can be broken down into four parts:

1. Key on/Engine off (KOEO): checks system inputs for hard faults and intermittent faults.

2. Computed ignition timing check: determines the ECA's ability to advance or retard ignition timing. This check is made while the self-test is activated and the engine is running.

3. Key on/Engine running segment (KOER): checks system output for hard faults only.

4. Continuous monitoring test (wiggle test): allows the technician to look for and set intermittent faults while the engine is running.

Within these four tests, there are six types of service codes. There are on-demand codes, keep-alive memory codes, separator codes, dynamic response codes, fast codes, and engine ID codes. To understand what you are dealing with when the codes start to display, a brief explanation of each type is necessary.

On-demand codes are used to identify hard faults. A hard fault means that a system failure has been detected and is still present at the time of testing. The term on-demand simply means a technician is asking the computer if a problem exists right now while it is running its self-test. Keep-alive memory codes mean that a malfunction was noted sometime during the last 20 vehicle warm-ups but is not present now. The continuous memory code comes on after an approximate ten second delay. Make the on-demand code repairs first. Once you have completed repairs, repeat the key on, engine off test. If all the parts are repaired correctly, a pass code of 11 should be received.

A separator code (10) indicates that the on-demand codes are over, and the memory codes are about to begin. When a code 10 appears during the engine running seg-

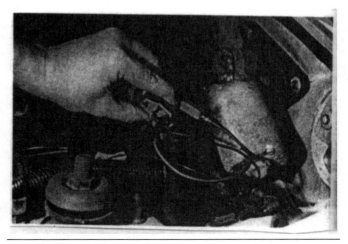

Figure 24-8 Ford's self-test connectors are used to connect test equipment to the system's computer.

ment of the self-test, it is referred to as a dynamic response code. The dynamic response code is a signal to the technician to goose the throttle momentarily so that the ECA can verify the operation of the throttle position (TP) and manifold absolute pressure (MAP) sensors. On some models, the technician must goose the throttle, step on the brake pedal, and turn the steering wheel 180 degrees. Failure to respond to the dynamic code within 15 seconds after it appears sets a code 77 (operator did not respond).

Fast codes are designed for factory use and are transmitted about 100 times faster than even a scan or Ford's Star tester can read. On a voltmeter, fast codes cause the needle to rapidly pulse between zero and three volts. On the scan tool, the LED light flickers. Although fast codes have no practical use in the service bay, pay attention to when they occur so you know what is coming next. Fast codes appear twice during the entire self-test: once at the very beginning of the key on/engine off test (right before the on-demand codes) and again after the dynamic response code (prior to hard fault transmission).

Engine identification codes are used to tell automated assembly line equipment how many cylinders the engine has. Two needle pulses indicate a four-cylinder, three pulses a six-cylinder, and four pulses identify the engine as an eight-cylinder model. Engine ID codes appear at the beginning of the engine-running segment only.

Nissan Systems

Late model Nissan vehicles are equipped with an OBD-II standardized diagnostic system. However, they have also retained the earlier self-diagnostic. Entering self-diagnostics without the use of a scan tool is accomplished by using a screwdriver activated switch in the computer. The computer is located under the front passenger seat on earlier systems. Later year models have the computer located behind the instrument panel near the glove box. Diagnostic codes are flashed by the use of an LED on the side of the computer. On later models, the check engine light is used to flash the codes.

EI System Service

Standard test procedures using an oscilloscope, ohmmeter, and timing light can be used to diagnose problems in distributorless ignition systems. Keep in mind, however, that problems involving one cylinder may also occur in its companion cylinder that fires off the same coil. Some oscilloscopes require their pickups placed on each pair of cylinders to view all patterns. Special adapters are available to make these hook-ups less troublesome. Many newer engine analyzers with an oscillo-

scope use a single adapter that allows viewing of all cylinder patterns at one time.

SERVICE TIP *According to the SAE J1930 standards, the term "distributor ignition" (DI) replaces all previous terms for distributor-type ignition systems that are electronically controlled. Also according to this mandate, the term "electronic ignition" (EI) replaces all previous terms for distributorless ignition systems.*

Follow the testing procedures outlined in the vehicle service manual for EI systems and other computer-controlled ignition systems. Specific computer-generated trouble codes are designed to help troubleshoot ignition problems on computer-controlled systems. The diagnostic procedure for EI systems varies depending on the vehicle make and model year. Always follow the procedure recommended in the vehicle's service manual.

On distributorless or EI systems, visually inspect the secondary wiring connections at the individual coil modules. If a plug wire is loose, inspect the terminal for signs of burning. Check for evidence of terminal resistance. Separate the coils and inspect the underside of the coil and the ignition module wires (Figure 24-9). A loose or damaged wire or a bad plug can lead to carbon tracking of the coil. If this condition exists, the coil must be replaced.

The coils in an EI system can be tested in much the same manner as conventional coils. When checking the resistance across the windings, pay particular attention to the meter reading. If the reading is out of specifications, even if it is only slightly out, the coil or coil assembly should be replaced. Steps for the removal and inspection of coil packs and the ignition module are illustrated in Photo Sequence 10, which is included in this chapter.

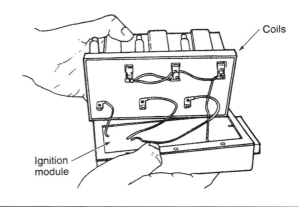

Figure 24-9 On distributorless dystems, check the wiring linking the ignition coils to the ignition module.

Removing and Replacing Various DIS Components

P9-1 Late-model engines are equipped with different components than older models and are not equipped with some that were used for many years.

P9-2 This coil pack replaces the ignition coil and distributor assembly.

P9-3 The coil pack is removed by first disconnecting the spark plug wires from the coils. Prior to doing this, make sure their exact location is marked.

P9-4 Unbolt the assembly from the engine block.

P9-5 The individual coils are bolted to the mounting plate that also houses the ignition module. If a coil or the module is to be replaced, unbolt it from the assembly.

 # Removing and Replacing Various DIS Components (cont'd)

P9-6 Often, service of the assembly is done with the unit totally disconnected from the engine. To do this, loosen the remaining bolt on the electrical connector and pull the connector off the assembly. It is now free from the engine.

P9-7 Another integral part of the DIS system is the throttle position sensor (TPS).

P9-8 To remove this sensor and many other similar components, disconnect the electrical connector and loosen the assembly retaining bolts.

P9-9 Spark plugs are removed in the normal manner with a spark plug socket and rachet.

The specifications for EI coils vary with manufacturer and design. For GM coils, the resistance in the primary of the coil should be between 0.35 to 1.50 ohms. A reading below this indicates a shorted winding and an infinite reading indicates an open. The resistance of the secondary windings varies with system design. The winding in type 1 systems should have 10,000 to 14,000 ohms resistance, whereas the windings in type 2 systems should have 5,000 to 7,000 ohms resistance.

Electromagnetic Interference

Electromagnetic interference (EMI) or radio frequency interference (RFI) can cause problems with the vehicle's onboard computer. Unfortunately, an automobile's spark plug wires, ignition coil, and generator all possess the ability to generate these radio waves. Under the right conditions, RFI can trigger sensors or actuators. The result may be an intermittent drivability problem that appears to be ignition system related.

To minimize the effects of RFI, make sure your visual inspection is thorough. Also check to make sure that sensor wires running to the computer are routed away from potential RFI sources. Rerouting a wire by no more than an inch or two may keep RFI from falsely triggering or interfering with computer operation. RFI can be present on a voltage signal or on a ground.

Most manufacturers shield their PCMs from EMI and RFI. However, this shielding will only work if the PCM is properly grounded. Always confirm a good ground before condemning the PCM. Some PCMs may have up to 5 grounds.

Testing Hall-effect Sensors

Most Hall-effect sensors can be tested by connecting the correct voltage across the plus (+) and minus (-) voltage terminals of the Hall layer, and a voltmeter across the minus (-) and signal voltage terminals.

With the voltmeter hooked up, insert a steel feeler gauge or knife blade between the Hall layer and magnet (Figure 24-10). If the sensor is good, the voltmeter should read within 0.5 volt of specifications when the feeler gauge or knife blade is inserted and touches the magnet. When the feeler gauge or blade is removed, the voltage should read less than 0.5 volt.

You can also watch the activity of a Hall-effect switch with a lab scope. This is especially handy because you do not need to run a separate power circuit to the pickup unit. With a lab scope, the unit can be checked while the engine is running. Set the scope on an AC voltage scale and connect the positive lead to the Hall signal

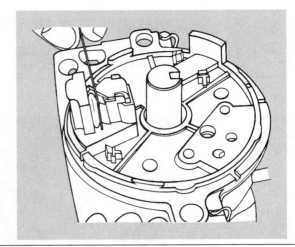

Figure 24-10 Test a Hall-effect switch with a steel feeler gauge.

lead. The negative lead should connect to ground or the ground terminal at the sensor's connector. With the engine running, the pattern should show a square wave pattern ranging from approximately 0 to 5 volts or 0 to 12 volts (Figure 24-11) depending on the system. If the range is out of specifications or distorted, replace the sensor.

On Chrysler optical distributors, the pickup voltage supply and ground wires may be tested at the four-wire connector near the distributor (Figure 24-12). With the ignition switch ON, connect the voltmeter from the orange voltage supply wire to ground. The reading should be 9.2 to 9.4 volts. Because the expected reading may vary with the model year, always refer to the manufacturer's service manual before conducting the test.

When the voltmeter is connected between the black/light blue ground wire and an engine ground, the reading should be less than 0.2 volt. When the voltmeter is connected from the grey/black reference pickup wire or the tan/yellow SYNC pickup wire to an engine ground, the voltmeter should cycle from nearly 0 volts to 5 volts while the engine is cranking. If the pickup signal is not within specifications, the pickup is defective. A defective SYNC pickup in the distributor should not cause a no-start problem but will affect engine performance.

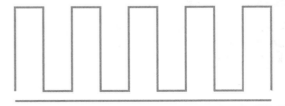

Figure 24-11 Square wave trace produced by a normal Hall-effect switch.

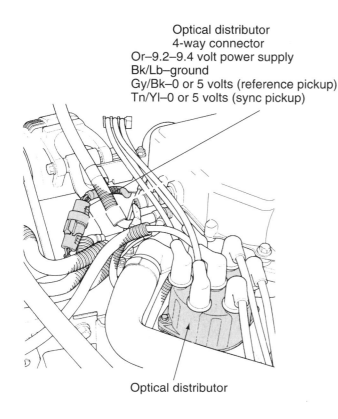

Optical distributor
4-way connector
Or–9.2–9.4 volt power supply
Bk/Lb–ground
Gy/Bk–0 or 5 volts (reference pickup)
Tn/Yl–0 or 5 volts (sync pickup)

Optical distributor

Figure 24-12 The four-wire connector for a Chrysler optical distributor. (Courtesy of Chrysler Corporation)

Computer-Controlled DI Ignition System Service and Diagnosis

Basic Ignition Timing Tests

CAUTION *To avoid timing light damage and personal injury, keep the timing light and the timing light lead wires away from rotating parts such as fan blades and belts while adjusting basic timing.*

CAUTION *To avoid personal injury, keep hands and clothing away from rotating parts such as fan blades and belts while adjusting basic timing. Remember that electric-drive cooling fans may start at any time.*

Prior to any basic timing check, the engine must be at normal operating temperature. The ignition timing specifications are provided on the underhood emission control information label. On some vehicles, such as General Motors, instructions regarding the procedure for checking ignition timing are also provided on the emission control information label. On computer-controlled distributor ignition (DI) systems, the ignition timing procedure varies depending on the vehicle make and model year. On some vehicles, the manufacturer may recommend disconnecting certain components while checking the basic ignition timing. Always follow the ignition timing procedure in the vehicle manufacturer's service manual.

When the basic timing is checked on most Chrysler fuel-injected engines, the computer system must be in the limp-in mode. The coolant temperature sensor may be disconnected to place the system in the limp-in mode, and then the timing may be checked with a timing light in the normal manner. On some Chrysler 4-cylinder engines, the timing window is in the top of the flywheel housing. On many engines, the timing mark or marks are on the crankshaft pulley, and the timing indicator is mounted above the pulley. If a timing adjustment is necessary, the distributor clamp bolt must be loosened, and the distributor rotated until the timing mark appears at the specified position on the timing indicator.

After a timing adjustment, the distributor clamp bolt must be tightened to the specified torque. Removal of the coolant temperature sensor wires places a fault code in the computer memory. This code should be erased following the timing adjustment.

WARNING *When disconnecting or reconnecting wires on a computer system, always be sure the ignition switch is off. Disconnecting computer system component wires with the ignition switch on may damage system components.*

On many DI systems, such as Ford and General Motors, a **timing connector** located in the engine compartment must be disconnected. The emission control information label usually provides the location of the timing connector on General Motors vehicles. When the timing connector is disconnected, the PCM cannot affect spark advance, and the pickup signal goes directly to the module. The distributor clamp bolt must be loosened and the distributor rotated to adjust the basic timing. After the timing adjustment is completed, the clamp bolt must be tightened and the timing connector reconnected.

CAUTION *On Chrysler magnum engines, adjusting the distributor does not alter ignition timing. Ignition timing is controlled by the PCM. Rotating the distributor housing on these engines changes the SYNC signal for fuel injection operation. This system is discussed in Chapter 25.*

No-Start Ignition Tests

The same no-start tests may be performed on conventional DI systems with centrifugal and vacuum advances, and DI systems with computer-controlled spark advance. These tests were explained previously.

Connect a 12-V test light from the coil tachometer (TACH) terminal to ground, and crank the engine. If the 12-V test lamp does not flutter while the engine is cranked, the pickup or ignition module is likely defective. Under this condition, always check the voltage supply to the positive primary coil terminal with the ignition switch on before the diagnosis is continued.

On most Chrysler fuel-injected engines, the voltage is supplied through the automatic shutdown (ASD) relay to the coil positive primary terminal and the electric fuel pump. Therefore, a defective ASD relay may cause 0V at the positive primary coil terminal. This relay is controlled by the PCM. On some Chrysler products, the relay closes when the ignition switch is turned on, whereas on other models it only closes while the engine is cranking or running. If the ASD relay closes with the ignition switch on and the engine not cranking or running, it only remains closed for about 1 second. This action shuts off the fuel pump and prevents any spark from the ignition system if the vehicle is involved in a collision with the ignition switch on and the engine stalls. A fault code should be present in the computer memory if the ASD relay is defective.

SERVICE TIP *If the engine dies after it has been run for an extended period of time, and will not restart until it has cooled down, use a heat gun to heat the ignition module and/or the PCM. Transistors do not operate well if they get hot. If heating the module causes the engine to die, replace the module.*

Pickup Tests

WARNING *Never short across or ground terminals or wires in a computer system unless instructed to do so in the vehicle manufacturer's service manual.*

If a magnetic-type pickup is used, the pickup may be checked for open circuits, shorts, and grounds with an ohmmeter. These tests are performed in the same way as the pickup tests on conventional distributors described previously. If the pickup coil tests are satisfactory, and the 12-V test light connected from the coil TACH terminal to ground does not flutter while cranking the engine, the ignition module is defective.

Prior to testing a Hall-effect pickup, an ohmmeter should be connected across each of the wires between the pickup and the computer with the ignition switch off. A computer terminal and pickup coil wiring diagram are essential for these tests. Satisfactory wires have nearly 0Ω resistance, while higher or infinite readings indicate defective wires. If the distributor has a Hall-effect pickup, the voltage supply wire and the ground wire should be checked before the pickup signal. In the following tests, the distributor connector is connected, and this connector may be backprobed to complete the necessary connections. With the ignition switch on, a voltmeter should be connected from the voltage input wire to ground, and the specified voltage must appear on the meter.

The ground wire should be tested with the ignition switch on and a voltmeter connected from the ground wire to a ground connection near the distributor. With this meter connection, the meter indicates the voltage drop across the ground wire, which should not exceed 0.2V if the wire has a normal resistance.

Connect a digital voltmeter from the pickup signal wire to ground. If the voltmeter reading does not fluctuate while cranking the engine, the pickup is defective. A voltmeter reading that fluctuates from nearly 0V to between 9V and 12V indicates a satisfactory pickup. During this test, the voltmeter reading may not be accurate because of the short duration of the voltage signal. If the Hall-effect pickup signal is satisfactory, and the 12-V test lamp does not flutter during the no-start test, the ignition module is probably defective. The ignition module is contained in the PCM on Chrysler products. On Chrysler fuel-injected engines, the reference pickup and the SYNC pickup should be tested. If either of these pickups is defective, a fault code may be stored in the computer memory.

Electronic Ignition (EI) System Diagnosis and Service

No-Start Ignition Diagnosis, Cam and Crank Sensors

The diagnostic procedure for EI systems varies depending on the vehicle make and model year. Always follow the procedure recommended in the vehicle manufacturer's service manual. The following procedure is based on Chrysler EI systems. The crankshaft timing sensor and camshaft reference sensor in these systems are modified Hall-effect switches.

If a crank or cam sensor fails, the engine will not start. Both of these sensor circuits can be checked with a voltmeter or lab scope. If the sensors are receiving the correct amount of voltage and have good low-resistance ground circuits, their output should be a digital signal or a pulsing voltmeter reading while the engine is cranking. If any of these conditions do not exist, the circuit needs to be repaired or the sensor needs to be replaced.

When the engine fails to start, follow these steps:

1. Check for fault codes 11 and 43. Code 11, "Ignition Reference Signal," could be caused by a defective camshaft reference signal or crankshaft timing sensor signal. Code 43 is caused by low primary current in coil number 1, 2, or 3.

2. With the engine cranking, check the voltage from the orange wire to ground on the crankshaft timing sensor and the camshaft reference sensor (Figure 24-13). Over 7V is satisfactory. If the voltage is less than specified, repeat the test with the voltmeter connected from PCM terminal 7 to ground. If the voltage is satisfactory at terminal 7 but low at the sensor orange wire, repair the open circuit or high resistance in the orange wire. If the voltage is low at terminal 7, the PCM may need replacement. Be sure 12V are supplied to PCM terminal 3 with the ignition switch off or on, and 12V must be supplied to PCM terminal 9 with the ignition switch on. Check PCM ground connections on terminals 11 and 12 before PCM replacement. See the note at the end of this section before replacing the PCM.

3. With the ignition switch on, check the voltage drop across the ground circuit (black/light blue wire) on the crankshaft timing sensor and the camshaft reference sensor. A reading below 0.2V is satisfactory.

4. If the readings in steps 2 and 3 are satisfactory, connect a lab scope or a digital voltmeter from the gray/black wire on the crankshaft timing sensor and the tan/yellow wire on the camshaft reference sensor to ground. When

the engine is cranking, a digital pattern should be displayed (Figure 24-14), or the voltmeter should cycle between 0 and 5 volts. If the voltage does not cycle, sensor replacement is required. Each sensor voltage signal should cycle from low voltage to high voltage as the engine is cranked.

SERVICE TIP *When using a digital voltmeter to check a crankshaft or camshaft sensor signal, crank the engine a very small amount at a time and observe the voltmeter. The voltmeter reading should cycle from almost 0 volts to a higher voltage of about 5 volts. Since digital voltmeters do not react instantly, it is difficult to see the change in voltmeter reading if the engine is cranked continually.*

A no-start condition can occur if the PCM "locks up". In step 2 above, if 0 volts is indicated the PCM may be faulty or it may be locked up. If the PCM is locked up it will not store a fault code for the reason. Basically, the PCM will lock up when it goes into a safeguard routine if the 9-volt or 5-volt reference voltage shorts to ground. This shuts down the PCM to protect it. Since it shuts down, no DTCs are stored. The engine will not start as long as the ground is present. An intermittent ground will cause the engine to stop running. Attempting to restart the engine without cycling the ignition switch to the full LOCK position will not start the engine, even if the ground is lifted. Cycle the ignition switch to the LOCK position and wait about 5 to 10 seconds. If the ground is lifted, the PCM will reset and the engine will start and run until the ground occurs again.

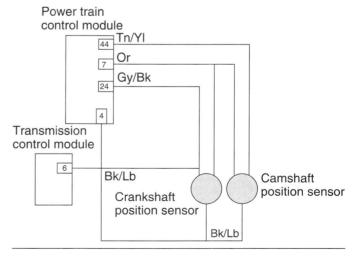

Figure 24-13 Crankshaft timing and camshaft reference sensor terminals. (Courtesy of Chrysler Corporation)

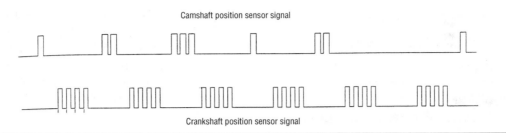

Figure 24-14 Lab scope patterns from the camshaft and crankshaft position sensors. (Courtesy of Chrysler Corporation.)

On 1996 and newer SBEC III and JTEC engine controllers, there are two 5-volt reference signals. The sensors that require 5 volts are separated, thus if this signal shorts to ground the engine will still stop running, but for the first time a DTC can be set.

Also note, if the 9-volt reference voltage is opened, there will be no DTC stored for the crankshaft or camshaft positions sensors. With an open circuit the PCM cannot tell if the engine is cranking or not. The diagnostic routine does not begin until the PCM senses engine cranking.

Sensor Replacement

If the crankshaft timing sensor or the camshaft reference sensor is removed, follow this procedure when the sensor is replaced:

1. Thoroughly clean the sensor tip and install a new spacer (part number 5252229) on the sensor tip. New sensors should be supplied with the spacer installed (Figure 24-15).

2. Install the sensor until the spacer lightly touches the sensor ring, and tighten the sensor mounting bolt to 105 in. lb.

WARNING *Improper sensor installation may cause sensor, rotating drive plate, or timing gear damage.*

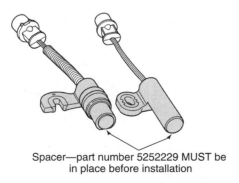

Spacer—part number 5252229 MUST be in place before installation

Figure 24-15 Spacer on crankshaft timing sensor and camshaft reference sensor tips. (Courtesy of Chrysler Corporation)

No-Start Ignition Diagnosis Coil and PCM Tests

If the sensor tests are satisfactory, proceed with these coil and PCM tests:

1. Check the spark plug wires with an ohmmeter as explained previously.

2. With the engine cranking, connect a voltmeter from the dark green/black wire on the coil to ground. If this reading is below 12V, check the automatic shutdown (ASD) relay circuit (Figure 24-16).

SERVICE TIP *Later model Chrysler vehicles have separate fuel pump and ASD relays. Always use the proper wiring diagram for the vehicle being tested.*

3. If the reading in step 2 is satisfactory, check the primary and secondary resistance in each coil with the ignition switch off. Primary resistance is 0.52 to 0.62 ohm, and secondary resistance is 11,000 to 15,000 ohms. If these ohm readings are not within specifications, replace the coil assembly.

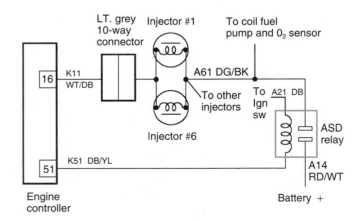

Figure 24-16 Automatic shutdown (ASD) relay circuit. (Courtesy of Chrysler Corporation)

4. With the ignition switch off, connect an ohmmeter across the three wires from the coil connector to PCM terminals 17, 18, and 19 (Figure 24-17). These terminals are connected from the coil primary terminals to the PCM. If an infinite ohmmeter reading is obtained on any of the wires, repair the open circuits.

5. Connect a 12-V test lamp from the dark blue/black wire, dark blue/gray wire, and black/gray wire on the coil assembly to ground while cranking the engine. If the test lamp does not flutter on any of the three wires, replace the PCM. Since the crankshaft and camshaft sensors, wires from the coils to the PCM, and voltage supply to the coils have been tested already, the ignition module must be defective. This module is an integral part of the PCM on Chrysler vehicles; thus, PCM replacement is necessary.

WARNING *Do not crank or run an EI-equipped engine with a spark plug wire completely removed from a spark plug. This action may cause leakage defects in the coils or spark plug wires.*

CAUTION *Since EI systems have more energy in the secondary circuit, electrical shocks from these systems should be avoided. The electrical shock may not injure the human body, but such a shock may cause you to jump and hit your head on the hood or push your hand into contact with a rotating cooling fan.*

6. If the tests in steps 1 through 5 are satisfactory, connect a test spark plug to each spark plug wire and ground and crank the engine. If any of the coils do not fire on the two spark plugs connected to the coil, replace the coil assembly.

Diagnosis and Service of Ford EEC Low Data and High Data Rate EI Systems

No-Start Diagnosis

The diagnostic procedure for the low data rate and high data rate systems varies depending on the system being diagnosed. The technician must have the proper wiring diagram for the system being diagnosed, and the procedure in the vehicle manufacturer's service manual must be followed. Ford Motor Company provides separate diagnostic harnesses for the low data rate and high data rate EI systems. The diagnostic harness has various leads that connect in series with each component in the ignition system (Figure 24-18). A large connector on the diagnostic harness is connected to the 60-pin breakout box (Figure 24-19). Overlays for the ignition system on each engine are available to fit on the breakout box terminals. These overlays identify the ignition terminals connected to the breakout box terminals.

The technician must follow the test procedures in the vehicle manufacturer's service manual, and measure the voltage or resistance at the specified breakout box terminals connected to the ignition system. If a diagnostic harness, overlay, and breakout box are not available, the technician must measure the voltage or resistance at the terminals on the individual ignition components. Following is a general no-start diagnostic procedure for a high data rate system:

1. Connect a test spark plug from each of the spark plug wires to ground and crank the engine. If the test spark plug does not fire on a pair of spark plugs connected to the same coil, test the spark plug wires connected to that coil. If these wires are satisfactory, that coil is

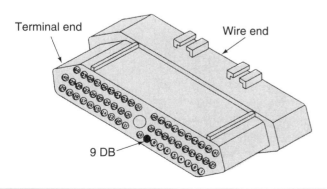

Figure 24-17 PCM terminal identification. (Courtesy of Chrysler Corporation)

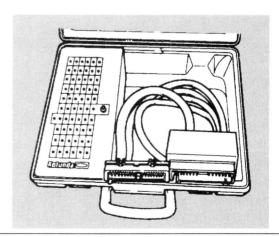

Figure 24-19 60-pin breakout box. (Reprinted with the permission of Ford Motor Company)

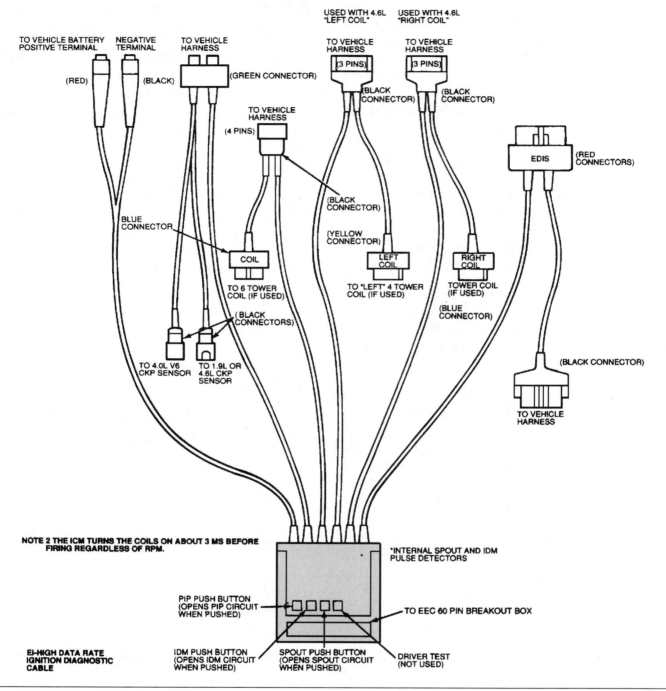

Figure 24-18 Diagnostic harness for a high data rate ignition system (Reprinted with the permission of Ford Motor Company)

probably defective. When the test spark plug does not fire on any spark plug, proceed to step 2.

2. Connect a digital voltmeter from terminal 2 on each coil pack to ground (Figure 24-20). With the ignition switch on, the voltmeter should indicate 12V. If the voltage is less than specified, test the wire from the ignition switch to the coils, and the ignition switch.

3. With the ignition switch off, connect the ohmmeter leads to the primary terminals in each coil pack. If the primary resistance is not within specifications, replace the coil pack.

4. Connect the ohmmeter leads to the secondary terminals in each coil pack. Replace the coil pack if the secondary resistance in any coil is not within specifications.

5. Connect the ohmmeter leads from each primary coil terminal to the terminal on the ignition control module (ICM) to which the primary terminal is connected. Each wire should read less than 0.5Ω. If any of these wires has more resistance than specified, repair the wire.

6. Connect a digital voltmeter from terminal 6 on the ICM to ground. With the ignition switch on, the voltmeter should indicate 12V. If the voltage is less than specified, test the wire from the power relay to terminal 6, and the power relay. Turn off the ignition switch.

7. Connect a digital voltmeter from terminal 10 on the ICM to ground. The voltmeter should indicate 0.5V or less with the ignition switch on. When the voltmeter reading is higher than specified, repair the ground wire.

8. Connect a digital voltmeter from terminal 7 on the ICM to ground. With the ignition switch on, the volt-

meter reading should be less than specified. If the voltmeter reading is more than specified, repair the ground wire.

9. Connect a digital voltmeter from terminals 8, 9, 11, and 12 on the ICM to ground and crank the engine. The voltmeter reading should fluctuate on each wire. If the voltmeter reading does not fluctuate on one of the wires, replace the ICM. When the voltmeter does not fluctuate on any of these primary wires, proceed to step 10.

10. With the ignition switch off and the ICM terminal disconnected, connect an ohmmeter to terminals 4 and 5 in the ICM wiring harness. If the ohmmeter reads 2,300 to 2,500 ohms, the crankshaft position (CKP) sensor and connecting wires are satisfactory. When the ohmmeter reading is not within specifications, repeat the test with the ohmmeter leads connected to the CKP

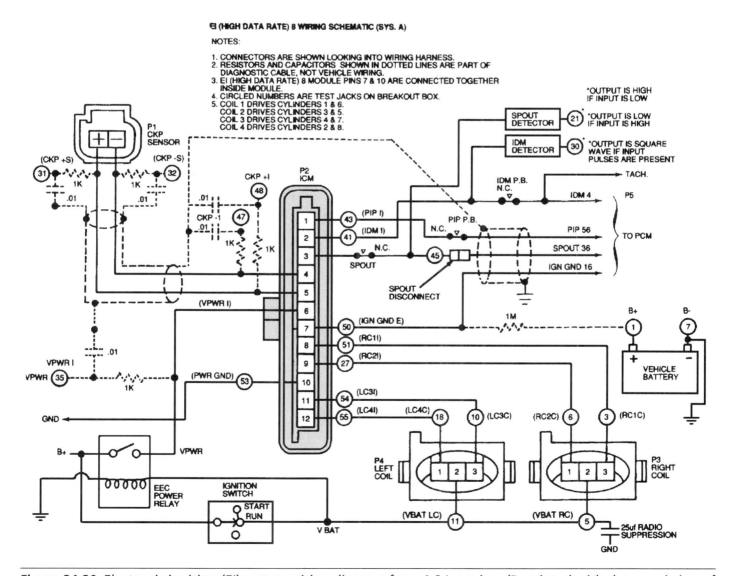

Figure 24-20 Electronic ignition (EI) system wiring diagram for a 4.6-L engine. (Reprinted with the permission of Ford Motor Company)

sensor terminals. If the resistance of the CKP sensor is not within specifications, replace the sensor. When the resistance reading is satisfactory at the CKP sensor terminals, but out of specifications at ICM terminals 4 and 5, repair the CKP wires.

11. Inspect the trigger wheel behind the crankshaft pulley and the CKP sensor for damage.

12. If the digital voltmeter readings do not fluctuate in step 9, but the CKP sensor test in step 10 is satisfactory, replace the ICM.

Engine Misfire Diagnosis

If the engine is misfiring, test for intake manifold vacuum leaks. Test the engine compression, spark plugs, and fuel injectors. Test the spark plug wires and coils as indicated in steps 1 through 4 of the no-start diagnosis. Timing adjustments are not possible in the low data rate or high data rate EI systems.

General Motors Electronic EI System Service and Diagnosis

Coil Winding Ohmmeter Tests

With the coil terminals disconnected, an ohmmeter calibrated on the X1 scale should be connected to the primary coil terminals to test the primary winding. The primary winding in any EI coil should have 0.35 to 1.50 ohms of resistance. An ohmmeter reading below the specified resistance indicates a shorted primary winding, and an infinite meter reading proves that the primary winding is open.

An ohmmeter calibrated on the X1,000 scale should be connected to each pair of secondary coil terminals to test the secondary windings. The coil secondary winding in EI type 1 systems should have 10,000 to 14,000 ohms resistance, whereas secondary windings in EI type 2 systems should have 5,000 to 7,000 ohms resistance. If the secondary winding is open, the ohmmeter reading is infinite. A shorted secondary winding provides an ohmmeter reading below the specified resistance.

Crankshaft Sensor Adjustment, 3.0-L, 3300, 3.8-L, and 3800 Engines

A basic timing adjustment is not possible on any EI system. However, if the gap between the blades and the crankshaft sensor is not correct, the engine may fail to start, or may stall, misfire, or hesitate on acceleration.

Follow these steps during the crankshaft sensor adjustment procedure:

1. Loosely install the sensor on the pedestal.
2. Position the sensor and pedestal on the J37089 adjusting tool.
3. Position the adjusting tool on the crankshaft surface (Figure 24-21).
4. Tighten the pedestal-to-block mounting bolts to 30 to 35 ft. lb. (20 to 40 Newton meters [Nm]).
5. Tighten the pinch bolt to 30 to 35 in. lb. (3 to 4 Nm).

The interrupter rings on the back of the crankshaft pulley should be checked for a bent condition. The same crankshaft sensor adjusting tool may be used to check these rings. Place the J37089 tool on the pulley extension surface and rotate the tool around the pulley (Figure 24-22). If any blade touches the tool, replace the pulley.

Figure 24-21 Crankshaft sensor adjustment for 3.0-L, 3300, 3.8-L, and 3800 engines. (Courtesy of Oldsmobile Division, General Motors Corporation)

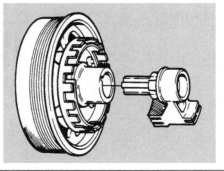

Figure 24-22 Interrupter ring checking procedure. (Courtesy of Oldsmobile Division, General Motors Corporation)

Crankshaft Sensor Adjustment, Single Slot Sensor

If a single slot crankshaft sensor requires adjustment, follow this procedure:

CAUTION *Always be sure the ignition switch is off before attempting to rotate the crankshaft with a socket and breaker bar. If the ignition switch is on, the engine may start suddenly and rotate the socket and breaker bar with tremendous force. This action may result in personal injury and vehicle damage.*

1. Be sure that the ignition switch is off. Then rotate the crankshaft with a pull handle and socket installed on the crankshaft pulley nut. Continue rotating the crankshaft until one of the interrupter blades is in the sensor and the edge of the interrupter window is at the edge of the defector on the pedestal.

2. Insert adjustment tool J36179 or its equivalent between each side of the blade and the sensor. If the tool does not fit between each side of the blade and the sensor, adjustment is required. The gap measurement should be repeated at all three blades.

3. If a sensor adjustment is necessary, loosen the pinch bolt and insert the adjusting tool between each side of the blade and the sensor. Move the sensor as required to insert the gauge.

4. Tighten the sensor pinch bolt to 30 in. lb. (3 to 4 Nm).

5. Rotate the crankshaft and recheck the gap at each blade.

No-Start Ignition Diagnosis for EI Type 1 and Type 2 Systems

If the engine fails to start, follow these steps for a no-start ignition diagnosis:

1. Connect a test spark plug from each spark plug wire to ground and crank the engine while observing the test spark plug.

2. If the test spark plug does not fire on any spark plug, check the 12-V supply wires to the coil module. Some coil modules have two fused 12-V supply wires. Consult the vehicle manufacturer's wiring diagrams for the car being tested to identify the proper coil module terminals.

3. If the test spark plug does not fire on a pair of spark plugs, the coil connected to that pair of spark plugs is probably defective.

4. If the test spark plug does not fire on any of the spark plugs and the 12-V supply circuits to the coil module are satisfactory, disconnect the crankshaft and camshaft sen-

sor connectors and connect short jumper wires between the sensor connector and the wiring harness connector. Be sure the jumper wire terminals fit securely to maintain electrical contact. Each sensor has a voltage supply wire, a ground wire, and a signal wire on 3.8-L engines. On the 3.3-L and 3300 engines, the dual crankshaft sensor has a voltage supply wire, ground wire, crank signal wire, and SYNC signal wire. Identify each of these wires on the wiring diagram for the system being tested.

5. Connect a digital voltmeter to each of the camshaft and crankshaft sensor black ground wires to an engine ground connection. With the ignition switch on, the voltmeter reading should be 0.2V or less. If the reading is above 0.2V, the sensor ground wires have excessive resistance.

6. With the ignition switch on, connect a digital voltmeter from the camshaft and crankshaft sensor white/red voltage supply wires to an engine ground (Figure 24-23). The voltmeter readings should be 5V to 11V. If the readings are below these values, check the voltage at the coil module terminals that are connected to the camshaft and crankshaft sensor voltage supply wires. When the sensor voltage supply readings are low at the coil module terminals, the coil module should be replaced. If the voltage supply readings are low at either sensor connector but satisfactory at the coil module terminal, the wire from the coil module to the sensor is defective. On the 3.3-L and 3300 engines, the crankshaft sensor ground wire and voltage supply wire are checked in the same way as explained in steps 5 and 6.

7. If the camshaft and crankshaft sensor ground and voltage supply wires are satisfactory, connect a digital voltmeter to each sensor signal wire and crank the engine. Each sensor should have a 5-V to 7-V fluctuating signal. On the 3.3-L and 3300 engines, test this voltage signal on the crank and SYNC signal wires at the crankshaft sensor. If the signal is less than specified, replace the sensor with the low signal.

8. When the camshaft and crankshaft sensor signals on 3.8-L engines or crank and SYNC signals on 3.3-L and 3300 engines are satisfactory and the test spark plug does not fire at any spark plug, the coil module is probably defective.

9. On 3.8-L engines where the coil assembly is easily accessible, the coil assembly screws may be removed and the coil lifted up from the module with the primary coil wires still connected. Connect a 12-V test lamp across each pair of coil primary wires and crank the engine. If the test lamp does not flutter on any of the coils, the coil module is defective, assuming that the crankshaft and camshaft sensor readings are satisfactory.

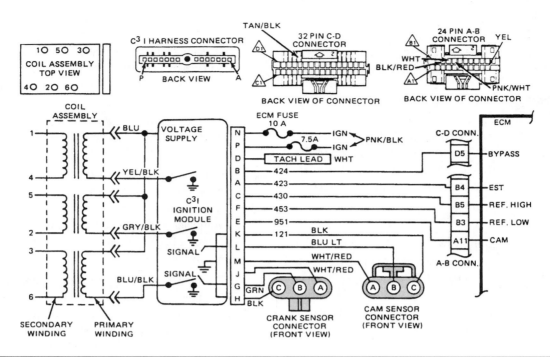

Figure 24-23 Crankshaft and camshaft sensor wiring connections for a 3.8-L engine. (Courtesy of Buick Motor Division, General Motors Corporation)

No-Start Ignition Diagnosis for EI Type 3 Fast-Start Systems

Complete steps 1, 2, and 3 in the No-Start Ignition Diagnosis for EI type 1 and type 2 systems, and then complete these steps:

1. If the 12-V supply circuits to the coil module are satisfactory, disconnect the crankshaft sensor connector and connect four short jumper wires between the sensor connector and the wiring harness connector.

2. Connect a digital voltmeter from the sensor ground wire to an engine ground. With the ignition switch on, the voltmeter should read 0.2V or less. A reading above this value indicates a defective ground wire.

3. Connect the voltmeter from the sensor voltage supply wire to an engine ground. With the ignition switch on, the voltmeter reading should be 8V to 10V. If the reading is lower than specified, check the voltage at coil module terminal N (Figure 24-24). When the voltage at terminal N is satisfactory and the reading at the sensor voltage supply wire is low, the wire from terminal N to the sensor is defective. A low voltage reading at terminal N indicates a defective coil module.

4. If the readings in steps 3 and 4 are satisfactory, connect a voltmeter from the 3X and 18X signal wires at the sensor connector to an engine ground and crank the engine. The voltmeter reading should fluctuate from 5V to 7V. The exact voltage may be difficult to read, especially on the 18X signal, but the reading must fluctuate. If the voltmeter reading is steady on either sensor signal, the sensor is defective.

5. Connect a digital voltmeter from the 18X and 3X signal wires at the coil module to an engine ground and crank the engine. The voltmeter readings should be the same as in step 4. If these voltage signals are satisfactory at the sensor terminals but low at the coil module, repair the wires between the coil module and the sensor.

6. If the 18X and 3X signals are satisfactory at the coil module terminals, remove the coil assembly-to-module screws and lift the coil assembly up from the module. Connect a 12-V test lamp across each pair of coil primary terminals and crank the engine. If the test lamp does not flash on any pair of terminals, the coil module is defective.

Cam Sensor Timing for 3.8-L Turbocharged Engines

If the cam sensor is removed from the engine on 3.8-L turbocharged engines, the sensor must be timed to the engine when the sensor is installed. The cam sensor gear has a dot that must be positioned opposite the sensor disc window prior to sensor installation. As the cam sensor is installed in the engine, the gear dot must face away from

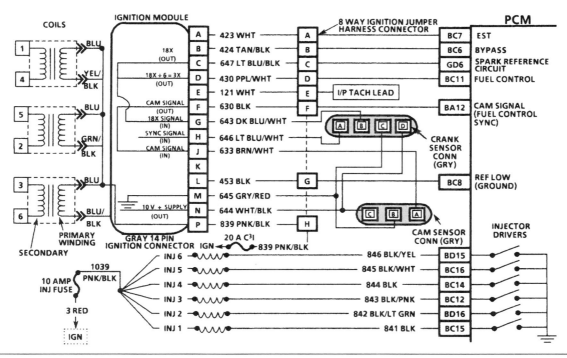

Figure 24-24 Terminal identification for an EI fast-start system 3800 engine. (Courtesy of Oldsmobile Division, General Motors Corporation)

the timing chain toward the passenger's side of the vehicle. When the cam sensor is installed in the engine, the sensor wiring harness must face toward the driver's side of the vehicle. Follow this procedure for cam sensor installation and timing:

1. Remove the spark plug wires from the coil assembly.

2. Remove the number 1 spark plug and crank the engine until compression is felt at the spark plug hole.

3. Slowly crank the engine until the timing mark lines up with the 0í position on the timing indicator.

4. Measure 1.47 to 1.5 in. (3.7 to 3.8cm) from the 0í position toward the after-TDC position on the crankshaft pulley, and mark the pulley at this location.

5. Slowly crank the engine until the mark placed on the pulley in step 4 is lined up with the 0í position on the timing indicator.

6. Use a weatherpack terminal removal tool to remove the center terminal B in the cam sensor connector, and connect a short jumper wire between this wire and the terminal in the connector. Terminal B is the cam sensor signal wire.

7. Connect a digital voltmeter from the cam sensor signal wire to an engine ground, and turn on the ignition switch.

8. Rotate the cam sensor until the voltmeter reading changes from high volts (5V to 12V) to low volts (0V to 2V).

9. Hold the cam sensor in this position and tighten the cam sensor-to-block retaining bolt.

Diagnosis of EI Systems with Magnetic Sensors

Magnetic Sensor Tests

With the wiring harness connector to the magnetic sensor disconnected and an ohmmeter calibrated on the X10 scale connected across the sensor terminals, the meter should read 900 to 1,200 ohms on 2.0-L, 2.8-L, and 3.1-L engines. The meter should indicate 500 to 900 ohms on a Quad 4 engine and 800 to 900 ohms on a 2.5-L engine (Figure 24-25).

Meter readings below the specified value indicate a shorted sensor winding, whereas infinite meter readings prove that the sensor winding is open. Since these sensors are mounted in the crankcase, they are continually splashed with engine oil. In some sensor failures, the engine oil enters the sensor and causes a shorted sensor winding. If the magnetic sensor is defective, the engine fails to start.

With the magnetic sensor wiring connector disconnected, an alternating current (AC) voltmeter may be connected across the sensor terminals to check the sensor signal while the engine is cranking. On 2.0-L, 2.8-L, and 3.1-L engines, the sensor signal should be 100 millivolts

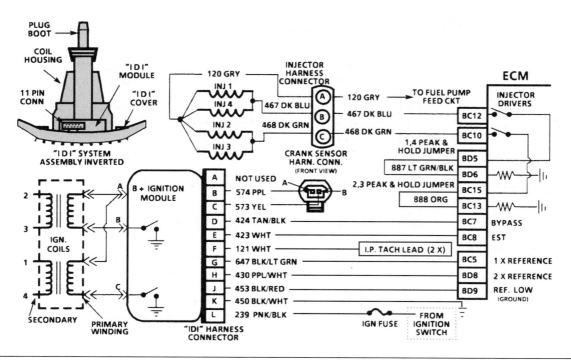

Figure 24-25 Terminal identification for an EI system 2.3-L Quad 4 engine. (Courtesy of Oldsmobile Division, General Motors Corporation)

(mV) AC. The sensor voltage on a Quad 4 engine should be 200 mV AC. When the sensor is removed from the engine block, a flat steel tool placed near the sensor should be attracted to the sensor if the sensor magnet is satisfactory.

No-Start Diagnosis for an EI System with a Magnetic Sensor

When an engine with an EI system and a magnetic sensor fails to start, complete steps 1, 2, and 3 of the No-Start Diagnosis for EI type 1 and type 2 systems, and then follow this procedure:

1. If the test spark plug did not fire on any of the spark plugs, check for 12V at the coil module voltage input terminals. Consult the wiring diagram for the system being tested for terminal identification.

2. If 12V are supplied to the appropriate coil module terminals, test the magnetic sensor as explained under Magnetic Sensor Tests.

3. When the magnetic sensor tests are satisfactory, the coil module is probably defective.

Some coil module changes were made on 1989 EI systems with magnetic sensors, and 1989 coil modules

will operate satisfactorily on 1988 EI systems. However, if a 1988 module is installed on a 1989 EI system, the malfunction indicator light (MIL) will come on and code 41 will be stored in the PCM memory.

SERVICE TIP *When diagnosing any computer system, never forget the basics. For example, always be sure the engine has satisfactory compression and ignition before attempting to diagnose the computer-controlled fuel injection.*

Engine Misfire Diagnosis

If the engine misfires all the time or on acceleration only, test the following components:

1. Engine compression

2. Spark plugs

3. Spark plug wires

4. Ignition coils test for firing voltage with a test spark plug

5. Crankshaft sensor

CUSTOMER CARE *Some intermittent automotive problems are difficult to diagnose unless we can catch the car in the act. In other words, the problem is hard to diagnose if the symptoms are not present while we are performing diagnostic tests. One solution to this problem is to have the customer leave the vehicle with the shop, and then drive the car under the conditions when the problem occurs. However, if the problem only appears once every week, this solution is not likely to work, because the problem will not occur in the short time we have to drive the car.*

6. Fuel injectors on multiport and sequential fuel injection systems

A second solution, especially in no-start situations, is to have the customer phone the shop immediately when the problem occurs, and send a technician to diagnose the problem. If this solution is attempted, inform the customer not to attempt starting the car until the technician arrives. This solution may be expensive, but in some cases, it may be the only way to diagnose the problem successfully. Always be willing to go the extra mile to diagnose and correct the customer's problem. By doing so, you will obtain a lot of satisfied, repeat customers.

Summary

❑ Regardless of the type of computerized ignition system used (DI or EI), determining the defective part or area of the system requires a thorough knowledge of how the system works and a logical troubleshooting process.

❑ There are many variations in the operation of electronic engine controls. Therefore, always check the appropriate factory service manual before beginning any diagnosis and repair.

❑ Five methods are used to check individual components: visual, ohmmeter, voltmeter, lab scope, and scan tool checks.

❑ In some cases, a final check on the computer can be made only by substitution.

❑ Most DI and EI ignition systems have onboard diagnostics. These systems test the input sensors and actuators for continuity and in some case for functionality.

❑ The MIL serves two purposes. First, it is a signal to the driver that a detectable system failure has occurred. Second, it can also be used by a technician as an aid to identify any problems in the system.

❑ If a 12-V test light connected from the negative primary coil terminal to ground does not flash when the engine is cranked, the pickup coil or module is defective.

❑ When a 12-V test light connected from the negative primary coil terminal to ground flashes while cranking the engine, but a test spark plug connected from the coil secondary wire to ground does not fire while cranking the engine, the coil is defective.

❑ When a test spark plug connected from the coil secondary wire to ground fires while cranking the engine, but fails to fire when connected from the spark plug wires to ground, the distributor cap or rotor is defective.

❑ The ohmmeter leads may be connected to the pickup coil leads to test the pickup coil for open and shorted circuits.

❑ The ohmmeter leads may be connected from one of the pickup leads to ground to check the pickup for a grounded condition.

❑ On many vehicles, such as General Motors and Ford vehicles, a timing connector must be disconnected while checking basic timing.

❑ On Chrysler vehicles the computer system must be in the limp-in mode while checking basic timing.

❑ The voltage supply and ground connection should be checked on Hall-effect pickups before the voltage signal from the pickup is checked with the engine cranking.

❑ After the voltage supply and ground wires have been checked on an optical pickup, the voltage signal from the pickup should fluctuate between 0V and 5V while cranking the engine.

❑ Prior to checking the voltage signal from a crankshaft or camshaft sensor on an EI system, the sensor voltage supply and ground wire should be checked.

❑ The voltage signal from crankshaft or camshaft sensors may be checked with a 12-V test light, a digital voltmeter, or a lab scope while cranking the engine.

Terms-To-Know

Automatic shutdown (ASD) relay   Onboard diagnostics   Timing connector

ASE Style Review Questions

1. Ignition coil ohmmeter testing is being discussed. Technician A says the ohmmeter should be placed on the X1,000 scale to test the secondary winding. Technician B says the ohmmeter tests on the coil check the condition of the winding insulation. Who is correct?
   a. Technician A only
   b. Technician B only
   c. Both technicians
   d. Neither technician

2. EI system coil test results show one primary winding has 0.53 resistance and the specified resistance is 13. Technician A says the primary winding in this coil is grounded. Technician B says the primary winding in this coil is shorted. Who is correct?
   a. Technician A only
   b. Technician B only
   c. Both technicians
   d. Neither technician

3. Basic ignition timing adjustment on vehicles with computer-controlled distributor ignition (DI) is being discussed. Technician A says on some DI systems, a timing connector must be disconnected. Technician B says the distributor must be rotated until the timing mark appears at the specified location on the timing indicator. Who is correct?
   a. Technician A only
   b. Technician B only
   c. Both technicians
   d. Neither technician

4. An electronic ignition (EI) system's crankshaft and camshaft sensor tests are satisfactory, but a test spark plug connected from the spark plug wires to ground does not fire. Technician A says the coil assembly may be defective. Technician B says the voltage supply wire to the coil assembly may be open. Who is correct?
   a. Technician A only
   b. Technician B only
   c. Both technicians
   d. Neither technician

5. EI service and diagnosis is being discussed. Technician A says the crankshaft sensor may be rotated to adjust the basic ignition timing. Technician B says the crankshaft sensor may be moved to adjust the clearance between the sensor and the rotating blades on some EI systems. Who is correct?
   a. Technician A only
   b. Technician B only
   c. Both technicians
   d. Neither technician

6. EI systems with crankshaft and camshaft sensors that require a paper spacer on the sensor tip prior to installation are being discussed. Technician A says the sensor should be installed so the paper spacer lightly touches the rotating sensor ring. Technician B says the sensor should be installed so the paper spacer lightly touches the rotating sensor ring and then pulled outward 0.125 inches. Who is correct?
   a. Technician A only
   b. Technician B only
   c. Both technicians
   d. Neither technician

7. Engine misfire diagnosis is being discussed. Technician A says a defective EI coil may cause cylinder misfiring. Technician B says the engine compression should be verified first if the engine is misfiring continually. Who is correct?
   a. Technician A only
   b. Technician B only
   c. Both technicians
   d. Neither technician

8. Possible causes for RFI are being discussed. Technician A says spark plug wires and the generator can be sources for RFI. Technician B says the ignition coil can be a source for RFI. Who is correct?
   a. Technician A only
   b. Technician B only
   c. Both technicians
   d. Neither technician

9. While discussing what will appear on a voltmeter when the engine is cranking and the meter is connected to the output of a properly working crankshaft sensor: Technician A says the meter will fluctuate between low and high readings. Technician B says the meter will show a constant low voltage. Who is correct?
   a. Technician A only
   b. Technician B only
   c. Both technicians
   d. Neither technician

10. A vehicle with DI ignition is towed into the shop due to a no-start condition. A spark tester connected to the plug wire does not spark when the engine is cranked. However, a spark appears at the spark tester when it is connected to the ignition coil wire. Technician A says the primary windings of the ignition coil may be open. Technician B says the distributor pickup coil may be faulty. Who is correct?
    a. Technician A only
    b. Technician B only
    c. Both technicians
    d. Neither technician

Electronic Fuel Control

Objective

Upon completion and review of this chapter, you should be able to:

- ❏ Explain what is meant by a speed density electronic fuel injection (EFI) system.
- ❏ Explain the function of an air density electronic fuel injection (EFI) system.
- ❏ Explain the operation of a fuel pump circuit in an EFI system when the ignition switch is on and the engine is not cranked.
- ❏ Describe the purpose of the inertia switch used in some fuel pump circuits.
- ❏ Explain the purpose of the oil pressure switch used in some fuel pump circuits.
- ❏ Describe the operation of the fuel pressure regulator.
- ❏ Explain the fuel strategies used during different modes of operation.
- ❏ Describe how an idle air control (IAC) motor controls idle speed.
- ❏ Describe the difference between a sequential fuel injection (SFI) system and a multipoint fuel injection (MPI) system.
- ❏ List and describe the functions of various input sensors used in fuel injection systems.
- ❏ Describe the meaning of pulse width.
- ❏ Explain the operation of a cold start injector.
- ❏ Describe the operation of the central injector and poppet nozzles in a central port injection (CPI) system.
- ❏ Explain the purpose of the intake manifold tuning valve (IMTV).

Introduction

During the 1980s, automotive engineers changed many engines from carburetors or computer-controlled carburetors to **electronic fuel injection (EFI)** systems. This action was taken to improve fuel economy, performance, and emission levels. Automotive manufacturers had to meet increasingly stringent **corporate average fuel economy (CAFE)** regulations and comply with emission standards at the same time. CAFE standards are federally imposed regulations requiring vehicle manufacturers to a meet a specific average fleet fuel mileage for all their vehicles.

The term electronic fuel injection was used by some manufacturers to describe their fuel system. Today this term is no longer used and has been replaced with more specific descriptions. In this book, the term EFI is applied to any electronic or computer-controlled fuel injection system.

Many of the early EFI systems were **throttle body injection (TBI)** systems in which the fuel is injected above the throttles. The TBI systems have been gradually changed to **port fuel injection (PFI)** systems with the injectors located in the intake ports. On carbureted and throttle body injected engines, some intake manifold heating was required to prevent fuel condensation on the intake manifold passages. When the injectors are positioned in the intake ports, intake manifold heating is not required. This provided engineers with increased intake manifold design flexibility. Intake manifolds could now be designed with longer, curved air passages, which increased air flow and improved torque and horsepower.

Since intake manifolds no longer require heating, they can now be made from plastic materials such as glass-fiber reinforced nylon resin. This material is considerably lighter than cast iron or even aluminum. Saving weight means an improvement in fuel economy. The plastic-type intake manifold does not transfer heat to the air and fuel vapor in the intake passages, which improves economy and hot start performance.

This chapter discusses the common components found in most EFI systems and the principles of various types of EFI. Typical EFI systems are covered to provide a basis for understanding the many different EFI systems when you encounter them.

Input Sensors

Most EFI systems use a combination of the following sensors. Some of these sensors were discussed in Chapter 22. However, they will be discussed again here in reference to the fuel system.

1. Crankshaft and camshaft position sensors
2. Manifold absolute pressure (MAP) sensor
3. Mass air flow (MAF) sensor
4. Barometric pressure (BARO) sensor
5. Throttle position sensor (TPS)
6. Engine coolant temperature (ECT) sensor
7. Intake air temperature (IAT) sensor
8. Oxygen sensor (O_2S)
9. Exhaust gas recirculation valve position (EVP) sensor
10. Vehicle speed sensor (VSS)
11. Neutral drive switch (NDS)
12. Idle switch

Crankshaft and Camshaft Position Sensors

As with the ignition system, the PCM needs these inputs to determine when to fire the injectors. If both sensors are used, they must be synchronized during engine starting before the injectors are energized. The crankshaft position sensor supplies engine speed information to the PCM. This information is required to determine if the engine is starting or in the run mode. In addition, the rpm

signal (used with other sensors) determines the volume of fuel entering the engine. All EFI systems use some method of determining the volume of air flow. Most systems will not start if this input is missing.

Manifold Absolute Pressure Sensor

The **manifold absolute pressure (MAP) sensor** is usually mounted in the engine compartment. A hose may be connected from the intake manifold to a vacuum inlet on the MAP sensor, and three wires are connected from the sensor to the computer. Some systems insert the map sensor directly into the intake manifold, thus eliminating the need for a hose. The computer supplies a constant 5-V reference voltage to the sensor. The other wires are a signal wire and a ground wire.

The computer uses the MAP sensor signal to determine the engine load and volume of air entering the engine. When the MAP sensor signal indicates wide-open throttle, heavy load conditions, the computer provides a richer air/fuel ratio. The computer supplies a leaner air/fuel ratio if the MAP sensor signal indicates light load, moderate cruising speed conditions.

When the ignition switch is turned on before the engine is started, many MAP sensors act as a barometric pressure sensor. Under this condition, the MAP sensor signal informs the computer regarding atmospheric pressure, which varies in relation to altitude and atmospheric conditions such as the humidity. Some applications have a separate barometric pressure sensor.

Many manufacturers use MAP sensors that contain a silicon diaphragm. When the manifold vacuum stresses the silicon diaphragm, an analog voltage signal is sent from the sensor to the computer in relation to the amount of vacuum. On a typical MAP sensor, the signal voltage changes from 1V to 1.5V at idle to 4.5V at wide-open throttle. The signal voltage may be checked with a digital voltmeter on MAP sensors with a silicon diaphragm.

Another variation of the MAP sensor is a capacitance discharge MAP. Instead of using a silicon diaphragm, this system uses a variable capacitor. In the capacitor capsule-type MAP sensor, two flexible alumina plates are separated by an insulating washer (Figure 25-1). A film electrode is deposited on the inside surface of each plate and a connecting lead is extended for external connections. The result of this construction is a parallel plate capacitor. This capsule is placed inside a sealed housing which is connected to intake manifold pressure. As manifold pressure increases (goes toward atmospheric), the alumina plates deflect inward, resulting in a decrease in the distance between the electrodes. The formula for capacitance is:

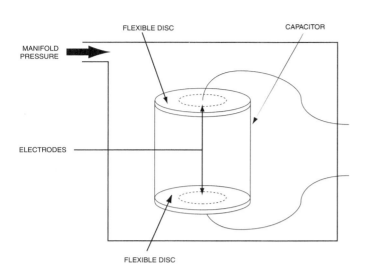

Figure 25-1 Capacitance discharge MAP sensor.

$$C = \frac{\varepsilon_0 A}{d}$$

ε_0 is the dielectric constant of air

A is the area of the film electrodes

d is the distance between electrodes

Since everything in the formula is constant except for the distance between the electrodes, a measure of capacitance constitutes a measurement of pressure.

Ford MAP sensors contain a pressure-sensitive disk capacitor (Figure 25-2). This type of MAP sensor changes manifold pressure to a digital voltage signal of varying frequency. The MAP sensor actually senses the difference between atmospheric pressure and manifold vacuum. When the engine is idling, high intake manifold vacuum is about 18 in. of mercury (Hg). Under this condition, the MAP sensor signal is approximately 95 hertz. Ford refers to this condition as low MAP, because there is a greater difference between atmospheric pressure and manifold vacuum.

If the engine is operating at, or near, wide-open throttle, the manifold vacuum may be 2 in. Hg, and the MAP sensor hertz approximates 160. This condition may be called high MAP, because the manifold vacuum is closer to atmospheric pressure.

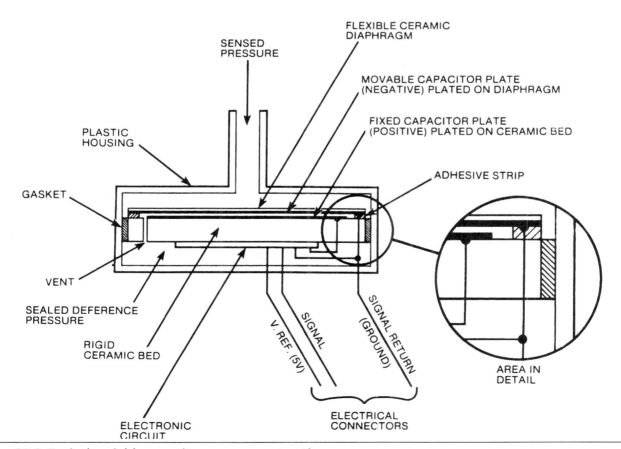

Figure 25-2 Typical variable capacitor sensor construction.

WARNING *Never connect any type of voltmeter directly to the voltage signal wire on a Ford MAP sensor. This action causes MAP sensor damage! Ford MAP sensor testers read the sensor hertz directly, and other Ford MAP sensor testers convert the hertz to a voltage signal.*

Most manufacturers will use a **limp-in** value if the MAP sensor signal is considered invalid by the PCM. Limp-in values are used to keep the engine running. However engine performance suffers greatly under these conditions. Usually the predetermination is based on throttle position sensor inputs.

Mass Air Flow Sensors

Mass air flow (MAF) sensors are used to determine the volume of air entering the engine. The MAF sensors may be classified as vane, heated grid, hot wire, pressure, or ultrasonic. The MAF sensor is often mounted in the hose between the air cleaner and the throttle body so that all the intake air must flow through the sensor.

Some vane-type MAF sensors are referred to as volume air flow meters (Figure 25-3). In a vane-type MAF sensor, a pivoted air-measuring plate is lightly spring-loaded in the closed position. As the intake air flows through the sensor, the air-measuring plate moves toward the open position. Plate movement is proportional to intake air flow (Figure 25-4).

A movable pointer is attached to the measuring plate shaft. The pointer contacts a resistor to form a potentiometer which sends a voltage signal to the PCM in relation to intake air flow. A thermistor in the air flow meter sends a signal to the PCM in relation to air intake temperature.

The heated resistor-type MAF sensor has a heated resistor mounted in the center of the air passage. In addition, an electronic module is mounted on the side of the sensor. When the ignition switch is turned on, voltage is supplied to the module. The module sends current through the resistor to maintain a specific resistor temperature.

If an engine is accelerated suddenly, the rush of air tries to cool the resistor. Under this condition, the module

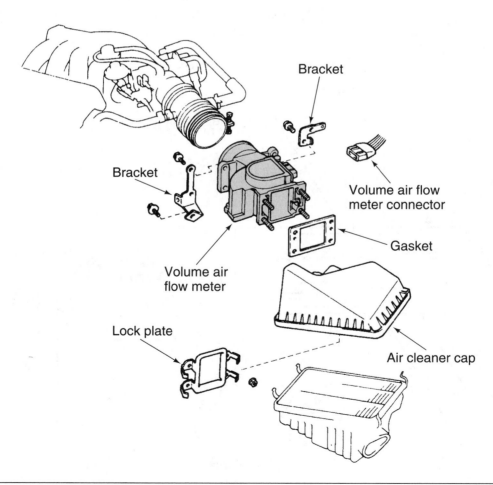

Figure 25-3 A vane-type mass air flow sensor used to measure the volume of air entering the engine.

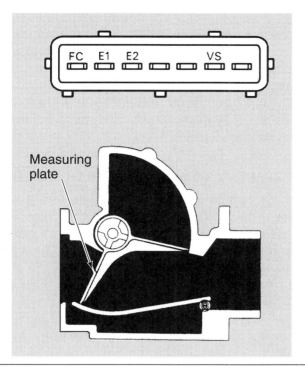

Figure 25-4 The measuring plate movement in a MAF sensor is proportional to intake air flow.

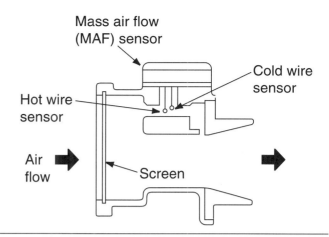

Figure 25-5 Hot wire-type MAF sensor. (Reprinted with permission of Ford Motor Company)

supplies more current to maintain the resistor temperature. The module sends the increasing current signal (proportional to the air flow entering the engine) to the PCM. When the PCM receives this increasing current flow signal it increases the amount of fuel delivery with the additional air flow entering the engine. The MAF module reacts in a few milliseconds to maintain the resistor temperature. Some MAF sensors have an electric grid in place of the resistor.

In a hot wire-type MAF sensor, a hot wire is positioned in the air stream through the sensor, and an ambient temperature sensor wire is located beside the hot wire. This ambient temperature sensor wire senses intake air temperature and may be referred to as a cold wire (Figure 25-5).

When the ignition switch is turned on, the MAF module sends enough current through the hot wire to maintain the temperature of this wire at 392°F (200°C) above the ambient temperature sensed by the cold wire. If the engine is accelerated suddenly, the rush of air tries to cool the hot wire. The module immediately sends more current through the wire to maintain the wire. The module sends the increasing current signal to the PCM that is directly proportional to the intake air flow. When this signal is received, the PCM supplies more fuel to go with the increased intake air flow. This action maintains the exact air/fuel ratio required by the engine. Some MAF sensors have a burn-off relay and related circuit. When

the ignition switch is turned off, after the engine has been running for a specific length of time, the computer closes the burn-off relay. This relay activates a burn-off circuit in the MAF which heats the hot wire to a very high temperature for a short time period. This action burns contaminants off the hot wire.

The pressure-type MAF sensor uses the Karman vortex phenomenon to determine the volume of air entering the engine. A vortex-generating column is located in the path of air flow (Figure 25-6). Vortexes are generated in proportion to air flow speed (air volume).

PRESSURE TYPE MAF

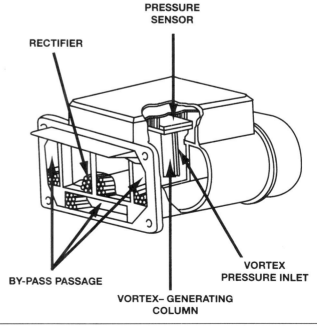

Figure 25-6 Pressure-type MAF sensor. (Courtesy of Chrysler Corporation)

A pressure inlet is positioned downstream of the vortex-generating column (Figure 25-7). The pressure inlet is connected to a pressure sensor (stress gauge). The vortexes generated by the column result in pressure variations that are detected by the pressure sensor. As the air flow through the MAF sensor increases, so does the number of pressure variations. The voltage changes resulting from the pressure variations are changed to a pulsed digital signal that is proportional to intake air flow. The digital signal is then sent to the PCM.

The ultrasonic MAF sensor also uses the Karman vortex phenomenon. The difference is this sensor measures air flow with ultrasonic waves. An ultrasonic transmitter is positioned downstream of the vortex-generating column. The transmitter is located across from an ultrasonic receiver (Figure 25-8). The sensor determines the volume of intake air by measuring the length of time required for the ultrasonic waves from the transmitter to reach the receiver.

When the ignition switch passes through the RUN position on the way to the START position, there is no air flow over the sensor. During this time a fixed reference time for the ultrasonic transmission is established. A clockwise rotating vortex speeds up the ultrasonic waves and they reach the receiver more quickly than the reference time. A counterclockwise vortex tends to slow down the ultrasonic waves.

The resulting signal is then converted into digital pulses and sent to the PCM. The greater number of pulses indicates a greater volume of air.

Barometric Pressure (BARO) Sensors

Many systems use a **barometric pressure (BARO) sensor** to determine the amount of fuel delivery needed to start the engine and for spark timing strategies. In some applications the BARO sensor is part of the MAF sensor. As its name implies, this sensor measures atmospheric pressure. At lower pressures, the engine requires more air for efficient combustion. At high pressures, it requires less air. Basically the BARO sensor operates the same as the MAP sensor discussed earlier, except there is no vacuum source attached to it.

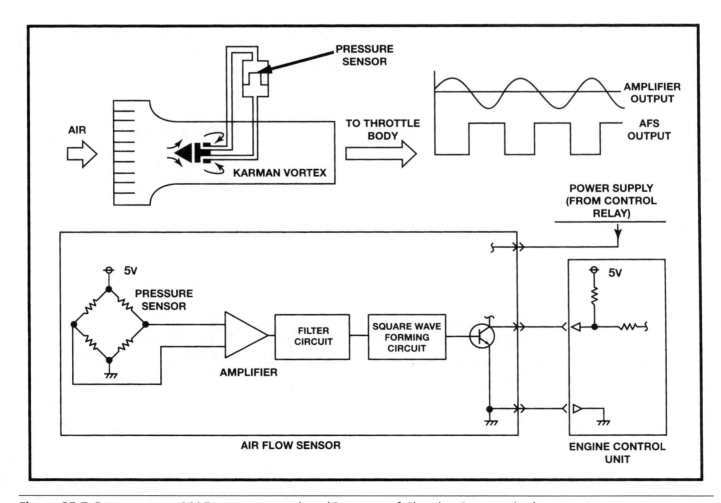

Figure 25-7 Pressure-type MAF sensor operation. (Courtesy of Chrysler Corporation)

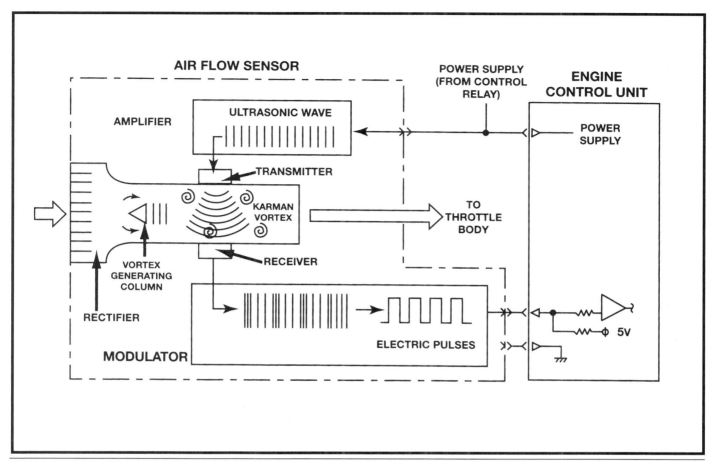

Figure 25-8 Ultrasonic mass air flow sensor. (Courtesy of Chrysler Corporation)

Throttle Position Sensor

The most common type of **throttle position sensor (TPS)** contains a potentiometer with a pointer that is rotated by the throttle shaft. Therefore, the TPS is mounted on the end of the throttle shaft in the throttle body.

Three wires are connected from the TPS to the computer. When the ignition switch is on, the computer supplies a constant 5-V reference voltage to the sensor on one of these wires. One of the other wires is a TPS signal wire from the sensor to the computer, and the third wire is a ground wire between these two components. Sensor ground wires are usually black, or black with a colored tracer. Some manufacturers will send a second 5 volts up the signal wire from the PCM (Figure 25-9). This 5 volts is sent through a pull-up resistor to the wiper of the potentiometer. The PCM measures the voltage drop over the pull-up resistor which changes as the resistance between the wiper and ground of the TPS changes. This method keeps the TPS sensor voltage extremely accurate through the temperature range the TPS is exposed to. In addition, it provides for a means for the PCM to determine if the TPS sensor is operating properly.

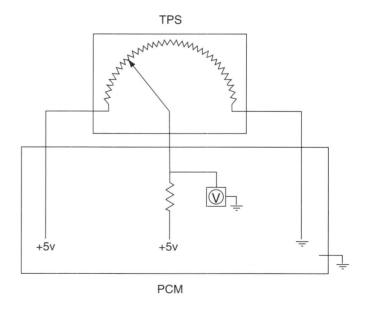

Figure 25-9 Some TPS sensors have 5 volts sent from the PCM to the wiper.

A typical TPS has 1,000Ω with the throttle in the idle position, and 4,000Ω at wide-open throttle. The voltage signal from a typical TPS is 0.5V to 1V at idle and 3.8V to 4.5V at wide-open throttle. This signal informs the computer regarding the exact throttle position.

The computer also knows how fast the throttle is opened from the TPS signal. When the engine is accelerated suddenly, a richer air/fuel ratio is required with the additional air flow into the engine. If the computer receives a TPS signal indicating sudden acceleration, the computer supplies the necessary richer air/fuel ratio. The computer also uses the TPS signal to control other outputs.

Elongated mounting holes on some TPSs provide a sensor adjustment. When a TPS adjustment is necessary, the mounting bolts are loosened and the sensor is rotated until the specified signal voltage is available with the engine idling. On many TPSs, there is no provision for sensor adjustment.

Most manufacturers will substitute a predetermined limp-in value if the TPS signal is considered invalid by the PCM. The predetermined value is usually based on rpm and/or MAP sensor inputs.

Engine Coolant Temperature Sensor

The **engine coolant temperature (ECT) sensor** sends an analog voltage signal to the computer in relation to coolant temperature. The engine coolant temperature (ECT) sensor is often threaded into the intake manifold with the lower tip immersed in engine coolant. Usually, the ECT sensor contains a thermistor which provides high resistance when cold and much lower resistance at higher temperatures (NTC thermistor). A typical ECT sensor may have 35,000Ω at -40°F (-40°C) and 1,200Ω at 248°F (120°C). The ECT sensor has two wires connected between the sensor and the computer. One of these wires is a voltage signal wire and the other wire provides a ground. The computer senses the voltage drop across the ECT sensor, and this voltage changes in relation to the coolant temperature and sensor resistance. For example, at low coolant temperature and high sensor resistance, the voltage drop across the sensor might be 4.5V, whereas at high coolant temperature and low sensor resistance, this voltage drop would be approximately 0.3V (Figure 25-10). Notice the PCM does not send a voltage to the ECT sensor and measure the voltage that is returned. Instead the reference voltage is sent through a fixed pull-up resistor that is in series with the ECT sensor. The PCM measures the voltage drop over the fixed resistor, which will change as the resistance of the thermistor changes. Most systems use a dampened response, negative temperature coefficient (NTC) thermistor-type sensor.

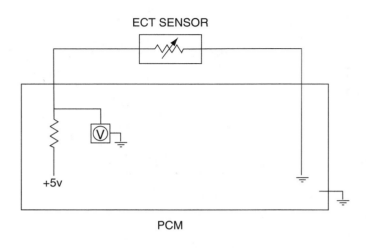

Figure 25-10 Typical ECT sensor circuit.

The computer must know the coolant temperature to complete many of the necessary decisions regarding output control functions. For example, the computer must provide a richer air/fuel ratio when the engine is cold and a leaner stoichiometric air/fuel ratio once the engine is at normal operating temperature. Therefore, the computer must know engine coolant temperature from the ECT signal to provide the correct air/fuel ratio.

A cold engine requires more spark advance for improved performance, whereas a hot engine needs less spark advance to prevent detonation. The computer must know engine coolant temperature and other inputs to provide the correct spark advance.

Some emission devices, such as the exhaust gas recirculation (EGR) valve, are not required on a cold engine, but this valve must be operational on a hot engine under certain operating conditions. The computer uses the ECT signal and other inputs to control the EGR valve.

A limp-in substitution value may be used by the PCM if the signal from the ECT becomes invalid. This value usually is about 115°F. This value will allow the engine to run with limited performance.

Dual Ramping Temperature Circuits

Many manufacturers use a **dual ramping ECT** sensor circuit (Figure 25-11). This system provides a more precise method of determining the engine temperature. Since an 8-bit computer is limited to 256 numbers within

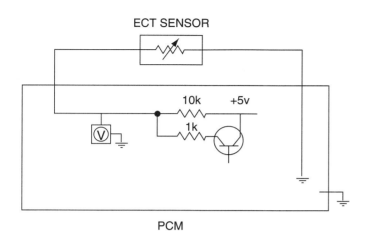

Figure 25-11 Typical dual ramping ECT sensor circuit.

one byte of information, only 256 temperature graduations are possible without additional computer calculations. By using a dual ramping system the number of temperature graduations are squared. This makes the sensor accurate at cold and hot temperatures. In figure 25-11, the 5-volt reference signal is sent through a $10,000\Omega$ resistor when the engine is cold. As the engine temperature increases, the voltmeter value will decrease. When the voltmeter reaches a predetermined value (1.25 volts for example), the PCM will turn on the transistor. This action puts a $1,000\Omega$ resistor in parallel to the $10,000\Omega$ resistor. The calculated resistance of the parallel circuit is now 909Ω. This causes the voltage reading to go high again and provides a second set of inputs (Figure 25-12). The dual ramping allows for particularly accurate high temperature readings with 50,000 possible divisions in the 110-250° range.

Intake Air Temperature Sensor

Some **intake air temperature (IAT) sensors** are threaded into the intake manifold with the lower end of the sensor protruding into one of the air passages. On some applications, the ACT sensor is mounted in the air cleaner and senses air intake temperature at this location. The intake air temperature sensor contains a thermistor, which has similar ohm and voltage drop readings compared to the ECT sensor (Figure 25-13). Usually the IAT input is used during engine start-up, colder ambient temperatures, and during wide-open throttle conditions.

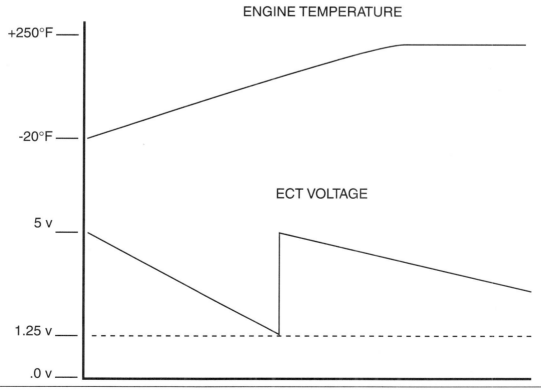

Figure 25-12 Engine coolant temperature will continue in a linear line while the voltage switches to start another set of binary numbers.

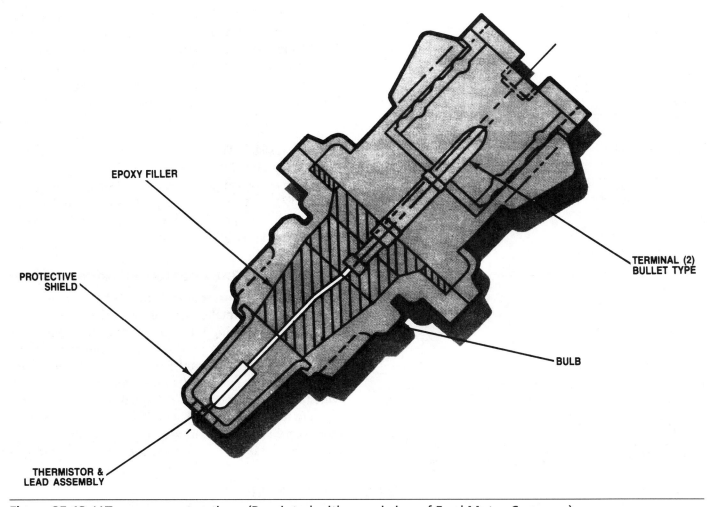

Figure 25-13 IAT sensor construction. (Reprinted with permission of Ford Motor Company)

A voltage signal wire and a ground wire are connected between the intake air temperature sensor and the computer. Cold intake air is denser; therefore, a richer air/fuel ratio is required. When the ACT sensor signal indicates colder intake air temperature, the computer provides a richer air/fuel ratio.

Some manufacturers use a dual ramping intake air temperature circuit. This system works identically as the dual ramping ECT system discussed earlier.

Oxygen Sensors

As discussed in Chapter 21, most oxygen sensors (O_2S) are voltage-generating devices consisting of a variable voltage source and a resistor wired in series (Figure 25-14). The voltage source produces an analog signal ranging between 0 and 1 volt. The oxygen sensor is threaded into the exhaust manifold or into the exhaust pipe near the engine. Some manufacturers refer to this sensor as an exhaust gas oxygen (EGO) sensor or a heated exhaust gas oxygen (HEGO) sensor.

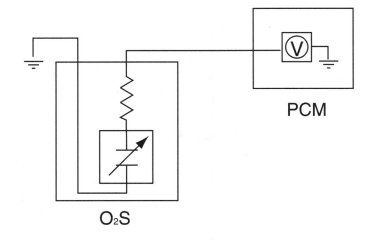

Figure 25-14 Typical O_2S circuit.

A steel cover or a neoprene boot is installed over the top of the sensor. On many oxygen sensors, the steel cover is loosely installed on the sensor, which allows a constant supply of oxygen from the atmosphere inside the oxygen-sensing element. If a neoprene boot is installed on the sensor, the boot has grooves cut on the inner surface to allow air inside the sensing element. On some later model oxygen sensors, the top of the sensor is tightly sealed, and oxygen enters the sensor through the signal wire.

A shield covers the bottom end of the sensor that protrudes into the exhaust manifold or pipe. Flutes in this shield help to continually swirl the exhaust gas around the sensor element when the engine is running.

The sensors detect exhaust oxygen content by a galvanic reaction within the sensor. Once the O_2S reaches about 600°F (315°C) the zirconium becomes active. Instead of dissimilar metals in the presence of electrolyte (as with the vehicle's battery) the O_2S uses dissimilar amounts of oxygen in the presence of electrolyte to generate the voltage. When the oxygen content of the exhaust is low (rich air/fuel) the difference is greater and the O_2S generates about 1 volt. If there is a high oxygen content in the exhaust, there is little potential and the voltage generated is about 0 volts.

A lean air/fuel ratio provides excess quantities of oxygen in the exhaust stream, because the mixture entering the cylinders has an excessive amount of air in relation to the amount of fuel. Therefore, air containing oxygen is left over after the combustion process. When the exhaust stream has high oxygen content, oxygen from the atmosphere is also present inside the O_2S element. When oxygen is present on both sides of the sensor element, the sensor produces a low voltage.

A rich air/fuel ratio contains excessive fuel in relation to the amount of air entering the cylinders. A rich air/fuel mixture produces very little oxygen in the exhaust stream because the oxygen in the air is all mixed with fuel, and excess fuel is left over after combustion is completed. When the exhaust stream with very low oxygen content strikes the O_2S, there is high oxygen content from the atmosphere inside the sensor element. With different oxygen levels on the inside and outside of the sensor element, the sensor produces up to 1V. As the air/fuel ratio cycles from lean to rich, the O_2S voltage changes in a few milliseconds.

In a gasoline fuel system, the stoichiometric, or ideal, air/fuel mixture is 14.7:1. This indicates that for every 14.7 pounds of air entering the air intake, the fuel injectors supply 1 pound of fuel. At the stoichiometric air/fuel ratio, combustion is most efficient, and nearly all the oxygen in the air is mixed with fuel and burned in the combustion chambers. Computer-controlled fuel injection systems maintain the air/fuel ratio at stoichiometric under most operating conditions. As the air/fuel ratio cycles slightly rich and lean from the stoichiometric ratio, the O_2S voltage cycles from high to low. Once the engine is at normal operating temperature, the O_2S voltage should vary between 0.3V and 0.8V if the air/fuel ratio is at, or near, stoichiometric.

WARNING *The use of some RTV sealants will contaminate the O_2S.*

WARNING *The use of leaded gasoline or coolant leaks into the combustion chamber will contaminate the O_2S.*

If leaded regular fuel is used in an engine with an O_2S, the sensor becomes lead-coated in a short time. Under this condition, the sensor signal is no longer satisfactory, and sensor replacement is likely required. Certain types of room temperature vulcanizing (RTV) sealant or antifreeze entering the combustion chambers will contaminate the O_2S. Always use the RTV sealant recommended by the vehicle manufacturer.

WARNING *The O_2S will be damaged if it is tested with an analog voltmeter. A digital meter must be used to test this sensor.*

An analog voltmeter must never be used to check the O_2S voltage. Use a 10M impedance digital meter or lab scope to check the O_2S. The sensor threads should be coated with an anti-seize compound prior to sensor installation, or sensor removal may be difficult the next time it is necessary.

Titania-type O_2S

Some vehicles are now equipped with a **titania-type O_2S**. This sensor design is similar to the zirconia-type sensor, but the sensing element is made from titania rather than zirconia.

The titania-type sensor modifies voltage, whereas the zirconia-type sensor generates voltage. The computer supplies battery voltage to the titania-type sensor, and this voltage is lowered by a resistor in the circuit. The resistance of the titania varies as the air/fuel ratio cycles from rich to lean. When the air/fuel ratio is rich, the titania resistance is low, and this provides a higher voltage signal to the computer. If the air/fuel ratio is lean, the titania resistance is high, and a lower voltage signal is sent to the computer.

The titania-type O_2S provides a satisfactory signal almost immediately after a cold engine is started. This action provides improved air/fuel ratio control during engine warm-up.

Exhaust Gas Recirculation Valve Position Sensor

CAUTION *The EGR valve and EVP sensor may be very hot if the engine has been running. Wear protective gloves to avoid burns if it is necessary to remove and handle these components after the engine has been running.*

The **exhaust gas recirculation valve position (EVP) sensor** is mounted on top of the EGR valve (Figure 25-15). A linear potentiometer with a stem extending to the top of the EGR valve is mounted in the EVP sensor. When the EGR valve opens, the potentiometer stem moves upward and a higher voltage signal is sent to the PCM.

The EVP sensor acts as a feedback signal to inform the computer regarding the EGR valve position. Resistance in the EVP linear potentiometer changes from $3,000\Omega$ with the EGR valve closed to $5,000\Omega$ when the EGR valve is open. The voltage signal from the EVP sensor changes from 0.8V with the EGR valve closed to 4.5V with the EGR valve open. A 5-V reference wire, a ground wire, and a signal wire are connected from the EVP sensor to the computer. The EVP sensor is used on some Ford products; however, other manufacturers have similar sensors.

Vehicle Speed Sensor

On many systems, the **vehicle speed sensor (VSS)** is connected in the speedometer cable or mounted in the transaxle opening where the speedometer cable was previously located (Figure 25-16). In the latter case, the speedometer cable is connected to the VSS. The speedometer cable is not required if the vehicle has an electronic speedometer.

In some VSSs, the speedometer drive rotates a magnet inside a coil of wire. This type of sensor produces an

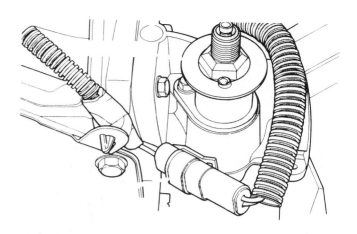

Figure 25-16 Vehicle speed sensor. (Courtesy of Chrysler Corporation)

AC voltage signal proportional to vehicle speed. In some other VSSs, a magnet with eight poles rotates past a set of reed points, and these rotating poles open the points eight times per sensor revolution. Each time the points open, a signal is sent to the computer. The computer uses the VSS signal for converter clutch lockup and cruise control operation. In some applications, the VSS is located in the speedometer head.

Most fuel systems do not require this signal for determining fuel strategies directly. Instead this input is used to determine which operating cell the system is in. Operating cells are discussed in the section on adaptive memory.

Neutral Drive Switch

The **neutral drive switch (NDS)** is operated by the transmission linkage (Figure 25-17). If the transmission is in park or neutral, the NDS is closed. When the transmission is placed in drive or reverse, the NDS opens. A closed NDS sends a voltage signal below 1V to the computer, whereas an open NDS provides a signal above 5V to the computer.

The NDS switch signal informs the computer regarding gear shift selector position, and the computer uses this signal to control idle speed. The NDS may be referred to as a neutral/park switch.

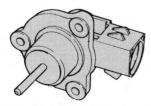

Figure 25-15 EVP sensor. (Reprinted with permission of Ford Motor Company)

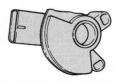

Figure 25-17 Neutral drive switch. (Reprinted with permission of Ford Motor Company)

Idle Switch

Some EFI systems use an **idle switch** to identify a closed throttle condition. This information is used to enable target idle control. This switch input is also used for decel fuel shut-off control.

The idle switch is closed (low voltage signal) when the throttle plates are in the idle position. If the idle switch input always shows a high (open) condition, the PCM will not perform decel fuel shut-off. In addition, the PCM will lose all idle speed control.

If the idle switch input always shows low (shorted), the engine may surge when the transmission is in gear and the vehicle is accelerating from a stop. For example, if the driver moves the accelerator pedal to increase engine speed (still below 2,000 rpm) as the engine speed increases the PCM starts to close the idle air control motor to decrease engine speed. This happens because the PCM is still trying to maintain target idle due to the closed idle switch input. In addition to trying to use the idle air control motor, the PCM will retard the spark timing in an attempt to reach target idle.

Outputs

Although there are several outputs controlled by the PCM in the fuel injection system, many of these operations have been covered in previous chapters. This includes such components as relays and solenoids. The following discussion covers some of the other types of output devices used by the fuel systems.

Injector Internal Design and Electrical Connections

The **fuel injector** is simply a fast acting solenoid. Whenever the injector is opened, it sprays a constant amount of fuel for a given amount of time. Because pressure drop across the injector is fixed, the fuel flow through the injector is constant. With the injector connected to a pressurized fuel supply, atomized fuel sprays from the injector nozzle behind the intake valve.

Inside the injector, the plunger and valve seat are held downward by a spring. In this position, the seat closes the metering orifices in the end of the injector. Openings in the sides of the injector allow fuel to enter the cavity surrounding the injector tip. A mesh screen filter inside the injector openings removes dirt particles from the fuel. In some injectors, a diaphragm is located between the valve seat and the housing (Figure 25-18). The tip of the injector may contain one to six metering orifices. Injector design varies depending on the manufacturer.

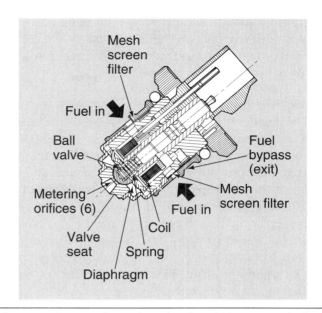

Figure 25-18 Throttle body injector design. (Reprinted with permission of Ford Motor Company)

Each injector contains two terminals, and an internal coil is connected across these terminals. The plunger is positioned in the center of the coil, and the lower end of the plunger has a tapered valve seat. When the ignition switch is turned on, 12V are supplied to one of the injector terminals, and the other injector terminal is connected to the computer. When the computer grounds this terminal, current flows through the injector coil to ground. When this action occurs, the injector coil magnetism moves the plunger and valve seat upward. Fuel, under pressure, sprays out the injector orifices into the air stream.

Injector Deposits

In some EFI systems, there have been problems with deposits on injector tips resulting from small quantities of gum present in gasoline. Injector deposits usually occur when this gum bakes onto the injector tips after a hot engine is shut off. Most oil companies have added detergents to their gasoline to help prevent injector tip deposits. Vehicle manufacturers and auto parts stores sell detergent additives to mix in the fuel tank to clean injector tips. Also, some vehicle manufacturers and parts suppliers have designed deposit-resistant injectors. These injectors have several different pintle tip and orifice designs to help prevent deposits. On one type of deposit-resistant injector, the pintle seat opens outward away from the injector body and more clearance is provided between the pintle and the body. Another type of deposit-resistant injector has four orifices in a metering plate rather than a single orifice. Some deposit-resistant injectors may be recognized by the injector body color. For example, conventional injectors supplied by Ford Motor Company are painted black, whereas their deposit-resistant injectors have tan or yellow bodies.

Pulse Width

The length of time the computer grounds the injector is referred to as pulse width. Under most operating conditions, the computer provides the correct injector pulse width to maintain the stoichiometric air/fuel ratio. For example, the computer might ground the injector for 2 milliseconds at idle speed and 7 milliseconds at a near wide-open throttle condition to provide the stoichiometric air/fuel ratio. In many TBI systems, the computer grounds an injector each time a signal is received from the distributor pickup. This type of TBI system is referred to as a **synchronized system**, because the injector pulses are synchronized with the pickup signals. In a dual injector throttle body assembly, the computer grounds the injectors alternately under most operating conditions.

Air/Fuel Ratio Enrichment

When the coolant temperature sensor signal to the computer indicates the engine coolant is cold, the computer increases the injector pulse width to provide a richer air/fuel ratio. This action eliminates the need for a conventional choke assembly. The PCM supplies the proper air/fuel ratio and engine rpm when starting a cold engine. This also eliminates the need for the driver to depress the accelerator pedal while starting the engine.

TRADE JARGON: Starting a fuel injected engine without depressing the accelerator pedal may be called no-touch starting.

When a EFI-equipped engine is cold, the computer provides a very rich air/fuel ratio for faster starting. However, if the engine does not start because of an ignition defect, the engine becomes flooded quickly. Under this condition, excessive fuel may run past the piston rings into the crankcase. Therefore, when a cold EFI engine does not start, periods of long cranking should be avoided.

If the driver suspects that the air/fuel ratio is extremely rich, he or she may push the accelerator pedal to the wide-open position while cranking the engine. Under these conditions, the computer program provides a very lean air/fuel ratio. Some manufacturers have programmed their PCMs to shut down injector operation all together during this time. This mode is referred to as a **clear flood**. However, under normal conditions, the driver should not push on the accelerator pedal at any time

when starting an engine with EFI. When the engine is decelerated, the computer reduces injector pulse width to provide a lean air/fuel ratio, which reduces emissions and improves fuel economy.

Idle Air Control Motors

Many TBI systems will use a DC reversible motor to control idle speed. In this system the pintle of the **idle speed control (ISC) motor** is located against a lever that is attached to the throttle plates (Figure 25-19). As the motor moves the pintle in and out the throttle plates move, allowing more or less air to enter the engine. For example, if idle speed needs to be increased, the PCM would activate the motor to move the pintle so the throttle plates would open slightly. At this time more air is allowed to enter the engine. The O_2S senses the high oxygen content and the PCM will increase the pulse width to maintain stoichiometric. With the increase of air and fuel, engine speed increases. The opposite will occur if idle speed needs to be decreased.

In most newer EFI systems idle speed is controlled by regulating air flow around the throttle plates. The **idle air control (IAC) stepper motor** is used to control the amount of air that will bypass the closed throttle plates (Figure 25-20).

Most IAC stepper motors use a permanent magnet and two to six electromagnets. Each command from the PCM results in one step movement. By switching which circuit supplies the current and which supplies ground (through the use of a H driver) the PCM can control the direction of motor rotation. To decrease engine speed, the controller will step the motor in to close off the air bypass. If engine speed needs to be increased, the PCM driver the stepper motor pintle out to increase the size of the bypass.

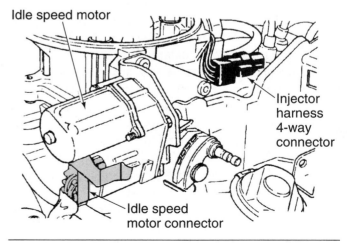

Figure 25-19 Idle speed control motor. (Courtesy of Chrysler Corporation)

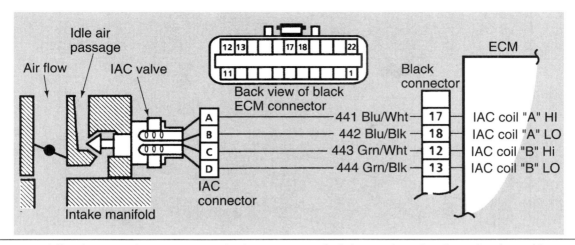

Figure 25-20 An idle air control motor controls the amount of air bypassing the throttle plate. (Courtesy of General Motors Corporation, Service Technology Group)

Open Loop vs Closed Loop

In a computer-controlled fuel system, **open loop** occurs when the engine coolant is cold. Under this condition, the O_2S is too cold to produce a satisfactory signal, and the computer program controls the air/fuel ratio without the O_2S input. During this open loop mode, the computer provides a richer air/fuel ratio.

As the engine approaches normal operating temperature, the computer enters the **closed loop** mode in which the computer uses the O_2S signal to control the air/fuel ratio. Closed loop in many computer-controlled fuel systems occurs when the engine coolant temperature (ECT) sensor signal informs the computer that the coolant temperature is approximately 175°F (79.4°C) and the O_2S signal is valid.

Therefore, the ECT sensor signal is very important because it determines the open or closed loop status. For example, if the engine thermostat is defective and the coolant temperature never reaches 175°F (79.4°C), the computer-controlled fuel system never enters closed loop. When this open loop condition occurs, the air/fuel ratio is continually rich, and fuel economy is reduced. Some systems go back into open loop during prolonged periods of idle operation when the O_2S cools down. Also, many systems revert to open loop at or near wide-open throttle to provide a richer air/fuel ratio.

To keep the computer in closed loop during idle, and to get the O_2S hotter faster, most newer O_2S are heated electrically (Figure 25-21). Battery voltage is sent to the heater element when the ignition switch is in the RUN position. The heater element has a positive temperature coefficient (PTC) thermistor which operates as a current regulator. When the O_2S is hot the PTC resistance increases, thus it causes a reduction in current flow to the heater element.

Computer Air/Fuel Ratio Strategy

In an EFI system, the computer must know the amount of air entering the engine so it can supply the stoichiometric air/fuel ratio. This air/fuel ratio is 14.7 to 1 (14.7 parts air to 1 part fuel). This ratio is ideal for both fuel efficiency and emission control. However, conditions inside the engine's combustion chamber are not ideal. Even with a stoichiometric ratio, the engine's exhaust gases contain a certain percentage of unburned fuel in the form of hydrocarbons (HC) and carbon monoxide (CO). In addition, at higher combustion temperatures some of the free oxygen and nitrogen gases combine, forming various oxides of nitrogen (NOx).

To adjust to the various operating conditions, vehicles must be equipped with fast responding onboard computer systems. The computer will use an oxygen sensor to measure the oxygen content of the exhaust. This provides information concerning how well the computer did in calculating the correct air/fuel mixture for the previous burn. The computer then works to maintain driveability while maintaining the 14.7 to 1 ratio as closely as possible.

The length of time the computer energizes the injector (pulse width) determines the duration of fuel flow and, therefore, the amount of fuel in the combustion chamber. This meets operating demands as closely as possible. There are three basic strategies, or methods, manufacturers use to maintain the stoichiometric ratio: speed density, air density, and density speed. Regardless of the system used, the computer must know the amount of air entering the engine in order to provide the proper pulse width.

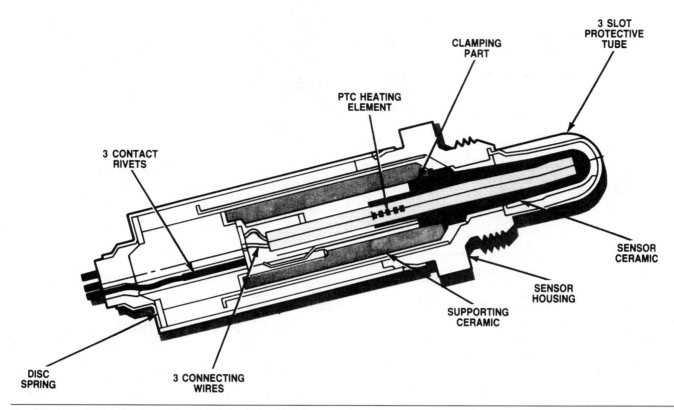

Figure 25-21 The heater element gets the O_2S hotter faster and keeps the PCM in closed loop during extended idle. (Reprinted with permission of Ford Motor Company)

Speed Density Systems

In EFI systems with a manifold absolute pressure (MAP) sensor, the computer program is designed to calculate the amount of air entering the engine from the MAP and rpm input signals. The distributor pickup or crank position sensor supplies an rpm signal to the computer. This type of EFI system is referred to as a **speed density system**, because the computer calculates the air intake flow from the engine rpm, or speed, input, and the density of intake manifold vacuum input. Therefore, the computer must have accurate signals from these inputs to maintain the stoichiometric air/fuel ratio. The other inputs are used by the computer to "fine tune" the air/fuel ratio. For example, if the TPS input indicates sudden acceleration, the computer momentarily supplies a richer air/fuel ratio.

As stated, in the speed density system, the two most important inputs are engine speed and MAP. All other inputs are used to modify the pulse width calculation. Below is the formula for a typical speed density pulse width calculation. The computer will assign a multiplicative value to the input signal, then determine the pulse width based on these values.

$$\frac{MAP}{RPM} \times TPS \times ETC \times IAT \times O_2S \times BATTERY \times$$

$$VOLTAGE \times ADAPTIVE\ MEMORY$$

Air Density Systems

EFI systems that use a mass air flow (MAF) sensor instead of a MAP sensor to determine air volume are referred to as **air density** or **mass air systems**. These systems are referred to as air density systems since the MAF sensor provides a direct input as to the amount of air entering the engine. The MAF sensor measures the temperature and volume of air entering the engine, as well as altitude for the PCM. However, in this system the primary input is the rpm signal. The following is a typical formula used by the computer to calculate pulse width in an air density system:

$$\frac{Air\ Flow\ Hz}{RPM} \times TPS \times ECT \times BARO \times IAT \times O_2S \times$$

Air Fuel Compensator $\times$ Adaptive Memory

There are a few manufacturers that use both MAF and MAP sensors. In this case, the MAP sensor is used mainly as a backup if the MAF sensor fails. In some 4-

cylinder applications using the air density strategy, the TPS sensor input is unique in that it does not affect the fuel injector base pulse width. Instead, the PCM (based on TPS input) will add four additional injector firings based on the rate of TPS increase. If the TPS sensor input identifies a rapid opening of the throttle plates, the PCM will add four additional infecter pulses (one to each cylinder). The extra pulses occur every 10 milliseconds as long as the TPS input increases at a rapid rate. The actual amount of fuel delivered by the extra pulses is determined by ECT inputs. However, the maximum pulse width of these pulses is 4 milliseconds. The effect of TPS inputs on fuel strategy is listed below:

TPS Voltage	Multiplicative Value
Steady	1.0 (no effect)
Rapidly increasing	5.0 (increase pulse width by 500%)
Rapidly decreasing	0.3 (decrease pulse width by 70%)

Usually the BARO sensor input has no effect on fuel strategies at sea level. The range of authority of a typical BARO sensor on fuel ranges from reducing pulse width by about 50% to 12 1/2% increase. This means at higher altitudes this sensor input can reduce the calculated base pulse width in half. If the engine is operating below sea level, more fuel will be added.

Density Speed Systems

In most **density speed** systems, the most important input is from the MAP sensor. In fact the engine will not run if this input is missing. The second important input is a combination TDC/Crankshaft position/Engine speed sensor. If this input is missing the engine will not run as well.

Modes of Operation

The following are typical modes of operation for most EFI systems. Variations of these modes are also used. For this reason, the correct service manual should be referenced when diagnosing a fuel system problem.

Key ON

Typically, when the ignition switch is turned to the ON or RUN position, the PCM receives a key "on" signal. The PCM will then gather the needed information in order to prepare the engine for starting. Usually the information required is:

- Barometric pressure
- Engine coolant temperature
- Throttle position
- Intake air temperature

Also during this time the fuel pump is turned on for a short time to prime the system.

Crank

When the ignition switch is located in the START position and the engine is cranking, the PCM receives crankshaft and camshaft position sensor signals. The PCM will then turn on the fuel pump. Depending on the system, the PCM may also energize all of the injectors. During the time the engine is being cranked, the PCM ignores the O_2S input.

The TPS allows the PCM to monitor the amount of air delivered to the engine. Some EFI systems also receive an injector sync signal at this point to inform the PCM which cylinder is approaching the compression stroke for proper fuel injector sequencing.

Open Loop

When the engine rpm increases to about 450 rpm, the PCM switches to "run" mode. Based on the input from the MAP or MAF sensor and engine speed inputs, the PCM activates the fuel injectors and generates ignition timing signals. Initially the O_2S input is ignored since the sensor is cold and a rich mixture was delivered during cranking to start the engine. The PCM delivers a predetermined air/fuel ratio and a higher idle speed to bring the engine up to operating temperature faster.

Closed Loop

Once the PCM has determined the O_2S is warm enough to send valid readings, it switches to closed loop operation. The PCM will now adjust the air/fuel mixture based on O_2S inputs.

Wide-Open Throttle

The wide-open throttle (WOT) signal is received by the PCM from the TPS. MAP sensor inputs also identify a WOT condition. Upon recognition of a WOT condition the PCM switches back to open loop operation and ignores the O_2S input. The main priority of the PCM during this mode is to increase the injector pulse width.

Adaptive Memory

The PCM's goal is to use the input information and control outputs to provide a constant stoichiometric ratio. The operating modes just discussed represent decision categories for the PCM. Within each of the modes of operation, smaller and more finite decisions are made.

To control air/fuel ratio feedback, the PCM uses **adaptive memory** correction. Within each cell, fuel pulse width calculations occur. Usually a cell is defined by the MAP and rpm signals (Figure 25-22). As discussed, a set formula determines pulse width. If the O_2S signals a rich or lean condition in a cell, the cell will update to aid in fuel control. There are different methods used by manufacturers to express the amount of correction occurring. A common method is a percentage value change to the formula.

For example, assume that the engine is operating in closed loop (adaptive memory only updates in closed loop) and there is a small vacuum leak. The O_2S reading will indicate a high oxygen content in the exhaust stream until the PCM increases the injector pulse width and the air/fuel mixture is returned to 14.7 to 1. The percentage of pulse width increase would be recorded by the PCM for that cell. Anytime the operating conditions re-enter that cell the PCM will use the percent of correction in its formula to determine pulse width.

Chrysler Adaptive Strategies

When the fuel system enters closed loop operation, there are two adaptive memory systems that begin to operate. The first system that becomes operational is **short term adaptive**. Short term adaptive memory corrects fuel delivery in direct proportion to the voltage signals from the O_2S (Figure 25-23). As the O_2S voltage switches in response to the air/fuel ratio changes, the PCM will adjust the amount of fuel delivered until the O_2S reaches its switch point. When the switch point is reached, short term adaptive begins with a quick kick.

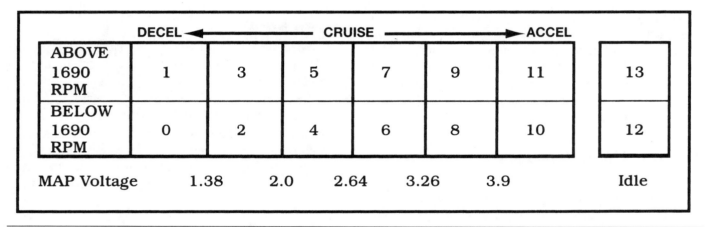

	DECEL ◄────		CRUISE ────		──► ACCEL		
ABOVE 1690 RPM	1	3	5	7	9	11	13
BELOW 1690 RPM	0	2	4	6	8	10	12
MAP Voltage	1.38	2.0	2.64	3.26	3.9		Idle

Figure 25-22 A example of memory cell construction based on engine rpm and MAP inputs. (Courtesy of Chrysler Corporation)

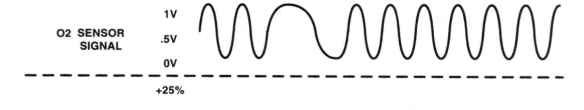

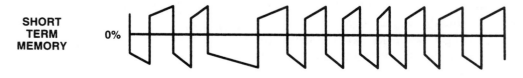

Figure 25-23 Short term memory is driven by O_2S inputs. (Courtesy of Chrysler Corporation)

Then the adaptive memory slowly ramps until the O_2S's output voltage indicates the switch point in the opposite direction. Short term adaptive will continue to increase and decrease the amount of fuel delivered based upon the O_2S input. For example, if the O_2S output voltage goes toward 0 volt (high oxygen), short term adaptive memory will start to add additional fuel until the O_2S begins switching again. The maximum range of authority for short term adaptive memory is ±25% of base pulse width.

Short term adaptive resets to a value of 0 when the engine is turned off. It will not update until the PCM enters closed loop. Normal short term adaptive activity is for the values to jump back and forth, crossing over 0.

The second system is called **long term adaptive** memory. This memory uses a cell structure to better maintain correct emission levels throughout all operating ranges. In this illustration there are 16 cells, however, different engines may have fewer cells. Two of the cells are used only in idle, based on TPS and Park/Neutral switch inputs. Two cells are used for deceleration, based on TPS, rpm, and vehicle speed. The other 12 cells are based on rpm and MAP inputs. The values shown in the previous illustration are examples only.

As the engine enters a cell the PCM monitors the short term adaptive value. The goal of long term adaptive is to keep short term at 0. If short term is correcting for a low oxygen content (for example), based on O_2S inputs, short term will begin to correct for it. After short term reaches a predetermined threshold, long term adaptive will start to correct in the same direction to bring short term back to 0.

Long term is retained in memory even after the engine is turned off. However, long term does not update until the engine reaches about 170 degrees F. Additionally, long term adaptive memory correction values are used in open loop operation. Long term can alter pulse width ±25% from base. Combined, short term and long term can correct pulse width 50% on either side.

General Motors Integrator and Block Learn

Pre-OBD II General Motors learning ability is called **block learn** and **integrator**. Integrator is the first correction. As a rich or lean condition is sensed by the O_2S, integrator will attempt to correct for it by adjusting the pulse width. If integrator is being observed on a scan tool, the value is displayed by a number. The number displayed by the scan tool is a base ten conversion of a binary code. If there is no correction then the number is 128 (half-way between 0 and 256). A number larger than 128 represents an increase of pulse width over base. A

lower number than 128 means pulse width is being reduced.

Block learn represents pulse width adjustments that have become a trend over an extended period of time. Once the engine has been running in closed loop long enough, the PCM will store the corrective factor (as a number) into memory. That number will be stored until additional correction needs to be made, or until the engine is turned off.

Block learn numbers are stored in 16 different cells (Figure 25-24). Each cell represents a different combination of engine speed and load.

Integrator and block learn work together to correct pulse width. For example, if block learn is above 128, but the integrator is at 128, the system has a lean tendency that is being corrected by block learn but the integrator is not having to make any additional changes. If block learn is at 128 but integrator is above 128, the system is attempting to run lean, but has not been doing so long enough for block learn to make any corrections.

Ford Adaptive Strategies

Inside of the PCM's ROM chip is a table that represents cells based on engine speed and load (Figure 25-25). Each cell is assigned a base value of 1. Based on rpm and engine load, the PCM will determine which cell it is operating in. The value in that cell then is used to

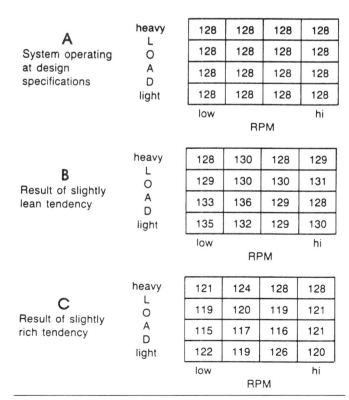

Figure 25-24 Block learn information.

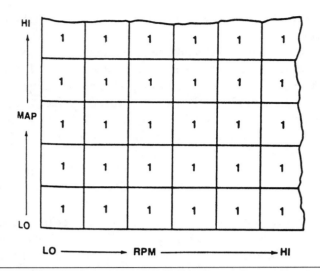

Figure 25-25 Adaptive fuel control table base values.

correct pulse width. The value is the multiplier used in pulse width calculations. A value of 1 will not change pulse width over base.

Notice this table is located in the computer's ROM chip. Since the values in ROM cannot be changed by the microprocessor there is a second copy of the table in volatile RAM. It is the second copy the computer uses for corrections (Figure 25-26). The adaptive control will add or subtract from 1. A number larger than 1 indicates an increase in pulse width over base while a number less than 1 represents a decrease in pulse width.

Fuel Pump Circuits

Regardless of the strategy used, all EFI systems use electric fuel pumps to supply fuel to the **injector rail**

(Figure 25-27). However, manufacturers differ on the control of the fuel pump.

Usually the fuel pump is located in the tank. This means the entire delivery system is pressurized, reducing the possibility of vapor bubbles forming. All fuel lines are special high pressure hose material. No other hose material should be used on EFI vehicles.

Most pumps are positive displacement, roller vane-type design with a permanent magnet motor. However a few use a two-stage gerotor design pump. Usually there are two internal check valves. One valve is located on the inlet side of the pump. This valve is used to regulate maximum fuel pump pressures (about 120 psi). If pressures exceed this amount, due to a restriction in the fuel line, the valve opens and dumps the fuel back into the tank.

The second check valve is located on the outlet side of the pump. This valve prevents any movement of fuel in either direction when the pump is not operating. This will lock a volume of fuel in the injector fuel rail for easier engine starting.

In addition some manufacturers equip their vehicles with a second booster fuel pump mounted on the frame rail under the vehicle. When the ignition switch is turned on, the PCM grounds the fuel pump relay winding and the relay points close. Usually this is done for about .5 to 2 seconds to prime the fuel rail and injectors for faster starting. This action supplies voltage through the relay points to the fuel pump, or pumps, and the pump supplies fuel to the injection system. The PCM then looks for a rpm signal over a predetermined value. Once this signal is seen, the fuel pump relay coil is energized again. This PCM and relay action prevents the fuel pump from running if the vehicle is in a collision and the fuel line is broken with the ignition switch on.

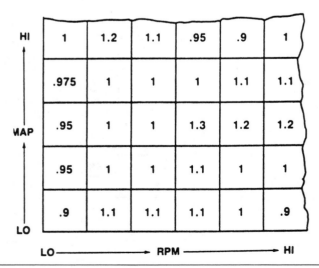

Figure 25-26 Adaptive fuel control table with adapted cell values.

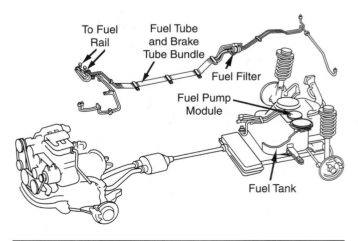

Figure 25-27 Fuel delivery system. (Courtesy of Chrysler Corporation)

Fuel Pump Circuit with Inertia Switch

On Ford products, an **inertia switch** is connected in series in the fuel pump circuit. A steel ball inside the inertia switch is held in place by a permanent magnet. If the vehicle is involved in a collision, this ball pulls away from the magnet and strikes a target plate to open the points in the switch (Figure 25-28). This inertia switch action opens the fuel pump circuit and prevents fuel pump operation.

A reset button on top of the inertia switch must be pressed to close the switch and restore fuel pump operation. The inertia switch is located in the trunk area on most Ford vehicles. A decal in the trunk of later model vehicles indicates the inertia switch location. On Ford EFI-equipped trucks, the inertia switch is located under the dash.

Fuel Pump Circuit with Oil Pressure Switch

General Motors supplies voltage to the fuel pump relay windings by the PCM when the ignition switch is turned on. This action closes the relay points, and voltage is supplied through the points to the in-tank fuel pump. The fuel pump relay remains closed while the engine is cranking or running. If the ignition switch is on for 2 seconds and the engine is not cranked, the PCM shuts off the voltage to the fuel pump relay, and the relay points open to stop the pump.

If the ignition switch is on and the fuel line is broken in a collision, this PCM and fuel pump relay action is a safety feature that prevents the fuel pump from pumping gasoline from the ruptured fuel line into a fire. An oil pressure switch is connected in parallel to the fuel pump

relay points. If the relay becomes defective, voltage is supplied through the oil pressure switch points to the fuel pump. This action keeps the fuel pump operating and the engine running, even though the fuel pump relay is defective. When the engine is cold, oil pressure is not available immediately, and the engine may be slow to start if the fuel pump relay is defective.

A fuel pump test connector is connected to a set of relay points that remain closed when the ignition switch is turned off. This test connector is not connected to any external circuit. It is located on the left fender shield on some vehicles. If 12 volts are supplied to this test connector with the ignition switch off, the fuel pump should run.

Fuel Pump Circuits with Automatic Shutdown (ASD) Relay

WARNING *Never turn on the ignition switch or crank the engine with a fuel line disconnected. This action may result in gasoline discharge from the fuel line and a fire, which could cause personal injury and/or property damage.*

The fuel pump relay in pre-1993 Chrysler EFI systems is referred to as an automatic shutdown (ASD) relay. In 1984 Chrysler EFI systems, the ASD relay is mounted under the right kick pad. The ASD relay in 1985 through 1987 Chrysler EFI systems is mounted in the power module in the engine compartment. This relay is not serviced separately. Chrysler PCMs from 1987 to present usually have the ASD relay located in the power distribution center (PDC) close to the battery (Figure 25-29). If the ignition switch is turned on, the PCM grounds the ASD relay winding and the relay points close. The ASD relay points

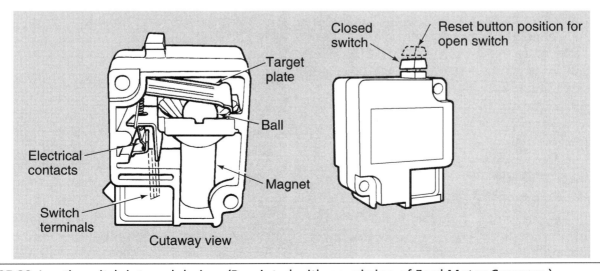

Figure 25-28 ■ Inertia switch internal design. (Reprinted with permission of Ford Motor Company)

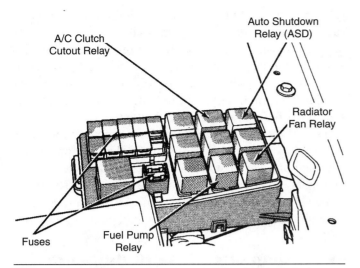

Figure 25-29 Power distribution center (PDC). (Courtesy of Chrysler Corporation)

supply voltage to the fuel pump, positive primary coil terminal, generator field windings, oxygen sensor heaters, and the fuel injectors in earlier systems.

On some Chrysler products with a power module and logic module, the engine has to be cranked before the power module grounds the ASD relay winding. The later model PCM grounds the ASD relay winding when the ignition switch is turned on, and the relay remains closed as long as a rpm signal is received by the PCM. If the ignition switch is on for .5 second and the engine is not cranked, the PCM opens the circuit from the ASD relay winding to ground. Under this condition, the ASD relay points open and voltage is no longer supplied to the fuel pump, positive primary coil terminal, injectors, and oxygen sensor heater.

Chrysler fuel pump circuits from 1993 and later have a separate ASD relay and a fuel pump relay. In these circuits, the fuel pump relay supplies voltage to the fuel pump and oxygen sensor heaters, and the ASD relay powers the positive primary coil terminal, injectors, and generator field windings. The ASD relay and the fuel pump relay operate the same as the previous ASD relay. The PCM grounds both relay windings through the same wire (Figure 25-30). In 1996 the circuits were changed so the PCM can control the fuel pump relay separate of the ASD relay. The oxygen sensor heater circuit was also moved to the ASD relay.

Fuel Pump Circuit with Circuit Opening Relay

On many Toyota vehicles, the fuel pump relay is called a **circuit opening relay**. This relay has dual windings; one of the circuit opening relay windings is connected between the starter relay points and ground,

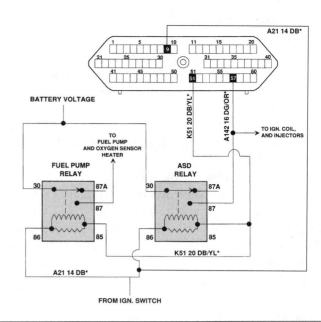

Figure 25-30 Chrysler fuel pump circuit with separate ASD and fuel pump relays. (Courtesy of Chrysler Corporation)

and the second relay winding is connected from the battery positive terminal to the PCM. When the engine is cranking and the starter relay points are closed, current flows through the starter relay points and the circuit opening relay winding to ground. This current flow creates a magnetic field around the circuit opening relay winding that closes the relay points. When these points close, current flows through the points to the fuel pump (Figure 25-31).

Once the engine starts, the starter relay is no longer energized, and current stops flowing through these relay points and the circuit opening relay winding. However, the PCM grounds the other circuit opening relay winding when the engine starts to keep the relay points closed while the engine is running.

Pressure Regulators

Since the fuel pump is a positive displacement pump, system pressures must be regulated. The **fuel pressure regulator** may be located on the end of the fuel rail (Figure 25-32). A diaphragm and valve assembly is positioned in the center of the regulator, and a diaphragm spring seats the valve on the fuel outlet (Figure 25-33). When fuel pressure reaches the regulator setting, the diaphragm moves against the spring tension, and the valve opens. This action allows fuel to flow through the return line to the fuel tank. The fuel pressure drops

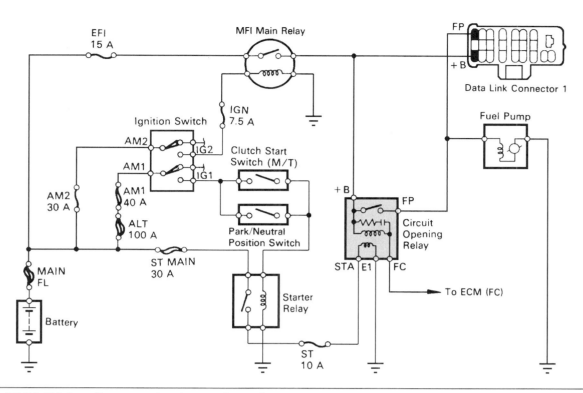

Figure 25-31 Wiring diagram, circuit opening relay.

slightly when the pressure regulator valve opens, and the spring closes the regulator valve. Under this condition, the fuel pressure increases and reopens the regulator valve. In many EFI systems, the pressure regulator maintains fuel pressure at 39 psi (269 kPa); however, the technician must have the fuel pressure specifications for each system when these systems are diagnosed.

A vacuum hose is connected from the intake manifold to the vacuum inlet on the pressure regulator. This hose supplies vacuum to the area where the diaphragm spring is located. This vacuum works with the fuel pres-

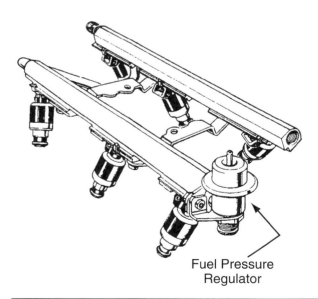

Figure 25-32 Fuel pressure regulator attached to the fuel rail. (Courtesy of General Motors Corporation, Service Technology Group)

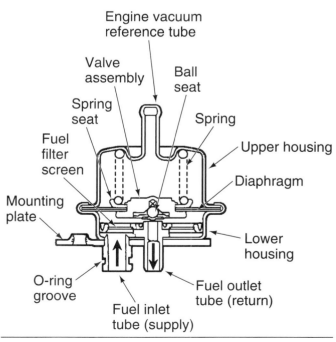

Figure 25-33 Internal pressure regulator construction. (Reprinted with permission of Ford Motor Company)

sure to move the diaphragm and open the valve. When the engine is running at idle speed, high manifold vacuum is supplied to the pressure regulator. Under this condition, 39 psi (269 kPa) fuel pressure opens the regulator valve. If the engine is operating at wide-open throttle, a very low manifold vacuum is supplied to the pressure regulator. When this condition is present, the vacuum does not help to open the regulator valve, and a higher fuel pressure of 49 psi (338 kPa) is required to open the valve.

When the engine is idling, higher manifold vacuum is supplied to the injector tips, and the injectors are discharging fuel into this vacuum. Under wide-open throttle conditions, very low manifold vacuum is supplied to the injector tips. When this condition is present, the injectors are actually discharging fuel into a higher pressure compared to idle speed conditions, because the very low manifold vacuum is closer to a positive pressure. If the fuel pressure remained constant at idle and wide-open throttle conditions, the injectors would discharge less fuel into the higher pressure in the intake manifold at wide-open throttle. The increase in fuel pressure supplied by the pressure regulator at wide-open throttle maintains the same pressure drop across the injectors at idle speed and wide-open throttle. When this same pressure drop is

maintained, the change in pressure at the injector tips does not affect the amount of fuel discharged by the injectors.

Returnless Fuel System Pressure Regulator

With the introduction of the Dodge Viper V-10 engine, Chrysler also started using a **returnless fuel system**. By 1997 most Chrysler-built vehicles used this system. In these systems, the fuel pressure regulator and filter are mounted in the top of the assembly containing the fuel pump and fuel gauge sending unit in the fuel tank. The fuel line from the fuel rail under the hood is connected to the filter with a quick-disconnect fitting. Fuel enters the filter through the fuel supply tube in the center of the regulator and filter assembly. Fuel pressure is applied against the regulator seat washer, which is seated by the seat control spring (Figure 25-34). At the specified regulator pressure (about 49 psi), the seat is forced downward against the spring, and fuel flows past the seat into the cavity around the seat control spring. Fuel returns from this cavity to the fuel tank. When the pressure drops slightly, the seat closes again. With the returnless fuel system, only the fuel needed by the engine is filtered, thus allowing the use of a smaller fuel filter.

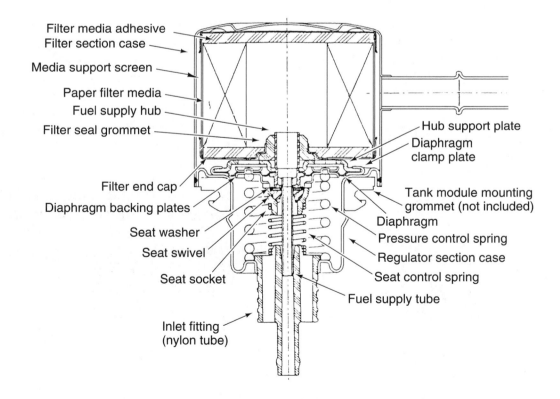

Figure 25-34 Pressure regulator and fuel filter used on returnless fuel systems. (Courtesy of Chrysler Corporation)

An additional benefit of the returnless system is the fuel being returned to the tank is not heated by under-hood temperatures, thus reducing hot fuel problems.

The fuel pressure regulators on returnless systems do not use a vacuum supply from the engine to maintain a constant pressure differential at the injector tip. The PCM has enhanced programming to compensate for this by increasing pulse width using MAP, TPS, and rpm inputs.

Results of Improper Fuel Pressure

Since the injectors in an EFI system are subjected to high temperatures, the EFI systems must have high fuel pressures to prevent fuel boiling in the rails and injectors. The computer program assumes the injector is discharging liquid fuel, and the computer provides the correct injector pulse width to supply the stoichiometric air/fuel ratio. If fuel boiling occurs in the fuel rails or injectors, some vapor is discharged and the air/fuel ratio is lean. This will result in reduced engine power, particularly on acceleration. The computer program also assumes the specified pressure is available in the fuel rails. Lower or higher than specified fuel pressures cause a different amount of fuel to be discharged than what the computer calculated.

Low fuel pressure may be caused by a worn fuel pump or a restricted fuel filter. If the fuel pressure is lower than specifications, the injectors do not discharge enough fuel, and the air/fuel ratio is lean.

High fuel pressure may be caused by a restricted return fuel line or a sticking pressure regulator. Under this condition, the injectors discharge excessive fuel, which increases fuel consumption and emission levels. When the injectors discharge excessive fuel, the air/fuel ratio is rich. This causes a rough running engine at low speed with excessive sulphur smell from the catalytic converter.

Throttle Body Injection Systems

CAUTION *Always relieve the fuel pressure before disconnecting a fuel system component to avoid gasoline spills that may cause a fire, resulting in personal injury and/or property damage.*

The **throttle body** assembly is mounted on top of the intake manifold where the carburetor was mounted on carbureted engines. Four-cylinder engines usually have a single throttle body assembly with one throttle, whereas V-6 and V-8 engines may be equipped with dual throttle bodies with two throttles on a common throttle shaft (Figure 25-35).

The throttle body assembly contans a pressure regulator, injector or injectors, TPS, idle speed control motor, and throttle shaft and linkage assembly. A fuel filter is located in the fuel line under the vehicle or in the engine compartment. When the engine is cranking or running, fuel is supplied from the fuel pump through the lines and filter to the throttle body assembly. A fuel return line connected from the throttle body to the fuel tank returns excess fuel to the fuel tank (Figure 25-36).

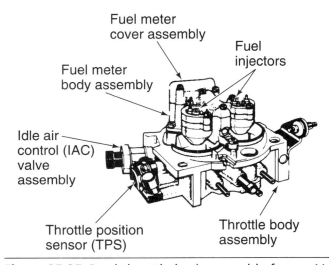

Figure 25-35 Dual throttle body assembly from a V-8 engine. (Courtesy of General Motors Corporation, Service Technology Group)

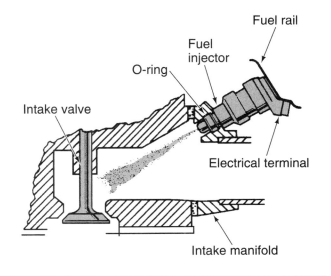

Figure 25-36 Fuel delivery system in a typical TBI system. (Courtesy of General Motors Corporation, Service Technology Group)

When fuel enters the throttle body fuel inlet, the fuel surrounds the injector or injectors at all times. Each injector is sealed into the throttle body with O-ring seals, which prevent fuel leakage around the injector at the top or bottom. Fuel is supplied from the injector through a passage to the pressure regulator. A diaphragm and valve assembly is mounted in this regulator, and a diaphragm spring holds the valve closed. At a specific fuel pressure, the regulator diaphragm is forced upward to open the valve, and some excess fuel is returned to the fuel tank.

In most TBI systems, the fuel pressure regulator controls fuel pressure at 10 psi to 25 psi (70 kPa to 172 kPa). The fuel pressure must be high enough to prevent fuel boiling in the TBI assembly. When the pressure on a liquid is increased, the boiling point is raised proportionally. If fuel boiling occurs in the TBI assembly, vapor and fuel are discharged from the injectors.

A **throttle body temperature senso**r is mounted in some Chrysler TBI assemblies (Figure 25-37). When the TBI assembly reaches the temperature at which some fuel boiling may occur, the throttle body temperature sensor signals the computer to provide a slightly richer air/fuel ratio. This action compensates for the vapor discharge from the injectors.

The fuel pressure is regulated at 14.5 psi (100 kPa) in Chrysler TBI assemblies with a temperature sensor. These systems were used from 1986 through 1990, and they are referred to as **low pressure TBI (LPTBI)**. A black plastic rivet is located on the top of the pressure regulator in a LPTBI system. In 1991, Chrysler installed a pressure regulator on their TBI systems that controls

the fuel pressure at 39.2 psi (270 kPa). These systems are called **high pressure TBI (HPTBI)**. A white plastic rivet is located on top of these pressure regulators, and the TBI temperature sensor is no longer required with the higher fuel pressure.

Port Fuel Injection Systems

Port fuel injection (PFI) is a term that may be applied to any fuel injection system with the injectors located in the intake ports. There are two basic types of PFI systems: multipoint fuel injection (MPI), which is also called multiport fuel injection, and sequential fuel injection (SFI). The **multipoint fuel injection (MPI or MFI)** system has one injector per cylinder. The injectors are fired in pairs, or in groups of three or four. Usually half of the fuel is delivered on each crankshaft revolution. In a **sequential fuel injection (SFI)** system each injector is fired individually, in ignition firing order prior to the intake valve opening.

There are many similarities in the MFI and SFI systems supplied by the various domestic and import vehicle manufacturers. For example, both MFI and SFI systems have injectors installed in the intake ports near the intake valve, and many of these systems share similar inputs and outputs. One of the major differences in MFI and SFI systems is the method of connecting the injectors to the computer. In SFI systems, each injector is connected individually into the computer, so the computer grounds one injector at a time. In MFI systems, the injectors are grouped together in pairs or groups. Each group of injectors shares a common wire to the computer so they are energized together. For example, on some four-cylinder engines, the injectors are connected in pairs, and each pair of injectors has a common connection to the computer. On some V-6 engines, each group of three injectors has a common ground wire connected to the computer. Groups of four injectors share a common ground connection to the computer on some V-8 engines.

Basically the same electric in-tank fuel pumps and fuel pump circuits are found on TBI, MFI, and SFI systems. These pumps and pump circuits were explained previously in this chapter. Some MFI and SFI systems such as those on Ford trucks have a booster fuel pump on the frame rail and the in-tank pump. A fuel filter is connected in the fuel line from the tank to the injectors. This filter may be under the vehicle or in the engine compartment. The fuel line from the filter is connected to a hollow fuel rail bolted to the intake manifold. The lower end of each port injector is sealed in the intake manifold with an O-ring seal, and a similar seal near the top of the injector seals the injector to the fuel rail. Most fuel rails

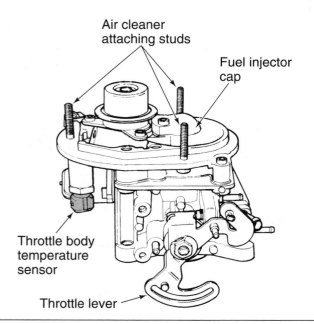

Air cleaner
attaching studs

Fuel injector
cap

Throttle body
temperature
sensor

Throttle lever

Figure 25-37 Throttle body temperature sensor used to prevent fuel boiling. (Courtesy of Chrysler Corporation)

have a Schrader valve for connecting a pressure gauge when testing fuel pressure.

The computer is programmed to ground the injectors well ahead of the actual intake valve openings so the intake ports are filled with fuel vapor before the intake valves open (Figure 25-38). In both SFI and MFI systems, the computer supplies the correct injector pulse width to provide the stoichiometric air/fuel ratio. The computer increases the injector pulse width to provide air/fuel ratio enrichment while starting a cold engine. A clear flood mode is also available in the computer in MFI and SFI systems. On some TBI, MFI, and SFI systems, if the ignition system is not firing, the computer

stops operating the injectors. This action prevents severe flooding from long cranking periods while starting a cold engine. On many MFI and SFI systems, the computer decreases injector pulse width while the engine is decelerating to provide improved emission levels and fuel economy. On some of these systems, the computer stops operating the injectors while the engine is decelerating in a certain rpm range.

On many V-6 and V-8 engines, there is a fuel rail on each side of the intake manifold. On these engines, the fuel is supplied to one rail and a connecting hose carries fuel to the second rail. A pressure regulator is connected to the end of the fuel rail, and excess fuel is returned from the regulator through a return line to the fuel tank. On some V-6 engines, the fuel rail is U-shaped, and each injector is mounted in this one-piece rail. Many fuel rails are made from steel or aluminum alloy. Plastic-type fuel rails have been installed recently on some engines. These fuel rails transfer less heat to the fuel and reduce the possibility of fuel boiling in the rail.

On some V-6 engines such as the General Motors 2.8-L and 3.1-L V-6, a one-piece fuel rail is mounted under the top of the intake manifold. In these MFI systems, the top of the intake must be removed to gain access to the fuel rail and injectors (Figure 25-39).

Cold Start Injector

Some engines supplied by domestic and import manufacturers utilize a **cold start injector** along with the normal fuel injectors. The cold start injector is mounted in the intake manifold. It is connected to the fuel rail by a pickup pipe.

Unlike the intake port injectors that are operated by the PCM, the cold start injector is operated by a thermo-

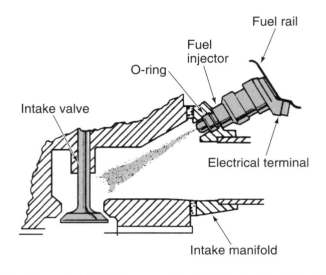

Figure 25-38 A port fuel injector sprays fuel into the intake manifold behind the intake valve. (Courtesy of General Motors Corporation, Service Technology Group)

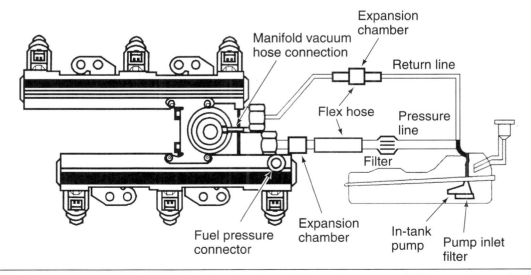

Figure 25-39 One-piece fuel rail mounted under the top of the intake manifold on a V6 engine. (Courtesy of General Motors Corporation, Service Technology Group)

time switch that senses coolant temperature. When the engine is cranked, voltage is supplied from the starter solenoid to one terminal on the cold start injector (Figure 25-40). If the coolant temperature is below 95 degrees F (35 degrees C), the thermo-time switch supplies the ground circuit for the cold start injector. Under this condition, the cold start injector is energized only while the engine is cranking. When the cold start injector is operating, it supplies additional fuel along with the normal fuel injectors.

A bimetal switch in the thermo-time switch is heated as current flows through the injector coil. The bimetal switch action opens the circuit through the thermo-time switch in a maximum of 8 seconds. The actual time that the thermo-time switch remains closed is determined by the coolant temperature. In this MFI system, the pulse width supplied by the PCM to the intake port injectors is programmed to operate with the cold start injector and supply the correct air/fuel ratio while cranking a cold engine. The cold start injector and thermo-time switch are also used on some General Motors V-8 engines and on MFI and SFI systems used by some import manufacturers.

Typical Domestic Sequential Fuel Injection System

In a Chrysler SFI system on a 3.5-L engine, each injector has a separate ground wire connected into the PCM (Figure 25-41). Many Chrysler engines previous to 1992 have multipoint fuel injection (MPI) systems with the injectors connected in pairs on the ground side. Each pair of injectors shares a common ground wire into the PCM. As mentioned previously in the discussion of the TBI systems, voltage is supplied through the ASD relay points to the injectors when the ignition switch is turned on, and a separate fuel pump relay supplies voltage to the fuel pump. This engine is equipped with an electronic ignition (EI) system, and the crank and cam sensors are inputs for this system. Since these inputs are connected to the PCM, the ignition module is contained in the PCM.

The inputs shown in the 3.5-L SFI system are:

1. Dual heated oxygen (O_2S) sensors
2. Crank sensor
3. Cam sensor
4. Throttle position sensor (TPS)
5. Manifold absolute pressure (MAP) sensor
6. Coolant temperature sensor
7. Charge temperature sensor
8. Electronic automatic transaxle (EATX) computer
9. Starter relay
10. Generator field
11. Dual knock sensors
12. Data links to other computers such as the EATX
13. Battery voltage
14. Power steering switch
15. Brake switch
16. Ignition switch

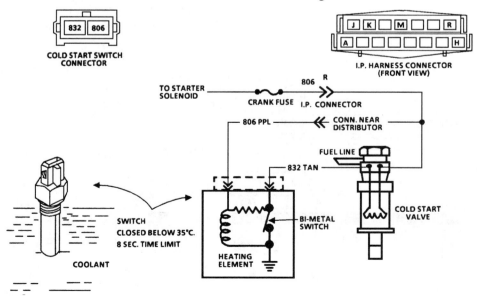

Figure 25-40 Cold start injector circuit. (Courtesy of General Motors Corporation, Service Technology Group)

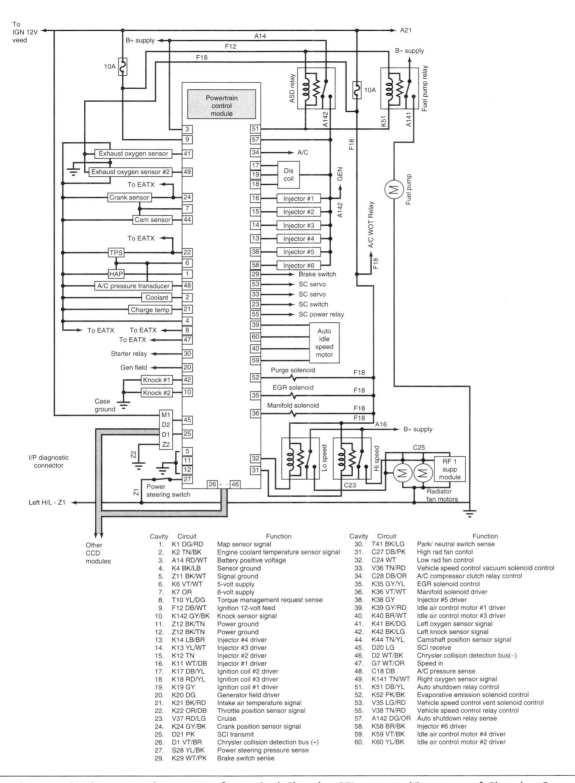

Figure 25-41 PCM inputs and outputs of a typical Chrysler SFI system. (Courtesy of Chrysler Corporation)

Cavity	Circuit	Function
1.	K1 DG/RD	Map sensor signal
2.	K2 TN/BK	Engine coolant temperature sensor signal
3.	A14 RD/WT	Battery positive voltage
4.	K4 BK/LB	Sensor ground
5.	Z11 BK/WT	Signal ground
6.	K6 VT/WT	5-volt supply
7.	K7 OR	8-volt supply
8.	T10 YL/DG	Torque management request sense
9.	F12 DB/WT	Ignition 12-volt feed
10.	K142 GY/BK	Knock sensor signal
11.	Z12 BK/TN	Power ground
12.	Z12 BK/TN	Power ground
13.	K14 LB/BR	Injector #4 driver
14.	K13 YL/WT	Injector #3 driver
15.	K12 TN	Injector #2 driver
16.	K11 WT/DB	Injector #1 driver
17.	K17 DB/YL	Ignition coil #2 driver
18.	K18 RD/YL	Ignition coil #3 driver
19.	K19 GY	Ignition coil #1 driver
20.	K20 DG	Generator field driver
21.	K21 BK/RD	Intake air temperature signal
22.	K22 OR/DB	Throttle position sensor signal
23.	V37 RD/LG	Cruise
24.	K24 GY/BK	Crank position sensor signal
25.	D21 PK	SCI transmit
26.	D1 VT/BR	Chrysler collision detection bus (+)
27.	S28 YL/BK	Power steering pressure sense
29.	K29 WT/PK	Brake switch sense

Cavity	Circuit	Function
30.	T41 BK/LG	Park/ neutral switch sense
31.	C27 DB/PK	High rad fan contol
32.	C24 WT	Low rad fan control
33.	V36 TN/RD	Vehicle speed control vacuum solenoid control
34.	C28 DB/OR	A/C compressor clutch relay control
35.	K35 GY/YL	EGR solenoid control
36.	K36 VT/WT	Manifold solenoid driver
38.	K38 GY	Injector #5 driver
39.	K39 GY/RD	Idle air control motor #1 driver
40.	K40 BR/WT	Idle air control motor #3 driver
41.	K41 BK/DG	Left oxygen sensor signal
42.	K42 BK/LG	Left knock sensor signal
44.	K44 TN/YL	Camshaft position sensor signal
45.	D20 LG	SCI receive
46.	D2 WT/BK	Chrysler collision detection bus(−)
47.	G7 WT/OR	Speed in
48.	C18 DB	A/C pressure sense
49.	K141 TN/WT	Right oxygen sensor signal
51.	K51 DB/YL	Auto shutdown relay control
52.	K52 PK/BK	Evaporative emission solenoid control
53.	V35 LG/RD	Vehicle speed control vent solenoid control
55.	V38 TN/RD	Vehicle speed control relay control
57.	A142 DG/OR	Auto shutdown relay sense
58.	K58 BR/BK	Injector #6 driver
59.	K59 VT/BK	Idle air control motor #4 driver
60.	K60 YL/BK	Idle air control motor #2 driver

17. Cruise control switches

18. Park/neutral switch, starter relay

Many V-6 and V-8 engines now have dual oxygen sensors, which provide improved control of the air/fuel ratio in each bank of cylinders. An oxygen sensor is usu-

ally mounted in each exhaust manifold. Dual knock sensors are located in many V-6 and V-8 engines to improve spark knock, or detonation, control. The knock sensors are positioned in each side of the block, or cylinder heads.

The outputs on the 3.5-L SFI system are:

1. Automatic shutdown (ASD) relay
2. Fuel pump relay
3. DIS coil
4. Spark advance
5. Injectors
6. Cruise control
7. Automatic idle speed motor
8. Purge solenoid
9. EGR solenoid
10. Manifold solenoid
11. Dual radiator fan relays

The Chrysler SFI system has many similarities to the Chrysler TBI system illustrated previously in this chapter. For example, the voltage regulator and the cruise control module are contained in the PCM board. The SFI system on the 3.5-L engine has a low-speed and a high-speed cooling fan relay. At a specific coolant temperature, the PCM grounds the low-speed relay winding, which closes the relay points and supplies voltage to the fan motor. If the engine coolant temperature continues to increase, the PCM grounds the high-speed cooling fan relay winding, which closes the fan relay points and supplies voltage to the high-speed fan motor.

The manifold solenoid controls the vacuum supplied to the intake manifold tuning valve. The manifold solenoid is mounted on the right shock tower, and the manifold tuning valve is positioned near the center of the intake manifold (Figure 25-42). The manifold contains a

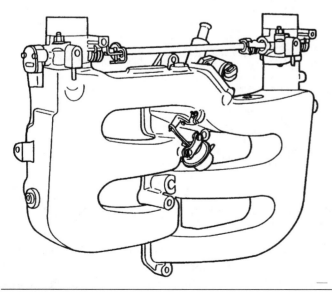

Figure 25-42 The manifold tuning valve is activated by a vacuum motor. (Courtesy of Chrysler Corporation)

pivoted butterfly valve that opens and closes to change the length of the intake manifold air passages. This butterfly valve is mounted on a shaft, and the outer end of the shaft is connected through a linkage to a diaphragm in a sealed vacuum chamber. A vacuum hose is connected from the outlet fitting on the manifold solenoid to the vacuum chamber in the manifold tuning valve. Another vacuum hose is connected from the inlet fitting on the manifold solenoid to the intake manifold.

One terminal on the manifold solenoid winding is connected to the ignition switch, and the other terminal is connected to the PCM. While the engine is running at lower rpm, the PCM opens the manifold solenoid circuit. Under this condition, the solenoid shuts off the manifold vacuum to the intake manifold tuning valve, and the butterfly valve closes some of the air passages inside the intake manifold.

At higher engine rpm, the PCM grounds the manifold solenoid winding and energizes the solenoid. This action opens the vacuum passage through the solenoid and supplies vacuum to the intake manifold tuning valve. Under this condition, the butterfly valve is moved so it opens additional air passages inside the intake manifold to improve air flow and increase engine horsepower and torque.

Crankshaft and Camshaft Position Sensor Input

As discussed, the crankshaft position sensor informs the PCM of engine speed and of two pistons approaching TDC. The camshaft position sensor informs the PCM which piston is approaching TDC compression stroke. The two sensors provide alternating high- and low-voltage signals to provide input for pulse width and ignition strategies.

The different slots on the crankshaft position sensor inform the PCM of needed functions as follows:

- 69° BTDC

  MAP reading is updated and a value is stored in RAM.

  Ignition period is determined by the time between two consecutive 69° edges.

  Detonation is mainly controlled on the 69° edge.

  Synchronized is mainly conducted on this edge along with camshaft interrupt.

  Ignition coil firing for the current cylinder, as well as the next cylinder is generated.

  Injection firing may occur based on operating conditions.

- 49° BTDC

Ignition coil firing adjustment may occur if engine speed is below a programmed value.

Injector firing may occur based on operating conditions.

- 29° BTDC

Ignition coil firing adjustment may occur if engine speed is below a programmed value.

Injector firing may occur based on operating conditions.

Adaptive knock is updated.

- 9° BTDC

If the engine is being started, the coil firing is done on this edge.

Cylinder indication is updated.

Adaptive dwell is updated.

MAP is read. The value is averaged between the reading at 69° BTDC and this reading.

TPS Programs

The PCM determines idle mode based upon inputs from the TPS. When the ignition switch is turned to the RUN position, during engine starting, the PCM is programmed to monitor the TPS input. Once the engine is started the PCM assumes the lowest value received from the TPS must be where the throttle plate is fully closed. Normally this voltage is between 0.5-1.0 volt. At the low-voltage position, the PCM records the signal as idle or **minimum TPS**.

If the sensed voltage from the TPS decreases lower than this value, the PCM will update its memory. However, the PCM will not make any corrections for an increase in voltage values. The PCM uses minimum TPS to determine other modes of operation.

If the throttle is opened, and the TPS moves from its minimum TPS value by approximately 0.06 volt, the PCM enters off-idle strategies. At this time spark advance is no longer used to control idle speed. Also, the idle air control motor is positioned to act as a dashpot if the throttle is released suddenly.

If the PCM receives a rapidly increasing voltage signal from the TPS, the PCM will enter acceleration mode. During this time the injector pulse width is increased based on the rate of TPS voltage increase. The PCM may activate three injectors: the injector that just closed, the injector being opened, and the one to be opened next.

Wide-open throttle (WOT) is determined by an increase in TPS volts of 2.608 above minimum TPS. The PCM enters open loop and increases the injector pulse width.

If a rapid deceleration is sensed, the PCM will lean out the air/fuel ratio. In some instances the injectors are turned off. This action reduces the amount of emissions.

If the throttle is at WOT during engine cranking, the injectors are turned off. This provides a clear flood mode. This program only occurs during cranking and when the TPS voltage exceeds 2.608 volts above minimum TPS.

Typical Domestic Multipoint Fuel Injection System

In many Ford MFI systems on V-8 engines, the injectors are connected in groups of four on the ground side. Each group shares a common ground wire to the powertrain control module (PCM). On a Ford 5.8-L V-8 engine, injectors 1, 4, 5, and 8 are connected to PCM terminal 58 with a common ground wire. The ground sides of injectors 2, 3, 6, and 7 are connected to PCM terminal 59 (Figure 25-43).

When the injectors are connected in groups of four on the ground side, a group of injectors is grounded by the computer every crankshaft revolution. The inputs are:

1. Heated exhaust gas oxygen (HEGO) sensor
2. Engine coolant temperature (ECT) sensor
3. Throttle position sensor (TPS)
4. Intake air temperature (IAT) sensor
5. Manifold absolute pressure (MAP) sensor
6. EGR valve position sensor
7. Brake on/off (BOO) switch
8. Manual lever position (MLP) sensor
9. Neutral/park switch
10. Self-test connector
11. Self-test input

The outputs on the 5.8-L V-8 MFI system are:

1. Injectors
2. Spark advance
3. Idle speed control (IAC) solenoid
4. EGR vacuum regulator (EVR) solenoid
5. Canister purge solenoid
6. Secondary air injection bypass (AIRB) solenoid
7. Secondary air injection diverter (AIRD) solenoid

Constant Control Relay Module

Some Ford products are equipped with a **constant control relay module** (Figure 25-44). This module con-

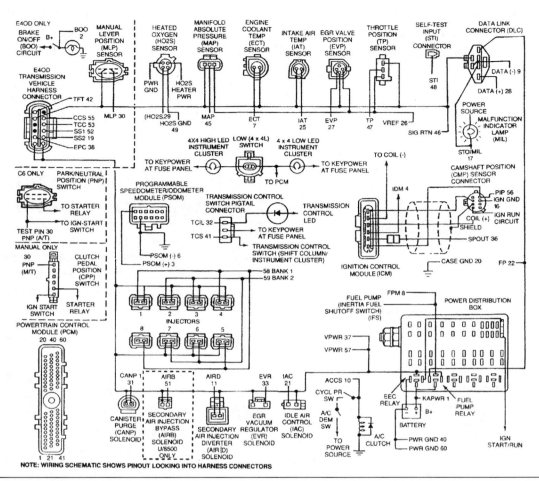

Figure 25-43 Inputs and outputs for 5.8 L multiport fuel injection system. (Reprinted with permission of Ford Motor Company)

tains several relays that are mounted separately on other models. These relays perform the same function whether they are mounted separately or located in the constant control relay module. When the ignition switch is turned on, voltage is supplied to the power relay winding. This action closes the power relay points, which supply voltage to the fuel pump relay winding, PCM terminals 37 and 57, the electric drive fan (EDF) relay winding, and the fuel pump relay winding.

The PCM grounds the fuel pump relay winding and closes these relay points under the same conditions as in other EEC IV systems. The PCM grounds the low fan control (LFC) relay winding when the engine coolant reaches a specific temperature. This action closes the LFC relay points, which supply voltage through a resistor to the cooling fan motor. Under this condition, the cooling fan motor runs at low speed. If the coolant temperature continues to increase, the PCM grounds the high fan control (HFC) relay winding, which closes these relay points and supplies voltage directly to the cooling fan motor. This action causes the cooling fan motor to run at high speed.

When the A/C mode is selected on the instrument panel controls, a signal is sent to the PCM and the solid-state A/C relay. When this signal is received, the A/C relay supplies voltage to the A/C compressor clutch. If the engine is operating at wide-open throttle, the PCM sends a signal from terminal 54 to the A/C solid-state relay. When this signal is received, the solid-state relay opens to disengage the compressor clutch.

A BIT OF HISTORY

Since the 1970s, fuel system technology has developed quickly from carburetors to computer-controlled carburetors and then to throttle body injection systems. Many of the throttle body injection systems have been replaced with multipoint fuel injection systems, and a significant number of multipoint injection systems have been changed to sequential fuel injection systems.

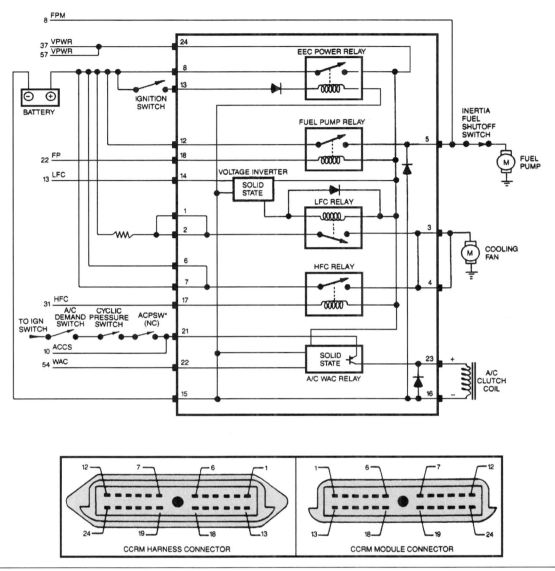

Figure 25-44 Wiring diagram for a constant control relay module. (Reprinted with permission of Ford Motor Company)

Central Port Injection

In a **central port injection (CPI)** system, a central port injector assembly is mounted in the lower half of the intake manifold. Fuel inlet and return lines are connected from the rear of the intake to the CPI assembly. A retaining clip attaches these lines to the CPI assembly. Small poppet nozzles are positioned in each intake port in the lower half of the intake. Nylon fuel lines connect these nozzles to the CPI assembly (Figure 25-45).

The pressure regulator is mounted with the central injector. Since this regulator is mounted inside the intake manifold, vacuum from the intake is supplied through an opening in the regulator cover to the regulator diaphragm. The regulator spring pushes downward on the diaphragm and closes the valve. Fuel pressure from the in-tank fuel pump pushes the diaphragm upward and opens the valve, which allows fuel to flow through this valve and the return line to the fuel tank (Figure 25-46).

The pressure regulator is designed to regulate fuel pressure to 54 to 64 psi (370 to 440 kPa), which is higher than many port fuel injection systems. Higher pressure is required in the CPI system to prevent fuel vaporization from the extra heat encountered with the CPI assembly, poppet nozzles, and lines mounted inside the intake manifold. The pressure regulator operates the same as the regulators explained previously in this chapter.

The central port injector contains a winding with two terminals extending from the ends of the winding through the top of the injector. When the ignition switch

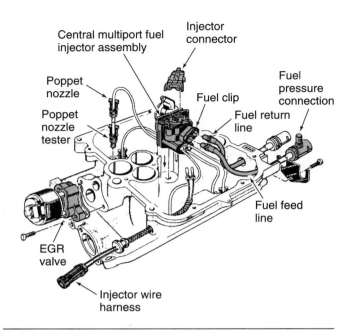

Figure 25-45 Central port injection components in the lower half of the intake manifold. (Courtesy of General Motors Corporation, Service Technology Group)

is on, voltage is supplied from the ignition switch to one of the injector terminals. The other injector terminal is connected to the PCM (Figure 25-47).

A pivoted armature is mounted under the injector winding in the central port injector. The lower side of this armature acts as a valve that covers the six outlet ports to the nylon tubes and poppet nozzles. A supply of fuel at a constant pressure surrounds the injector arma-

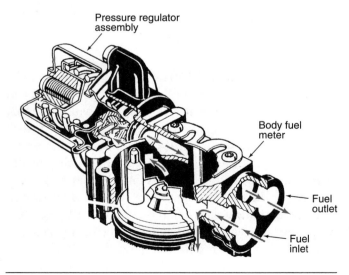

Figure 25-46 CPI system pressure regulator. (Courtesy of General Motors Corporation, Service Technology Group)

ture while the ignition switch is on. Each time the PCM grounds the injector winding, the armature is lifted upward, opening the injector ports. Under this condition, fuel is forced from the nylon tubes to the poppet nozzles (Figure 25-48).

The amount of fuel delivered by the central injector is determined by the length of time the PCM keeps the injector winding grounded. When the PCM opens the injector ground circuit, the injector spring pushes the armature downward and closes the injector ports. The injector winding has 1.5Ω resistance, and the PCM operates the injector with a **peak-and-hold current**. When the PCM grounds the injector winding, the current flow in this circuit increases rapidly to 4 amperes. When the current flow reaches this value, a current-limiting circuit in the PCM limits the current flow to 1 ampere for the remainder of the injector pulse width. The peak-and-hold function provides faster injector armature opening and closing.

The **poppet nozzles** are snapped into openings in the lower half of the intake manifold, and the tip of each nozzle directs fuel into an intake port. Each poppet nozzle contains a valve with a check ball seat in the tip of the nozzle. A spring holds the valve and check ball seat in the closed position. When fuel pressure is applied from the central injector through the nylon lines to the poppet nozzles, this pressure forces the valve and check ball seat open against spring pressure. The poppet nozzles open when the fuel pressure exceeds 37 to 43 psi (254 to 296 kPa), to spray fuel from these nozzles into the intake ports (Figure 25-49).

When fuel pressure drops below this value, the poppet nozzles close. Under this condition, approximately 40 psi (276 kPa) fuel pressure remains in the nylon lines and poppet nozzles. This pressure prevents fuel vaporization in the nylon lines and nozzles during hot engine operation or hot soak periods. If a leak occurs in a nylon line or other CPI component, fuel drains from the bottom of the intake manifold through two drain holes to the center cylinder intake ports. The in-tank fuel pump, fuel filter, lines, and fuel pump circuit used with the CPI system are similar to those used with SFI and MFI systems.

The two-piece aluminum intake manifold has an integral throttle body. An **intake manifold tuning valve (IMTV)** assembly is mounted in the top of the intake manifold. The IMTV assembly contains an electric solenoid that operates a rectangular-shaped valve inside the manifold. Two zip tubes are connected from the throttle body to dual plenums in the upper half of the intake manifold.

The IMTV rectangular valve is mounted in an opening between the two halves of the upper intake manifold. While the engine is operating at lower rpm, the IMTV

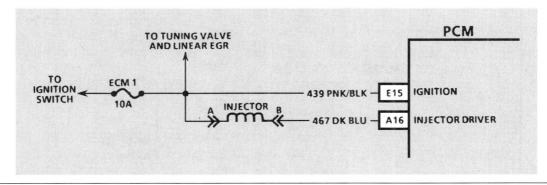

Figure 25-47 Wiring diagram for CPI system. (Courtesy of General Motors Corporation, Service Technology Group)

remains closed, and air flow in the two halves of the upper intake is separated. When the engine rpm reaches 3,025 to 4,650 and the throttle opening exceeds 36%, the PCM energizes the IMTV solenoid. This action opens the rectangular valve, and the two halves of the intake are now joined through this valve opening. Under this condition, air flow in the intake increases to improve horsepower and torque.

One end of the IMTV relay winding is connected to the ignition switch and the other end of this winding is connected to the PCM. When the engine operating conditions require IMTV operation, the PCM grounds the IMTV relay winding. This action closes the relay points. Current flows through the IMTV relay points to the IMTV solenoid winding in the intake manifold to open the rectangular valve (Figure 25-50).

Electronic Continuous Injection System (CIS-E)

The **electronic continuous injection system (CIS-E)** is manufactured by Robert Bosch Corporation. This system is used on many European imported vehicles. Robert Bosch Corporation manufactures a variety of SFI and MFI systems that have components similar to the ones described previously in this chapter. Early model CIS systems did not have electronic control.

In the electronic continuous injection system (CIS-E), the injectors in each intake port are injecting fuel continually while the engine is running. These injectors do not open and close while the engine is running. Fuel is supplied from a central **fuel distributor** to all the injectors. An electric fuel pump supplies fuel through an accumulator and filter to the fuel distributor. The accumulator pre-

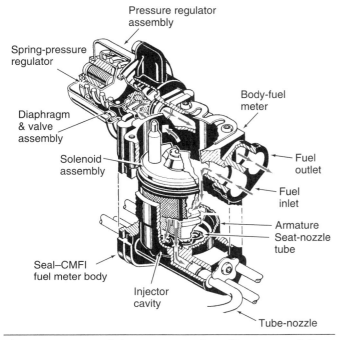

Figure 25-48 CPI injector operation. (Courtesy of General Motors Corporation, Service Technology Group)

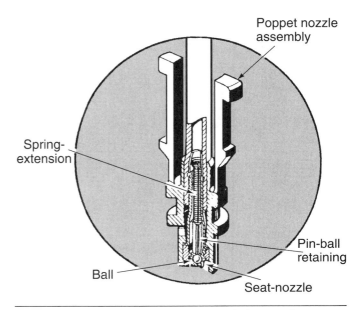

Figure 25-49 Poppet nozzle internal design. (Courtesy of General Motors Corporation, Service Technology Group)

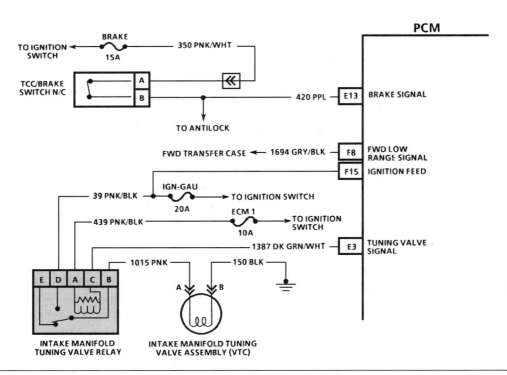

Figure 25-50 IMTV relay and solenoid wiring diagram. (Courtesy of General Motors Corporation, Service Technology Group)

vents fuel pressure pulses. Fuel is returned from the fuel distributor and the pressure regulator to the fuel tank.

A pivoted air flow sensor plate is positioned in the air intake. With the engine not running, the air flow sensor plate closes the air passage in the air intake. A control plunger in the fuel distributor rests against the air flow sensor lever (Figure 25-51).

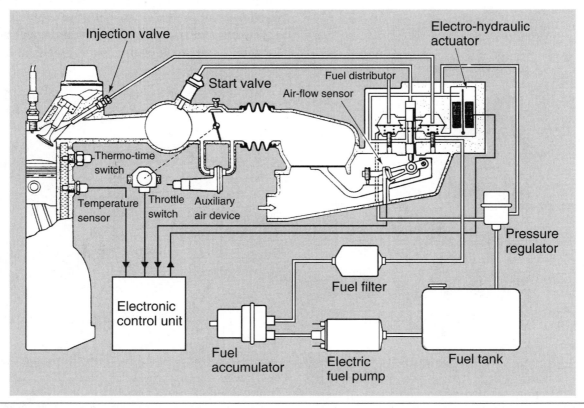

Figure 25-51 Air flow sensor plate and control plunger in CIS-E system. (Courtesy of SAE International)

When engine speed increases, the air velocity in the air intake gradually opens the air flow sensor plate. This plate movement lifts the control plunger, which precisely meters the fuel to the injectors. The differential pressure regulator winding is energized by the electronic control unit. The **differential pressure regulator** winding is cycled on and off by the electronic control unit. This action moves the plunger up and down in the differential pressure regulator, which controls fuel pressure in the lower chambers of the fuel distributor and provides precise control of the fuel delivery to the injectors and the air/fuel ratio.

The input sensors connected to the electronic control unit vary depending on the vehicle, but these sensors include an oxygen sensor and coolant temperature sensor. Some CIS-E systems are complete engine management systems in which the electronic control unit provides fully integrated control of the air/fuel ratio, spark advance, emission control devices, and idle speed.

The **cold start valve** is operated by a thermo-time switch. This system is very similar to the cold start injector explained before in this chapter. The auxiliary air device opens when the engine is cold and allows a small amount of air to bypass the throttle to increase idle speed.

Summary

❑ TBI, MFI, and SFI systems provide improved fuel economy, engine performance, and emission levels compared to carbureted engines.

❑ In a typical speed density TBI, MFI, or SFI system, the computer uses the MAP and engine rpm inputs to calculate the amount of air entering the engine. The computer then calculates the required amount of fuel to go with the air entering the engine.

❑ In a typical air density system, the PCM uses the MAF and engine speed to determine the pulse width.

❑ An inertia switch in the fuel pump circuit opens the fuel pump circuit immediately if the vehicle is involved in a collision.

❑ The oil pressure switch connected in parallel with the fuel pump relay operates the fuel pump if the fuel pump relay is defective.

❑ In some TBI, MFI, and SFI systems, the ASD relay supplies voltage to the fuel pump, positive primary coil terminal, oxygen sensor heater, and injectors.

❑ In any TBI, MFI, or SFI system, the fuel pressure must be high enough to prevent fuel boiling.

❑ In a TBI, MFI, or SFI system, the computer supplies the proper air/fuel ratio by controlling injector pulse width.

❑ In a TBI, MFI, or SFI system, the computer increases injector pulse width to provide air/fuel ratio enrichment while starting a cold engine.

❑ Most computers provide a clear flood mode if a cold engine becomes flooded. This mode is activated by pressing the gas pedal to the floor while cranking the cold engine.

❑ In an SFI system, each injector has an individual ground wire connected to the computer.

❑ In an MFI system, the injectors are connected together in pairs or groups on the ground side.

❑ The cold start injector is operated by a thermo-time switch, and this injector sprays additional fuel into the intake manifold while cranking a cold engine.

❑ The pressure regulator maintains the specified fuel system pressure and returns excess fuel to the fuel tank.

❑ In a returnless fuel system, the pressure regulator and filter assembly is mounted with the fuel pump and gauge sending unit assembly on top of the fuel tank. This pressure regulator returns fuel directly into the fuel tank.

❑ A central port injection system has one central injector and a poppet nozzle in each intake port. The central injector is operated by the PCM, and the poppet nozzles are operated by fuel pressure.

❑ An intake manifold tuning valve (IMTV) controls the air passages inside the intake manifold to provide improved air flow in the manifold.

❑ In a continuous injection system, the injectors spray fuel continually into the intake manifold while the engine is cranking or running. The injectors are supplied with fuel from a central fuel distributor.

❑ In a continuous injection system, fuel is metered to the injectors by a control plunger in the fuel distributor which is operated by an air flow sensor plate. In these injection systems, the computer controls a differential pressure regulator that controls pressure in the lower chambers of the fuel distributor and provides precise control of the fuel supplied to the injectors.

Terms-To-Know

Adaptive memory	(EVP) sensor	Neutral drive switch (NDS)
Air density system	Fuel distributor	Nonsynchronized mode
Barometric pressure (BARO) sensor	Fuel injector	Open loop
Block learn	Fuel pressure regulator	Peak-and-hold current
Central port injection (CPI)	High pressure TBI (HPTBI)	Poppet nozzles
Circuit opening relay	Idle air control (IAC) motor	Port fuel injection (PFI)
Clear flood	Idle speed control (ISC) motor	Returnless fuel system
Closed loop	Idle switch	Semi-closed loop
Cold start injector	Inertia switch	Sequential fuel injection (SFI)
Cold start valve	Injector rail	Short term adaptive
Constant control relay module	Intake air temperature (IAT) sensor	Speed density system
Corporate average fuel economy (CAFE)	Intake manifold tuning valve (IMTV)	Synchronized mode
	Integrator	Synchronized system
Density speed system	Limp-in	Throttle body
Differential pressure regulator	Long term adaptive	Throttle body backup (TBB)
Dual ramping	Low pressure TBI (LPTBI)	Throttle body injection (TBI)
Electronic continuous injection system (CIS-E)	Manifold absolute pressure (MAP) sensor	Throttle body temperature sensor
Electronic fuel injection (EFI)	Mass air flow (MAF) sensor	Throttle position sensor (TPS)
Engine coolant temperature (ECT) sensor	Mass air systems	Titania-type O_2S
Exhaust gas recirculation valve position	Minimum TPS	Vehicle speed sensor (VSS)
	Multipoint fuel injection (MPI)	

Review Questions

Short Answer Essays

1. Explain the purpose of the TPS input in a speed-density fuel injection system.

2. Describe the advantages of MFI and SFI systems compared to carburetor fuel systems.

3. Explain the purpose of the inertia switch in a fuel pump circuit, including the switch reset procedure.

4. When an oil pressure switch is connected in parallel to the fuel pump relay, describe the engine operating problem caused by a defective fuel pump relay.

5. Describe the operation of the central injector and poppet nozzles in a central port injection (CPI) system.

6. Describe the operation of a typical fuel pressure regulator.

7. Describe the purpose of a throttle body temperature signal on a TBI system.

8. Explain how the computer controls the air/fuel ratio on a TBI, MFI, or SFI system.

9. Describe how an idle air control (IAC) motor controls idle speed.

10. Describe the difference between a sequential fuel injection (SFI) system and a multipoint fuel injection (MPI) system.

Fill-in-the-Blanks

1. The computer determines the air entering the engine from the _____ and _____ input signals in a speed-density system.

2. The length of time the computer grounds the injector is referred to as _____ _____ .

3. In TBI, MFI, and SFI systems, the fuel pressure must be high enough to prevent _____ _____ .

4. The computer program assumes that only liquid fuel is available at the injector and that a specified _____ _____ is available at the injector.

5. On an SFI system, each injector has an individual _____ _____ connected to the computer.

6. If the injector pulse width is increased, the air/fuel ratio becomes _____.

7. When the engine is idling, the fuel pressure regulator provides _____ fuel pressure compared to the fuel pressure at wide-open throttle.

8. A plugged return fuel line on a TBI, MFI, or SFI system causes _____ fuel pressure and a _____ air/fuel ratio.

9. In a central port fuel injection (CPI) system, the air/fuel ratio is determined by the pulse width on the _____ _____ _____ .

10. _____ loop means the O_2S input is ignored by the PCM.

ASE Style Review Questions

1. While discussing electronic fuel injection principles:
   Technician A says the computer uses the TPS and ECT signals to determine the air entering the engine in a speed-density system.
   Technician B says the computer uses the TPS and oxygen sensor signals to determine the air entering the engine in a air-density system.
   Who is correct?
   a. A only
   b. B only
   c. Both A and B
   d. Neither A nor B

2. While discussing TBI and MFI systems:
   Technician A says that in a TBI or MFI system higher than normal fuel pressure causes a lean air/fuel ratio.
   Technician B says that in these systems lower than normal fuel pressure causes a rich air/fuel ratio.
   Who is correct?
   a. A only
   b. B only
   c. Both A and B
   d. Neither A nor B

3. While discussing TBI, MFI, and SFI systems:
Technician A says the PCM provides the proper air/fuel ratio by controlling fuel pressure.
Technician B says the PCM provides the proper air/fuel ratio by controlling injector pulse width.
Who is correct?
   a. A only
   b. B only
   c. Both A and B
   d. Neither A nor B

4. While discussing cold start injector systems:
Technician A says the cold start injector is operated by a thermo-time switch.
Technician B says the cold start injector is operated only while the engine is being cranked.
Who is correct?
   a. A only
   b. B only
   c. Both A and B
   d. Neither A nor B

5. While discussing fuel pressure regulators:
Technician A says that the pressure regulator in a returnless SFI system maintains the same fuel pressure regardless of throttle opening.
Technician B says that the manifold vacuum connection to the pressure regulator in a SFI system causes higher fuel pressure at wide-open throttle.
Who is correct?
   a. A only
   b. B only
   c. Both A and B
   d. Neither A nor B

6. While discussing returnless fuel systems:
Technician A says in a returnless fuel system the pressure regulator is mounted on the fuel rail.
Technician B says in this type of fuel system the fuel has a tendency to get hotter than in return type systems.
Who is correct?
   a. A only
   b. B only
   c. Both A and B
   d. Neither A nor B

7. Input sensors are being discussed. Technician A says the manifold absolute pressure (MAP) sensor is used to determine engine load. Technician B says the MAP sensor is used to determine the volume of air entering the engine. Who is correct?
   a. A only
   b. B only
   c. Both A and B
   d. Neither A nor B

8. While discussing fuel boiling in the fuel rail:
Technician A says fuel boiling in the fuel rail causes a lean air/fuel ratio.
Technician B says the computer calculations for pulse width assume proper fuel pressure.
Who is correct?
   a. A only
   b. B only
   c. Both A and B
   d. Neither A nor B

9. While discussing central port injection (CPI):
Technician A says the poppet nozzles are opened by the computer during engine starting.
Technician B says the poppet nozzles are opened by fuel pressure when the engine is running.
Who is correct?
   a. A only
   b. B only
   c. Both A and B
   d. Neither A nor B

10. While discussing electronic continuous injection systems:
Technician A says the air/fuel ratio delivered is controlled by the fuel pressure in the lower chambers of the fuel distributor, which is controlled by the computer and the differential pressure regulator.
Technician B says the air/fuel ratio is determined by the position of the air flow sensor plate and control plunger.
Who is correct?
   a. A only
   b. B only
   c. Both A and B
   d. Neither A nor B

26 Electronic Fuel Injection Diagnosis and Service

Objective

Upon completion and review of this chapter, you should be able to:

❏ Perform a preliminary diagnostic procedure on a throttle body injection (TBI), multiport fuel injection (MFI), or sequential fuel injection (SFI) system.

❏ Properly relieve fuel pressure.

❏ Test fuel pump pressure and properly interpret the results.

❏ Test the fuel injector and determine needed repairs.

❏ Test the cold start injector and properly interpret the results.

❏ Perform a minimum idle speed adjustment.

❏ Perform a flash code diagnosis on various vehicles.

❏ Erase fault codes.

❏ Perform a scan tester diagnosis on various vehicles.

❏ Diagnose and determine needed repairs of common input sensors used in fuel injection systems.

❏ Diagnose and determine needed repairs of common outputs used in fuel injection systems.

❏ Interpret data recordings.

❏ Determine the root cause of computer system failures.

❏ Determine the root cause of repeated component failures.

❏ Determine the root cause of multiple component failures.

Introduction

When engine performance or economy complaints occur on fuel injected vehicles, the tendency of many technicians is to think the problem is in the fuel injection and computer system. However, many other defects can affect the engine and fuel injection system operation. For example, an intake manifold vacuum leak causes a rough idle condition and engine surging at low speed. If the engine has a MAF sensor, an intake manifold vacuum leak allows additional air into the intake, and this air does not flow through the MAF sensor. Therefore, the MAF sensor signal indicates to the powertrain control module (PCM) less air flow is entering the engine in relation to the throttle opening. Under this condition, the PCM supplies less fuel through the injectors. This creates a lean air/fuel ratio, which results in engine surging and acceleration stumbles. Expensive hours of diagnostic time may be saved if these items are proven to be satisfactory before the fuel injection system is diagnosed:

1. Intake manifold vacuum leaks

2. Emission devices such as the EGR valve and related controls

3. Ignition system condition

4. Engine compression

5. Battery fully charged

6. Engine at normal operating temperature

7. All accessories turned off

After all other considerations are made, then the fuel injection system is suspect. This chapter details common diagnostic procedures used to test the inputs and outputs of typical fuel injection systems.

Service Precautions

These precautions must be observed when TBI, MFI, and SFI systems are diagnosed and serviced:

1. Always relieve the fuel pressure before disconnecting any component in the fuel system.

2. Never turn the ignition switch to the RUN position when any fuel system component is disconnected.

3. Use only the test equipment recommended by the vehicle manufacturer.

4. Always turn off the ignition switch before connecting or disconnecting any system component or test equipment.

5. Never allow electrical system voltage to exceed 16V. This could be done by disconnecting the circuit between the alternator and the battery with the engine running.

6. Avoid static electric discharges when handling computers, modules, and computer chips.

Preliminary Inspection

A complete visual inspection will often reveal the cause of a driveability complaint. Connections must be tight and clean. In addition, all wires must be checked for signs of damage. While inspecting the wires also look for signs of tampering and modifications.

Along with inspection of wires, check the vacuum hoses. Use the under-hood sticker to assure the vacuum hose routing is correct. Check the hoses for signs of cracking, loose connections, and restrictions.

Include in the inspection all emission control devices. If the engine is equipped with an air pump, check the tension and condition of the belt. Also, check the condition of the diverter valve and all hoses connected to the system. Make sure the EGR valve and the canister purge system is functioning properly.

Locating Service Information

The first step in accurate fuel injection system diagnosis (as with all other diagnosing) is to locate the proper service information for the vehicle. The technician must have the correct specifications, wiring diagrams, and test procedures.

The service manual is one of the most important tools for today's technician. Manuals provide information concerning engine identification, service procedures, specifications, and diagnostic information. In addition, the service manual provides information concerning wiring harness connections and routing, component location, and fluid capacities. Service manuals can be supplied by the vehicle manufacturer or through aftermarket suppliers.

To obtain the correct information, you must be able to identify the engine you are working on. This may involve using the vehicle identification number (VIN). This number will have a code for model year and engine. Which numbers are used will vary between manufacturers but the service manual will provide instructions for proper VIN number usage (Figure 26-1). The service manual may also assist you in engine identification through the interpretation of casting numbers and marks on the block or cylinder head.

With the engine properly identified the required information can be retrieved. In the past, each manufacturer and manual publisher used its own method of organizing its manuals. Recent guidelines now require manufacturer service manuals to have a standard organization (Figure 26-2).

Procedural information provides the steps necessary to perform the task. Most service manuals provide illustrations to guide the technician through the task. To get the most out of the service manual, you must use the correct manual for the vehicle and system being worked on, and follow each step in order. Some technicians lead themselves down the wrong trail by making assumptions and skipping steps.

Diagnostic procedures are often presented in a chart form or a tree (Figure 26-3). The tree guides you through the process as system tests are performed. The result of a test then directs you to a branch. Keep following the steps until the problem is isolated.

Service and parts information can also be provided through computer services. This is a popular method since libraries of service manuals can take up a lot of room. Computerized systems can have the information stored onto disks, or the computer can be connected to a central database. The computer system helps the techni-

VEHICLE IDENTIFICATION NUMBERS

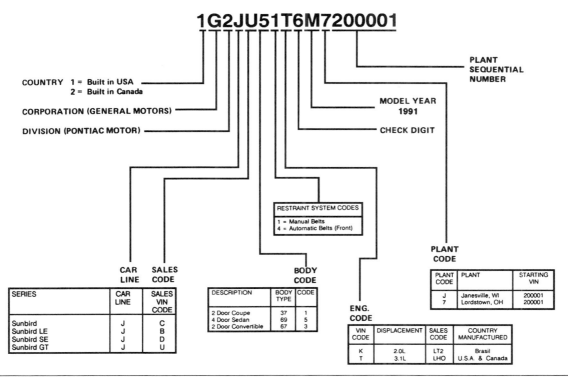

Figure 26-1 The service manual provides information on how to use the VIN number. (Courtesy of General Motors Corporation, Service Technology Group)

cian find the required information more quickly and easier than in a book-type manual. Using the computer keyboard, light pen, touch sensitive screen, or mouse the

technician makes choices from a series of menus on the monitor. If needed, the information can be printed to paper.

Service procedures may change in mid-year production or problem components may be replaced with new designs, etc. **Service bulletins** provide up-to-date corrections, additions, and information concerning common problems and their solutions.

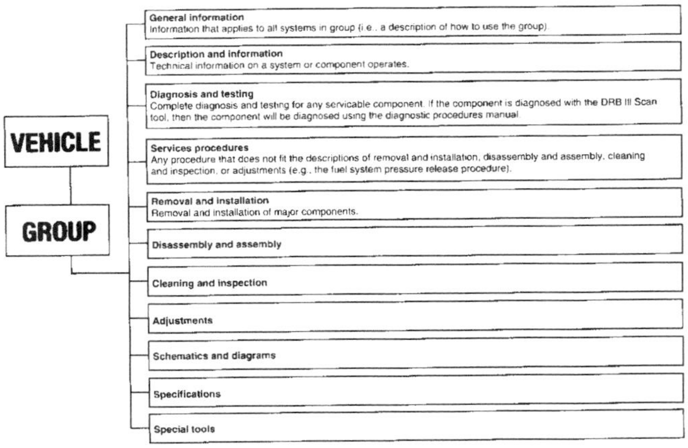

Disconnecting Battery Cables

During the diagnosis of TBI, MFI, and SFI systems, many procedures require the disconnection of the battery. Disconnecting the battery will erase the adaptive strategies stored in the PCM. If the adaptive strategy in the computer is erased, the engine operation may be rough at low speeds when the engine is restarted. This is because the computer must relearn the adaptive values. After disconnecting the battery and performing repairs to the fuel or ignition systems the vehicle should be road tested for at least 5 minutes with the engine at normal operating temperature. Some manufacturers recommend a 12-V dry cell battery be connected from the positive battery

Figure 26-2 Uniform service manual layout. (Courtesy of Chrysler Corporation)

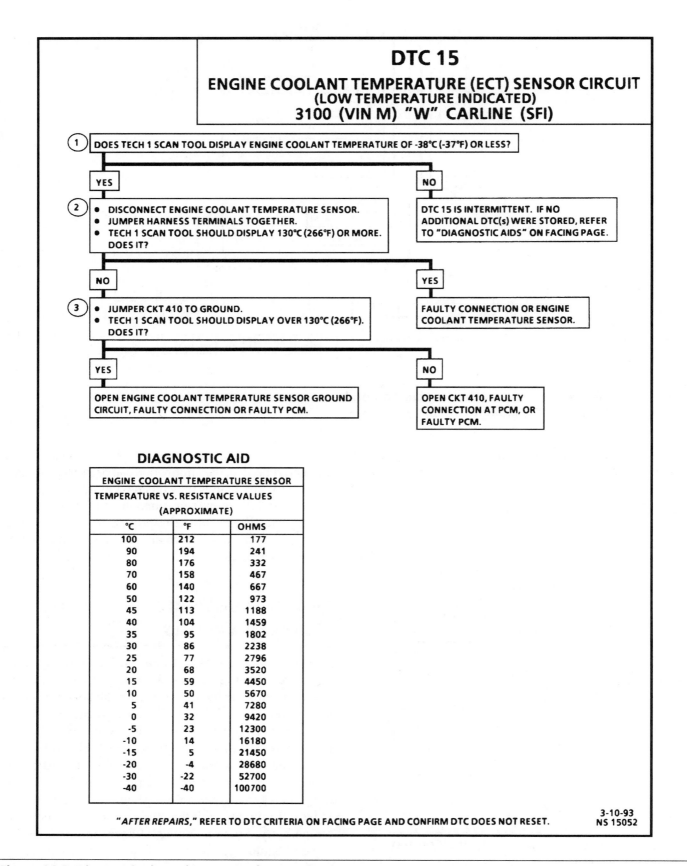

Figure 26-3 Diagnostic chart. (Courtesy of General Motors Corporation, Service Technology Group)

cable to ground if the battery is disconnected. The 12V supplied by the dry cell prevents deprogramming and memory erasing. Some 12-V sources for this purpose are designed to plug into the cigarette lighter socket.

Also, if diagnosis showed a higher than normal adaptive correction value and a repair to the system is performed, the adaptives should be reset to zero. This can be done with the scan tool, by disconnecting the battery, or pulling the fuse for the circuit that supplies battery voltage to the PCM. The vehicle should be road tested so the PCM can relearn the new values. For example, if the adaptive values were showing a 25% decrease in base pulse width due to a dirty air cleaner (causing a rich air/fuel ratio) and the air cleaner is replaced, the PCM will still take away the 25% from base pulse width. This will make the air/fuel ratio too lean. The PCM will learn the new values over time and correct the adaptive, however, it is best to start at zero. In some systems there are different adaptive strategies. Chrysler uses long term and short term adaptive memories. Short term is reset to zero every time the ignition switch is turned off, but long term is kept in RAM. In addition, long term is used in open loop fuel calculations and does not update until engine coolant temperature reaches about 170° F. If the memory is not erased, this value will be used until the engine temperature is increased and the vehicle is driven in that cell for a particular length of time. If the vehicle is returned to the customer without erasing the memory and test driving the vehicle, they may experience driveability problems.

Fuel Pressure Testing

SERVICE TIP *Remember, many Ford products have an inertia switch in the fuel pump circuit. If there is no fuel pump pressure, always push the inertia switch reset button first.*

SERVICE TIP *Many fuel pump circuits are connected through a fuse in the fuse panel or a fusible link. If there is no fuel pump pressure, always check the fuel pump fuse or fusible link first.*

When tests are performed to diagnose any automotive problem, always start with the tests that are completed quickly and easily. The fuel pressure test is usually one of the first tests to consider when TBI, MFI, and SFI systems are diagnosed. Remember low fuel pressure may cause lack of power, acceleration stumbles, engine surg-

ing, and limited top speed, whereas high fuel pressure results in excessive fuel consumption, rough idle, engine stalling, and excessive sulphur smell from the catalytic converter.

In some cases, in-tank fuel pumps have the specified pressure when the ignition switch is turned on or when the engine is idling, but the fuel pump cannot meet the engine demand for fuel at or near wide-open throttle. Therefore, if the customer complains about the engine cutting out or stalling at higher speeds, the fuel pump pressure should be tested at higher speeds during a road test.

Relieving Fuel Pressure

Prior to pressure gauge connection on TBI, MFI, or SFI systems, the fuel pressure must be relieved. There are several methods that can be used to relieve fuel pressure. Always refer to the correct service manual to determine the manufacturer's recommend procedure.

One method used is to momentarily supply 12V to one injector terminal and ground the other injector terminal. This action lifts the injector plunger and the fuel discharges from the injector to relieve the fuel pressure. Do not supply 12V and a ground to an injector for more than 5 seconds unless a vehicle manufacturer's recommended procedure specifies a longer time period. A variation of this is to remove the fuel pump relay, turn the ignition switch to the RUN position, and use the scan tool actuator tests to energize the injector.

On vehicles with a **Schrader valve** in the fuel line, the manufacturer may recommend a drain hose be connected to the valve with the open end of the hose placed into an approved container (Figure 26-4).

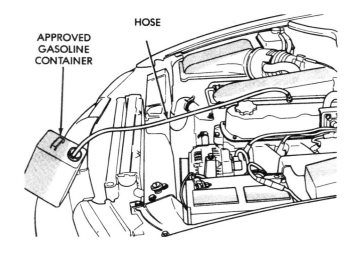

Figure 26-4 Relieving fuel pressure with a drain hose and approved container. (Courtesy of Chrysler Corporation)

Another method is to disable the fuel pump by removing the fuel pump relay, disconnecting the fuel pump, or pulling the fuel pump fuse. With the fuel pump disabled, start the engine and allow it to die. Then crank the engine a few times to remove all fuel from the rails.

Regardless of the method used, always remove the gas tank filler cap first. This will relieve any pressure in the tank and prevent siphoning.

Connecting Fuel Pressure Gauge

Connection of the fuel pressure gauge depends on system design. Again, the service manual must be referenced during this procedure. Common methods are discussed here.

In some TBI systems, the inlet fuel line at the throttle body assembly must be removed and the pressure gauge hose installed in series between the inlet line and the inlet fitting (Figure 26-5). On other TBI systems, the vehicle manufacturer recommends connecting the fuel pressure gauge at the fuel filter inlet. Use new gaskets on the union bolt when the pressure gauge is connected at this location.

CAUTION *Never turn the ignition switch to the RUN position or crank the engine with a fuel line disconnected. This action causes the fuel pump to discharge fuel from the disconnected line, which may result in a fire, causing personal injury and/or property damage.*

In many MFI or SFI systems a Schrader valve on the fuel rail is used to connect the fuel pressure gauge. On some SFI systems, the vehicle manufacturer recommends connecting the fuel pressure gauge to the cold start injector fuel line (Figure 26-6). Install new gaskets on the union

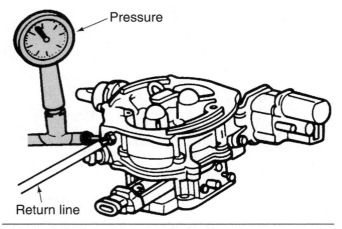

Figure 26-5 Connecting a fuel pressure gauge to the throttle body assembly. (Courtesy of Chrysler Corporation)

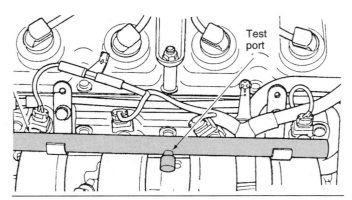

Figure 26-6 Connecting a fuel pressure gauge to the Schrader valve. (Courtesy of Chrysler Corporation)

bolt when the pressure gauge is installed at this location. If the system does not have either a Schrader valve or cold start injector, a special adaptor is connected at the end of the fuel rail.

Select the proper gauge for the system pressures being tested. The recommended pressures should be in the middle of the gauge's operating range. For example, if the specification calls for 49 psi, do not use a gauge that reads up to 200 psi or one that has a maximum reading of 55 psi.

Operating the Fuel Pump to Test Pressure

The technician must have pressure specifications for the make and model year of the vehicle being tested. Once the pressure gauge is connected, the ignition switch may be cycled several times to read the fuel pressure, or the pressure may be read with the engine idling. In cases where the engine will not start or when further diagnosis of the fuel pump circuit is required, it may be helpful to operate the fuel pump continually. Many fuel pump circuits have a provision for operating the fuel pump for testing fuel pump pressure if the engine will not run. Some manufacturers, such as Toyota, recommend operating the fuel pump with a jumper wire connected across the appropriate terminals in the DLC. On many Toyota products, the jumper wire must be connected across the B+ and FP terminals in the DLC with the ignition switch in the RUN position (Figure 26-7).

Fuel pump test procedures vary depending on the year and make of the vehicle. Always follow the recommended procedure in the vehicle manufacturer's service manual. Following is a typical fuel pump test procedure on a Toyota vehicle:

1. Connect a 12-V power supply to the cigarette lighter socket and disconnect the negative battery cable. If the vehicle is equipped with an air bag, wait one minute.

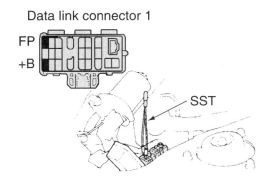

Figure 26-7 Connecting a fuel pressure gauge to the cold start injector fuel line.

2. Bleed pressure from the fuel system as mentioned previously.

3. Connect the fuel pressure gauge as outlined previously. Use a shop towel to wipe up any spilled gasoline.

4. Connect the jumper wire across the B+ and FP terminals in the DLC.

5. Reconnect the battery negative cable, and turn the ignition switch in the RUN position.

6. Observe the fuel pressure on the gauge.

7. Disconnect the jumper wire from the DLC terminals.

8. Disconnect and plug the vacuum hose from the pressure regulator, and start the engine.

9. Observe the fuel pressure on the gauge with the engine idling.

10. Reconnect the vacuum hose to the pressure regulator and observe the fuel pressure on the gauge.

The fuel pump pressure must equal the manufacturer's specifications under all conditions. This pressure is usually about 10 psi (70 kPa) higher with the vacuum hose removed from the pressure regulator than when this vacuum hose is connected. If the pressure is higher than specified, check the return fuel line and pressure regulator.

When there is no fuel pump pressure, check the fusible link, fuses, SFI main relay, fuel pump, PCM, and wiring connections. If the fuel pump pressure is lower than specified, check the fuel lines and hoses, fuel pump, fuel filter, pressure regulator, and cold start injector.

On many Ford products, a self-test connector is located in the engine compartment. This connector is tapered on both ends, but one tapered end is longer than the other end. A wire is connected from the fuel pump relay to the outer terminal in the short tapered end of the DLC.

WARNING *When instructed to ground a wire for diagnostic purposes, always be sure you are grounding the proper wire under the specified conditions. Improper grounding of computer system terminals may damage computer system components.*

The PCM normally grounds this wire to close the fuel pump relay points. If this wire is grounded with a jumper wire, when the ignition switch is on, the fuel pump runs continually for diagnostic purposes.

On many Chrysler products, a square DLC is located in the engine compartment. This connector has a notch in one corner (Figure 26-8). The terminal in the corner of the diagnostic connector directly opposite the notch may be grounded with a 12-V test lamp to operate the fuel pump continually. This terminal could be grounded with a jumper wire, but there is a 12-V power wire in one of the other diagnostic connector terminals. If this power wire is accidentally grounded with a jumper wire, severe computer and wiring harness damage may result. On some Chrysler products, the fuel pump test wire is discontinued. Always check the wiring diagram for the vehicle being diagnosed.

On many General Motors products, a 12-terminal DLC is located under the instrument panel. In most of these connectors, the terminals are lettered A to F across the top row, and G to M across the bottom row (Figure 26-9).

On some General Motors vehicles, a fuel pump test connector is located in terminal G on the DLC. On other General Motors vehicles, this fuel pump test wire is located in the engine compartment. If 12V are supplied to the fuel pump test wire with the ignition switch off, voltage is supplied through a pair of fuel pump relay points to the fuel pump. The technician may observe the

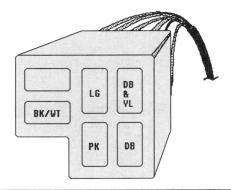

Figure 26-8 On many Chrysler vehicles, the terminal directly opposite the notch in the corner of the DLC may be grounded with a 12-volt test light to operate the fuel pump. (Courtesy of Chrysler Corporation)

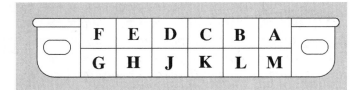

Figure 26-9 Data link connector (DLC).

fuel pump pressure under this condition or listen at the fuel tank filler neck for the fuel pump running. If the fuel pump operates satisfactorily under this test condition, the fuel pump and the wire from the relay to the pump are satisfactory. If the fuel pump does not run when the ignition switch is turned on, the fuel pump relay, PCM, or connecting wires are defective.

After performing the fuel pressure test, turn the ignition switch off and wait 10 minutes. Recheck the gauge reading with the ignition switch still off. Pressures should be the same as when the engine was turned off. If the pressure is lower now, the check valve in the pump may be defective. An injector pintle that is stuck open will also cause pressures to drop. It is normal for fuel pressures to drop after the vehicle has been sitting for a few hours. However, fuel volume in the rail should not drop. Turn the ignition switch to the RUN position; pressure should be at specifications.

Causes of Low Fuel Pump Pressure

If the fuel pressure is low, always check the filter and fuel lines for restrictions before the fuel pump is diagnosed as the cause of the problem. In some cases, dirt or sediment in the fuel tank covers and plugs the pickup sock on the in-tank fuel pump. This action restricts the fuel supply to the pump. Usually this will cause the engine to cut out or stall at highway speeds.

Also, if pressures are low, gently pinch the fuel return line while observing the pressure gauge. If the pressures come up, the fuel pressure regulator is defective.

Fuel Pump Volume Testing

Some manufacturers recommend a fuel volume test be performed if the fuel pressure test passes. Always refer to the service manual for the recommended procedure. The following is a typical procedure:

1. Relieve the fuel pressure as discussed previously.
2. Disconnect the return line from the fuel rail.

3. Connect a hose from the fuel rail return connection to a graduated container with a 2-quart capacity. Properly secure the container to prevent spilling. Make sure the hose is routed away from moving parts and the exhaust manifold.

4. Energize the fuel pump for thirty seconds with a scan tool or by the specified procedure in the service manual.

5. Measure the quantity of fuel in the container.

If fuel pump volume is less than specifications, there may be a restriction in the fuel lines. Also, air leaks in the lines will cause low volume. Check all connections carefully.

If the fuel lines are good, check for applied voltage at the fuel pump while it is energized. The voltage should be within .5 volt of battery voltage. If there is sufficient voltage, check the fuel pump ground. If everything checks good, replace the fuel pump.

Diagnosis of Computer Voltage Supply and Ground Wires

SERVICE TIP *Never replace a computer unless the ground wires and voltage supply wires are proven to be in satisfactory condition.*

A computer can not operate properly unless it has proper ground connections and satisfactory voltage supply at the required terminals. A computer wiring diagram for the vehicle being tested must be available for these tests. Backprobe the BATT terminal on the computer and connect a pair of digital voltmeter leads from this terminal to ground. The voltage at this terminal should be .5 volt of battery voltage with the ignition switch off. If the proper voltage is not present at this terminal, check the computer fuse and related circuit. Turn the ignition switch to the RUN position and check applied voltage to all battery and ignition feed terminals of the PCM. The voltage measured at these terminals should be within .5 volt of battery voltage with the ignition switch in the RUN position. When the specified voltage is not available, test the voltage supply wires to these terminals. These terminals may be connected through fuses, fuse links, or relays. Always refer to the vehicle manufacturer's wiring diagram for the vehicle being tested.

Computer ground wires usually extend from the computer to a ground connection on the engine or battery. In addition, there may be a case ground provided by a mounting screw. With the ignition switch in the RUN position, perform a voltage drop test across the ground wires. More than 0.2 volt indicates excessive resistance in the ground circuit.

SERVICE TIP *When diagnosing computer problems, it is usually helpful to ask the customer about service work that has been performed lately on the vehicle. If service work has been performed in the engine compartment, it is possible that a computer ground wire may be loose or disconnected.*

Input Sensor Diagnosis and Service

Most input voltage values can be checked using a scan tool. Often this is the easiest way to get this information since it does not require access to the component or PCM. However, remember the scan tool only gives an interpretation of these values. Sometimes the scan tool may display a value that is within specifications but actual DMM testing of the sensor proves the component is defective. The following examples of input sensor testing are provided if a scan tool is not available or is suspected of erroneous display.

Oxygen Sensor Diagnosis

The engine must be at normal operating temperature before the oxygen (O_2S) sensor is tested. Always follow the test procedure in the vehicle manufacturer's service manual, and use the specifications supplied by the manufacturer.

WARNING *An oxygen sensor must be tested with a 10M impedance digital voltmeter. If an analog meter is used for this purpose, the sensor may be damaged*

SERVICE TIP *A contaminated oxygen sensor may provide a continually high voltage reading because the oxygen in the exhaust stream does not contact the sensor.*

A digital voltmeter connected from the O_2S wire to ground can be used to test this sensor. Use an electric probe to backprobe the connector near the O_2S to connect the voltmeter to the sensor signal wire. If possible, avoid probing the insulation to connect a meter to a wire. With the engine operating at normal temperature and idle

speed, in closed loop, the O_2S voltage should be cycling from low voltage to high voltage. A typical O_2S cycles from 0.3V to 0.8V.

If the voltage is continually high, the air/fuel ratio may be rich, or the sensor may be contaminated. The O_2S may be contaminated with room temperature vulcanizing (RTV) sealant, antifreeze, or lead from leaded gasoline. A rich mixture can be caused by a faulty input sensor (such as ECT or TPS), a dirty air cleaner element, a leaking injector, and so forth.

When the O_2S voltage is continually low, the air/fuel ratio may be lean, due to a vacuum leak, stuck or plugged injector, open injector winding, or open injector circuit. Also the sensor may be defective, or the wire between the sensor and the computer may have high resistance. A low voltage O_2S output can also be the result of a faulty spark plug or spark plug cable. Since none of the oxygen is used during the combustion process, a high oxygen content passes the O_2S. Remember the O_2S does not measure hydrocarbons, only oxygen content.

If the O_2S voltage signal remains in a mid-range position, the computer may be in open loop, the sensor may be defective, or the sensor circuit is open.

When the O_2S is removed from the engine, a digital voltmeter may be connected from the signal wire to the sensor case, and the sensor element may be heated in the flame from a propane torch. The propane flame keeps the oxygen in the air away from the sensor element, causing the sensor to produce voltage. While the sensor element is in the flame, the voltage should be nearly 1V, and the voltage should drop to zero immediately when the flame is removed from the sensor. If the sensor does not produce the specified voltage, it should be replaced.

If a defect in the O_2S signal wire is suspected, backprobe the sensor signal wire at the computer and connect a digital voltmeter from the signal wire to ground with the engine idling. The difference between the voltage readings at the sensor and at the computer should not exceed the vehicle manufacturer's specifications.

With the engine idling, connect a digital voltmeter from the sensor case to the sensor ground wire on the computer. A typical maximum voltage drop reading across the sensor ground circuit is 0.2V. Always use the vehicle manufacturer's specifications. If the voltage drop across the sensor ground exceeds specifications, repair the ground wire or the sensor ground in the exhaust manifold.

WARNING *When using a digital storage oscilloscope (DSO), always follow the DSO manufacturer's recommended operating procedures. Improper DSO use may result in tester damage.*

Oxygen Sensor Testing With DSO

The DSO provides more accurate diagnosis of automotive computer systems because it has a much faster signal sampling speed than most other types of test equipment. Before diagnosing an oxygen sensor, the engine must be at normal operating temperature and the computer must be in closed loop. When diagnosing input sensor waveforms, adjust the horizontal time base on the DSO so three waveforms appear on the screen (Figure 26-10). When testing any input sensor, the DSO leads must be connected from the sensor signal wire to ground. The O_2S voltage should be cycling continually in the 0 to 1-V range (Figure 26-11). O_2S voltage signal cross counts are the number of times the O_2S voltage signal changes above or below 0.45V per second. If the O_2S cross counts are below specifications, the sensor may be defective or partially contaminated. When the voltage remains about 0.5V with very little fluctuation, the signal wire or sensor may have an open circuit. If the voltage remains below 0.5V with very little fluctuation, the intake manifold may have a vacuum leak.

Propane enrichment may be used to force the fuel system into a very rich condition, and verify the O_2S operation. A propane cylinder with a precision metering valve must be used for this test. Always maintain the propane cylinder in an upright position, and position the cylinder and hose away from rotating components. To verify rich-to-lean O_2S switching, connect the propane hose to a large intake manifold port such as the brake booster hose. Be sure there are no leaks between the

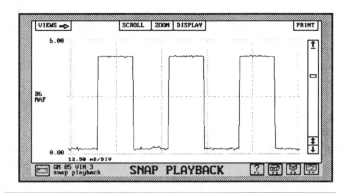

Figure 26-11 O_2S voltage signal cross counts. (Courtesy of OTC-SPX Corp. Aftermarket Tool & Equipment Group)

propane hose and the brake booster hose. Open the propane cylinder valve and introduce propane into the intake manifold. If necessary, hold the throttle open to maintain engine speed. The O_2S signal should become a flat line above 600mV. Quickly disconnect the propane hose and expose the brake booster hose to atmosphere. The O_2S signal should drop from 600mV to 300mV in less than 125ms (Figure 26-12).

Hold the throttle open to maintain a reasonable engine rpm with the brake booster hose exposed to atmosphere. Quickly install the propane hose into the brake booster hose and introduce propane into the intake manifold. Under this condition the O_2S signal should increase from 300mV to 600mV in 100ms (Figure 26-13). If the O_2S voltage signal does not switch from rich-to-lean or lean-to-rich in the specified time, the sensor is defective.

Oxygen Sensor Heater Diagnosis

If the O_2S heater is not operating properly, the sensor warm-up time is extended. Since most PCMs now have a close-loop timer that puts them into closed after a prede-

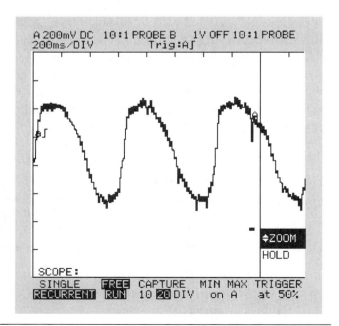

Figure 26-10 Normal O_2S activity. (Reproduced with permission of Fluke Corporation)

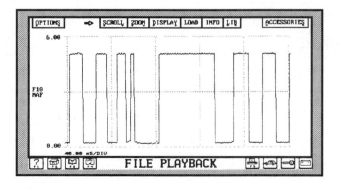

Figure 26-12 Rich-to-lean O_2S switching. (Reproduced with permission of Fluke Corporation)

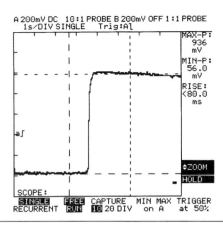

A 200mV DC 10:1 PROBE B 200mV OFF 1:1 PROBE
1s/DIV SINGLE   Trig:A1

MAX-P:
936
mV
MIN-P:
56.0
mV
RISE:
<80.0
ms

⇕ZOOM
HOLD

SCOPE:
SINGLE **FREE** CAPTURE MIN MAX TRIGGER
RECURRENT **RUN** **10** 20 DIV on A at 50%

Figure 26-13 Lean-to-rich O₂S switch. (Reproduced with permission of Fluke Corporation)

termined length of time, if the heater fails the computer supplies a richer air/fuel ratio. To test the heater, disconnect the O₂S connector and connect a digital voltmeter from the heater voltage supply wire and ground. With the ignition switch in the RUN position, 12V should be supplied on this wire. If the voltage is less than 12V, repair the fuse in this voltage supply wire or the wire itself.

On some vehicles the heaters are supplied voltage from a relay, such as the fuel pump relay or ASD relay. If all other circuits that receive voltage from this relay function properly, the relay is not at fault. However, for these relays to stay energized the engine must be running. Voltage will be present at the heater terminal at the O₂S only momentarily if checked with the ignition switch in the RUN position.

With the O₂S wire disconnected, connect an ohmmeter across the heater terminals in the sensor connector. If the heater does not have the specified resistance, replace the sensor.

Engine Coolant Temperature (ECT) Ohmmeter Diagnosis

WARNING *Never apply an open flame to an engine coolant temperature (ECT) sensor or intake air temperature (IAT) sensor for test purposes. This action will damage the sensor.*

A defective ECT sensor may cause some of the following problems:

1. Hard engine starting
2. Rich or lean air/fuel ratio
3. Improper operation of emission devices
4. Reduced fuel economy
5. Improper converter clutch lockup
6. Hesitation on acceleration
7. Engine stalling

With the sensor installed in the engine, the sensor terminals may be backprobed to connect a digital voltmeter to the sensor terminals. The sensor should provide the specified voltage drop at any coolant temperature (Figure 26-14).

Some computers use dual ramping in the ECT sensor circuit. The computer switches these resistors at approximately 120°F (49°C). This resistance change inside the computer causes a significant change in voltage drop across the sensor as indicated in the specifications. This is a normal condition on any computer with this feature. This change in voltage drop is always evident in the vehicle manufacturer's specifications.

The ECT sensor may be removed and placed in a container of water with an ohmmeter connected across the sensor terminals. A thermometer is also placed in the water. When the water is heated, the sensor should have the specified resistance at any temperature (Figure 26-15). Always use the vehicle manufacturer's specifications. If the sensor does not have the specified resistance, replace the sensor.

WARNING *Before disconnecting any computer system component, be sure the ignition switch is turned off. Disconnecting components may cause high induced voltages and computer damage.*

COLD CURVE 10,000-OHM RESISTOR USED		HOT CURVE CALCULATED RESISTANCE OF 909 OHMS USED	
−20°F	4.70 V	110°F	4.20 V
−10°F	4.57 V	120°F	4.00 V
0°F	4.45 V	130°F	3.77 V
10°F	4.30 V	140°F	3.60 V
20°F	4.10 V	150°F	3.40 V
30°F	3.90 V	160°F	3.20 V
40°F	3.60 V	170°F	3.02 V
50°F	3.30 V	180°F	2.80 V
60°F	3.00 V	190°F	2.60 V
70°F	2.75 V	200°F	2.40 V
80°F	2.44 V	210°F	2.20 V
90°F	2.15 V	220°F	2.00 V
100°F	1.83 V	230°F	1.80 V
110°F	1.57 V	240°F	1.62 V
120°F	1.25 V	250°F	1.45 V

Figure 26-14 Voltage specifications for the engine coolant temperature (ECT) sensor. (Courtesy of Chrysler Corporation)

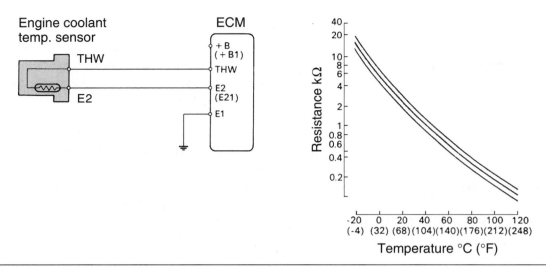

Figure 26-15 ECT sensor wiring and specifications. (Courtesy of Toyota Motor Corp.)

To diagnosis the ECT sensor circuit begin by disconnecting the ECT sensor and the computer connectors. Connect an ohmmeter from each sensor terminal to the computer terminal to which the wire is connected. Both sensor wires should indicate less resistance than specified by the vehicle manufacturer. If the wires have higher resistance than specified, the wires or wiring connectors must be repaired.

Intake Air Temperature (IAT) Sensor Diagnosis

The results of a defective intake air temperature (IAT) sensor may vary depending on the vehicle make and year. A defective IAC sensor may cause the following problems:

1. Rich or lean air/fuel ratio

2. Hard engine starting

3. Engine stalling or surging

4. Acceleration stumbles

5. Excessive fuel consumption

The IAT sensor may be removed from the engine and placed in a container of water with a thermometer. When a pair of ohmmeter leads is connected to the sensor terminals and the water in the container is heated, the sensor should have the specified resistance at any temperature (Figure 26-16). If the sensor does not have the specified resistance, sensor replacement is required.

With the IAT sensor installed in the engine, the sensor terminals may be backprobed and a voltmeter connected across the sensor terminals. The sensor should have the specified voltage drop at any temperature. The wires between the intake air temperature sensor and the computer may be tested in the same way as the IAT wires.

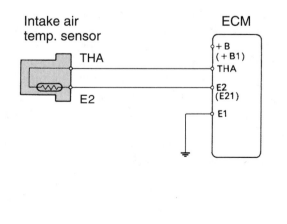
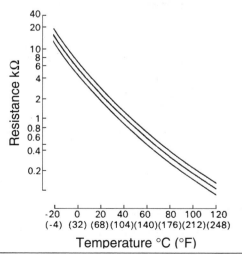

Figure 26-16 IAT sensor circuit and specifications. (Courtesy of Toyota Motor Corp.)

Throttle Position Sensor Diagnosis, Three-Wire Sensor

A defective throttle position sensor (TPS) may cause acceleration stumbles, engine stalling, and improper idle speed. Backprobe the sensor terminals to complete the meter connections. With the ignition switch in the RUN position, connect a voltmeter from the 5-V reference wire to ground (Figure 26-17). The voltage reading on this wire should be approximately 5V. Always refer to the vehicle manufacturer's specifications.

If the reference wire is not supplying the specified voltage, check the voltage on this wire at the computer terminal. If the voltage is within specifications at the computer but low at the sensor, repair the 5-V reference wire. When this voltage is low at the computer, check the voltage supply wires and ground wires on the computer. If these wires are satisfactory, replace the computer.

With the ignition switch in the RUN position, connect the voltmeter from the sensor ground wire to the battery ground. If the voltage drop across this circuit exceeds specifications, repair the ground wire from the sensor to the computer.

SERVICE TIP *When testing the throttle position sensor voltage signal, use an analog voltmeter, because the gradual voltage increase on this wire is quite visible on the meter pointer. If the sensor voltage increase is erratic, the voltmeter pointer fluctuates.*

SERVICE TIP *When the throttle is opened gradually to check the throttle position sensor voltage signal, tap the sensor lightly and watch for fluctuations on the voltmeter pointer indicating a defective sensor.*

With the ignition switch in the RUN position, connect a voltmeter from the sensor signal wire to ground. Slowly open the throttle and observe the voltmeter. The voltmeter reading should increase smoothly and gradually. Typical TPS voltage readings are 0.5V to 1V with the throttle in the idle position, and 3.5 to 4.5 volts at wide-open throttle. Always refer to the vehicle manufacturer's specifications. If the TPS does not have the specified voltage or if the voltage signal is erratic, replace the sensor.

Throttle Position Sensor Diagnosis, Four-Wire Sensor

Some TPSs contain an idle switch that is connected to the computer. These sensors have the same wires as a three-wire TPS, and an extra wire for the idle switch.

The four-wire TPS is tested with an ohmmeter connected from the sensor ground terminal to each of the other terminals (Figure 26-18). A specified feeler gauge is placed between the throttle lever and the stop for some of the ohmmeter tests. When the ohmmeter is connected from the ground terminal to the VTA terminal, the throttle must be held in the wide-open position (Figure 26-19).

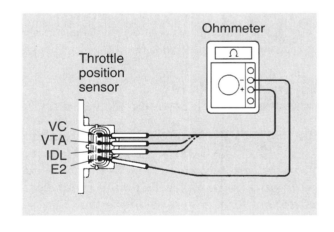

Figure 26-18 Ohmmeter test connections to test the TPS.

CLEARANCE BETWEEN LEVER AND STOP SCREW	BETWEEN TERMINALS	RESISTANCE
0 mm (0 in.)	VTA – E2	0.28 – 6.4 kΩ
0.35 mm (0.014 in.)	IDL – E2	0.5 kΩ or less
0.70 mm (0.028 in.)	IDL – E2	Infinity
Throttle valve fully open	VTA – E2	2.0 – 11.6 kΩ
–	VC – E2	2.7 – 7.7 kΩ

Figure 26-19 Specifications for TPS ohmmeter tests.

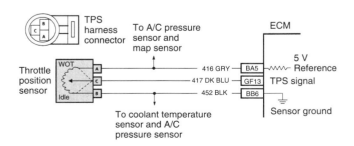

Figure 26-17 TPS sensor circuit. (Courtesy of General Motors Corporation, Service Technology Group)

DSO Testing of the TPS

Each time the throttle is opened, the TPS sensor should provide a smooth analog voltage signal. If the TPS sensor is defective, glitches will appear in the sensor signal as the throttle is opened (Figure 26-20).

Throttle Position Sensor Adjustment

A TPS adjustment may be performed on some vehicles, but this adjustment is not possible on many newer applications. Check the vehicle manufacturer's service manual for the TPS adjustment procedure. An improper TPS adjustment may cause inaccurate idle speed, engine stalling, and acceleration stumbles. Follow these steps for a typical TPS adjustment:

1. Backprobe the TPS signal wire and connect a voltmeter from this wire to ground.

2. Turn the ignition switch to the RUN position and observe the voltmeter reading with the throttle in the idle position.

3. If the TPS does not provide the specified signal voltage, loosen the TPS mounting bolts and rotate the sensor housing until the specified voltage is indicated on the voltmeter.

4. Hold the sensor in this position and tighten the mounting bolts to the specified torque.

Manifold Absolute Pressure (MAP) Sensor Diagnosis

SERVICE TIP *Manifold absolute pressure sensors have a much different calibration on turbocharged engines compared to non-turbocharged engines. Be sure to use the proper specifications for the sensor being tested.*

A defective manifold absolute pressure sensor may cause a rich or lean air/fuel ratio, excessive fuel consumption, and engine surging. In some systems the engine will not run if the MAP sensor is defective. Diagnosis of MAP sensors differ between types.

To test a MAP sensor that produces an analog signal, place the ignition switch in the RUN position and backprobe the 5-V reference wire with a voltmeter (Figure 26-21). If the reference wire is not supplying the specified voltage, check the voltage on this wire at the computer. If the voltage is within specifications at the computer, but low at the sensor, repair the 5-V reference wire. When this voltage is low at the computer, check the voltage supply wires and ground wires on the computer. If these wires are satisfactory, replace the computer.

With the ignition switch in the RUN position, connect the voltmeter from the sensor ground wire to the battery ground. If the voltage drop across this circuit exceeds specifications, repair the ground wire from the sensor to the computer.

Backprobe the MAP sensor signal wire and connect a voltmeter from this wire to ground with the ignition switch in the RUN position. The voltage reading indicates the barometric pressure signal from the MAP sensor to the computer. Many MAP sensors send a barometric pressure signal to the computer each time the ignition switch is turned on and when the throttle is in the wide-open position. If the voltage supplied by the barometric pressure signal in the MAP sensor does not equal the vehicle manufacturer's specifications, replace the MAP sensor.

The barometric pressure voltage signal varies depending on altitude and atmospheric conditions. Follow this calculation to obtain an accurate barometric pressure reading:

1. Determine present barometric pressure; e.g., 29.85 inches.

2. Multiply your altitude by 0.001; e.g., 600 feet times 0.001 = 0.6.

3. Subtract the altitude correction from the present barometric pressure; e.g., 29.85 - 0.6 = 29.79.

Check the vehicle manufacturer's specifications to obtain the proper barometric pressure voltage signal in relation to the present barometric pressure.

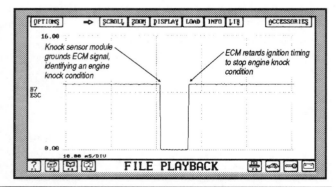

Figure 26-20 Defective TPS waveform. (Courtesy of Edge Diagnostic Systems)

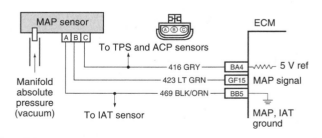

Figure 26-21 MAP sensor circuit. (Courtesy of General Motors Corporation, Service Technology Group)

To diagnose the MAP sensor voltage signal, leave the ignition switch in the RUN position and the voltmeter connected to the MAP sensor signal wire. Connect a vacuum hand pump to the MAP sensor vacuum connection and apply 5 inches of vacuum to the sensor. On some MAP sensors, the sensor voltage signal should change 0.7V to 1.0V for every 5 inches of vacuum change applied to the sensor. Always use the vehicle manufacturer's specifications. If the barometric pressure voltage signal was 4.5V, with 5 inches of vacuum applied to the MAP sensor, the voltage should be 3.5V to 3.8V. When 10 inches of vacuum is applied to the sensor, the voltage signal should be 2.5V to 3.1V. Check the MAP sensor voltage at 5-inch intervals from 0 to 25 inches. If the MAP sensor voltage is not within specifications at any vacuum, replace the sensor.

If the MAP sensor produces a digital voltage signal of varying frequency, check the 5-V reference wire and the ground wire with the same procedure used on other MAP sensors. This sensor diagnosis is based on the use of a MAP sensor tester that changes the MAP sensor varying frequency voltage to an analog voltage. Follow the steps in Photo Sequence 13 to test the MAP sensor.

DSO Testing of the MAP Sensor

When the engine is accelerated and returned to idle, the MAP sensor voltage signal should increase and decrease (Figure 26-22). If the engine is accelerated and the MAP sensor voltage signal does not rise and fall, or the signal is erratic, the sensor or connecting wires may be defective. Remember the MAP output volts change as pressure in the intake manifold changes. Internal engine problems will cause improper MAP voltage output.

A Ford MAP sensor produces a frequency-type signal, and the frequency should change as the engine speed increases. If the frequency is erratic or the frequency does not change in relation to engine speed, the sensor is defective.

Mass Air Flow Sensor Diagnosis

To diagnose a vane-type mass air flow sensor with a voltmeter, always check the voltage supply wire and the ground wire to the MAF module before checking the sensor voltage signal. Always follow the recommended test procedure in the manufacturer's service manual and use the specifications supplied by the manufacturer. The following procedure is based on the use of a Fluke multimeter. Follow these steps to measure the MAF sensor voltage signal:

1. Set the multimeter on the Vdc scale and connect the red meter lead to the MAF signal wire with a special

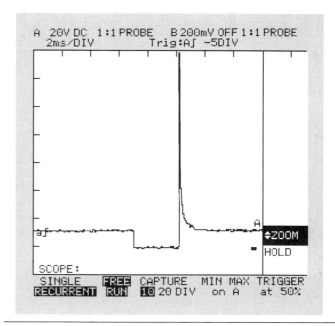

Figure 26-22 Normal MAP sensor waveform. (Courtesy of Edge Diagnostic Systems)

piercing probe, and connect the black meter lead to ground (Figure 26-23).

2. Turn the ignition switch to the RUN position and press the min/max button to activate the min/max feature.

WARNING *While pushing the mass air flow sensor vane open and closed, be careful not to mark or damage the vane or sensor housing.*

3. Slowly push the MAF vane from the closed to the wide-open position, and allow the vane to slowly return to the closed position.

4. Touch the min/max button once to read the maximum voltage signal recorded, and press this button again to read the minimum voltage signal. If the minimum voltage signal is zero, there may be an open circuit in the MAF sensor variable resistor. When the voltage signal is not within the manufacturer's specifications, replace the sensor.

Some vehicle manufacturers specify ohmmeter tests for the MAF sensor. The following is a typical test procedure using a Toyota MAF sensor. With the MAF sensor removed, connect the ohmmeter to the E2 and VS MAF sensor terminals (Figure 26-24). The resistance at these terminals should be 200Ω to 600Ω (Figure 26-25). Connect the ohmmeter leads to the other recommended MAF sensor terminals as shown in the table, and

MAP Sensor Testing

P10-1 Remove the MAP sensor connector and vacuum hose.

P10-2 Connect the MAP sensor tester to the MAP sensor.

P10-3 Connect the digital voltmeter leads from the proper MAP sensor tester lead to ground.

P10-4 Connect the MAP sensor tester leads to the battery terminals with the proper polarity.

P10-5 Observe the MAP sensor barometric pressure (BARO) voltage reading on the voltmeter and compare this reading to specifications.

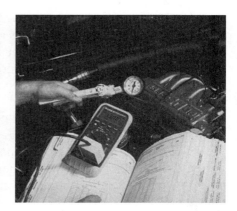

P10-6 Connect a hand vacuum pump to the MAP sensor and supply 5 in. Hg. Observe the MAP sensor voltage and compare to specifications.

P10-7 Increase the vacuum applied to the MAP sensor to 10 in. Hg. and observe the voltmeter reading. Compare this reading to specifications.

P10-8 Apply 15 in. Hg. to the MAP sensor and compare the voltmeter reading to specifications.

P10-9 Apply 20 in. Hg. to the MAP sensor and compare the reading to specifications. If any of the MAP sensor readings do not meet specifications, replace the MAP sensor.

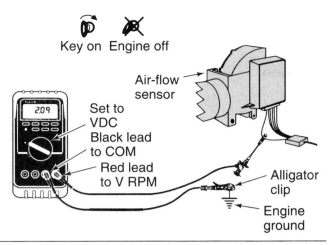

Figure 26-23 Voltmeter connected to measure MAF voltage signal. (Reproduced with permission of Fluke Corporation)

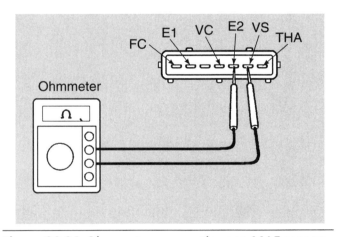

Figure 26-24 Ohmmeter connections to MAF sensor.

BETWEEN TERMINALS	RESISTANCE (Ω)	TEMP. °C (°F)
VS – E2	200 – 600	—
VC – E2	200 – 400	—
THA – E2	10,000 – 20,000 4,000 – 7,000	-20 (-4) 0(32)
FC – E1	Infinity	—

Figure 26-25 MAF sensor ohm specifications between terminals.

record the resistance readings. Since the THA and E2 sensor terminals are connected internally to the thermistor, temperature affects the ohm readings at these terminals as indicated in the specifications. If the specified resistance is not present in any of the test connections, replace the MAF sensor.

Connect the ohmmeter leads to the specified MAF sensor terminals (Figure 26-26), and move the vane from the fully closed to the fully open position. With each specified meter connection and vane position, the ohmmeter should indicate the specified resistance. When the ohmmeter leads are connected to the E2 and VS sensor terminals, the ohm reading should increase smoothly as the sensor vane is opened and closed.

The test procedure for heated resistor and hot wire MAF sensors varies depending on the vehicle make and year. Always follow the test procedure in the vehicle manufacturer's service manual to determine if the MAF sensor produces a frequency signal or a DC voltage signal. Always test the MAF voltage supply and ground wires first.

A frequency test may be performed on some MAF sensors, such as the AC Delco MAF on some General Motors products. The following test procedure is based on the use of a Fluke multimeter. Follow these steps to check the MAF sensor voltage signal and frequency:

1. Place the multimeter on the V/rpm scale and connect the meter leads from the MAF voltage signal wire to the ground wire (Figure 26-27).

2. Start the engine and observe the voltmeter reading. On some MAF sensors, this reading should be 2.5V. Always refer to the manufacturer's specifications.

3. Lightly tap the MAF sensor housing with a screwdriver handle and watch the voltmeter pointer. If the pointer fluctuates or the engine misfires, replace the MAF sensor. Some MAF sensors have loose internal connections, which cause erratic voltage signals and engine misfiring and surging.

4. Be sure the meter dial is on DC volts, and press the rpm button three times so the meter displays voltage frequency. The meter should indicate about 30 hertz (Hz) with the engine idling.

Between terminals	Resistance (Ω)	Measuring plate opening
FC – E1	Infinity	Fully closed
FC – E1	Zero	Other than closed
VS – E2	200 – 600	Fully closed
VS – E2	20 – 1,200	Fully open

Figure 26-26 MAF sensor ohm specifications at various vane positions.

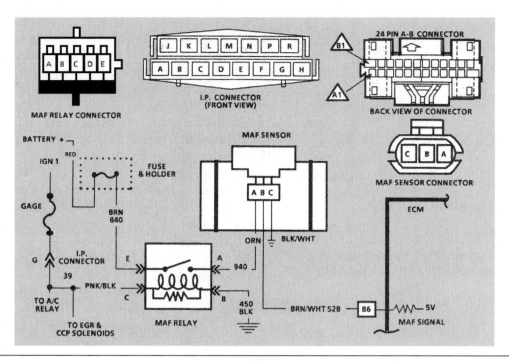

Figure 26-27 MAF sensor circuit. (Courtesy of General Motors Corporation, Service Technology Group)

5. Increase the engine speed, and record the meter reading at various speeds.

6. Graph the frequency readings. The MAF sensor frequency should increase smoothly and gradually in relation to engine speed. If the MAF sensor frequency reading is erratic, replace the sensor (Figure 26-28).

If the MAF sensor being tested produces a DC voltage signal, connect a digital voltmeter from the signal wire to ground. Test the MAF voltage signal at various engine speeds. If the MAF sensor voltage signal at any rpm does not match the specified voltage signal, replace the sensor. A scan tester displays the grams per second of air flow through MAF sensors that provide a frequency or DC voltage signal.

DSO Testing of the MAF Sensor

The DSO leads must be connected from the MAF sensor signal wire to ground. On frequency-type MAF sensors such as General Motors and Ford, the waveform should appear as a series of digital signals. When engine speed and intake air flow are increased, the frequency of the MAF sensor signals increases proportionally. If the MAF or connecting wires are defective, there is an intermittent change in signal frequency (Figure 26-29).

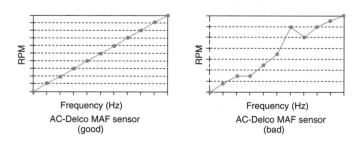

Figure 26-28 Satisfactory and unsatisfactory MAF sensor frequency readings. (Reproduced with permission of Fluke Corporation)

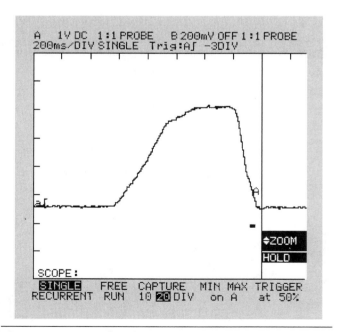

Figure 26-29 Defective frequency-type MAF sensor waveform. (Courtesy of Edge Diagnostic Systems)

SERVICE TIP *Tap the MAF sensor while observing the waveform. If this action causes an erratic waveform, the sensor is defective or the connecting wires are defective.*

A vane-type MAF sensor should provide an analog voltage signal when the engine is accelerated. A defective vane-type MAF sensor waveform displays sudden and erratic voltage changes (Figure 26-30).

Exhaust Gas Recirculation Valve Position Sensor Diagnosis

CAUTION *The EGR valve and EVP sensor may be very hot if the engine has been running. Wear protective gloves if it is necessary to service these components.*

Many exhaust gas recirculation valve position (EVP) sensors have a 5-V reference wire, a voltage signal wire, and a ground wire. The 5-V reference wire and the ground wire may be checked using the same procedure explained previously on TPS and MAP sensors. Connect the voltmeter leads from the voltage signal wire to ground, and turn the ignition switch to the RUN position. The voltage signal should be approximately 0.8V. Connect a vacuum hand pump to the vacuum fitting on the EGR valve and slowly increase the vacuum from 0 to 20 in. Hg. The EVP sensor voltage signal should gradually increase to 4.5V at 20 in. Hg. Always use the EVP test procedure and specifications supplied by the vehicle manufacturer. If the EVP sensor does not have the specified voltage, replace the sensor.

Vehicle Speed Sensor Diagnosis

A defective vehicle speed sensor may cause different problems depending on the computer output control functions. A defective vehicle speed sensor (VSS) may cause the following problems:

1. Improper converter clutch lockup
2. Improper cruise control operation
3. Inaccurate speedometer operation
4. Overspeed cut-out too early

Prior to VSS diagnosis, the vehicle should be lifted on a hoist so the drive wheels are free to rotate. In a typical General Motors system, backprobe the VSS yellow wire, and connect the voltmeter leads from this VSS wire to ground. Then start the engine.

Place the transaxle in drive and allow the drive wheels to rotate. If the VSS voltage signal is not 0.5V or more, replace the sensor. When the VSS provides the specified voltage signal, backprobe the GD 14 PCM terminal and repeat the voltage signal test with the drive wheels rotating. If 0.5V is available at this terminal, the trouble may be in the PCM.

If 0.5V is not available at this terminal, turn off the ignition switch and disconnect the VSS terminal and the PCM terminals. Connect the ohmmeter leads from the 400 VSS terminal to the GD 14 PCM terminal. The meter should read 0 ohms. Repeat the test with the ohmmeter leads connected to the 401 VSS terminal and the GD 13 PCM terminal. This wire should also have 0 ohms resistance. If the resistance in these wires is more than specified, repair the wires.

Park/Neutral Switch Diagnosis

A defective park/neutral switch may cause improper idle speed and failure of the starting motor circuit. Always follow the test procedure in the vehicle manufacturer's service manual. Disconnect the park/neutral switch wiring connector and connect a pair of ohmmeter leads from the B terminal in the wiring connector to ground (Figure 26-31). If the ohmmeter does not indicate less than 0.5Ω, repair the ground wire.

With the wiring harness connected to the switch, backprobe terminal A on the park/neutral switch and connect a pair of voltmeter leads from this terminal to ground. Turn the ignition switch to the RUN position, and move the gear selector through all positions. The voltmeter should read over 5V in all gear selector positions except neutral or park.

If the voltmeter does not indicate the specified voltage, backprobe the ECM terminal B10 and connect the meter from this terminal to ground. When the specified

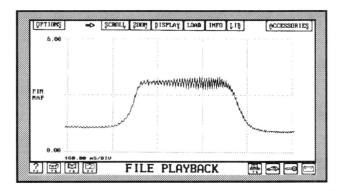

Figure 26-30 Defective vane-type MAF sensor waveform. (Reproduced with permission of Fluke Corporation)

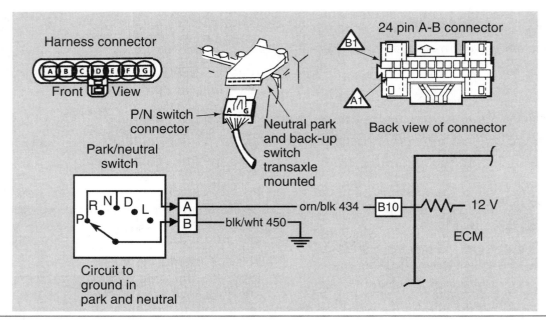

Figure 26-31 Park/neutral switch circuit. (Courtesy of General Motors Corporation, Service Technology Group)

voltage is available at this terminal, repair the wire from the ECM to the park/neutral switch. If the specified voltage is not available, replace the ECM.

When the gear selector is placed in neutral or park, the voltmeter reading should be less than 0.5V. If the specified voltage is not available in these gear selector positions, replace the park/neutral switch.

Injector Testing

Since injectors on MFI and SFI systems are subject to more heat than TBI injectors, port injectors have more problems with tip deposits. The symptoms of restricted injectors are:

1. Lean surge at low speeds
2. Acceleration stumbles
3. Hard starting
4. Acceleration sag, cold engine
5. Engine misfiring
6. Rough engine idle
7. Lack of engine power
8. Slow starting when cold
9. Stalling after a cold start

An **injector balance test** may be performed to diagnose restricted injectors on MFI and SFI systems. A fuel pressure gauge and an injector balance tester are required for this test. Fuel pressure should be checked before the injector balance test is performed. The injector balance tester contains a timer circuit which energizes each injector for an exact time period when the timer button is pressed. When the injector balance test is performed, follow these steps:

1. Connect the fuel pressure gauge to the Schrader valve on the fuel rail.
2. Connect the injector tester leads to the battery terminals with the correct polarity. Remove one of the injector wiring connectors and install the tester lead to the injector terminals (Figure 26-32).
3. Cycle the ignition switch on and off until the specified fuel pressure appears on the fuel gauge. Many fuel pressure gauges have an air bleed button that must be

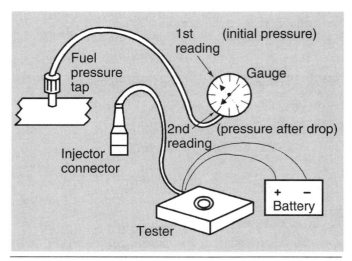

Figure 26-32 Injector balance tester and gauge connections. (Courtesy of General Motors Corporation, Service Technology Group)

pressed to bleed air from the gauge. Cycle the ignition switch or start the engine to obtain the specified pressure on the fuel gauge, and then leave the ignition switch off.

4. Push the timer button on the tester and record the gauge reading. When the timer energizes the injector, fuel is discharged from the injector into the intake port, and the fuel pressure drops in the fuel rail.

5. Repeat steps 2, 3, and 4 on each injector, and record the fuel pressure after each injector is energized by the timer.

6. Compare the gauge readings on each injector. When the injectors are in satisfactory condition, the fuel pressure will be the same after each injector is energized by the timer. If an injector orifice or tip is restricted, the fuel pressure does not drop as much when the injector is energized by the timer. When an injector plunger is sticking in the open position, the fuel pressure drop is excessive. If the fuel pressure on an injector is 1.4 psi (10 kPa) below or above the average pressure when the injectors are energized by the timer, the injector is defective (Figure 26-33).

In addition to this method, many scan tools provide for an **injector kill test** on SFI systems. With the engine running and rpms stabilized, the scan tool will kill one injector at a time. Engine rpm should drop equally for each injector killed. With this test it is important to remember other factors may cause a cylinder to drop very few rpms than just the injector. Low compression, fouled or faulty spark plugs, worn spark plug cable, or open injector circuit will also cause a low rpm drop.

If the injector balance test indicates that some of the injectors are restricted, the injectors may be cleaned. Tool manufacturers market a variety of injector cleaning equipment. The injector cleaning solution is poured into a canister on some injector cleaners, and the shop air supply is used to pressurize the canister to the specified pressure. The injector cleaning solution contains unleaded fuel mixed with injector cleaner. The container hose is connected to the Schrader valve on the fuel rail.

Automotive parts stores usually sell a sealed pressurized container of injector cleaner with a hose for Schrader valve attachment. During the cleaning process, the engine is operated on the pressurized container of unleaded fuel and injector cleaner. The fuel pump operation must be stopped to prevent the pump from forcing fuel up to the fuel rail, and the fuel return line must be plugged to prevent the solution in the cleaning container from flowing through the return line into the fuel tank. Follow these steps for the injector cleaning procedure:

1. Disconnect the wires from the in-tank fuel pump or the fuel pump relay to disable the fuel pump. If you disconnect the fuel pump relay on General Motors products, the oil pressure switch in the fuel pump circuit must also be disconnected to prevent current flow through this switch to the fuel pump.

2. Plug the fuel return line from the fuel rail to the tank.

3. Connect a can of injector cleaner to the Schrader valve on the fuel rail and run the engine for about 20 minutes on the injector solution.

After the injectors are cleaned or replaced, rough engine idle may still be present. This problem occurs because the adaptive memory in the computer has learned previously about the restricted injectors. If the injectors were supplying a lean air/fuel ratio, the computer increased the pulse width to try to bring the air/fuel ratio back to stoichiometric. With the cleaned or replaced injectors, the adaptive computer memory is still supplying the increased pulse width. This action makes the air/fuel ratio too rich now that the restricted injector problem does not exist. With the engine at normal operating temperature, drive the vehicle for at least 5 minutes to allow the adaptive computer memory to learn about the cleaned or replaced injectors. After this time, the computer should supply the correct injector pulse width and the engine should run smoothly. This same problem may occur when any defective computer system component is replaced.

CYLINDER	1	2	3	4	5	6
HIGH READING	225	225	225	225	225	225
LOW READING	100	100	100	90	100	115
AMOUNT OF DROP	125	125	125	135	125	110
	OK	OK	OK	FAULTY, RICH (TOO MUCH) (FUEL DROP)	OK	FAULTY, LEAN (TOO LITTLE) (FUEL DROP)

Figure 26-33 Pressure readings from injector balance test indicating defective injectors. (Courtesy of General Motors Corporation, Service Technology Group)

Injector Sound Test

The injector sound test is a method of quickly checking the operation of the pintle on engines where the injectors are accessible. A port injector that is not functioning may cause a cylinder misfire at low engine speeds. With the engine idling, a stethoscope pickup may be placed on the side of the injector body (Figure 26-34). Each injector should produce the same clicking noise. If an injector does not produce any clicking noise, the injector, connecting wires, or PCM may be defective. When the injector clicking noise is erratic, the injector plunger may be sticking. If there is no injector clicking noise, proceed with the injector ohms test and noid light test to locate the cause of the problem.

Injector Ohmmeter Test

An ohmmeter may be connected across the injector terminals to check the injector winding after the injector wires are disconnected. If the ohmmeter reading is infinite, the injector winding is open. An ohmmeter reading below the specified value indicates the injector winding is shorted. A satisfactory injector winding has the amount of resistance specified by the manufacturer. Injector replacement is necessary if the injector winding does not have the specified resistance.

CAUTION *The injectors used may change in mid-model year. Also replace the injector with the correct unit. Usually the connector end of the injector is color coded by the manufacturer.*

Noid Light Test

Some manufacturers of automotive test equipment market **noid lights**, which have terminals designed to plug into most injector wiring connectors after these connectors are disconnected from the injector. When the engine is cranked, the noid light flashes as the computer cycles the injector on and off. If the light is not flashing, the computer or connecting wires are defective.

DSO Testing of the Injector

The DSO leads must be connected from the switched side of the injector to ground to obtain an injector waveform. When the PCM grounds the injector, the voltage should decrease. The voltage should remain low while the PCM keeps the injector grounded. When the PCM opens the injector ground and the injector shuts off, a voltage spike should occur from the induced voltage in the injector winding. When the voltage spike is lower on one injector compared to the other injectors, the injector with the low spike probably has a shorted winding. If the PCM operates the injectors with a peak-and-hold current, the waveform has a voltage spike when the PCM reduces the injector current and a second spike when the injector shuts off (Figure 26-35). When the waveform is aligned properly with the horizontal grids on the screen, the injector pulse width can be measured (Figure 26-36). When the injector waveform is ragged, the injector driver in the PCM may be degraded. In some fuel systems, the PCM grounds the injector winding and then uses a pulse width modulated (PWM) signal to reduce injector current.

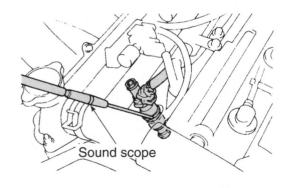

Figure 26-34 Performing an injector sound test.

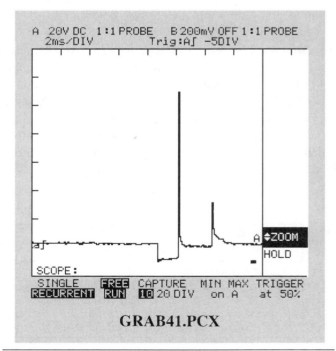

Figure 26-35 Normal peak-and-hold injector waveform. (Reproduced with permission of Fluke Corporation)

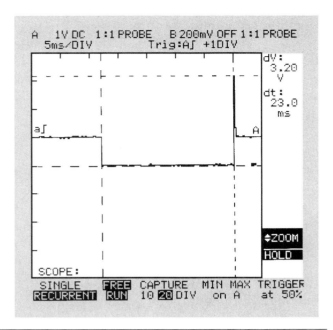

Figure 26-36 Measuring injector pulse width with a DSO. (Reproduced with permission of Fluke Corporation)

Injector Flow Testing

Some vehicle manufacturers recommend an **injector flow test** rather than the balance test. Follow these steps to perform an injector flow test:

1. Connect a 12-V power supply to the cigarette lighter socket and disconnect the negative battery cable. If the vehicle is equipped with an air bag, wait one minute.

2. Remove the injectors and fuel rail and place the tip of the injector to be tested in a calibrated container. Leave the injectors in the fuel rail.

3. Connect a jumper wire between the B+ and FP terminals in the DLC as in the fuel pump pressure test.

4. Turn the ignition switch to the RUN position.

5. Connect a special jumper wire from the terminals of the injector being tested to the battery terminals.

6. Disconnect the jumper wire from the negative battery cable after 15 seconds.

7. Record the amount of fuel in the calibrated container.

8. Repeat the procedure on each injector. If the volume of fuel discharged from any injector varies more than 0.3 cubic inch (5 cc) from the specifications, the injector should be replaced.

9. Connect the negative battery cable and disconnect the 12-V power supply.

Cold Start Injector Diagnosis

WARNING *Energizing the cold start injector for more than five seconds may damage the injector winding.*

The cold start injector diagnosis procedure varies depending on the vehicle. A typical cold start injector removal and testing procedure follows:

1. Bleed the pressure from the fuel system.

2. Connect a 12-V power supply to the cigarette lighter socket and disconnect the negative battery cable. If the vehicle is equipped with an air bag, wait one minute.

3. Wipe excess dirt from the cold start injector with a shop towel.

4. Remove the electrical connector from the cold start injector.

5. Remove the union bolt and the cold start injector fuel line (Figure 26-37).

6. Remove the cold start injector retaining bolts and remove the cold start injector.

7. Connect an ohmmeter across the cold start injector terminals. If the resistance is more or less than specified, replace the injector.

8. Connect the fuel line and union bolt to the cold start injector and place the injector tip in a container.

9. Connect a jumper wire to the B+ and FP terminals in the data link connector (DLC) and turn the ignition switch to the RUN position.

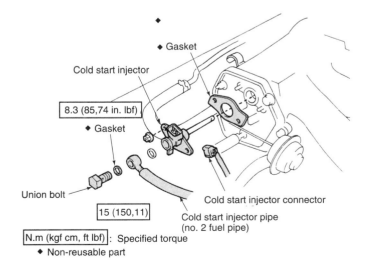

Figure 26-37 Removing the cold start injector and fuel line.

10. Connect a special jumper wire from the cold start injector terminals to the battery terminals.

11. Check the fuel spray pattern from the injector. This pattern should be as illustrated in Figure 26-38. If the pattern is not as shown, replace the injector. Do not energize the cold start injector for more than five seconds.

Minimum Idle Speed Adjustment

The **minimum idle speed** adjustment may be performed on some MFI or SFI systems with a minimum idle speed screw in the throttle body. This screw is factory-adjusted and the head of the screw is covered with a plug. This adjustment should only be required if throttle body parts are replaced. If the minimum idle speed adjustment is not adjusted properly, engine stalling may result. The procedure for performing a minimum idle speed adjustment varies considerably depending on the vehicle. Always follow the vehicle manufacturer's recommended procedure in the service manual. Following is a typical minimum idle speed adjustment procedure for a General Motors vehicle:

1. Be sure the engine is at normal operating temperature, and turn off the ignition switch.

2. Connect terminals A and B in the DLC, and connect a tachometer from the ignition tach terminal to ground.

3. Turn the ignition switch to the RUN position and wait 30 seconds. Under this condition, the idle air control (IAC) motor is driven completely inward by the PCM.

4. Disconnect the idle air control (IAC) motor connector.

5. Remove the connection between terminals A and B in the DLC and start the engine.

6. Place the transmission selector in drive with an automatic transmission or neutral with a manual transmission.

7. Adjust the idle stop screw if necessary to obtain 500 to 600 rpm with an automatic transmission, or 550 to 650 rpm with a manual transmission. A plug must be removed to access the idle stop screw (Figure 26-39).

8. Turn off the ignition switch and reconnect the IAC motor connector.

9. Turn the ignition switch to the RUN position and connect a digital voltmeter from the TPS signal wire to ground. If the voltmeter does not indicate the specified voltage of 0.55V, loosen the TPS mounting screws and rotate the sensor until this voltage reading is obtained. Hold the TPS in this position and tighten the mounting screws.

Minimum Idle Speed Adjustment, Throttle Body Injection

The minimum idle speed adjustment procedure varies depending on the year and make of vehicle. Always follow the adjustment procedure in the vehicle manufacturer's service manual. The minimum idle speed adjustment is only required if the TBI assembly or TBI assembly components are replaced. If the minimum idle speed adjustment is not adjusted properly, engine stalling

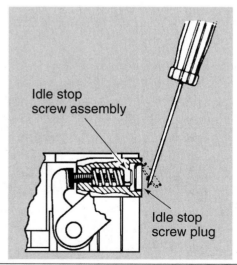

Figure 26-38 Adjusting the idle stop screw for minimum air adjustment. (Courtesy of General Motors Corporation, Service Technology Group)

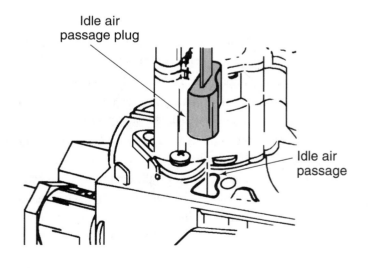

Figure 26-39 Special tool to plug the air passage to the IAC motor while checking minimum idle speed adjustment on a TBI system. (Courtesy of General Motors Corporation, Service Technology Group)

may result. Proceed as follows for a typical minimum idle speed adjustment on a General Motors TBI system:

1. Be sure that the engine is at normal operating temperature and remove the air cleaner and TBI-to-air cleaner gasket. Plug the air cleaner vacuum hose inlet to the intake manifold.

2. Disconnect the throttle valve (TV) cable to gain access to the minimum air adjustment screw. A tamper-resistant plug in the TBI assembly must be removed to access this screw.

3. Connect a tachometer from the ignition tach terminal to ground, and disconnect the IAC motor connector.

4. Start the engine and place the transmission in park with an automatic transmission, or neutral with a manual transmission.

5. Plug the air intake passage to the IAC motor. Tool J33047 is available for this purpose (Figure 26-40).

6. On 2.5-L four-cylinder engines, use the appropriate torx bit to rotate the minimum air adjustment screw until the idle speed on the tachometer is 475 to 525 rpm with an automatic transaxle, or 750 to 800 rpm with a manual transaxle.

7. Stop the engine and remove the plug from the idle air passage. Cover the minimum air adjustment screw opening with silicone sealant, reconnect the TV cable, and install the TBI gasket and air cleaner.

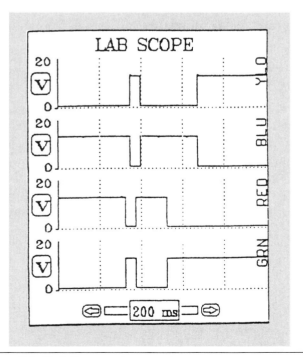

Figure 26-40 Idle speed control stepper motor waveforms. (Courtesy of OTC-SPX Corp. Aftermarket Tool & Equipment Group)

Idle Air Control Motor Testing

Some EFI systems use a stepper motor to control idle speed. This motor contains two windings. The PCM energizes one stepper motor winding and de-energizes the other winding to extend the motor pintle and close the air passage to decrease idle speed. The stepper motor pintle is retracted to open the air passage and increase idle speed when the PCM reverses the energizing and de-energizing of the stepper motor windings. The PCM operates the stepper motor windings in a series of steps. Each step is displayed on the DSO screen as an opposed square wave when the motor advances or retracts a step (Figure 26-41). When the ignition switch is shut off, the PCM positions the stepper motor pintle in the half-open position for restarting.

The DSO must be capable of displaying four waveforms simultaneously to display the four square waves from the idle speed control stepper motor. Backprobe the four stepper motor connections, and connect the four DSO leads to these connections. When the idle speed changes, or when the ignition is turned off, the four square waves from the idle speed control stepper motor should be displayed on the DSO screen.

If the signals from the PCM are satisfactory, but there is no change in rpm, the IAC motor is defective or the motor air passages are plugged. When the signal from the PCM is erratic, the PCM is defective or the connecting wires are defective.

Some IAC motors have a single winding and the PCM operates the winding with a duty cycle to pulse the IAC motor plunger and control the amount of air bypassing the throttle. In some applications, this voltage signal from the PCM varies between battery voltage and 5V,

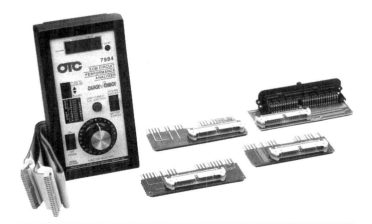

Figure 26-41 Duty cycle signal from the PCM to the IAC motor. (Courtesy of Edge Diagnostic Systems)

and the PCM never decreases this voltage to 0V (Figure 26-42). If the A/C clutch is engaged, the duty cycle should change on the IAC motor. An increase in duty cycle causes a corresponding increase in engine rpm. If the duty cycle remains fixed, the PCM may be in the **failure mode effects management (FMEM)**.

Flash Code Retrieval

Flash codes are provided to help the technician diagnose the circuit that has been detected as faulty by the PCM. Remember, the code displayed may not necessitate the replacement of the component. For example, if the fault code indicates a high TPS voltage, the fault may not be the throttle position sensor itself, rather an open in the circuit. A low O_2S fault may not be a faulty O_2S, rather a vacuum leak that the O_2S is sensing.

There are several different methods used by manufacturers to display flash codes. For this reason, always refer to the correct service manual. In addition, scan tools are available that will display the diagnostic trouble codes (DTC). The flash code provides a means of retrieving DTCs when a scan tool is not available.

Chrysler Flash Code Diagnosis

If a TBI, MFI, or SFI system is working normally, the **malfunction indicator light (MIL)** is illuminated when the ignition switch is turned on, and this light goes out a few seconds after the engine is started. The MIL light should remain off while the engine is running.

If a defect occurs and a diagnostic trouble code (DTC) is set in the computer memory, the computer may enter a limp-in mode. In this mode, the malfunction indicator light (MIL) or check engine light is on, the air/fuel ratio is rich, and the spark advance is fixed, but the vehicle can be driven to an automotive service center. When a vehicle is operating in the limp-in mode, fuel consump-

tion and emission levels increase, and engine performance may decrease.

Prior to any DTC diagnosis, the Preliminary Diagnostic Procedure mentioned previously in this chapter must be completed, and the engine must be at normal operating temperature. If the engine is not at normal operating temperature, the computer may provide erroneous DTCs. The vehicle's battery must be fully charged prior to DTC diagnosis.

Follow these steps to read the DTCs from the flashes of the MIL light on most Chrysler products:

1. Cycle the ignition switch on and off, on and off, and on in a five-second interval.

2. Observe the MIL lamp flashes to read the DTCs. Two quick flashes followed by a brief pause and three quick flashes indicates code 23. The DTCs are flashed once in numerical order.

3. When code 55 is flashed, the DTC sequence is completed. The ignition switch must be turned off, and steps 1 and 2 repeated to read the DTCs a second time.

On any TBI or PFI system, a DTC indicates a defect in a specific area. For example, a TPS code indicates a defective TPS, defective wires between the TPS and the computer, or the computer may be unable to receive the TPS signal. Specific ohmmeter or voltmeter tests may be necessary to locate the exact cause of the fault code. On logic module and power module systems, disconnect the quick-disconnect connector at the positive battery cable for 10 seconds with the ignition switch off to erase DTCs. On later module PCMs, this connector must be disconnected for 30 minutes to erase fault codes.

Toyota Flash Code Diagnosis

Prior to the flash code output, the Preliminary Diagnostic Procedure must be performed as mentioned at the beginning of this chapter. Follow these steps for DTC diagnosis:

1. Turn the ignition switch to the RUN position and connect a jumper wire between terminals E1 and TE1 in the data link connector (DLC). Some round DLCs are located under the instrument panel, while other rectangular-shaped DLCs are positioned in the engine compartment (Figure 26-42).

2. Observe the MIL light flashes. If the light flashes on and off at 0.26-second intervals, there are no DTCs in the computer memory (Figure 26-43).

3. If there are DTCs in the computer memory, the MIL light flashes out the DTCs in numerical order. For example, one flash followed by a pause and three flashes is code 13, and three flashes followed by a

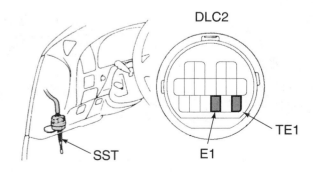

Figure 26-42 E1 and TE1 terminals in round DLC located under the instrument panel.

pause and one flash represents code 31 (Figure 26-44). The codes will be repeated as long as terminals E1 and TE1 are connected and the ignition switch is on.

4. Remove the jumper wire from the DLC.

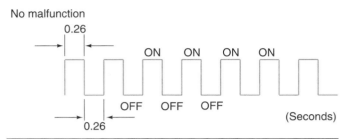

Figure 26-43 If the MIL light flashes at 0.26 second intervals, there are no DTCs.

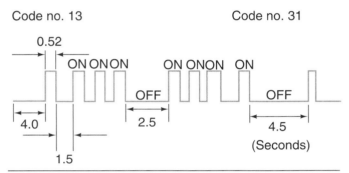

Figure 26-44 DTCs 13 and 31.

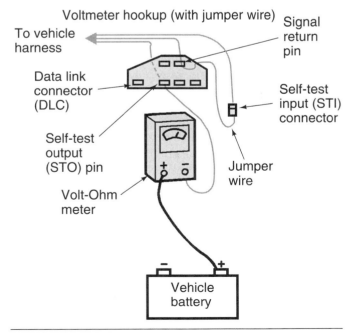

Figure 26-45 Jumper wire and voltmeter connection to Ford DLC. (Reprinted with permission of Ford Motor Company)

Follow this procedure to obtain fault codes during a **driving test mode**:

1. Turn the ignition switch to the RUN position and connect terminals E1 and TE2 in the DLC.

2. Start the engine and drive the vehicle at speeds above 6 mph (10 km/h). Simulate the conditions when the problem occurs.

3. Connect a jumper wire between terminals E1 and TE1 on the DLC.

4. Observe the flashes of the MIL light to read the DTCs, and remove the jumper wire from the DLC.

Ford Flash Code Diagnosis

Most Ford vehicles have a MIL light on the instrument panel, and this light flashes the DTCs in the diagnostic mode. When a defect occurs in a major sensor, the PCM illuminates the MIL light and enters the limp-in mode in which the air/fuel ratio is rich and the spark advance is fixed. In this mode, engine performance decreases, and fuel consumption and emission levels increase.

Prior to any fault code diagnosis the engine must be at normal operating temperature and the Preliminary Diagnostic Procedure mentioned previously in the chapter must be completed. A jumper wire must be connected from the self-test input wire to the appropriate DLC terminal to enter the self-test mode. When the ignition switch is turned on after this jumper wire connection, the MIL light begins to flash any DTCs in the PCM memory.

Optional Voltmeter Connection

If the vehicle does not have a check engine light (MIL), a voltmeter may be connected from the positive battery terminal to the proper DLC terminal (Figure 26-46). The voltmeter must be connected with the correct polarity as indicated in the figure.

When the ignition switch is turned on after the jumper wire and voltmeter connections are completed, the DTCs may be read from the sweeps of the voltmeter pointer or the flashes of the MIL. For example, if three upward sweeps of the voltmeter pointer are followed by a pause and then four upward sweeps, code 34 is displayed.

Key On, Engine Off (KOEO) Test

Follow these steps for the **key on, engine off (KOEO)** fault code diagnostic procedure:

1. With the ignition switch off, connect the jumper wire to the self-test input wire and the appropriate terminal in the DLC.

2. If the vehicle does not have a MIL light, connect the

Digit pulses are 1/2 second "on" and 1/2 second "off"

Analog meter:
Each pulse equals
1 meter sweep

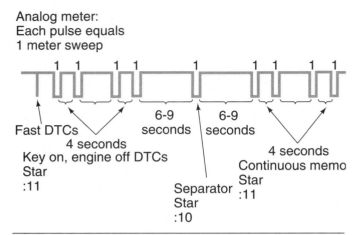

Fast DTCs
4 seconds
Key on, engine off DTCs
Star
:11

6-9 seconds
6-9 seconds

Separator
Star
:10

4 seconds
Continuous memo
Star
:11

Figure 26-46 Key on, engine off (KOEO) test procedure. (Reprinted with permission of Ford Motor Company)

voltmeter to the positive battery terminal and the appropriate DLC terminal.

3. Turn the ignition switch to the RUN position and observe the MIL light or voltmeter. **Hard fault** DTCs are displayed, followed by a separator code 10 and continuous **memory faults** (Figure 26-47).

Hard fault DTCs are present at the time of testing, whereas memory DTCs represent intermittent faults that occurred some time ago and are set in the computer memory. Separator code 10 is displayed as one flash of the MIL light or one sweep of the voltmeter pointer. Each fault DTC is displayed twice and provided in numerical order. If there are no DTCs, system pass code 11 is displayed. If the technician wants to repeat the test or proceed to another test, the ignition switch must be turned off for 10 seconds.

Digit pulses are 1/2 second "on" and 1/2 second "off"

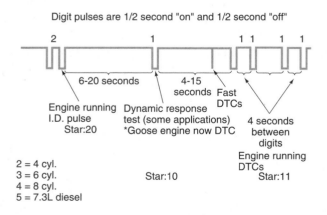

2 = 4 cyl.
3 = 6 cyl.
4 = 8 cyl.
5 = 7.3L diesel

6-20 seconds
Engine running
I.D. pulse
Star:20

4-15 seconds
Dynamic response
test (some applications)
*Goose engine now DTC
Star:10

Fast DTCs
4 seconds
between digits
Engine running DTCs
Star:11

Figure 26-47 Key on, engine running (KOER) test. (Reprinted with permission of Ford Motor Company)

Key On, Engine Running (KOER) Test

Follow these steps to obtain the fault codes in the **Key on, engine running (KOER) test** sequence:

1. Connect the jumper wire and the voltmeter as explained in steps 1 and 2 of the KOEO test.
2. Start the engine and observe the MIL lamp or the voltmeter. The engine identification code is followed by the separator code 10 and hard fault codes (Figure 26-48).

The engine identification (ID) code represents half of the engine cylinders. On a V-8 engine, the MIL light flashes four times or the voltmeter pointer sweeps upward four times during the engine ID display.

On some Ford products, the brake on/off (BOO) switch and the power steering pressure switch (PSPS) must be activated after the engine ID code or DTCs 52 and 74, representing these switches, are present. Step on the brake pedal and turn the steering wheel to activate these switches immediately after the engine ID display.

Separator code 10 is presented during the KOER test on many Ford products. When this code is displayed, the throttle must be pushed momentarily to the wide-open position. The best way to provide a wide-open throttle is to push the gas pedal to the floor momentarily. On some Ford products, the separator code 10 is not displayed during the KOER test, and this throttle action is not required.

Hard fault DTCs are displayed twice in numerical order, and if no faults are present, system pass code 11 is given.

Fault Code Erasing Procedure

DTCs may be erased by entering the KOEO test procedure and disconnecting the jumper wire between the self-test input wire and the DLC during the code display.

General Motors Flash Code Testing

When a fault occurs in a major sensor, the PCM illuminates the MIL light, and a fault code is set in the PCM memory. Once this action takes place, the PCM is usually operating in a limp-in mode, and the air/fuel ratio is rich with a fixed spark advance. In this mode, driveabil-

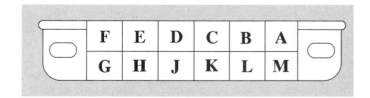

Figure 26-48 DLC terminals.

ity is adversely affected and fuel consumption and emission levels increase. Prior to any fault code diagnosis, the engine must be at normal operating temperature and the Preliminary Diagnostic Procedure mentioned previously must be completed. Follow these steps to obtain the DTCs with the MIL light flashes:

1. With the ignition switch off, connect a jumper wire between terminals A and B in the DLC under the instrument panel. A special tool that has two lugs that fit between these terminals is available. Usually terminals A and B are located at the top right corner of the DLC, but some DLCs are mounted upside down or vertically. Always consult the vehicle manufacturer's service manual for exact terminal location.

2. Turn the ignition switch to the RUN position, and observe the MIL lamp.

3. One lamp flash followed by a brief pause and two more flashes indicates code 12. This code indicates that the PCM is capable of diagnosis. Each code is flashed three times, and codes are given in numerical order. If there are no DTCs in the PCM, only code 12 is provided. The code sequence keeps repeating until the ignition switch is turned off.

DTC Erasing Procedure

The fault codes may be erased by disconnecting the quick-disconnect connector at the positive battery terminal for 10 seconds with the ignition switch off. If the vehicle does not have a quick-disconnect connector, the PCM B fuse may be disconnected to erase fault codes. On later model General Motors vehicles with P4 PCMs, the quick-disconnect, or PCM B, fuse may have to be disconnected for a longer time to erase codes.

If DTCs are left in a computer after the defect is corrected, the codes are erased automatically when the engine is stopped and started 30 to 50 times. This applies to most computer-equipped vehicles.

Field Service Mode

If the A and B terminals are connected in the DLC and the engine is started, the PCM enters the **field service mode**. In this mode, the speed of the MIL lamp flashes indicate whether the system is in open loop or closed loop. If the system is in open loop, the MIL lamp flashes quickly. When the system enters closed loop, the MIL lamp flashes at half the speed of the open loop flashes.

Nissan Flash Code Testing

In some Nissan electronic concentrated engine control systems (ECCS), the PCM has two light emitting diodes (LEDs), which flash a fault code if a defect occurs in the system. One of these LEDs is red, and the second LED is green. The technician observes the flashing pattern of the two LEDs to determine the DTC. If there are no DTCs in the ECCS, the LEDs flash a system pass code. The flash code procedure varies depending on the year and model of the vehicle, and the procedure in the manufacturer's service manual must be followed. Later model Nissan engine computers have a five-mode diagnostic procedure. Be sure the engine is at normal operating temperature and complete the Preliminary Diagnostic Procedure explained previously in this chapter. Turn the diagnosis mode selector in the PCM to obtain the diagnostic modes (Figure 26-49).

These diagnostic modes are available on some Nissan products:

Mode 1 This mode checks the oxygen sensor signal. With the system in closed loop and the engine idling, the green light should flash on each time the oxygen sensor detects a lean condition. This light goes out when the oxygen sensor detects a rich condition. After 5–10 seconds, the PCM clamps on the ideal air/fuel ratio and pulse width, and the green light may be on or off. This PCM clamping of the pulse width only occurs at idle speed.

Mode 2 In this mode, the green light comes on each time the oxygen sensor detects a lean mixture, and the red light comes on when the PCM receives this signal and makes the necessary correction in pulse width.

Mode 3 This mode provides DTCs representing various defects in the system.

Mode 4 Switch inputs to the PCM are tested in this mode. Mode 4 cancels codes available in mode 3.

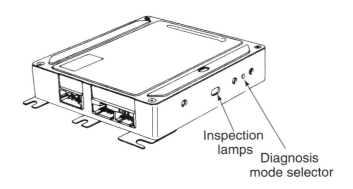

Figure 26-49 Diagnosis mode selector on the PCM. (Courtesy of Nissan Corporation)

Mode 5 This mode increases the diagnostic sensitivity of the PCM for diagnosing intermittent faults while the vehicle is driven on the road.

After the defect is corrected, turn off the ignition switch, rotate the diagnosis mode selector counterclockwise, and install the PCM securely in the original position.

Scan Tester Diagnosis

Several makes of scan testers are available to read the fault codes and perform other diagnostic functions. The exact tester buttons and test procedures vary on these testers, but many of the same basic diagnostic functions are completed regardless of the tester make. When test procedures are performed with a scan tester, these precautions must be observed:

1. Always follow the directions in the manual supplied by the scan tester manufacturer.

2. Do not connect or disconnect any connectors or components with the ignition switch in the RUN position. This includes the scan tester power wires and the connection from the tester to the vehicle diagnostic connector.

3. Never short across or ground any terminals in the electronic system except those recommended by the vehicle manufacturer.

4. If the computer terminals must be removed, disconnect the scan tester diagnostic connector first.

5. Observe the service precautions listed previously in this chapter.

A typical scan tester keypad contains these buttons (Figure 26-50):

1. Numbered keys — digits 0 through 9.

2. UP/DOWN arrow keys — allow the technician to move back and forward through test modes and menus.

3. ENTER key — enters information into the tester.

4. MODE key — allows the technician to interrupt the current procedure and go back to the previous modes.

5. F1 and F2 keys — allow the technician to perform special functions described in the scan tester manufacturer's manuals.

Scan tester operation varies depending on the make of the tester, but a typical example of initial entries follows:

1. Be sure the engine is at normal operating temperature and the ignition switch is off. With the correct module in the scan tester, connect the power cord to the vehicle battery.

2. Enter the vehicle year. The tester displays enter 199X, and the technician presses the correct single digit and the ENTER key.

3. Enter VIN code. This is usually a two-digit code based on the model year and engine type. These codes are listed in the scan tester operator's manual. The techni-

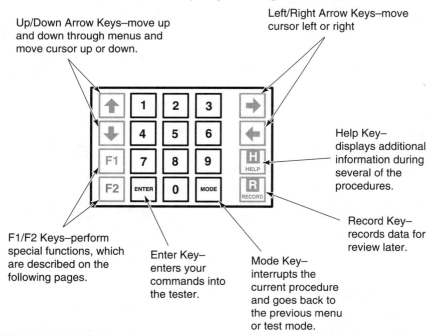

Figure 26-50 Scan tester features. (Courtesy of OTC-SPX Corp. Aftermarket Tool & Equipment Group)

cian enters the appropriate two-digit code and presses the ENTER key.

4. Connect the scan tester adaptor cord to the diagnostic connector on the vehicle being tested.

When the technician has programmed the scan tester by performing the initial entries, some entry options appear on the screen. These entry options vary depending on the scan tester and the vehicle being tested. Following is a typical list of initial entry options:

1. Engine
2. Antilock brake system (ABS)
3. Suspension
4. Transmission
5. Data line
6. Deluxe CDR
7. Test mode

The technician presses the number beside the desired selection to proceed with the test procedure. In the first four selections, the tester is asking the technician to select the computer system to be tested. If data line is selected, the scan tester provides a voltage reading from each input sensor in the system. If the technician selects Chrysler digital readout (CDR) on a Chrysler product, some of the next group of selections appear on the screen. When the technician selects the test mode number on a General Motors product, each test mode provides voltage or status readings from specific sensors or components.

When the technician makes a selection from the initial test selection, the scan tester moves on to the actual test selections. These selections vary depending on the scan tester and the vehicle being tested. The following list includes many of the possible test selections and a brief explanation:

1. Faults codes — displays fault codes on the scan tester display.

2. Switch tests — allows the technician to operate switch inputs such as the brake switch to the PCM. Each switch input should change the reading on the tester.

3. ATM tests — forces the PCM to cycle all the solenoids and relays in the system for 5 minutes or until the ignition switch is turned off.

4. Sensor tests — provides a voltage reading from each sensor.

5. Automatic idle speed (AIS) motor — forces the PCM to operate the AIS motor when the up and down arrows are pressed. The engine speed should increase 100 rpm each time the up arrow is touched. RPM is limited to 1,500 or 2,000 rpm.

6. Solenoid state tests or output state tests — displays the on or off status of each solenoid in the system.

7. Emission maintenance reminder (EMR) test — allows the technician to reset the EMR module. The EMR light reminds the driver when emission maintenance is required.

8. Wiggle test — allows the technician to wiggle solenoid and relay connections. An audible beep is heard from the scan tester if a loose connection is present.

9. Key on, engine off (KOEO) test — allows the technician to perform this test with the scan tester on Ford products.

10. Computed timing check — forces the PCM to move the spark advance 20° ahead of the initial timing setting on Ford products.

11. Key on, engine running (KOER) test — allows the technician to perform this test with the scan tester on Ford products.

12. Clear memory or erase codes — quickly erases fault codes in the PCM memory.

13. Code library — reviews fault codes.

14. Basic test — allows the technician to perform a faster test procedure without prompts.

15. Cruise control test — allows the technician to test the cruise control switch inputs if the cruise module is in the PCM.

Many scan testers have **snapshot** capabilities on some vehicles, which allow the technician to operate the vehicle under the exact condition when a certain problem occurs and freeze the sensor voltage readings into the tester memory. The vehicle may be driven back to the shop, and the technician can play back the recorded sensor readings. During the play back, the technician watches closely for a momentary change in any sensor reading, which indicates a defective sensor or wire. This action is similar to taking a series of sensor reading "snapshots," and then reviewing the pictures later. The snapshot test procedure may be performed on most vehicles with a data line from the computer to the DLC.

Mass Air Flow Sensor Testing, General Motors

While diagnosing a General Motors vehicle, one test mode displays grams per second from the MAF sensor. This mode provides an accurate test of the MAF sensor. The grams per second reading should be 4 to 7 with the engine idling. This reading should gradually increase as the engine speed increases. When the engine speed is constant, the grams-per-second reading should remain constant. If the grams-per-second reading is erratic at a constant engine speed or if this reading varies when the sensor is tapped lightly, the sensor is defective. A MAF sensor fault code may not be present with an erratic grams-per-second reading, but the erratic reading indicates a defective sensor.

Block Learn and Integrator Diagnosis, General Motors

While using a scan tester on General Motors vehicles, one test mode displays block learn and integrator. A scale of 0 to 255 is used for both of these displays and a mid-range reading of 128 is preferred. The oxygen sensor signal is sent to the integrator chip and then to the pulse-width calculation chip in the PCM, and the block learn chip is connected parallel to the integrator chip. If the O_2S voltage changes once, the integrator chip and the pulse width calculation chip change the injector pulse width. If the O_2S provides four continually high or low voltage signals, the block learn chip makes a further injector pulse width change. When the integrator, or block learn, numbers are considerably above 128, the PCM is continually attempting to increase fuel; therefore, the O_2S voltage signal must be continually low, or lean. If the integrator, or block learn, numbers are considerably below 128, the PCM is continually decreasing fuel, which indicates that the O_2S voltage must be always high, or rich.

Cylinder Output Test, Ford

A **cylinder output test** may be performed on electronic engine control IV (EEC IV) SFI engines at the end of the KOER test. The cylinder output test is available on many SFI Ford products regardless of whether the KOER test was completed with a scan tester or with the flash code method.

When the KOER test is completed, momentarily push the throttle wide open to start the cylinder output test. After this throttle action, the PCM may require up to 2 minutes to enter the cylinder output test. In the cylinder output test, the PCM stops grounding each injector for about 20 seconds, and this action causes each cylinder to misfire.

While the cylinder is misfiring, the PCM looks at the rpm that the engine slows down. If the engine does not slow down, there is a problem in the injector, ignition system, engine compression, or a vacuum leak. If the engine does not slow down as much on one cylinder, a fault code is set in the PCM memory. For example, code 50 indicates a problem in number 5 cylinder. The correct DTC list must be used for each model year.

Ford Breakout Box Testing

On Chrysler and General Motors products, a data line is connected from the computer to the DLC, and the sensor voltage signals are transmitted on this data line and read on the scan tester. If these sensor voltage signals are normal on the scan tester, the technician knows that these signals are reaching the computer.

In 1989, Ford introduced data links on some Lincoln Continental models. Since that time, Ford has gradually installed data links on their engine computers. If a Ford vehicle has data links, there are two extra pairs of wires in the DLC. Ford has introduced a new generation star (NGS) tester that has the capability to read data on some Ford engine computers. This technology is now available on other scan testers.

Since most Ford products previous to 1990 do not have a data line from the computer to the DLC, the sensor voltage signals cannot be displayed on the scan tester. Therefore, some other method must be used to prove that the sensor voltage signals are received by the computer. A breakout box is available from Ford Motor Company and some other suppliers (Figure 26-51).

Two large wiring connectors on the breakout box allow the box to be connected in series with the PCM wiring. Once the breakout box is connected, each PCM terminal is connected to a corresponding numbered breakout box terminal. These terminals match the terminals on Ford PCM wiring diagrams. For example, on many Ford EEC IV systems, PCM terminals 37 and 57 are 12-V supply terminals from the power relay to the PCM. Therefore, with the ignition switch in the RUN position and the power relay closed, a digital voltmeter may be connected from breakout box terminals 37 and 57 to ground, and 12V should be available at these terminals. The wiring diagram for the model year and system being tested must be used to identify the breakout box terminals.

Fuel Injection Data Recording Interpretation

As stated earlier, many scan tools can be used to take **data recordings** (Figure 26-52). This function is known by several names such as copilot, snapshots, and so forth. The main design purpose of this feature is to record the vehicle in an actual driving mode that cannot be duplicated in the shop. This is ideal for diagnosing intermittent problems. In addition, some systems come with a smaller version of the scan tool that can be placed in the vehicle and operated by the customer. When the problem

Figure 26-51 A breakout box may be connected in series with the PCM terminals on Ford vehicles to allow meter test connections. (Courtesy of OTC-SPX Corp. Aftermarket Tool & Equipment Group)

is encountered, the customer pushes a button and the tool will record the event. The customer then returns to the shop and the technician retrieves the recording.

Although setup and actual operation differs between scan tool manufacturers, proper interpretation of the recordings is the critical step to diagnosing the problem. Usually the data recording is about 45 seconds in length,

35 seconds before the trigger and 10 seconds after. Many scanners allow for this default time after trigger to be changed. Once the recording is displayed, there are usually means to manipulate the graphs. Use the operator's manual that came with the scan tool for information on how to record, retrieve, and manipulate the recordings.

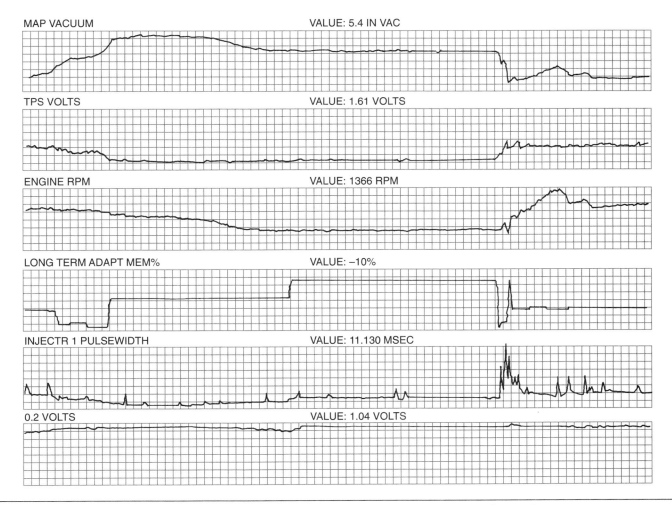

Figure 26-52 A data recording provides a method to capture the data stream information.

Method of Approach to Interpreting Data Recordings

There are several things to consider when looking at a data recording. These include:

1. Can the condition shown really happen? For example, is it possible for the engine coolant temperature reading to go from 102°F to 1°F in 5 seconds (Figure 26-53)?

2. If a sudden change in input voltages or output activation can happen, prove it by finding two additional supported indicators. For example, if the TPS is reading 3.80 volts the engine should be operating at wide-open throttle. If this is really the case then MAP and engine rpm should support this.

3. Study the line characteristics. Since the graph line is a representation of values over time, vertical changes indicate changes that may be too fast to be physically possible. In addition, notice the pattern the line makes when it changes direction. Also note if the line pattern is consistent or erratic.

Examine the data recording shown in Figure 26-54. Notice the MAP vacuum dropping suddenly and the MAP voltage increaseing suddenly. Is it possible for the MAP vacuum to drop that much as a normal condition? If so, what will support this conclusion? Obviously MAP vacuum cannot drop that fast with a wide-open throttle condition. The engine rpm graph does not indicate any increase in engine speed. In fact, it shows a decrease due to the increased injector pulse width. Injector pulse width is increased because the MAP indicates the engine is under load. Notice the TPS value does not support a low vacuum condition either. This graph indicates an open in the MAP circuit. Notice the second time MAP volts goes to 5 volts, the MAP vacuum does not drop as much. This

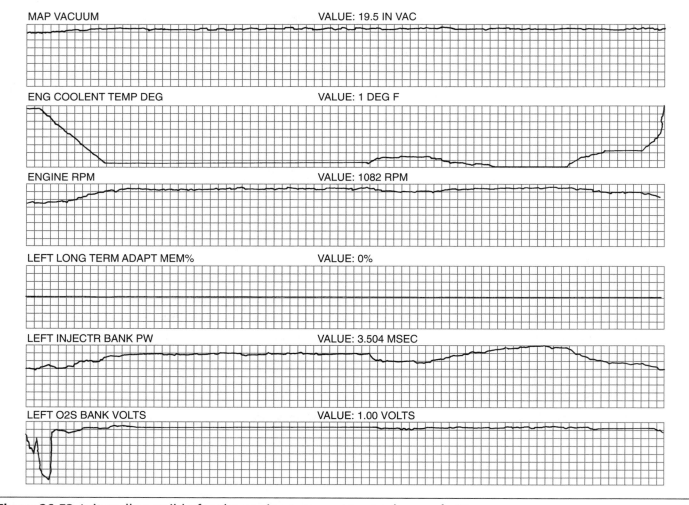

MAP VACUUM   VALUE: 19.5 IN VAC

ENG COOLENT TEMP DEG   VALUE: 1 DEG F

ENGINE RPM   VALUE: 1082 RPM

LEFT LONG TERM ADAPT MEM%   VALUE: 0%

LEFT INJECTR BANK PW   VALUE: 3.504 MSEC

LEFT O2S BANK VOLTS   VALUE: 1.00 VOLTS

Figure 26-53 Is it really possible for the engine temperature to drop so fast?

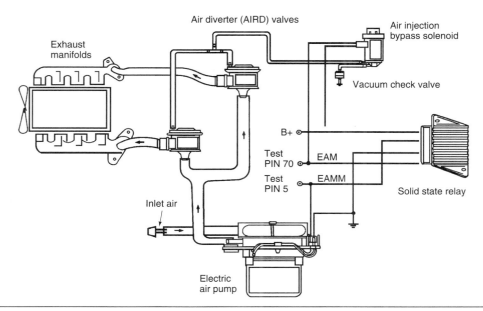

Figure 27-36 Some Ford vehicles are equipped with an electric air pump (EAP) system. (Reprinted with permission of Ford Motor Company)

function properly. The length of EAP operation after the engine is started depends on engine temperature. Cold engine temperatures provide longer EAP operation. Once the catalytic converter is operating properly, the PCM signals the SSR to shut off the EAP. The PCM also de-energizes the air-injection bypass solenoid, which allows the air diverter valves to close.

The PCM monitors the SSR and the EAP to determine if secondary air is present. This PCM monitor for the EAP system functions once per drive cycle. When a malfunction occurs in the EAP system on two consecutive drive cycles, a DTC is stored in the PCM memory and the MIL is turned on. If this malfunction corrects itself, the MIL is turned off after three consecutive drive cycles in which the fault is not present.

Comprehensive Monitor

The comprehensive component monitor uses two strategies to monitor inputs and two strategies to monitor outputs. One strategy for monitoring inputs involves checking certain inputs for electrical defects and out-of-range values by checking the input signals at the analog digital (A/D) converter. The input signals monitored in this way are:

1. Rear HO_2S inputs
2. HO_2S inputs
3. Mass air flow (MAF) sensor
4. Manual lever position (MLP) sensor
5. Throttle position (TP) sensor
6. Engine coolant temperature (ECT) sensor
7. Intake air temperature (IAT) sensor

The comprehensive component monitor (CCM) checks frequency signal inputs by performing rationality checks. During a rationality check, the monitor uses other sensor readings and calculations to determine if a sensor reading is proper for the present conditions. The CCM checks these inputs with rationality checks:

1. Profile ignition pickup (PIP)
2. Output shaft speed (OSS) sensor
3. Ignition diagnostic monitor (IDM)
4. Cylinder identification (CID) sensor
5. Vehicle speed sensor (VSS)

The PCM output that controls the idle air control (IAC) motor is monitored by checking the idle speed demanded by the inputs against the closed loop idle speed correction supplied by the PCM to the IAC motor.

The output state monitor in the CCM checks most of the outputs by monitoring the voltage of each output solenoid, relay, or actuator at the output driver in the PCM. If the output is off, this voltage should be high. This voltage is pulled low when the output is on.

Monitored outputs include:

1. Wide-open throttle A/C cutoff (WAC)
2. Shift solenoid 1 (SS1)
3. Shift solenoid 2 (SS2)
4. Torque converter clutch (TCC) solenoid
5. HO_2S heaters
6. High fan control (HFC)
7. Fan control (FC)
8. Electronic pressure control (EPC) solenoid

Summary

❏ An OBD II system has six monitors to check system operation, and the MIL light is illuminated if vehicle emissions exceed 1.5 times the allowable standard for that model year.

❏ Compared to previous systems, the main difference in an OBD II system is in the software contained in the PCM.

❏ OBD II systems must have a 16-terminal DLC with 12 terminals defined by SAE.

❏ An OBD II system has six monitors to check system operation, including catalyst efficiency, engine misfire, fuel system, heated exhaust gas oxygen sensors, EGR, and comprehensive component monitors.

❏ To complete an OBD II drive cycle, the engine must be started and the vehicle driven until the five monitors are completed, followed by the catalyst efficiency monitor.

Terms-To-Know

Active testing

Adaptive numerator

Alternate good trip

Big slope

Calibrated frequency

California Air Resources Board (CARB)

Catalyst monitor

Catalyst monitor sensor (CMS)

Class 2 data links

Comprehensive components

Conflict

Continuous DTCs

Delta pressure feedback EGR sensor

Diagnostic executive

Diagnostic management system

Diagnostic switch

Diagnostic weight

Drive cycle

Electric air pump (EAP) system

Enhanced EVAP

Excess vacuum test

Exhaust gas recirculation (EGR)

Federal test procedure (FTP)

Freeze frame

Fuel system good trip

Fuel system monitor

Fuel tank pressure sensor

Global good trip

Half cycle counter

International Standards Organization (ISO)

Intrusive test

Leak detection pump (LDP)

Loaded canister test

Maturing code

Misfire

Misfire good trip

Misfire monitor

Non-enhanced EVAP

Onboard diagnostics second generation (OBD II)

One trip monitors

Passive testing

Pended

Pending DTC

Post-catalyst oxygen sensor

Power-up vacuum test

Pressure check mode

Pump mode

Punch through

Purge flow sensor (PFS)

Purge free cells

Purge solenoid leak test

Readiness indicator

Response time test

Rotational velocity fluctuations

Secondary air injection (AIR)

Sensor voltage test

Similar conditions window

Small leak test

Standard corporate protocol (SCP)

Steady-state catalyst efficiency monitor

Stricter EVAP

Suspending

Switch points

Switch ratio

Task manager

Test fail actions

Test frequency

Test mode

Time to activity test

Trip

Two trip monitors

Universal asynchronous receive and transmit (UART)

Vapor management valve (VMV)

Warm-up cycle

Weak vacuum test

Review Questions

Short Answer Essay

1. Briefly describe the six monitors in an OBD II system.

2. Describe the main hardware differences between OBD I and OBD II systems.

3. Describe an OBD II warm-up cycle.

4. Explain trip and drive cycle in an OBD II system.

5. Describe how engine misfire is detected in an OBD II system.

6. What priority does a matured fuel control monitor DTC have in freeze frame?

7. Describe the half cycle counter.

8. Explain the term "Adaptive Numerator".

9. Define what a rationality test is.

10. What is an intrusive test?

Fill-In-The-Blanks

1. In an OBD II system, the MIL light is illuminated if the emission levels exceed _____ times the standard for that model year.

2. The downstream HO_2S monitor _____ efficiency.

3. Type A engine misfires are excessive if misfiring exceeds _____ to _____ percent in a _____ rpm period.

4. OBD II vehicles must use a _____ terminal DLC.

5. The _____ _____ detects the catalyst's ability to store and give off oxygen through the use of a _____ _____ oxygen sensor.

6. A _____ _____ is a specific driving method to verify a symptom, or the repair for the symptom, and to begin and complete a specific OBD II monitor.

7. The minimum requirement for a _____ is that an ignition switch-off period must precede an OBD II trip. Following an ignition off period, the engine must be started and the vehicle driven.

8. A _____ _____ is defined by vehicle operation after an engine shutdown period.

9. All OBD II monitored systems provide _____ _____ data on the vehicle's operating conditions when a maturing code was set.

10. If a DTC is stored due to _____ or _____ system related problems, the PCM requires the engine be returned to a similar conditions window.

ASE Style Review Questions

1. OBD II systems are being discussed. Technician A says these systems have two heated oxygen sensors downstream from the catalytic converters to monitor converter operation. Technician B says the PCM checks exhaust temperature to monitor ignition misfiring. Who is correct?
   a. Technician A only
   b. Technician B only
   c. Both technicians
   d. Neither technician

2. Technician A says a maturing fault code is a one trip failure of a two trip monitor. Technician B says a freeze frame is stored when a maturing fault is detected. Who is correct?
   a. Technician A only
   b. Technician B only
   c. Both technicians
   d. Neither technician

3. OBD II systems are being discussed. Technician A says the PCM illuminates the MIL if a defect causes emission levels to exceed 2.5 times the emission standards for that model year. Technician B says if a misfire condition threatens engine or catalyst damage, the PCM flashes the MIL. Who is correct?
   a. Technician A only
   b. Technician B only
   c. Both technicians
   d. Neither technician

4. The catalyst efficiency monitor is being discussed. Technician A says if the catalytic converter is not reducing emissions properly, the voltage frequency increases on the downstream HO2S. Technician B says if a fault occurs in the catalyst monitor system on three drive cycles, the MIL is illuminated. Who is correct?
   a. Technician A only
   b. Technician B only
   c. Both technicians
   d. Neither technician

5. The misfire monitor is being discussed. Technician A says while detecting type A misfires, the monitor checks cylinder misfiring over a 500-rpm period. Technician B says while detecting type B misfires, the monitor checks cylinder misfiring over a 1,000-rpm period. Who is correct?
   a. Technician A only
   b. Technician B only
   c. Both technicians
   d. Neither technician

6. The EGR system and monitor is being discussed. Technician A says two hoses are connected from the delta pressure feedback EGR (DPFE) sensor to an orifice under the EGR valve. Technician B says when the EGR valve is open, the pressure should be the same on both sides of the orifice under the EGR valve. Who is correct?
   a. Technician A only
   b. Technician B only
   c. Both technicians
   d. Neither technician

7. A technician is using a DSO to diagnose an OBD II system. The downstream HO$_2$S has the same voltage waveform as the upstream HO$_2$S. Technician A says the catalytic converter is defective. Technician B says the air/fuel ratio is too rich. Who is correct?
   a. Technician A only
   b. Technician B only
   c. Both technicians
   d. Neither technician

8. Technician A says in an enhanced EVAP system there is no leak detection pump. Technician B says in a non-enhanced EVAP system there is a leak detection pump. Who is correct?
   a. Technician A only
   b. Technician B only
   c. Both technicians
   d. Neither technician

9. Technician A says a drive cycle is a specific driving method to verify a symptom or the repair for the symptom. Technician B says it takes three good trips within one drive cycle to extinguish the MIL. Who is correct?
   a. Technician A only
   b. Technician B only
   c. Both technicians
   d. Neither technician

10. Technician A says the purpose of the warm-up cycle is to turn off the MIL. Technician B says the purpose of a trip is to erase the fault code. Who is correct?
    a. Technician A only
    b. Technician B only
    c. Both technicians
    d. Neither technician

28 OBD II Diagnosis

Objectives

Upon completion and review of this chapter, you should be able to:

❏ Describe the differences between diagnosing OBD I and OBD II vehicles.

❏ Determine the enabling conditions for OBD II monitors.

❏ Run and evaluate a comprehensive component monitor.

❏ Locate and interpret freeze-frame data.

❏ Perform a good trip after a comprehensive component monitor failure.

❏ Run and evaluate a fuel system lean monitor.

❏ Locate and interpret similar conditions window information.

❏ Perform a good trip after a fuel system monitor failure within the similar conditions window.

Introduction

Because all manufacturers have continually updated, expanded, and improved their engine control systems, there are now hundreds of different domestic and import systems on the road. Methods of reading on board diagnostic data and troubleshooting the systems can vary greatly between manufacturers and between model lines and years of the same manufacturer. This is why it is extremely important the service manual be referred to during diagnostics.

Fortunately, laws have been passed requiring a standardized test procedure for all electronic control systems. As a result of this mandate, known as OBD II, manufacturers will use the same terms, acronyms, and definitions to describe their components (commonly referred to as the SAE J1930 standards). They will also have the same type of diagnostic connector, the same basic test sequences, and will display the same trouble codes. OBD II began in 1994 and is required to be on all vehicles sold in North America after 1997 model year.

The purpose of OBD II was to provide standard service procedures without the use of dedicated special tools. To accomplish this goal, manufacturers needed to change many aspects of their electronic control systems. According to the guidelines of OBD II, all vehicles will have:

- A universal diagnostic test connector, known as the data link connector (DLC), with dedicated pin assignments.

- A standard location for the DLC. It must be under the dash on the driver's side of the vehicle.

- A standard list of diagnostic trouble codes (DTCs).

- Vehicle identification that is automatically transmitted to the scan tool.

- Stored trouble codes that can be cleared from the computer's memory with the scan tool.

- The ability to record, and store in memory, a snapshot of the operating conditions that existed when a fault occurred.

- The ability to store a code whenever something goes wrong and affects exhaust quality.

- A standard glossary of terms, acronyms, and definitions used for all components in the electronic control systems.

The standard DLC (Figure 28-2) is a 16-pin, D-shaped connector with guide keys which allow the scan tool to be installed only one way. Using a standard connector design and designating the pins allows data retrieval with any scan tool designed for OBD II.

The DLC pins are arranged in two rows and are numbered consecutively. Seven of the sixteen pins have been assigned by the OBD II standard. The remaining nine pins can be used by the individual manufacturers to meet their needs. The MIL is only used to inform the driver that the vehicle should be serviced soon and to inform the technician that the computer has set a trouble code.

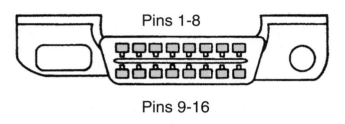

Pins 1-8

Pins 9-16

Figure 28-1 Standard OBD II data link connector. (Reprinted with permission of Ford Motor Company)

Many of the trouble codes from an OBD II system will mean the same thing regardless of manufacturer. However, some of the codes will pertain only to a particular system or will mean something different with each system. The DTC is a five-character code with both letters and numbers (Figure 28-2). This is called the alphanumeric system.

OBD I vs OBD II Diagnosis

In OBD I equipped vehicles, the MIL light would come on if a fault was detected. If the fault was not found the next time the PCM checked the system, the MIL would go out. If the light came on for a few seconds then never came back on, most drivers would not bother bringing the vehicle to the shop to be diagnosed. With OBD II there will not be any intermittent MIL lights. Once the MIL is turned on it takes a series of trips without a fault to turn the light off. Because of this, customers will be bringing their vehicles in for diagnosis.

Once the vehicle is in the shop, other differences exists between OBD I and OBD II. On OBD I systems it

The SAE J2012 standards specify that all DTCs will have a five-digit alphanumeric numbering and lettering system. The following prefixes indicate the general area to which the DTC belongs:

1. P — power train
2. B — body
3. C — chassis

The first number in the DTC indicates who is responsible for the DTC definition.

1. 0 — SAE
2. 1 — manufacturer

The third digit in the DTC indicates the subgroup to which the DTC belongs. The possible subgroups are:

0 — Total system
1 — Fuel-air control
2 — Fuel-air control
3 — Ignition system misfire
4 — Auxiliary emission controls
5 — Idle speed control
6 — PCM and I/O
7 — Transmission
8 — Non-EEC power train

The fourth and fifth digits indicate the specific area where the trouble exists. Code P1711 has this interpretation:

P — Power train DTC
1 — Manufacturer-defined code
7 — Transmission subgroup
11 — Transmission oil temperature (TOT) sensor and related circuit

Figure 28-2 OBD II trouble code formate.

is a common practice to read all fault codes, record them, then erase them. After the DTCs are erased the vehicle is driven to see which DTCs reappear. These codes are then diagnosed.

On OBD II equipped vehicles, the last step in the diagnosis process is to erase the fault codes. If the fault codes are erased on OBD II vehicles, all OBD II data is cleared. This includes freeze frame and similar condition windows. The following is a typical procedure when diagnosing an OBD II vehicle that came in with the MIL light on:

1. Read and record all DTCs (Figure 28-3).

2. Examine the freeze-frame data and determine which DTC corresponds with the freeze frame (Figure 28-4).

3. If applicable, examine the similar conditions window (Figure 28-5).

4. Determine the cause of the fault using the diagnostic procedures outlined in the service manual.

5. Perform the needed repairs.

6. Test drive the vehicle while staying in the similar conditions window (if applicable) or in the same conditions as shown in the freeze frame.

7. Observe the good trip counter. If it increments by one (Figure 28-6), pull over to the side of the road and shut the engine off.

8. Restart the engine and drive back to the shop. During this time the good trip counter should increment to 2 good trips.

9. At this time you have confirmed the repair was successful. If no good trips were incremented, there is still a failure in the system.

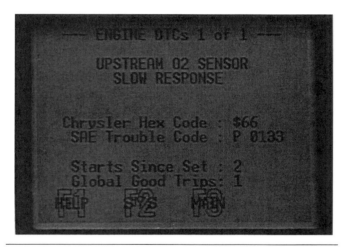

Figure 28-3 Scan tool display of DTCs. (Courtesy of Chrysler Corporation)

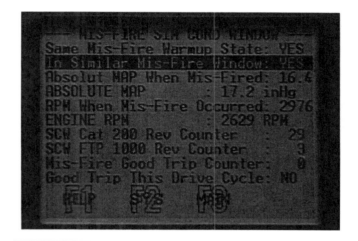

Figure 28-5 Similar conditions window. (Courtesy of Chrysler Corporation)

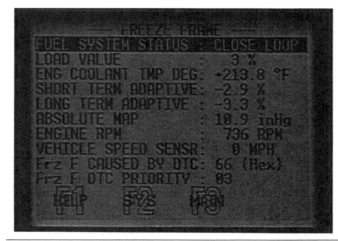

Figure 28-4 Freeze frame data is stored when the PCM first detects a failure. (Courtesy of Chrysler Corporation)

Figure 28-6 Good trip counter. (Courtesy of Chrysler Corporation)

10. If the repair is confirmed as successful, erase the fault codes and return the vehicle to the customer.

Also, keep in mind the freeze-frame data and similar conditions window is recorded when the PCM first detected the failure. The driving conditions explained by the customer may not match those found in these two recordings. If the scan tool has the capabilities, it is a good practice to read any one-trip failures that are recorded.

Summary

❏ As a result of OBD II, manufacturers will use the same terms, acronyms, and definitions to describe their components (commonly referred to as the SAE J1930 standards).

❏ The standard DLC is a 16-pin, D-shaped connector with guide keys which allow the scan tool to be installed only one way.

❏ The DTC is a five-character code with both letters and numbers. This is called the alphanumeric system.

❏ With OBD II there will not be any intermittent MIL lights. Once the MIL is turned on it takes a series of trips without a fault to turn the light off. Because of this, customers will be bringing their vehicles in for diagnosis.

❏ On OBD II equipped vehicles, the last step in the diagnosis process is to erase the fault codes. If the fault codes are erased on OBD II vehicles, all OBD II data is cleared. This includes freeze frame and similar condition windows.

❏ Freeze-frame data and similar conditions window information is recorded when the PCM first detected the failure.

ASE Style Review Questions

1. Technician A says a rationality check is when the PCM compares two or more inputs to determine if they make sense under the current operating conditions. Technician B says a functionality check is when the PCM is commanding a certain output then monitors related inputs to determine if the command was executed. Who is correct?
   a. Technician A
   b. Technician B
   c. Both technicians
   d. Neither technician

2. Technician A says a freeze frame is set for all emission related faults. Technician B says a similar conditions window is recorded for all emission related faults. Who is correct?
   a. Technician A
   b. Technician B
   c. Both technicians
   d. Neither technician

3. Technician A says when diagnosing an OBD II vehicle, the first step is to erase the fault codes. Technician B says erasing fault codes will also erase freeze-frame data. Who is correct?
   a. Technician A
   b. Technician B
   c. Both technicians
   d. Neither technician

4. Technician A says the last DTC will always overwrite the current freeze frame. Technician B says the freeze-frame data is for when the fault was first detected. Who is correct?
   a. Technician A
   b. Technician B
   c. Both technicians
   d. Neither technician

5. Technician A says for a fuel system lean monitor to mature and turn on the MIL it must fail in the same conditions recorded in the similar conditions window. Technician B says it takes three good trips in the similar conditions window in order to extinguish the MIL for a fuel system lean failure. Who is correct?
   a. Technician A
   b. Technician B
   c. Both technicians
   d. Neither technician

6. Technician A says an ECT sensor failure will set a higher priority DTC than a matured misfire monitor failure. Technician B says a monitor may be pended if there is a comprehensive component failure. Who is correct?
   a. Technician A
   b. Technician B
   c. Both technicians
   d. Neither technician

7. Technician A says some systems cannot run a misfire monitor until the PCM learns the manufacturing tolerances of the crankshaft position sensor inputs. Technician B says a type A 10–12% of misfire in 200 revolutions. Who is correct?
   a. Technician A
   b. Technician B
   c. Both technicians
   d. Neither technician

8. Technician A says a drive cycle is a specific driving method to verify a symptom or the repair for the symptom. Technician B says a drive cycle is a specific driving method to begin and complete a specific OBD II monitor. Who is correct?
   a. Technician A
   b. Technician B
   c. Both technicians
   d. Neither technician

9. Technician A says a warm-up cycle is defined by vehicle operation after an engine shut down period when the engine coolant temperature must rise at least 10 degrees F, and must reach at least 80 degrees F. Technician B says most DTCs are erased after 40 warm-up cycles if the problem does not reoccur after the MIL light is turned off for that problem. Who is correct?
   a. Technician A
   b. Technician B
   c. Both technicians
   d. Neither technician

10. Technician A says the "P" in the uniform Diagnostic Trouble Code format means the failure is within the PCM. Technician B says the first digit of the code indicates if the DTC is generic or manufacturer specific. Who is correct?
    a. Technician A
    b. Technician B
    c. Both technicians
    d. Neither technician

29 Advanced Lighting Circuits and Electronic Instrumentation

Objective

Upon completion and review of this chapter, you should be able to:

❑ Explain the operation of the most common types of automatic headlight dimming systems.

❑ Describe the operation of twilight systems that are body computer controlled.

❑ Explain the use and function of fiber optics.

❑ Describe the different methods used to provide lamp outage indicators.

❑ Describe the operating principles of the digital speedometer.

❑ Explain the operation of IC chip and stepper motor odometers.

❑ Describe the operation of electronic fuel, temperature, oil, and voltmeter gauges.

❑ Explain the use and operation of light emitting diodes, liquid crystal, vacuum fluorescent, and CRT displays in electronic instrument clusters.

❑ Describe the operation of quartz analog instrumentation.

❑ Explain the operation of body computer-controlled instrument panel illumination light dimming.

Introduction

With the addition of solid-state circuitry in the automobile, manufacturers have been able to incorporate several different lighting circuits or modify the existing ones. Some of the refinements to the lighting system include automatic headlight washers, automatic headlight dimming, automatic on/off with timed delay headlights, and illuminated entry systems. Some of these systems use sophisticated body computer-controlled circuitry and fiber optics.

Some manufacturers have included such basic circuits as turn signals into their body computer to provide for pulse width dimming in place of a flasher unit. The body computer can also be used to control instrument panel lighting. By using the body computer to control many of the lighting circuits, the amount of wiring has been reduced. In addition, the use of computer control has provided a means of self-diagnosis in some applications.

Computer-Controlled Concealed Headlights

The body computer (BCM) has been utilized by some manufacturers to operate the concealed headlight system. The BCM will receive inputs from the headlight and flash-to-pass switches (Figure 29-1). When the headlight switch is turned on, the BCM receives a signal that the headlights are being activated. To open the headlight doors, the BCM energizes the door open relay. The contacts of the open relay are closed and current is applied to the door motor (Figure 29-2). In this example, battery voltage to operate the door motor is supplied from the 30-ampere circuit breaker. The computer energizes the door open relay through circuit L50, which moves the normally open relay contact arm. Ground is provided through the door close relay contact.

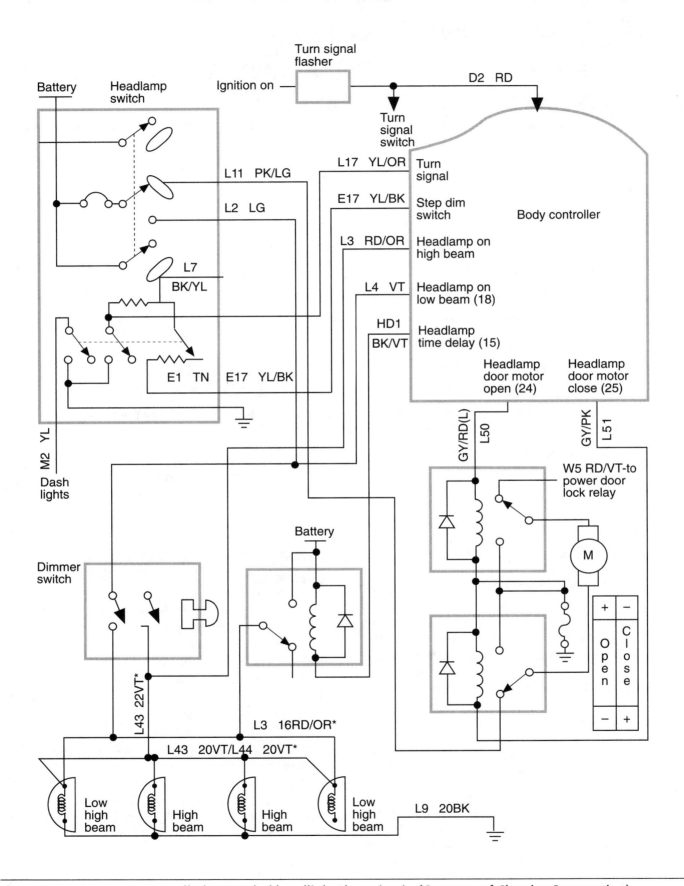

Figure 29-1 Computer-controlled concealed headlight door circuit. (Courtesy of Chrysler Corporation)

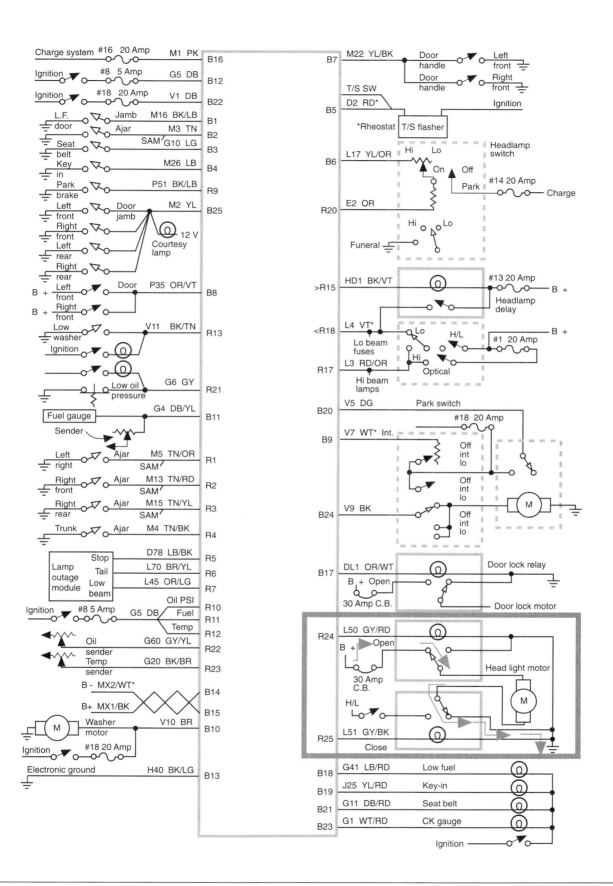

Figure 29-2 Concealed headlight door circuit operation with doors opening. (Courtesy of Chrysler Corporation)

When the headlight switch is turned off, the computer energizes the door close relay through circuit L51. With the door close relay energized, the contacts provide current to the door motor. The ground is supplied through the door open relay. Reversing the polarity through the door motor closes the door.

If the flash-to-pass option is activated, the body computer receives a high (on) signal and energizes the door open relay. When the switch is released, the computer receives a low (off) signal and activates the door close relay. The computer delays the activation of the door close relay for 3 seconds.

Automatic Headlight Dimming

Automatic headlight dimming automatically switches the headlights from high beams to low beams under two different conditions: when light from oncoming vehicles strikes the photocell-amplifier, or light from the taillights of a vehicle being passed strikes the photocell-amplifier. Modern automatic headlight dimming systems use solid-state circuitry and electromagnetic relays to control the beam switching. Most systems consist of the following major components:

1. Light sensitive photocell and amplifier unit
2. High-low beam relay
3. Sensitivity control
4. Dimmer switch
5. Flash-to-pass relay
6. Wiring harness

The **photocell** is a variable resistor that uses light to change resistance. The photocell-amplifier is usually mounted behind the front grill (ahead of the radiator), or in the rearview mirror support. The **sensitivity control** sets the intensity level at which the photocell-amplifier will energize. This control is set by the driver and is located next to, or is a part of, the headlight switch assembly. The sensitivity control is a potentiometer that allows the driver to adjust the sensitivity of the automatic dimmer system to surrounding ambient light conditions. The driver is able to adjust the sensitivity level of the system by rotating the control knob. An increase in the sensitivity level will make the headlights switch to the low beams sooner (approaching vehicle is further away). A decrease in the sensitivity level will switch the headlights to low beams when the approaching vehicle is closer. If the knob is rotated to the full counterclockwise position, the system enters manual override.

The high-low relay is a single-pole, double-throw unit that provides the switching of the headlight beams (Figure 29-3). The dimmer switch is usually a flash-to-pass design. If the turn signal lever is pulled part-way up, the flash-to-pass relay is energized. The high beams will stay on as long as the lever is held in this position, even if the headlights are off. In addition, the driver can select either low beams or automatic operation through the dimmer switch.

Although the components are similar in most systems, there are differences in system operations. Systems differ in how the manufacturer uses the relay to do the switching from high beams to low beams. The system can use either an energized relay to activate the high beams or an energized relay to activate the low beams. If the system uses an energized relay to activate the high beams, the relay control circuit is opened when the dimmer switch is placed in the low beam position or the driver manually overrides the system. With the headlight switch in the ON position and the dimmer switch in the low beam position, battery voltage is applied to the relay coil through circuit 221. The relay coil is not energized because there is no ground provided. The automatic feature is bypassed when the dimmer switch is in the low beam position.

With the dimmer switch in the automatic position, ground is provided for the relay coil through the sensor-amplifier. The energized coil closes the relay contacts to the high beams and current is available to the headlamps. When the photocell sensor receives enough light to overcome the sensitivity setting, the amplifier opens the relay's circuit to ground. This de-energizes the relay coil and switches current flow from the high beam to the low beam position.

If the system uses an energized relay to switch to low beams, placing the dimmer switch in the low beam position will energize the relay (Figure 29-4). With the headlights turned on and the dimmer switch in the automatic position, battery voltage is applied to the photocell-amplifier, one terminal of the high-low control, and through the relay contacts to the high beams. The voltage drop through the high-low control is an input to the photocell-amplifier. When enough light strikes the photocell-amplifier to overcome the sensitivity setting, the amplifier allows current to flow through the high-low relay coil, closing the contact points to the low beams. Once the light has passed, the photocell-amplifier opens battery voltage to the relay coil and the contacts close to the high beams.

When flash-to-pass is activated, the switch closes to ground. This bypasses the sensitivity control and de-energizes the relay to switch from low beams to high beams.

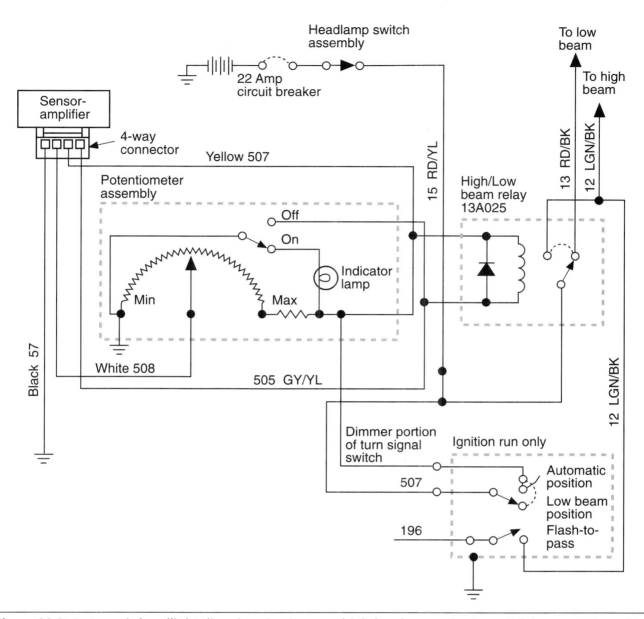

Figure 29-3 Automatic headlight dimming circuit uses a high-low beam relay to switch beam settings. (Reprinted with permission of Ford Motor Company)

Automatic On/Off with Time Delay

The **automatic on/off with time delay** feature has two functions: to turn on the headlights automatically when ambient light decreases to a predetermined level, and to allow the headlights to remain on for a certain amount of time after the vehicle has been turned off. This system can be used in combination with automatic dimming systems. The common components of the automatic on/off with time delay include:

1. Photocell and amplifier

2. Power relay

3. Timer control

In a typical system, a photocell is located inside the vehicle's dash to sense outside light. In most systems, the headlight switch must be in the OFF or AUTO position to activate the automatic mode (Figure 29-5). Battery voltage is applied to the normally open headlight contacts of the relay through the headlight switch. Battery voltage is also supplied to the normally open exterior light contacts through the fuse panel.

To activate the automatic on/off feature, the photocell and amplifier must receive current from the ignition switch (circuit 640). As the ambient light level decreases, the internal resistance of the photocell

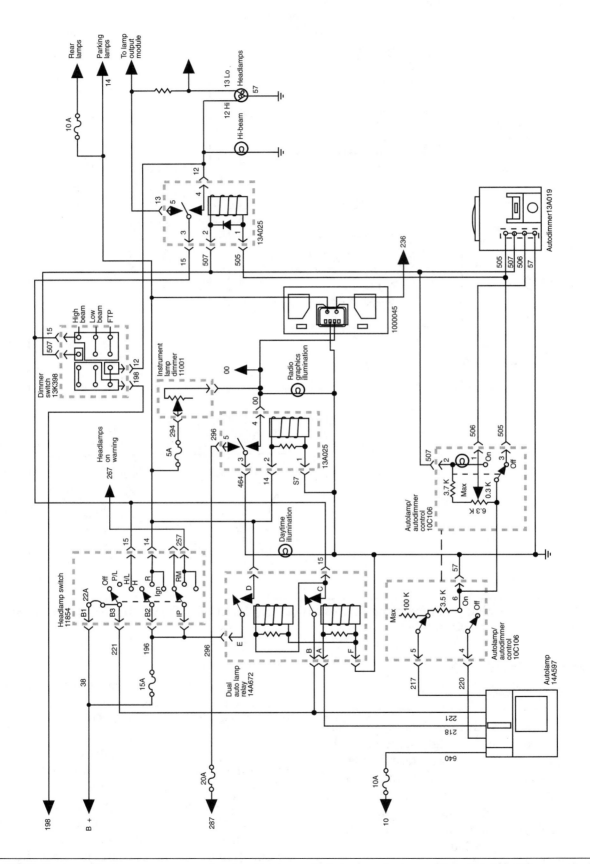

Figure 29-4 System that uses an energized relay to switch to low beams. (Reprinted with permission of Ford Motor Company)

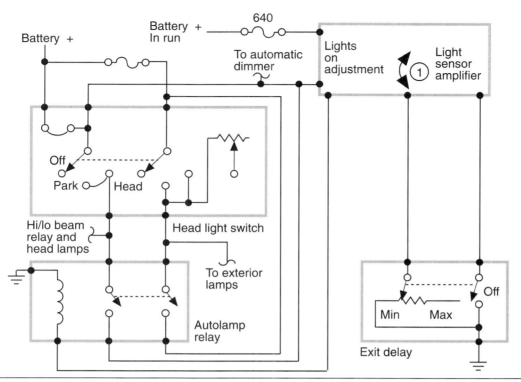

Figure 29-5 Schematic of automatic headlight on/off with time delay system.

increases. When the resistance value reaches a predetermined value, the photocell and amplifier trigger the sensor-amplifier module. The sensor-amplifier module energizes the relay, turning on the headlights and exterior parking lights.

Some systems provide a time delay feature that allows drivers to set a timer circuit to control how long the headlights remain on after they leave the vehicle. The **timer control** is a potentiometer that is a part of the headlight switch in most systems. The timer control unit controls the automatic operation of the system and the length of time the headlights stay on after the ignition switch is turned off. The timer control signals the sensor-amplifier module to energize the relay for the requested amount of time.

If the headlights are on when the ignition switch is turned off, the photocell and amplifier's circuit (640) is opened. This activates a timer circuit in the amplifier. The amplifier still receives current from the headlight switch, and uses this current to keep the relay energized for the requested time interval. When the preset length of time has passed, the amplifier module removes power to the relay and the headlights (and exterior lights) turn off.

The driver can override the automatic on/off feature by placing the headlight switch in the ON position. This bypasses the relay and sends current directly to the headlight circuit.

Depending on model application, General Motors' Twilight Sentinel System can use the body computer to control system operation (Figure 29-6). The BCM senses the voltage drop across the photocell and the delay control switch. The delay control switch resistance is wired in series with the photocell. If the ambient light level drops below a specific value, the BCM grounds the headlamp and parklamp relay coils. The BCM also keeps the headlights on for a specific length of time after the ignition switch is turned off.

Daytime Running Lamps

All late-model Canadian vehicles and some newer vehicles sold in the United States are equipped with **daytime running lamps (DRL)**. The basic idea behind daytime running lights is dimly lit headlamps during the day. This allows other drivers to see the vehicle from a distance. Manufacturers have taken many different approaches to achieve this lighting. Most have a control module or relay (Figure 29-7) that turns the lights on when the engine is running and allows normal headlamp operation when the driver turns on the headlights.

The dimmer headlights can result from headlight current passing through a resistor during daylight hours. The resistor reduces the available voltage and current to the headlights. The resistor is bypassed during normal headlamp operation.

Some systems use a control module, which uses pulsed current to the high-beam lamps. Pulse width mod-

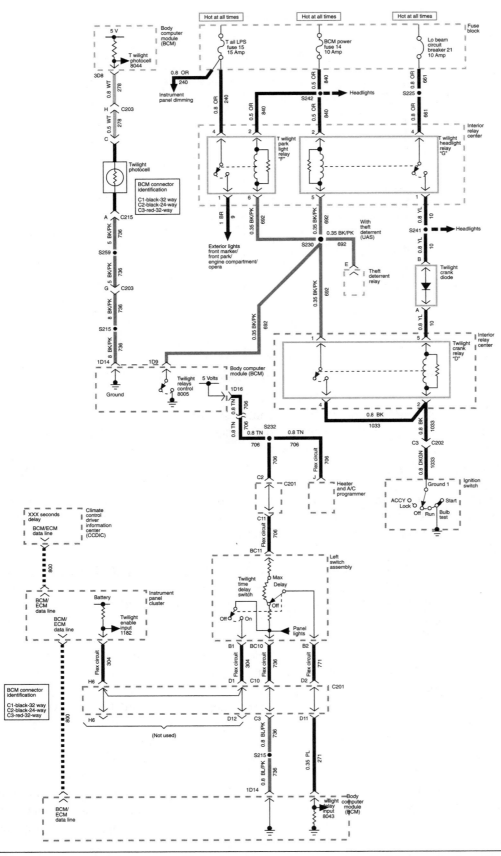

Figure 29-6 Some General Motors' Twilight Sentinel systems use the BCM to sense inputs from the photocell and delay control switch. (Courtesy of Chilton Book Company)

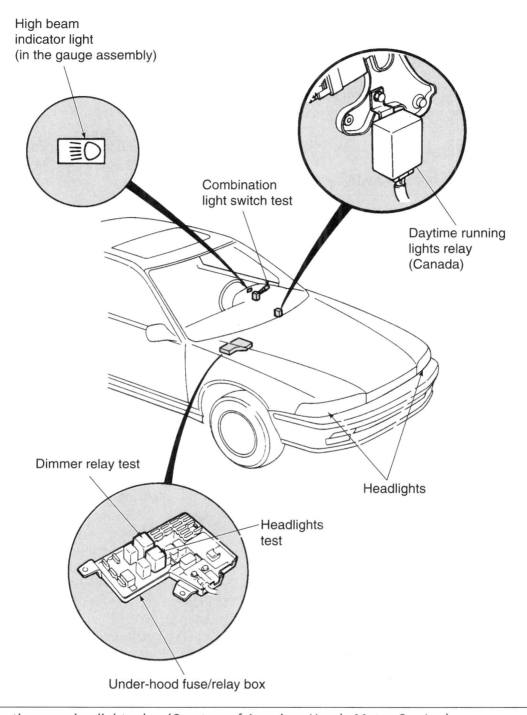

High beam
indicator light
(in the gauge assembly)

Combination
light switch test

Daytime running
lights relay
(Canada)

Dimmer relay test

Headlights
test

Headlights

Under-hood fuse/relay box

Figure 29-7 A daytime running light relay. (Courtesy of American Honda Motor Co., Inc.)

ulation reduces the illumination level of the high beams to less than a normally operating low-beam light.

GM's daytime running lamp system includes a solid-state control module assembly, a relay, and an ambient light sensor assembly. The system lights the low beam headlights at a reduced intensity when the ignition switch is in the RUN position during daylight. The daytime running lamp system is designed to light the low

beam headlamps at full intensity when low light conditions exist.

As the intensity of the light reaching the ambient light sensor increases, the electrical resistance of the sensor assembly decreases. When the DRL control module assembly senses the low resistance, the module allows voltage to be applied to the DRL diode assembly and then to the low beam headlamps. Because of the voltage drop across the diode assembly, the low beam headlamps are on with a low intensity.

As the intensity of the light reaching the ambient light sensor decreases, the electrical resistance of the sensors increases. When the DRL module assembly senses high resistance in the sensor, the module closes an internal relay that allows the low beam headlamps to illuminate with full intensity.

Illuminated Entry Systems

The **illuminated entry system** provides for activation of the courtesy lights before the doors are opened. Most modern illuminated entry systems incorporate solid-state circuitry that includes an illuminated entry actuator and side door switches in the door handles. The **illuminated entry actuator** contains a printed circuit, and a relay. Illumination of the door lock tumblers can be provided by the use of fiber optics or light emitting diodes.

When either of the front door handles are lifted, a switch in the handle will close the ground path from the actuator. This signals the logic module to energize the relay (Figure 29-8). With the relay energized, the contacts close and the interior and door lock lights come on. A timer circuit is incorporated to turn off the lights after 25 to 30 seconds. If the ignition switch is placed in the RUN position before the timer circuit turns off the interior lights, the timer sequence is interrupted and the interior lights turn off.

Some manufacturers have incorporated the illuminated entry actuator into their body computer. Activation of the system is identical as discussed. The signal from the door handle switch can also be used as a "wake-up"

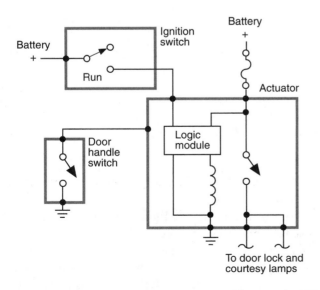

Figure 29-8 Illuminated entry actuator circuit.

signal to the body computer. A **wake-up signal** is used to notify the BCM that an engine start and operation of accessories is going to be initiated soon. This signal is used to warm-up the circuits that will be processing information (Figure 29-9).

The signal from the door handle switch informs the body computer to activate the courtesy light relay control circuit. Some systems use a pair of door jam switches to signal the body computer to keep the courtesy lights on when the door is open. When the door is closed, and the ignition switch is in the RUN position, the lights are turned off.

Some manufacturers use the twilight photocell to inform the body computer of ambient light conditions. If the ambient light is bright, the photocell signals the body computer that courtesy lights are not required.

Instrument Panel Dimming

In the computer-controlled **instrument panel dimming** system, the headlight switch dimming control is used as an input to the computer instead of having direct control of the illumination lights. The body computer uses inputs from the panel dimming control and photocell to determine the illumination level of the instrument panel lights (Figure 29-10). With the ignition switch in the RUN position, a 5-volt signal is supplied to the panel dimming control potentiometer. The wiper of the potentiometer returns the signal to the body computer.

When the dimmer control is moved toward the dimmer positions, the increased resistance results in a decreased voltage signal to the BCM. By measuring the feedback voltage, the BCM is able to determine the resistance value of the potentiometer. The body computer controls the intensity level of the illumination lamps by pulse width modulation.

Some manufacturers send battery voltage to the dimming rheostat then use the voltage to the BCM as an input for illumination level. An input of approximately 1 volt indicates a request for high illumination level and higher voltage level inputs indicate less intensity.

Fiber Optics

Fiber optics is the transmission of light through **polymethylmethacrylate plastic** that keeps the light rays parallel even if extreme bends are in the plastic. The invention of fiber optics has provided a means of illuminating several objects with a single light source (Figure 29-11). Plastic fiber optic strands are used to transmit light from the source to the object to be illuminated. The

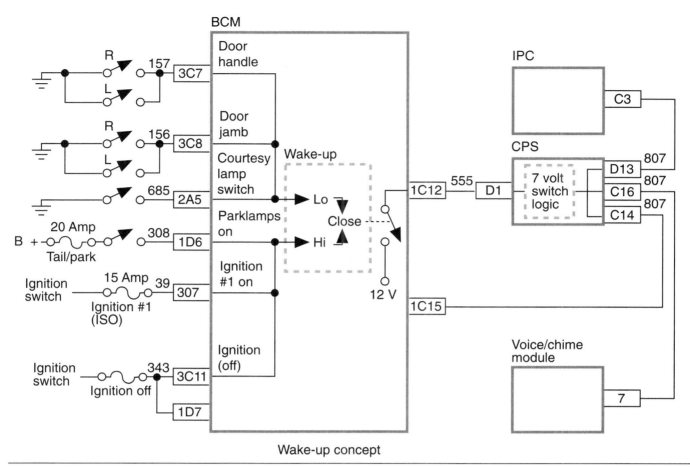

Figure 29-9 Body computer control of the illuminated entry system. (Courtesy of Chilton Book Company)

strands of plastic are sheathed by a polymer that insulates the light rays as they travel within the strands. The light rays travel through the strands by means of internal reflections.

Fiber optics are commonly used as indicator lights to show the driver that certain lights are functioning. Many vehicles with fender-mounted turn signal indicators use fiber optics from the turn signal light to the indicator. The indicator will only show light if the turn signal light is on and working properly.

The advantage of fiber optics is it can be used to provide light in areas where bulbs would be inaccessible for service. Other uses of fiber optics include:

- Lighting ash trays
- Illuminating instrument panels
- Dash lighting over switches
- Ignition key "halo" light

Lamp Outage Indicators

The most common **lamp outage indicator** uses a translucent drawing of the vehicle (Figure 29-12). If one

of the monitored systems fails, or is in need of driver attention, the graphic display illuminates a light to indicate the location of the problem.

The basic lamp outage indicator system is used to monitor the stop light circuit. This system consists of a reed switch and opposing electromagnetic coils (Figure 29-13). Opposing magnetic fields are created since the coils are wound in opposite directions. When the ignition switch is turned to the RUN position, battery voltage is applied to the normally open reed switch. When the brake light switch is closed, current flows through the coils on the way to the stop light bulbs. If both bulbs are operating properly, the coils create opposing magnetic fields, keeping the reed switch in the open position. If one of the stop light bulbs burns out, current will only flow through one of the coils which attracts the reed switch contacts and closes them. This completes the stop light warning circuit and illuminates the warning light on the dash. The warning light will remain on as long as the stop light switch is closed.

Some manufacturers use a **lamp outage module** either as a stand-alone module or in conjunction with the BCM. A lamp outage module is a current measuring sen-

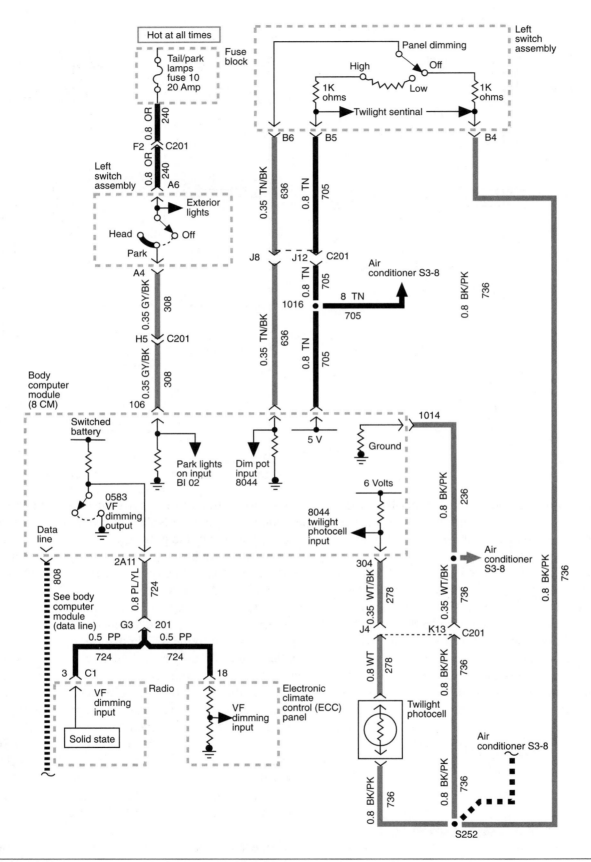

Figure 29-10 The dimming control and photocell are inputs to the BCM to control instrument panel dimming. (Courtesy of General Motors Corporation, Service Technology Group)

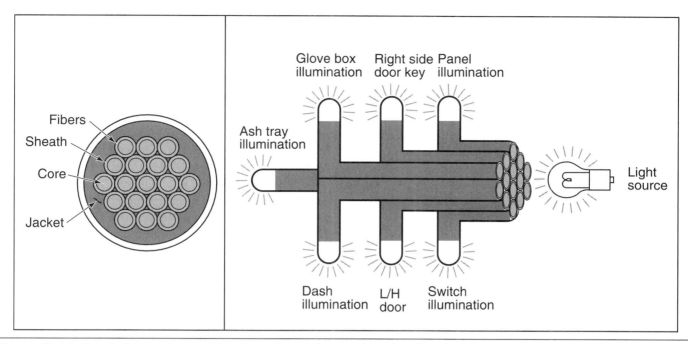

Figure 29-11 One light source can illuminate several areas by using fiber optics.

sor containing a set of resistors, wired in series with the power supply to the headlights, taillights, and stop lights. If the module is a "stand-alone" unit, it will operate the warning light directly. The module monitors the voltage drop of the resistors. If the circuits are operating prop-

erly, there is a 0.5 volt input signal to the module. If one of the monitored bulbs burns out, the voltage input signal drops to about 0.25 volt. The module completes the ground circuit to the warning light to alert the driver of a bulb malfunction. The module is capable of monitoring several different light circuits. However, the bulbs are monitored only when current is supplied to them.

Many vehicles today use a computer-driven information center to keep the driver informed of the condition of monitored circuits. The vehicle information center usually receives its signals from the BCM (Figure 29-14). In this system the lamp outage module is used to

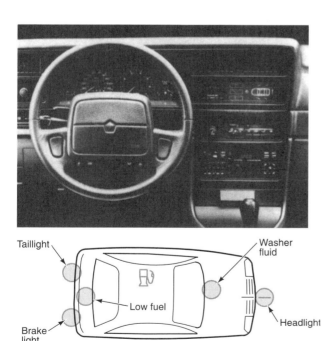

Figure 29-12 Many vehicle information systems may use a graphic display to indicate warning areas to the driver.   (Courtesy of Chrysler Corporation)

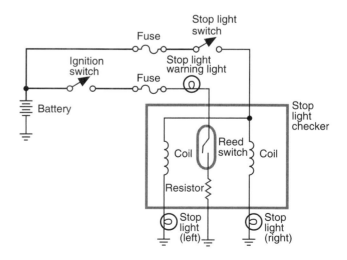

Figure 29-13 Stop light lamp outage indicator circuit. (Courtesy of General Motors Corporation, Service Technology Group)

send signals to the BCM. The BCM will either illuminate a warning light, give a digital message, and/or activate a voice warning device to alert the driver of a light bulb malfunction.

A burned out light bulb means there is a loss of current flow in one of the resistors of the lamp outage module. A monitoring chip in the module compares the voltage drop across the resistor. If there is no voltage drop across the resistor, there is an open in the circuit (burned out light bulb). When the chip measures no voltage drop across the resistor, it signals the body computer, which then gives the necessary message to the vehicle information center (VIC).

General Motors uses the lamp monitor module to connect the light circuits to ground (Figure 29-15). When the circuits are operating properly, the ground connection in the module causes a low circuit voltage. Input from the lamp circuits are through two equal resistance wires.

The module output to the bulbs is from the same module terminals as the inputs.

If a bulb burns out, the voltage at the lamp monitor module terminal will increase. The module will open the appropriate circuit from the BCM, signaling the BCM to send a communication to the IPC computer, which displays the message in the information center.

Electronic Instrumentation Introduction

Computer-driven instruments are becoming increasingly popular on today's vehicle. These instruments provide far more accurate readings than their conventional analog counterparts. Today's technician will be required to service these systems on a more frequent basis as they grow in popularity. This section introduces you to the

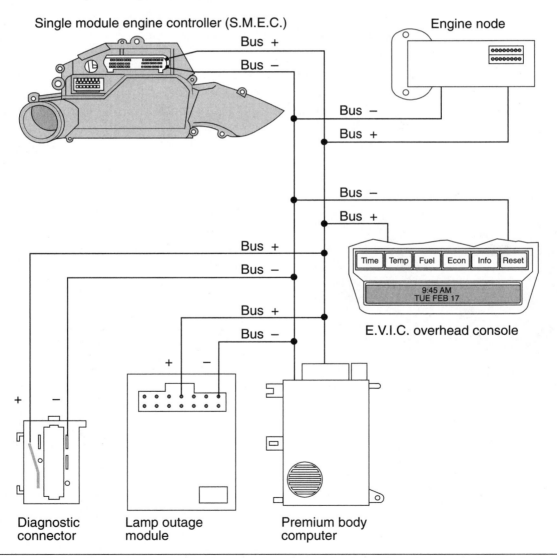

Figure 29-14 The body computer can be used to receive signals from various inputs and give signals to control the information center. (Courtesy of Chrysler Corporation)

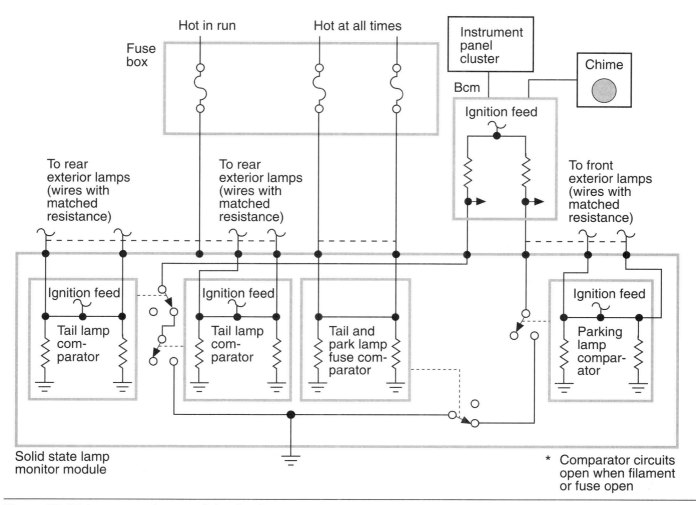

Figure 29-15 Lamp monitor module circuit.

most commonly used computer-driven instrumentation systems. These systems include the speedometer, odometer, fuel, oil, and temperature gauges.

The computer-driven instrument panel uses a microprocessor to process information from various sensors and to control the gauge display. Depending on the manufacturer, the microprocessor can be a separate computer that receives direct information from the sensors and makes the calculations, or it may use the body computer to perform all functions. The illustration (Figure 29-16) shows the different inputs and outputs to the computer-driven instrument panel. The computer may control a digital instrument or an analog cluster.

In addition, there are many types of information systems used today. These systems are used to keep the driver informed of a variety of monitored conditions, including vehicle maintenance, trip information, and navigation.

Digital Instrumentation

Digital instrument clusters use digital and linear displays to notify the driver of monitored system conditions (Figure 29-17). Digital instrumentation is far more precise than conventional analog gauges. Analog gauges display an average of the readings received from the sensor; a digital display will present exact readings. In some systems the information to the gauge is updated as often as 16 times per second.

Most digital instrument panels provide for display in English or metric values. Also, many gauges are a part of a multigauge system. Drivers select which gauges they wish to have displayed. Most of these systems will automatically display the gauge to indicate a potentially dangerous situation. For example, if the driver has chosen the oil pressure gauge to be displayed and the engine temperature increases above set limits, the temperature gauge will automatically be displayed to warn the driver. A warning light and/or a chime will also activate to get the driver's attention.

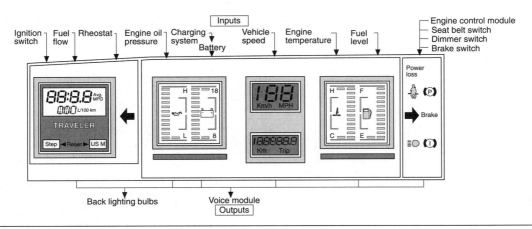

Figure 29-16 Inputs and outputs of the electronic instrument panel. (Courtesy of Chrysler Corporation)

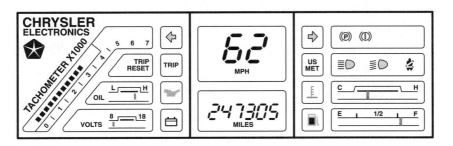

Figure 29-17 Digital instrument cluster. (Courtesy of Chrysler Corporation)

Most electronic instrument panels have self-diagnostic capabilities. The tests are initiated through a scan tool or by pushing selected buttons on the instrument panel. The instrument panel cluster also initiates a self-test every time the ignition switch is turned to ACC or RUN. Usually the entire dash is illuminated and every segment of the display is lighted. International Standards Organization (ISO) symbols are used to represent the gauge function (Figure 29-18). The ISO symbols generally flash during this test. At the completion of the test, all gauges will display current readings. A code is displayed to alert the driver if a fault is found.

Speedometers

The electronic speedometer receives voltage signals directly from the vehicle speed sensor (VSS) or over a data bus from the BCM. The VSS can be either a PM generator, Hall-effect switch, or an optical sensor. Ford, GM, and Toyota have used optical vehicle speed sensors. The Ford and Toyota optical sensors are operated from

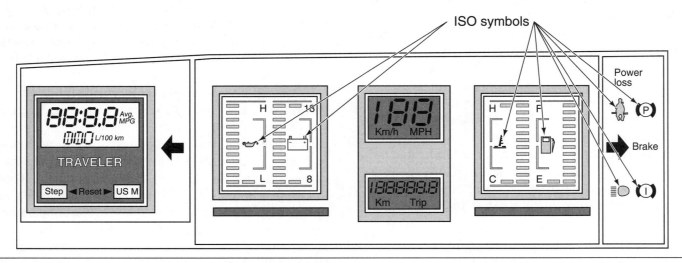

Figure 29-18 A few of the ISO symbols used to identify the gauge. (Courtesy of Chrysler Corporation)

the conventional speedometer cable. The cable rotates a slotted wheel between an LED and a phototransistor (Figure 29-19). As the slots in the wheel break the light, the transistor conducts an electronic pulse signal to the speedometer. An integrated circuit rectifies the analog input signal from the optical sensor and counts the pulses per second. The value is calculated into mph and displayed in the digital readout. The display is updated every half-second. If the driver selected the readout to be in kilometers per hour, the computer makes an additional calculation to convert the readout. These systems may use a conventional gear-driven odometer.

The early style of GM speed sensor also operated from the conventional speedometer cable. The LED directs its light onto the back of the speedometer cup. The cup is painted black, and the drive magnet has a reflective surface applied to it. As the drive magnet rotates in front of the LED, its light is reflected back to a phototransistor. A small voltage is created every time the phototransistor is hit with the reflective light.

The illustration (Figure 29-20) is a schematic of an instrument panel cluster using a PM generator for the VSS. As the PM generator is rotated, it causes a small AC voltage to be induced in its coil. This AC voltage signal is sent to the engine control module (ECM) and is shared with the BCM. The signal is rectified into a digital signal used to control the output to the instrument panel cluster (IPC) module. The BCM calculates the vehicle speed and provides this information to the IPC module through the serial data link. The IPC module turns on the proper display.

The microprocessor will initiate a self-check of the electronic instrument cluster anytime the ignition switch is placed in the ACC or RUN position. The self-check usually runs for about 3 seconds. The most common sequence for the self-check is as follows:

1. All display segments are illuminated.

2. All displays go blank.

3. 0 mph or 0 km/h is displayed.

In addition to the above methods of sensing speed, Hall-effect switches are also used. The sensor is attached to a gear driven wheel that rotates a trigger wheel. The gear is determined by tire size and the final drive gear ratio of the vehicle. As the trigger wheel rotates it will cause the Hall-effect switch to switch voltages high and low at set amounts and times each revolution. The amount of switches per revolution remains constant, regardless of vehicle speed. Once the control module receives a programmed number of switches (8,000 for example) it knows it has traveled one mile.

Odometers

If the speedometer uses an optical sensor, the odometer may be of conventional design. Two other types of odometers are used with electronic displays: the electromechanical type with a stepper motor, and the electronic design using an IC chip.

The electromechanical odometer uses a DC stepper motor that receives control signals from the speedometer circuit (Figure 29-21). The digital signal impulses from the speedometer are processed through a circuit that will halve the signal. The stepper motor receives one-half of the VSS signals sent to the instrument panel cluster. As the stepper motor is activated, the rollers are rotated to accurately display accumulated mileage.

General Motors controls the stepper motor through the same impulses that are sent to the speedometer (Figure 29-22). The stepper motor uses these signals to turn the odometer drive IC on and off. An H-gate arrangement of transistors is used to drive the stepper motor by alternately activating a pair of its coils. An **H-gate** is a set of transistors used to reverse current to the motor windings to control the direction of rotation. The H-gate is constantly reversing system polarity, causing the permanent magnet poles to rotate in the same direction.

The IC chip-type odometer uses a nonvolatile RAM that receives distance information from the speedometer circuit or from the PCM. The controller can update the odometer display every half-second.

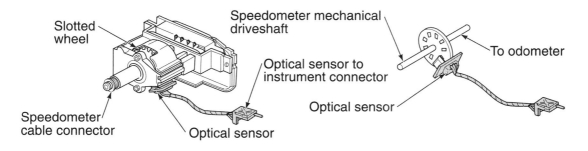

Figure 29-19 Optical speed sensor. (Courtesy of Chilton Book Company)

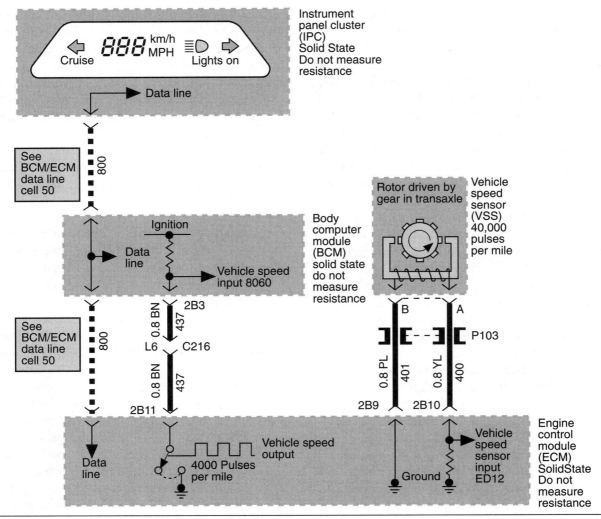

Figure 29-20 The instrument panel cluster module receives its instructions from the BCM which shares the signals from the VSS with the ECM .(Courtesy of General Motors Corporation, Service Technology Group)

Many instrument panel clusters cannot display both trip mileage and odometer readings at the same time. Drivers must select which function they wish to have displayed (Figure 29-23). By depressing the trip reset button, a ground is applied as an input to the microprocessor. The microprocessor clears the trip odometer readings from memory and returns the display to zero. The trip odometer will continue to store trip mileage even if this function is not selected for display.

If the IC chip fails, some manufacturers provide for replacement of the chip. Depending on the manufacturer, the new chip may be programmed to display the last odometer reading. Most replacement chips will display an X, S, or * to indicate the odometer has been changed. If the odometer IC chip cannot be programmed to display correct accumulated mileage, a door sticker must be installed to indicate the odometer has been replaced. IC odometer reading corrections or changes can only be performed in the first 10 miles of operation.

If an error occurs in the odometer circuit, the display will change to notify the driver. The form of error message differs between manufacturers. In some systems the word "ERROR" is displayed.

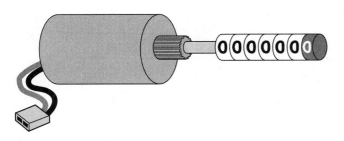

Figure 29-21 Stepper motor used to rotate the odometer dial.

Tachometers

The tachometer can be a separate function that is displayed at all times, or a part of a multigauge display.

Ford uses a multigauge that sequentially changes the gauge between four different gauges (Figure 29-24). The tachometer receives its voltage signals from the ignition system and displays the readout in a bar graph. The Ford multigauge has a built in power supply that provides a 5-volt reference signal to the other monitored systems for the gauge. Also, the gauge has a "watchdog" circuit incorporated in it. The power on/off **watchdog circuit** supplies a reset voltage to the microprocessor in the event pulsating output signals from the microprocessor are interrupted.

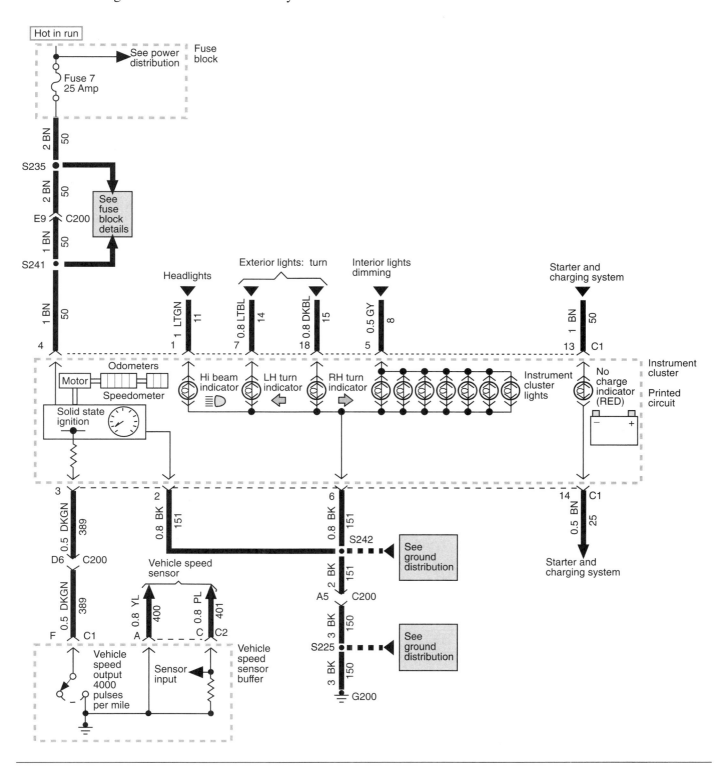

Figure 29-22 The stepper motor receives its control signals through the speedometer circuit. (Courtesy of General Motors Corporation, Service Technology Group)

General Motors uses the signal from the direct ignition system (DIS) module to the PCM. The signal is sent from the PCM to the BCM on the serial data link. The IPC does not receive this signal directly, but it eavesdrops on the data link communications. The IPC uses this reference signal to calculate and display the tachometer reading.

Electronic Fuel Gauges

Most digital fuel gauges use a fuel level sender that decreases resistance value as the fuel level decreases. This resistance value is converted to voltage values by the microprocessor. A voltage-controlled **oscillator** changes the signal into a cycles-per-second signal. An oscillator creates a rapid back and forth movement of voltage. The microprocessor counts the cycles and sends the appropriate signal to operate the digital display.

An F is displayed when the tank is full and an E is displayed when less than 1 gallon is remaining in the tank. Other warning signals include incandescent lamps, a symbol on the dash, or flashing of the fuel ISO symbol. If the warning is displayed by a bulb, usually a switch is located in the sending unit that closes the circuit. Flashing digital displays are usually controlled by the microprocessor.

WARNING *Not all sending units used for bar graph and digital displays are the same. Some sending units used for bar graph displays increase resistance as the fuel level is decreased.*

The bar graph-style gauge uses segments to represent the amount of fuel remaining in the tank (Figure 29-25). The segments divide the tank into equal levels. The display will also include the F, 1/2, and E symbols along with the ISO fuel symbol. A warning to the driver is displayed when only one bar is lit. The gauge will also alert the driver to problems in the circuit. A common method

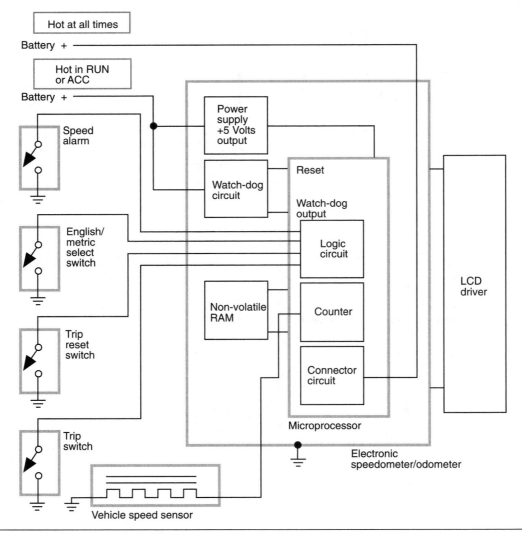

Figure 29-23 The trip reset button provides a ground signal to the logic circuit, which is programmed to erase the trip odometer memory while retaining total accumulated mileage in the odometer.

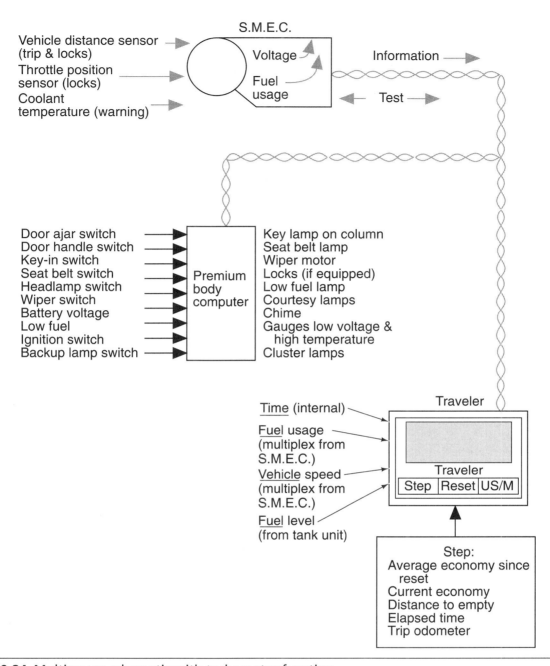

Figure 29-24 Multigauge schematic with tachometer function.

of indicating an open or short is to flash the F, 1/2, and E symbols while the gauge reads empty.

Other Digital Gauges

Most of the gauges used to display temperature, oil pressure, and charging voltage are of bar graph design. Another popular method is to use a floating pointer.

The temperature gauge will usually receive its input from an NTC thermistor. When the engine is cold, the resistance value of the thermistor is high, resulting in a low voltage input to the microprocessor. This input sig-

nal is translated into a low temperature reading on the gauge. As the engine coolant warms, the resistance value drops. At a predetermined resistance level, the microprocessor will activate an alert function to warn the driver of excessive engine temperature.

The voltmeter calculates charging voltage by comparing the voltage supplied to the instrument panel module to a reference voltage signal.

The oil pressure gauge uses a piezoresistive sensor that operates like those used for conventional analog gauges.

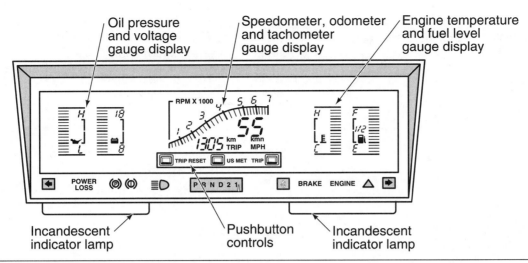

Figure 29-25 Bar graph style of electronic instrumentation. Each segment represents a different value. (Courtesy of Chrysler Corporation)

Digital gauges perform self-tests. If a fault is found, a warning signal will be displayed to the driver. A "CO" indicates the circuit is open, a "CS" indicates the circuit is shorted. The gauge will continue to display these messages until the problem is corrected.

Digital Displays

The display can be by means of four common methods: light emitting diodes (LEDs), liquid crystal display (LCDs), vacuum fluorescent display (VFD), and cathode ray tube (CRT).

LED Digital Displays

When first used in the instrument panel, the LED would indicate an on/off status. Chrysler used the LED in conjunction with its gauges to alert the driver of conditions requiring immediate attention. Some manufacturers are using LEDs in bar graph displays (Figure 29-26). Also, the LEDs can be combined to display alphanumeric characters. There are two common methods of using the LED for digital display: (1) seven-segment display (Figure 29-27), and (2) **dot matrix display** (Figure 29-28). A matrix is a group of elements that are arranged in columns and rows.

By activating selected segments or dots, any number or letter can be displayed. LEDs generally are designed to produce a red, green, or yellow light. They work very well in low light conditions. However, they are difficult to see in bright light. Although some manufacturers use LED readout instrument panels, their use is limited due to their comparatively high power requirements.

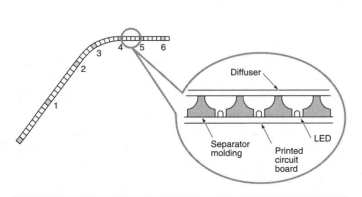

Figure 29-26 LEDs arranged to create a bar graph display.

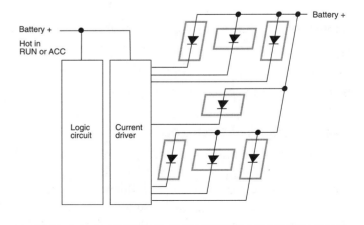

Figure 29-27 Seven segment display panel.

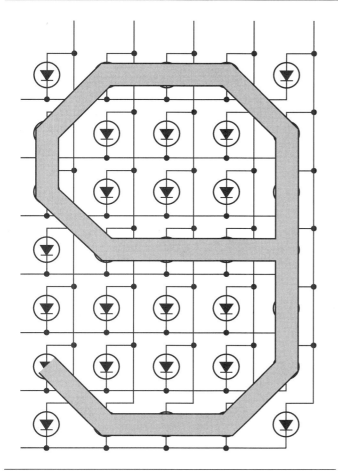

Figure 29-28 Dot matrix display panel.

Liquid Crystal Displays

Liquid crystal displays (LCDs) require an external light source (Figure 29-29). The external light source can be supplied by either ambient or artificial light. In day-light the segments are activated from the front; the artificial light activates the LCD from the back. The artificial light can be controlled by the headlight switch or turned on whenever the ignition switch is placed in the RUN or ACC position.

The LCD construction consists of a twisted **nematic fluid** sandwiched between two polarized glass sheets (Figure 29-30). The nematic fluid is a liquid crystal that has a thread-like form. It has light slots that can be rearranged by applying small amounts of voltage.

The front polarizer is a vertical polarizer and the rear polarizer is a horizontal polarizer. The display is viewed through the vertical polarizer. The **polarizers** make light

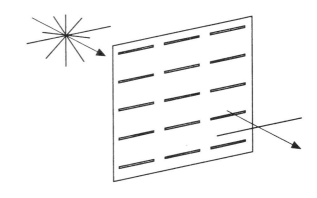

Figure 29-30 The polarizer makes the light waves vibrate in only one direction.

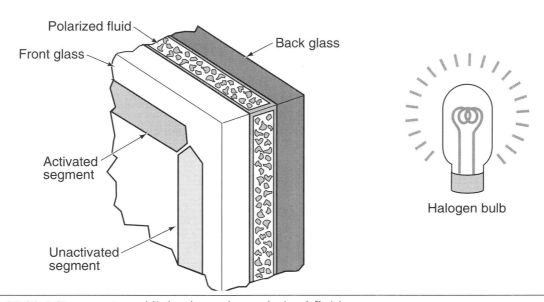

Figure 29-29 LC's use external light through a polarized fluid.

waves vibrate in only one direction. Light is composed of waves that vibrate in several different directions. The polarizer converts the light into polarized light.

The nematic fluid's molecules are arranged in such a manner that they rotate the light from the vertical polarizer 90 degrees (Figure 29-31). The light leaves the fluid in a horizontal waveform. The light continues to pass through the horizontal polarizer to the reflector. The light is then reflected back through the horizontal polarizer to the fluid. The fluid once again rotates the light wave into a vertical position and out the vertical polarizer. When light passes through the LCD in this manner, the display appears light and no pattern is seen.

When a small square wave voltage is applied to the fluid its light slots are rearranged. The fluid will no longer rotate the light waves. The light waves leave the fluid in a vertical plane and cannot pass through the horizontal polarizer to get to the reflector. Because the light cannot be reflected back, the display appears dark. Characters are displayed by controlling which segments are dark and which segments remain light.

In most instrument panels, the LCD cluster is constantly backlit with halogen lights, but a slightly different principle is used. When the segment is not activated, the light is unable to transmit through the opaque fluid and the segment appears dark. When voltage is applied to the fluid, its light slots align and allow the light to pass to the segment. The intensity of the halogen lights is controlled through pulse width modulation to provide the correct illumination levels for the LCD under different ambient light conditions. A photocell is used to sense the amount of light intensity inside the vehicle. This information is processed by the microprocessor to determine the correct

light intensity for the display. Additional control of light intensity is provided by the driver through the headlight switch rheostat.

The voltage to the fluid is provided through the polarizers that have contacts with each segment in the display. The front glass has metallized shapes where the characters will be displayed. The back glass also is metallized. The color of the display is determined by filters placed in front of the display.

WARNING *LCDs operate off waveform alternating current. Do not apply DC current or the LCD will be destroyed.*

Vacuum Fluorescent Displays

The **vacuum fluorescent display (VFD)** is constructed of a hot cathode of tungsten filaments, a grid, and a phosphorescent screen that is the anode (Figure 29-32). The components are sealed in a flat glass envelope that has been evacuated of oxygen and filled with argon or neon gas.

A constant voltage is applied to the hot cathode, which results in tungsten electrons being released from the filament wires. The grid is at a higher positive voltage than the cathode. The freed tungsten electrons are accelerated by the positive grid wires and pass through the grid to the anode. The grid ensures that the tungsten electrons will strike the anode uniformly.

The anode is at a higher positive voltage than the grid. The phosphorescent coated anode (screen) will be luminescent when the tungsten electrons strike it. The display is controlled by which segments of the screen are activated by a digital circuit. If the segment is activated,

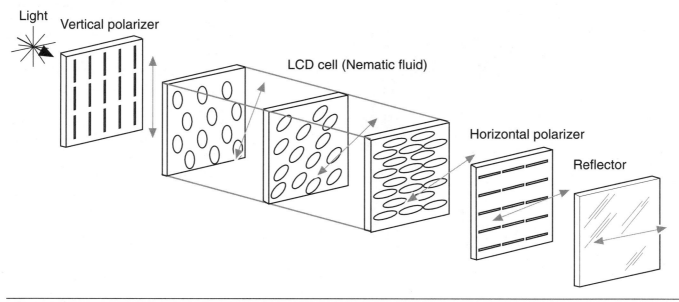

Figure 29-31 The nematic fluid rotates the polarized light wave 90 degrees.

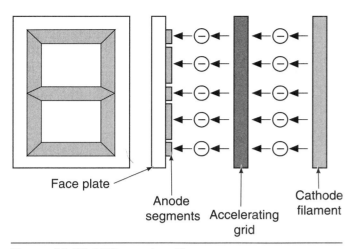

Figure 29-32 VFD construction.

the screen will illuminate. If the segment is not activated, the electrons striking the screen have no effect on the phosphors and the screen remains off.

The segments of a VFD can be arranged in several different patterns. The most common are seven- or fourteen-segment patterns. The computer selects the sets of segments that are to emit light for the message.

The VFD display is very bright. Most manufacturers will dim the intensity of the VFD to 75% brightness whenever the headlights are turned on. To provide sufficient brightness in the daylight, with the headlights on, the headlight switch rheostat may have an additional detent to allow bright illumination of the VFD.

CRT Displays

A **cathode ray tube (CRT)** display was first offered as standard equipment on the 1986 Buick Riviera. A cathode ray tube is similar to a television tube. It contains a cathode that emits electrons and an anode that attracts them. The screen will glow at the points being hit by the electrons. Control plates control the direction of the electrons. The screen of the CRT is touch sensitive. By touching the button on the screen, the menu can be changed to display different information. The menu-driven instrumentation brings up a screen with a "menu" of items. The driver can select a particular area of vehicle operation. The menu of items includes the radio, climate control, trip computer, and dash instrument information. The technician can access diagnostics through the CRT.

The CRT receives information from the BCM and ECM. It also provides inputs to the BCM and ECM in the form of driver commands to control the various functions (Figure 29-33).

Quartz Analog Instrumentation

Computer-driven **quartz swing needle displays** are similar in design to the air core electromagnetic gauges used in conventional analog instrument panels (Figure 29-34). A common application for this type of gauge is for the speedometer. However, it may be used for any gauge.

In most speedometer gauge systems, a permanent magnet generator sensor is installed in the transaxle or transmission. The AC signal from the PM generator is sent to a **buffer**, then to the processing unit. The buffer circuit changes the AC voltage from the PM generator into a digital signal. Next, the signal is passed to a quartz clock circuit, a gain selector circuit, and a driver circuit. The driver circuit sends voltage pulses to the coils of the gauge which moves the needle in the same manner as conventional air core gauges. The "A" coil is connected to system voltage and the "B" coil receives a voltage that is proportional to input frequency. The magnetic armature reacts to the changing magnetic fields.

WARNING *The signal produced by the speed sensor can also be used by the antilock brake and the cruise control systems. If the tire size or gear ratios are changed, the operation of these systems will be affected. The speedometer must be accurately recalibrated if tires or gear ratios are changed from OE.*

Chrysler vehicles equipped with the 41TE or 42LE electronic shift transaxles, beginning in the 1993 model year, use a method called **"pinion factor"** to determine vehicle speed. The output speed sensor, located at the rear of the transaxle's output shaft, is a PM generator. The AC signal it generates is the result of a rotating 24-tooth trigger wheel mounted on the rear planetary unit. This AC signal is sent to the transmission control module (TCM). The TCM then conditions the signal and sends 8,000 pulses per mile to the PCM. The PCM then sends the signal over the CCD bus circuit to the BCM. Finally, the BCM sends the message to the mechanical instrument cluster (MIC). The MIC then sets the needle to read the vehicle speed. Even though the MIC is on the bus system, it does not respond to the vehicle speed signal sent by the PCM. It is programmed to accept messages only from the BCM.

The signal produced by the output speed sensor is not a true representation of vehicle speed since it is before the final drive, and tire size has not been taken into consideration. The TCM must apply pinion factor to this signal. The pinion factor tells the TCM what the gear ratio

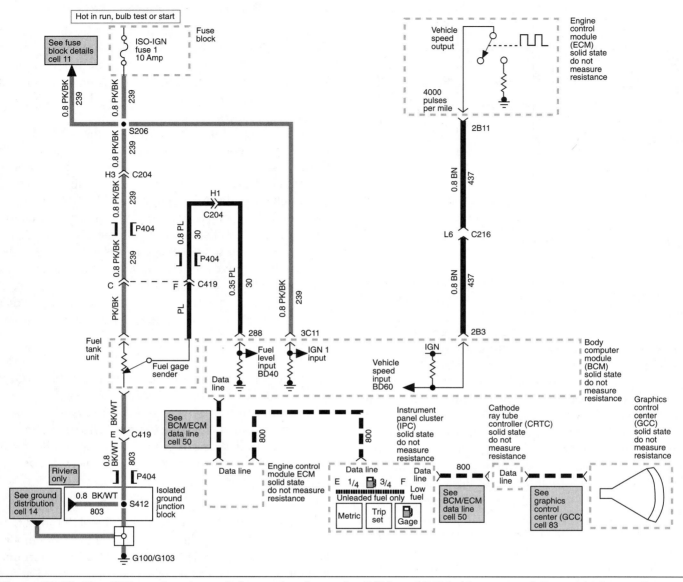

Figure 29-33 The CRT receives input information from the ECM and BCM. (Courtesy of General Motors Corporation, Service Technology Group)

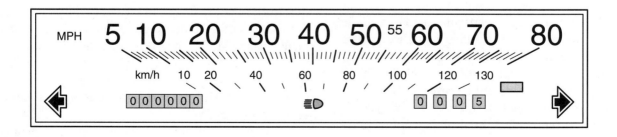

Figure 29-34 Electronic-controlled swing needle speedometer. (Courtesy of General Motors Corporation, Service Technology Group)

and tire size is. This information is set into the TCM at the factory. If the TCM is replaced in the field the scan tool must be used to program in the tire size on the vehicle. In early versions the gear ratio also had to be programmed. In later versions the TCM would learn the gear ratio from the VIN number transmitted by the PCM, however tire size must still be programmed. This is true only for the 41TE. The 42LE uses only one ratio and both tire sizes available have the same rolling diameter. If the pinion factor is not programmed into the TCM, the speedometer and cruise control systems will not function.

Digital Instrument Panel Dimming

Many conventional and computer-driven analog instrument panels use a direct input from the headlight switch rheostat to control the brightness of the display. Other systems use the rheostat as an input to the BCM which in turn controls instrument cluster illumination by pulse width dimming. Depending on the design and type of digital instrument panel, the display may require constant changes in its intensity so the driver will be able to read the display under different lighting conditions.

If the vehicle is equipped with a minimum of digital displays, and an analog speedometer, the brightness level of the displays can be controlled by one input from the rheostat (Figure 29-35). In this system the body controller monitors the voltage level supplied by the head-light switch rheostat. Based on the input levels, the pulse width dimming module controls the illumination level.

Some digital instrument panel modules also use an ambient light sensor in addition to the rheostat. The ambient sensor will control the display brightness over a 35 to 1 range and the rheostat will control over a 30 to 1 range. When the headlights are turned on, the module compares the values from both inputs and determines the illumination level. When the headlights are off, the module uses only the ambient light sensor for its input.

Head-up Display

Some manufacturers have equipped selected models with a **head-up display (HUD)** feature. This system displays visual images onto the inside of the windshield in the driver's field of vision (Figure 29-36). The images are projected onto the windshield from a vacuum fluorescent light source, much like a movie projector.

The head-up control module is mounted in the top of the instrument panel. This module contains a computer and an optical system that projects images to a **holographic combiner** integrated into the windshield above the module. The holographic combiner projects the images in the driver's view just above the front end of the hood. The HUD contains the following displays and warnings:

1. Speedometer reading with USC/metric indicator

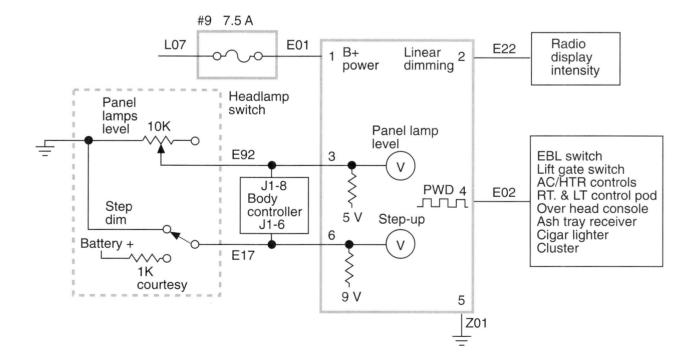

Figure 29-35 Pulse width dimming module. (Courtesy of Chrysler Corporation)

Figure 29-36 The HUD displays various information onto the inside of the windshield. (Courtesy of General Motors Corporation, Service Technology Group)

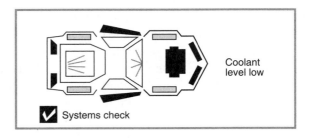

Figure 29-37 The readout panel displays a written message to alert the driver of problems. (Courtesy of Chrysler Corporation)

2. Turn signal indicators
3. High beam indicator
4. Low fuel indicator
5. Check gauges indicator

The head-up control switch contains a head-up display on/off switch, USC/metric switch, and a head-up dimming switch. The head-up dimming switch is a rheostat that sends an input signal to the head-up module. The vertical position of the head-up display may be moved with one of the switches in the control switch assembly. This switch is connected through a mechanical cable-drive system to the head-up module. Moving the vertical position switch moves the position of the head-up module.

Voice Warning Systems

Some warning systems use a **voice synthesizer** to alert the driver of monitored conditions. A voice synthesizer uses a computer-controlled phoneme generator to reproduce the phonemes used for basic speech. The computer puts the phonemes into the right combination to create words and sentences. The voice warning system can be a basic system that alerts the driver of about six conditions. Or, it may be very complex and monitor several functions.

Chrysler used a 24-function monitor with voice alert. This system supplements the warning indicators on the instrument panel and consists of the following components:

1. An alphanumeric readout panel.
2. A car graphic condition/location indicator.
3. An electronic voice alert module.

The alphanumeric readout panel provides a warning message to be displayed (Figure 29-37). The message is displayed until the condition is corrected. The car graphic

indicator is a vehicle silhouette that is displayed when the ignition switch is in the RUN position. When a condition occurs that requires the driver's attention, a colored indicator will be lighted and remain on until the condition is corrected. The electronic voice alert delivers a verbal message if a new warning condition is detected. The illustration shows the monitored conditions of this system.

Sensors are placed throughout the vehicle to supply information to the microprocessor (Figure 29-38). Four different types of sensors supply information to the microprocessor:

1. Modules that monitor headlights, taillights, and stop lights for proper operation.
2. A thermistor to monitor oil levels in the engine. When the temperature increases to a predetermined value due to lack of oil cooling, the microprocessor is sent a warning signal.
3. A voltage sensor to measure charging system output.
4. Normally open switches that provide ground to the microprocessor when there is a component failure or a hazardous condition. These switches are used to monitor door ajar, brake and coolant fluid levels, oil pressure, and brake pad wear.

Additional information is supplied from an internal clock, the coolant temperature sensor, the vehicle speed sensor, and the ignition system for determining engine speed.

Some faults require more than one input to trigger the voice alert. When an unsafe or harmful condition is detected by the sensors, a warning is activated. A warning message is displayed on the readout panel and a tone sounds to alert the driver. The voice alert module delivers a verbal message by interrupting the radio (if on) and delivers the message through the radio speaker closest to the driver. Some systems use a speaker that is built into the module.

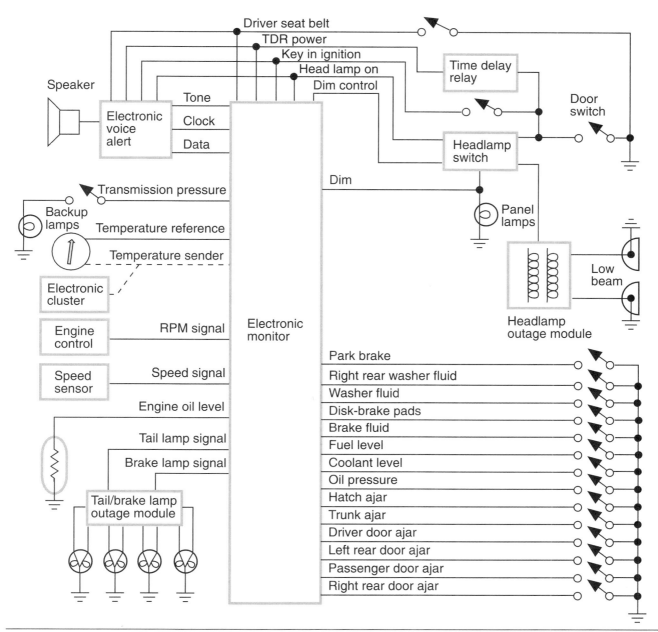

Figure 29-38 Circuit diagram of Chrysler's 24-function voice warning alert system. (Courtesy of Chrysler Corporation)

Travel Information Systems

The travel information system can be a simple calculator that computes fuel economy, distance to empty, and remaining fuel. Other systems provide a much larger range of functions.

Fuel data centers display the amount of fuel remaining in the tank and provide additional information for the driver (Figure 29-39). By depressing the RANGE button, the body computer calculates the distance until the tank is empty by using the amount of fuel remaining and the average fuel economy. When the INST button is depressed, the fuel data center displays instantaneous fuel economy. The display is updated every half-second and is computed by the BCM.

Depressing the AVG button displays average fuel economy for the total distance traveled since the reset button was last pushed. FUEL USED displays the amount of fuel used since the last time this function was reset. The RESET button resets the average fuel economy and fuel used calculations. The function to be reset must be displayed on the fuel data center.

The system illustrated (Figure 29-40) uses the fuel sender unit and ignition voltage references as inputs for

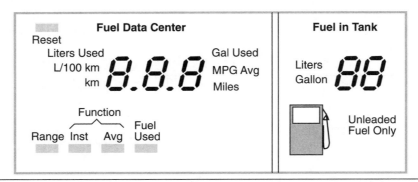

Figure 29-39 Fuel data center. (Courtesy of General Motors Corporation, Service Technology Group)

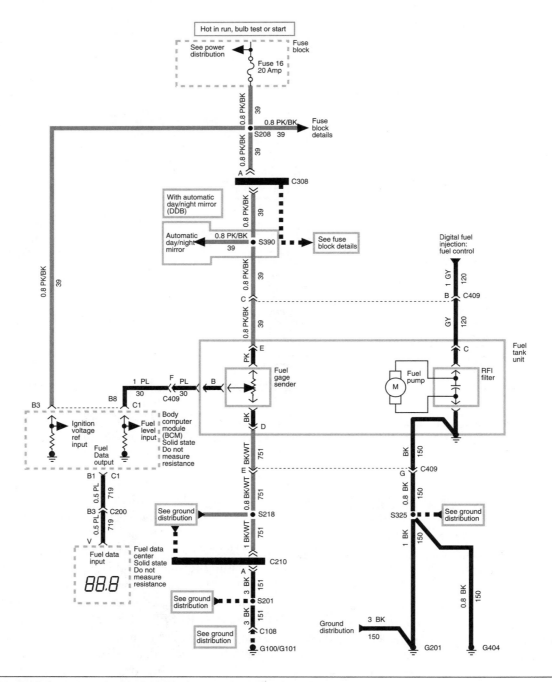

Figure 29-40 Fuel data circuit schematic. (Courtesy of General Motors Corporation, Service Technology Group)

calculating fuel data functions. Other inputs can include speed and fuel flow sensors.

Deluxe systems may incorporate additional features such as outside temperature, compass, elapsed time, estimated time of arrival, distance to destination, day of the week, time, and average speed. The illustration shows the inputs used to determine many of these functions. Fuel system calculations can use injector on time and vehicle speed pulses to determine the amount of fuel flow. Some manufacturers use a fuel flow sensor that provides pulse information to the microprocessor concerning fuel consumption.

Many vehicles are equipped with navigation systems (Figure 29-41). These computerized route guidance systems help drivers to their destinations. Using information broadcast by satellites, these systems display precise road maps and directions to get anywhere on earth. The display unit is mounted to the instrument panel. The information received from the satellites is combined with information stored on a compact disc by the navigational computer. After the driver inputs the intended destination, the system displays the immediate route to get there (Figure 29-42).

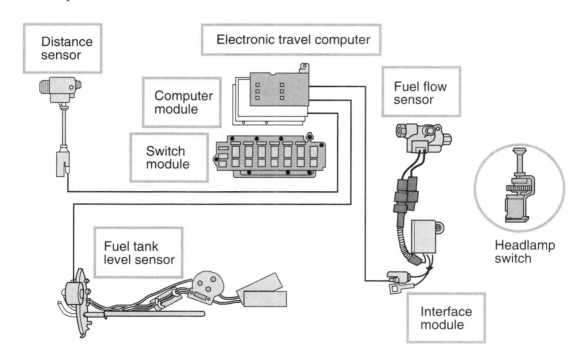

Figure 29-41 Ford's Route Guidance System. (Reprinted with permission of Ford Motor Company)

Figure 29-42 Immediate directions are displayed by Ford's Route Guidance System. (Reprinted with permission of Ford Motor Company)

Summary

❑ Most automatic headlight dimming systems consist of a light sensitive photocell and amplifier unit, high-low beam relay, sensitivity control, dimmer switch, flash-to-pass relay, and a wiring harness.

❑ The automatic on/off with time delay has two functions: to turn on the headlights automatically when ambient light decreases to a predetermined level and to allow the headlights to remain on for a certain amount of time after the vehicle has been turned off.

❑ Fiber optics is the transmission of light through polymethylmethacrylate plastic that keeps the light rays parallel even if there are extreme bends in the plastic.

❑ The lamp outage indicator alerts the driver, through an information center on the dash or console, that a light bulb has burned out.

❑ Digital instrument clusters use digital and linear displays to notify the driver of monitored system conditions.

❑ The most common types of displays used on electronic instrument panels are light emitting diodes (LEDs), liquid crystal displays (LCDs), vacuum fluorescent displays (VFDs), and a cathode ray tube (CRT).

❑ Computer-driven quartz swing needle displays are similar in design to the air core electromagnetic gauges used in conventional analog instrument panels.

❑ A head-up display system displays visual images onto the inside of the windshield in the driver's field of vision.

❑ A voice synthesizer uses a computer-controlled phoneme generator that is capable of reproducing the phonemes used for basic speech.

Terms-To-Know

Automatic headlight dimming	Illuminated entry actuator	Pinion factor
Automatic on/off with time delay	Illuminated entry system	Polarizers
Buffer	Instrument panel dimming	Polymethylmethacrylate plastic
Cathode ray tube (CRT)	Lamp outage indicator	Quartz swing needle display
Daytime running lamps (DRL)	Lamp outage module	Sensitivity control
Digital instrument clusters	Liquid crystal displays (LCDs)	Timer control
Fiber optics	Matrix	Vacuum fluorescent display (VFD)
H-gate	Nematic fluid	Voice synthesizer
Head-up display (HUD)	Oscillator	Wake-up signal
Holographic combiner	Photocell	Watchdog circuit

Review Questions

Short Answer Essays

1. Describe the operation of computer-controlled concealed headlight systems.

2. List the common components of the automatic headlight dimming system.

3. Explain the operation of body computer-controlled instrument panel illumination dimming.

4. What is the function of the sensitivity control in the automatic dimmer system?

5. What is the basic operation of the illuminated entry system?

6. Describe the operating principles of the digital speedometer.

7. Explain the operation of IC chip-type odometers.

8. Describe the operation of the electronic fuel gauge.

9. Describe the operation of quartz analog speedometers.

10. What is meant by pulse width dimming?

Fill-in-the-Blanks

1. With body computer-controlled concealed headlights, the computer receives inputs from the _____ and _____ switches.

2. The sensitivity control used with automatic dimming sets the sensitivity at which the photocell and amplifier are _____ .

3. The photocell will have _____ resistance as the ambient light level increases.

4. In some illuminated entry systems, _____ _____ signals the body computer that the courtesy lights are not required.

5. The body computer uses inputs from the _____ _____ _____ and _____ to determine the illumination level of the instrument panel lights.

6. The body computer dims the illumination lamps by using a _____ _____ _____ signal to the panel lights.

7. Digital instrument clusters use _____ and _____ displays to notify the driver of monitored system conditions.

8. Some digital instrument panel modules also use an _____ _____ sensor in addition to the rheostat.

9. Most digital fuel gauges use a fuel level sender that _____ resistance value as the fuel level decreases.

10. Computer-driven quartz swing needle displays are similar in design to the _____ _____ electromagnetic gauges used in conventional analog instrument panels.

ASE Style Review Questions

1. The sensitivity control of the automatic headlight dimming system is being discussed. Technician A says decreasing the sensitivity means the headlights will switch to the low beams when the approaching vehicle is farther away. Technician B says increasing the sensitivity means the headlights will switch to the low beams when the approaching vehicle is closer. Who is correct?
   a. A only
   b. B only
   c. Both A and B
   d. Neither A nor B

2. Illuminated entry is being discussed. Technician A says when either of the front door handles are lifted a switch in the handle will close the ground path from the actuator. Technician B says if the ignition switch is placed in the RUN position before the timer circuit turns off the interior lights, the timer sequence is shut off and the interior lights turn off. Who is correct?
   a. A only
   b. B only
   c. Both A and B
   d. Neither A nor B

3. Computer-controlled instrument panel dimming is being discussed. Technician A says the body computer dims the illumination lamps by varying resistance through a rheostat that is wired in series to the lights. Technician B says the body computer can use inputs from the panel dimming control and photocell to determine the illumination level of the instrument panel lights on certain systems. Who is correct?
   a. A only
   b. B only
   c. Both A and B
   d. Neither A nor B

4. Fiber optic applications are being discussed. Technician A says fiber optics is the transmission of light through several plastic strands that are sheathed by a polymer. Technician B says fiber optics are used only in external lighting applications. Who is correct?
   a. A only
   b. B only
   c. Both A and B
   d. Neither A nor B

5. Computer-driven instrumentation is being discussed. Technician A says a computer-driven instrument panel uses a microprocessor to process information from various sensors and to control the gauge display. Technician B says some manufacturers use the body computer to perform all functions. Who is correct?
   a. A only
   b. B only
   c. Both A and B
   d. Neither A nor B

6. The IC chip odometer is being discussed. Technician A says if the chip fails, some manufacturers provide for replacement of the chip. Technician B says depending on the manufacturer, the new chip may be programmed to display the last odometer reading. Who is correct?
   a. A only
   b. B only
   c. Both A and B
   d. Neither A nor B

7. Computer-driven quartz swing needle displays are being discussed. Technician A says the "A" coil is connected to system voltage and the "B" coil receives a voltage that is proportional to input frequency. Technician B says the quartz swing needle display is similar to air core electromagnetic gauges. Who is correct?
   a. A only
   b. B only
   c. Both A and B
   d. Neither A nor B

8. Technician A says digital instrumentation displays an average of the readings received from the sensor. Technician B says conventional analog instrumentation gives more accurate readings but is not as decorative. Who is correct?
   a. A only
   b. B only
   c. Both A and B
   d. Neither A nor B

9. The microprocessor-initiated self-check of the electrical instrument cluster is being discussed. Technician A says during the first portion of the self-test all segments of the speedometer display are lit. Technician B says the display should not go blank during any part of the self-test. Who is correct?
   a. A only
   b. B only
   c. Both A and B
   d. Neither A nor B

10. Technician A says bar graph-style gauges do not provide for self-tests. Technician B says the digital instrument panel will display "CO" to indicate the circuit is shorted. Who is correct?
    a. A only
    b. B only
    c. Both A and B
    d. Neither A nor B

30 Advanced Lighting Systems and Electronic Instrumentation Diagnosis and Repair

Objective

Upon completion and review of this chapter, you should be able to:

❑ Diagnose computer-controlled concealed headlight systems.

❑ Perform a functional test of the automatic headlight system.

❑ Diagnose the automatic headlight system.

❑ Test the automatic headlight system's photocell.

❑ Replace the photocell assembly.

❑ Test the automatic headlight system's amplifier.

❑ Diagnose Ford's illuminated entry system as an example of control module systems.

❑ Diagnose Chrysler's illuminated entry system as an example of BCM-controlled systems.

❑ Diagnose fiber optic systems.

❑ Enter diagnostic mode and retrieve trouble codes from BCM-controlled electronic instrument clusters.

❑ Perform self-diagnostic tests on the electronic instrumentation system.

❑ Determine faults as indicated by the self-test.

❑ Diagnose speedometer and odometer malfunctions.

❑ Test magnetic pickup speed sensors.

❑ Test optical-type speed sensors.

❑ Determine the cause of constant low gauge readings.

❑ Diagnose and locate the cause for constantly high gauge readings.

Introduction

Diagnosis of computer-controlled lighting systems is designed to be as easy as possible. The controller may provide trouble codes to assist the technician in diagnosis. Most manufacturers provide a detailed diagnostic chart for the most common symptoms. The most important thing to remember when diagnosing these systems is to follow the diagnostic procedures in order. Do not attempt to get ahead of the chart by assuming the outcome of a test. This will lead to replacement of good parts and lost time.

In this chapter you will perform selected service samples on the computer-controlled concealed headlight, automatic headlight, and illuminated entry systems. It is out of the scope of this manual to provide service procedures for the different manufacturers that use these systems. Technicians must follow the procedure in the service manual for the vehicle they are diagnosing.

The second section of this chapter covers diagnosis and service of electronic instrumentation.

Computer-Controlled Concealed Headlight Diagnosis

Customer concerns associated with the concealed headlight system may include:

1. Headlight doors will not open.

2. Headlight doors will not close.

3. Headlight doors will not open for flash-to-pass.

4. Headlight doors do not operate by headlight switch.

5. Headlights do not turn off.

The illustration (Figure 30-1) is a schematic of the concealed headlight system used on some Chrysler vehicles. Study the schematic of the system you are diagnosing until you understand how it is supposed to operate. Confirm the complaint by trying to operate the system. Once the complaint is confirmed, make a complete visual inspection of the system. Look for loose connections, broken wires, damaged components, and so on. If the visual inspection does not reveal an obvious fault, refer to the service manual for the specific tests to be performed as determined by the symptoms.

The following is a diagnostic service sample of a customer concern that the headlight doors will not open. Always refer to the correct service manual procedures for the vehicle you are diagnosing.

Connect the scan tool to the diagnostic connector then turn the ignition switch to the RUN position. Follow the menus on the scan tool screen to access the body controller. Next select the HEADLIGHT DOOR TEST option.

The headlight doors should open and close as the scan tool operates the system. If the doors do not operate, go to step 1. If the doors operate, begin at step 4.

CAUTION *Do not place your hands close to the headlight doors when the test is activated.*

1. Check the relay fuse. A blown fuse indicates a possible problem in the circuit to the relay module and door motors. If the fuse is blown, go to step 2. If the fuse is not blown, go to step 3.

2. Disconnect the headlight door relay module (Figure 30-2). Use an ohmmeter to measure the resistance between the connector red wire and ground (Figure 30-3). The resistance should be higher than 5 ohms. If it is not, there is a short to ground in the red wire. If the resistance is above 5 ohms, disconnect the headlight

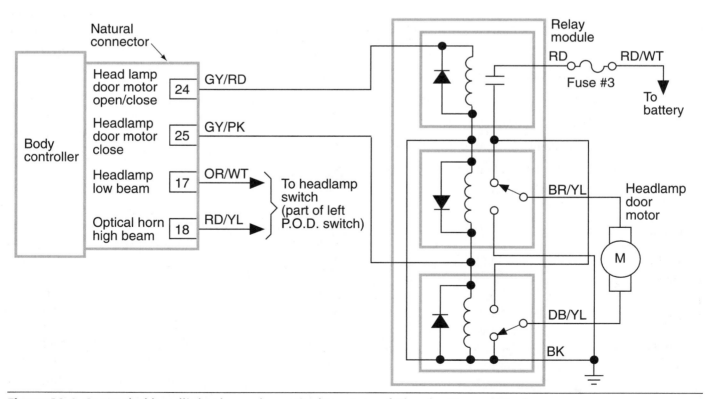

Figure 30-1 Concealed headlight door schematic. (Courtesy of Chrysler Corporation)

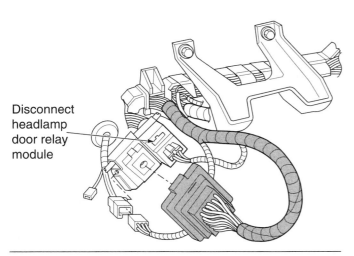

Figure 30-2 Headlight door relay module location. (Courtesy of Chrysler Corporation)

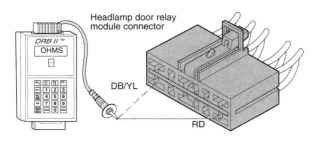

Figure 30-3 Door relay module connection test locations. (Courtesy of Chrysler Corporation)

door connector and measure the resistance between the relay module connector DB/YL wire.

A. If the resistance is lower than 5 ohms, the DB/YL wire has a short to ground between the relay module and the door motor.

B. If it is higher than 5 ohms, reconnect the relay module. Use a DMM to test for voltage across the headlight door motor connector (Figure 30-4). Voltage will be present only for a few seconds when the headlight switch is turned on. If the voltage is 10

volts or higher, replace the door motor. If it is less than 10 volts, replace the door relay module.

3. Use a voltmeter to test for applied voltage to the fuse. No voltage at this point indicates there is an open or short in the circuit between the battery and the fuse box. If there is over 10 volts applied to the fuse box, disconnect the headlight door motor connector. Reinstall the fuse and measure the voltage across the connector terminals when the headlight switch is turned on. The voltage will only be present for a few seconds.

A. If the voltage is higher than 10 volts, replace the motor if a good ground is confirmed.

B. If the voltage is less than 10 volts, turn off the headlights and back probe the BR/YL wire at the relay module using the voltmeter (Figure 30-5). Voltage should be present for a few seconds when the headlight switch is turned on.

   a. If the voltage is higher than 10 volts, there is an open in the BR/YL wire between the relay module and the motor.

   b. If the voltage is less than 10 volts, backprobe the RD wire at the relay module connection.

      (1) If the voltage is lower than 10 volts, there is an open in the RD wire between the relay module and the relay fuse.

      (2) If the voltage is over 10 volts, turn the ignition switch off and use an ohmmeter to test the resistance at the relay module connector BK wire (Figure 30-6). A resistance reading higher than 5 ohms means there is an open in the BK wire from the connector to ground. If it is lower than 5 ohms, connect a jumper wire from the battery positive terminal to the relay module GY/RD wire. The doors should open. If not, replace the relay module. If the doors open, use a jumper

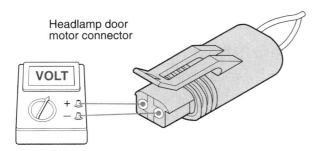

Figure 30-4 Connect a voltmeter across the door motor connector. (Courtesy of Chrysler Corporation)

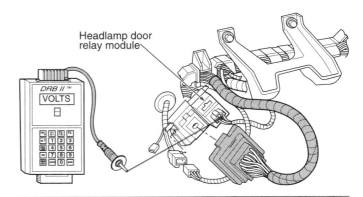

Figure 30-5 Backprobing the door relay module for voltage with the scan tool in voltmeter mode. (Courtesy of Chrysler Corporation)

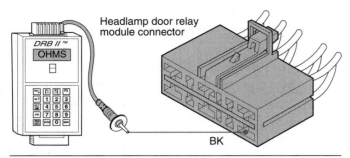

Figure 30-6 Using the scan tool ohmmeter function to test the resistance in the circuit to the door relay module connector. (Courtesy of Chrysler Corporation)

wire to the body controller cavity 24 and ground (Figure 30-7). If the doors open, replace the BCM. If they do not open, there is an open or short in the circuit between the BCM and relay module.

4. If the doors operated properly when the scan tool performed the headlight door test, use the scan tool to access the switch test mode and select headlight switch from the menu.

5. While observing the scan tool, turn on the headlight switch. If the scan tool indicates the circuit is closed, replace the BCM.

6. If the scan tool displays the circuit is open, disconnect the left POD switch connector. Use a jumper wire connected between the OR/WT wire terminal and ground.

7. Observe the scan tool. If it reads that the circuit is open, there is an open in the OR/WT wire between the POD and the BCM terminal 18. If a closed circuit is indicated, use an ohmmeter to measure the resistance to ground at the POD switch connector BK/OR wire.

A. If resistance is lower than 5 ohms, replace the POD switch.

B. A reading higher than 5 ohms indicates an open in the circuit to ground.

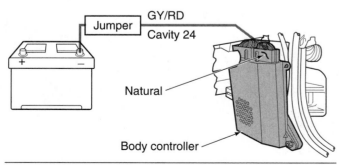

Figure 30-7 Using a jumper wire to ground circuit cavity 24 to the BCM. (Courtesy of Chrysler Corporation)

Automatic Headlight System Diagnosis

If the headlights do not turn on, check the regular headlight system first before condemning the automatic headlight system. If the headlights do not illuminate when the automatic system is turned off, the problem is in the basic circuit.

To perform a **functional test** of the automatic headlight system, activate the automatic headlight system. Cover the photocell and turn the ignition switch to the RUN position. The headlights should turn on within 10 seconds. Remove the cover from the photocell and shine a bright light onto it. The lights should turn off after 10 seconds but within 60 seconds. Cover the photocell again. When the lights turn on, wait 15 seconds. Then turn off the ignition switch. The headlights should turn off after the selected amount of time delay.

Once it is determined the fault is within the automatic headlight system, make a few quick checks of the system. Inspect the photocell lens for obstructions. Check all connections from the headlight switch, as well as all fuses used in the system.

The following are some of the most common concerns resulting from problems in the automatic headlight:

1. Lights turning on and off at wrong light levels.

2. Lights that do not turn on in darkness.

3. Lights not turning off in bright light.

4. Lights not staying on for an adjustable time after the ignition switch is turned off.

5. Lights do not turn off after the ignition switch is turned off.

These problems can be caused by faults in the headlight switch, ignition switch, amplifier, or photocell. To locate the problem, perform the following test series. As a service sampler of automatic headlight system diagnosis, refer to the schematic (Figure 30-8) of a GM Twilight Sentinel System. The following test connections will refer to those used by this system. Use the correct service manual for the system you are diagnosing.

Photocell Test

To perform a **photocell test**:

1. Separate the photocell connector.

2. Turn the ignition switch to the RUN position.

3. Turn the Twilight Sentinel control to ON.

4. Turn the headlight switch OFF.

5. If the lights turn on within 60 seconds, replace the photocell.

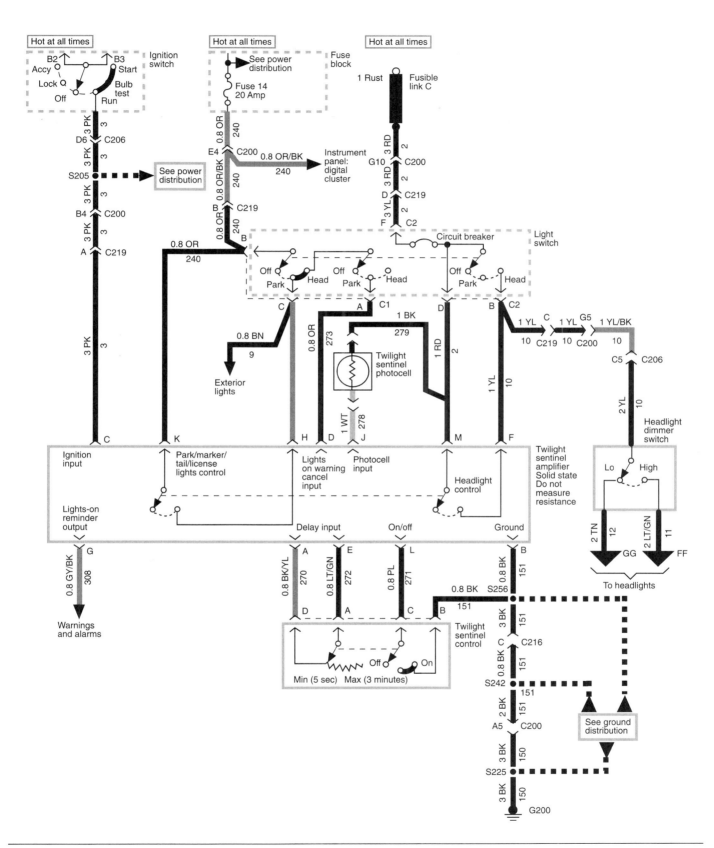

Figure 30-8 General Motors' Twilight Sentinel schematic. (Courtesy of General Motors Corporation, Service Technology Group)

6. If the lights do not turn on within 60 seconds, test the amplifier.

Photo Sequence 14 illustrates the procedure for replacing the photocell.

Amplifier Test

To perform the **amplifier test**:

1. Disconnect the wire connector to the amplifier.

2. Turn the headlight switch to the OFF position.

3. Disconnect the negative battery terminal.

4. Turn the control switch to the ON position.

5. Use an ohmmeter to measure the resistance between the wire connector terminal L and ground. There should be zero ohms of resistance. If there is more than zero ohms of resistance, check the wire from amplifier terminal L to control switch terminal C for opens. Also check circuit 151 from amplifier terminal B for opens. If the ohmmeter indicated zero ohms of resistance, continue testing.

6. Turn the control switch to the OFF position.

7. The ohmmeter should read infinite when connected between the L terminal and ground. If there is low resistance, check the control switch and circuit 271 for a short. If the ohmmeter indicated infinite resistance, continue testing.

8. Connect the ohmmeter between terminals H and D. There should be zero ohms of resistance. If not, check circuits 273 and 274 for an open. If the ohmmeter indicated zero ohms of resistance, continue testing.

9. Place the control switch in the middle of MIN and MAX settings.

10. Connect the ohmmeter between terminals A and E. There should be between 500 and 250,000 ohms of resistance. If the resistance value is not within this range, check circuits 270 and 272 between the amplifier and the control switch for opens.

    A. If the circuits are good, observe the ohmmeter as the control switch is moved from the MIN position to the MAX position. The resistance value should change smoothly and consistently from one position to the next. If the ohmmeter indicates a resistance value out of limits or an erratic reading, replace the control switch.

11. If all of the resistance values tested are correct, continue testing.

12. Turn off all switches and reconnect the amplifier wire connector. Reconnect the battery negative terminal.

13. Turn the ignition switch to the RUN position. Turn the headlight switch to the OFF position and place the control switch to ON.

14. Connect a fused jumper wire between amplifier terminals J and M. The headlights should go off within 60 seconds. If the headlights go off within 60 seconds, yet they do not go off normally in bright light, check circuits 278 and 279 for opens. If these circuits are good, replace the photocell.

15. Turn off all switches.

16. Disconnect the wire connector to the amplifier.

17. Turn the ignition switch to RUN and turn the headlight switch to OFF.

18. Connect a test light between terminal M and ground. The test light should light. If not, there is a short or open in circuit 2 between the battery and the amplifier or in circuit 279 between the amplifier and the photocell. If the test light illuminates, continue testing.

19. Connect the test light between terminals M and B. The test light should light. If not, check circuit 151 for an open. If the test light illuminates, continue testing.

20. Connect the test light between terminal K and ground. The test light should light. If not, check the fuse and circuit 240 for an open. If the test light illuminates, continue testing.

21. Connect the test light between terminal C and ground. The test light should light. If not, check circuit 3 for an open. If this circuit is good, check the ignition switch. If the test light illuminates, continue testing.

22. Turn the ignition switch OFF.

23. With the test light connected as in step 21, the test light should turn off. If not, replace the ignition switch. If the test light turns off, continue testing.

24. Connect the test light between terminal H and ground. If the test light is illuminated, replace the light switch. If the test light is off, continue testing.

25. Connect the test light between terminal F and ground. If the test light is illuminated, replace the light switch. If the test light remains off, continue testing.

26. With the test light connected as in step 25, place the headlight switch in the HEAD position. If the test light does not turn on, check circuit 10 and the headlight switch for opens. If the test light lights, and all other tests have the correct results, replace the amplifier.

WARNING *Skipping any of the tests will result in replacement of the amplifier, even if it is good.*

PHOTO SEQUENCE 11 Typical Procedure for Replacing a Photocell Assembly

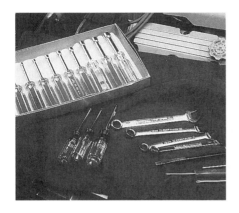

P11-1 Tools required to replace the photocell assembly include nut driver set, thin flat blade screw driver, phillips screw driver set, battery terminal pullers, battery pliers, box end wrenches, safety glasses, seat covers, and fender covers.

P11-2 Place the fender covers over the vehicle's fenders and disconnect the battery negative terminal.

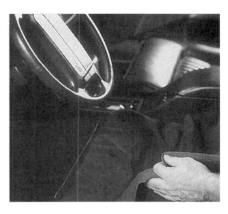

P11-3 Protect the seats by placing the seat covers over them.

P11-4 Remove the grill that covers the photocell.

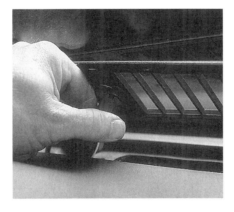

P11-5 Twist the socket to free the photocell from the grill.

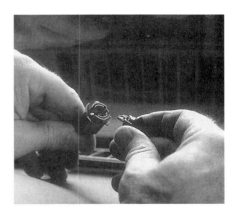

P11-6 Gently work the photocell from its socket. Replace the photocell and replace the socket. Next, re-install the grill.

Resistance Assembly Test

The **resistance assembly test** is performed when the lights turn on and off at the wrong light levels. This test requires the construction of a **photocell resistance assembly** (Figure 30-9). This assembly is a test tool that replaces the photocell to produce predictable results.

Replace the photocell with the resistance assembly. Turn the resistance assembly switch to the OFF (open) position and place the ignition switch in the RUN position. If the lights do not come on within 60 seconds, replace the amplifier.

If the lights turn on, wait 30 seconds and turn the resistance assembly switch on. If the lights turn off within 60 seconds, replace the photocell. If the lights do not turn off within 60 seconds, replace the amplifier.

Illuminated Entry System Diagnosis

The diagnostic procedures for testing the illuminated entry system are different depending upon manufacturer. Always refer to the service manual for the vehicle you are working on. Always perform a visual inspection before performing any tests. First check the fuse. Then check to make sure all connections are tight and clean. Inspect the ground wires for good connections. Check all visible wires for fraying or damaged insulation, especially where they go through body parts. Make sure all doors are closed properly and the headlight switch is in the detent position.

The following are typical procedures for diagnosing the illuminated entry system.

Ford Illuminated Entry Diagnosis

Activation of the system is by lifting the outside door handle. This momentarily closes a switch that completes the ground circuit of the actuator module. The module activates the interior lights for 25 seconds or until the ignition switch is placed in the RUN or ACC position.

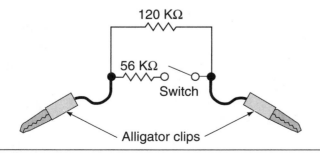

120 KΩ

56 KΩ

Switch

Alligator clips

Figure 30-9 Photocell resistance assembly.

Ford has built in a logic circuit to prevent battery drain if the door handle is held up for longer than 25 seconds. The system will operate as normal until the 25 seconds have elapsed. The module will then turn off the lights. The lights will remain off and cannot be reactivated until the handle is returned to the released position.

The system has four main components (Figure 30-10). The door lock cylinder uses an LED to provide the illumination of the cylinder. The lens of the LED is built into the cylinder.

WARNING *The normal operating voltage for the LED is 3 volts. The circuit uses a dropping resistor to protect the LED. Do not apply 12 volts to the LED circuit ahead of the resistor. If the resistor is bypassed, the LED will be destroyed. When applying voltage to test the LED, apply it only to the connector terminals.*

To test the system, disconnect the actuator harness from the actuator. Connect the test light between terminal 8 and ground (Figure 30-11). The test light should illuminate with the ignition switch in the OFF or RUN position. If the test light fails to come on, trace circuit 54 back to the fuse box to locate the problem.

NOTICE: On some systems, terminal 4 is used. On these systems jumping between terminals 6 and 8 will illuminate the courtesy lights. Jumping between terminals 4 and 8 will light the door lock cylinders only. If they do not, trace circuit 464.

Connect the test light between terminal 7 and ground to test circuit 296. The test light should not glow with the ignition switch in the OFF position. When the ignition switch is turned to the RUN or ACC position, the test light should come on. If the test light does not turn on and off as the ignition switch is turned, trace circuit 296 to the fuse box to locate the problem.

Connect a jumper wire between connector terminals 6 and 8 (circuits 54 and 53). Make sure all doors are closed. The courtesy lights and door lock cylinders should be illuminated. If the lights did not operate, trace circuit 53 to locate the problem.

Connect an ohmmeter between connector terminal 2 (circuit 465) and ground. The ohmmeter should indicate an infinite reading. However, a minimum of 10,000 ohms is acceptable. Lift up on each of the outside door handles to close the latch switch. Hold the handle up while observing the ohmmeter. The ohmmeter should indicate a resistance value of 50 ohms maximum. If either of the ohmmeter readings are out of specifications,

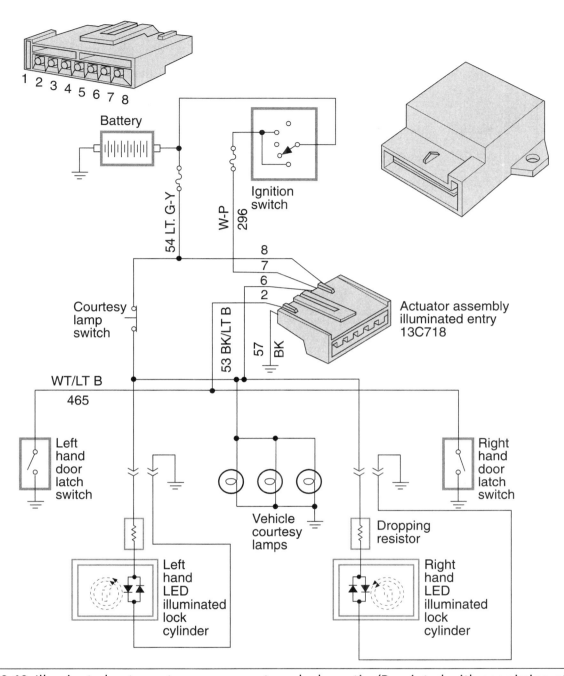

Figure 30-10 Illuminated entry system components and schematic. (Reprinted with permission of Ford Motor Company)

trace circuit 465 to the latch switch. Also test the latch switches for correct operation.

With the test light connected between connector terminals 1 and 8, the light should be on (Figure 30-12). If the test light fails to come on, trace circuit 57 to ground.

If the preceding tests did not indicate any problems, the actuator module assembly is faulty. The module must be replaced.

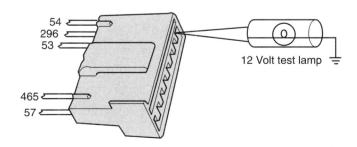

Figure 30-11 Testing circuit 54 for opens. (Reprinted with permission of Ford Motor Company)

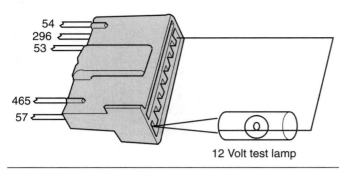

Figure 30-12 Testing ground circuit 57 with a test light. (Reprinted with permission of Ford Motor Company)

Chrysler's BCM-Controlled Illuminated Entry Diagnosis

Refer to the schematic (Figure 30-13) of a Chrysler illuminated entry system. Follow these steps to diagnose the system:

1. Move the dimmer control to the center position.

2. Open the driver's side door to activate the courtesy lights. If none of the courtesy lights turn on, continue testing. If only one bulb is inoperative, check its circuit and the bulb.

3. Lower the driver side window and close all doors. Manually lock the driver's door. Wait 30 seconds with the ignition switch off.

4. Activate the illuminated system by lifting the driver's door handle. If the lights come on, repeat the test for the right side door. If the system does not operate when the right side door handle switch is closed, refer to the service manual for the circuits to be tested. The procedure will be the same as when the left door is inoperative. However, the circuit designations are different.

5. If the lights do not turn on when the door handles are lifted, connect the scan tool to the diagnostic connector then turn the ignition switch to RUN. Cycle through the menus to get select BODY in the system menu. Then select INPUTS/OUTPUTS from the menu. Scroll through the list until door handle switch statue is displayed.

6. With the ignition switch on, observe the scan tool while lifting the door handle. The display should indicate the switch is closed when the handle was lifted. If the switch operated correctly, go to step 7. If the display did not indicate proper switch operation, connect a jumper wire from BCM terminal 7 to ground. Observe the scan tool. If it indicates the circuit is closed, follow

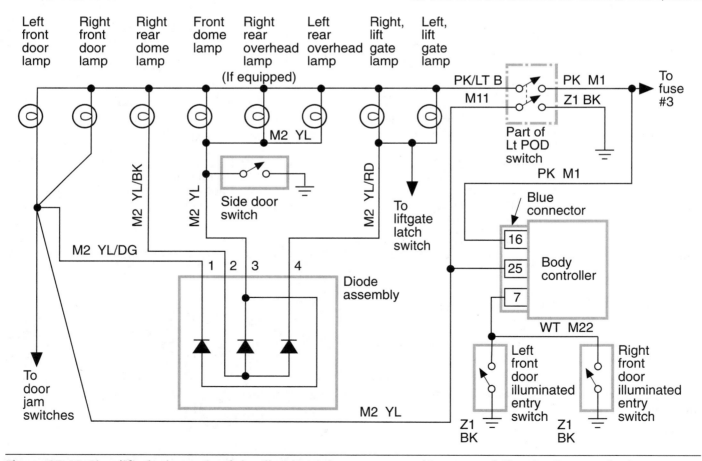

Figure 30-13 Simplified schematic of the illuminated entry system. (Courtesy of Chrysler Corporation)

the service manual procedure for testing the door switch. If the display indicates the circuit is open, replace the BCM.

WARNING *The terminal to jump from the controller is different between years and models. Be sure to refer to the service manual for the correct terminals.*

7. Open the driver's side door. If the courtesy lights do not turn on, go to step 8. Close all doors and jumper the terminals of the body computer as identified in the service manual. If the lights do not turn on, there is an open circuit between the controller and the lamps. If the lights turned on when the BCM was jumped, read the scan tool display. If it reads the door handle switch is open, replace the BCM. If it reads the switch is closed, follow the service manual procedure for testing the switch.

8. With the ignition switch in the OFF position, close all doors and wait 60 seconds. After the waiting period, turn the ignition switch to RUN while observing the scan tool display. If the display reads the circuit is closed, check the system fuse for applied voltage. If good, repair the wiring from the fuse to the lamps. If the fuse is blown, locate the circuit defect and repair.

9. In step 8, if the scan tool reads the circuit is closed, gain access to the driver's side door switch harness and disconnect it (Figure 30-14). Observe the scan tool display. If it reads circuit open, replace the door switch. If the display does not show the circuit as being open, gain access to the passenger's side door switch connector and disconnect it. Again, observe the scan tool display. If it displays the circuit is open, replace the

passenger's door switch. If the scan tool displays the circuit is closed, there is a short to ground from the door switches to the controller or the BCM is faulty.

Fiber Optics Diagnosis

If the fiber optics do not illuminate, most likely the light source has failed. Check the operation of the bulb and its circuit. If the bulb illuminates, check that the fiber optic lead is connected to the light source and to the lens. If these are good, the only other cause is that the cable is cut. It will need to be replaced.

Diagnosing and Servicing Electronic Instrumentation

This section introduces some of the service procedures used to diagnose and repair computer-driven instrumentation systems. These systems include the speedometer, odometer, fuel, and engine instrumentation. It is out of the scope of any textbook to cover service procedures for every type of electronic instrument cluster. To illustrate, in one model year alone Chrysler offered two different electronic clusters (Huntsville and Motorola). The Huntsville cluster was available in four different variations: one with a message center, one with a trip computer, and two options with tachometers. In addition, the system could also have a 24-voice alert function. In the same year, Ford offered five different electronic clusters, and each division of General Motors offered its own electronic instrument cluster. Add to this the many different types of import vehicles that use their own systems.

This section provides a familiarization with general procedures. However, it is important to remember that each system uses its own diagnostic procedures.

Usually the technician will be required to isolate the faulty component and replace it. Most instrument panel components are not repaired or serviced in the shop. They are sent back to the manufacturer or specialty shop for rebuilding.

BCM Diagnostics

In some systems the BCM is capable of running diagnostic checks of the electronic instrument cluster to determine if a fault is present. If the values received from monitored functions are outside of programmed parameters, a trouble code is set. This code can be retrieved by the technician to aid in troubleshooting. Depending upon

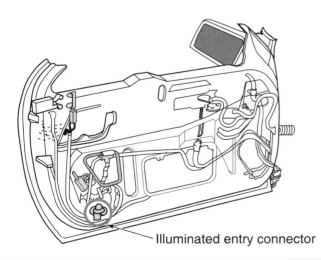

Illuminated entry connector

Figure 30-14 Door switch connector. (Courtesy of Chrysler Corporation)

the vehicle, code retrieval is done through the ECC, IPC, jumping terminals in the DLC, or by a scan tool.

To retrieve diagnostic codes in most General Motors vehicles with an electronic climate control (ECC) display, turn the ignition switch to the RUN position and simultaneously press the OFF and WARMER buttons on the climate control panel. ECM codes will be displayed first, followed by BCM codes. The system will display codes twice. The first pass are all codes in memory. Codes included in the first set but not in the second are history codes. All codes displayed during the second pass are current codes.

Record all trouble codes displayed. If a vehicle speed sensor code is present, perform the steps listed in the diagnostic chart to determine if the ECM, BCM, or speed sensor is at fault (Figure 30-15).

After the trouble codes have been retrieved from memory, refer to the proper diagnostic chart to isolate the fault. It is possible for a problem to exist that does not set a trouble code. In these instances, use the symptom or troubleshooting chart in the service manual to locate the fault (Figure 30-16).

Some manufacturers provide a means of overriding the instrument cluster display and change the parameters to allow for testing. By changing the parameters in this test mode, the gauge will change its indicated reading. If the gauge changes its readings correctly, the fault is in the control module.

Self-Diagnostics

Most instrument panel display modules have a **diagnostic mode** within their programming. The diagnostic mode allows the module to isolate any faults within the instrument panel cluster. In most systems, if the module is not able to complete its self-diagnostic test, the fault is within the module and it must be replaced. Successful completion of the self-diagnostic test indicates the problem is not the module. The following are examples of self-diagnostic procedures.

Diagnosis of a Typical Electronic Instrument Cluster

WARNING *VFD displays are easily damaged by physical shock. When handling EICs, do not drop or jar them.*

WARNING *When servicing EICs, follow all service precautions related to static discharge in the vehicle manufacturer's service manual to avoid EIC damage.*

All electronic instrument clusters (EICs) are sensitive to static electricity damage, and EIC cartons usually have a static electricity warning label. When servicing EICs:

1. Do not open the EIC carton until you are ready to install the component.
2. Ground the carton to a known good ground before opening the package.
3. Always touch a known good ground before handling the component.
4. Do not touch EIC terminals with your fingers.
5. Follow all service precautions and procedures in the vehicle manufacturer's service manual.

Prove-Out Display

Most electronic instrument clusters have a **prove-out display** each time the ignition switch is turned on. During this display, all the EIC segments are illuminated momentarily and then turned off momentarily (Figure 30-17). The EIC returns to normal displays after the prove-out display. If the EIC is not illuminated during the prove-out display, check the power supply and grounds to the EIC. If these are good, replace the EIC.

If, during the prove-out display, some of the segments do not illuminate, the EIC is defective and must be replaced. During the prove-out mode, the turn signal indicators and high beam indicator are not illuminated. Other indicator lights remain on when the EIC display is turned off momentarily in the prove-out mode. These indicator lights go out shortly after the EIC returns to normal displays after the prove-out mode is completed.

Function Diagnostic Mode

The diagnostic procedure for electronic instrument clusters varies depending on the vehicle make and model year. Always follow the diagnostic procedures in the vehicle manufacturer's service manual. Some EICs have a **function diagnostic mode** which provides diagnostic information in the display readings if certain defects occur in the system. If the coolant temperature sender has a shorted circuit (for example), the two top and bottom bars are illuminated in the temperature gauge, and the ISO symbol is extinguished (Figure 30-18). If the engine coolant never reaches normal operating temperature or the coolant temperature sender circuit has an open circuit, the bottom bar in the temperature gauge is illuminated with the ISO symbol.

If the fuel gauge sender develops a shorted or open circuit, the two top and bottom bars in the fuel gauge are illuminated, and the ISO symbol is not illuminated. A shorted fuel gauge sender causes CS to be displayed in the fuel remaining or distance to empty displays. If the

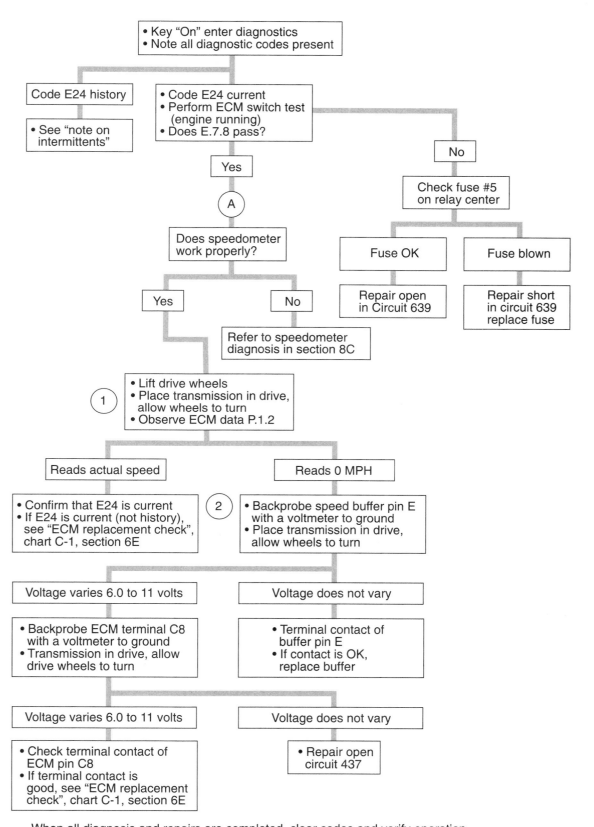

Figure 30-15 Diagnostic chart for speed sensor circuit failure. (Courtesy of General Motors Corporation, Service Technology Group)

SYMPTOM TABLE

SYMPTOM	ACTION
Entire display does not light up	See Test B. Check pins M, J, C and N
All segments light up constantly	Replace Digital Cluster (see section 8C)
Display has only partial segments lit	Replace Digital Cluster (see section 8C)
Display does not change from English to metric	Replace Digital Cluster (see section 8C)
Display does not change brightness when dimming rheostat is turned	See Test B. Check pins G and H
Speedometer does not operate, Odometers OK	Replace Instrument Cluster (see section 8C)
An Odometer does not operate, Speedometer OK	Replace Instrument Cluster (see section 8C)
Speedometer and/or Odometers are inaccurate	Do Test C
Speedometer and both Odometers do not operate	Do Test C
Trip Odometer does not reset to zero	Replace Digital Cluster (see section 8C)
Fuel Data Center does not work properly	See Instrument Panel: Standard Cluster, Section 8A-80
SERVICE VEHICLE SOON Indicator inoperative	See Section 8D
SERVICE ENGINE SOON Indicator inoperative	See Section 8D
COOLANT TEMP FAN Indicator inoperative	See Section 8D
STOP ENGINE TEMP Indicator inoperative	See Warnings and Alarms, section 8A-75
BRAKE Indicator inoperative	See Brake Warning System, section 8A-41
Turn Indicators inoperative	See Exterior Lights, section 8A-110
NO CHARGE Indicator lights	See Starter and Charging System, section 8A-110
SECURITY Indicator inoperative	See Theft Deterrent System, section 8A-133
ANTI-LOCK Indicator inoperative	See Antilock Brake System, section 8A-44
LOW WASHER FLUID Indicator inoperative	See Wiper/Washer, section 8A-90 or 8A-91
FASTEN BELTS Indicator inoperative	See Warnings and Alarms, section 8A-75
CAR IS LEVELING Indicator inoperative	See Electronic Level Control, section 8A-42
SERVICE AIR CONDITIONER Indicator inoperative	See Instrument Panel: Standard Cluster section 8A-80
TRUNK WARNING Indicator inoperative	See Trunk Pull-Down, section 8A-135
HI BEAM Indicator inoperative	See Headlights, sections 8A-100 to 8A-103
STOP ENGINE OIL does not work properly	See Instrument Panel: Standard Cluster, section 8A-80

Figure 30-16 Symptom tables assist the technician in locating the fault when no trouble codes are present. (Courtesy of General Motors Corporation, Service Technology Group)

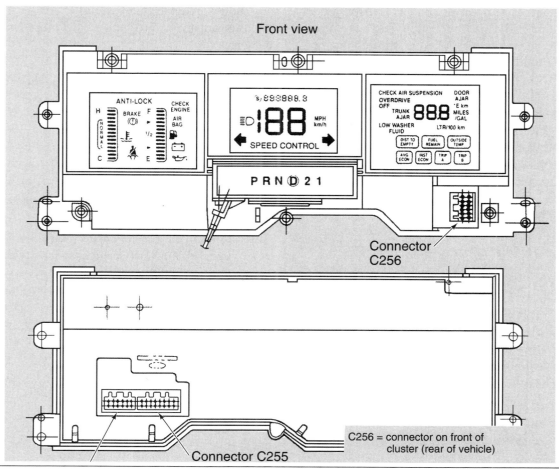

Figure 30-17 All EIC segments are illuminated during the prove-out display. (Reprinted with permission of Ford Motor Company)

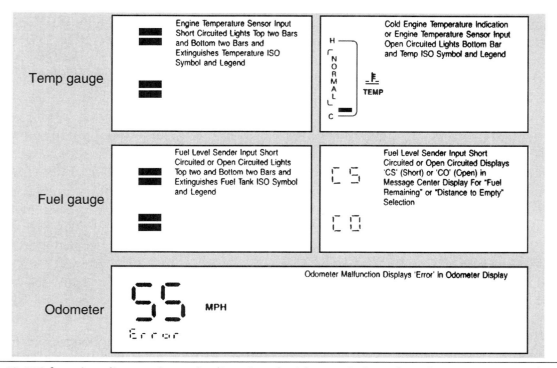

Figure 30-18 EIC function diagnostic mode. (Reprinted with permission of Ford Motor Company)

fuel gauge sender has an open circuit, CO is displayed in the fuel remaining and distance to empty displays. When the function diagnostic mode indicates shorted or open circuits in the inputs, the cause of the problem must be located by performing voltmeter and ohmmeter tests in the circuit with the indicated problem. These voltmeter and ohmmeter tests are included in the vehicle manufacturer's service information.

When the word ERROR appears in the odometer display, the EIC computer cannot read valid odometer information from the nonvolatile memory chip.

Special Test Mode

Most EICs have a **self-test** procedure to determine if the display is working properly. To enter the special test mode, press the E/M and SELECT buttons simultaneously, and turn the ignition switch from the off to the run position. When this action is completed, a number appears in the speedometer display, and two numbers are illuminated in the odometer display. The gauges and message center displays are not illuminated. If any of the numbers are flashing in the speedometer or odometer displays, the EIC is defective and must be replaced.

Diagnosis of a Typical Import Electronic Instrument Cluster

The **display check** tests for an open circuit in each display segment and shorts between the segments. Press and hold reset switch A, and turn the ignition switch from OFF to RUN to initiate the display check (Figure 30-19). After this action is taken, all the display segments should illuminate, one after the other. If any segment is not illuminated, the EIC must be replaced.

The **preprogrammed signal check** tests for defects in various displays. To complete the preprogrammed signal check:

1. Disconnect the negative battery cable. If the vehicle is equipped with an air bag, wait the specified time recommended by the vehicle manufacturer.

2. Remove the EIC power unit.

3. Remove the retaining nuts on the EIC switches and remove the EIC switches.

4. Remove the cluster lid, then remove the EIC assembly.

5. Connect the special self-checking wiring harness to the EIC terminals (Figure 30-20).

6. Connect the negative battery cable, turn the ignition switch to the RUN position, and observe the EIC displays. Each display should change to a specific reading (Figure 30-21). If each display changes as specified by the vehicle manufacturer, the EIC is satisfactory. When some of the displays do not change as specified, voltmeter and ohmmeter tests are required to locate the exact cause of the problem.

After completing the test procedure, turn off the ignition switch, and disconnect the negative battery cable. If the vehicle is equipped with an air bag, wait the length of time specified by the vehicle manufacturer. Disconnect the special self-checking wiring harness, and connect all EIC connectors securely. Complete reinstallation of the EIC, cluster lid, and switches.

CAUTION *If the odometer has been repaired or replaced, and the odometer cannot indicate the same mileage indicated before it was removed, the law in most areas requires that an odometer mileage label must be attached to the left front door frame. Failure to comply with this procedure could lead to court action.*

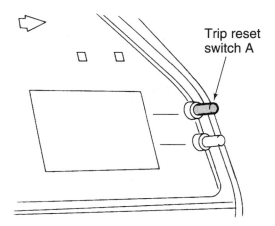

Figure 30-19 EIC reset button. (Courtesy of Nissan Corportion)

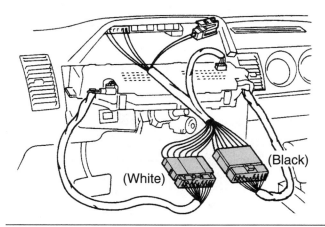

Figure 30-20 Special self-checking wiring harness connected to EIC terminal. (Courtesy of Nissan Corporation)

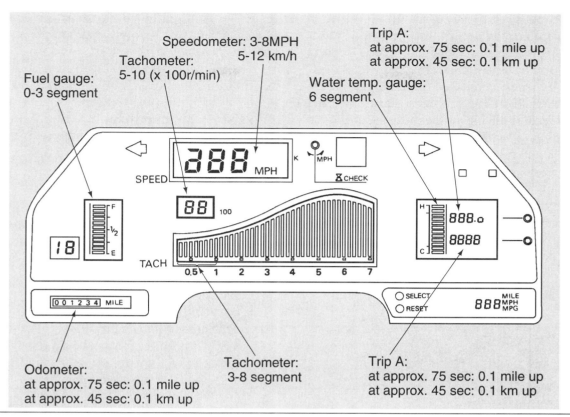

Figure 30-21 Display changes during the preprogrammed signal check. (Courtesy of Nissan Corporation)

A defective power unit may cause the EIC displays to be inoperative. The power unit supplies different voltages to various EIC displays. Therefore, it is possible for a defective power unit to cause the failure of specific EIC displays to illuminate.

To perform the **power unit test**, begin by removing the power unit and leaving the wiring harness connected to the unit (Figure 30-22). With the ignition switch on, test the voltage at the power unit terminals. Each power unit terminal should have the voltage specified by the vehicle manufacturer (Figure 30-23). The power unit ground wire is connected from terminal 9 to ground.

With the ignition switch in the OFF position, connect a pair of ohmmeter leads from power unit terminal 9 to ground. If the meter reading is above 0.5Ω, repair the ground wire. If the power unit does not have the specified voltage at some of the terminals, replace the unit.

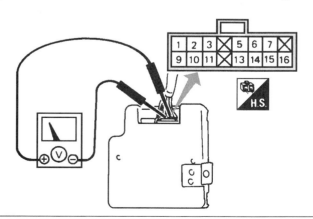

Figure 30-22 Power unit with wiring harness connected. (Courtesy of Nissan Corporation)

Voltmeter terminal		Voltage [V]	Remarks
⊕	⊖		
②	⑨	Approx. 12	Check when no display appears.
③		Approx. 0	
⑤		Approx. 22	
⑥		Approx. 26	
⑨	⑦	Approx. 23	
	⑬	Approx. 14	For speedometer, fuel, information, tachometer
	⑭		
	⑮	Approx. 19	For temp., trip
	⑯		

Figure 30-23 Voltage at various power unit terminals. (Courtesy of Nissan Corporation)

A defective speed sensor may cause an inoperative or erratic speedometer reading. To test the speed sensor begin by removing the cluster lid to gain access. Connect a pair of voltmeter leads to terminals 11 and 1 on the EIC with the wiring harness connected (Figure 30-24). Turn the ignition switch to the RUN position while observing the voltmeter reading. If the voltage is zero, check the power unit. If battery voltage is observed, turn off the ignition switch and disconnect the speedometer cable from the speed sensor. Remove the wiring harness connector containing terminals 1 and 12 from the EIC, and connect analog voltmeter leads to terminals 1 and 12. Use a small screwdriver to slowly rotate the speed sensor (Figure 30-25). If the voltmeter pointer does not deflect, the speed sensor or the connecting wires are defective.

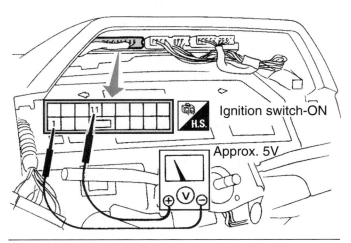

Figure 30-24 Voltmeter connections to EIC terminals. (Courtesy of Nissan Corporation)

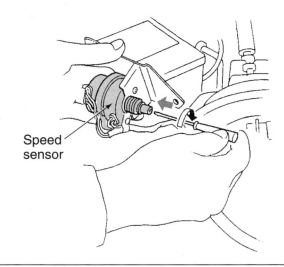

Figure 30-25 Testing the speed sensor. (Courtesy of Nissan Corporation)

SERVICE TIP *While testing the speed sensor, slowly turn the sensor with a small screwdriver. The sensor produces 24 signals per revolution, which are difficult to read during fast sensor rotation.*

Chrysler Electronic Instrument Cluster Diagnostics

The electronic cluster has two forms of diagnostic routines that will isolate faults within the cluster: two-button diagnostics and the scan tool.

If a scan tool is used, plug its cord into the diagnostic connector and follow the instructions to select electronic cluster from the "Select Module" menu. Follow the function flow chart as instructed. If the cluster illuminated when the ignition switch was placed in the RUN position, conduct the self-test.

If a scan tool is not available, **two-button diagnostics** can be used to initiate the self-test. Two-button diagnostics refers to the buttons on the instrument panel that, when pressed in the correct combination, place the module into self-test. Before performing the self-test, turn the ignition switch to the RUN position. The instrument cluster panel should illuminate. If not, check the fuses and wiring to the cluster. Also test for battery voltage to the cluster. If all tests are good, replace the cluster.

WARNING *Do not attempt to perform the self-test if the cluster did not illuminate or if any of the segments failed to light. Replace the cluster before continuing. Otherwise, incorrect diagnostic procedures and codes will result in replacement of good components.*

When entering the self-test mode, if either code 1, 2, or 3 appears and the word FAIL is in the odometer display, the cluster must be replaced before continuing. If a code 4 appears and the word FAIL is displayed in the odometer, replace the odometer chip. The chip can be located in the cluster or in the BCM, depending upon year of manufacture.

WARNING *Follow all state and federal regulations associated with odometer replacement.*

When the first parts of the self-test are complete, a 5 will appear in the speedometer display. These tests illuminate all of the digital displays. If unable to check all displays during this first test, repeat the illumination sequence by pressing the TRIP RESET button after a 5-second wait.

When satisfied that all segments are operating properly, continue to test six by pressing the U.S./METRIC button. This will illuminate all individual display segments in sequence. If any two segments in the same digit appear at the same time, or if more than one image appears in the gauge sequence at once, replace the cluster. A 6 will appear to indicate the test is completed. The test can be repeated by pressing the TRIP RESET button.

To begin test seven, press the U.S./METRIC button. The gauge scales should illuminate and sequence the speedometer and odometer. If this test fails, replace the cluster.

If test seven passes, press the U.S./METRIC button to begin test eight. On clusters having warning lights, the battery, temperature, and fuel lights should turn on in sequence, followed by an 8 in the speedometer window. If any or all of the lights fail to illuminate, check the bulbs. Replace any faulty bulbs and repeat the test. If the cluster still fails the test, replace it. On clusters not using warning lights, an 8 should appear immediately in the speedometer window. If the 8 does not appear, replace the cluster. An 8 indicates the test is completed and the cluster is in good operating condition. Next test the wiring and sensors.

Chrysler Electromechanical Cluster Self-Test

Beginning in 1993 many of Chrysler's electromechanical clusters (MIC) can be tested with or without a scan tool. Two-button diagnostics will activate a self-test of the cluster. This test does not check any of the inputs, only the cluster. As an example of this system a 1997 NS body mini-van is used. However, diagnostics of most MICs are similar.

To enter self-diagnostics the ignition key must be in the LOCK position. Push in and hold the TRIP and RESET buttons on the cluster at the same time. While continuing to hold these buttons, turn the ignition switch to the RUN position. Note that the cluster will illuminate in the UNLOCK position but it will not activate self-diagnosis. Continue to hold the two buttons until CODE is displayed in the odometer. Release the buttons. If there are any fault codes they will be displayed in the odometer. A code 999 means there are no faults. If fault codes are present, use the correct diagnostic manual to diagnose the system.

After the codes are displayed the cluster will go through a series of tests as follows:

Check 0. This tests all of the VF display segments in the odometer and PRND3L. All segments should be illuminated.

Check 1. This tests the operation of all gauges. The gauge swing needles will move to programmed values.

Check 2. Each odometer VF segment is illuminated individually.

Check 3. Tests the PRND3L display.

Check 4. Illuminates all of the warning lamps that are controlled by the MIC.

Observe operation of the MIC during each test. If any of these tests fail proper operation the MIC must be replaced.

On this cluster it is possible for the technician to calibrate the gauges. This requires the use of a DRBIII scan tool. The following steps provide a guide to recalibrating the speedometer. All gauges are calibrated in the same manner.

1. Plug the DRBIII cable into the DLC. The ignition switch does not need to be in the RUN position, however, the bus must be active. Opening a door will awaken the BCM.

2. Once the DRBIII powers up and displays the MAIN MENU select '94–'97 DIAGNOSTICS.

3. From the SELECT SYSTEMS menu select BODY.

4. Select ELECTRO/MECH CLUSTER from the BODY menu.

5. Select MISCELLANEOUS.

6. Select CALIBRATE GAUGES.

7. The DRBIII will then ask if the cluster has a tachometer. Answer with the YES or NO key.

8. The DRBIII will then ask if the vehicle is a diesel. Answer with the YES or NO key.

9. The DRBIII will then ask if the cluster units are in MPH. Answer with the YES or NO key.

10. Place the ignition switch into the UNLOCK position.

This will place the DRBIII in the mode to run gauge calibration. The first gauge to be calibrated will be the speedometer. The DRBIII screen will display that it is sending a signal to the MIC to set the mph at 0. If the needle is not aligned with the 0 then use the up or down arrow keys to move the needle until it is aligned. Once the gauge is calibrated to 0, press the enter key and the DRBIII will move to the next calibration unit. The next unit is at 20 mph. Follow the same procedure to align the needle with the 20 mph mark on the cluster. After each calibration press the enter key to move to the next unit. The other calibration units are at 55 and 75 mph.

The tachometer, fuel, and temperature gauges are calibrated in the same manner. Once all of the gauges are calibrated the DRBIII will instruct the MIC to write the new values to memory.

Ford Electronic Cluster Self-Diagnostic Test

The electronic cluster is capable of indicating a fault and providing an explanation of the cause. Use the illustration (Figure 30-26) as a guide to the function of the gauges when in diagnostic mode. Use the gauge display to determine the nature of the fault. Then refer to the service manual for diagnostic charts to locate the problem.

Electronic Speedometers and Odometers

Vehicles equipped with conventional electromagnetic gauges and a swing quartz speedometer use normal procedures for testing the conventional instrument panel (Figure 30-27). When diagnosing quartz swing or digital speedometer assemblies, the following is a common diagnostic approach to locating the fault.

On most systems, the odometer and speedometer receive their input from the vehicle speed sensor (Figure 30-28). If there is a fault with the speed sensor, other systems (such as cruise control) will also be affected. When test driving the vehicle, attempt to activate the cruise control system to determine if it is operating properly. If the cruise control system fails, the problem can be the speed sensor, its circuit, BCM, or the ECM.

Some systems may provide for replacement of the stepper motor or odometer chip separate of the cluster. Always refer to the service manual for the vehicle being diagnosed.

Generally, if the BCM and/or cluster module pass their self-diagnostic tests, the fault will be in the speed sensor circuit. Common test procedures for the speed sensor are presented later in this chapter.

If the speedometer is not operating, but the odometer works properly, the fault is in the instrument cluster. Also, if the speedometer operates, but the odometer fails, the fault is in the cluster. In either case the cluster must be replaced. If the speedometer and/or odometer are inaccurate, or both do not operate, check the following items:

1. If the system uses an optical vehicle speed sensor, check the speedometer cable for kinks, twists, or other defects that will cause an inaccurate reading.

2. Check for shorts in the wiring circuit of the speed sensor.

3. Make sure of proper gear ratio and tire sizes. Both of these items will affect correct speedometer and odometer operation.

If the display illuminates but remains at zero (or any other digit), or operates erratically or intermittently,

check the connector at the speed sensor for proper installation and corrosion. Next, check the wiring circuit for any shorts or opens. If these tests do not isolate the problem, the speed sensor should be tested. Testing of the speed sensor will depend on the type of sensor used.

To test magnetic pickup speed sensors, disconnect the connector. With the ignition switch placed in the RUN position, use a jumper wire to make and break the connection between the two wires (Figure 30-29). This should cause the speedometer display to change. Change the rate of speed at which you make and break the connection and the display should indicate the changes in speed. The faster you make and break the connection, the higher the speedometer reading.

If there is no change in the speedometer display, check for opens and shorts in the sensor circuit. If the cluster passed its self-test (and you did not skip any steps) this is the only area in which the fault can be located.

If the speedometer changed speeds, the problem is in the speed sensor. To test the speed sensor, remove it from the transaxle. Connect an ohmmeter to the connector terminals of the sensor and select the lowest scale. Rotate the sensor gear while observing the ohmmeter. Distinct pulses should be detected on the ohmmeter. Compare the number of pulses per revolution with specifications. Also, compare the resistance value with specifications. If the number of pulses and resistance values are within specifications, the sensor is good.

WARNING *Do not use a test light to test the sensor. Damage to the unit may result.*

To test optical speed sensors, disconnect the speedometer cable at the transaxle and rotate the cable in its housing as fast as possible. If the speedometer display operates properly, check the speedometer pinion and drive gear for damage.

SERVICE TIP *A reversible, variable speed drill can be used to rotate the speedometer cable if you are sure there are no twists or kinks in the cable.*

If there is no speedometer operation, check for a broken speedometer cable. This can usually be determined by feeling for resistance while turning the cable by hand. Little resistance indicates a broken cable. Excessive resistance indicates a damaged cable or sensor head.

If the cable is good, the problem is in the sensor or in the wiring between the sensor and the speedometer. Fol-

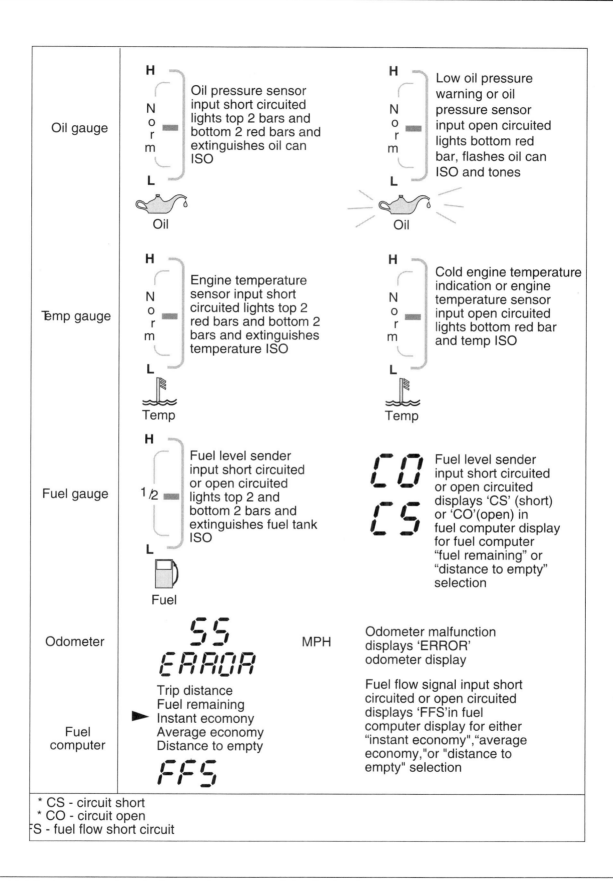

Figure 30-26 Gauge readout indicates the nature of the fault when in the diagnostic mode. (Reprinted with permission of Ford Motor Company)

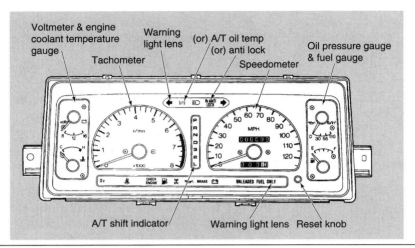

Figure 30-27 Quartz swing needle speedometer used with conventional gauges. (Courtesy of American Isuzu Motors Inc.)

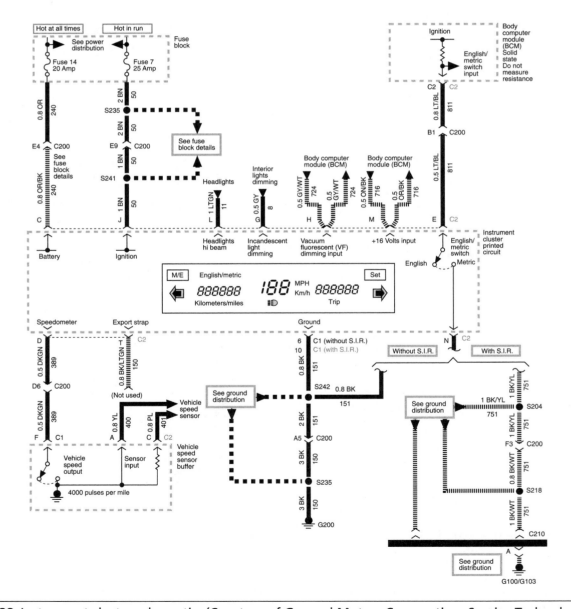

Figure 30-28 Instrument cluster schematic. (Courtesy of General Motors Corporation, Service Technology Group)

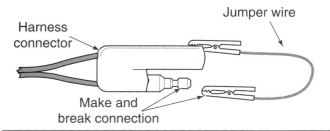

Figure 30-29 Testing the magnetic pickup speed sensor circuit. Making and breaking the connection should produce a reading in the speedometer window. (Courtesy of General Motors Corporation, Service Technology Group)

low the manufacturer's procedure for removing the instrument cluster. Connect a DMM to read the pulsed speed signal from the sensor (Figure 30-30). Rotate the speedometer cable while observing the voltmeter. Compare the pulses per revolution and pulse output values with specifications. Replace the sensor if the values are not within specifications.

Electronic Gauges

The following are guidelines for testing the individual gauges for proper operation. These tests are to be conducted after the self-diagnostic test indicates there is no problem with the cluster or the module. Test procedures will vary between manufacturers. To properly troubleshoot the gauges you will need the manufacturer's diagnostic procedure, specifications, and circuit diagram.

Gauge Reads Low Constantly

A gauge that constantly reads low when the ignition switch is in the RUN position indicates an open in the gauge circuit. To locate the open, follow these steps:

1. Disconnect the wire harness from the sending unit.
2. Connect a jumper wire between the wire circuit from the gauge and ground.
3. Turn the ignition switch to the RUN position. The gauge should indicate maximum.

If the gauge reads high, check the sending unit ground connection. If the ground is good, the sending unit is faulty and must be replaced.

SERVICE TIP *Although some service manual procedures do not require it, it is a good practice to connect a 10-ohm resistor into the jumper wire when performing these tests. This prevents a nonresistive short to ground, yet does not noticeably affect gauge operation.*

WARNING *Most electronic instrument cluster fuel gauges operate with sensors that decrease resistance values as the fuel level decreases. Jumping the wire to ground would indicate a low reading in these systems. Before faulting the sending unit, refer to the service manual for correct test results for the vehicle you are diagnosing. Do not leave the ignition switch in the RUN position for longer than 30 seconds. This is all the time required to test gauge operation.*

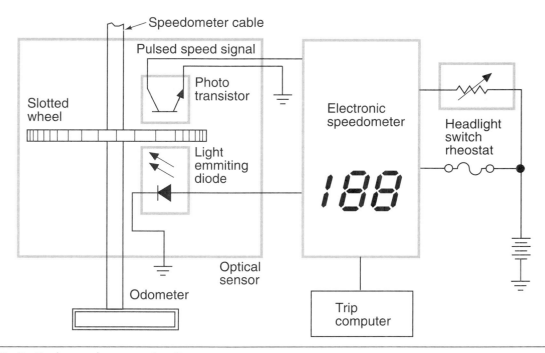

Figure 30-30 Optical speed sensor circuit.

WARNING *Some fuel gauge systems determine fuel level by sending a reference voltage from the module to the sending unit, then measuring the input voltage back from the sending unit. Do not directly ground this system. Use a voltmeter to backprobe the connector terminals for the proper voltage values.*

If the gauge continues to read low, follow the circuit diagram for the vehicle being serviced to test for opens in the wire from the sending unit. If the circuit is good, test the control module following recommended diagnostic procedures.

Gauge Reads High Constantly

A gauge that reads high when the ignition switch is placed in the RUN position indicates there is a short to ground in the circuit. To test the circuit, disconnect the wire harness at the gauge sending unit. Place the ignition switch in the RUN position while observing the gauge. If the gauge reads low, the sending unit is faulty and needs replacement.

If the gauge continues to read high, use the circuit diagram to test for shorts to ground in the circuit from the sending unit. If the circuit is good, test the control module following recommended diagnostic procedures.

WARNING *Most electronic instrument cluster fuel gauges operate with sensors that decrease resistance values as the fuel level decreases. Opening the wire would indicate a high reading in these systems. Before faulting the sending unit, refer to the service manual for correct test results for the vehicle you are diagnosing.*

Inaccurate Gauge Readings

Inaccurate gauge readings are usually caused by faulty sending units. To test the operation of the gauge you will need the manufacturer's specifications concerning resistance values as they relate to gauge readings. Gauge testers are available to test the units as different resistance values are changed.

SERVICE TIP *If a gauge tester is not available, you can substitute a rheostat of correct resistance range or place different resistors into a jumper wire. Connect the rheostat or resistor between the sending unit wire and ground. For example, if the gauge is designed to read high at zero ohms resistance and low at 90 ohms, placing 45 ohms of resistance in the circuit should produce a reading of midpoint.*

Other reasons for inaccurate gauge readings include poor connections, resistive shorts, and poor grounds. Also look for damage around the sending unit. For example, a damaged fuel tank can result in inaccurate gauge readings.

Trip Computers

Simple trip computers use inputs from the speed sensor and the fuel gauge to perform their functions. Like most electronic instrument clusters, trip computers will usually have a self-diagnostic test procedure. Follow the manufacturer's procedure for initiating this test. If the trip computer passes the self-test, check the fuel gauge and speed sensor inputs.

Complex trip computers may receive inputs from several different areas of the vehicle (Figure 30-31). These systems will require the use of specific manufacturer diagnostic charts, specifications, and procedures. Using the skills you have acquired, you should be able to perform these tests competently.

Overhead Travel Information System (OTIS) Self-Diagnostic Procedure

These steps may be followed to perform an overhead travel information system (OTIS) self-diagnostic procedure:

1. Place the ignition switch in the OFF position. Press the OTIS C/T and STEP buttons simultaneously.

2. All OTIS segments must be illuminated for 2 to 4 seconds.

3. If the OTIS displays PASS after the segment illumination, the OTIS module is satisfactory.

4. If the OTIS displays FAIL after the segment illumination, the OTIS module must be replaced.

5. When the OTIS displays CCD after the segment illumination, there is an open or shorted circuit in the C2D bus.

6. Press the C/T and STEP buttons simultaneously to exit the diagnostic mode.

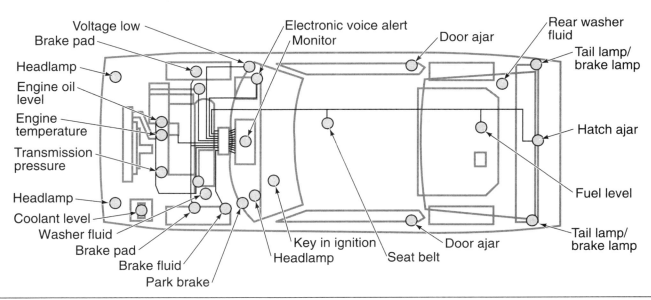

Figure 30-31 Complex information centers use many inputs to monitor the vehicle's subsystems. (Courtesy of Chrysler Corporation)

Adjusting Compass Variance

The compass in the OTIS is self-calibrating and does not require any adjustment. The word CAL may be displayed in the OTIS, indicating the compass is in a fast calibration mode. This display is turned off when the vehicle has gone in three complete circles without stopping in an area free of magnetic disturbance. However, the **variance** between the magnetic north pole and geographic north pole may cause inaccurate compass readings depending on the area in which the vehicle is driven. The procedure for setting the compass variance follows:

1. Select the proper variance number from the variance chart (Figure 30-32).

2. Turn the ignition switch to the RUN position. Press the C/T button to select the compass/temperature display.

3. Press and hold the reset button for 5 seconds. OTIS now displays the present variance zone entered in the OTIS module with the letters VAR.

4. Press the STEP button until the proper variance number appears in the display.

5. Press the RESET button to set this proper number in the OTIS memory.

Head-Up Display (HUD) Diagnosis

To check the HUD display, turn the HUD brilliance control switch to maximum intensity, and turn the igni-

tion switch from the OFF to the RUN position. When this action is taken, all segments and indicators in the HUD display should be illuminated for 4 seconds. After this time, the speedometer should read 0 with the mph or km/h indicator illuminated, and other indicators should remain off.

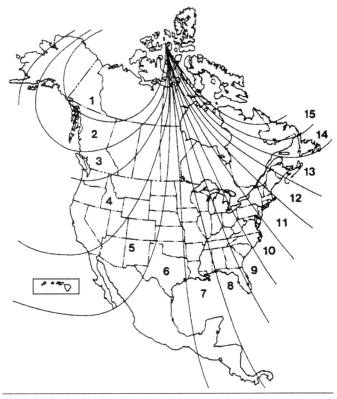

Figure 30-32 Compass variance zone numbers. (Courtesy of Chrysler Corporation)

HUD Window and Windshield Cleaning

WARNING *Do not spray glass cleaner directly on the HUD window, because this action may damage HUD module internal components.*

If the display is not sharp and clear, clean the windshield and the module window. Also, check for anything that may be obstructing the HUD module window or the windshield. Spray glass cleaner on a soft clean cloth, and then use this cloth to clean the HUD module window. Do not spray glass cleaner directly on the HUD module window. The windshield may be cleaned with a chamois and a soft clean cloth.

HUD Diagnosis

If the HUD display is inoperative, follow these steps to diagnose the system:

1. Disconnect the HUD control switch connector and connect a pair of digital voltmeter leads from terminal 5 in the wiring harness connector to ground (Figure 30-33).

2. Turn the ignition switch to the RUN position. If the voltage is less than 11 volts, check for an open between the ignition switch and the control switch. If the voltage is above 11 volts, proceed to step 3.

3. Turn off the ignition switch, and check for loose wiring connections on the HUD module. Disconnect the HUD module connector.

4. Turn the ignition switch to the RUN position and connect a DMM voltmeter between terminal 11 of the HUD control module and ground. If the voltage is less than 11 volts, repair the open circuit between the ignition switch and the module. If the voltage is above 11 volts, proceed to step 5.

5. Connect a DMM voltmeter across terminals 11 and 12 of the HUD module connector. If the reading is less than 11 volts, repair the open circuit in the HUD ground wire. If the reading is over 11 volts, replace the HUD module.

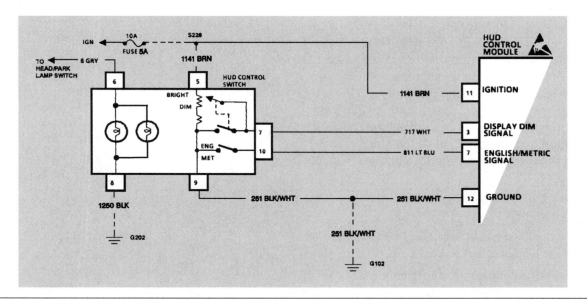

Figure 30-33 HUD wiring diagram. (Courtesy of General Motors Corporation, Service Technology Group)

Summary

❏ Diagnosis of computer-controlled lighting systems is designed to be as easy as possible. The controller may provide trouble codes to assist the technician in diagnosis.

❏ Study the schematic of the system you are diagnosing until you understand how it is supposed to operate.

❏ If the headlights do not turn on, check the regular headlight system first before condemning the automatic headlight system.

❏ To perform a functional test of the automatic headlight system, activate the automatic headlight system. Note the operation of the system as the photocell is covered and exposed to light.

❑ The resistance assembly test is performed when the lights turn on and off at the wrong light levels.

❑ Most illuminated entry systems can be diagnosed with a test light, voltmeter, and an ohmmeter.

❑ If the fiber optics do not illuminate, most likely the light source has failed.

❑ In some systems the BCM is capable of running diagnostic checks of the electronic instrument cluster to determine if a fault is present.

❑ Most instrument panel display modules have a diagnostic mode within their programming to allow the module to isolate any faults within the instrument panel cluster.

❑ Most EICs have a self-test procedure to determine if the display is working properly.

❑ Vehicles equipped with conventional electromagnetic gauges and a swing quartz speedometer use normal procedures for testing the conventional instrument panel.

❑ A gauge that constantly reads low when the ignition switch is in the RUN position indicates an open in the gauge circuit.

❑ A gauge that reads high when the ignition switch is placed in the RUN position indicates there is a short to ground in the circuit.

❑ Inaccurate gauge readings are usually caused by faulty sending units.

Terms-To-Know

Amplifier test	Photocell resistance assembly	Resistance assembly test
Diagnostic mode	Photocell test	Self-test
Display check	Power unit test	Two-button diagnostics
Functional test	Preprogrammed signal check	Variance
Function diagnostic mode	Prove-out display	

ASE Style Review Questions

1. The results of a functional test on the automatic headlight system are being discussed. Technician A says when the photocell is covered and the ignition switch is in the RUN position, the headlights should turn on within 10 seconds. Technician B says when a bright light is shone onto the photocell the lights should turn off after the selected amount of time delay. Who is correct?
   a. A only
   b. B only
   c. Both A and B
   d. Neither A nor B

2. The results of the photocell resistance test are being discussed. Technician A says if the lights do not turn on within 60 seconds when the resistance assembly switch is in the OFF (open) position and the ignition switch is in the RUN position, the photocell should be replaced. Technician B says if the lights turn off within 60 seconds after the resistance assembly switch is turned on, the photocell should be replaced. Who is correct?
   a. A only
   b. B only
   c. Both A and B
   d. Neither A nor B

3. Technician A says problems with the automatic headlight system can be the fault of the headlight switch. Technician B says the fault may be in the ignition switch. Who is correct?
   a. A only
   b. B only
   c. Both A and B
   d. Neither A nor B

4. The fiber optic indicator is not operating. Technician A says the cable can have a bend in it. Technician B says the light source may not be operating. Who is correct?
   a. A only
   b. B only
   c. Both A and B
   d. Neither A nor B

5. The photocell resistance assembly is being discussed. Technician A says it is a technician-made test tool consisting of resistors and a switch. Technician B says it is a replacement photocell that is known good. Who is correct?
   a. A only
   b. B only
   c. Both A and B
   d. Neither A nor B

6. A digital speedometer constantly reads 0 mph. Technician A says the speed sensor may be faulty. Technician B says the throttle position sensor may have an open. Who is correct?
   a. A only
   b. B only
   c. Both A and B
   d. Neither A nor B

7. All gauges read low. Technician A says the connector to the cluster may be loose or off. Technician B says the cluster module may be at fault. Who is correct?
   a. A only
   b. B only
   c. Both A and B
   d. Neither A nor B

8. Two-button diagnostics has been entered on a Chrysler vehicle. Technician A says the 5 being displayed in the speedometer window is a trouble code. Technician B says the last test can be repeated by pressing the TRIP RESET button. Who is correct?
   a. A only
   b. B only
   c. Both A and B
   d. Neither A nor B

9. Technician A says if the BCM and/or cluster module pass their self-diagnostic tests, the fault is probably in the vehicle speed sensor circuit. Technician B says if the speedometer works but the odometer does not, the problem is in the vehicle speed sensor. Who is correct?
   a. A only
   b. B only
   c. Both A and B
   d. Neither A nor B

10. Ford diagnostic mode displays are being discussed. Technician A says a CO displayed in the fuel gauge window indicates that the fuel level is near empty. Technician B says a CS indicates an open in the sender circuit. Who is correct?
    a. A only
    b. B only
    c. Both A and B
    d. Neither A nor B

31 Electronic Climate Control

Objective

Upon completion and review of this chapter, you should be able to:

❏ Explain the purpose of the automatic temperature control.

❏ Explain the purpose and operation of the control assembly in the SATC and EATC systems.

❏ List and describe the types of sensors used in SATC and EATC systems.

❏ Explain the differences in operation between semiautomatic temperature control and electronic automatic temperature control systems.

❏ Explain the design and operation of the in-vehicle sensor.

❏ Describe the design and operation of the ambient sensor.

❏ Explain the design and operation of the sunload sensor.

❏ Describe the design and operation of the evaporator temperature sensor.

❏ Explain the design and purpose of the mode door feedback inputs.

Introduction

In this chapter you will learn the operation of the semiautomatic and automatic temperature control systems. Air conditioning systems have not escaped the electronics revolution. Many cars now have computer-controlled air conditioning systems. This chapter discusses some of these computer-controlled systems.

Introduction to Semiautomatic and Electronic Automatic Temperature Control

Automatic air conditioning systems operate with the same basic components as the conventional systems. The major difference is the **automatic temperature control (ATC)** system is capable of maintaining a preset level of comfort as selected by the driver. Sensors are used to determine the present temperatures and the system can adjust the level of heating or cooling as required.

The system uses actuators to control the position of air-blend doors to achieve the desired in-vehicle temperature. Some systems will control fan motor speeds to keep the temperature very close to that requested by the driver.

There are two types of automatic temperature control: **semiautomatic temperature control (SATC)** and **electronic automatic temperature control (EATC)**. The SATC system usually requires the driver to select the mode position and blower speed. The EATC system will automatically select mode position and blower speed based on the temperature selected. In addition, most SATC systems do not provide for storing of trouble codes. EATC systems monitor system operation and set codes in a RAM module for diagnostic use. Other differences include actuator types and the number of sensors used.

Though the systems differ in methods of operation, they are all designed to provide in-vehicle temperatures and humidity conditions at a preset level. The in-vehicle humidity and temperature levels are maintained, regardless of the climate conditions outside the vehicle. The in-vehicle humidity level is maintained at 45 to 55%.

A BIT OF HISTORY

In 1939, Packard introduced a car with air conditioning. It used refrigeration coils in an air duct behind the rear seat.

Semiautomatic Temperature Control

Basic SATC systems are not much different than manual systems. The primary difference is in the use of a programmer, electric servomotor, and/or control module to operate the actuators. The SATC system maintains a driver-selected comfort level by sensing air temperature and A/C door positions through the programmer. The driver selects the operating mode and blower speeds manually from the control assembly.

Common Components

Not all systems will have all of the components described here, but most will have a combination of several of them.

The **control assembly** (or control panel) provides for driver input into the temperature control microprocessor. The control assembly used in SATC systems is similar to those used in manual systems. The main difference is the

temperature control has a temperature range imprinted under it or displayed (Figure 31-1). The control assembly is located in the instrument panel and provides the means for driver input to the climate control system. The temperature control lever is a part of a sliding resistor that mechanically converts the lever setting into an electrical resistance. The driver selects operating modes (A/C, heat, defogger, and vent) and fan motor speeds through push button selection on the control assembly.

The **programmer** (Figure 31-2) controls the blower speed, air mix doors, and vacuum motors of the SATC system. Depending on manufacturer, they are also called servo assemblies. The push buttons of the control assembly will input the programmer which controls the actuators at the air distribution doors. The programmer also receives electrical input from the in-vehicle and exterior ambient temperature sensors. Based on the inputs from these sensors and the control assembly, the programmer provides output signals to turn the compressor clutch on/off, open/close the heater water valve, and position the mode doors.

There can be several different sensors used on the SATC system. The most common are the in-vehicle temperature, ambient temperature sensors, sunload, and evaporator temperature sensors.

The **in-vehicle sensor** contains a temperature sensing NTC thermistor to measure the average temperature inside the vehicle (Figure 31-3). The in-vehicle sensor is located in the **aspirator** unit, which is usually mounted in the dashboard (Figure 31-4). An aspirator tube is connected from the in-vehicle sensor to the heater and air conditioning duct (Figure 31-5). The rush of air past the aspirator tube in the heater duct creates a slight vacuum,

Figure 31-1 Semiautomatic temperature control assembly. (Courtesy of Chrysler Corporation)

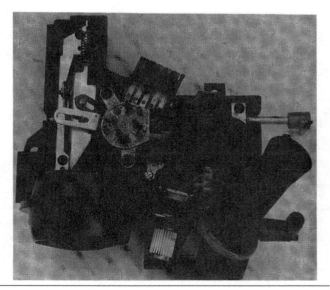

Figure 31-2 A programmer. (Courtesy of BET, Inc.)

Figure 31-3 The in-car sensor is a NTC thermistor located in the aspirator unit.

which pulls a small amount of air through the in-vehicle sensor and aspirator tube. Some A/C systems have a small electric fan motor to move air through the in-vehicle sensor.

The in-vehicle sensor is a NTC thermistor electrically connected to the programmer. As the resistance of the in-vehicle sensor changes, the programmer senses the change in voltage drop across the sensor.

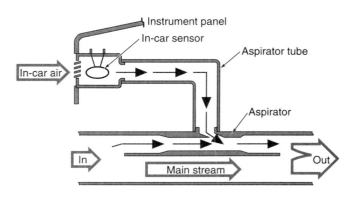

Figure 31-5 A typical aspirator. The main air stream creates a low pressure at the inlet of the aspirator drawing in-car air over the sensor.

The **ambient sensor** is also an NTC thermistor used to measure the temperature outside the vehicle (Figure 31-6). The programmer senses the change in voltage drop across the sensor terminals as the resistance changes in relation to temperature. The ambient sensor is usually located behind the grill. Due to its location, and possible influence by engine temperatures, the sensor circuit has several memory features that prevent false input.

Some systems use a **sunload sensor** (Figure 31-7). The sunload sensor is a **photovoltaic diode** that sends signals to the programmer concerning the extra generation of heat as the sun beats through the windshield. Photovoltaic diodes are capable of producing a voltage when exposed to radiant energy. The sensor converts the light signal into a current value that is sent to the programmer or computer (Figure 31-8). This sensor is usually located on the dash next to a speaker grill (Figure 31-9).

The **evaporator temperature sensor** (or FIN sensor) is mounted in the cooling unit. Many evaporator sensors have a pickup mounted in the evaporator fins so the sensor accurately measures evaporator temperature (Figure 31-10). Some evaporator sensors are mounted so they

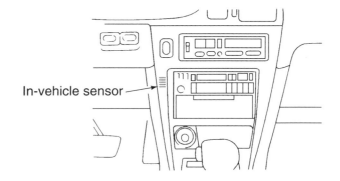

Figure 31-4 In-vehicle sensor. (Courtesy of Nissan Corporation)

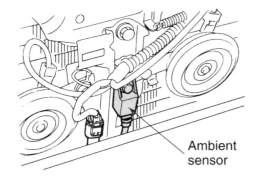

Figure 31-6 Ambient temperature sensor located behind the grill. (Courtesy of Nissan Corporation)

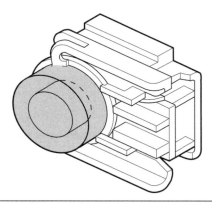

Figure 31-7 The sunload sensor produces a signal proportional to the heat intensity of the sun's heat through the windshield.

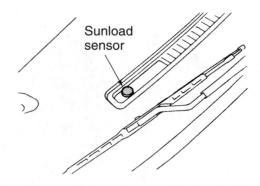

Figure 31-9 Sunload sensor mounted in one of the defroster ducts. (Courtesy of Nissan Corporation)

measure air temperature as it leaves the evaporator. These sensors may be called intake sensors. The thermistor in the evaporator temperature sensor is usually a NTC.

Common outputs of the system include the blend air door actuator, heater core flow control valve and solenoid, mode door actuator, recirc/air inlet door actuator, and the compressor clutch control.

The system may have several actuators to perform the commands inputted by the driver. Actuators can be either vacuum motors, electrically controlled vacuum sole-noids, or electrical motors. Those discussed here are electrically controlled.

The **blend air door actuator** is an electric motor that controls the position of the blend air door to supply the temperature in the vehicle selected by the driver (Figure 31-11). For example, if the driver selects 72°F (22°C) on the A/C temperature control and the in-vehicle temperature is 32°F (0°C), the programmer operates the blend air door motor to move the door so it blocks airflow through the evaporator and allows air flow through the heater core into the vehicle interior. This action brings the in-vehicle temperature up to the selected temperature as quickly as possible.

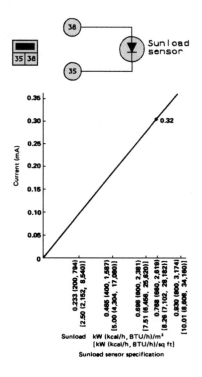

Figure 31-8 The conversion of sunlight intensity to a current value used by the programmer. (Courtesy of Nissan Corporation)

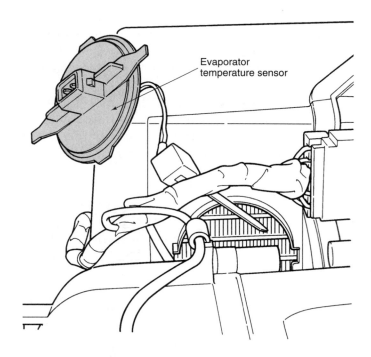

Figure 31-10 Evaporator temperature sensor with pickup mounted in the evaporator fins. (Courtesy of Chrysler Corporation)

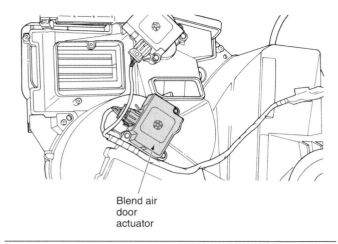

Figure 31-11 Blend airdoor actuator. (Courtesy of Chrysler Corporation)

When the selected temperature is 72°F (22°C) and the in-vehicle temperature is 98°F (37°C), the programmer operates the blend air door motor to move the door so it blocks air flow through the heater core and moves air through the evaporator into the vehicle interior. This action quickly cools down the vehicle interior to supply the selected temperature.

Under most operating conditions, the actuator motor positions the blend air door to allow a blend of warm air through the heater core and cold air through the evaporator to supply the in-vehicle temperature selected by the driver.

Some systems use a **heater core flow valve** and solenoid to shut off the coolant flow through the heater core (Figure 31-12). When the A/C system is in the max air mode, the programmer grounds the coolant control solenoid winding. This action moves the solenoid plunger

and supplies vacuum through the solenoid to the vacuum diaphragm connected to the valve in the heater hose. Under this condition, the valve closes to stop the coolant flow through the heater core. In other operating modes, the programmer opens the coolant control solenoid ground circuit. Under this condition, the solenoid shuts off vacuum to the diaphragm, and the coolant flow valve in the heater hose is open.

The **mode door actuator** is an electric motor that is linked to the mode door (Figure 31-13). The programmer operates the mode door to supply air flow to the floor ducts, A/C panel ducts, or defrost ducts. In the bi-level mode, the mode door is positioned to supply air to the floor and A/C panel ducts. The mix mode supplies air flow from the defrost and floor ducts. The programmer positions the mode door to supply the air flow selected

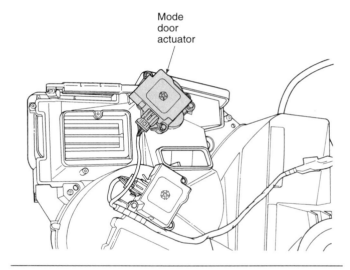

Figure 31-13 Mode door actuator. (Courtesy of Chrysler Corporation)

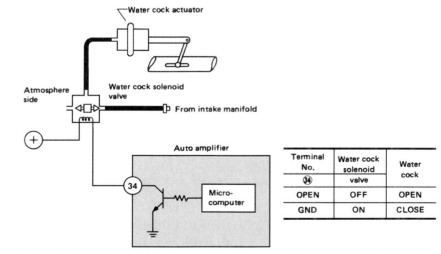

Terminal No.	Water cock solenoid valve	Water cock
㉞		
OPEN	OFF	OPEN
GND	ON	CLOSE

Figure 31-12 Solenoid and coolant control valve in the heater hose. (Courtesy of Nissan Corporation)

by the driver. For example, if the driver presses the defrost button on the A/C controls, the programmer commands the mode door actuator to supply air flow from the defrost ducts.

In some systems, if the auto button is pressed, the mode door is positioned automatically. For example, if A/C is selected, the programmer operates the mode door so air flow is directed from the panel ducts in the dashboard. In these systems, the driver can press the A/C control buttons and override the auto function. The auto button may be pressed to return the system to automatic operation. When the system is in the auto mode, auto fan mode is displayed in the A/C control display (Figure 31-14). If the temperature control is placed in the full hot or full cold manual override position, the system will not enter the auto mode. The control head or programmer illuminates an LED in each A/C control button to indicate the operating mode to the driver (Figure 31-15).

The **recirc/air inlet door actuator** is also an electric motor that is linked to the recirc door (Figure 31-16). The programmer operates the recirc/air inlet door actuator to position the door to move outside air or in-vehicle air into the A/C heater case. If the A/C system is in the max air mode, the programmer operates the recirc/air door actuator to position the door so in-vehicle air is

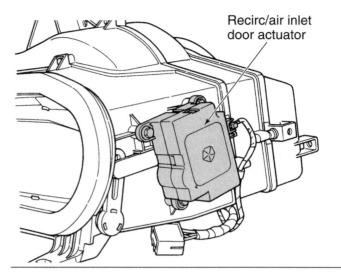

Figure 31-16 Recirc/air inlet door actuator. (Courtesy of Chrysler Corporation)

moved into the A/C heater case. This action provides faster cooling of the passenger compartment. In most other modes, the programmer positions the recirc/air door so outside air is moved into the A/C heater case.

When the driver presses the recirc button on the A/C control, the system enters a manual recirc mode, and the programmer positions the recirc door so in-vehicle air is moved into the A/C heater case.

When the A/C mode is selected by the driver, the programmer grounds the compressor clutch relay winding. On some A/C systems, this winding is grounded by the PCM (Figure 31-17). When the A/C compressor clutch relay winding is grounded, current flows through the closed relay contacts to the compressor clutch winding. This action energizes the compressor clutch, and the drive belt begins turning the compressor shaft.

On many A/C systems, the A/C compressor clutch is not energized if the ambient temperature sensor signal indicates the temperature is below a preset value.

Chrysler SATC

The Chrysler SATC system contains the following components:

- Control assembly
- Ambient sensor
- Servomotor
- Aspirator

The control assembly consists of a temperature control level, a push-button mode switch and a 4-speed blower motor switch. Mode selection is through push-

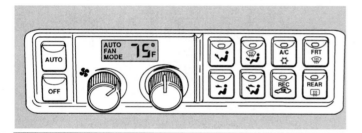

Figure 31-14 A/C control head. (Courtesy of Chrysler Corporation)

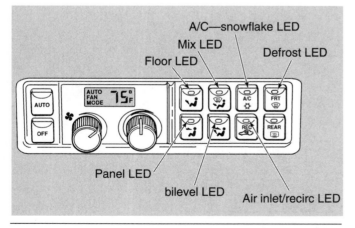

Figure 31-15 LEDs in the A/C control buttons indicate the operating mode to the driver. (Courtesy of Chrysler Corporation)

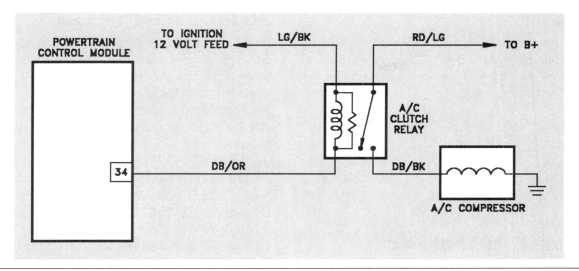

Figure 31-17 Compressor clutch circuit. (Courtesy of Chrysler Corporation)

button operated vacuum and electric switches. The vacuum switches control water valve operation and position of the mode doors. The blend air door is controlled by the servomotor.

The in-car and ambient temperature sensor resistance values are measured by the servomotor assembly. The temperature control lever of the control assembly is connected to a sliding resistor that is used by the servomotor to determine blend air door position requirements.

High sensor or control head resistance results in the servomotor positioning the blend air door to allow additional heat to enter the passenger compartment. When the resistance value lowers, as a result of the increased flow of warm air, the servomotor positions the door to allow cooler air to enter. The servomotor provides a feedback in the form of the in-car sensor that measures the temperature changes.

This system of temperature control is performed in all modes except for MAX A/C. In this position, the water valve is closed by vacuum from the control assembly. This restricts the flow of engine coolant through the heater core, reducing the temperature of the heater core. The air is recirculated inside of the passenger compartment for maximum cooling.

GM C65 System

SERVICE TIP *The C65 system is capable of maintaining the temperature selected by the driver to +/- 4 degrees. An adjustment screw is provided to correct the resistance values of a potentiometer in the programmer in the event this tolerance is no longer maintained.*

The operation principles of the General Motors C65 Tempmatic system are similar to the Chrysler system just discussed. The difference is in the programmers used. The programmer contains a servomotor that is connected to the air mix valve, and four vacuum solenoids.

Engine vacuum is always present to the programmer. The vacuum solenoids are held closed by gravity, and open when the programmer energizes the solenoid. When the driver uses the control assembly to direct air flow out of the floor, dash, or defogger vents, the programmer energizes the appropriate solenoid to allow vacuum to the mode door.

CAUTION *There are differences in operation between models of General Motors vehicles equipped with C65 systems. Always refer to the correct service manual.*

GM C61 System

The C61 programmer is an integral component of the control assembly. A rotary switch provides an electrical path to the A/C compressor clutch based on mode lever position. The temperature dial on the control assembly changes the resistance value of a rheostat based on driver selection. The value of the sensor resistance is added to the resistance of the rheostat and used by the programmer to make system operation determinations.

The programmer also incorporates an amplifier to increase the voltage signals from the sensors, a **transducer**, and a feedback potentiometer. A transducer is a vacuum valve that changes a voltage signal into a vacuum signal.

Ford ATC

Temperature control of Ford's ATC system is determined by vacuum modulation produced by a bi-metallic sensor. The sensor controls the servo vacuum motor. The sensor and modulator assembly is located within the aspirator.

The sensor is connected to an internal vacuum modulator. As air is drawn over the bi-metallic sensor, the sensor mechanically controls and regulates a vacuum that is proportional to the air temperature. The vacuum motor positions the blend door according to the regulated vacuum.

Electronic Automatic Temperature Control

The determining factor separating electronic automatic temperature control (EATC) systems from SATC systems is the ability to perform self-diagnostics. The body control module (BCM) or automatic temperature control (ATC) computer will set trouble codes. Also, the EATC system provides a continuously variable blower speed signal and readjusts interior temperature several times a second.

In addition to the sensors used in the SATC system, the EATC system may also use engine coolant temperature, vehicle speed, and throttle position sensors as inputs. If these input sensors indicate a wide-open throttle condition or high engine rpm, the computer does not energize the compressor clutch. On many A/C systems, the compressor clutch is energized in the defrost mode. Since the A/C system removes moisture from the air, this action helps to reduce windshield fogging.

On some A/C systems, the driver's and passenger's door ajar switches send input signals to the A/C computer. When either door is initially opened after the car has been sitting, the door ajar switch signal to the A/C computer causes this computer to turn on the aspirator motor and flush hot air out of the in-vehicle sensor. This action only occurs above a specific in-vehicle temperature.

On many systems, some input signals are sent to the PCM and then relayed to the BCM. On some vehicles, these input signals are transmitted on data links between the PCM and BCM.

The EATC system may also incorporate a **cold engine lock-out switch** to signal the BCM or A/C controller to prevent blower motor operation until the air entering the passenger compartment reaches a specified temperature.

There are two categories of EATC systems: BCM-controlled systems and stand-alone systems that use their own computer.

BCM-Controlled Systems

Climate control is one of the primary functions of the BCM on many vehicles. The following discussion uses a common system used by General Motors where the body computer module has central control over the EATC system. The BCM monitors all system sensors and switches, compares the data with programmed instructions, and commands the actuators to provide accurate control of the system.

The **climate control panel (CCP)** contains a circuit board that translates driver inputs into electrical signals (Figure 31-18). The CCP and BCM communicate with each other over a data circuit. The information sent to the CCP can be displayed by the vacuum fluorescent display.

The driver uses the "Warmer" and "Cooler" buttons to change the desired temperature control by one degree increments between a range of 65°F (18°C) and 85°F (29°C). If the buttons are held down, the set temperature will change until the button is released or at the end of the scale.

The driver can select temperature settings of 60°F (16°C) or 90°F (32°C) that will override the automatic temperature control. The "Outside Temp" button will display the outside temperature as sensed from the ambient temperature sensor. The other buttons enable operating mode selection. The display will indicate which mode has been selected.

The ambient temperature sensor input is controlled by special BCM programming. During idle and low speed conditions, the sensor may be influenced by engine heat, and the CCP will display higher than actual outside temperatures. At vehicle speeds less than 20 mph, the BCM will limit the update of the CCP displays to once every 100 seconds. Between 20 and 45 mph, the display can update only after a 2-minute delay. At speeds above 45 mph, the display can be updated quickly.

The BCM monitors the air conditioning system through several sensors. High side temperature is moni-

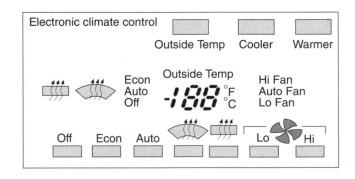

Figure 31-18 Climate control panel. (Courtesy of General Motors Corporation, Service Technology Group)

tored through a sensor in the pressure line. By monitoring the high side temperature, the BCM is capable of making calculations that translate into pressure. The calculations are based on the pressure-temperature relationship of R-12 or R-134a. The BCM will also monitor low side pressure in the same manner through a low side temperature sensor. The BCM also receives a signal from the low pressure switch in the accumulator. The BCM will shut down the compressor clutch if system operation is not within set parameters.

The BCM uses a bi-directional motor to adjust the blend door. A potentiometer feedback signal is used by the BCM to determine blend door position. Feedback inputs are sent from the blend air mode door, mode door, and recirc door to the BCM (Figure 31-19). A potentiometer is mounted in each door motor (Figure 31-20). Each door motor potentiometer may be called a **potentio balance resistor (PBR)**. These potentiometers send a signal to the BCM in relation to door position so the computer knows the exact door position. The resistance of the potentiometer in each door changes in relation to

the door opening. As the resistance of the potentiometer changes, the BCM senses the voltage drop across the potentiometer.

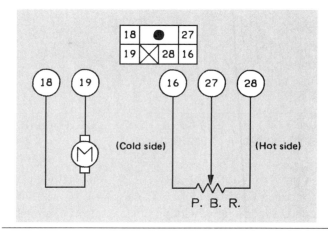

Figure 31-20 The potentiometer in each mode door signals the exact position of the door. (Courtesy of Nissan Corporation)

BLOCK DIAGRAM

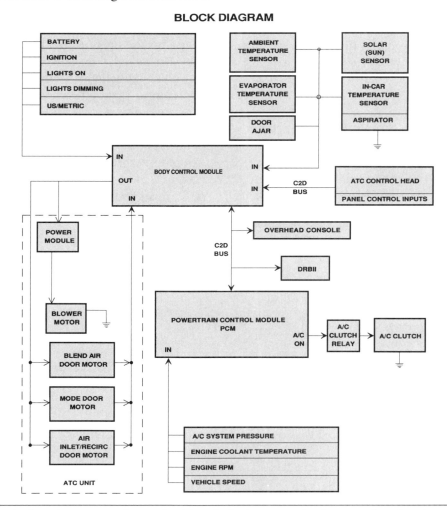

Figure 31-19 Feedback signals are sent from the blend air door, mode door, and recirc door motors to the BCM. (Courtesy of Chrysler Corporation)

Figure 31-21 Power module amplifies BCM signals to control the blower motor speed.

A power module (Figure 31-21) controls the blower motor based on drive signals from the BCM. The body computer sends a pulse width modulated (PWM) signal to the power module, and the power module amplifies the signals to provide variable fan blower speeds (Figure 31-22). If the auto button is pressed on the A/C control head, the blower speed is automatically controlled by the body computer depending on the temperature setting, ambient temperature, in-vehicle temperature, amount of sunload, evaporator temperature, and blend door position. The blower speed may be manually controlled by pressing the blower speed switch in the A/C control head.

Operation

The BCM calculates a **program number,** which represents the amount of heating or cooling required to obtain the temperature set by the driver. This number is based on inputs from the control assembly (driver input), ambient temperature, and in-vehicle temperature. Based on this number air delivery mode, fan blower speed, and blend door positioning are determined. A program number of 0 represents maximum cooling, 100 represents maximum heating. The program number can be observed while in the diagnostic mode.

To provide the proper mix of inside air temperature entering the passenger compartment, the BCM monitors the ambient and in-vehicle temperatures, the average low side temperature, and the coolant temperature. These inputs are combined with the program number. The BCM commands the programmer to position the blend door for the correct temperature of incoming air. The programmer feedback potentiometer is also monitored by the BCM.

Blower speed is determined by a combination of the program number and driver input temperature. The CCP will signal the BCM over data line 718. The signal is sent from the BCM to the power module from terminal A9. This signal is a constantly variable voltage proportional to blower speed. The power module will amplify the signal then apply it to the blower motor through circuit 65. The BCM monitors the blower speed through a feedback voltage from the motor on circuit 761.

Solenoid valves in the programmer control the air valves that are operated by mechanical and vacuum controls. The air-inlet, up-down, and A/C defrost valves are controlled individually. Commands from the BCM to the programmer control the operation of the solenoids.

Air conditioner compressor clutch control is performed by the powertrain control module (PCM) through inputs from the BCM (Figure 31-23). The PCM cycles the compressor on and off based on input signals from the high side and low side temperature sensors to the BCM. In addition, the PCM will anticipate clutch cycling and will adjust engine idle speed accordingly.

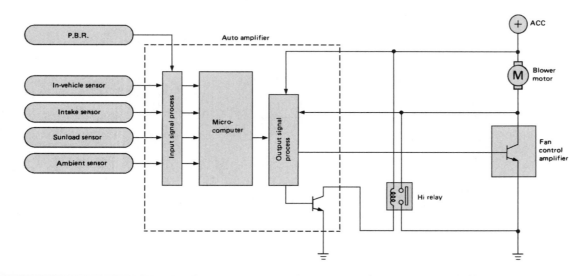

Figure 31-22 Blower motor circuit with fan control amplifier and high relay. (Courtesy of Nissan Corporation)

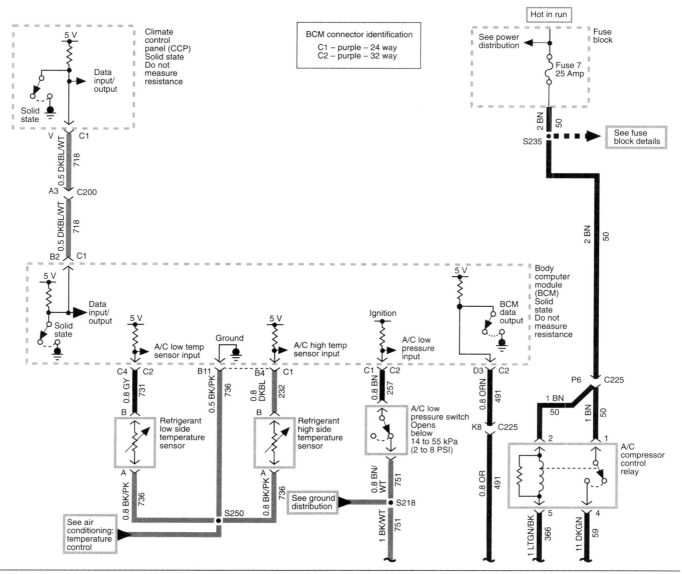

Figure 31-23 Air conditioning clutch controls. (Courtesy of General Motors Corporation, Service Technology Group)

The BCM monitors system inputs and feedback signals. If these voltage signals fall outside programmed parameters, the BCM will turn on the "Service Air Cond" indicator lamp.

Separate Computer-Controlled Systems

Several EATC systems use a separate computer for the sole purpose of climate control. The systems described here are representative examples of these types of systems.

GM Electronic Touch Climate Control

General Motors electronic touch climate control (ETCC) systems are fully automatic and regulate fan speed, air inlet, and outlet positions. The control head contains the microprocessor that will execute desired selection and remember the last selection (Figure 31-24). The control head receives inputs from ambient temperature, in-vehicle temperature, and setting selection to determine position of the blend door and fan blower speed.

Depending on the model of vehicle the system is installed on, the mode door and water flow actuators can be controlled by either vacuum or electric servomotors. All systems use an electrically operated blend door.

The control head signals the blower and A/C clutch control module to provide for clutch cycling and appropriate blower speeds (Figure 31-25). The blower motor provides up to 256 different speeds.

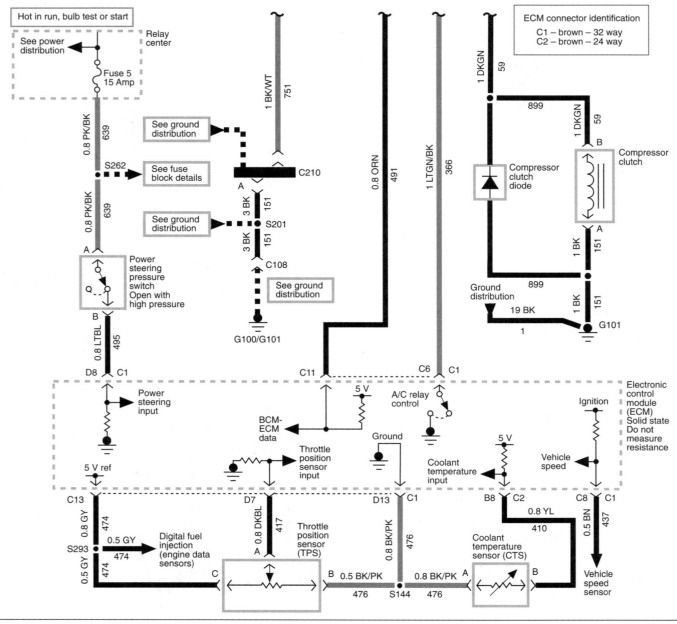

Figure 31-23 (Continued)

Figure 31-24 Electronic touch climate control panel.

Ford EATC

The Ford EATC system uses a microprocessor that is built into the control assembly. The control unit controls four DC motor rotary actuators to operate each air distribution door. The sensors and inputs are the same as in previously discussed systems. The control assembly sends continuously variable voltage signals to the **blower motor speed controller (BMSC)**. The BMSC module amplifies the signal to control motor operation. Figure 31-28 is a schematic of one EATC system used by Ford.

Chrysler EATC

The Chrysler EATC system uses an individual computer to regulate incoming air temperature and adjust the

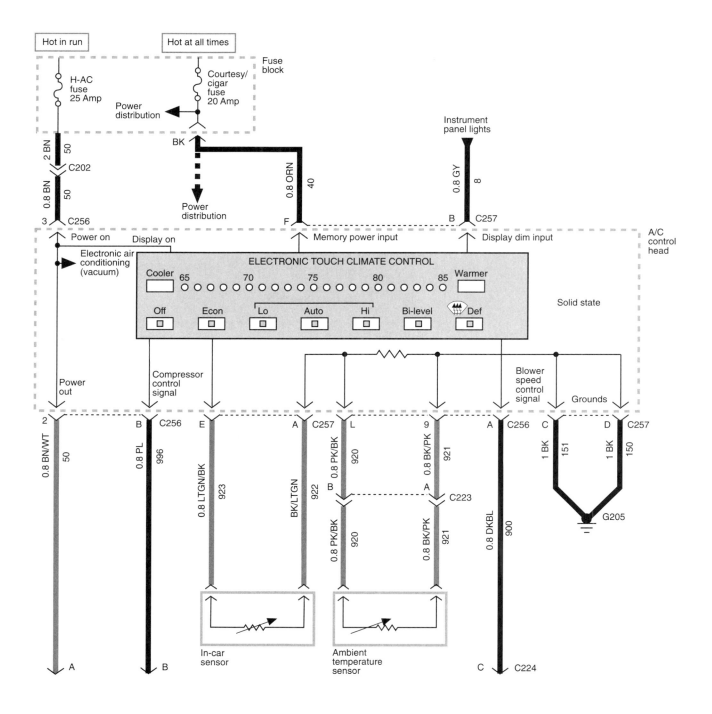

Figure 31-25 EATC wiring schematic. (Courtesy of General Motors Corporation, Service Technology Group)

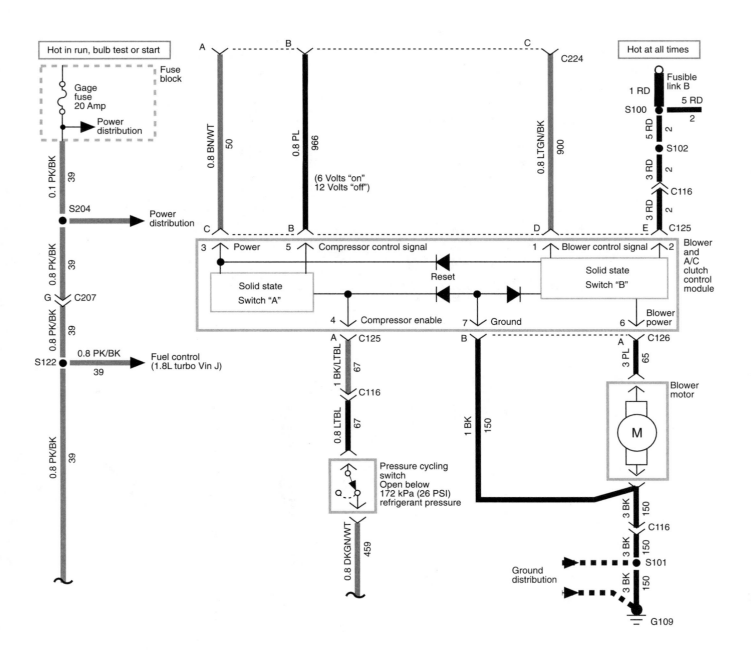

Figure 31-25 (Continued)

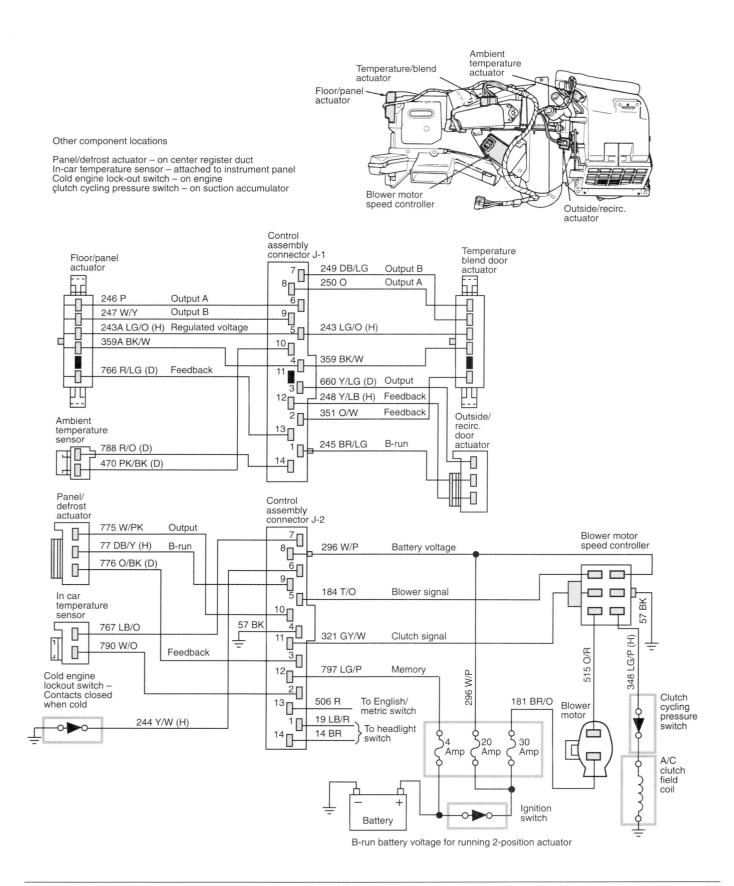

system as required every 7 seconds (Figure 31-27). The **power-vacuum module (PVM)** uses logic signals from the control assembly microprocessor to send a variable voltage signal to the blower motor. In addition, the PVM controls A/C compressor clutch and a voltage signal to the blend door actuator, and vacuum to all other actuators (Figure 31-28).

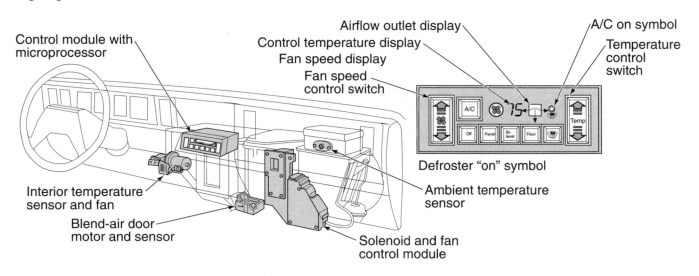

Figure 31-27 Chrysler's EATC system components and control assembly. (Courtesy of Chrysler Corporation)

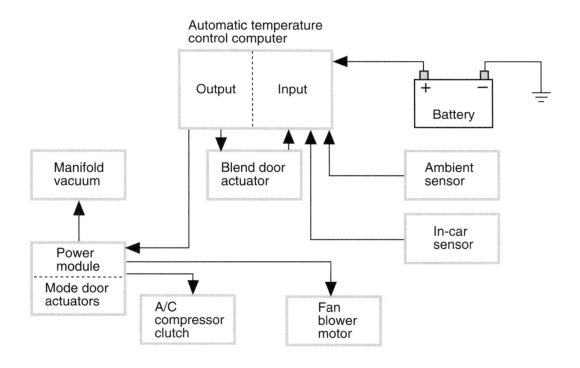

Figure 31-28 Block diagram of Chrysler's EATC system. (Courtesy of Chrysler Corporation)

Summary

❑ The in-vehicle sensor is a thermistor mounted in the instrument panel. This sensor sends a voltage signal to the A/C computer in relation to in-vehicle air temperature.

❑ An aspirator tube draws a small amount of air past the in-vehicle sensor. In some systems, a small electric motor and fan move air past the in-vehicle sensor.

❑ The ambient temperature sensor is a thermistor that sends a voltage signal to the A/C computer in relation to ambient air temperature.

❑ The sunload sensor contains a photo diode that senses the amount of sunlight striking the sensor and converts this light signal to a current value that is sent to the A/C computer.

❑ The evaporator temperature sensor is a thermistor that sends a voltage signal to the A/C computer in relation to the evaporator temperature.

❑ On some vehicles, the driver's and passenger's door ajar switch inputs cause the A/C computer to turn on the aspirator motor and flush hot in-vehicle air out of the in-vehicle sensor.

❑ In many computer-controlled A/C systems, potentiometers on the mode door actuators send feedback signals to the BCM. From these feedback signals, the BCM knows the exact position of the door actuators.

❑ In some computer-controlled A/C systems, the BCM sends a pulse width modulated signal to the power module, and this module sends a varying voltage signal to the blower motor to control blower speed.

❑ The blend air door controls hot air flow from the heater core and cold air flow from the evaporator into the passenger compartment.

❑ The BCM operates the mode door actuator to position the mode door to deliver air flow from the floor ducts, panel ducts, or defrost ducts.

Terms-To-Know

Ambient sensor

Aspirator

Automatic temperature control (ATC)

Blend air door actuator

Blower motor speed controller (BMSC)

Cold engine lock-out switch

Control assembly

Climate control panel (CCP)

Electronic automatic temperature control (EATC)

Evaporator temperature sensor

Heater core flow valve

In-vehicle sensor

Mode door actuator

Photovoltaic diode

Potentio balance resistor (PBR)

Power-vacuum module (PVM)

Programmer

Program number

Recirc/air inlet door actuator

Semiautomatic temperature control (SATC)

Sunload sensor

Transducer

Review Questions

Short Answer Essays

1. Describe the design and operation of the in-vehicle sensor and aspirator.

2. Explain the location, design, and operation of the ambient sensor.

3. Describe the location, design, and operation of the sunload sensor.

4. Explain the location, design, and operation of the evaporator temperature sensor.

5. Describe the purpose of the potentiometers in the mode door actuator motors.

6. Explain how the BCM and power module control blower speed.

7. What is meant by a program number?

8. Explain how the blend air actuator motor controls in-vehicle temperature.

9. Describe the operation of the heater core flow control valve and solenoid.

10. Explain the operation of the mode door actuator.

Fill-in-the-Blanks

1. When the temperature of the in-vehicle sensor decreases, the sensor resistance _____ .

2. The sunload sensor contains a _____ _____ .

3. When a signal is received from the driver's door ajar switch on some systems, the A/C computer turns on the _____ motor.

4. In some computer controlled A/C systems, the BCN sends a _____ _____ _____ signal to the power module, and this module sends a _____ voltage signal to the blower motor to control blower speed.

5. The ATC system uses _____ that will open and close air-blend doors to achieve the desired in-vehicle temperature.

6. The heater core flow control valve is closed when the A/C system is in the _____ _____ mode.

7. In some systems, the mix mode delivers air flow from the _____ and _____ ducts.

8. The computer will not energize the compressor clutch at _____ ambient temperatures.

9. The recirc/air inlet door actuator delivers _____ air or _____ air to the A/C heater case.

10. The sunload sensor is a _____ diode that sends signals to the programmer concerning the extra generation of heat as the sun beats through the windshield.

ASE Style Review Questions

1. The in-vehicle sensor is being discussed. Technician A says the resistance of the sensor increases as the temperature increases. Technician B says the A/C computer senses the voltage drop across this sensor as the resistance of the sensor changes. Who is correct?
   a. A only
   b. B only
   c. Both A and B
   d. Neither A nor B

2. The ambient sensor is being discussed. Technician A says the ambient sensor contains a thermistor. Technician B says the resistance of this sensor increases as the sensor becomes colder. Who is correct?
   a. A only
   b. B only
   c. Both A and B
   d. Neither A nor B

3. The sunload sensor is being discussed. Technician A says this sensor contains a light emitting diode. Technician B says this sensor contains a thermistor. Who is correct?
   a. A only
   b. B only
   c. Both A and B
   d. Neither A nor B

4. The evaporator temperature sensor is being discussed. Technician A says some evaporator temperature sensors have a pickup mounted in the evaporator fins. Technician B says some systems have an evaporator sensor that senses the air temperature leaving the evaporator. Who is correct?
   a. A only
   b. B only
   c. Both A and B
   d. Neither A nor B

5. Door ajar switch inputs as they relate to A/C operation are being discussed. Technician A says when the A/C computer receives a door ajar switch input, it energizes the compressor clutch. Technician B says when the A/C computer receives a door ajar switch input, it moves the blend air door to the max air position. Who is correct?
   a. A only
   b. B only
   c. Both A and B
   d. Neither A nor B

6. Feedback inputs from the mode doors are being discussed. Technician A says the blend air door, mode door, and recirc/air inlet door motors contain feedback potentiometers. Technician B says only the blend air door motor contains a feedback potentiometer. Who is correct?
   a. A only
   b. B only
   c. Both A and B
   d. Neither A nor B

7. Control assemblies are being discussed. Technician A say that the control assembly is located in the instrument panel and provides the means for driver input to the climate control system. Technician B says that the temperature control lever is a part of a sliding resistor that mechanically converts the lever setting into an electrical resistance. Who is correct?
   a. A only
   b. B only
   c. Both A and B
   d. Neither A nor B

8. Blend air door actuator operation is being discussed. Technician A says in the max A/C mode, the blend air door is positioned so air flow into the passenger compartment is a blend of cold air through the evaporator and warm air through the heater core. Technician B says in the max A/C mode, the blend air door is positioned so all the air flow entering the passenger compartment comes through the evaporator core. Who is correct?
   a. A only
   b. B only
   c. Both A and B
   d. Neither A nor B

9. Mode door actuator operation is being discussed. Technician A says in the bi-level mode, air flow is directed from the floor and defrost ducts. Technician B says in the mix mode, air flow is directed from the floor and defrost ducts. Who is correct?
   a. A only
   b. B only
   c. Both A and B
   d. Neither A nor B

10. Compressor clutch operation is being discussed. Technician A says in many systems, the compressor clutch is not energized if the throttle is wide open. Technician B says on some systems, the compressor clutch is not engaged if the ambient temperature is below a preset value. Who is correct?
   a. A only
   b. B only
   c. Both A and B
   d. Neither A nor B

32 Diagnosis of Electronic Climate Control Systems

Objective

Upon completion and review of this chapter, you should be able to:

- ❏ Perform a preliminary inspection on a computer-controlled air conditioning system.
- ❏ Complete an A/C performance test on a computer-controlled air conditioning system.
- ❏ Diagnose common SATC systems.
- ❏ Test the aspirator assembly for proper operation.
- ❏ Test the sensors used on SATC and EATC systems.
- ❏ Test the servomotor for proper operation.
- ❏ Test the control assembly sliding resistor.
- ❏ Perform a functional test of the SATC system.
- ❏ Diagnose common EATC systems.
- ❏ Enter self-diagnostic tests of common EATC systems.
- ❏ Perform self-diagnostic tests to illuminate the A/C control head displays.
- ❏ Perform control rod adjustments on the door actuator motors.
- ❏ Complete a scan tester diagnosis of computer-controlled air conditioning systems.

Introduction

Precise control of the interior temperature is made possible through the use of electronic components that control vacuum or electric actuators. These components include microprocessors, thermistors, and potentiometers. A failure in any of these components, or the actuators, will result in customer complaints ranging from inaccurate temperature control to no operation. Today's technician must possess a basic knowledge and understanding of the operating principles of both the SATC and EATC systems. In addition, today's technician must be proficient at diagnosing and servicing these systems. In this chapter you will learn the common diagnostic procedures used with both systems.

Preliminary Inspection

In this chapter, we discuss mainly the electronic diagnosing and servicing of computer-controlled air conditioning systems. It is assumed that the student has already studied refrigeration systems and manual air conditioning. Before an air conditioning system is diagnosed, a preliminary A/C system inspection must be completed. The procedure for this inspection follows:

1. Visually inspect the air passages in the condenser and radiator for restrictions (Figure 32-1).
2. Visually inspect the cooling system for leaks and proper operation. Tape a thermometer to the top radiator hose and be sure the cooling system is operating at the proper temperature.

3. Check the condition and tension of the compressor and water pump belts.

4. Check all fuses in the A/C system, including the compressor clutch fuse. Some compressor clutches have a thermal fuse.

5. Visually inspect all wiring connections for loose or corroded terminals. Inspect all wiring harnesses for damaged wires.

6. Check for compressor operation when the system is in the A/C mode.

7. Be sure the refrigerant system has the proper refrigerant charge and pressures.

Semiautomatic Temperature Control System Diagnosis

When diagnosing the system, remember the SATC system is a control system and faults within the evaporator and heater systems will have an effect on its operation. You have learned how to test motors and electromagnetic components in previous chapters. Use these same procedures to check the motor and compressor clutch circuits of the SATC system. You will also need the circuit schematic and specifications for the vehicle you are diagnosing.

The air delivery control systems of SATC systems differ between manufacturers and require specific diagnostic procedures. Most systems can be tested with specific tests designed to troubleshoot that particular system.

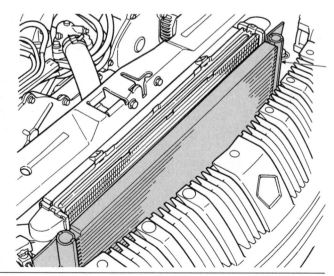

Figure 32-1 Visually inspect the condenser air passages for restrictions. (Courtesy of Chrysler Corporation)

SATC systems, even those of different model lines made by the same manufacturer, will require specific troubleshooting procedures found in that model's service manual. Examples of such systems and their procedures are described here.

Component Testing

The troubleshooting chart will direct the technician to test certain components based on the symptom reported by the vehicle owner. The next section outlines common methods of testing the Chrysler SATC system.

A paper test is a quick method of testing the aspirator assembly to verify air flow to the in-car sensor. Set the controls for high blower speed while in the heat mode of operation. Use a piece of paper large enough to cover the aspirator air inlet (Figure 32-2). The suction of the aspirator assembly should be sufficient to hold the paper against the inlet grill. If the paper is not held in place, refer to the aspirator system diagnostic chart (Figure 32-3).

To test the ambient sensor, remove it from its socket and measure its resistance when the temperature is between 70°F and 80°F (21°C and 27°C). Because current flow through the sensor and body heat will affect the readings, do not hold the sensor in your hand or leave the ohmmeter connected for longer than 5 seconds (Figure 32-4). The resistance through the sensor at the previous temperature should be between 225 and 335 ohms. If the resistance levels fall out of the specification range, replace the ambient sensor.

To test the in-vehicle sensor, disconnect its electrical connector. Do not disconnect the aspirator tubes or remove the sensor from the panel. Place the test thermometer stem into the air inlet grill near the in-car sensor. With the blower motor speed set on the second notch, depress then pull out the MAX-A/C button. This

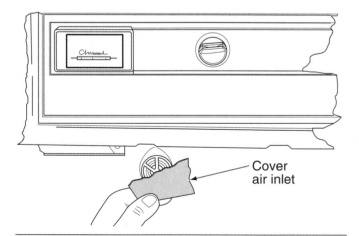

Figure 32-2 When performing the aspirator paper test, the vacuum should hold the paper over the grill. (Courtesy of Chrysler Corporation)

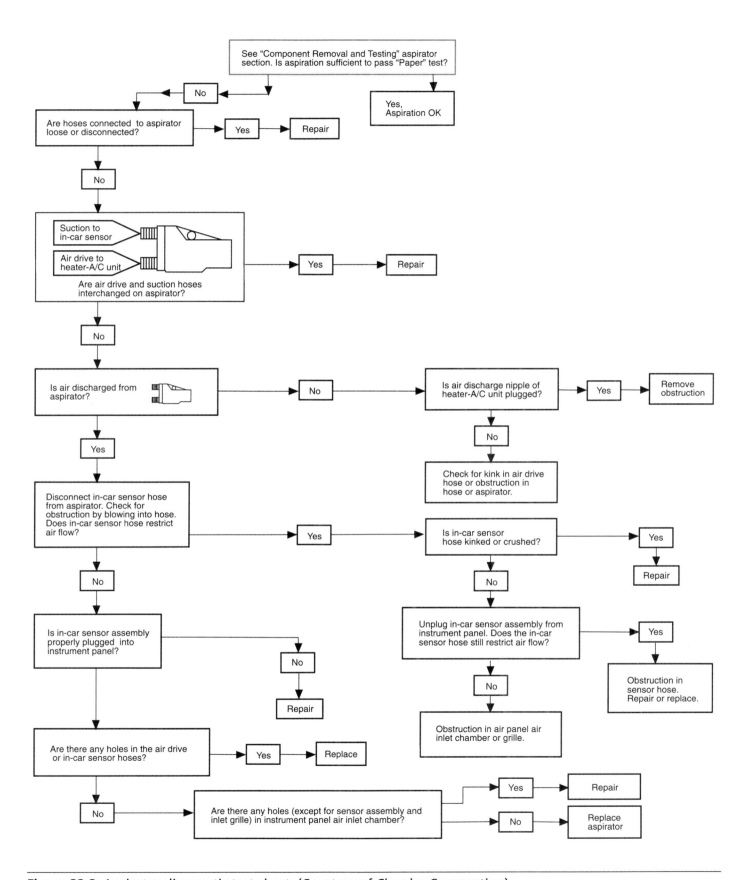

Figure 32-3 Aspirator diagnostic test chart. (Courtesy of Chrysler Corporation)

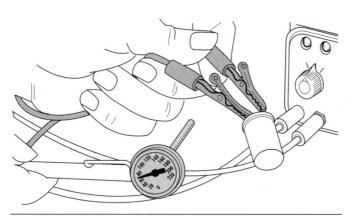

Figure 32-4 Test the ambient temperature sensor at room temperature and avoid touching the sensor during testing. Also, do not connect the ohmmeter for longer than 5 seconds. (Courtesy of Chrysler Corporation)

will turn off the compressor and close the water valve. Operate the blower while quickly measuring the resistance of the sensor. Do not leave the ohmmeter connected to the sensor terminals for longer than 5 seconds or inaccurate readings will result from the sensor warming. The resistance of the in-car sensor should be between 1,100 and 1,800 ohms when the thermometer reads between 70°F and 80°F (21°C and 27°C). Replace the sensor if the resistance is not within specifications.

Perform the servomotor test only when it has been determined the aspirator assembly is functioning correctly. To test the servomotor, it must be removed from the vehicle. Follow the service manual procedures for removing the motor. Use the following steps to test the motor:

1. Use a blend door crank from a manual A/C unit to check for free door movement. If the door binds, attempt to locate the cause and repair the unit, if the cause is in the motor then the servo will have to be replaced.

2. Connect the ambient sensor across the two servomotor terminals on the left side of the connector.

3. Connect a test light between the right side terminal and battery positive, and the jumper wire from battery negative to the middle terminal.

4. When the connections are made, the motor should run in the clockwise direction while the test light glows dimly.

5. When the motor reaches its internal stop, the test light should glow brightly.

6. To operate the motor in the counterclockwise direction, remove the ambient sensor.

While operating the motor, check for smooth operation and observe the test light. If the motor encounters a brief jam, the test light illumination level will increase. A test light that flickers while the motor is operating indicates the motor is not moving in a smooth fashion. If the cause cannot be repaired, the servo will need to be replaced. For the motor to be considered acceptable and placed back into service it must move to the full clockwise and full counterclockwise positions.

The control assembly resistor can be tested either in or out of the vehicle. Disconnect the electrical connector to the sliding resistor and connect the ohmmeter test leads across the terminals (Figure 32-5). Set the comfort control lever to the 65 selection. The resistance value in this position should be less than 390 ohms.

Slowly move the comfort control lever toward the right while observing the ohmmeter. When the ohmmeter indicates 930 ohms, the lever should be near the 75 setting. Also, the ohmmeter should indicate a smooth increase in resistance. Next, move the lever to the 85 setting. The resistance value should increase smoothly to at least 1,500 ohms. If the resistance values are not within specifications, or the increase in resistance is not smooth, the control assembly must be replaced.

GM C65 Troubleshooting

The C65 Tempmatic system is used in several versions of General Motors vehicles. Each division of GM uses a different version of the C65 system requiring the use of test procedures for the specific system being serviced. The following test procedure is typical. Perform the functional test to check the operation of the system; the other tests are performed as directed in the functional test.

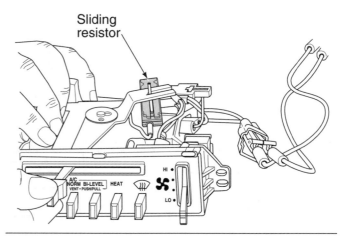

Figure 32-5 Testing the control assembly sliding resistor. Resistor values should be within specifications and should change smoothly. (Courtesy of Chrysler Corporation)

CAUTION *Connect the vehicle exhaust system to the shop ventilation system before starting the engine.*

To perform the functional test, the engine must be warmed to normal operating temperature. Leave the engine running and place the control assembly in the OFF position and the temperature control to 90. Observe the control assembly lamps. Continue the test depending on the attitude of the lamps; if they do not light begin with step 1, if they light begin at step 2.

1. Check the fuse and power supply to the control assembly. If the fuse is good and 12 volts is applied to the control assembly, go to step 2. If the fuse is blown, check the circuit for the cause. If the problem is not in the circuit, replace the control assembly.

2. Place the control assembly in the heat and high fan positions. Check for air coming from the heater and defogger outlets. The air exiting the vents should be hot. If there is no air flow, check the operation of the blower motor circuit. If the air is too cold, go to the TEMPERATURE SELECTED IS ALWAYS COLD TEST.

3. Set the control assembly to the DEFOG position and check for an increase in air flow at the defogger outlets. If there is no increase, perform the STRONG AIR FLOW FROM DEFOG OUTLETS WITH HEAT SELECTED TEST.

4. Set the control assembly to the VENT and 60 positions. Check for air flow from the vent outlets. The air temperature should decrease. If there is no air flow through the vent outlets, perform the NO AIR FROM A/C OUTLETS TEST. If the air is not cooler, perform the TEMPERATURE SELECTED IS ALWAYS HOT TEST.

5. Place the control assembly in the BI-LEVEL position and check for a decrease in air temperature from the heater outlets. If there is no decrease in air temperature, check the compressor clutch circuit.

6. Place the control assembly to the NORMAL position and check for cool air from the vent and heater outlets. There should not be any air flow from the heat outlets. If there is, perform the STRONG AIR FROM HEATER OUTLETS WITH VENT SELECTED.

7. Place the control assembly to the MAX position and check for an increase in air flow from the vent outlets. Also listen for an increase in blower speed. If there is no increase in air flow or blower speed, perform the NO MAXIMUM A/C TEST.

8. Turn the blower motor off and place the control assembly in the OFF position. If the blower motor does not turn off there is a short in the motor circuit. If the motor turns off, the system is operating properly.

Strong Air Flow From Defog Outlets With Heat Selected Test

If the function test directed you to perform this test, follow the service manual procedures for removing the control assembly from the instrument panel. Leave the wire harness connected to the control assembly. Place the controls in the HEAT and 60 settings. Connect a voltmeter between control assembly terminal 4 and case ground. There should be less than 1 volt between these connections. If battery voltage is present, replace the control assembly.

Leave the voltmeter connected as above, and place the control assembly in the DEFOG position. If battery voltage is present, check the A/C programmer for a poor connection of the red vacuum hose and binding of the vacuum motor and its linkage. If there is less than 1 volt indicated, check for an open in circuit 994 to the electronic climate control (ECC) programmer. If there are no circuit defects, replace the programmer.

No Air From A/C Outlets Test

If the function test directed you to perform this test, follow the service manual procedures for removing the control assembly from the instrument panel. Leave the wire harness connected to the control assembly. Place the controls in the A/C NORM and 60 settings. Connect a voltmeter between control assembly terminal 2 and case ground. There should be less than 1 volt between these connections. If battery voltage is present, replace the control assembly.

Leave the voltmeter connected as above, and place the control assembly in the DEFOG position. If battery voltage is present, check the A/C programmer for a poor connection of the blue vacuum hose and binding of the defog door vacuum motor and its linkage. If there is less than 1 volt indicated, check for an open in circuit 996 to the ECC programmer. If there are no circuit defects, replace the programmer.

Strong Air From Heater Outlets With Vent Selected Test

If the function test directed you to perform this test, follow the service manual procedures for removing the control assembly from the instrument panel. Leave the wire harness connected to the control assembly. Place the controls in the A/C NORM and 60 settings. Connect a voltmeter between control assembly terminal 3 and case ground. There should be less than 1 volt between these connections. If battery voltage is present, replace the control assembly.

Leave the voltmeter connected as above, and place the control assembly in the HEAT position. If battery voltage is present, check the A/C programmer for a poor connection of the yellow vacuum hose and binding of the bi-directional door vacuum motor and its linkage. If there is less than 1 volt indicated, check for an open in circuit 995 to the ECC programmer. If there are no circuit defects, replace the programmer.

No MAX A/C

If the function test directed you to perform this test, follow the service manual procedures for removing the control assembly from the instrument panel. Leave the wire harness connected to the control assembly. Place the controls in the A/C NORM and 60 settings. Connect a voltmeter between control assembly terminal 6 and case ground. There should be less than 1 volt between these connections. If battery voltage is present, replace the control assembly.

Leave the voltmeter connected as above, and place the control assembly in the DEFOG position. If battery voltage is present, check the A/C programmer for a poor connection of the orange vacuum hose and binding of the inlet door vacuum motor and its linkage. If there is less than 1 volt indicated, check for an open in circuit 997 to the ECC programmer. If there are no circuit defects, replace the programmer.

Temperature Selected Is Always Hot Test

If the function test directed you to perform this test, follow the service manual procedures for removing the control assembly from the instrument panel. Leave the wire harness connected to the control assembly. Place the controls in the A/C NORM and 60 settings. Connect a jumper wire between terminal A and B of connector C212 to bypass the in-car sensor. The air temperature should decrease. If not, connect a jumper wire between terminal N of the ECC programmer and a good ground. If the temperature decreases, check for an open in circuit 198.

If the temperature does not change when grounding terminal N, check for binding linkage. If the linkage is operating properly, replace the programmer.

If the temperature decreased when the in-car sensor was bypassed, connect a voltmeter between control assembly terminal A and case ground. The voltmeter should read between 4 and 5 volts. Observe the meter reading while changing the temperature setting from 60 to 90. If there is no change in voltage, replace the control assembly. If the voltage reading increases, replace the in-car sensor.

Temperature Selected Is Always Cold

If the function test directed you to perform this test, follow the service manual procedures for removing the control assembly from the instrument panel. Leave the wire harness connected to the control assembly. Place the controls in the HEAT and 90 settings. Unplug connector C212 while observing the temperature of the air flow. If the temperature increases, replace the in-car sensor. If the air temperature does not change, check for a short in circuit 198. If there is no short in the circuit then replace the programmer.

Temperature Is Controlled But Is Inaccurate Test

The C65 system is designed to maintain a temperature within 4 degrees of that selected by the driver. If the temperature is controlled but it is inaccurate, perform this test to isolate the fault.

If the temperature inside of the vehicle seems to be warmer on hot days, check the ambient sensor for a short. Also check circuit 61 for a short to ground. If the temperature inside of the vehicle feels cooler on cold days, check the ambient sensor and circuit 61 for an open.

If the temperature inside of the vehicle is always lower or higher than that selected by the driver, adjust the programmer. The programmer is adjusted by the plastic head screw located on the side of the programmer. Turn the adjusting screw 1/8 of a turn at a time until the temperature control works properly.

Electronic Automatic Temperature Control System Diagnosis

Diagnostics of electronic automatic temperature control (EATC) systems depend on system design. There are two basic system designs: those that use their own microprocessor and those that incorporate the controls of the system into the BCM.

Separate Microprocessor-Controlled Systems

Most EATC systems that use a separate microprocessor have it contained in the control assembly (Figure 32-6). A majority of these systems provide a means of self-diagnostics and have a method of retrieving trouble codes.

Chrysler EATC System Troubleshooting

Before entering self-diagnostics, start the vehicle and allow it to reach normal operating temperature. Make

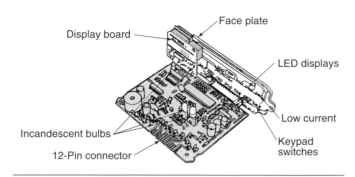

Figure 32-6 Most EATC systems use a separate microprocessor that is located in the control assembly. (Courtesy of Chrysler Corporation)

sure all exterior lights are off. Then press the PANEL button. If the display illuminates, self-diagnostic mode can be entered. If the display does not illuminate, check the fuses and circuits to the control assembly. If the circuits are good, replace the ATC computer.

The self-diagnostic mode is entered by pressing the BI-LEVEL, FLOOR, and DEFROST buttons at the same time (Figure 32-7). If no trouble codes are present, the self-test program will be completed within 90 seconds and display a 75.

During the process of running the self-diagnostic tests, the technician must make four observations that the computer is not able to make by itself. When the test is first initiated, all of the display symbols and indicators should illuminate. The blower motor should operate at its highest speed. Also, air should flow through the panel outlets, and the air temperature should become hot then cycle to cold. The diagnostic flow chart will direct the technician to the correct test to perform if any of the previous functions fail (Figure 32-8). The proper procedures for an observed failure are found in Figure 32-9.

If a fault is detected in the system, a trouble code will be flashed on the display panel. To resume the test, record the trouble code then press the PANEL button.

Refer to the service manual to diagnose the trouble codes received.

Ford EATC System Troubleshooting

To correctly diagnose Ford's EATC system you will need the exact system description as well as the exact procedures for trouble code retrieval. Ford uses different versions of EATC systems that have different diagnostic capabilities. The following is a typical service example of performing the self-test. However, this is not the same procedure used on all Ford systems.

1. Turn the ignition switch to the RUN position.

2. Place the temperature selector to the 90 setting and select the OFF mode.

3. Wait 40 seconds while observing the display panel. If the VFD display begins to flash, there is a malfunction in the blend actuator circuit, the actuator, or the control assembly. If the LED light begins to flash, this indicates there is a malfunction in one of the other actuator circuits, the actuators, or the control assembly.

4. If no flashing of displays occurs, place the temperature selection to 60 and select the DEF mode.

5. Wait 40 seconds while observing the VFD and LED displays. If there are no malfunctions in the actuator drive or feedback circuits, the displays will not flash.

6. Regardless of whether or not flashing displays were indicated, continue with self- diagnostics. Press the OFF and DEFROST buttons at the same time.

7. Within 2 seconds, press the AUTO button.

Once you have entered self-diagnostics, if an 88 is displayed, there are no trouble codes present. If there are any trouble codes retrieved, they will be displayed in sequence until the COOLER button is pressed. Always exit self-test mode by pressing the COOLER button before turning the ignition switch to the OFF position. Refer to the trouble code chart (Figure 32-10).

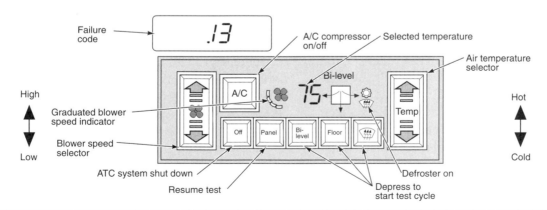

Figure 32-7 Use the panel buttons to enter diagnostics. (Courtesy of Chrysler Corporation)

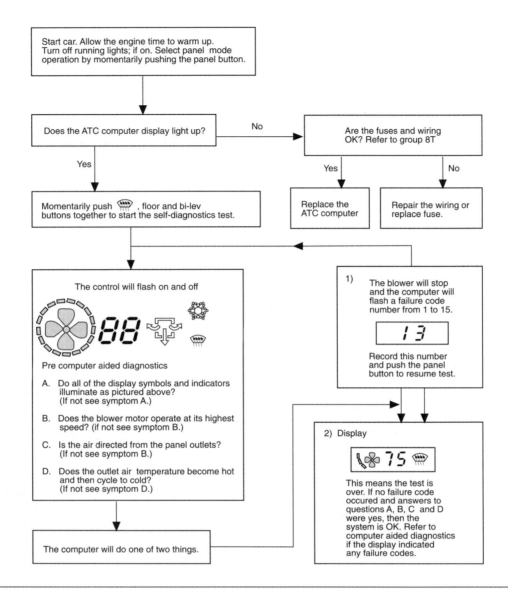

Figure 32-8 Chrysler EATC diagnostic flow chart. While the self-test is being performed, the technician must answer four questions. (Courtesy of Chrysler Corporation)

When service repairs have been performed on the system, rerun the self-test to confirm all faults have been corrected.

GM ETCC System Troubleshooting

General Motors uses several different versions of the microprocessor-controlled electronic touch climate control (ETCC) system. Depending on the GM division, and system design, the door controls can be by either vacuum or electric servomotors. Methods of entering diagnostics also vary between divisions and models. For this reason you will need to have the correct service manual for the system being serviced in order to perform correct diagnostic procedures. Knowledge of one ETCC system type is no guarantee of being able to service other ETCC sys-

tems without the use of the service manual. This is true even between models of the same GM division.

Even though the methods and procedures are different between the various system designs, you are able to rely on your knowledge of electrical and electronic component diagnosis to follow the service manual steps and find the cause of the fault. Once a trouble code is retrieved, refer to the diagnostic chart for directions in performing the tests required to isolate the fault.

Many GM EATC systems can be checked for proper operation by using a functional chart (Figure 32-11). In addition, troubleshooting charts that correspond with fault symptoms are a great help to the technician (Figure 32-12).

"NO" ANSWERS	PROBABLE CAUSE	PROCEDURE
A	1) Control	a Replace the Control Module
B	1) Wiring problem 2) Power/Vacuum Module	a) CAUTION: STAY CLEAR OF THE BLOWER MOTOR WHEEL. POWER/VACUUM HEAT SINK IS HOT (12 VOLTS), DO NOT RUN THE POWER/VACUUM MODULE EXCESSIVELY (10 MINUTES) WITH THE UNIT REMOVED FROM THE A/C HEATER HOUSING. b) Check to see if connections are made at the blower motor and at Power/Vacuum Module. c) Did the diagnostic test give an error code of 8 or 12? If yes, refer to Fault Code Page. If no error code then check 30 AMP fuse for blower motor (fuse #4). d) Disconnect blower motor, check voltage at connector (green is +, black is −). Reading should be 3 to 12 volts for 1 to 8 bar segment on the display. If correct then problem is the motor. e) If blower voltage is not correct, then measure volts at green wire to vehicle ground (not blower ground) should be 12 volts (key on). If the voltage is OK, then replace the P/V Module.
C	1) Vacuum Leakage 2) Power/Vacuum Leakage	a) Service, if any codes are found. b) Check all connectors. c) Disconnect 7 port vacuum connector and connect it to a "manual control" and test each mode. Test Check Valve selecting the Panel Mode and disconnecting engine vacuum to see if mode changes quickly. d) Try a new Power/Vacuum Mod.
D	1) Refrigeration System 2) Heater System 3) Blend-Air Door	a) Complete diagnostics test, refer to Fault Code Page if error occurs. b) If a temperature difference of at least 40 Fahrenheit degrees is felt during the diagnostic test, then the Blend-Air door is engaged in the Servo Motor Actuator. If temp difference is less than 40 degrees, then a possible problem is the Blend-Air Door operation. c) Check heater system. 85 temp setting is full heat and 65 is full cool. d) Check refrigeration system. NOTE: Panel, A/C and holding down the bottom of the TEMP button for 4 seconds once 65 is obtained will cause Max A/C.

Figure 32-9 If the technician answers "NO" to any of the questions asked during the self-test, this diagnostic chart is used to isolate the fault. (Courtesy of Chrysler Corporation)

CODE	SYMPTOM	POSSIBLE SOURCE
1	Blend actuator is out of position. VFD flashes.	• Open circuit in one or more of the actuator leads • Actuator output arm jammed. • Actuator inoperative. • Control assembly inoperative.
2	Mode actuator is out of position. LED flashes.	• Same as 1.
3	Pan/Def actuator is out of position. LED flashes.	• Same as 1.
4	Fresh air/recirculator actuator is out of position. LED flashes.	• Same as 1.
1, 5	Blend actuator output shorted. VFD flashes.	• Outputs A or B shorted to ground, or to the supply voltage or together. • Actuator inoperative. • Control assembly inoperative.
2, 6	Mode actuator output shorted. LED flashes.	• Same as 1, 5.
3, 7	Pan/Def actuator output shorted. LED flashes.	• The actuator output is shorted to the supply voltage. • Actuator inoperative. • Control assembly inoperative.
4, 8	Fresh air/recirculator actuator output shorted. LED flashes.	• Same as 3, 7.
9	No failures found. See Supplemental Diagnosis.	
10, 11	A/C clutch never on.	• Circuit 321 open. • BSC inoperative. • Control assembly inoperative.
10, 11	A/C clutch always on.	• Circuit 321 shorted to ground. • BSC inoperative. • Control assembly inoperative.
12	System stays in full heat. In-car temperature must be stabilized above 60°F for this test to be valid.	• Circuit 788, 470, 767, or 790 is open. • The ambient or in-car temperature sensor is inoperative.
13	System stays in full A/C.	• Remove control assembly connectors. Measure the resistance between pin 10 of connector #1 and pin 2 of connector #2. • If the resistance is less than 3K ohms, check the wiring and in-car and ambient temperature sensors. • If the resistance is greater than 3K ohms, replace the control assembly.
14	Blower always at maximum speed.	• Turn off ignition. Remove connector #2 from control assembly. Using a small screwdriver remove terminal #5 from the connector. Replace connector, tape terminal end and turn on ignition. • If blower still at maximum speed, then check circuit 184 and the BSC. • If the blower stops then the control assembly is inoperative.
15	Blower never runs.	• Circuit 184 shorted to the power supply. • BSC inoperative. • Control assembly inoperative.

Figure 32-10 Trouble code chart for Ford EATC system. (Reprinted with permission of Ford Motor Company)

ELECTRONIC CLIMATE CONTROL FUNCTIONAL TEST

A LOGICAL SERVICE PROCEDURE IS AVAILABLE TO ISOLATE AN ECC PROBLEM. ALL SYSTEM DIAGNOSIS SHOULD BEGIN BY CONDUCTING THE FUNCTIONAL TEST IN THE EXACT ORDER IN WHICH IT IS PRESENTED. IF THE ANSWER TO ANY FUNCTIONAL TEST QUESTION IS NO, THEN TURN TO THE SPECIFIC TROUBLE TREE FOR ANALYSIS OF THE MALFUNCTION. IT IS ESSENTIAL THAT THE FUNCTIONAL TEST BE PERFORMED WITHOUT OMITTING STEPS BECAUSE OMISSIONS MAY RESULT IN THE INABILITY TO ISOLATE SPECIFIC MALFUNCTIONS.

IT IS NOT NECESSARY TO PERFORM THE FUNCTIONAL TEST AT THE BEGINNING OF EACH TROUBLE TREE IF THE FUNCTIONAL TEST HAS BEEN UTILIZED TO SELECT THE PROPER TROUBLE TREE. WHEN A MALFUNCTION HAS BEEN ISOLATED AND REPAIRED, THE TECHNICIAN SHOULD EXIT FROM THE TROUBLE TREE AT THE POINT OF REPAIR.

PROCEED AS FOLLOWS:

1. CHECK FUSES (CHECK STOP HAZARD OPERATION TO VERIFY STOP HAZARD FUSE)
2. WARM ENGINE BEFORE TESTS
3. CHECK LED ABOVE EACH BUTTON AS TEST IS PERFORMED.

SYSTEM CHECKS	CONTROL SETTING	TROUBLE TREES
1. DO THE MPG AND CONTROL HEAD DISPLAY? (MPG IS OPTIONAL)	ALL	1
2. DO COOLER AND WARMER PUSHBUTTONS OPERATE?	ALL	1
3. SET AT 60°		
A. DOES THE BLOWER OPERATE?	LO, AUTO, HI	2A
B. IS THERE LO BLOWER?	LO	2B
C. IS THERE HI BLOWER?	HI	2C
D. IS THERE AIR FROM A/C OUTLETS?	ECON, LO, AUTO, HI	3
E. DOES COMP CLUTCH ENGAGE?	LO, AUTO, HI	4
F. DOES COMP CLUTCH DISENGAGE?	OFF, ECON	5
G. IS THERE COLD AIR FROM A/C OUTLETS? IF NOT:	LO, AUTO, HI	6
1) IS THERE HEAT ONLY?		
2) IS THERE INSUFFICIENT COOLING?		7
H. *DOES RECIRC DOOR OPEN FULLY?*	AUTO, HI	8
4. *SET AT 90°		
A. IS THERE SUFFICIENT HEAT FROM HEATER OUTLETS?	AUTO	9
5. *SET AT 85°		
A. IS THERE WARM OR HOT AIR AT HEATER OUTLET?	LO, AUTO	10
6. DOES FRONT DEFROSTER OPERATE?	FRT DEF	11
7. DOES REAR DEFROSTER OPERATE?	RR DEF	12
8. DOES REAR DEFROSTER TURN OFF?	RR DEF OFF	13

*WAIT ONE OR TWO MINUTES FOR SYSTEM TO CHANGE.

Figure 32-11 GM EATC function test. (Courtesy of General Motors Corporation, Service Technology Group)

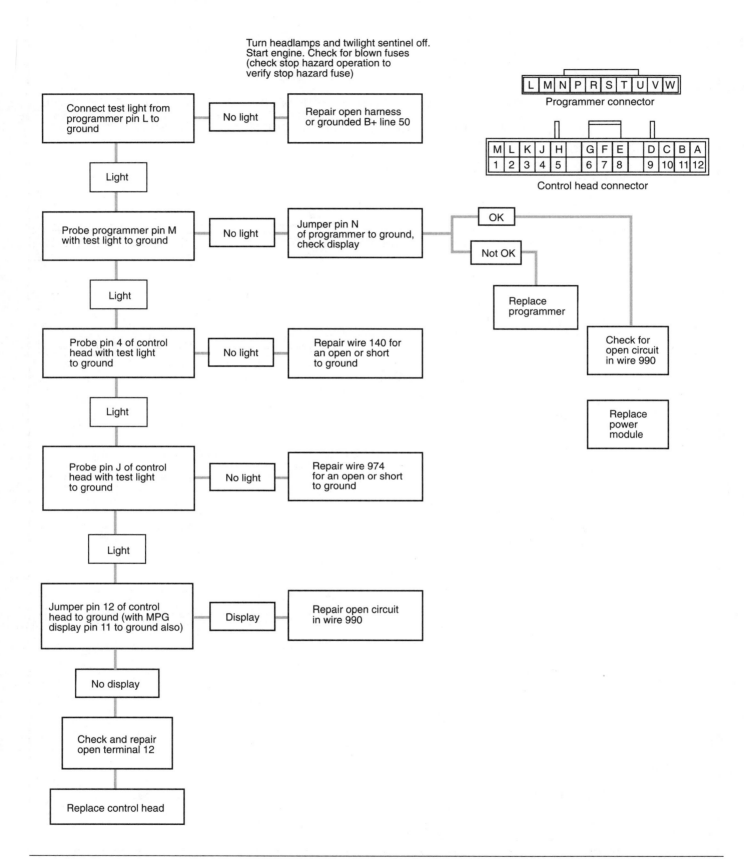

Figure 32-12 GM EATC troubleshooting chart. (Courtesy of General Motors Corporation, Service Technology Group)

Nissan EATC Diagnostics

Self-diagnostic mode of the EATC system used on Nissan Maxima is entered by starting the engine and pressing the off button in the A/C control head for 5 seconds. The off button must be pressed within 10 seconds after the engine is started, and the fresh vent lever must be in the off position. The self-diagnostic mode may be canceled by pressing the auto button or turning off the ignition switch.

The self-diagnostic tests are completed in five steps. The up arrow for temperature setting on the A/C control head is pressed to move to the next step. When the down arrow for temperature setting is pressed, the diagnostic system returns to the previous step (Figure 32-13).

Step 1. During step 1, all the LEDs and display segments are illuminated. If some segments or LEDs are not illuminated, these components are defective.

Step 2. When the up temperature control arrow is pressed to move to step 2, the input sensors are tested by the A/C computer. If there are no defects in the input sensors, 20 is displayed in the A/C control head. If any of the input sensors are defective, a code for that sensor is displayed in the control head (Figure 32-14).

Step 3. When the up temperature arrow is pressed to move to step 3, the mode door position switches are tested. If all these position switches are in satisfactory condition, 30 is displayed in the A/C control head. A defect in one of the position switches results in a code display in the A/C control head. These codes range from 31 to 36 (Figure 32-15).

Step 4. If the up temperature arrow is pressed to move to step 4, the A/C control head displays 41. In this mode, the A/C computer positions the mode door in the vent position. The intake door is in the recirculation (REC) position, and the air mix door is in the full cold position. The defrost (DEF) button is pressed to move to the next mode, which displays 42 in the A/C control head. There are six modes in step 4 and each mode is represented by a number in the A/C control head. These numbers range from 41 to 46. The DEF button is used to select the next mode. In each mode, the A/C computer commands specific door positions, blower motor voltage, and compressor clutch operation (Figure 32-16). Door operation may be checked by the air discharge from the various ducts.

Step 5. When the up temperature arrow is pressed to select step 5, the temperature detected by each input sensor is displayed. After this mode is entered, 5 is displayed in the A/C control head. If the DEF button is pressed, the temperature sensed by the ambient sensor is displayed in the A/C control head. Pressing the DEF button a second time displays the temperature detected by the in-vehicle sensor. If the DEF button is pressed again, the temperature detected by the intake sensor is displayed (Figure 32-17). This sensor is mounted near the evaporator outlet. When the temperature displayed varies significantly from the actual temperature, the sensor and connecting wires should be tested with an ohmmeter and a voltmeter.

At the end of step 5, the blower speed button may be pressed to enter the auxiliary mode. In this mode, the temperature on the display may be adjusted so it is the same as the in-vehicle temperature felt by the driver. After the auxiliary mode is entered, press the up or down temperature control buttons until the A/C control head displays the same temperature as the temperature inside the vehicle.

BCM EATC-Controlled Systems

Because the BCM-controlled EATC system incorporates many different microprocessors within its system, diagnostics can be very complex (Figure 32-18). Faults that seem to be unrelated to the EATC system may cause the system to malfunction. Since it was first introduced in 1986, BCM-controlled EATC systems have become increasingly popular on many vehicles. Each model year brings forth revisions and improvements in the system that also require different diagnostic procedures.

It would be impossible to describe the diagnostic procedures required to service the many systems in use. For this reason, the technician must have the correct service manual for the system they are diagnosing. There are several different methods used to retrieve trouble codes. Be sure to follow the correct procedure.

Once the codes have been retrieved, refer to the correct diagnostic chart. This pinpoint test will lead to the fault in a logical manner. After all repairs to the system are complete, follow the service manual procedure for erasing codes and for resetting the system. Rerun the diagnostic test to confirm the system is operating properly.

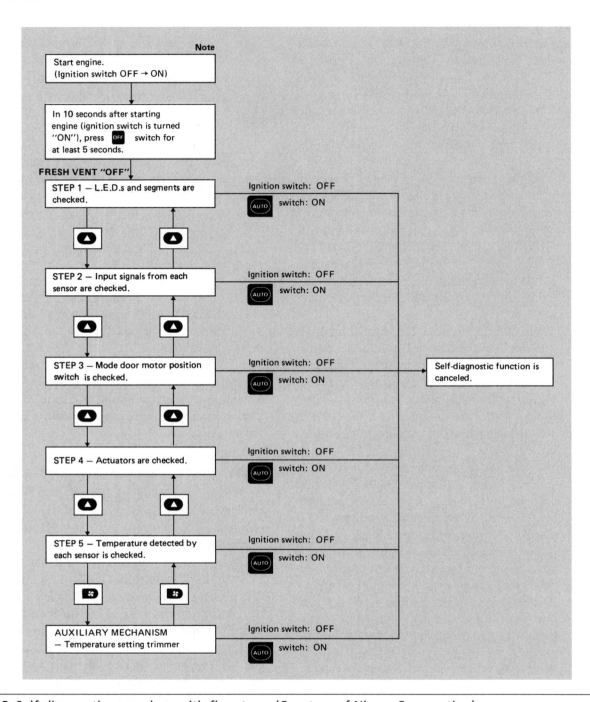

Figure 32-13 Self-diagnostic procedure with five steps. (Courtesy of Nissan Corporation)

Self-Diagnostic Tests on Automatic Temperature Control Systems

Many BCM computer-controlled air conditioning systems have self-diagnostic capabilities. These self-diagnostic tests vary depending on the vehicle make and model year. Always follow the instructions in the vehicle manufacturer's service manual. On some vehicles, such as Chrysler LH and LHS cars, the self-diagnostic mode for the automatic temperature control (ATC) system is entered by pressing the floor, mix, and defrost buttons simultaneously with the engine running and the vehicle not moving (Figure 32-19).

When the diagnostic mode is entered, the digital display in the A/C control head begins blinking. The display continues blinking until self-diagnostic tests are completed. While the display is blinking, the door actuator motors are calibrated to the unit on which they are installed. Any diagnostic trouble codes are displayed. When one DTC is displayed, the PANEL button may be pressed to scroll through any other DTCs. DTC numbers range from 23 to 36 (Figure 32-20). If there are no DTCs, the system returns to normal operation indicated by the temperature display. When DTCs are displayed, voltmeter or ohmmeter tests are usually required to locate the exact cause of the problem. DTCs may be

erased with a scan tester or by disconnecting battery voltage from the body computer for 10 minutes.

Scan Tester Diagnosis of BCM-Controlled Air Conditioning Systems

WARNING *Do not disconnect the wiring connector from any computer-controlled A/C system component with the ignition switch on. This action may result in damage to the A/C computer.*

WARNING *Do not connect or disconnect the scan tester to or from the DLC with the ignition switch on. This action may damage the A/C computer.*

Code No.	Malfunctioning sensor (including circuits)
21	Ambient sensor
22	In-vehicle sensor
24	Intake sensor
25	Sunload sensor*1
26	P.B.R.

Figure 32-14 In step 2, a defective input sensor causes a code to be displayed. (Courtesy of Nissan Corporation)

Code No.	Malfunctioning mode door motor position switch (including circuits)
31	VENT
32	B/L
33	B/L
34	FOOT/DEF 1
35	FOOT/DEF 2
36	DEF

Figure 32-15 In step 3, a defective mode door switch causes a code to be displayed. (Courtesy of Nissan Corporation)

Code No.	Actuators test pattern				
	Mode door	Intake door	Air mix door	Blower motor	Compressor
41	VENT	REC	Full Cold	4 - 5V	ON
42	B/L	REC	Full Cold	9 - 11V	ON
43	B/L	20% FRE	Full Hot	7 - 9V	ON
44	D/F 1	FRE	Full Hot	7 - 9V	OFF
45	D/F 2	FRE	Full Hot	7 - 9V	OFF
46	DEF	FRE	Full Hot	10 - 12V	ON

Figure 32-16 Step 4 contains six modes. In each mode the computer commands a specific door position, blower motor voltage, and compressor clutch operation (Courtesy of Nissan Corporation)

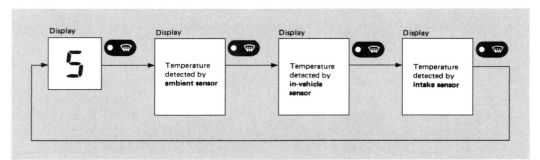

Figure 32-17 Step 5 displays the temperature detected by each sensor. (Courtesy of Nissan Corporation)

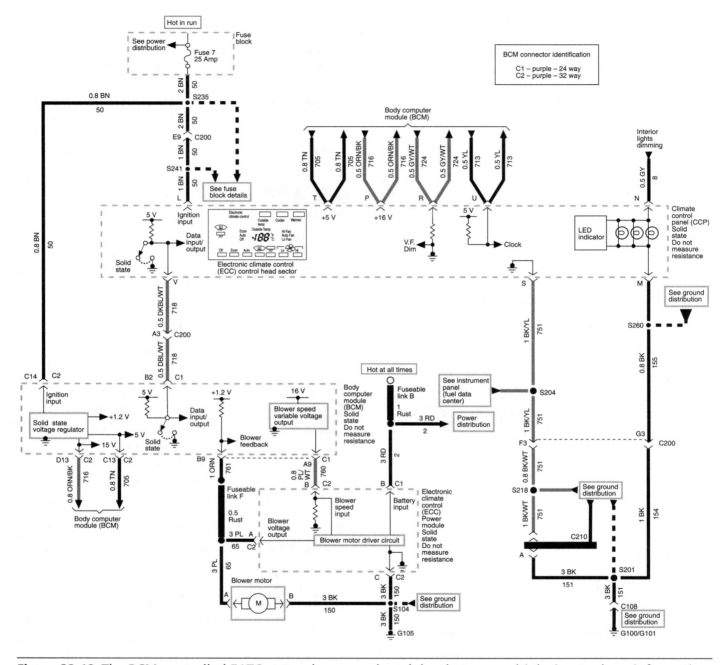

Figure 32-18 The BCM-controlled EATC system has several modules that use multiplexing to share information. (Courtesy of General Motors Corporation, Service Technology Group)

The following is a typical procedure for using a scan tool to diagnose the ATC system. Refer to Photo Sequence 15 for use of the DRB III scan tool.

The scan tester diagnosis of computer-controlled air conditioning systems varies depending on the vehicle make and model year. Always use the scan tester diagnostic procedures in the vehicle manufacturer's service manual or in the scan tester manufacturer's manual. The proper module for the vehicle make, model, and year must

be installed in the scan tester. The scan tester must be connected to the data link connector. Usually the scan tool is capable of displaying ATC fault codes and the status of system switches and sensors. Also, it is able to activate relays, motors, and other actuators used in the system.

When the initial menu appears on the scan tester, select "Climate Control." On the first climate control menu, press the proper number to select the desired function (Figure 32-21).

Typical Procedure for Performing a Scan Tester Diagnosis of an Automatic Temperature Control System

P12-1 Insert the proper module (if needed) for the vehicle being serviced into the scan tester.

P12-2 With the ignition switch in the LOCK position, connect the scan tester power lead (if needed) to the cigarette lighter socket.

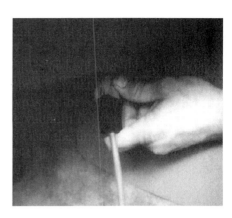

P12-3 Connect the scan tester data lead to the DLC.

P12-4 Program the scan tester for the vehicle being serviced. Select climate control on the SELECT SYSTEM screen.

P12-5 Select SYSTEM TEST, READ FAULTS, SENSOR DISPLAY, ACTUATOR TESTS, or ADJUSTMENTS from the climate control menu screen.

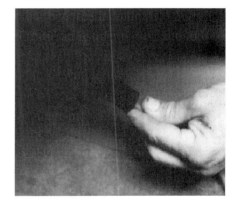

P12-6 When completed, turn the ignition switch to the LOCK position, then disconnect the data lead from the DLC and the scan tester power cord (if used).

Summary

❑ Before an air conditioning system is diagnosed, a preliminary A/C system inspection must be completed.

❑ When diagnosing the system, remember the SATC system is a control system and faults within the evaporator and heater systems will have an effect on its operation.

❑ The troubleshooting chart will direct the technician to test certain components based on the symptom reported by the vehicle owner.

❑ A paper test is a quick method of testing the aspirator assembly to verify air flow to the in-car sensor.

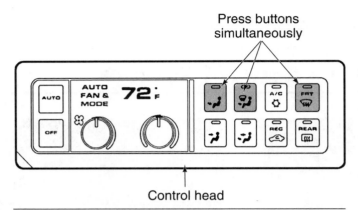

Press buttons
simultaneously

Control head

Figure 32-19 The self-diagnostic mode is entered by pressing the floor, mix, and defrost buttons simultaneously with the engine running and the vehicle not moving. (Courtesy of Chrysler Corporation)

DIAGNOSTIC TROUBLE CODE CHART

CODE	DESCRIPTION
23	ATC Blend Door Feedback Failure
24	ATC Mode Door Feedback Failure
25	Ambient Sensor
26	ATC In-Car Sensor
27	Sun Sensor Failure
31	ATC Recirculation Door Stall Failure
32	ATC Blend Door Stall Failure
33	ATC Mode Door Stall Failure
34	Engine Temperature Message not Received
35	Evaporator Sensor Failure
36	ATC Head Communication Failure

Figure 32-20 Diagnostic trouble codes related to the ATC system. (Courtesy of Chrysler Corporation)

❑ To test the ambient sensor, remove it from its socket and measure its resistance when the temperature is between 70°F and 80°F (21°C and 27°C).

❑ The control assembly resistor can be tested using an ohmmeter to measure changes in resistance.

❑ Many systems require a function test be performed as the first diagnostic test. This test is also performed to confirm repairs.

❑ Most EATC systems provide a method of retrieving DTCs, either with or without a scan tool.

❑ Usually the scan tool is capable of displaying ATC fault codes and the statues of system switches and sensors. Also, it is able to activate relays, motors, and other actuators used in the system.

ASE Style Review Questions

1. The service manual has instructed the technician to test the ambient temperature sensor. Technician A says to measure its resistance when the temperature is between 70 and 80 degrees F. Technician B says connecting the ohmmeter for longer than 5 seconds may result in false readings. Who is correct?
   a. A only
   b. B only
   c. Both A and B
   d. Neither A nor B

2. A customer says there is no air flow from the A/C outlets. Technician A says the control assembly could be faulty. Technician B says the ambient temperature sensor is open. Who is correct?
   a. A only
   b. B only
   c. Both A and B
   d. Neither A nor B

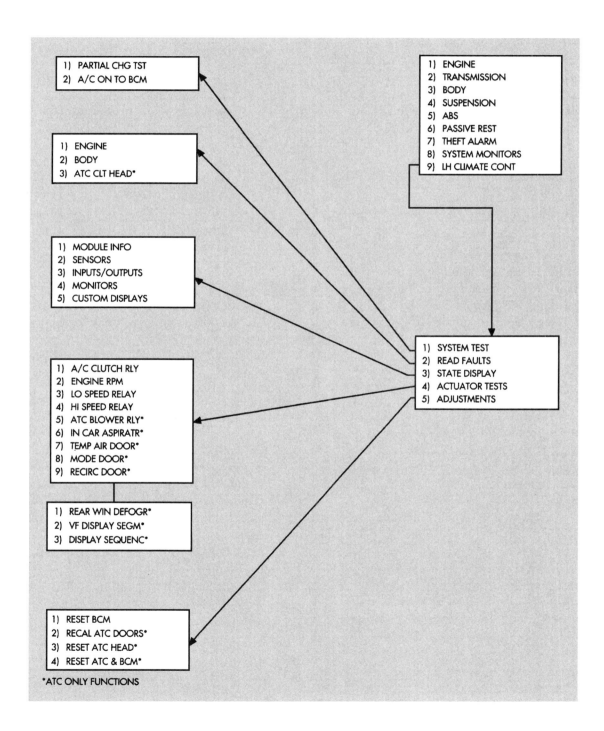

Figure 32-21 Scan tester diagnosis of ATC system. (Courtesy of Chrysler Corporation)

3. The customer says the air entering the vehicle is always hotter than that selected. Technician A says the fault could be a faulty in-vehicle sensor. Technician B says the programmer is defective. Who is correct?
   a. A only
   b. B only
   c. Both A and B
   d. Neither A nor B

4. Diagnosis of EATC systems using a separate microprocessor is being discussed. Technician A says to enter the self-test, most systems require that a series of panel buttons be pressed. Technician B says it may be required to observe certain functions during the diagnostic test. Who is correct?
   a. A only
   b. B only
   c. Both A and B
   d. Neither A nor B

5. Diagnosis of the Ford EATC system is being discussed. Technician A says all Ford systems retrieve trouble codes in the same manner. Technician B says on some systems, codes are received after pressing the OFF and DEFROST buttons at the same time then the AUTO button. Who is correct?
   a. A only
   b. B only
   c. Both A and B
   d. Neither A nor B

6. While discussing the self-diagnostic mode: Technician A says on Chrysler LH and LHS cars, this mode is entered with the ignition switch on and the engine not running. Technician B says on these cars, the self-diagnostic mode may be entered while driving the vehicle. Who is correct?
   a. A only
   b. B only
   c. Both A and B
   d. Neither A nor B

7. The self-diagnostic mode is being discussed. Technician A says on some automatic A/C systems, such as those on the Nissan Maxima, the self-diagnostic mode is entered by pressing the A/C off button for 5 seconds within 10 seconds after the engine is started. Technician B says on these automatic air conditioning systems, the self-diagnostic mode is entered by pressing the defrost and off buttons at the same time. Who is correct?
   a. A only
   b. B only
   c. Both A and B
   d. Neither A nor B

8. Self-diagnostic mode on the automatic A/C system in a Nissan Maxima is being discussed. Technician A says there are seven steps in the self-diagnostic mode. Technician B says the auto button is pressed to move to the next mode in the sequence. Who is correct?
   a. A only
   b. B only
   c. Both A and B
   d. Neither A nor B

9. The self-diagnostic mode on an automatic A/C system in a Nissan Maxima is being discussed. Technician A says step 1 displays faults in the input sensors. Technician B says step 2 illuminates all the A/C display segments and LEDs. Who is correct?
   a. A only
   b. B only
   c. Both A and B
   d. Neither A nor B

10. Scan tester diagnosis of computer-controlled A/C systems is being discussed. Technician A says the scanner is not able to show the status of the A/C switch and sensors. Technician B says activations tests are available. Who is correct?
    a. A only
    b. B only
    c. Both A and B
    d. Neither A nor B

33 Vehicle Accessories

Objective

Upon completion and review of this chapter, you should be able to:

❏ Define the purpose of the cruise control system.

❏ Explain the operation of the components of the cruise control system.

❏ Explain the basic operation of electromechanical cruise control systems.

❏ Explain the operating principles of the electronic cruise control system.

❏ List and describe the safety modes incorporated into electronic cruise control systems.

❏ Define integrated and nonintegrated cruise control systems and explain the difference.

❏ Explain the operating principles of the memory seat feature.

❏ Describe the control concepts of electronically controlled sunroofs.

❏ Detail the operation of common anti-theft systems.

❏ Explain the function of the pass-key security system.

❏ Explain the purpose and operation of automatic door lock systems.

❏ Detail the operation of the keyless entry system.

❏ Explain the operating principles of Ford's and GM's heated windshield systems.

❏ Describe the purpose and configuration of vehicle sound systems.

Introduction

Most vehicle accessories are electrically or electronically controlled. These systems are designed to make the driver more comfortable, make the vehicle safer to drive, and make operation of the vehicle easier. There are numerous accessories that may be installed on today's vehicles. This chapter details some of the most common vehicle accessories.

Introduction to Electronic Cruise Control Systems

Cruise control is a system that allows the vehicle to maintain a preset speed with the driver's foot off the accelerator pedal. Cruise control was first introduced in the 1960s for the purpose of reducing driver fatigue. When engaged, the cruise control system sets the throttle position to maintain the desired vehicle speed.

Most cruise control systems are a combination of electrical and mechanical components. The components used depend on manufacturer and system design. However, the operating principles are similar.

Electromechanical Systems

A review of **electromechanical** systems is helpful when trying to understand the operation of electronic cruise control. Electromechanical cruise control receives its name because of the two subsystems: the electrical portion, and the mechanical portion. The illustration (Figure 33-1) shows the location of the main components of the electromechanical control system. These components include:

1. Cruise control switch: The control switch is located on the turn signal stock or in the steering wheel (Figure 33-2). The switch assembly is actually a set of driver-operated switches. On most systems, the switches are ON, SET/ACCEL, COAST, and RESUME.

2. Transducer: The transducer receives vehicle speed signals through the speedometer cable. Electrical signals from the control switch, brake switch, or clutch switch are sent to the transducer. In addition, the transducer receives engine manifold vacuum. It regulates the vacuum to the servo through the electrical signals received.

3. Servo: The servo controls throttle plate position. It is connected to the throttle plate by a rod, bead chain, or Bowden cable. The servo maintains the set speed by receiving a controlled amount of vacuum from the transducer. When vacuum is applied to the servo, the spring is compressed and the throttle plate is moved to increase speed (Figure 33-3). When the vacuum is released, the spring returns the throttle plate to reduce engine speed.

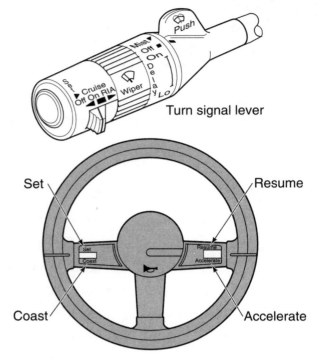

Mode control, steering wheel and on-off rocker switch, floor console

Figure 33-2 The control switch can be mounted on the turn signal stock or into the steering wheel. The switch is used to provide driver inputs for the system. (Courtesy of General Motors Corporation, Service Technology Group)

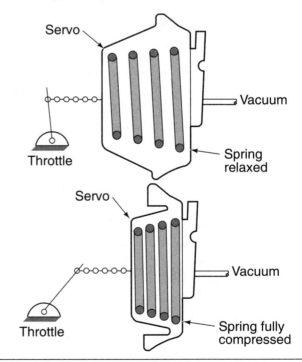

Figure 33-3 Cutaway view of the servo. Vacuum is used to compress the spring and open the throttle. (Courtesy of General Motors Corporation, Service Technology Group)

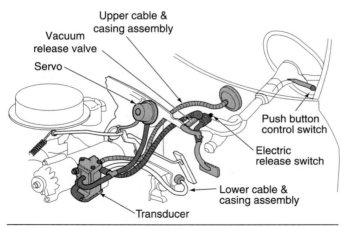

Figure 33-1 Components of typical electromechanical cruise control system. (Courtesy of General Motors Corporation, Service Technology Group)

4. Safety switches: When the brake pedal is depressed, the cruise control system is disengaged through electrical and vacuum switches (Figure 33-4). The switches are usually located on the brake pedal bracket. The two switches provide a fail-safe means of assuring that the cruise control is disengaged when the brakes are applied. If one of the switches fails, the other will still be able to return vehicle speed control over to the driver. Vehicles equipped with manual transmissions may use switches on the clutch pedal to disengage the system whenever the clutch pedal is depressed.

Some manufacturers combine the transducer and servo into one unit. They usually refer to this unit as a **servomotor**.

When the cruise control switch is in the "at rest" position, battery current flows through the switch to the resistance wire to the hold terminal of the transducer (Figure 33-5). Voltage to the hold terminal is too low to activate the solenoid coil because the resistor drops the voltage.

The transducer contains a rubber clutch with an operating arm. At speeds below 30 mph (48 km/h) the rubber clutch arm holds the low speed switch open. When speeds exceed 30 mph (48 km/h), the clutch arm rotates. The rotation of the arm allows the switch to close and the system can be engaged. By pushing the momentary contact switch in the SET position, the current flow is through the engage terminal of the transducer, through the low speed switch, to the solenoid coil. The resistor is bypassed and sufficient current is applied to the solenoid to activate it. When the momentary switch is released, the current flow is returned through the hold terminal. The current applied to the solenoid coil is sufficient to hold the solenoid in the activated position.

When the solenoid is activated, the vacuum valve opens to allow engine vacuum to the servo and the brake release valve. The air control valve is a variable orifice and is the control mechanism that adjusts vacuum level to the system. At lower speed settings, the air control valve bleeds off vacuum so less is sent to the servo. At higher set speeds, less vacuum is bled to allow for more throttle plate opening.

If the brake pedal is depressed, with the cruise control system engaged, the brake switch provides an alternate path to ground and bypasses the solenoid. With the solenoid deactivated, the vacuum valve closes and returns the throttle control over to the driver. At the same time, the resume solenoid is closed and the vacuum release valve opens to release the vacuum in the system.

Electronic Cruise Control

The electronic cruise control system uses an electronic module to operate the actuators that control throttle position (Figure 33-6). Electronic cruise control offers more precise speed control than the electromechanical system. In addition, other benefits include:

- More frequent throttle adjustments per second.
- More consistent speed increase/decrease when using the tap-up/tap-down feature.
- Greater correction of speed variation under loads.
- Rapid deceleration cut-off when deceleration rate exceeds programmed rates.
- Wheelspin cut-off when acceleration rate exceeds programmed parameters.
- System malfunction cut-off when the module determines there is a fault in the system.

Common Components

Common components of the electronic cruise control system include:

1. The control module: The module can be a separate cruise control module, the engine control module, or the body control module. The operation of the systems are similar regardless of the module used.

2. The control switch: Depending on system design, the control switch contacts apply the ground circuit through resistors. Because each resistor has a different value, a different voltage is applied to the control module. In some systems the control switch will send a 12-volt signal to different terminals of the control module.

3. The brake or clutch switch.

4. Vacuum release switch.

5. Servo: The servo operates on vacuum that is controlled by supply and vent valves. These operate from controller signals to solenoids.

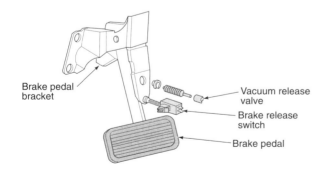

Figure 33-4 The brake release switch and the vacuum release switch work together to disengage the cruise control switch when the brake pedal is depressed. (Courtesy of General Motors Corporation, Service Technology Group)

Depending on system design, the sensors used as inputs to the control module include the vehicle speed sensor, servo position sensor, and throttle position sensor. Other inputs are provided by the brake switch, instrument panel switch, control switch, and the park-neutral switch.

The control module receives signals from the speed sensor and the control switch. When the vehicle speed is fast enough to allow cruise control operation and the driver pushes the SET button on the control switch, an electrical signal is sent to the controller. The voltage level received by the controller is set in memory. This signal is

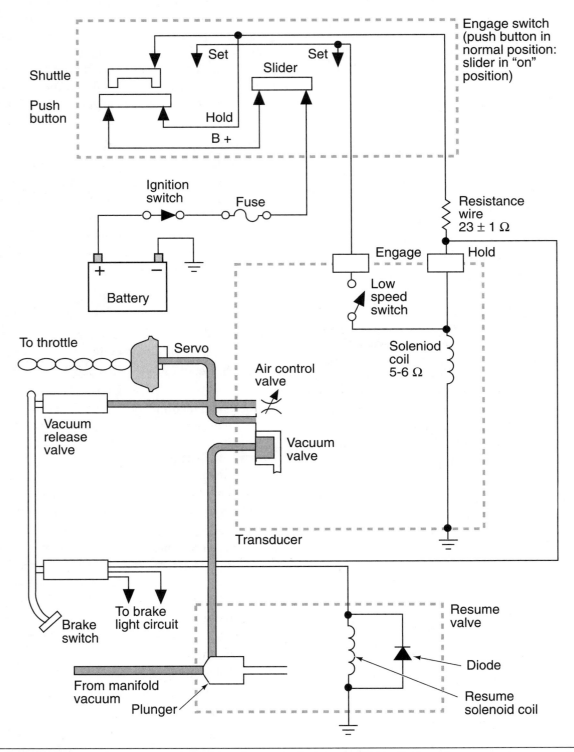

Figure 33-5 Electrical and vacuum schematic of a typical electromechanical cruise control system. (Courtesy of General Motors Corporation, Service Technology Group)

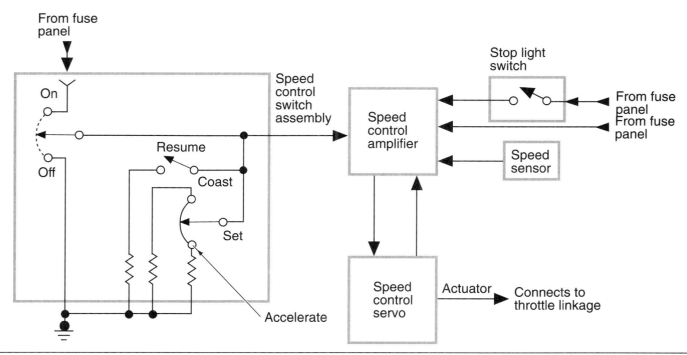

Figure 33-6 Block diagram of electronic cruise control system. (Courtesy of General Motors Corporation, Service Technology Group)

used to create two additional signals. The two signal values are set at 1/4 mph above and below the set speed. The module uses the comparator values to change vacuum levels at the servo to maintain set vehicle speed.

Three safety modes are operated by the control module:

1. **Rapid deceleration cutoff**: If the module determines that deceleration rate is greater than programmed values, it will disengage the cruise control system and return operation back over to the driver.

2. **Wheelspin cutoff**: If the control module determines that the acceleration rate is greater than programmed values, it will disengage the system.

3. **System malfunction cutoff**: The module checks the operation of the switches and circuits. If it determines there is a fault, it will disable the system.

The vacuum-modulated servo is the primary actuator. Vacuum to the servo is controlled by two solenoid valves: supply and vent. The vent valve is a normally open valve and the supply valve is normally closed (Figure 33-7). The servo receives signals from the controller to operate the solenoid valves to maintain a preset throttle position.

Principles of Operation

When the driver sends a SET signal to the controller it sets the voltage signals received from the vehicle speed sensor (VSS) into memory. It then determines the high

and low **comparators**. The signals that are created above and below the set speed are called high and low comparators.

The controller energizes the supply and vent valves to allow manifold vacuum or atmospheric pressure to enter the servo. The servo uses the vacuum and pressure to move the throttle and maintain the set speed. The vehicle speed is maintained by balancing the vacuum in the servo.

If the voltage signal from the VSS drops below the low comparator value, the control module energizes the supply valve solenoid to allow more vacuum into the servo and increases the throttle opening. When the VSS signal returns to a value within the comparator levels, the supply valve solenoid is de-energized.

If the VSS signal is greater than the high comparator value, the control module de-energizes the vent solenoid valve to release vacuum in the servo. The vehicle speed is reduced until the VSS signals are between the comparator values, at which time the control module will energize the vent valve solenoid again. This constant modulation of the supply and vent valves maintains vehicle speed.

During steady cruise conditions, both valves are closed and a constant vacuum is maintained in the servo.

System Examples

The following are some system examples that will illustrate some of the differences between manufacturers.

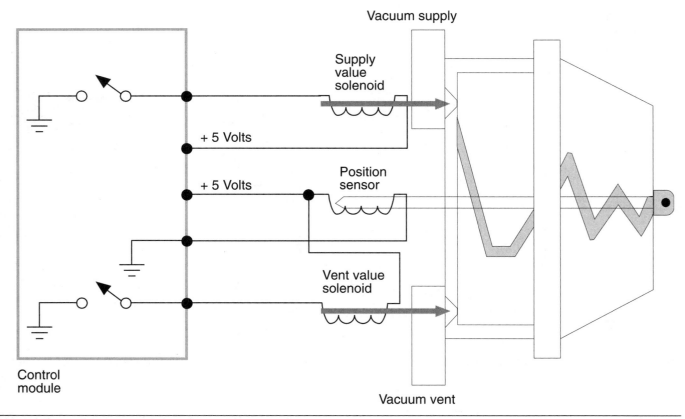

Figure 33-7 Servo valve operation in electronic control system. The servo position sensor informs the controller of servo operation and position.

The operating principles just discussed are basic to most systems. The difference is mainly in control module capabilities and accuracy of control.

General Motors Custom Cruise III Systems

The Custom Cruise III system is a common system used by General Motors. In this system the control module can be integrated or nonintegrated. In the **nonintegrated cruise control** system, the module can be a self-contained unit or built into the instrument cluster module. In 1987 the control module function was included into the PCM on some models. This system is referred to as an **integrated cruise control**. The system provides for tap-up and tap-down control that changes set vehicle speed in 1 mph increments.

The cruise control operation is one function of the PCM. Input signals from the control switches, the brake release switch, and speed sensor are received by the PCM. The PCM processes this information, in combination with engine control information, and commands the servo to open or close its valves to control vehicle speed. When the desired speed is selected by the driver through the SET switch, the PCM records the value in its mem-

ory. The PCM will pulse voltage signals to the vacuum and vent valves to maintain set vehicle speed.

The throttle position sensor informs the PCM of the position and motion of the servo. The feedback signal provides smooth throttle changes while the cruise control is engaged.

When the cruise control switch is turned on, a 12-volt signal is sent to the PCM that alerts the module to operate the cruise control system. When the SET switch is pressed, a 12-volt signal is delivered to the PCM. Depending on the mode of operation when this switch is pressed, the PCM will engage and maintain the current vehicle speed or it will allow the vehicle speed to coast down.

By momentarily tapping the button, the vehicle speed set can be decreased by 1 mph increments. The desired speed set by the driver is maintained in memory. If the brake is applied, the set speed can be resumed by applying the resume button. A 12-volt signal is delivered to the PCM and it operates the servo to obtain the previously set speed. Resume will operate as long as the set speed has not been erased from memory. The memory is erased when the ignition switch is turned off or vehicle speed falls below 30 mph.

If the RESUME/ACCEL switch is moved toward the accelerate position, the PCM will open the throttle and accelerate the vehicle until the switch is released. When the switch is released, it sets the new speed into memory.

By momentarily tapping the RESUME/ACCEL switch the set speed can be increased by 1 mph increments.

GM DFI Engine Cruise Control

This system does not use vent and supply solenoids, instead it uses a cruise vacuum solenoid and a cruise power solenoid.

When the SET button is pushed, the PCM sets the value in memory. The PCM then grounds the vacuum solenoid coil to allow vacuum to the cruise control servo. The amount of vacuum that is applied to the servo is controlled by the power solenoid. The PCM controls the power solenoid through pulse width modulation. As the power solenoid is cycled on and off, vacuum and atmospheric pressure is alternately exposed to the servo.

Ford Cruise Control

Ford uses both nonintegrated and integrated systems. The integrated system uses the EEC-IV engine control module to operate the cruise control system. In the integrated system, the EEC-IV computer is capable of detecting system faults that can be retrieved through diagnostic procedures. The integrated system controls two solenoid valves in the servo: vent solenoid and vacuum solenoid. The EEC-IV computer controls the ground circuit of the solenoid coils to activate the valves.

In the nonintegrated system, the control switch signals the controller (amplifier) as to the position that the driver has selected through a resistor. The resistor values are different for each position, and the voltage drop over the resistors sends a different voltage signal to the controller. When the SET signal is impressed upon the controller, it will send commands to the servo to maintain that speed. The servo unit uses a modulating supply solenoid to control throttle position. The position of the modulating valve depends upon which coil is energized by the controller.

Memory Seats

The **memory seat** feature is an addition to the basic power seat system that allows the driver to program different seat positions. These seat positions can be recalled at the push of a button. Most memory seat systems share the same basic operating principles. The difference is in programming methods and number of positions that can be programmed.

The power seat system may operate in any gear position. However, the memory seat function will only operate when the transmission is in the PARK position or if there is no vehicle speed input. The purpose of the memory disable feature is to prevent accidental seat movement while the vehicle is being driven. In the PARK position (or during no speed input), the seat memory module will receive a 12-volt signal that will enable memory operation. In any other gear selection, the 12-volt signal is removed and the memory function is disabled. This signal can come from the gear selector switch or the neutral safety switch (Figure 33-8).

Most systems provide for two seat positions to be stored in memory. Ford allows for three positions to be stored by pushing both position 1 and 2 buttons together. With the seat in the desired position, depressing the set memory button and moving the memory select switch to either the memory 1 or 2 position will store the seat position into the module's memory (Figure 33-9).

When the seat is moved from its memory position, the seat memory module transmits the voltage applied from the switch to the motors. The module counts the pulses produced by motor operation, and then stores the number of pulses and direction of movement in memory. When the memory switch is closed, the module will operate the seat motors until it counts down to the preset number of pulses. Ford (and some other manufacturers) use a variable resistance sensors to monitor seat position instead of counting pulses.

Some systems offer an **easy exit** feature. This additional function of the memory seat feature provides for easier entrance and exit of the vehicle by moving the seat all the way back and down. Some systems will also move the steering column up and to full retract. When the easy exit switch is closed, voltage is applied to both memory 1 and 2 inputs of the module. This signal is interrupted by the module to move the seat to its full down and full back position. As the seat moves to the easy exit position it counts the pulses and stores this information in memory. In some systems, the easy exit feature is activated when the door is opened.

Memory is not lost when the ignition switch is turned off. However, it is lost if the battery is disconnected. If memory is lost, the position of the seat at the time power is restored becomes set in memory for both positions.

Electronic Sunroof Concepts

Many manufacturers have introduced electronic control of their electric sunroofs. These systems incorporate a pair of relay circuits and a timer function into the control module. Although there are variations between manufacturers, the systems discussed here provide a study of the two basic types of systems.

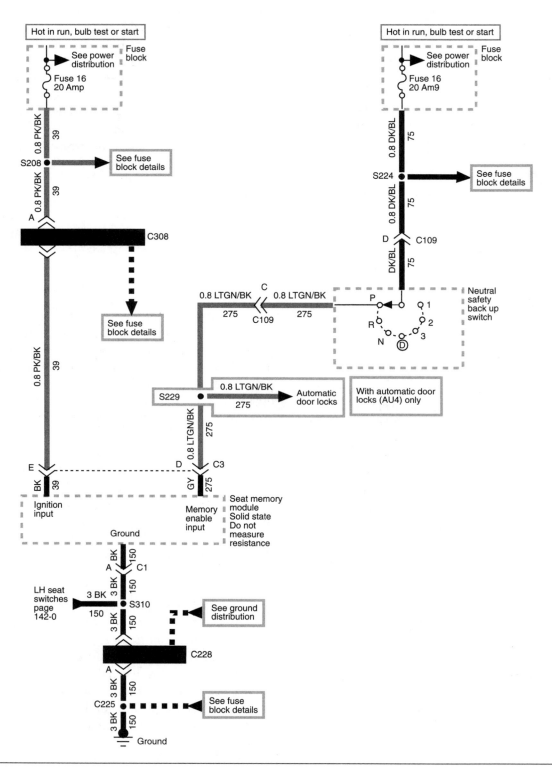

Figure 33-8 The neutral safety back-up switch signals the seat memory module when the transmission is in park. (Courtesy of General Motors Corporation, Service Technology Group)

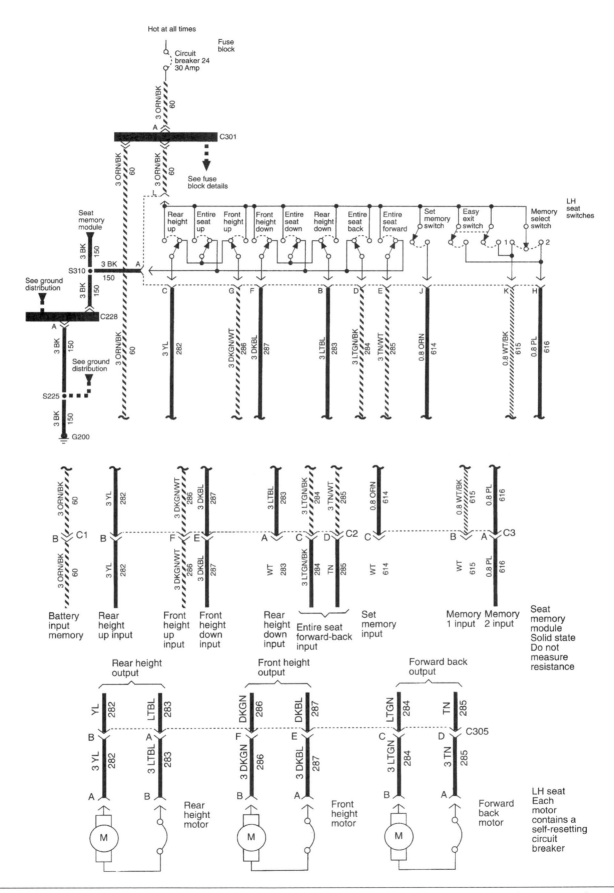

Figure 33-9 Memory seat circuit. (Courtesy of General Motors Corporation, Service Technology Group)

Electronic-Controlled Toyota Sunroof

Refer to the schematic (Figure 33-10) of a typical sunroof control circuit used by Toyota. Note the use of logic gates. If needed, refer back to the chapter on logic gates to review their operation. The movement of the sunroof is controlled by the motor that operates a drive gear. The drive gear either pushes or pulls the connecting cable to move the sunroof.

Motor rotation is controlled by relays that are activated according to signals received from the slide, tilt, and limit switches. The limit switches are operated by a cam on the motor.

The logic gates of this system operate on the principle of **negative logic**. Negative logic defines the most negative voltage as a logical 1 in the binary code. When the slide switch is moved to the OPEN position, either limit switch 1 or both limit switches are closed. Limit switches 1 and 2 provide a negative side signal to the OR gate labeled F. The output from gate F is sent to gate A. Gate A is an AND gate, requiring input from gate F and the open slide switch. The output signal from gate A is used to turn on TR2. This provides a ground path for the coil in relay 2. Battery voltage is applied to the motor through relay 2; the ground path is provided through the de-energized relay 1. Current is sent to the motor as long as the OPEN switch is depressed. If the OPEN switch is held in this position too long, a clutch in the motor disengages the motor from the drive gear.

Operation of the system during closing depends on how far the sunroof is open. If the sunroof is open more than 7.5 inches and the slide contact is moved to the CLOSE position, an input signal is sent to gate E. The other input signal required at gate E is received from the limit switches. The limit switch 1 signal passes through the OR gate G to the AND gate D. Limit switch 2 provides the second signal required by gate D. The output signal from D is the second input signal required by gate E. The output signal from E turns on TR1. This energizes relay 1 and reverses the current flow through the motor. The motor will operate until the slide switch is opened or limit switch 2 opens.

If the sunroof is open less than 7.5 inches and the slide switch is placed in the CLOSE position, the timer circuit is activated. The CLOSE switch signals the timer and provides an input signal to gate E. Limit switch 1 is open when the sunroof is opened less than 7.5 inches. The second input signal required by gate D is provided by the timer. The timer is activated for .5 second. This turns on TR1 and operates the motor for .5 second, or long enough for rotation of the motor to close limit switch 1. When limit switch 1 is closed, the operation is the same as described when the sunroof is closed when it is more than 7.5 inches open.

When the tilt switch is located in the UP position, a signal is imposed on gate B. This signal is inverted by the NOT gate and is equal to the value received from the opened number 2 limit switch. The output signal from

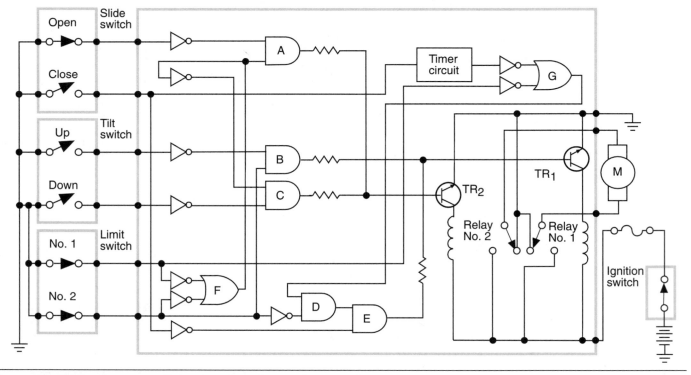

Figure 33-10 Toyota sunroof circuit using electronic controls.

gate B turns on TR1, which energizes relay 1 to turn on the motor. The motor clutch will disengage if the switch is held in the closed position longer than needed.

When the tilt switch is placed in the DOWN position, a signal is imposed on gate C. The second signal to gate C is received from the limit switches (both are open) through gate F. The signal from gate F is inverted by the NOT gate and is equal to that from the DOWN switch. The output signal from gate C turns on TR2 and energizes relay 2 to lower the sunroof. If the DOWN switch is held longer than necessary, limit switch 1 closes. When this switch is closed, the signals received by gate F are not opposite. This results in a mixed input to gate C and turns off the transistor.

Electronically Controlled General Motors Sunroof

See the schematic (Figure 33-11) of the sunroof system used on some GM model vehicles. The timing module uses inputs from the control switch and the limit switches to direct current flow to the motor. Depending on the inputs, the relays will be energized to rotate the motor in the proper direction. When the switch is located in the OPEN position, the open relay is energized sending current to the motor. The sunroof will continue to retract as long as the switch is held in the OPEN position. When the sunroof reaches its full open position, the limit switch will open and break the circuit to the open relay.

Placing the switch in the CLOSE position will energize the close relay. The current sent to the motor is in the opposite direction to close the sunroof. If the close

switch is held until the sunroof reaches the full closed position, the limit switch will open.

Anti-theft Systems

A vehicle is stolen in the United States every 26 seconds. In response to this problem, vehicle manufacturers are offering **vehicle theft security systems (VTSS)** as optional or standard equipment. Anti-theft systems are deterrent systems designed to scare off would-be thieves by sounding alarms and/or disabling the ignition system. The illustration (Figure 33-12) shows many of the common components used in an anti-theft system. These components include:

1. An electronic control module
2. Door switches at all doors
3. Trunk key cylinder switch
4. Hood switch
5. Starter inhibitor relay
6. Horn relay
7. Alarm

In addition, many systems incorporate the exterior lights into the system. The lights are flashed if the system is activated.

For the system to operate, it must first be **armed**. This puts the system in readiness to detect an illegal entry. This is done when the ignition switch is turned off and the doors are locked electrically by either the door switches or the remote keyless system. When the driver's door is shut, a security light will illuminate for approxi-

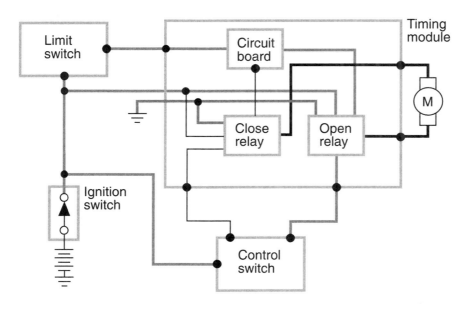

Figure 33-11 Block diagram of the GM sunroof.

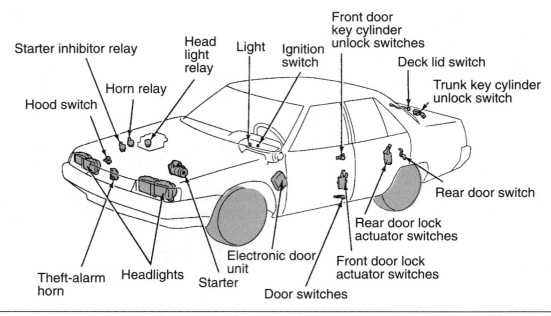

Figure 33-12 Typical components of an anti-theft system. (Courtesy of Mitsubishi Motor Sales of America, Inc.)

mately 30 seconds to indicate that the system is armed and ready to function. If any other door is open, the system will not arm until it is closed. The system can be bypassed by manually locking the doors.

The control module monitors the switches. If the doors or trunk are opened, or the key cylinders are rotated, the module will activate the system. The control module will sound the alarm and flash the lights until the timer circuit has counted down. At the end of the timer function, the system will automatically rearm itself.

Some systems use ultrasonic sensors to signal the control module if someone attempts to enter the vehicle through the door or window. The sensors can be placed to sense the perimeter of the vehicle and sound the alarm if someone enters within the protected perimeter distance.

The system can also use current sensitive sensors that will activate the alarm if there is a change in the vehicle's electrical system. The change can occur if the courtesy lights come on or if an attempt is made to start the engine.

The following systems are provided to give you a sample of the types of anti-theft systems used.

Ford Anti-theft Protection System

If the system is triggered, it will sound the horn; flash the low-beam headlights, taillights, and parking lamps; and disable the ignition system. See the schematic (Figure 33-13) of the system.

The arming process is started when the ignition switch is turned off. Voltage provided to the module at

terminal K is removed. When the door is opened, a voltage is applied to the courtesy lamp circuit 24 through the closed switch. This voltage energizes the inverter relay and provides a ground for module terminal J. This signal is used by the control module to provide an alternating ground at terminal D, causing the indicator lamp to blink. The flashing indicator light alerts the driver that the system is not armed. When the door lock switch is placed in the LOCK position, battery voltage is applied to terminal G of the module. The module uses this signal to apply a steady ground at terminal D, causing the indicator light to stay on continuously. When the door is closed, the door switch is opened. The opened door switch de-energizes the inverter relay coil. Terminal J is no longer grounded and the indicator light goes out after a couple of seconds.

To disarm the system, one of the front doors must be opened with a key or by pressing the correct code into the keyless entry keypad. Unlocking the door closes the lock cylinder switch and grounds terminal H of the module. This signal disarms the system.

Once the system is armed, if terminal J and C receive a ground signal, the control module will trigger the alarm. Terminal C is grounded if the trunk tamper switch contacts close. Terminal J is grounded when the inverter relay contacts are closed. The inverter relay is controlled by the doorjamb switches. If one of the doors is opened, the switch closes and energizes the relay coil. The contacts close and ground is provided to terminal J.

When the alarm is activated, a pulsating ground is provided at module terminal F. This pulsating ground energizes and de-energizes the alarm relay. As the relay

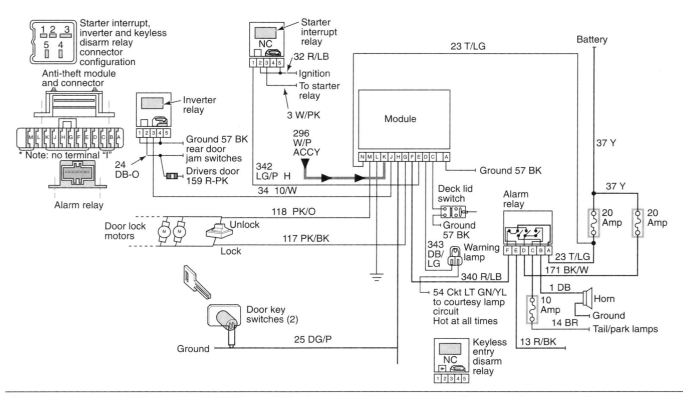

Figure 33-13 Circuit schematic of Ford's anti-theft system. (Courtesy of Chilton Book Company)

contacts open and close, a pulsating voltage is sent to the horns and exterior lights.

At the same time, the start interrupt circuit is activated. The start interrupt relay receives battery voltage from the ignition switch when it is in the START position. When the alarm is activated, the module provides a ground through terminal E, causing the relay coil to be energized. The energized relay opens the circuit to the starter system, preventing starter operation.

Ford's new ignition-disabling anti-theft system is based on the use of a key that contains a passive transponder and integrated circuit. When the key is inserted into the lock cylinder and switched ON, a radio transmitter in the lock cylinder sends out a low-power signal. This signal energizers the key circuit, which responds with a code. If the code from the key matches the code from the transmitter, the engine starts. If the code does not match, the engine will quit running after one second.

GM Pass-key Anti-Theft System

The name **"Pass-key"** is derived from **personal automotive security system**. The basic operation of the GM system is similar to that of the Ford system. An additional feature that GM offers can be used as a stand-alone anti-theft system or in combination with other systems. This system acts as an engine disable system by

using a pass-key arrangement. The ignition key has an electronic pellet that has a coded resistance value (Figure 33-14). Each of the different pellets used has a specific resistance value that ranges between 380 ohms and 12,300 ohms. The ignition key must be the correct cut to operate the lock and the correct electrical code to close the starter circuit.

When the ignition key is inserted into the cylinder, the pellet makes contact with the resistor sensing contact. When the cylinder is rotated to the START position, battery voltage is sent to a decoder module. In addition, the resistance value of the key pellet is sent to the decoder module. The resistance value is compared to memory and if they match, the starter enable relay is energized. This completes the starter circuit and signals the PCM to start fuel delivery.

If the key pellet resistance does not match, the decoder will prevent starting of the engine for two to four minutes. Even if the lock cylinder is removed, the engine will not start since the start enable relay will not energize.

Two anti-theft devices were introduced in 1996. One is GM's new low-cost pass-lock system, which works like the pass-key system, but without the chip in the key. Pass-lock systems have a Hall-effect sensor in the key cylinder that measures the magnetic properties of the key as it is inserted into the cylinder. The cut pattern of every

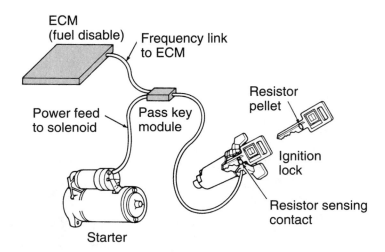

Figure 33-14 Basic components of the pass-key system. (Courtesy of General Motors Corporation, Service Technology Group)

key has its own magnetic identity. If the wrong key is inserted into the lock cylinder, the car will not start, even if the lock cylinder turns.

Chrysler VTSS

Chrysler vehicles can be equipped with a stand-alone **security alarm module (SAM)** or incorporate this function within the BCM. If a BCM is used, it functions in

the same manner as the SAM. The following is an explanation of a stand-alone system.

The VTSS module receives inputs from the following components (Figure 33-15):

1. Ignition switch
2. Trunk cylinder sense switch
3. Door key cylinder switch
4. Door ajar switch

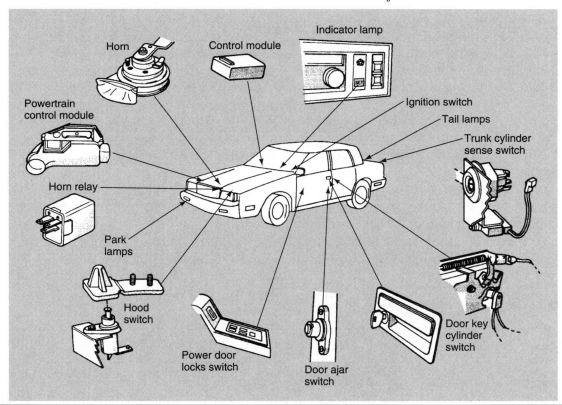

Figure 33-15 Vehicle theft security system. (Courtesy of Chrysler Corporation)

5. Power door locks switch

6. Hood switch

The VTSS module operates the following outputs (Figure 33-16):

1. Horn relay and horn

2. PCM engine no-start feature

3. Park and tail lamps

4. Set lamp

The vehicle theft security system inputs, outputs, and operation may vary depending on make and model year. Always consult the vehicle manufacturer's service information. The system is armed during a normal vehicle exit in which the VTSS module receives the following inputs after the ignition switch is turned off and the key is removed:

1. Door ajar switch, door opened

2. Power door locks activated with switch or remote control

3. Door ajar switch, door closed

The arming sequence does not have to be completed in a specific order. The set light in the instrument panel flashes quickly for 15 seconds during the arming process. When the arming process is complete, the set light flashes slower. On some systems, the set light goes out once the system is armed. Passive disarming of the system occurs when either door is unlocked with the key or the remote control. This disarming will also halt the alarms once they are activated.

The alarms are triggered by any of the following events:

1. Opening any door without using the key in one of the front doors

2. Removing the trunk lock cylinder

3. Turning the ignition switch to the on position

4. Opening the hood

Once the alarm is triggered, the module flashes the park and tail lamps, sounds the horn, and signals the PCM to prevent injector operation. During the first 3 minutes after the alarm is triggered, the horn sounds and the park and taillights flash. After 3 minutes the horn is shut off, but the lights continue flashing for another 15 minutes. Once the alarm is triggered, the system can only be disarmed by unlocking either of the front doors with the key.

WARNING *On some Chrysler vehicles, if either of the battery terminals is disconnected, the no-start feature in the theft security system is activated. The key must be used to lock and unlock one of the front doors to allow the engine to start.*

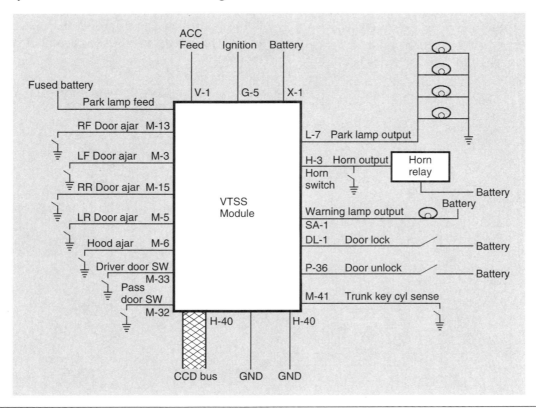

Figure 33-16 Vehicle theft security system wiring diagram. (Courtesy of Chrysler Corporation)

When the alarm has been triggered and the 18-minute alarm mode completed, if the driver enters the car in the normal manner, the module provides three horn pulses to alert the driver regarding the alarm activation.

Chrysler requires the engine controller to have sensed at least 20 engine cranks before the theft alarm becomes operational. Anytime the engine controller is replaced, the anti-theft system will not operate until the 20 cranks have been obtained.

Automatic Door Locks

Many automobile manufacturers are incorporating a passive system to lock all doors when the required conditions are met. The **automatic door lock (ADL)** system is an additional safety and convenience system. Most systems lock the doors when the gear selector is placed in drive, the ignition switch is in RUN, and all doors are shut. Some systems will lock the doors when the gear shift selector is passed through the reverse position; others do not lock the doors unless the vehicle is moving 15 mph or faster.

The system may use the body computer to control the door lock relays (Figure 33-17), or a separate controller. The controller (or body computer) takes the place of the door lock switches for automatic operation. In order for the door lock controller to lock the doors, the following conditions must be met:

1. Ignition switch is in the RUN position.
2. Seat switch is closed by the driver.
3. All doors are closed (switches are open).
4. Gear selection is not in PARK.
5. Courtesy light switch is off.

When all of the door jam switches are open (doors closed), the ground is removed from the WHT wire to the controller (Figure 33-18). This signals the controller to enable the lock circuit.

When the gear selection is moved from the PARK position, the neutral safety switch removes the power signal from the controller. The controller sends voltage through the LH seat switch to the lock relay coil. Current is sent through the motors to lock all doors.

When the gear selector is returned to the PARK position, voltage is applied through the neutral safety switch to the controller. The controller then sends power to the unlock relay coil to reverse current flow through the motors.

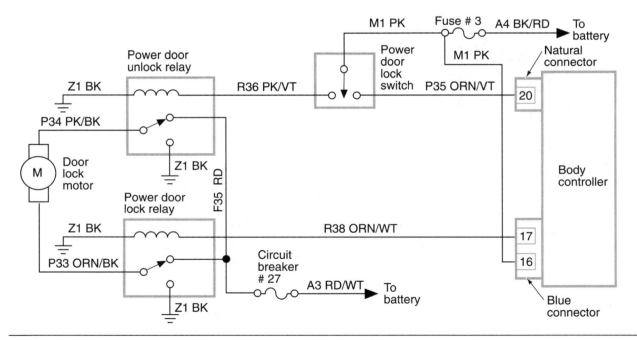

Figure 33-17 Automatic door lock system utilizing the body computer. (Courtesy of Chrysler Corporation)

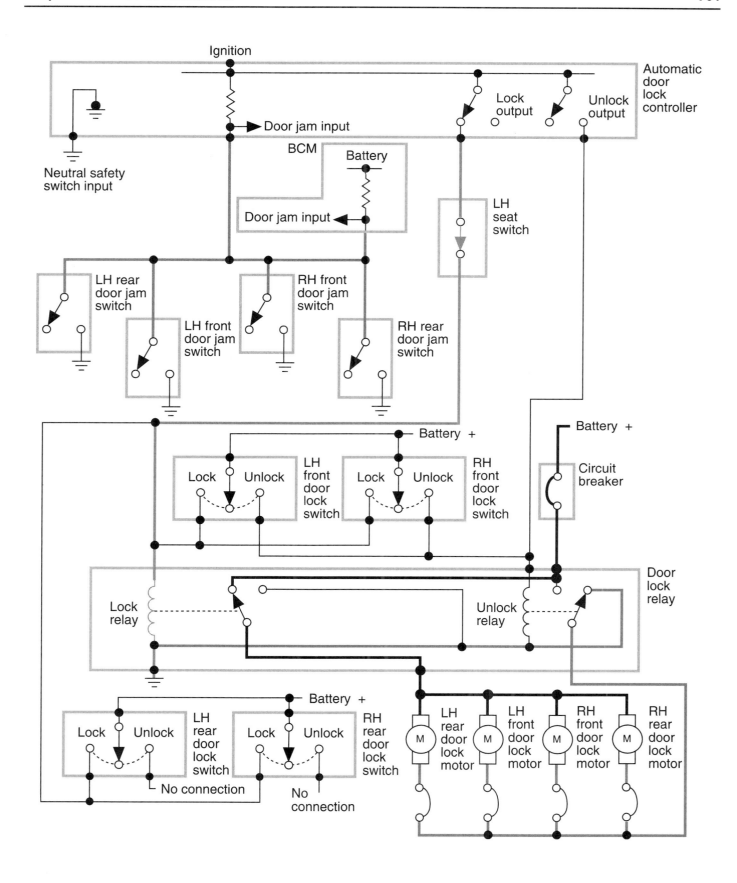

Figure 33-18 Automatic door lock system circuit schematic indicating operation during the lock procedure.

Keyless Entry

The **keyless entry** system allows the driver to unlock the doors or the deck lid (trunk) from outside of the vehicle without the use of a key. The main components of the keyless entry system are the control module, a coded-button keypad located on the driver's door, and the door lock motors.

The **keypad** consists of five normally open, single-pole, single-throw switches. Each switch represents two numbers: 1-2, 3-4, 5-6, 7-8, and 9-0 (Figure 33-19). The keypad is wired into the circuit to provide input to the control module (Figure 33-20). The control module is programmed to lock the doors when the 7-8 and 9-0 switches are closed at the same time. The driver's door can be unlocked by entering a five-digit code through the keypad. The unlock code is programmed into the con-

Figure 33-19 Keyless entry system keypad.

troller at the factory. However, the driver may enter a second code. Either code will operate the system.

In addition to the aforementioned functions, the keyless entry system also:

1. Unlocks all doors when the 3-4 button is pressed within 5 seconds after the five-digit code has been entered.

2. Releases the deck lid lock if the 5-6 button is pressed within 5 seconds of code entry.

3. Activates the illuminated entry system if one of the buttons is pressed.

4. Operates in conjunction with the automatic door lock system and may share the same control module.

See the schematic (Figure 33-21) of the keyless entry system used by Ford. When the 7-8 and 9-0 buttons on the keypad are pressed, the controller applies battery voltage to all motors through the lock switch.

When the five-digit code is entered, the controller closes the driver's switch to apply voltage in the opposite direction to the driver's door motor. If the driver presses the 3-4 button, the controller will apply reverse voltage to all motors to unlock the rest of the doors.

Remote Keyless Entry

Many new vehicles are equipped with a **remote keyless entry** system that is used to lock and unlock the doors, turn on the interior lights, and release the trunk latch. A small receiver is installed in the vehicle. The transmitter assembly is a hand-held item attached to the

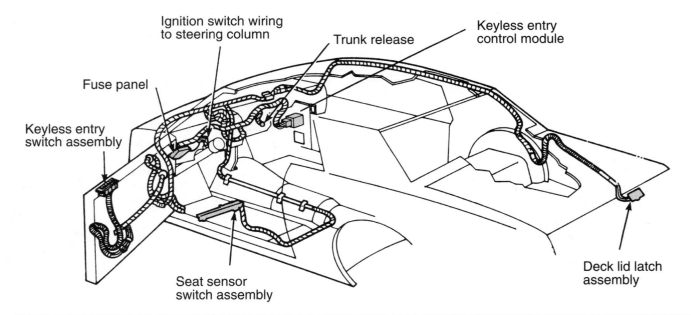

Figure 33-20 Wiring harness and components of a typical keyless entry system. (Reprinted with permission of Ford Motor Company)

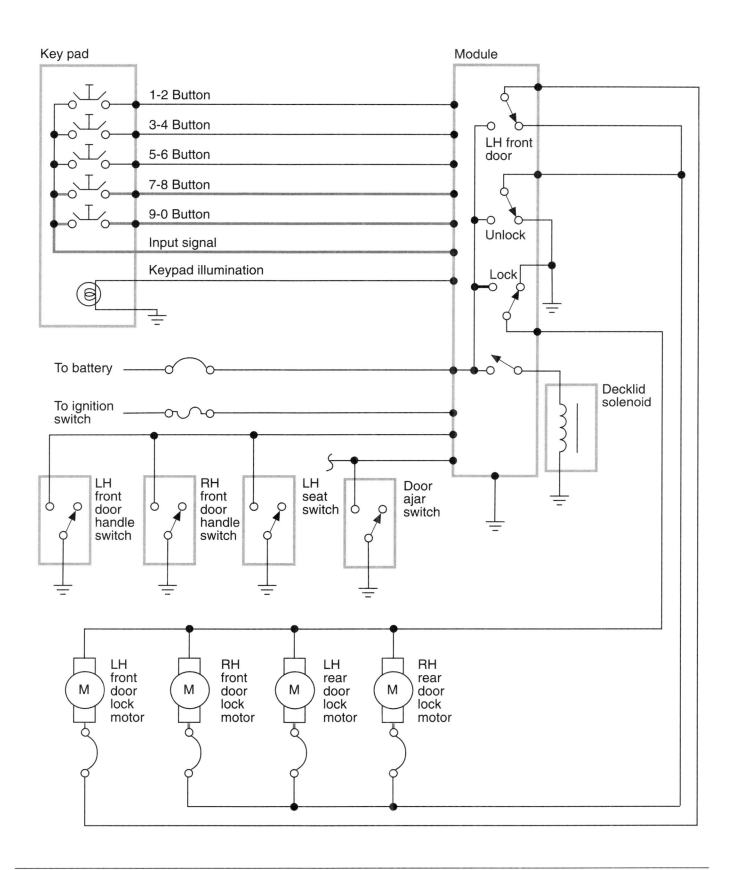

Figure 33-21 Simplified keyless entry system schematic.

key ring. Pressing a button on a hand-held transmitter will allow operation of the system from distances of 25 to 50 feet (Figure 33-22). When the unlock button is pressed, the driver's door unlocks and the interior lights are illuminated. If a theft deterrent system is installed on the vehicle, it is also disarmed when the unlock button is pressed. A driver exiting the vehicle can activate the door locks and arm the security system by pressing the lock button. Many transmitters also have a third button for opening the deck lid.

The system operates at a fixed radio frequency. If the unit does not work from a normal distance, check for two conditions: weak batteries in the remote transmitter or a stronger radio transmitter close by (radio station, airport transmitter, etc.).

Electronic Heated Windshield

The heated windshield system is designed to melt ice and frost from the windshield three to five times faster than conventional defroster systems (Figure 33-23). The windshield undergoes a special process during manufacturing to allow for current flow through the glass without interfering with the driver's vision.

There are two basic methods used to make the heated windshield:

1. Use a layer of plastic laminate that is between two layers of glass. The back of the outer layer is fused with a silver and zinc oxide coating. The coating carries the electrical current. Bus bars are attached to the coating at the top and bottom of the windshield (Figure 33-24). A sensor is used to check the condition of the windshield coating. If the windshield has a crack or chip that will affect heating (Figure 33-25), the voltage drop across the resistor will indicate this condition to the control module. If the windshield is damaged, the controller will not allow heated windshield operation.

Figure 33-22 Remote keyless entry system transmitter. (Courtesy of General Motors Corporation, Service Technology Group)

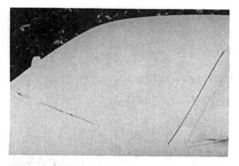

Figure 33-23 The heated windshield removes ice and frost from the windshield in just a few minutes. (Reprinted with permission of Ford Motor Company)

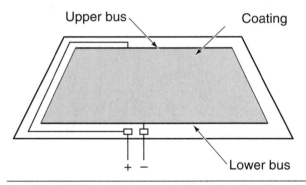

Figure 33-24 The power and ground circuits are connected to the silver and zinc coating through busbars. (Reprinted with permission of Ford Motor Company)

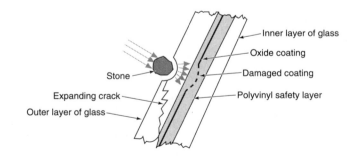

Figure 33-25 An open in the circuit can be caused by a chip or crack in the windshield. A sensor is used to prevent operation if the windshield is damaged. (Reprinted with permission of Ford Motor Company)

2. Use a layer of resistive coating sprayed between the inner and outer windshield layers. The coating is transparent and does not provide any tint. A sensor is used to indicate if the coating has been damaged. If a chip or crack is not deep enough to penetrate the coating, it will not affect the system operation.

The two systems discussed here are representative of the methods used to heat the windshield.

General Motors' Heated Windshield

General Motors' heated windshield consists of the following components:

1. The heated windshield: Contains a transparent internal resistive coating that heats when current is applied to it.

2. Special CS 144 generator: There are three special phase terminals to provide AC power to the system's power module (Figure 33-26). The generator can continue to supply its normal DC voltage while AC power is being supplied.

3. The power module: Converts the AC voltage from the generator to a higher DC voltage for use by the windshield.

4. The control module: Controls the heating cycle and provides automatic shut-down at the end of the time cycle, or if a fault is detected in the system.

5. The control switch.

See the schematic of the GM heated windshield. When the system is activated by the driver, the control module starts its turn-on sequence. First it checks that there is more than 11.2 volts present at terminal B6. This assures there will be sufficient voltage to operate other circuits.

The second step for the control module is to check the vehicle's inside temperature. For the system to operate, the temperature must be below 65°F (18°C). Next, the controller checks the windshield sensor to see if there is any damage to the film coating.

If all of these conditions are met, the control module sends a signal to the BCM to increase the engine speed. The BCM passes the request on to the PCM. If the gear selector is in PARK or NEUTRAL, the PCM will increase the idle speed to approximately 1,400 rpm. The PCM will send a signal back to the BCM to indicate that the speed has been increased. When this feedback signal is received, the control module will turn on the power module relays. The power module will draw AC current from the generator. The current is amplified and rectified by the power module, then sent to the windshield. Voltage at the windshield is between 50 and 90 volts.

Incorporated into the control module is a timer circuit. When the activation switch is turned on for the first time, the control module will operate the system for 3 minutes. If the switch is pressed again, at the end of the first cycle, it will result in a 1-minute cycle. If the switch

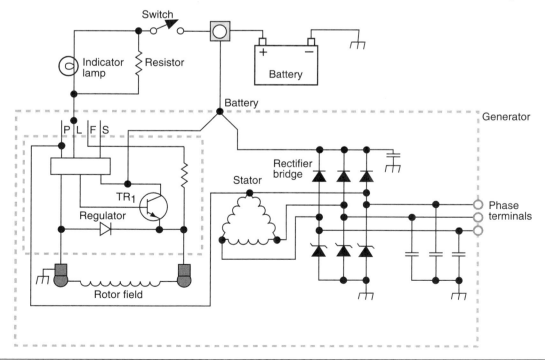

Figure 33-26 Schematic of CS 144 alternator used on vehicles equipped with heated windshields. The three terminals provide AC current to the power module.

is pressed while the cycle is still in operation, the system is turned off.

Ford's Heated Windshield System

The illustration (Figure 33-27) shows the major components of the Ford heated windshield system. For the system to be activated, the engine must be running and inside temperature must be 40°F or less. When the driver activates the system the control module shuts off the voltage regulator and energizes the generator output control relay. This switches the generator output from the electrical system to the windshield circuit (Figure 33-28). After the switch has been completed, the control module turns on the voltage regulator to restore generator output.

With the generator output disconnected from the battery, battery voltage drops below 12 volts. The voltage regulator attempts to charge the battery by full fielding the generator. Because the battery does not receive the generator output, full field voltage reaches 30 to 70 volts. All of the full field power is sent to the windshield.

The control module will monitor the battery voltage and generator output. It will prevent the output from increasing over 70 volts to protect the system. To prevent damage to the battery, if its voltage drops below 11 volts the control module reconnects the electrical system to the generator.

When the system is activated, the control module sends a signal to the EEC controller to increase the idle speed to about 1,400 rpm. If the transmission is placed into a gear selection other than PARK or NEUTRAL, the EEC will return the idle speed to the normal setting.

Intelligent Windshield Wipers

To avoid making the driver select the correct speed of the windshield wipers according to the amount of rain, manufacturers have developed **intelligent wiper systems**. Two intelligent wiper systems will be discussed here: one senses the amount of rainfall and the other adjusts wiper speed according to vehicle speed.

Cadillac's Rainsense system automatically selects the wiper speed needed to keep the windshield clear by sensing the presence and amount of rain on the windshield. The system relies on a series of eight LEDs that shine at an angle onto the inside of the windshield glass and an equal number of light collectors. The outer surface of a dry windshield will reflect the lights from the LEDs back into a series of collectors. The presence of water on the windshield will refract some of the light away from the collectors. When this happens, the wipers are turned on.

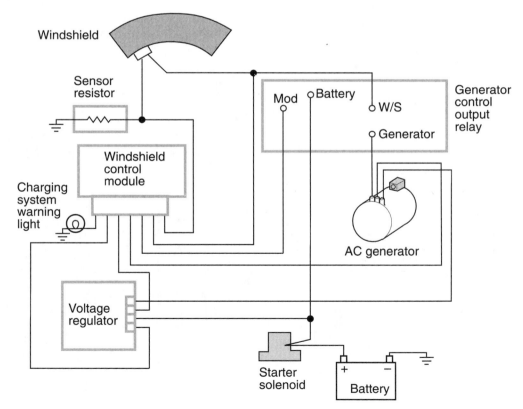

Figure 33-27 Components of Ford's heated windshield system.

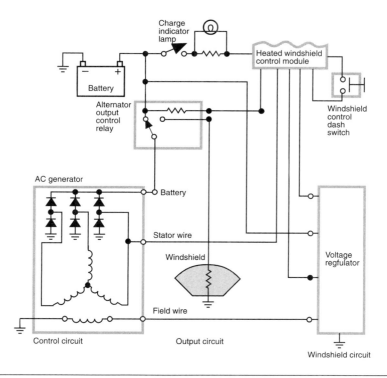

Figure 33-28 Simplified circuit schematic of Ford's heated windshield system.

If the water is not cleared by one complete travel of the wipers, the wipers will operate again. Therefore if the rainfall is heavy, the wipers will operate quickly and often.

One of the functions of Ford Motor Company's Generic Electronic Module (GEM) system is front wiper control and a speed-dependent wiper system. Speed-dependent wipers compensate for extra moisture that normally accumulates on the windshield at higher speeds in the rain. At higher speeds, the delay between wipers shortens when the wipers are operating in the interval mode. This delay is automatically adjusted at speeds between 10 and 65 miles per hour. Basically this system functions according to the input the computer receives about vehicle speed.

Vehicle Audio Entertainment Systems

When radios were first introduced to the automotive market, they produced little more than "tinny" noise and static. Today's audio sound systems produce music and sound that rivals the best that home sound systems can produce. And with nearly the same or even greater amounts of volume or sound power!

The most common sound system configuration is the all-in-one unit more commonly called a receiver. Housed in this unit is the radio tuner, amplifier, tone controls, and unit controls for all functions. These units may also include internal capabilities such as cassette players, compact disc players, digital audiotape players, and/or graphic equalizers. Most will be electronically tuned with a display that shows all functions being accessed/controlled and digital clock functions.

Recent developments by the manufacturers have been made to take individual functions (tape, disc, equalizer, control head, tuner, amplifier, etc.) and put them in individual boxes and call them components. This would allow owners greater flexibility in selecting options to suit their needs and tastes (Figure 33-29). Componentizing has allowed greater dash design flexibility. Some components, such as multiple CD changers, can be remotely mounted in a trunk area for greater security.

Wiring diagrams for component systems will be more complex (Figure 33-30). In addition to power and audio signal wires, note that some systems will have a serial data wire for microprocessor communication between components for the controlling of unit functions. Some functions are shared and integrated with factory-installed cellular phones, such as radio mute. Some systems allow remote control of functions through a control assembly

Figure 33-29 Components, like those shown here, allow for a more compact and optimized mounting location within the vehicle.

mounted in the steering wheel or alternate passenger compartment location.

A BIT OF HISTORY

Radios were introduced in cars by Daimler in 1922. Cars were equipped with Marconi wireless receivers.

Summary

❏ Cruise control is a system that allows the vehicle to maintain a preset speed with the driver's foot off the accelerator.

❏ Safety switches return control of vehicle speed to the driver when the brake pedal is depressed.

❏ The controller energizes the supply and vent valves to allow manifold vacuum to enter the servo. The servo moves the throttle to maintain the set speed. The vehicle speed is maintained by balancing the vacuum in the servo.

❏ Electronic cruise control systems can be either integrated or nonintegrated design.

❏ The memory seat feature allows the driver to program different seat positions that can be recalled at the push of a button.

❏ The easy exit feature is an additional function of the memory seat that provides for easier entrance and exit of the vehicle by moving the seat all the way back and down.

❏ Anti-theft systems are deterrent systems designed to scare off would-be thieves by sounding alarms and/or disabling the ignition system.

❏ The pass-key system acts as an engine disable system by using an ignition key that has an electronic pellet containing a coded resistance value.

❏ Automatic door locks is a passive system used to lock all doors when the required conditions are met. Many automobile manufacturers are incorporating the system as an additional safety and convenience feature.

❏ The keyless entry system allows the driver to unlock the doors or the deck lid from outside the vehicle without the use of a key.

❏ The heated windshield system is designed to melt ice and frost from the windshield three to five times faster than conventional defroster systems.

❏ Vehicle audio entertainment systems are generally only a single component (receiver). Recently manufacturers have been developing multiple components (tuner, amplifier, control head, etc.) to allow for greater flexibility.

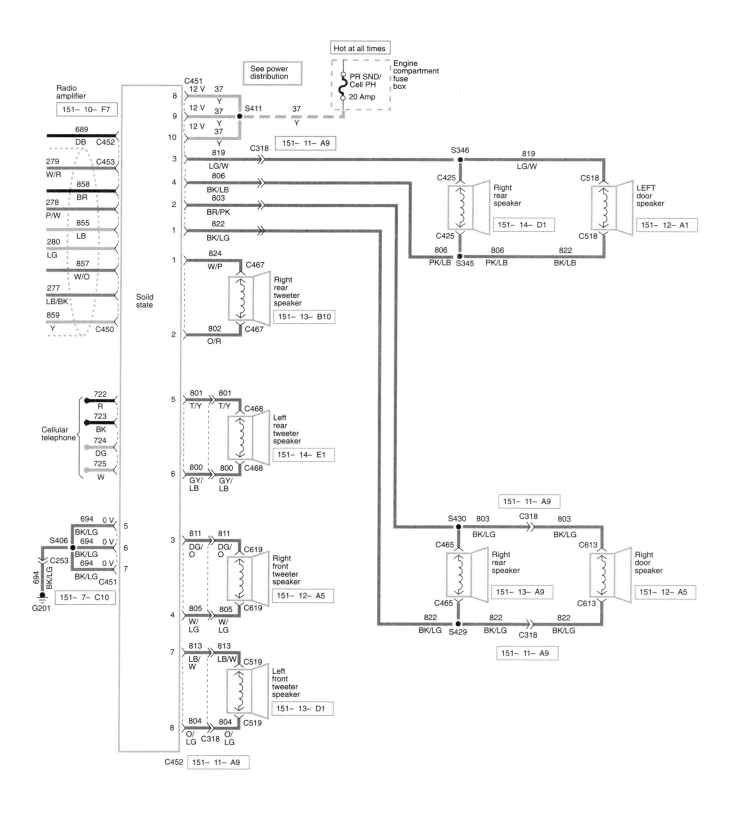

Figure 33-30 Schematic diagram showing the wiring hook-ups for remote-mounted components. (Reprinted with permission of Ford Motor Company)

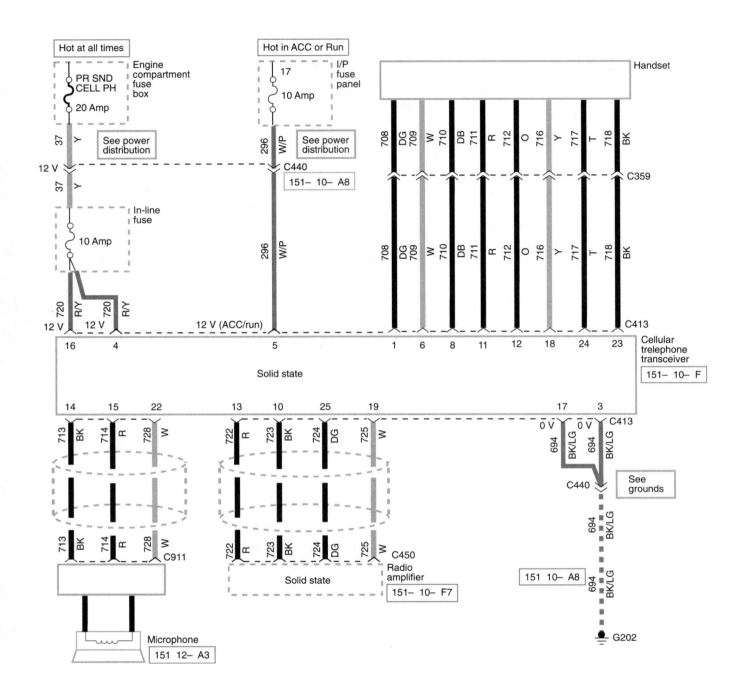

Figure 33-30 (Continued)

Terms To Know

Armed	Intelligent wiper systems	Rapid deceleration cutoff
Automatic door lock (ADL)	Keyless entry	Remote keyless entry
Comparators	Keypad	Security alarm module (SAM)
Cruise control	Memory seat	Servomotor
Electromechanical	Negative logic	System malfunction cutoff
Electronic cruise control	Nonintegrated cruise control	Vehicle theft security systems (VTSS)
Easy exit	Pass-key	Wheelspin cutoff
Integrated cruise control	Personal automotive security system	

Review Questions

Short Answer Essay

1. What is the purpose of the cruise control system?

2. Explain the operating principles of the electromechanical cruise control system.

3. Explain the basic operating principles of the electronic cruise control system.

4. Explain the purpose and operation of the safety switches used on electromechanical and electronic cruise control systems.

5. List and describe the safety modes incorporated into electronic cruise control systems.

6. What is meant by "comparator"?

7. Explain two methods used by the memory seat control module to determine seat position.

8. List the main components of common anti-theft systems.

9. Describe the basic operating principles of the automatic door lock system.

10. Explain the difference between a radio receiver unit system and a component radio system.

Fill-in-the-Blanks

1. When the brake pedal is depressed, the cruise control system is disengaged through _____ and _____ switches.

2. When the driver sends a SET signal to the controller, through the control switch, it sets the voltage signals received from the _____ _____ into memory.

3. If the VSS signal is greater than the high comparator value, the control module de-energizes the _____ solenoid valve.

4. Ford integrated systems use the EEC-IV engine control module to control _____ solenoid valves in the servo.

5. In some electronic sunroofs, the _____ module uses inputs from the control switch and the limit switches to direct current flow to the motor.

6. Radio receiver units usually contain at least the following: an _____ , a _____ , and function _____ .

7. The _____ _____ feature is an additional function of the memory seat that provides for easier entrance and exit of the vehicle.

8. In the pass-key system, the ignition key must be the correct _____ to operate the lock and the correct _____ _____ to close the starter circuit.

9. The generator in the General Motors heated windshield system provides _____ power to the system's power module.

10. In the Ford system for heated windshield operation, the voltage regulator _____ _____ the generator to supply voltage to the windshield.

ASE Style Review Questions

1. A typical electronic cruise control system is being discussed. Technician A says the system utilizes vent and supply solenoids. Technician B says the amount of vacuum applied to the servo is controlled by the power solenoid. Who is correct?
   a. A only
   b. B only
   c. Both A and B
   d. Neither A nor B

2. Electronic cruise control systems are being discussed. Technician A says if the voltage signal from the VSS drops below the low comparator value, the control module energizes the vent valve solenoid. Technician B says if the VSS signal is greater than the high comparator value, the control module energizes the supply solenoid valve. Who is correct?
   a. A only
   b. B only
   c. Both A and B
   d. Neither A nor B

3. The heated windshield system is being discussed. Technician A says if the windshield has a crack or chip that will affect heating, the voltage drop across the resistor will indicate this condition to the control module. Technician B says if the windshield is damaged, the controller reduces the voltage to the windshield to 20 volts. Who is correct?
   a. A only
   b. B only
   c. Both A and B
   d. Neither A nor B

4. Technician A says the DFI system controls the power solenoid through the transducer. Technician B say that the servo is a stepper motor that mechanically changes throttle position. Who is correct?
   a. A only
   b. B only
   c. Both A and B
   d. Neither A nor B

5. Ford cruise control systems are being discussed. Technician A says the integrated system uses the EEC-IV engine control module to operate the cruise control system. Technician B says the EEC-IV computer is capable of detecting system faults that can be retrieved through diagnostic procedures. Who is correct?
   a. A only
   b. B only
   c. Both A and B
   d. Neither A nor B

6. The safety switches of a typical cruise control system are being discussed. Technician A says when the brake pedal is depressed, the cruise control system maintains preset vehicle speeds. Technician B says the two switches provide a fail-safe means of assuring that the cruise control is disengaged when the brakes are applied. Who is correct?
   a. A only
   b. B only
   c. Both A and B
   d. Neither A nor B

7. Technician A says that the electronic cruise control system offers more precise speed control than the electromechanical system. Technician B says that the throttle position sensor is used to provide smooth throttle changes while the cruise control is engaged. Who is correct?
   a. A only
   b. B only
   c. Both A and B
   d. Neither A nor B

8. Memory seats are being discussed. Technician A says the power seat and memory seat functions can only be operated when the transmission is in the PARK position. Technician B says when the seat is moved from its memory position, the module stores the number of pulses and direction of movement in memory. Who is correct?
   a. A only
   b. B only
   c. Both A and B
   d. Neither A nor B

9. The operation of Toyota electronic controlled sunroof is being discussed. Technician A says that it is not necessary to understand logic gate operation to understand the control of the sunroof. Technician B says the movement of the sunroof is controlled by a motor that operates a drive gear. Who is correct?
   a. A only
   b. B only
   c. Both A and B
   d. Neither A nor B

10. Electromechanical cruise control systems are being discussed. Technician A says the transducer receives vacuum from the servo to position the throttle. Technician B says that the vehicle must be traveling at speeds over 45 mph before the low speed switch closes. Who is correct?
    a. A only
    b. B only
    c. Both A and B
    d. Neither A nor B

34 Diagnosing and Servicing Vehicle Accessories

Objective

Upon completion and review of this chapter, you should be able to:

❑ Perform self-diagnostic procedures on electronic cruise control systems.

❑ Diagnose causes for no, intermittent, and erratic cruise control operation.

❑ Replace the servo assembly and adjust actuator cable.

❑ Replace the switch assembly.

❑ Determine the cause of poor, intermittent, and no memory seat operation.

❑ Properly determine the circuit fault causing a malfunction in sunroof operation.

❑ Diagnose vehicle alarm systems that sound for no apparent reason.

❑ Perform tests on the anti-theft controller to determine proper operation.

❑ Test the anti-theft relay for proper operation.

❑ Use self-diagnostic tests on alarm systems that provide this feature.

❑ Diagnose automatic door lock systems for poor, intermittent, or no operation.

❑ Diagnose keyless entry systems.

❑ Determine the causes of heated windshield malfunctions.

❑ Diagnose vehicle radio/stereo audio systems.

Introduction

In this chapter you will learn how to service electronic accessories designed to increase passenger comfort, provide ease of operation, and increase passenger safety. These systems include memory seats, sunroof, automatic door locks, keyless entry, heated windshields, anti-theft systems, and vehicle radio/stereo audio systems.

Most of these systems are additions to existing systems. For example, the memory seat feature is an addition to the conventional power seat system. As vehicles and accessories become more sophisticated, these "luxury" features will become more commonplace. Today's technician is expected to accurately and quickly diagnose malfunctions in these systems.

Diagnosis and Service of Electronic Cruise Control Systems

The cruise control system is one of the most popular electronic accessories installed on today's vehicles. Problems with the system can vary from no operation, to intermittent operation, to not disengaging. To diagnose these system complaints, today's technicians must be able to rely on their knowledge and diagnostic capabilities. Most of the system is tested using familiar diagnostic procedures. Build on this knowledge and ability to diagnose cruise control problems. Use system schematics, troubleshooting diagnostic charts, and switch continuity tables to assist in isolating the cause of the fault.

CAUTION *When servicing and testing the cruise control system you will be working close to the air bag and the antilock brake systems. The service manual will instruct you when to disarm and/or depressurize these systems. Failure to follow these procedures can result in injury and additional costly repairs to the vehicle.*

Self-Diagnostics

Most vehicle manufacturers have incorporated **self-diagnostics** into their cruise control system. This allows some means of retrieving trouble codes to assist the technician in locating system faults. The following are some common methods of retrieving diagnostic trouble codes (DTCs).

On any vehicle, perform a visual inspection of the system. Check the vacuum hoses for disconnects,

pinches, loose connections, and so forth. Inspect all wiring for tight, clean connections. Also look for good insulation and proper wire routing. Check the fuses for opens and replace as needed. Check and adjust linkage cables or chains, if needed. Some manufacturers will require additional preliminary checks before entering diagnostics. In addition, perform a road test (or simulated road test) in compliance with the service manual to confirm the complaint.

General Motors

General Motors has four different types of electronic cruise control systems. The common procedure for performing self-diagnostics and retrieving codes for system types 1, 3, and 4 follows. See the application chart (Figure 34-1) which indicates the type of system installed in a particular vehicle. Type 2 systems do not provide for code retrieval.

APPLICATION CHART

Year	Model	Type
BUICK		
Century	1990–92	2
Electra	1990	2
Estate Wagon	1990	2
LeSabre	1990–91	2
LeSabre	1992	3
Park Avenue	1990	2
Park Avenue	1991–92	3
Reatta	1990–91	3
Regal	1990–92	2
Riviera	1990–92	3
Roadmaster	1991–92	2
Skylark	1990–92	2
CADILLAC		
Brougham	1990–92	1
DeVille	1990–92	1
Eldorado	1990–92	1
Fleetwood	1990–92	1
Seville	1990–92	1
CHEVROLET		
Beretta	1990–92	2
Camaro	1990–92	2
Caprice	1990–92	2
Cavalier	1990–92	2
Celebrity	1990	2
Corsica	1990–92	2

Year	Model	Type
CHEVROLET—Continued		
Corvette	1990–92	2
Lumina	1990–92	2
GEO		
Prizm	1990–92	4
OLDSMOBILE		
Achieva	1992	2
Custom Cruiser	1990–92	2
Cutlass Calais	1990–91	2
Cutlass Ciera	1990–92	2
Cutlass Cruiser	1990–92	2
Cutlass Supreme	1990–92	2
Toronado/Trofeo	1990–92	3
88	1990–91	2
88	1992	3
98	1990	2
98	1991–92	3
PONTIAC		
Bonneville	1990–91	2
Bonneville	1992	3
Firebird	1990–92	2
Grand Am	1990–91	2
Grand Prix	1990–92	2
LeMans	1990–92	2
Sunbird	1990–92	2
6000	1990–91	2

Figure 34-1 Cruise control application chart. (Courtesy of General Motors Corporation, Service Technology Group)

Type 1 Diagnostics

CAUTION *Additional codes, other than those described, may be stored in the module memory. It may be necessary to identify these codes and diagnose the cause to assure proper cruise control operation.*

Brougham models with a type 1 system do not provide for trouble code retrieval, instead a special tool called a **"Quick Checker"** is used to test the system. On other vehicles equipped with this system, to enter self-diagnostics place the ignition switch in the RUN position and press the OFF and WARMER buttons on the climate control panel (CCP) simultaneously. All of the panel segments should light. If not, the affected panel must be replaced before continuing.

All PCM trouble codes will be displayed, followed by all BCM codes. Engine controller codes are prefixed with an "E", BCM codes are prefixed with an "F". On Eldorado and Seville models, "Current" or "History" will accompany the codes. On DeVille and Fleetwood models, trouble codes that are displayed on the first pass but not on the second represent history codes. History codes are intermittent faults that have occurred in the past, but are not present at the last self-test.

Type 3 Diagnostics

Code accessing can be done through the use of a scanner, flashing the "Service Engine Soon" light, or through the ECC panel. The method used depends on the model of vehicle being serviced.

To flash the service light, ground terminal B of the data link connector with the ignition switch in the RUN position. Any stored memory codes will be flashed by the light, also all PCM-controlled relays and solenoids will be energized.

The indicator light will flash a code 12 (one flash followed by two more flashes), to indicate the system is operating properly. If there are any trouble codes, they are each displayed three times. At the end of the trouble codes a code 12 will be flashed to indicate code resequencing has begun. If a scan tool is used to retrieve the codes, connect it to the DLC and follow the instructions to retrieve the codes.

Toronado, Reatta, and Riviera models provide for diagnostics through the ECC or CRT panel. Trouble codes are accessed by placing the ignition switch in the RUN position then pressing the OFF and WARMER buttons on the ECC/CRT panel at the same time. On models equipped with CRT displays, depress the OFF hardkey and the WARM softkey. On Reatta and Riviera models the WARMER button is identified as TEMP-UP.

After the **diagnostic service mode** is selected, any trouble codes in memory will be displayed. Engine controller codes are prefixed with an "E", BCM codes are prefixed with a "B", instrument panel cluster (IPC) codes are prefixed with an "I", and supplemental restraint system (SRS) codes are prefixed with a "R".

Type 4 Diagnostics

Use the chart (Figure 34-2) to determine the type of code to read for the symptom. If the chart instructs you to read a type A code, refer to the other chart (Figure 34-3) and follow the procedure below:

WARNING *Checking of number 4 code is performed with the drive wheels lifted from the floor and the engine idling.*

1. Turn the ignition switch to the RUN position.
2. Turn the SET/COAST switch on.
3. Push the main switch on.
4. Turn the SET/Coast switch off.
5. Perform the condition requirements listed in the chart.

The diagnostic code is read on the main switch indicator by counting the flashes.

If instructed to read type B codes, refer to the proper chart (Figure 34-4) and follow the procedure below:

1. Road test the vehicle.
2. If the system cancels because of a malfunction in the actuator or speed sensor, the indicator light will blink 5 times. If a malfunction occurs, do not turn off the ignition switch or the control switch. Inspect the system with the switches on. Turning off the switches will erase the codes from memory.
3. While driving at a speed less than 10 mph, press the SET/COAST button three times within two seconds. Any codes will be displayed through the indicator light.

If there are no codes displayed, refer to the service manual to perform diagnostic tests.

Ford IVSC System Diagnostics

Ford's integrated vehicle speed control (IVSC) system has self-test capabilities that are contained within the KOEO and KOER routine of the PCM. The KOEO routine is a static test of the IVSC inputs and outputs. The KOER routine is a dynamic check of the engine in operation. Testing of the IVSC system is broken down into two divisions: quick tests and pinpoint test.

The quick test is a functional test of the system. This test will check the operation and function of all system

Symptom	Read	Result	Circuits
• Cruise control switch indicator blinks five times or • Cruise control system does not set or • Cruise control system does not operate.	Type B Codes	11	• Actuator • Cruise Control Module
		21	• Speed Sensor • Cruise Control Module
		23	• Check speedometer cable operation • Actuator • Speed Sensor • Vacuum Pump and Hose • Vacuum Switch • Cruise Control Module
		31	• Engage Switch • Cruise Control Module
		33	• Engage Switch • Cruise Control Module
	Type A Code 5	No Code	• Speed Sensor • Cruise Control Module
		OK	• Cruise Control Switch • Engage Switch • Stoplamp Switch • Clutch or Neutral Start Switch • Parking Brake Switch • Check speedometer cable operation • Actuator • Cruise Control Module • Vacuum Hose and Brake Fluid
Setting speed deviated on high or low speed.	Type A Code 3	OK	• Check speedometer cable operation • Speed Sensor • Vacuum Pump • Vacuum Switch • Actuator • Cruise Control Module
		No Code	• Speed Sensor
Vehicle speed fluctuates when control switch is turned to "SET/COAST."			• Speed Sensor • Check speedometer cable operation • Actuator • Cruise Control Module
Setting speed does not cancel when the brake pedal is depressed.	Type A Code 4	OK	• Actuator • Stoplamp Switch • Cruise Control Module
		No Code	• Stoplamp Switch • Cruise Control Module
Setting speed does not cancel when the parking brake is applied.	Type A Code 4	OK	• Actuator • Cruise Control Module
		No Code	• Parking Brake Switch • Cruise Control Module

Figure 34-2 General Motors type 4 troubleshooting chart. (Courtesy of General Motors Corporation, Service Technology Group)

Symptom	Read	Result	Circuits
Setting speed does not cancel when shifted to "N" range. (A/T)	Type A Code 4	OK	• Actuator • Cruise Control Module
		No Code	• Neutral Start Switch • Cruise Control Module
Setting speed does not cancel when the clutch pedal is depressed. (M/T)	Type A Code 4	OK	• Actuator • Cruise Control Module
		No Code	• Clutch Switch • Cruise Control Module
Vehicle speed does not decrease when the engage switch is turned to SET/COAST.	Type A Code 1	OK	• Actuator • Check speedometer cable operation • Cruise Control Module
		No Code	• Engage Switch • Cruise Control Module
Vehicle does not accelerate when the engage switch is turned to "RESUME/ACCEL."	Type A Code 2	OK	• Actuator • Check speedometer cable operation • Cruise Control Module
		No Code	• Engage Switch • Cruise Control Module
Vehicle speed does not return to set speed when engage switch is turned to "RESUME/ACCEL."	Type A Code 2	OK	• Actuator • Check speedometer cable operation • Cruise Control Module
		No Code	• Engage Switch • Cruise Control Module
Speed can be set below approximately 30 Km/hr (19 mph)	Type A Code 5	OK	• Actuator • Cruise Control Module
		No Code	• Speed Sensor • Cruise Control Module
Cruise control will not disengage at approximately 30 Km/hr (19 mph)	Type A Code 5	OK	• Actuator • Cruise Control Module
		No Code	• Speed Sensor • Check speedometer cable operation • Cruise Control Module
Acceleration is sluggish when the engage switch is turned to "RESUME/ACCEL."	Type A Code 3	OK	• Check speedometer cable operation • Vacuum Pump • Actuator • Cruise Control Module • Check Vacuum Hose
		No Code	• Vacuum Switch • Vacuum Pump

Figure 34-2 (Continued)

components except the vehicle speed sensor. The quick test is performed first. Then, if any failure codes are displayed, the pinpoint test is performed. Pinpoint tests are specific component tests. If there is a complaint with the cruise control system, and the quick test does not indicate any faults, test the speed sensor. Perform the quick test after the system has been serviced to verify proper operation.

The processor stores the self-test program within its memory. When this test is activated, the processor initi-

ates a function test of the IVSC system to verify the sensors and actuators are connected and operating properly. The quick test will detect faults that are present at the time of the test. It will not store history codes.

The quick test can be performed with a STAR tester or an analog voltmeter. If the STAR tester is being used, connect it to the service connectors (Figure 34-5).

No.	Conditions	Indicator code	Diagnosis
1	"SET/CAST" on	ON OFF — 0.25S ← ← 1.0S → ← 0.25S	"SET/COAST" circuit is normal.
2	"RESUME/ACCEL" on	ON OFF	"RESUME/ACCEL" circuit is normal.
3	Vacuum switch on	ON OFF	Vacuum switch circuit is normal.
4	Each cancel switch on (Stoplamp switch, parking brake switch, clutch switch, neutral start switch)	ON OFF	Each cancel switch circuit is normal.
5	Drive 40 km/h (25 mph) or over	ON OFF	Speed sensor circuit is normal.
6	Drive 30 km/h (19 mph) or over	ON OFF	Speed sensor circuit is normal.

Figure 34-3 Type A code chart. (Courtesy of General Motors Corporation, Service Technology Group)

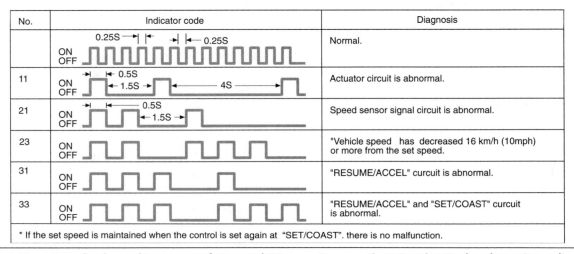

No.	Indicator code	Diagnosis
	ON OFF — 0.25S → ← ← 0.25S	Normal.
11	ON OFF → ← 0.5S ← 1.5S → ← 4S →	Actuator circuit is abnormal.
21	ON OFF → ← 0.5S ← 1.5S →	Speed sensor signal circuit is abnormal.
23	ON OFF	"Vehicle speed has decreased 16 km/h (10mph) or more from the set speed.
31	ON OFF	"RESUME/ACCEL" curcuit is abnormal.
33	ON OFF	"RESUME/ACCEL" and "SET/COAST" curcuit is abnormal.
* If the set speed is maintained when the control is set again at "SET/COAST". there is no malfunction.		

Figure 34-4 Type B code chart. (Courtesy of General Motors Corporation, Service Technology Group)

To use an analog voltmeter, place the ignition switch in the OFF position. Connect a jumper wire from the **self-test input (STI)** terminal to pin 2 (single return) of the self-test connector (Figure 34-6). The STI is the sin-

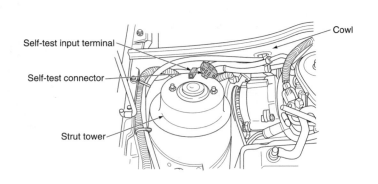

Figure 34-5 Self-test connector and STI location. (Reprinted with permission of Ford Motor Company)

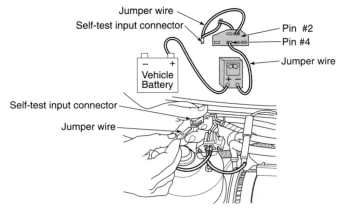

Figure 34-6 Voltmeter connections to the self-test connector and STI. (Reprinted with permission of Ford Motor Company)

gle pigtail connector located next to the self-test connector. Set the analog voltmeter on the DC 15-volt scale. Then connect the positive voltmeter lead to battery positive and the negative lead to pin 4 (self-test output) of the self-test connector.

Perform the KOEO, KOER, and intermittent (wiggle) test procedures to fully check the cruise control system.

KOEO Test

If you are using the STAR tester, leave the STI connector unplugged. Turn on the tester. The numbers 88 should flash followed by a steady 00. Press the push button to latch it in position. A colon should appear in front of the 00.

WARNING *During the KOEO test do not depress the throttle.*

Turn the ignition switch to the RUN position and turn on the speed control switch. A code 10 should appear to indicate the IVSC test is in progress. Press speed control OFF, COAST, ACCEL, RESUME buttons; tap brake pedal once; and depress clutch pedal once (if equipped with MTX). Observe the STAR tester display to retrieve any codes.

A code 11 indicates the system passed the KOEO test. If any other codes are present refer to the service manual to determine the action to take for the code(s) received. If no codes are received, repeat the KOEO test and if no codes are outputted perform pinpoint test Q1 in the service manual.

If an analog voltmeter is used, observe the sweeps of the needle. Count the sweeps to determine the code, for example, a code 23 is displayed by two sweeps followed by a two-second delay then three more sweeps.

KOER Test

The engine should be warm before performing this test. If it is not, start the engine and run it until the upper radiator hose is warm to the touch. The throttle must be off fast idle and the idle speed stabilized. Shut the engine off and follow the steps below.

CAUTION *When performing the KOER test, place the transmission in PARK (neutral for MTX).*

If you are using a STAR tester, connect it to the self-test connector and the STI connector. Start the engine and turn the START tester on. Within 30 seconds of starting the engine, turn on the speed control switch. Wait 15 seconds and press the STAR pushbutton.

CAUTION *Connect the engine exhaust to the shop's ventilation system before starting the engine.*

WARNING *Do not depress the throttle or brake pedal during this test. Doing so will result in incorrect codes or aborting of the test. At the conclusion of the test the engine may stall. If it does, turn off the ignition switch before the system enters EEC-IV KOEO self-test.*

A code 10 should be displayed to indicate the test has been initiated. The code 10 will then be followed by the service codes. A code 11 indicates the system passed the KOER test. If the complaint condition is still present and you received a code 11, refer to the symptom chart in the service manual.

If any other codes are present, refer to the service manual for the correct action to take.

If you are using an analog voltmeter to retrieve codes, follow the same procedure found in KOEO testing to read the code.

Intermittent Problem Testing

Intermittent problems can be detected by performing the KOEO and KOER quick test and wiggling the wires, connectors, and vacuum hoses while the self-test is being conducted. If a code is set when the wiggle test is being performed, but not when the quick tests were done the first time, check all connects and wires in the identified circuit.

Chrysler

Chrysler has four different types of electronic cruise control systems (Figure 34-7). The type 3 system does not provide trouble code diagnostics.

Type 1 and 4 Diagnostics

Trouble codes can be retrieved using a scanner or by reading the flashes from the "Check Engine" light. To use the scanner, connect the tester to the engine diagnostic connector with the ignition switch in the RUN position. Select the correct year of the vehicle you are servicing. When asked to select the system, choose "Engine." Next, select "Read Fault Data" and read the fault messages. Depending on the fault message received, go to the correct test chart to locate the fault.

If a scan tester is not available, trouble codes can be retrieved through the "Check Engine" light by cycling the ignition from RUN to OFF three times within 5 seconds.

If fault code 34 is displayed, perform the speed control system test located in the service manual. This series of tests will check the electrical condition of the system, including switches and the servo.

CAUTION *Before conducting the system test, disarm the air bag system as described in the service manual. Failure to disarm the air bag may result in deployment and personal injury. After all service is complete, rearm the air bag according to the service manual procedures.*

If the indicator light flashes a code 15, the distance sensor needs to be tested. A code 77 indicates that a speed control relay test must be performed.

If the cruise control system is inoperative, and no trouble codes are displayed, replace the engine controller after confirming all electrical connections and grounds.

Type 2 Diagnostics

There are up to six trouble codes that can be displayed. One of these codes indicates normal operation. To access the codes, connect a voltmeter between the

APPLICATION CHART

Year	Model	Type	Year	Model	Type
1990	Acclaim	1	1991—Cont'd	Monaco	1
	Colt	2		New Yorker Salon	1
	Colt Vista	2		Premier	1
	Colt Wagon	2		Shadow	1
	Daytona	1		Spirit	1
	Dynasty	1		Stealth	2
	Fifth Avenue	1		Summit	2
	Horizon	1		Sundance	1
	Imperial	1		Talon	2
	Laser	2	1992	Acclaim	1
	LeBaron Convertible	1		Colt	2
	LeBaron Coupe	1		Colt Vista	2
	LeBaron Landau	1		Daytona	1
	Monaco	3		Dynasty	1
	New Yorker Landau	1		Fifth Avenue	1
	New Yorker Salon	1		Imperial	1
	Omni	1		Laser	2
	Premier	3		LeBaron Convertible	1
	Shadow	1		LeBaron Coupe	1
	Spirit	1		LeBaron Sedan	1
	Summit	2		Monaco	1
	Sundance	1		New Yorker Salon	1
	Talon	2		Premier	1
1991	Acclaim	1		Shadow	1
	Colt	2		Spirit	1
	Colt Vista	2		Stealth	2
	Daytona	1		Summit	2
	Dynasty	1		Summit Wagon	2
	Fifth Avenue	1		Sundance	1
	Imperial	1		Talon	2
	Laser	2		Concorde	4
	LeBaron Convertible	1		Intrepid	4
	LeBaron Coupe	1	1993	Vision	4
	LeBaron Sedan	1			

Figure 34-7 Chrysler cruise control application chart. (Courtesy of Chrysler Corporation)

ground and auto-cruise control terminals of the diagnostic connector. The connector is located on the lower left side of the instrument panel. With the ignition switch located in the RUN position, read the needle sweeps to determine the code. Once the code is determined, refer to the diagnostic display pattern chart for the vehicle being diagnosed (Figure 34-8). This chart will refer you to the correct check chart to locate the fault.

WARNING *The same trouble code between models and years of manufacture may have different diagnostic charts.*

Diagnosing Systems without Trouble Codes

Systems that do not provide for trouble code diagnostics require the technician to perform a series of diagnostic tests. The test performed will depend on the symptom. The following sections discuss areas of generic troubleshooting procedures for all types of systems.

Simulated Road Test

The **simulated road test** will allow the technician to perform a road test without leaving the shop. Before performing this test, connect the shop's ventilation system to the vehicle's exhaust pipe. Lift the drive wheels from the floor and place jack stands under the vehicle. If the vehicle is equipped with CV joint shafts, place the jack stands under the lower control arms so the shafts are in their normal drive position. If the vehicle is rear wheel drive with a solid axle, place the jack stands under the axle.

CAUTION *Block the wheels that are to remain on the ground. Leave the wheels blocked throughout the test.*

Start the engine and place the transmission into drive. Turn the speed control switch into the ON position.

CAUTION *During the process of this test it is possible the engine will overspeed. If the system should appear to go out of control, the technician must be ready to turn it off. This can be done by turning off the ignition or turning off the speed control switch.*

Accelerate and hold the speed at 35 mph. Press and release the SET ACCEL button. Maintain a slight foot pressure on the accelerator. The speed should be maintained at 35 mph for a short period of time, then gradually start to surge. The engine surge is the result of operating the system while there is no load on the engine, and is normal.

Press the OFF button and the engine should decelerate to an idle speed. Stop the drive wheels by lightly applying the brakes.

WARNING *At any time during the test, do not attempt to place the transmission back into PARK without first stopping the drive wheels with the brakes. Doing so may result in damage to the transmission.*

Press the ON button and accelerate to 35 mph. Press and hold the SET/ACCEL button and gradually remove your foot from the accelerator pedal. Engine rpm should begin to increase. Continue to hold the SET ACCEL button until the indicated speed reaches 50 mph, then

Code No.	Display patterns (output codes) (use with voltmeter)	Probable cause	Check chart No.
1	The same pattern repeatly displayed	Vacuum pump assembly drive output system out of order	5
2		Vehicle speed signal system out of order	4
3		Control switch out of order (When SET or "RESUME" switch is kept ON state continuously for more than 60 seconds)	2
4		Control unit out of order	–
5		Throttle position sensor or idle switch out of order	9

Figure 34-8 Trouble code display pattern chart. (Courtesy of Chrysler Corporation)

release the button. Vehicle speed should remain at 50 mph for a short period of time, then the engine will start to surge.

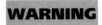 *Do not exceed 50 mph or damage to the differential assembly may result.*

Press the COAST button and hold it. Engine rpm should return to idle speed. Allow the indicated speed to slow to 35 mph without applying the brakes. When the speed is returned to 35 mph, release the COAST button. The speed should be held at 35 mph for a short period of time, then the engine will begin to surge.

Tap the brake pedal and the speed control system should shut off and engine speed return to idle. Set the indicated speed to 50 mph, then use the brakes to slow to 35 mph. Maintain 35 mph using the accelerator. Depress the RESUME button and the speed should climb to 50 mph.

Diagnosis of No Operation

The first step in a verified no-operation complaint is to check all fuses. Next visually inspect the system for any obvious problems. If the visual inspection does not pinpoint the problem, perform the following steps:

1. Apply the brake pedal to observe proper brake light operation. If the brake lights do not operate, check the switch and circuit.

2. If the vehicle is equipped with a manual transmission, check to assure that the clutch deactivator switch is operating properly. Use an ohmmeter or voltmeter to test its operation.

3. Check for proper operation of the actuator lever and throttle linkage.

4. Disconnect the vacuum hose between the check valve and the servo (on the servo side of the check valve). Apply 18 inches of vacuum to the open end of the hose to test the check valve. It should hold the vacuum. If not, replace the check valve.

5. Check the vacuum dump valve for proper operation.

6. Test control switches and circuits following the procedure already learned. Use the circuit diagram and switch continuity charts to aid in testing.

7. Test servo operation.

8. Test speed sensor operation.

9. If all tests indicate proper operation, yet the system is not operational, replace the amplifier (controller).

Diagnosing Continuously Changing Speeds

If the vehicle speed changes up and down while the cruise control is on, use the following steps to locate the problem:

1. Check the actuator linkage for smooth operation.

2. Check the speedometer for proper routing and to make sure there are no kinks in the cable.

3. Test the servo.

4. Check the speed sensor.

5. Check the operation of the vacuum dump valve.

6. Check all electrical connections.

7. If none of these tests locate the fault, replace the amplifier (controller).

Some computer-controlled systems have adaptive memory functions. This allows the control module to learn variances in cables between vehicles during manufacturing. Usually this does not cause any problems. However, if the driver lifts their foot from the accelerator just before pushing the set button (or accelerates above the desired speed then coasts before pushing the set button), the amount of throttle movement required to maintain the vehicle speed is considered slack by the control module. This slack value is stored in adaptive memory and is used whenever the cruise control system is activated. This erroneous adaptive factor may cause the system to under or over shoot the set speed. To clear the memory, drive the vehicle and activate the cruise control while maintaining a steady throttle setting and push the set button. Allow cruise to operate for at least ten seconds then push the cancel button or tap the brakes to disengage the system. Repeat this series ten to fifteen times.

Diagnosis of Intermittent Operation

Intermittent operation is usually caused by loose electrical or vacuum connections. If a visual inspection fails to locate the fault, test drive the vehicle and identify when the intermittent problem occurs. If the problem occurs during normal cruising, begin at step 1. If the problem occurs when operating the control buttons, or when the steering wheel is rotated, begin with step 3.

1. Connect the vacuum gauge to the hose entering the servo. There should be at least 2.5 inches Hg of vacuum.

2. Test the servo assembly.

3. Use the service manual's switch continuity chart and system schematic to test switch operation. Turn the steering wheel through its full range while testing the switches. For example, using the Ford system shown

(Figure 34-9), this test would be conducted by disconnecting the connector at the amplifier and connecting an ohmmeter between the terminal for circuit 151 and ground (with the ignition switch off). While rotating the steering wheel throughout its full range, make the following checks:

Depress the OFF button; the ohmmeter reading should read between zero and 1 ohm.

Depress the SET/ACCEL button and check for a reading between 646 and 714 ohms.

Depress the COAST button and the ohmmeter should read between 126 and 114 ohms.

When the RESUME button is depressed, the reading should be between 2,310 and 2,090 ohms.

If the resistance values fluctuate while the steering wheel is being turned, the most likely cause is contamination on the slip rings. Remove the steering wheel and clean the brushes. Apply a light coat of lubricant to the brushes using an approved lubricant. If the resistance values are above specifications, check the switches and ground circuit.

If the preceding tests (or the road test) fail to identify the fault, conduct a simulated road test while wiggling the electrical and vacuum connections.

Component Testing

Testing of the safety switches and circuits is performed using normal testing procedures you have already learned. Testing of the servo assembly, dump valve, and speed sensor is included to familiarize you with these procedures.

Servo Assembly Test

Actuator tests vary depending on design. Some manufacturers use vacuum servos and others use stepper motors. Be sure to follow the service manual procedures for the vehicle you are diagnosing. The following servo assembly test is a common test for Ford's cruise control system used on Continental, Cougar, and Thunderbird models. Use the schematic (Figure 34-9) to perform the following test.

Disconnect the eight-pin connector to the amplifier. Connect an ohmmeter between circuits 144 and 145. The resistance value should be between 40 and 125 ohms. Move the lead from circuit 145 to circuit 146. The resistance value should read between 60 and 190 ohms. If the resistance levels are out of specifications, check and repair the wiring between the amplifier and the servo. If the wiring is good, replace the servo.

If the resistance values are within specifications, leave the amplifier disconnected and start the engine.

Jump 12 volts to circuit 144 and jump circuit 146 to ground. Momentarily jump circuit 145 to ground. The servo actuator arm should pull in and the engine speed should increase.

WARNING *Be sure to have the transmission in PARK or NEUTRAL. Block the wheels and set the parking brake before performing the servo test.*

WARNING *Be ready to abort the test by turning off the ignition if engine rpm should rise to a level where internal damage may result.*

Remove the jumper to ground on circuit 146. The servo should release and engine speed should return to idle. The servo must be replaced if it does not operate as described.

WARNING *Do not short the jumper wires from circuit 144 to circuits 145 or 146. Damage to the amplifier will result if the amplifier is connected while this is done.*

Dump Valve Test

A dump valve that is stuck open or leaks will cause a no-operation or erratic operation complaint. Failure of the dump valve to release vacuum may not, by itself, be noticed by the driver. It is part of a fail-safe system. If the dump valve does not release, the electrical switch signal is also used to disengage the cruise control system when the brakes are applied. It is good practice to test the dump valve any time the vehicle is in the shop for cruise control service.

To test the dump valve, disconnect the vacuum hose from the servo assembly to the dump valve. Connect a hand vacuum pump to the hose and apply vacuum to the dump valve. If vacuum cannot be applied, either the hose or the dump valve is defective.

If the valve holds vacuum, press the brake pedal. The vacuum should be released. If not, adjust the dump valve according to the service manual procedures. If the dump valve fails to release vacuum when the brake is applied, and it is properly adjusted, it must be replaced.

Speed Sensor Test

Disconnect the six-pin connector from the amplifier (Figure 34-9). Connect an ohmmeter between circuits 150 and 57A. The resistance should be approximately 200 ohms. If the resistance value is less than 200 ohms, check for a short in the circuits between the amplifier and the speed sensor. If there is no problem in the wiring, the coil in the sensor is shorted. If the resistance value is infinite, there is an open in the wires or in the sensor coil.

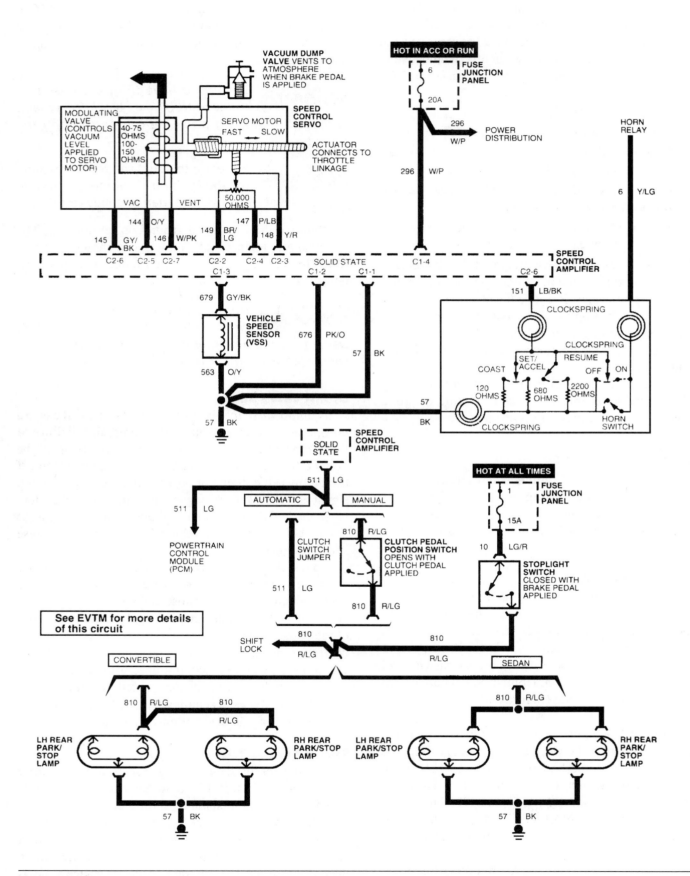

Figure 34-9 Speed control electrical schematic. (Reprinted with permission of Ford Motor Company)

Typical Procedure for Replacing the Servo Assembly

P13-1 Tools required to replace the servo assembly: fender covers, screwdriver set, combination wrench set, ratchet and socket set.

P13-2 Remove the retaining screws attaching the speed control actuator cable to the accelerator cable bracket and intake manifold support bracket.

P13-3 Disconnect the cable from the brackets.

P13-4 Disconnect the speed control cable from the accelerator cable and the electrical connection to the servo assembly.

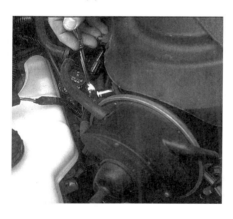

P13-5 Remove the two retaining bolts attaching the servo assembly bracket to the shock tower.

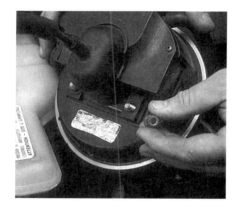

P13-6 Remove the two bolts that attach the servo assembly to the bracket and remove the servo and cable assembly.

P13-7 Remove the two cable covers to servo assembly retaining bolts and pull off the cover. Next remove the cable from the servo assembly.

To test the sensor separate from the wiring harness, disconnect the wire connector from the sensor and connect the ohmmeter between the two terminals. This test should be used after testing at the amplifier connector to determine if there is a fault in the entire circuit.

Component Replacement

The two most common components to be replaced in the cruise control system are the servo and the switches. The following section covers replacement of these units.

Servo Assembly Replacement

Follow Photo Sequence 16 to replace the servo assembly. Reverse the procedure to install the servo assembly. To adjust the actuator cable, leave the cable adjusting clip off and pull the cable until all slack is removed. Maintain light pressure on the cable and install the adjusting clip. The clip must snap into place.

Switch Replacement

Switch removal differs depending on location. If the switch is a part of the multiple switch assembly on the turn signal stock, refer to the service manual section for removing this switch. The following is a common method of switch replacement for switches contained in the steering wheel.

With the air bag system properly disarmed, remove

CAUTION *Follow the service manual procedure for disarming the air bag system before performing this task. Failure to disarm the air bag system may result in accidental deployment and personal injury.*

the air bag module. Disconnect the electrical connections to the switch assembly. Remove the screws that attach the switch assembly to the steering wheel. Then remove the switch.

To install the new switch assembly, position it into the steering wheel pad cover and attach the retaining screws. If the horn connectors had to be disconnected, attach them to the pad cover. Reinstall the air bag module and rearm the air bag system.

Memory Seat Diagnosis

If the seat motors fail to operate under any condition, test the motors and switches as outlined in an earlier chapter. This section relates only to that portion of the system that operates the memory function.

Using the illustration (Figure 34-10) of a memory seat circuit, this system would be diagnosed as follows. All tests are performed at memory seat module connectors C1 and C3. The connectors are disconnected from the module to perform the tests. Place the ignition switch in the RUN position with the gear selector in the PARK position.

With the test light connected between C1 connector terminal B and ground, the lamp should illuminate. If the light does not come on, check for a circuit fault in wire 60. Connect the test light between terminals A and B at the C1 connector. If the test light fails to illuminate, check circuit 150 for an open. Move the test light between circuit 39 and ground. The light should turn on. If not, there is an open in circuit 39.

Connect the test light between terminal D of connector C3 and ground. If the light does not turn on, there is a problem in the neutral safety switch circuit. Check the adjustment of the neutral safety switch and circuits 75 and 275 for opens. With the test light connected between terminal B of the C3 connector and ground, the test light should remain off. An illuminated light indicates the left-hand seat switch assembly must be replaced.

Leave the test light connected between terminal B and ground. Place the memory select switch in position 1. If the test light does not illuminate, check circuit 615 for an open. If the wire is good, use an ohmmeter to test the memory select switch for an open. Release the memory select switch and press the exit button. The test light should light. If the light fails to illuminate, there is a fault in the left-hand seat switch assembly. It therefore must be replaced.

Move the probe of the test light to terminal A and press the exit button. If the light fails to illuminate, the problem is in circuit 616 or in the exit switch. With the test light still connected to the A terminal of the C3 connector, release the exit button. The test light should turn off. If the test light remains illuminated, replace the left-hand seat switch assembly.

Continue to leave the test light connected to terminal A. Place the memory select switch in the number 2 position. The test light should light. If not, replace the left-hand seat switch assembly. With the test light connected between C3 connector terminal C and ground and the memory select switch released, the test light should be off. If it remains on, the seat switch assembly is defective. Press the set memory switch. The test light should illuminate. If not, check the set memory switch and circuit 614 for an open.

If all the test results were correct, the fault is in the control module. This module must be replaced.

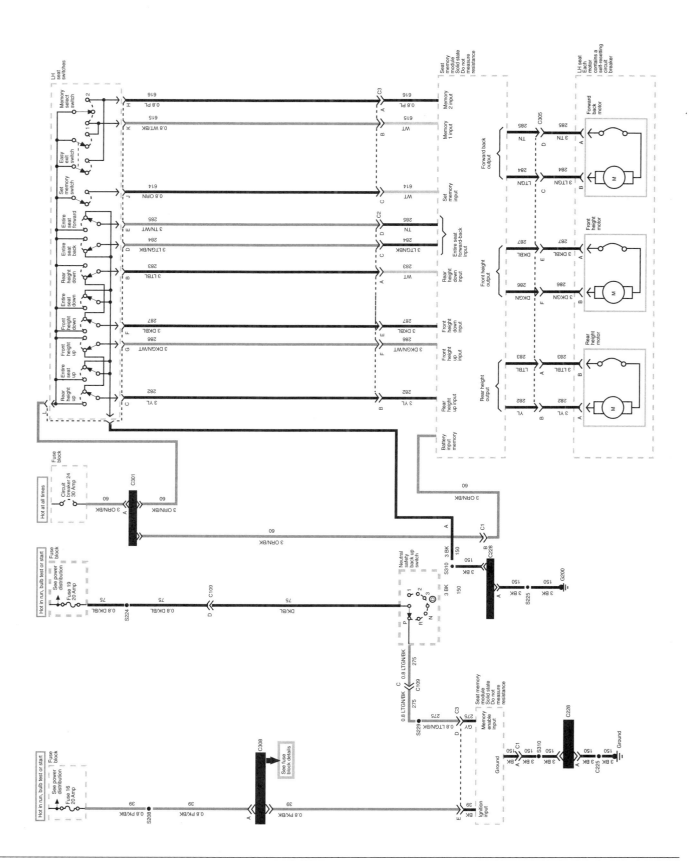

Figure 34-10 To diagnose the memory seat feature, a circuit schematic is required. (Courtesy of General Motors Corporation, Service Technology Group)

Electronic Sunroof Diagnosis

Troubleshooting the causes of slow, intermittent, or no sunroof operation is a relatively simple procedure. Unlike many systems, the sunroof operation is not integrated with other systems. Because the system stands alone, diagnostics are generally performed in the same manner as testing any other motor driven accessory. The following is the diagnostic procedure used to troubleshoot the GM electronic sunroof system shown (Figure 34-11). In addition, refer to the troubleshooting chart (Figure 34-12) to determine the causes of other system malfunctions.

Slow sunroof operation may be caused by excessive resistance in the circuit or motor. Excessive resistance can be determined by performing a voltage drop test. Obtain the correct schematic for the vehicle being diagnosed and follow through the circuit to locate the cause of the excessive resistance. Resistance can also occur inside the motor as a result of brush wear, bushing wear, and corroded connections.

Shorted armature or field coils can also result in slow motor operation. An ammeter can be used to test the current draw of the circuit to determine motor condition.

Intermittent problems may be the result of loose or corroded connections. To locate the cause of an intermittent fault, operate the system while wiggling the wires. This will assist in isolating the location of the poor connection. Some systems also use a circuit breaker to protect the motor. It may be overheating and tripping prematurely, or there may be resistance to window movement in the rails. The circuit breaker will trip, then cool

down and reset. Check that the glass is able to move easily in the rails. In addition, many intermittent problems are caused by a faulty control switch. Operate the control switch several times while performing the circuit test. Replace the switch if it fails at any time during the test.

Follow the procedures listed in your service manual to locate an intermittent or no-operation fault within the circuit. Perform the usual visual inspections of the circuit before continuing. Be sure to check for proper system grounds.

Anti-theft System Troubleshooting

As with many electrical systems, there are many different approaches taken by manufacturers in designing their anti-theft system. Most of the testing of relays, switches, and circuits require only basic electrical troubleshooting capabilities. Use the troubleshooting chart (Figure 34-13) as a guide in locating the fault. Refer to the service manual for the correct procedure of arming the system you are diagnosing.

Self-Diagnostic Systems

Some anti-theft systems offer self-diagnostic capabilities. Follow the service manual procedures for the proper method of entering diagnostics for the vehicle you are diagnosing. The following is a typical example of entering diagnosis.

Some vehicle theft security systems enter the diagnostic mode when the ignition switch is cycled three times from the off to the accessory position. When the

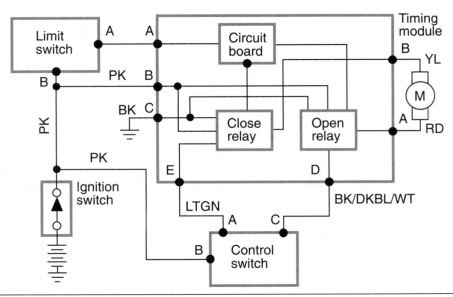

Figure 34-11 General Motors electronic sunroof schematic.

CONDITION	POSSIBLE CAUSE	CORRECTION
1. Sunroof fails to rise or close completely.	1. Weak battery.	a. Start car motor to get proper battery voltage and activate system.
	2. Panel mispositioned in opening.	a. Loosen 8 nuts (with panel in closed position) and position panel. b. Loosen front guide stop and adjust fore or aft.
	3. Panel not flush to roof surface.	a. Loosen guide assembly and adjust guide up or down. (Cable lifter links must be vertical for proper sealing.)
	4. Cable front guides and front guide stop(s) misaligned.	a. Loosen cable guide adjusting screws and raise or lower guide. b. Loosen front guide stop and adjust fore or aft.
	5. Cable assembly lifter link not positioned identically (right to left), out of synchronization.	a. Loosen sliding panel front support(s) and pivot support(s) forward or rearward and retighten. b. Remove actuator and adjust cable cams by sliding cables forward or rearward. (Cams must be in same position on each side.)
	6. Actuator slippage. Motor buzzing with no cable movement.	a. Tighten bolt in bottom of actuator. b. Replace actuator.
	7. Cable slippage, clicking or ratcheting sound.	a. Remove actuator and center guide and replace guide or actuator gear if damaged. b. Check cable at center guide area for wear or stripping; replace cable if damaged.
	8. Wrong sliding panel weatherstrip.	a. Replace weatherstrip.
2. Sunroof panel jammed in roof opening.	1. Broken or stripped cable.	a. Replace cable.
	2. Foreign material in guide.	a. Remove foreign material from guide.
	3. Cable jammed in conduit.	a. Replace cable and conduit.
	4. Side guide(s) out of adjustment.	a. Adjust guide(s).
3. Sunroof actuator inoperative (ignition switch on).	1. Short or open within sunroof wiring.	
	2. Faulty switch.	a. Replace switch.
	3. Defective actuator.	a. Replace actuator.
4. Sunroof panel snaps or clicks when closing.	1. Rear cable support(s) mispositioned to panel.	a. Loosen support(s) and adjust.

Figure 34-12 Sunroof diagnostic troubleshooting chart. (Courtesy of Chilton Book Company)

COMPREHENSIVE TROUBLESHOOTING GUIDE

System Won't Disarm
- CHECK LOCK CYLINDER SWITCHES FOR LOOSENESS.
- TRY TO DISARM THE SYSTEM USING A KEY IN THE PASSENGER DOOR LOCK CYLINDER. (THIS WILL TELL IF THERE IS A PROBLEM ON THE DRIVER'S SIDE.)
- IF SYSTEM DISARMS, CHECK FOR AN OPEN IN THE LIGHT-GREEN WIRE LEADING TO THE LOCK CYLINDER IN THE DRIVER'S SIDE DOOR.
- IF NO OPEN IN THE WIRE, REPLACE THE LOCK CYLINDER SWITCH ASSEMBLY.

System Won't Disarm from Either Door
- HOLD THE LOCK CYLINDER IN THE UNLOCK POSITION WITH THE DOOR KEY.
- CHECK FOR AN OPEN IN THE LIGHT-GREEN WIRE COMING FROM THE CONTROLLER, TERMINAL H.

System Goes Off by Itself
- CHECK THE DIODE (LOCATED UNDER THE DASH IN THE LT. BLUE WIRE FROM TERMINAL J OF CONTROLLER TO DRIVER'S FRONT DOOR JAMB SWITCH.).
- ON CARS EQUIPPED WITH AUTOMATIC DOOR LOCKS (ADL), CHECK FOR DIODE IN YELLOW WIRE FROM DOOR UNLOCK RELAY TO DRIVER'S DOOR LOCK CYLINDER SWITCH.
- CHECK JAMB SWITCHES AND GROUND WIRES FOR CORROSION AND CLEAN OR REPLACE AS NECESSARY.
- CHECK LOCK CYLINDERS, TAMPER SWITCHES, AND THE DOOR JAMB SWITCH WIRE FOR LOOSENESS.

Security Light Inoperative
- CHECK 20 AMP FUSES (SEE ELECTRICAL DIAGNOSIS SECTION).
- CHECK SECURITY LIGHT BULB.
- CHECK FOR A BREAK IN THE ORANGE WIRE LEADING TO THE BULB.
- CHECK THE DIODE (LOCATED UNDER THE DASH IN THE LT. BLUE WIRE FROM TERMINAL J OF CONTROLLER TO DRIVER'S FRONT DOOR JAMB SWITCH.

Security Light Glows But System Won't Disarm
- CHECK TO SEE IF EITHER DOOR OR TRUNK CYLINDER SWITCHES HAVE BEEN TAMPERED WITH.
- CHECK TAMPER SWITCHES FOR LOOSENESS.
- CHECK FOR FRAYS OR PINCHED WIRES TO THE CYLINDER AND JAMB SWITCHES.
- CHECK JAMB SWITCHES FOR PROPER ADJUSTMENT.
- CHECK CIRCUITS EXTERNAL TO THE CONTROLLER. (PLEASE SEE WIRING DIAGRAMS.)
- IF ALL CIRCUITS EXTERNAL TO THE CONTROLLER ARE COMPLETE, SEND THE CONTROLLER AND RELAY TO AN AUTHORIZED REPAIR STATION FOR SERVICE.

Security Light Blinks On and Off
- CHECK FOR LOOSE TAMPER SWITCHES.
- CHECK WIRE BETWEEN DOOR JAMB SWITCH AND CONTROLLER FOR PINCH OR FRAY.
- CHECK WIRE LEADING TO TAMPER SWITCHES.
- CHECK TO SEE THAT LIGHT BLUE WIRE IS PROTECTED BY PLASTIC CONDUIT.
- CHECK DOOR LOCKED SWITCHES AND WIRING.

Figure 34-13 Anti-theft troubleshooting chart. (Courtesy of Chilton Book Company)

vehicle theft security system is in the diagnostic mode, the horn should sound twice and the park and tail lamps should flash. If the horn does not sound or the lights do not flash, voltmeter and ohmmeter tests are required to locate the cause of the problem.

The scan tester may be used to diagnose some vehicle theft alarm systems. Follow the scan tester manufacturer's recommended procedure to enter the vehicle theft alarm system diagnostic mode. When this diagnostic mode is entered, the horn sounds twice to indicate the trunk lock cylinder is in its proper position. When the key is placed in the ignition switch, the park and tail lamps should begin flashing.

The following procedures should cause the horn to sound once if the system is operating normally:

1. Activate the power door locks to the locked and unlocked positions.

2. Use the key to lock and unlock each front door.

3. Turn on the ignition switch.

When the ignition switch is turned on in step 3, the diagnostic mode is exited.

CUSTOMER CARE *Always check the indicator lights in a customer's vehicle. These indicator lights may be indicating a dangerous situation, but the customer may not have noticed the indicator light. For example, the vehicle theft security system set light may not be flashing when the normal system arming procedure is followed. This indicates an inoperative security system, and someone could break into the car without triggering the alarms. The customer paid a considerable amount of money to have this system on the car, and it should be working. If this defect is brought to the customer's attention, he will probably have you repair the system and will appreciate your interest in the vehicle.*

Alarm Sounds for No Apparent Reason

Mechanical and corrosion factors on the cylinder tamper switches can cause the system to activate for no apparent reason. If the customer's vehicle is experiencing this condition, check the lock cylinder for looseness. Any looseness of the cylinder can cause the switch to activate the alarm.

Other causes of alarm system activation include loose, corroded, or improperly adjusted jamb switches. The switches should be adjusted to assure they remain in the OFF position when the doors are fully closed. The switch is adjusted by a nut located at the base of the switch.

Controller Test

To test the controller used in the illustration (Figure 34-14), disconnect the harness from the controller and the harness to the relay. Connect the test light between the N terminal and ground. The test light should illuminate to indicate voltage to the horns and controller.

WARNING *Failure to disconnect both harnesses will lead to false test indications.*

Move the test light probe to terminal M. The light should turn on only when the electrical door lock switch is moved to the UNLOCK position. Next, connect the test light between terminal B and a 12-volt source. The light should light only if the doors are locked. The light should go out if any doors are unlocked.

Probe terminal K for voltage. The test light should illuminate only when the ignition switch is placed in the RUN position. Check for a blown fuse or an open circuit if it does not light.

To test whether the cylinders are operating properly and have not been tampered with, connect the test light between terminal J and a 12-volt source. The test light should light only if a door is open. If it glows with the doors closed, inspect the lock cylinders for damage.

With the test light connected between terminal H and a 12-volt source, the test light should be on only when the outside door key is turned to the unlock position. Move the test light between terminal G and a good ground. The light should illuminate when the electric door lock switch is operated. This indicates there is electrical power to the switch. The test light should light in the LOCK position and go out in the UNLOCK position.

Reconnect the relay harness. When the test light is connected between the F terminal of the controller connector and ground, the horn should sound and the lights should turn on. This indicates that the relay coil is functioning.

Next, turn the ignition switch to the RUN position with the test light connected between the E terminal of the controller connector and ground. Use a voltmeter to measure voltage to the starter. There should be zero volts. This indicates the starter interrupt relay is opening to prevent engine starting.

SERVICE TIP *In some instances it may be easier to attempt starting the engine than to check for voltage at the starter. If the relay is working properly, the engine will not start.*

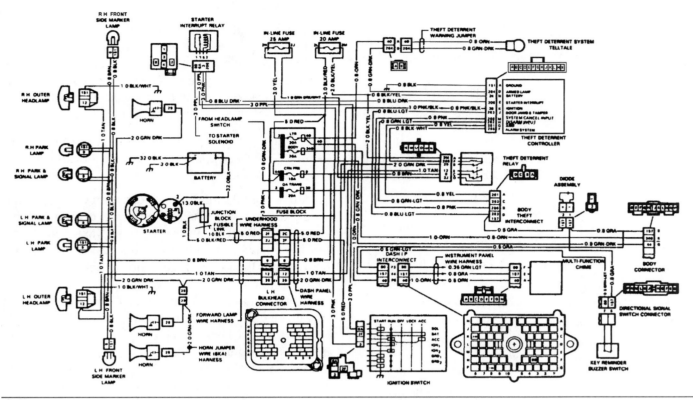

Figure 34-14 Circuit schematic of anti-theft system. (Courtesy of Chilton Book Company)

Connect a jumper wire between terminal D and a good ground while observing the security warning light. The warning light should be on. Connect a test light between terminal A of the controller connector and a 12-volt power supply. The test light should light. If not, there is a problem in the ground circuit.

Relay Test

Faulty relays are a leading cause of anti-theft system malfunction. Testing the relay is a simple matter of using a jumper wire to bypass the relay. If the circuits operate with the relay bypassed, but not with the relay connected, the relay is probably at fault. However, do not replace the relay until you have tested for the proper amount of applied voltage to the relay and for proper ground switching of the controller.

Automatic Door Lock System Troubleshooting

Some systems offer self-diagnostics through the body computer. The service manual will provide the steps required to enter diagnostics on these vehicles. The following is an example of locating the fault in vehicles that

do not provide this feature when the door locks work, but they do not lock or unlock automatically. As with any electrical diagnosing, you will need the circuit diagram for the system you are working on. The following steps relate to the system shown (Figure 34-15).

1. Locate the controller and backprobe for voltage at the power input terminal D with the ignition switch in the RUN position. If there is no voltage present, there is an open in circuit 39.

2. Backprobe for voltage between terminals A and D. If there is no voltage, check for an open in circuit 150.

3. Make sure the courtesy lights are off and all doors are closed. With the gear selector in the PARK position, turn the ignition switch to the RUN position.

4. Connect a test light between controller terminal B and a good ground. If the neutral safety switch circuit is operating properly, the test light will light.

5. With the test light connected as in step 4, move the gear selector to any other position. The test light should go out. If the light does not go out, check the neutral safety switch. It may be out of adjustment or faulty.

6. Leave the gear selector as in step 5, and connect the test light between terminals C and D. The test light should not illuminate. If it does, check circuit 156 and the light switch and door jamb switches.

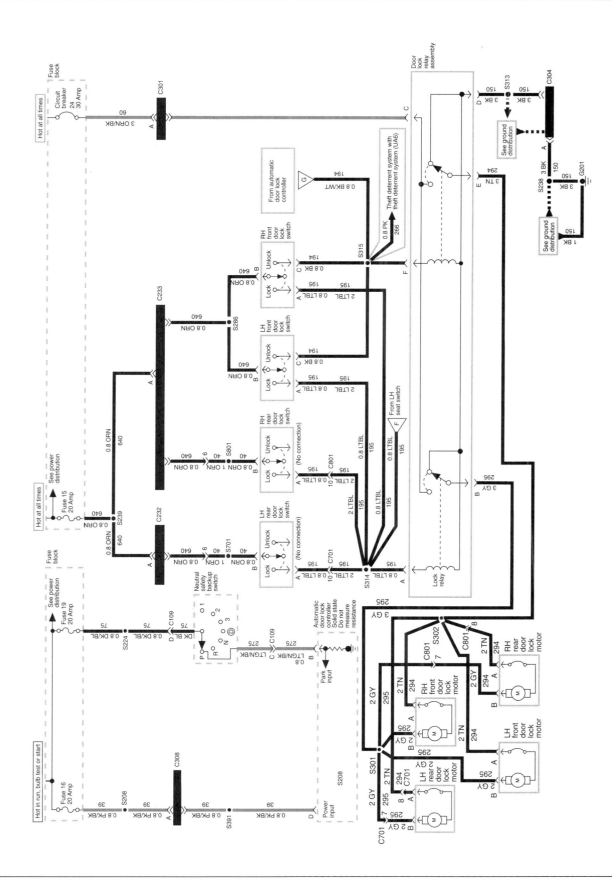

Figure 34-15 Automatic door lock schematic. (Courtesy of General Motors Corporation, Service Technology Group)

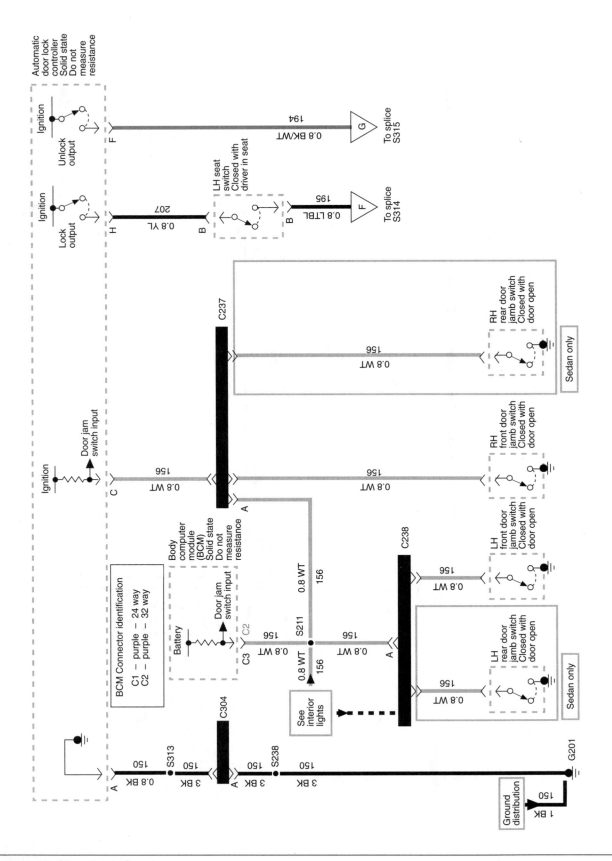

Figure 34-15 (Continued)

7. Return the gear selector to the PARK position. Connect the test light between terminal H and ground.

8. Observe the test light while the gear selector is moved from PARK to REVERSE. The test light should flash once. If not, replace the controller.

If the circuits passed all tests, check circuits 207 and 195, and the left-hand switch assembly for opens.

Keyless Entry Diagnosis

As an example of troubleshooting the keyless entry system, refer to the simplified schematic (Figure 34-16) of a Ford system. Poor, intermittent, or no-operation complaints can be verified using a quick test procedure. If the quick test indicates a fault, refer to the pinpoint test in the service manual to isolate the problem.

Before testing the system, make sure the battery is properly charged. Disconnect the battery and reconnect it to reset the system. Use the door lock switch to cycle between lock and unlock. If any or all of the doors fail to operate correctly, the fault is within that subsystem and not with the keyless entry system. Check the normal power lock system before continuing. If all door locks operate correctly, continue to test the system as follows:

WARNING *If the keyless entry system is not operating, be very careful not to lock the keys in the vehicle while performing this test. Make sure you have the keys in hand whenever the doors are to be locked with no one inside the vehicle.*

1. Depress one keypad button. The keypad light should turn on. If the light does not turn on, go to step 3.

2. After the light turns off again, press another button. Repeat this step for each keypad button, observing the light operation.

3. If the lights operate normally, go to step 4. If none of the lights turned on, refer to the service manual for performing pinpoint test B. If one or more lights fail to illuminate, refer to the service manual for performing pinpoint test C.

4. With the ignition switch in the OFF position and the key in the lock cylinder, close all doors including the trunk lid. Make sure all doors are unlocked. Press the 7-8 and 9-0 buttons simultaneously. All doors should lock. If one or more doors fail to lock, replace the control module. If all the doors fail to lock, refer to the service manual for pinpoint test C procedures.

5. Enter the code sequence located on the control module. The driver side door should unlock. If not, go to pinpoint test C.

6. When the keypad light turns off, reenter the code sequence and press the 3-4 button. (If the keypad light is still on, just press the 3-4 button.) The passenger side door should unlock. If not, perform pinpoint test C.

7. Repeat step 6 using the 5-6 button. The deck (trunk) lid should unlock. If not, refer to the service manual for pinpoint test F procedures.

8. Sit in the driver's seat with the gear selector in the PARK position. Close all doors, leaving them in the UNLOCK position.

9. Turn the ignition switch to the RUN position. Place the gear selector into REVERSE and then into DRIVE. All doors should lock. If any or all of the doors fail to lock, refer to the service manual for the procedure to perform pinpoint test A.

10. While still sitting in the driver's seat, open the driver's door. Depress the door jamb courtesy light switch to turn the lights off. The driver side door lock should not activate. Go to step 14 if the lock activates when the interior lights are turned off.

11. Return the gear selector to the PARK position and the ignition switch to the OFF position. Remove the key and exit the vehicle.

12. Press any of the keypad buttons while observing the interior lights. The courtesy and illuminated entry lights should turn on. If the system fails to operate, perform pinpoint test D.

13. When the interior lights go out, press any button to activate them again. Then press 7-8 and 9-0 simultaneously. The interior lights should turn off. If they do not, perform pinpoint test D.

14. Open the driver's side door and enter the vehicle. Turn the ignition switch to the RUN position. The interior lights should turn on when the door handle is lifted to open the door and turn off when the ignition switch is placed in the RUN position. If the interior lights did not illuminate, perform pinpoint test D. If the interior lights did not turn off when the ignition switch was placed in the RUN position, replace the control module and repeat the test sequence.

15. Enter the code sequence, then press the 1-2 button. This signals the control module that an alternate code is going to be entered. Depress the buttons in the following order: 9-0, 7-8, 5-6, 3-4, 1-2. At the completion of entering the alternate code the driver side door lock should unlock. If the door does not unlock, replace the control module and repeat the test sequence.

16. Disable the alternate code by reentering the code sequence located on the control module. Then depress the 1-2 button. Wait for the keypad light to turn off. Close and lock the driver's door. Then repeat the code

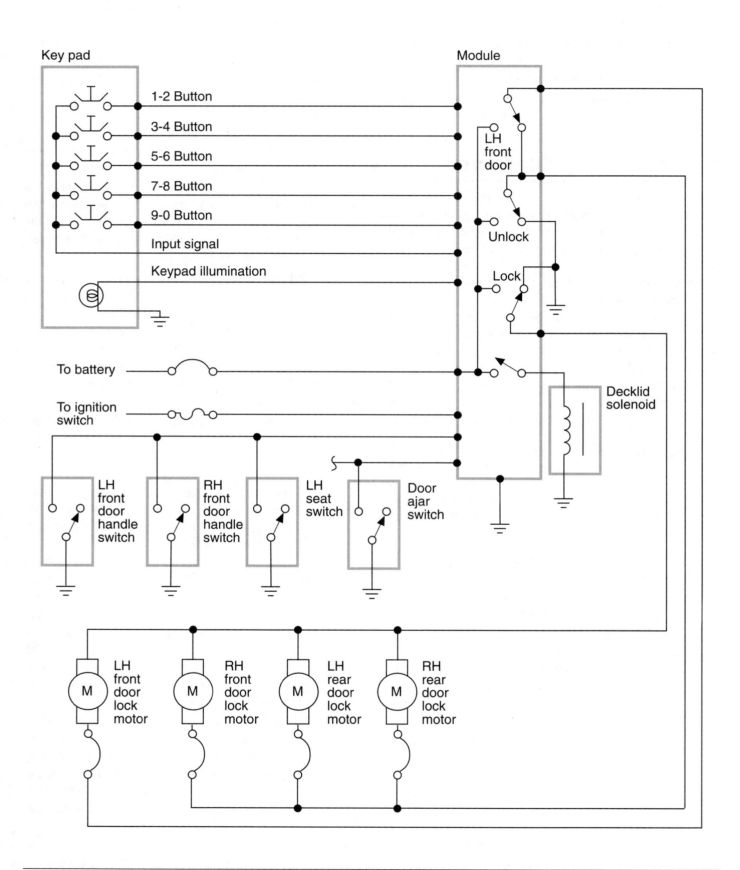

Figure 34-16 Simplified keyless entry system schematic.

installed in step 15. The door should remain locked. If not, replace the control module and repeat the test sequence.

Remote Keyless Entry

Many new vehicles are equipped with a remote keyless entry system that is used to lock and unlock the doors, turn on the interior lights, and release the trunk latch. A small receiver is installed in the vehicle. The transmitter assembly is a hand-held item attached to the key ring (Figure 34-17). It has three buttons that control the functions of the system.

The system operates at a fixed radio frequency. If the unit does not work from a normal distance, check for two conditions: weak batteries in the remote transmitter or presence of a stronger radio transmitter close by (radio station, airport transmitter, etc.)

If the system has other problems, make sure the door locks, trunk latch, and interior lamps work normally when manually activated. If these systems check out fine, detailed diagnosis of the remote system is necessary. Follow the manufacturers' recommendations for doing this.

Electronically Heated Windshield Service

The two basic styles of heated windshields require different approaches to diagnostics. However, in either system, check the operation of the alternator, alternator belt condition, and the windshield for damage before beginning service. Also, the system will not operate if the battery state of charge is low.

During the course of diagnosing the heated windshield system it will become necessary to override the temperature sensor. The Ford system will not activate unless the interior temperatures are less than 40°F. To override the system, connect a jumper wire between the black test lead pigtail and ground. The test lead is usually located in the engine compartment close to the wiper motor.

For the GM system to turn on, inside vehicle temperatures must be below 65°F (18°C). To override the internal thermistor, ground terminal C of the data link connector (Figure 34-18).

WARNING *Perform the diagnostic test as fast as possible. Prolonged operation of the system at temperatures above 65°F may cause permanent optical damage to the windshield.*

The following is a service sample of the test procedures used to diagnose the GM-style system.

If the customer states that the system does not turn on, verify this by starting the engine and pressing the activation switch. Observe the LED in the switch. The test procedure is determined by the attitude of the LED.

If the LED comes on for longer than half a second, but goes off again within 3 seconds, use the schematic (Figure 34-19) and follow these steps.

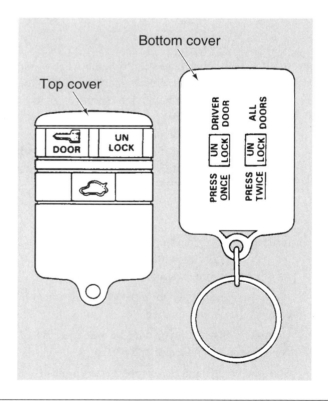

Figure 34-17 Typical door lock control transmitter assembly. (Courtesy of General Motors Corporation, Service Technology Group)

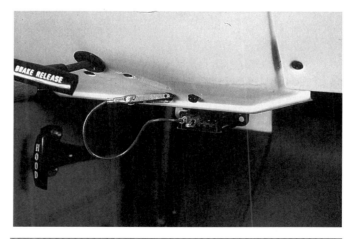

Figure 34-18 Jump the C terminal of the DLC to ground to bypass the thermistor.

1. Measure the voltage at the data link connector terminal C. If it is within 2 volts of battery voltage, check for a short to battery voltage in the yellow wire. If the wire is good, replace the control module. If the voltage is not within 2 volts of battery voltage, continue testing.

2. Ground data link connector terminal C and start the engine. Press the activation switch to turn on the system. If the LED lights and the windshield heats, replace the control module. The internal thermistor is bad. However, it is not serviced separately from the control module.

WARNING *Check the surface temperature of the windshield during testing. Turn off the system immediately if it gets too hot.*

3. Turn the ignition switch to the OFF position. Then return it to the RUN position. Measure the voltage from terminal B6 of the control module. If the measured voltage is less than 11.2 volts, there is an open or short in circuit 2.

4. With the ignition switch still in the RUN position, measure voltage at terminal A6. If battery voltage is not present, there is an open or short in circuit 50.

5. Measure the voltage between terminals A6 and A8 of the control module. If battery voltage is not present, measure the resistance between circuit 155 and ground. It must be less than 0.5 ohm.

6. With the ignition switch in the RUN position, measure voltage between terminals A6 and A3 while repeatedly pressing the activation switch. Zero volts should be indicated when the switch is pressed and 9.1 volts when it is released. Check circuit 648 for an open if the measured voltage is different than these values. If circuit 648 is good, check the continuity of the activation switch when it is in the released position. Also check for a good ground connection at terminal C of the switch.

7. Turn the ignition switch to the OFF position and disconnect the windshield connector. Measure the resistance of the windshield at the connector terminals. The resistance between terminals A and B should be less than 10 ohms. It should be less than 6 ohms between B and C. If the measured resistance values are different, inspect the connector. If the connector is good, replace the windshield.

WARNING *Measure the resistance on the windshield side of the connector. Measuring resistance on the controller side of the connector may damage the controller.*

8. Leave the windshield disconnected and measure the resistance of the windshield between each terminal of the connector to ground. All terminals should indicate 10,000 ohms or greater. If less than 10,000 ohms, check for shorts to ground between the windshield and the body.

9. Check circuit 475 for continuity between terminals D and A2 of the control module. Also check for continuity of circuit 378 between the windshield harness side connector terminal A and module terminal B4. If there is not continuity in either one of the circuits, repair the opens. If circuits 475 and 378 are good, replace the control module.

When confirming the complaint, if the LED remains on but the windshield does not heat, test the system as follows:

1. Start the engine and activate the system by pressing the switch.

2. Measure the three-phase voltage at the three posts on the back of the generator. Do not disconnect the connector at this time. Measure the voltage by backprobing the connector. The voltmeter must be on a scale higher than 20 volts AC.

3. Refer to the diagram (Figure 34-20) and measure the voltage as follows:

   X to Y

   X to Z

   Y to Z

4. In all cases the voltage should be between 9 and 14 volts. If the voltage is within specifications, go to step 6. If the voltage is not within these limits, turn the ignition switch to the OFF position and disconnect the three wires from the generator. Repeat the test again. Replace the generator if the voltage is still not within the limits. If the voltage is between 9 and 14 volts, check the wires from the generator to the power module.

5. Disconnect the windshield connector and measure the resistance between terminals B and C. If the resistance is less than 3 ohms, replace the power module. If the resistance is more than 3 ohms, replace the windshield.

6. Place the ignition switch in the OFF position and disconnect the control module.

7. Connect a fused jumper wire between pin B1 of the control module connector and battery positive.

8. Start the engine and backprobe between pins B and C of the connector with a voltmeter. Use 100-volt or greater DC scale. If the voltage is between 50 and 85 volts, replace the control module.

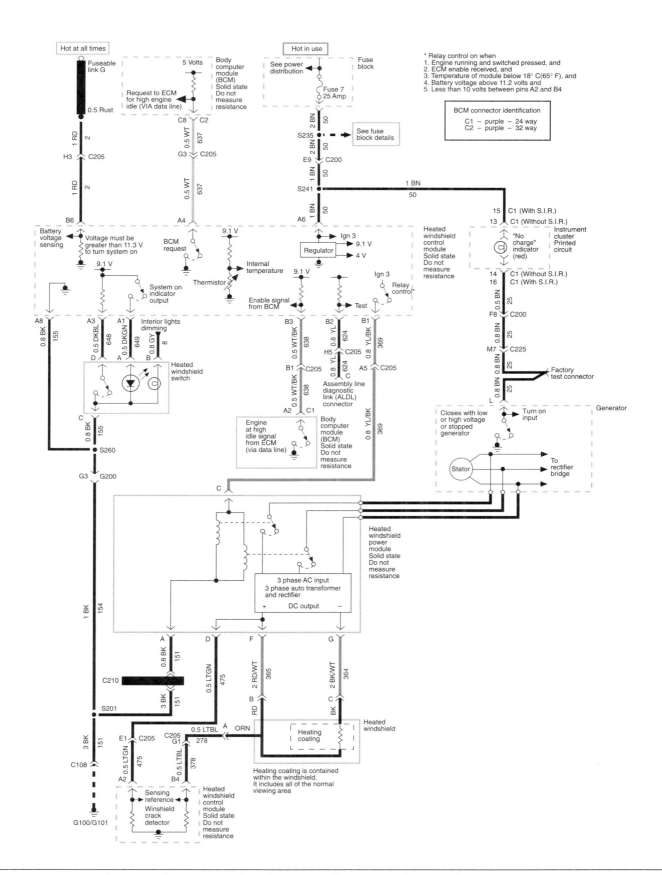

Figure 34-19 General Motors' heated windshield system. (Courtesy of General Motors Corporation, Service Technology Group)

9. With the seven-pin connector to the control module disconnected, check to see if circuit 369 has continuity between the control module and terminal B1. Also check that power module terminal C is free of shorts and grounds.

10. Check that circuit 151 has continuity between power module terminal A and ground.

11. If either circuit 369 or 151 have an open, repair as necessary.

12. If both circuits are good, measure the resistance between terminals G and F of the power module connector. If the resistance is less than 6 ohms, replace the power module. If the resistance is greater than 6 ohms, there is an open in circuit 364 or 365.

When confirming the customer complaint, if the LED illuminates for about 3 seconds then turns off, test as follows:

WARNING *It is normal for the LED to turn on and then off after 3 seconds if the ignition switch is in the RUN position but the engine is not started. Be sure to verify the complaint with the engine running.*

1. Start the engine and allow it to warm (engine off high idle).

2. Activate the system. If internal temperatures are above 65°F, jump DLC terminal C to ground.

3. The engine idle speed should increase within a few seconds of activating the system. Go to step 4 if the engine does not increase speed. Go to step 7 if the engine speed increases.

4. Turn the ignition switch off and disconnect the control module.

5. Start the engine. Ground terminal A4 of the module connector with a fused jumper wire.

6. If the engine speed increases, there is a problem with the BCM or ECM. Follow the service manual procedures for diagnosing these units. If the engine speed

does not increase, replace the control module.

7. With the engine running, backprobe terminal B3 of the control module connector with a voltmeter.

8. Activate the system while observing the voltmeter. The voltage should drop from battery voltage to less than 4 volts in approximately 3 seconds. If the voltage fails to decrease, check the wire between A2 of the BCM and B3 of the control module. If the wire is good, the BCM or ECM is defective.

9. If the voltage drops according to specifications, replace the control module.

Radio-Stereo Sound Systems

Because automotive technicians do not repair radios or system component units, there is a tendency to remove the unit when the customer has described having a particular problem before performing a thorough pre-diagnosis. In many cases, the units show "NO TROUBLE FOUND" and are sent back to the dealership or garage. Most of the problems could have been solved without taking the radio out of the dash.

It is important to try to avoid depriving customers of the use of their radio. Before removing the radio/component, do these simple checks to quickly determine whether the system problems are external:

- Test the vehicle's radio system outside, not inside a building. Make sure the hood is down.

- Most noise can be located on weak AM stations at the low frequency end of the tuning band.

- Ignition noise on FM usually indicates a problem in the ignition system.

- If a test antenna is going to be used, the base must be grounded to the vehicle's body. DON'T HOLD THE MAST.

- Ninety percent of radio noise enters by way of the antenna.

- Most "rubber" hoses (vacuum, coolant, etc.) are electrically conductive, unless they have a white stripe.

- When shielding hoses, wires, the dash, etc., use foil or screening material. Be sure to ground the material.

- A weak or fading AM signal is normally caused by an improperly adjusted antenna trimmer (when used).

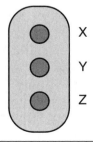

Figure 34-20 Generator connector terminal identification.

The technician must determine the exact nature of the problem when performing a diagnosis. Determining whether the problem is intermittent or constant or whether the problem occurs when the vehicle is moving or stationary will help pinpoint the nature of the problem. Use the chart (Figure 34-21) as an aid to diagnosing system problems.

A "diagnostic RF sniffer" tool can be made from an old piece of antenna lead-in from a mast or power antenna (Figure 34-22). This sniffer can be used, along with the radio, to locate "hot spots" that are generating radio frequency interference (RFI) noise. The noise can be found in wiring harnesses, in the upper part of the dash, or even between the hood and the windshield. When checking for noise on a wire, it is best to hold the sniffer parallel and close to the wire.

Consult your service manual for additional steps and procedures for further system diagnosis. The manufacturers usually have their own specialized service manuals that pertain to radio system problems and repair procedures.

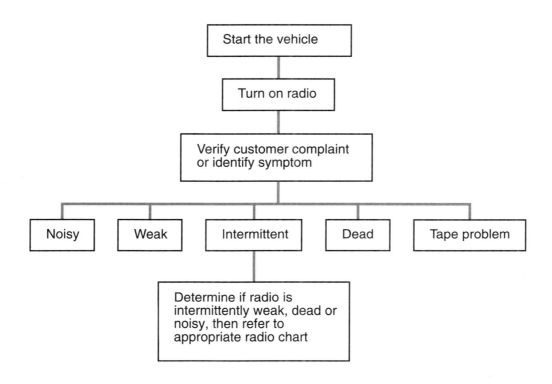

Figure 34-21 System diagnosis and analysis chart. (Courtesy of General Motors Corporation, Service Technology Group)

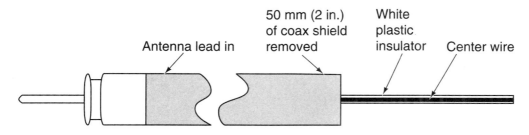

Figure 34-22 RF sniffer tool. (Courtesy of General Motors Corporation, Service Technology Group)

Summary

❑ Most vehicle manufacturers have incorporated self-diagnostics into their cruise control system. This allows some means of retrieving trouble codes to assist the technician in locating system faults.

❑ A "quick checker" is used to test some cruise control systems, however, a scan tool or panel displays can also be used.

❑ Ford's integrated vehicle speed control (IVSC) system has self-test capabilities that are contained within the KOEO and KOER routine of the PCM.

❑ The quick test is a functional test of the cruise control system. This test will check the operation and function of all system components except the vehicle speed sensor.

❑ The simulated road test will allow the technician to perform a road test without leaving the shop.

❑ Troubleshooting the causes of slow, intermittent, or no sunroof operation are generally performed in the same manner as testing any other motor driven accessory.

❑ Some vehicle theft security systems enter the diagnostic mode when the ignition switch is cycled three times from the off to the accessory position. When the vehicle theft security system is in the diagnostic mode, the horn should sound twice and the park and tail lamps should flash.

❑ The scan tester may be used to diagnose some vehicle theft alarm systems. Follow the scan tester manufacturer's recommended procedure to enter the vehicle theft alarm system diagnostic mode.

❑ Mechanical and corrosion factors on the cylinder tamper switches can cause the system to activate for no apparent reason. Other causes of alarm system activation include loose, corroded, or improperly adjusted jamb switches.

❑ Faulty relays are a leading cause of anti-theft system malfunction. Testing the relay is a simple matter of using a jumper wire to bypass the relay. If the circuits operate with the relay bypassed, but not with the relay connected, the relay is probably at fault.

❑ When troubleshooting the keyless entry system, poor, intermittent, or no-operation complaints can be verified using a quick test procedure. If the quick test indicates a fault, refer to the pinpoint test in the service manual to isolate the problem.

❑ If the remote keyless entry system does not work from a normal distance, check for two conditions: weak batteries in the remote transmitter or a stronger radio transmitter close by (radio station, airport transmitter, etc.).

❑ The two basic styles of heated windshields require different approaches to diagnostics. However, in either system, check the operation of the alternator, alternator belt condition, and the windshield for damage before beginning service.

❑ A "diagnostic RF sniffer" tool can be made from an old piece of antenna lead-in from a mast or power antenna. This sniffer can be used, along with the radio, to locate "hot spots" that are generating radio frequency interference (RFI) noise.

Terms-To-Know

Diagnostic service mode	Self-diagnostics	Simulated road test
Quick Checker	Self-test input (STI)	

ASE Style Review Questions

1. A customer states that the cruise control system is inoperative. Technician A says that the dump valve may be stuck open. Technician B says the servo may be faulty. Who is correct?
   a. A only
   b. B only
   c. Both A and B
   d. Neither A nor B

2. Diagnosis of Ford's IVSC system is being discussed. Technician A says the quick test will check the operation and function of all system components except the vehicle speed sensor. Technician B says the pinpoint test is performed first, then the quick test. Who is correct?
   a. A only
   b. B only
   c. Both A and B
   d. Neither A nor B

3. A customer says the power seats are not operating. Technician A says the problem is in the memory seat feature. Technician B says the switch assembly may be faulty. Who is correct?
   a. A only
   b. B only
   c. Both A and B
   d. Neither A nor B

4. Slow sunroof operation is being discussed. Technician A says the circuit breaker may be the wrong rating. Technician B says a shorted armature or field coils can be the cause. Who is correct?
   a. A only
   b. B only
   c. Both A and B
   d. Neither A nor B

5. Technician A says weak and fading AM signals could be the result of a bad or ungrounded antenna. Technician B says weak and fading AM signals could be the result of a misadjusted antenna trimmer. Who is correct?
   a. A only
   b. B only
   c. Both A and B
   d. Neither A nor B

6 A customer states that the vehicle alarm will trip when there is no apparent attempt of entry. Technician A says the fault may be a loose lock cylinder. Technician B says a misadjusted jamb switch may be the cause. Who is correct?
   a. A only
   b. B only
   c. Both A and B
   d. Neither A nor B

7. Testing of the alarm relay is being discussed. Technician A says if the system fails to operate with the relay bypassed, the relay is defective. Technician B says the relay is generally tested using an ammeter. Who is correct?
   a. A only
   b. B only
   c. Both A and B
   d. Neither A nor B

8. Automatic door lock (ADL) system diagnosis is being discussed. Technician A says some systems offer self-diagnostics through the body computer. Technician B says when the door locks work, but do not lock or unlock automatically, the neutral safety switch may be out of adjustment. Who is correct?
   a. A only
   b. B only
   c. Both A and B
   d. Neither A nor B

9. A keyless entry system is being diagnosed using the quick test sequence. Technician A says before testing the system, make sure the battery is properly charged. Technician B says to reset the system by disconnecting and reconnecting the battery. Who is correct?
   a. A only
   b. B only
   c. Both A and B
   d. Neither A nor B

10. Electronically heated windshield service is being discussed. Technician A says to override the GM system temperature sensor, ground terminal C of the data link connector. Technician B says the internal thermistor is not serviced separately from the control module. Who is correct?
    a. A only
    b. B only
    c. Both A and B
    d. Neither A nor B

35 Passive Restraint Systems

Objective

Upon completion and review of this chapter, you should be able to:

❏ Explain the basic operating principles of the emergency tensioning retractor.

❏ Explain the purpose of passive restraint systems.

❏ Describe the basic operation of automatic seatbelts.

❏ Describe the operation of air bag system sensors.

❏ List the components of the air bag module, and explain their function.

❏ Explain the functions of the diagnostic module used in air bag systems.

❏ List the sequence of events that occur during air bag deployment.

❏ Describe the function of the clock spring.

❏ Explain the basic operating principles of the emergency tensioning retractor.

❏ Describe normal operation of the air bag system warning light.

❏ Describe the operation of a hybrid inflator module and explain the advantages of this type of module.

❏ Explain the location and purpose of shorting bars.

Introduction

Federal regulations have mandated the use of automatic **passive restraint** systems in all vehicles sold in the United States after 1990. Two-point or 3-point automatic seatbelt and air bag systems are currently offered as a means of meeting this requirement. Passive restraints operate automatically with no action required on the part of the driver or occupant.

In this chapter you will learn the operation of the automatic passive restraint systems and of the air bag system. The safety of the driver and/or passengers depends on the technician properly diagnosing and repairing these systems. As with all electrical systems, the technician must have a basic understanding of the operation of the restraint system before attempting to perform any service.

There are many safety cautions associated with working on air bag systems. Safe service procedures are accomplished through proper use of the service manual and by understanding the operating principles of these systems.

Passive Seatbelt Systems

The **passive seatbelt system** automatically puts the shoulder and/or lap belt around the driver or occupant (Figure 35-1). The automatic seatbelt system uses reversible DC motors to move the belts by means of carriers on tracks (Figure 35-2).

One end of the seatbelt is attached to the carrier, the other end is connected to the inertia lock retractors. **Inertia lock retractors** use a pendulum mechanism to lock the belt tightly during sudden movement (Figure 35-3).

Figure 35-1 Passive automatic seatbelt system operation. (Reprinted with permission of Ford Motor Company)

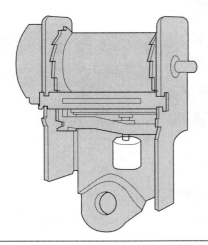

Figure 35-3 Inertia lock seatbelt retractor. (Reprinted with permission of Ford Motor Company)

When the door is opened, the outer end of the shoulder harness moves forward (to the A-pillar) to allow for easy entry or exit (Figure 35-4). When the door is closed and the ignition switch is placed in the RUN position, the motor moves the outer end of the harness to the locked position in the B-pillar (Figure 35-5).

The automatic seatbelt system uses a control module to monitor operation (Figure 35-6). The monitor receives inputs from door ajar switches, limit switches, and the emergency release switch.

The door ajar switches signal the position of the door to the module. The switch is open when the door is closed. This signal is used by the control module to activate the motor and move the harness to the lock point behind the occupant's shoulders. If the module receives a signal door open signal, regardless of ignition switch position, it will activate the motor to move the harness to the forward position.

The limit switches inform the module of the position of the harness. When the harness is moved from the FORWARD position, the front limit switch (limit A) closes. When the harness is located in the LOCK position, the rear limit switch (limit B) opens and the module turns off the current to the motor. When the door is opened, the module reverses the current flow to the motor until the A switch is opened.

An **emergency release mechanism** is provided in the event the system fails to operate. The normally closed

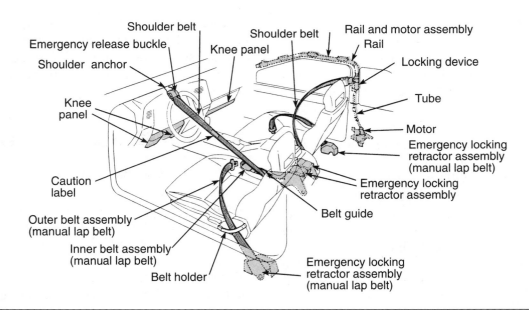

Figure 35-2 Passive seatbelt restraint system uses a motor to put the shoulder harness around the occupant. (Reprinted with permission of Ford Motor Company)

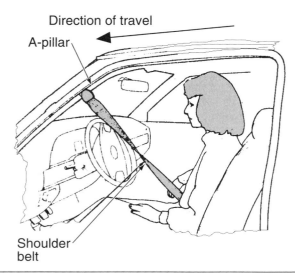

Figure 35-4 When the door is opened the motor pulls the harness to the A-pillar. (Reprinted with permission of Ford Motor Company)

(NC) emergency release switch is opened whenever the release lever is pulled. The module will turn on the warning lamp in the instrument panel and sound a chime to alert the driver. The opened switch also prevents the harness retractors from locking.

Ford incorporates the fuel pump inertia switch into the automatic seatbelt system. If the module receives a signal that the switch is open, it prevents the harness from moving to the forward position if the door opens.

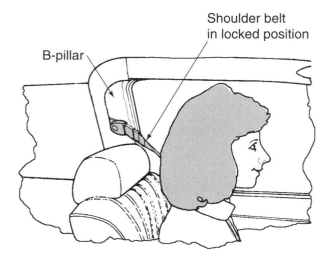

Figure 35-5 When the door is closed and the ignition switch is in the RUN position, the motor draws the harness to its lock position. (Reprinted with permission of Ford Motor Company)

Air Bag Systems

The need to supplement the existing restraint system during frontal collisions has led to the development of the **supplemental inflatable restraint (SIR)** or **air bag systems** (Figure 35-7). Vehicles with air bag systems also have seatbelts that must be worn at all times. The air bag is a supplementary restraint while the seatbelt is the primary restraint system. Seatbelts must be worn in an air bag-equipped vehicle for the following reasons:

1. Seatbelts hold the occupants in the proper position when the air bag inflates.

2. Seatbelts reduce the risk of injury in less severe accidents in which the air bag does not deploy.

3. Seatbelts reduce the risk of occupant ejection from the vehicle, and thus reduce the possibility of injury.

4. Because most air bag systems are not designed to deploy during side or rear collisions.

The air bag system contains an inflatable air bag module located in the steering wheel (Figure 35-8). If the vehicle is involved in a frontal collision, the air bag inflates rapidly to keep the driver's body from flying ahead and hitting the steering wheel or windshield. The frontal impact must be within 30° of the vehicle centerline to deploy the air bag (Figure 35-9). The air bag system helps to prevent head and chest injuries during a collision.

WARNING *Always refer to the specific manufacturer's recommendations. Each system has different safety requirements.*

A typical air bag system consists of sensors, a diagnostic module, a clock spring, and an air bag module. The illustration (Figure 35-10) shows the typical location of the common components of the SIR system.

Air Bag Module

The **air bag module** contains the air bag, air bag container, retainer and base plate, inflator, and trim cover. The retainer and base plate are made from stainless steel and riveted to the inflator module. The air bag is made from porous nylon. Some air bags have a neoprene coating. The purpose of the air bag module is to inflate the air bag in a few milliseconds when the vehicle is involved in a frontal collision. A typical fully inflated driver's side air bag has a volume of 2.3 cu. ft.

The air bag module uses pyrotechnology (explosives) to inflate the air bag. The igniter is an integral compo-

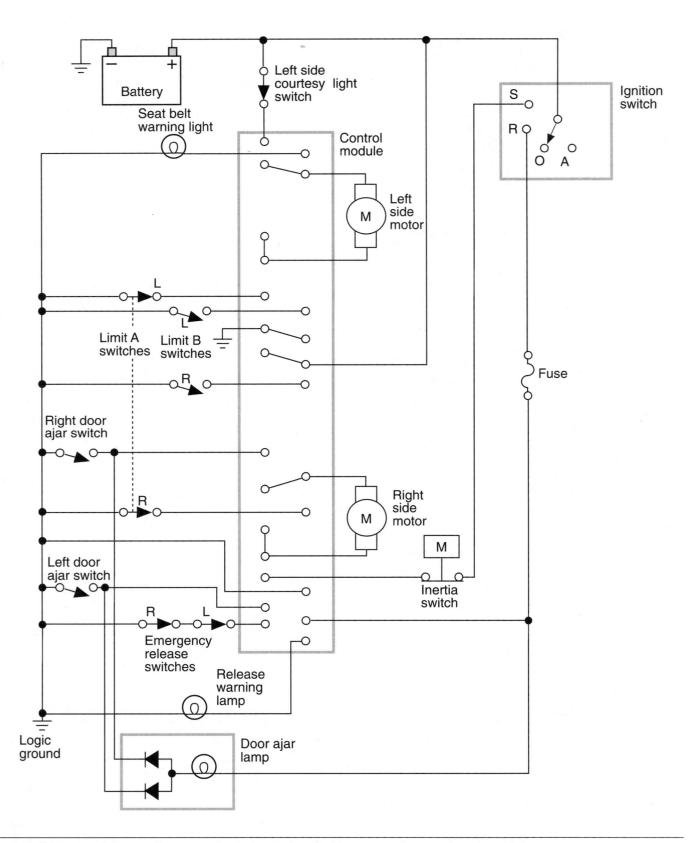

Figure 35-6 Typical circuit diagram of automatic seatbelt system using a control module.

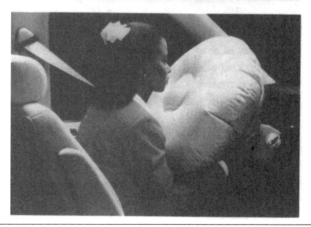

Figure 35-7 Air bag sequence. (Courtesy of Chrysler Corporation)

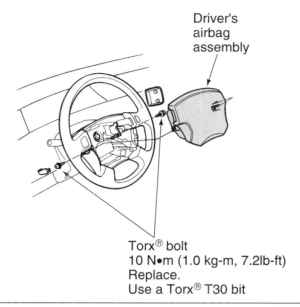

Driver's airbag assembly

Torx® bolt
10 N•m (1.0 kg-m, 7.2lb-ft)
Replace.
Use a Torx® T30 bit

Figure 35-8 Inflator assembly in steering wheel. (Courtesy of American Honda Motor Co., Inc.)

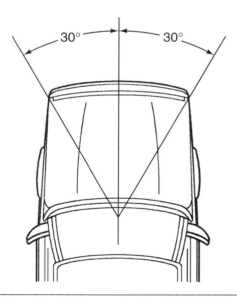

30°   30°

Figure 35-9 The air bag is deployed when the vehicle is involved in a frontal collision within 30° of the vehicle centerline. (Courtesy of General Motors Corporation, Service Technology Group)

nent of the **inflator assembly**. The **igniter** is a combustible device that converts electric energy into thermal energy to ignite the inflator propellant. It starts a chemical reaction to inflate the air bag (Figure 35-11).

At the center of the igniter assembly is the **squib** which contains zirconic potassium percolate (ZPP). The squib is similar to a blasting cap. When 1.75 amperes is supplied through the squib, the air bag deploys.

Three components are required to create an explosion: fuel, oxygen and heat. The squib and the igniter

charge of barium potassium nitrate (a very fast reacting explosive) provide the heat necessary for inflator module explosion. Fuel is supplied by the generant, which contains sodium azide and cupric oxide. The sodium azide provides hydrogen, and the cupric oxide provides oxygen. When the chemicals in the inflator module explode, large quantities of hot, expanding nitrogen gas are produced very quickly. This expanding nitrogen gas flows through the igniter assembly **diffuser**, where it is filtered

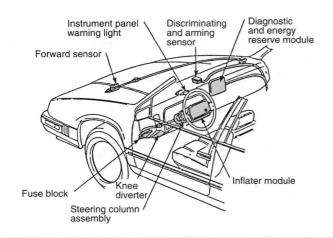

Figure 35-10 Typical location of components of the air bag system. (Courtesy of Chrysler Corporation)

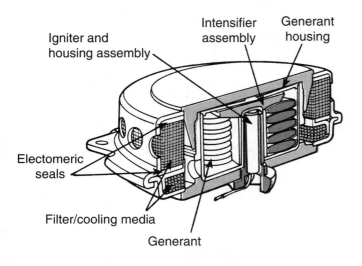

Figure 35-11 Igniter assembly.

and cooled before inflating the air bag. Four layers of screen are positioned on each side of the ceramic in the filter. Sodium oxide dust is trapped by the filter. Sodium hydroxide is an irritating caustic; therefore, automotive technicians are always warned to wear safety goggles and protective gloves when servicing deployed air bags. Within seconds after air bag deployment, the sodium hydroxide changes to sodium carbonate.

Tear seams in the steering wheel cover and in the instrument panel cover above the passenger's side air bag split easily and allow the air bags to exit from the module. Large openings under the air bag allow it to deflate in 1 second so it does not block the driver's view or cause a smothering condition.

Combustion temperature in the inflator module reaches about 2,500°F, but the air bag remains at approx-

imately room temperature. Typical by-products from inflator module combustion are:

1. Nitrogen — 99.2%
2. Water — 0.6%
3. Hydrogen — 0.1%
4. Sodium oxide — less than 1/10 of 1 part per million (ppm)
5. Sodium hydroxide — very minute quantity

Many air bags are packed with corn starch in the inflator module. This along with other combustion by-products may appear as a white dust during and after air bag deployment.

CAUTION *Wear gloves and eye protection when handling a deployed air bag module. Sodium hydroxide residue may remain on the bag. If this comes in contact with the skin, it can cause irritation.*

Clock Spring

The **clock spring** conducts electrical signals to the air bag module while allowing steering wheel rotation (Figure 35- 12). The clock spring is located between the column and the steering wheel. The clock spring electrical connector contains a long conductive ribbon. The wires from the air bag electrical system are connected through the underside of the clock spring electrical connector to one end of the conductive ribbon. The other end of the conductive ribbon is connected through wires on the top side of the clock spring electrical connector to the air bag module. When the steering wheel is rotated, the conductive ribbon winds and unwinds, allowing steering wheel rotation while completing electrical contact between the system and the air bag module.

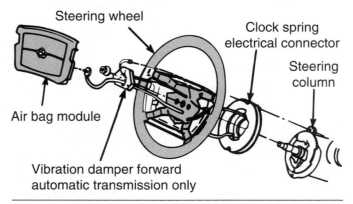

Figure 35-12 The clock spring provides for electrical continuity in all steering wheel positions. (Courtesy of Chrysler Corporation)

CAUTION *Whenever the air bag is deployed, the heat generated may damage the clock spring. The clock spring should be replaced whenever the air bag is deployed.*

Diagnostic Module

The **air bag diagnostic module (ASDM)** constantly monitors the readiness of the air SIR electrical system. If the module determines there is a fault in the system, it will illuminate the indicator light and store a diagnostic trouble code. Depending on the fault, the SIR system may be disarmed until the fault is repaired.

The diagnostic module also supplies back-up power to the air bag module in the event the battery or cables are damaged during an accident. The stored charge can last for up to 30 minutes after the battery is disconnected.

A typical diagnostic module performs the following functions:

1. Controls the instrument panel warning lamp.

2. Continuously monitors all air bag system components.

3. Controls air bag system diagnostic functions.

4. Provides an energy reserve to deploy the air bag if battery voltage is lost during a collision.

5. On some systems, the ASDM is responsible for deploying the air bag or bags when appropriate signals are received from the sensors.

Most air bag sensors contain a resistor connected in parallel with the sensor contacts. When the ignition switch is in the RUN position, the diagnostic module supplies a small amount of current through these resistors to monitor the system. If a short, ground, or open circuit occurs in the wiring or sensors, the current flow changes. When the diagnostic module senses this condition, it illuminates the air bag warning light in the instrument panel.

CAUTION *Before servicing the air bag system, the back-up power supply energy must be depleted. Disconnect the battery and isolate the cable terminal. Wait 5 to 30 minutes before servicing. Refer to the manufacturer's specifications for the recommended amount of time to wait.*

Sensors

To prevent accidental deployment of the air bag, most systems require at least two sensor switches to be closed in order to deploy the air bag (Figure 35-13). The number of sensors used depends on system design. Some systems use only a single sensor and others use up to five. The name used to identify the different sensors also varies between manufacturers.

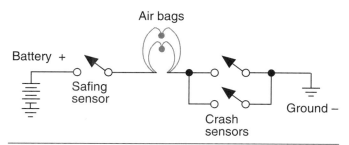

Figure 35-13 Typical sensor wiring circuit diagram.

On most three-sensor systems the **crash sensors** are normally open electrical switches designed to close when subjected to a predetermined impact. These sensors are usually located in the engine compartment or below the headlights. A single **safing sensor** is usually located on the centerline of the vehicle in the passenger compartment (usually under the center console). The safing sensor determines if the collision is severe enough to inflate the air bag.

When one of the crash sensors and the safing sensor closes, the electrical circuit to the igniter is complete. The igniter starts the chemical chain reaction that produces heat. The heat causes the generant to produce nitrogen gas, which fills the air bag.

There are several different types of sensors used. Common sensors include mass-type, roller-type, and accelerometers.

Mass-Type Sensors

The mass-type sensor contains a normally open set of gold-plated switch contacts and a gold-plated ball that acts as a sensing mass (Figure 35-14). The gold-plated ball is mounted in a cylinder coated with stainless steel. A magnet holds the ball about 1/8 in. away from the contacts. When the vehicle is involved in a frontal collision at 10 mph (16 km/h) or more, the sensing mass (ball) moves forward in the sensor and closes the switch contacts. The contacts will remain closed for 3 ms.

For proper operation, sensors must be mounted with the forward marking on the sensor facing toward the front of the vehicle, and they must be mounted in the original position designed by the manufacturer. Sensor brackets must not be bent or distorted.

Some mass-type air bag sensors contain a pivoted weight connected to a moving contact. When the vehicle is involved in a frontal collision with sufficient impact to deploy the air bag, the sensor weight moves in a circular path until the moving contact touches a fixed contact (Figure 35-15).

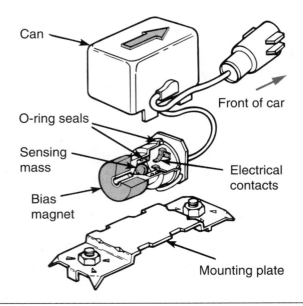

Figure 35-14 Some crash sensors hold the sensing mass by magnetic force. If the impact is severe enough to break the ball free, it will travel forward and close the electrical contacts. (Courtesy of Chrysler Corporation)

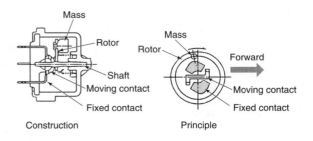

Figure 35-15 Mass-type air bag sensor with pivoted weight connected to a moving contact. (Courtesy of Chrysler Corporation)

Roller-Type Sensors

The **roller-type sensor** has a roller mass mounted on a ramp (Figure 35-16). One sensor terminal is connected to the ramp. The second sensor terminal is connected to a spring contact extending through an opening in the ramp without contacting the ramp. A 10,000Ω resistor is connected in parallel to the sensor contacts. The roller is held against a stop by small retractable springs on each side. These springs are similar to a retractable tape measure. If the vehicle is involved in a frontal collision at a high enough speed to deploy the air bag, the roller moves up the ramp and strikes the spring contact. In this position, the roller completes the circuit between the ramp and the spring contact.

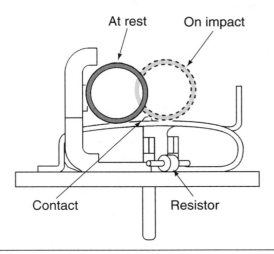

Figure 35-16 Roller-type air bag sensor. (Courtesy of Chrysler Corporation)

Accelerometers

In some air bag systems, **accelerometers** are used for air bag sensors. The accelerometer contains a piezoelectric element that is distorted during a collision. This element generates an analog voltage in relation to the impact force (Figure 35-17). The accelerometer-type sensor is usually an internal part of the air bag computer. The analog voltage from the piezoelectric element is sent to a collision-judging circuit in the air bag computer. The accelerometer is capable of determining the direction and severity of impact. If the collision impact is great enough, the computer deploys the air bag.

WARNING *SIR system sensors must be installed and tightened with the proper amount of torque and with the arrow pointing to the front of the vehicle for proper operation.*

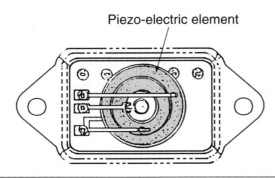

Figure 35-17 Accelerometer air bag sensor with piezoelectric element. (Courtesy of Chrysler Corporation)

Shorting Bars

The SIR electrical system is a dedicated system that is not interconnected with other electrical systems on the vehicle. All the wiring harness connectors in the system are the same color for easy identification.

Shorting bars are located in some of the component wiring harness connectors in the air bag system such as the inflator module connector at the steering column base. The shorting bars connect terminals together when the wiring connectors are disconnected (Figure 35-18). Since the terminals are shorted there is no way to have electrical potential, preventing accidental air bag deployment. If the ASDM connector is disconnected on some General Motors vehicles, shorting bars in the connector illuminate the air bag light.

Air Bag Deployment

The sequence of events occurring during an impact of a vehicle traveling at 30 mph is as follows:

1. When an accident occurs, the arming sensor is the first to close. It will close due to sudden deceleration caused by braking, or immediately upon impact. One of the discriminating (crash) sensors will then close. The amount of time required to close the switches is within 15 milliseconds.

2. Within 40 milliseconds, the igniter module burns the propellant and generates the gas to completely fill the air bag.

3. Within 100 milliseconds, the driver's body has stopped forward movement and the air bag starts to deflate. The air bag deflates by venting the nitrogen gas through holes in the back of the bag.

4. Within 2 seconds the air bag is completely deflated.

Air Bag Warning Lamp

The air bag system warning lamp indicates the system condition to the driver. The warning lamp is operated by the diagnostic module. Ignition on and crank signals are received by the diagnostic module. When the ignition switch is placed in the RUN position, the air bag warning lamp should illuminate for a bulb check. In some systems the lamp will flash 7 to 9 times, and then remain steadily illuminated while the engine is cranking. Once the engine starts, the air bag warning lamp should be extinguished. An air bag system failure may be indicated by any of the following warning lamp conditions:

1. If the lamp remains on but does not flash when the ignition is turned on.

2. If the lamp flashes 7 to 9 times and then remains on when the ignition is turned on.

3. If the lamp comes on when the engine is running.

4. If the lamp does not come on at any time.

5. If the lamp does not come on steadily while the engine is cranking.

If any of these lamp conditions are present, the driver should have the air bag system checked. When the technician places the air bag system in the diagnostic mode, the warning lamp flashes the trouble codes.

General Motors' SIR

General Motors has used several different versions of SIR systems. The version discussed here is representative of the system. Changes between models and years

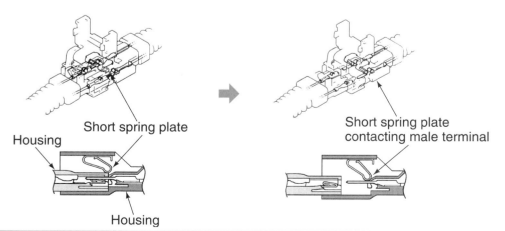

Figure 35-18 Shorting bars on air bag system wiring connectors.

require the technician to follow the service manual procedures for the vehicle they are working on.

Refer to the illustration (Figure 35-19) for the location of the components used on the SIR system. The major portions of the SIR system are the deployment loop and the **diagnostic energy reserve module (DERM)**.

The deployment loop supplies current to the inflator module in the steering wheel (Figure 35-20). The components of the deployment loop include the:

- arming sensor
- inflator module
- coil assembly
- discriminating sensors

The **arming sensor** switches power to the inflator module on the insulated side of the loop circuit. The arming sensor is calibrated to close at low level velocity changes. Either of the **discriminating sensors** can supply the ground. The discriminating sensors are calibrated to close with velocity changes that are severe enough to warrant air bag deployment (velocity changes higher than that of the arming sensor). For the inflator module to ignite, the arming sensor and at least one discriminating sensor must close simultaneously.

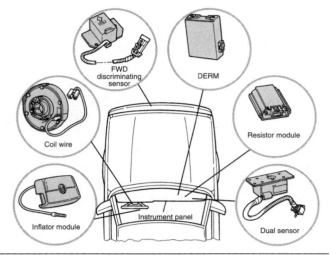

Figure 35-19 Component location of the GM SIR system. (Courtesy of General Motors Corporation, Service Technology Group)

There are two discriminating sensors used. One is located in front of the radiator, the other is part of the dual sensor located behind the instrument panel. The **dual sensor** is a combination of the arming and passenger compartment discriminating sensor.

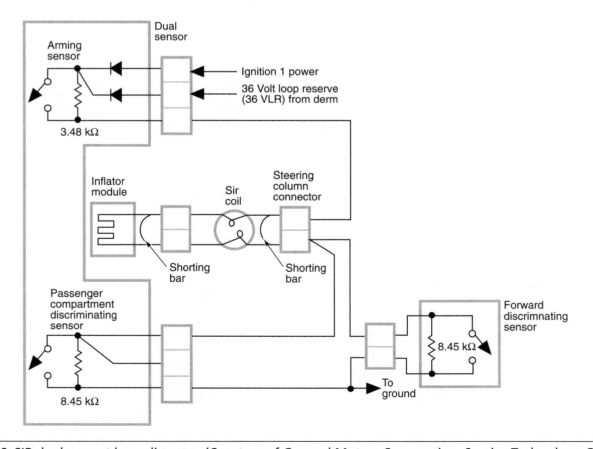

Figure 35-20 SIR deployment loop diagram. (Courtesy of General Motors Corporation, Service Technology Group)

The DERM is designed to provide an energy reserve of 36 volts to assure deployment for a few seconds if vehicle voltage is low or lost. The DERM also maintains constant diagnostic monitoring of the electrical system. It will store a code if a fault is found and provide driver notification by illuminating the warning light. On some early General Motors air bag systems, a **resistor module**, contained in a yellow plastic housing, is positioned under the left side of the instrument panel. Resistors in the module plus the resistors in the sensors assist the DERM in monitoring the system (Figure 35-21). On later model General Motors systems, the resistor module is discontinued and all the monitoring resistors are contained in the sensors (Figure 35-22).

If the vehicle is involved in a collision, the DERM detects the collision. Most DERMs are capable of recording four crashes, except the Corvette DERM, which only has enough memory to record one crash. When the DERM memory is full, a permanent fault code is set. On many air bag systems, this code is 52. If this code or code 71 is present, DERM replacement is required. On a Corvette air bag system, if code 52, 53, or 54 is present, DERM replacement is necessary. Once the vehicle is involved in a frontal collision and air bag deployment occurs, another permanent fault code is set in the DERM memory.

Mercedes-Benz

The supplemental restraint system (SRS) used on some Mercedes-Benz vehicles combines the air bag with a three-point seatbelt equipped with an **emergency tensioning retractor (ETR)**. The air bag operates similar to that discussed. However, it uses only one sensor. The SRS sensor incorporates two integrated chips (ICs) and an **acceleration pickup** to determine the degree and direction of impact. The SRS sensor uses a mercury switch to disconnect it from the circuit during normal driving. When the longitudinal deceleration is sufficient, the sensor sends a voltage through a bypass filter to an amplifier. The amplifier provides the voltage required to ignite the inflator module.

A voltage convertor is used to keep a constant 12 volts applied to the sensor and energy accumulator. The convertor is capable of maintaining 12 volts even though battery voltage may drop to as low as 4 volts. The **energy accumulator** operates like a capacitor to provide back-up current in the event that battery voltage is totally lost.

The ETR system is installed on the passenger side. The ETR contains an igniter device that produces a high pressure gas. The gas is used to operate a pulley that pulls the seatbelt harness snugly against the passenger.

CAUTION *Mercedes-Benz, Toyota, and some other manufacturers use mercury in their sensors. If these sensors require replacement, the old sensor must be treated as toxic waste material.*

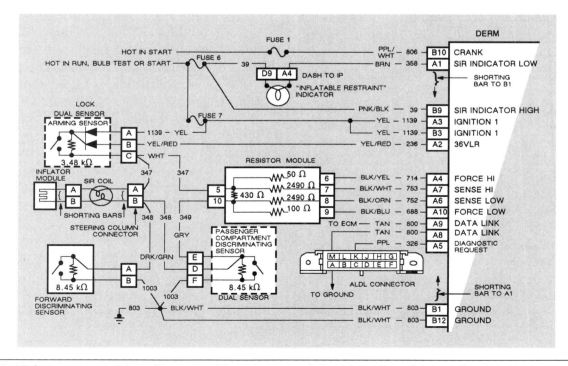

Figure 35-21 Air bag system wiring diagram with resistor module. (Courtesy of General Motors Corporation, Service Technology Group)

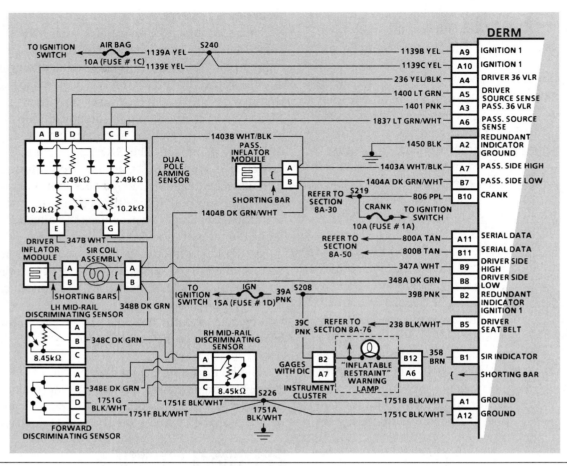

Figure 35-22 Air bag system wiring diagram with monitoring resistors in the sensors only. (Courtesy of Pontiac Division, General Motors Corporation)

Passenger Side Air Bags

Federal law expanded to require all passenger vehicles produced after 1995 be equipped with passenger side air bags. This feature has an additional air bag assembly located in the dash on the passenger's side (Figure 35-23). Since there is a greater distance between the passenger and the instrument panel compared to the distance between the driver and the steering wheel, the passenger's side air bag is much larger. A typical passenger's side air bag has a fully inflated volume of 7 cubic feet.

Passenger side air bags function in the same way as driver side air bags. However, not all air bag systems use nitrogen gas to inflate the bag; some (such as Chrysler) use a **hybrid inflator module**. The hybrid inflator module contains an initiator similar to the squib in other inflator modules. However, the hybrid inflator module also has a container of pressurized argon gas (Figure 35-

24). The same method is used to energize the initiator in the hybrid inflator. When the initiator is energized, the propellant surrounding the initiator explodes and pierces the propellant container. The exploding propellant heats the argon gas which escapes rapidly through the propellant container into the air bag (Figure 35-25).

The early version of the hybrid system used a pressure sensor mounted in the opposite end of the argon gas chamber from the initiator and propellant. This sensor sends a signal to the module in relation to the argon gas pressure. If the gas pressure decreases below a preset value, the module illuminates the air bag warning light.

CAUTION *Never place a rearward facing child safety seat in the front seat with an air bag. The deployment of the air bag can cause serious injury or death when it strikes the seat.*

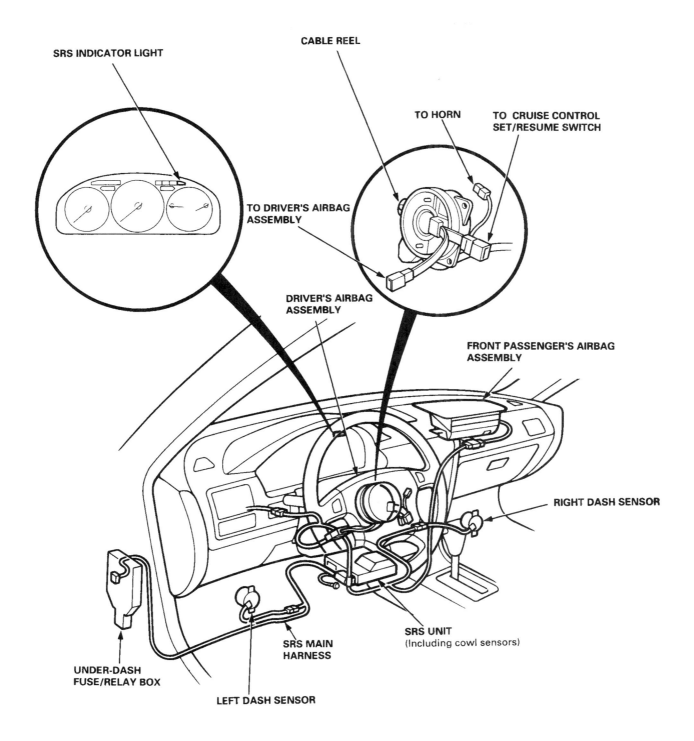

Figure 35-23 Component location for Honda's driver and passenger air bag system. (Courtesy of American Honda Motor Co., Inc.)

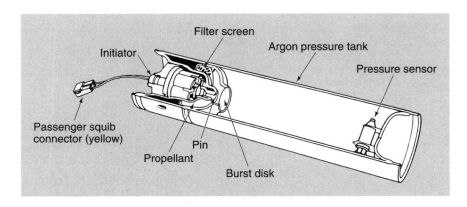

Figure 35-24 Hybrid inflator module with argon gas pressure chamber. (Courtesy of Chrysler Corporation)

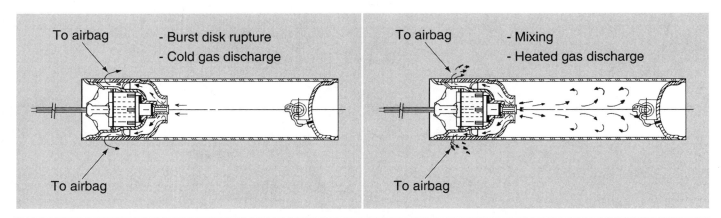

Figure 35-25 When the initiator is energized the propellant explodes and punctures the container and allows pressurized argon gas to fill the air bag.

Summary

❑ Passive restraints operate automatically with no action required on the part of the driver or occupant.

❑ The automatic seatbelt system uses a control module to monitor operation by receiving inputs from door ajar switches, limit switches, and the emergency release switch.

❑ An emergency release mechanism is provided in the event the passive seat belt system fails to operate.

❑ The air bag is a supplement. The seatbelt is the primary restraint system.

❑ The air bag module is composed of the air bag and inflator assembly. It is packaged in a single module and is mounted in the center of the steering wheel.

❑ The diagnostic module constantly monitors the readiness of the SIR electrical system. It supplies back-up power to the air bag module in the event the battery or cables are damaged during an accident.

❑ The igniter is a combustible device that converts electric energy into thermal energy to ignite the inflator propellant.

❑ Air bags will deploy if the vehicle is involved in a frontal collision of sufficient impact and the collision force is within 30° on either side of the vehicle centerline.

❑ Total air bag deployment time from the instant of impact until the air bag is inflated is less than 100 ms.

❑ An accelerometer-type air bag sensor generates an analog voltage in relation to the severity of impact force. The accelerometer also senses the direction of impact force.

❏ The clock spring electrical connector maintains electrical contact between the inflator module and the air bag electrical system.

❏ The air bag warning light indicates an inoperative air bag system.

❏ Safety goggles and protective gloves must be worn when handling deployed air bags.

❏ A hybrid inflator module contains a pressurized argon gas cylinder, which is punctured by the exploding propellant to inflate the air bag.

❏ Shorting bars connect some air bag system terminals together when the terminal is disconnected to help prevent accidental air bag deployment.

Terms To Know

Acceleration pickup	Diffuser	Mass-type sensor
Accelerometers	Discriminating sensors	Passive restraint
Air bag diagnostic module (ASDM)	Dual sensor	Passive seatbelt system
Air bag module	Emergency release mechanism	Resistor module
Air bag systems	Emergency tensioning retractor (ETR)	Roller-type sensor
Arming sensor	Energy accumulator	Safing sensor
Clock spring	Hybrid inflator module	Shorting bars
Crash sensors	Igniter	Squib
Diagnostic energy reserve module (DERM)	Inertia lock retractors	Supplemental inflatable restraint (SIR)
	Inflator assembly	

Review Questions

Short Answer Essay

1. Define the term passive restraint.

2. Describe the basic operation of automatic seatbelts.

3. List and describe the design and operation of three different types of air bag system sensors.

4. List the components of the air bag module, and explain their function.

5. List and explain two of the functions of the diagnostic module used in air bag systems.

6. List the sequence of events that occur during air bag deployment.

7. What is the purpose of the clock spring?

8. What is the difference between the arming sensor and the discriminating sensors?

9. Describe normal operation of an air bag warning light.

10. Describe the deployment of the hybrid inflator module.

Fill-In-The-Blanks

1. Safe service procedures of air bag systems are accomplished through proper use of the _____ _____ and by understanding the _____ principles of these systems.

2. The _____ _____ conducts electrical signals to the module while permitting steering wheel rotation.

3. In the automatic seatbelt system, the _____ switches inform the module of the position of the harness.

4. The _____ is a combustible device that converts electric energy into thermal energy to ignite the inflator propellant.

5. The diagnostic module supplies _____ _____ to the air bag _____ in the event that the battery or cables are damaged during an accident.

6. To prevent accidental deployment of the air bag, most systems require that at least _____ sensor switches be closed to deploy the air bag.

7. The frontal collision force must be within _____ degrees of the vehicle centerline to deploy the air bag.

8. An accelerometer-type air bag sensor produces an analog voltage in relation to _____ _____ .

9. The current flow through the squib required to deploy the air bag is approximately _____ amperes.

10. A hybrid inflator module contains a cylinder filled with compressed _____ gas.

ASE Style Review Questions

1. The input signals to the control module of the automatic seatbelt system are being discussed. Technician A says the door ajar switches signal the position of the harness. Technician B says that the limit switches signal when the emergency release switch is opened. Who is correct?
   a. A only
   b. B only
   c. Both A and B
   d. Neither A nor B

2. The air bag module is being discussed. Technician A says the module supplies emergency current in the event that battery voltage is lost. Technician B says the module is composed of the air bag and inflator assembly. Who is correct?
   a. A only
   b. B only
   c. Both A and B
   d. Neither A nor B

3. Technician A says that the igniter is a combustible device that converts electric energy into thermal energy. Technician B says that the inflation of the air bag is through an explosive release of compressed air. Who is correct?
   a. A only
   b. B only
   c. Both A and B
   d. Neither A nor B

4. Technician A says that the clock spring is located at the bottom of the steering column. Technician B says the clock spring conducts electrical signals to the module while permitting steering wheel rotation. Who is correct?
   a. A only
   b. B only
   c. Both A and B
   d. Neither A nor B

5. Technician A says the diagnostic module constantly monitors the readiness of the air SIR electrical system. Technician B says that a typical crash sensor is composed of a gold-plated ball held in place by a magnet. Who is correct?
   a. A only
   b. B only
   c. Both A and B
   d. Neither A nor B

6. While discussing air bag sensors, Technician A says the arrow on each sensor must face toward the rear of the vehicle. Technician B says air bag sensor brackets must not be bent or distorted. Who is correct?
   a. A only
   b. B only
   c. Both A and B
   d. Neither A nor B

7. While discussing accelerometer-type air bag sensors, Technician A says an accelerometer senses collision force and direction. Technician B says an accelerometer produces a digital voltage. Who is correct?
   a. A only
   b. B only
   c. Both A and B
   d. Neither A nor B

8. While discussing the air bag system warning light, Technician A says on some systems this warning light should flash 7 to 9 times after the engine is started and then go out. Technician B says on some systems this warning light should be illuminated while cranking the engine. Who is correct?
   a. A only
   b. B only
   c. Both A and B
   d. Neither A nor B

9. While discussing the air bag deployment loop, Technician A says if the arming sensor contacts close, this sensor completes the circuit from the inflator module to ground. Technician B says if the contacts close in two discriminating sensors, the air bag is deployed. Who is correct?
   a. A only
   b. B only
   c. Both A and B
   d. Neither A nor B

10. While discussing hybrid inflator modules, Technician A says a pressure sensor on the outside of the argon gas cylinder sends a signal to the ASDM in relation to gas pressure in the cylinder. Technician B says when the initiator is energized, the propellant explodes and pierces the propellant container, allowing the pressurized argon gas to escape into the air bag. Who is correct?
   a. A only
   b. B only
   c. Both A and B
   d. Neither A nor B

36 Diagnosing Passive Restraint Systems

Objective

Upon completion of this chapter, you will be able to:

❏ Diagnose automatic seatbelt systems.

❏ Replace the automatic seatbelt drive motor assembly.

❏ Replace the drive belt in an automatic seatbelt system.

❏ Enter air bag system self-diagnostics.

❏ Use a scan tool to properly retrieve air bag system trouble codes.

❏ Use the flashing warning light to retrieve air bag system trouble codes.

❏ Enter air bag diagnostics through the electronic climate control panel of GM vehicles equipped with digital instrument panels.

❏ Properly replace the air bag module according to manufacturer's service manual standards.

❏ Properly replace the clock spring assembly according to manufacturer's service manual standards.

❏ Safely service the air bag system.

Introduction

In this chapter you will learn how to properly and safely service automatic passive restraint systems. Federal mandates concerning the equipping of these systems has assured that today's technician will be required to service them. The safety of the driver and/or passengers depends on the ability of the technician to properly diagnose and repair these systems.

Passive restraint systems can be either automatic seatbelts, air bags, or a combination of both. In this chapter you will learn how to diagnose automatic seatbelt systems and to replace main components. You will also learn how to enter self-diagnostics and to retrieve trouble codes in the air bag system. Included in this chapter are procedures for replacing the air bag module and the clock spring.

Automatic Seatbelt Service

Even though the components used in the automatic seatbelt system vary according to manufacturer, the basic principles of locating and repairing the cause of a problem are similar. Refer to the service manual to obtain a circuit diagram of the system (Figure 36-1). The circuit schematic, troubleshooting charts, and diagnostic charts will assist you in finding the fault.

In addition, most manufacturers provide a troubleshooting chart (Figure 36-2) for the automatic seatbelt system. Troubleshooting the circuits, switches, lamps, and motors as indicated in this chart requires only the skills described in previous chapters. In addition to the troubleshooting chart, an operational logic chart will provide helpful information when testing switches (Figure 36-3). Use a diagnostic chart to test the

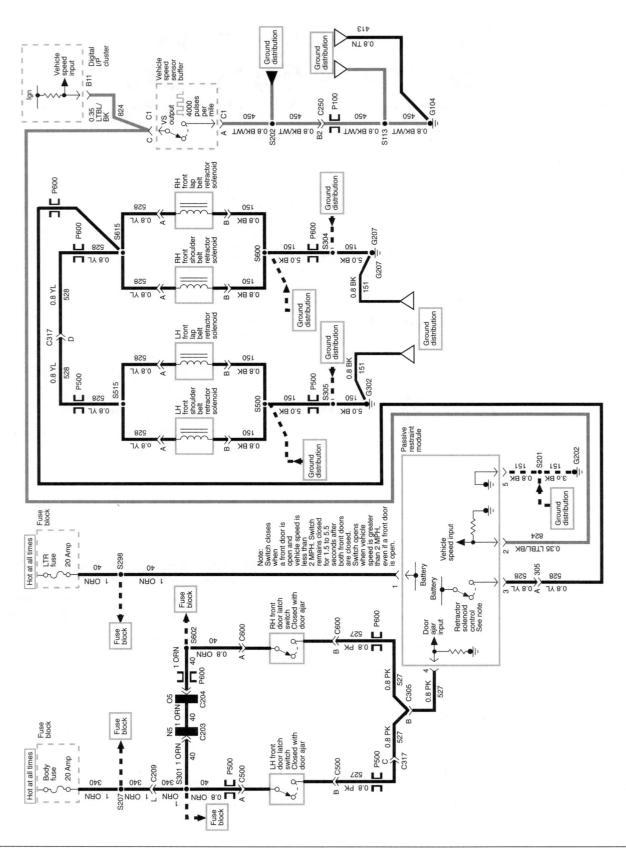

Figure 36-1 The circuit schematic is a vital tool to diagnose the automatic seatbelt system. (Courtesy of General Motors Corporation, Service Technology Group)

CONDITION	POSSIBLE SOURCE	ACTION
• Belt Appears to Run Fine But Motor Stalls at A-pillar. Motor Could Get Warm.	• "A" limit switch wire shorted to ground. "A" limit switch not opening.	• Check for pinched wire along A-pillar, or behind LH or RH side cowl panel. Check that carrier reaches "A" limit switch.
• Fasten Belt Indicator (IP) Remains Lit After 7-10 Seconds of Running to B-pillar. Belt Appears to Run Fine But Motor Stalls at B-pillar. Motor Could Get Warm.	• "B" limit switch wire is shorted to ground. • Obstruction.	• Check for pinched wire under guide attachment screws. • Check for trim screw, obstruction in track or track seal that prevents the carrier from reaching the "B" limit switch.
• Belt Will Not Run. Module May Be Burned.	• Motor wire shorted to ground.	• Check for short to ground on the motor circuit breaker terminals.
• Belt Will Not Run.	• Inertia switch tripped. • Motor wire not connected.	• Reset (depress) fuel pump inertia switch button. • Check for unconnected connectors on the motor, or behind LH and RH side cowl panels.
• Belt Runs to A-pillar Only. Belt Will Not Run Back to B-pillar.	• Door switch wire shorted to ground. • "B" limit switch wire not connected. • "B" limit switch plunger stuck in depressed position.	• Check for pinched wire behind door trim panel. • Check for unconnected connectors near motor, behind LH and RH side cowl panel, or on "B" limit switch. • Correct jammed switch or replace "B" limit switch.
• Belt Runs to B-pillar only. Belt Will Will Not Run Forward to A-pillar.	• Door switch wire not connected. • "A" limit switch wire not connected. • Obstruction.	• Check for unconnected connectors on door latch switch, or behind LH and RH side cowl panel. • Check for unconnected connectors behind LH and RH side cowl panel, or on "A" limit switch. • Check for a trim screw, track seal or jammed track locking pawl that prevents the carrier from reaching the "A" limit switch.
• Opening/Closing Door Causes Both Belts to Move.	• Damaged Door Ajar lamp assembly.	• Replace Door Ajar lamp assembly.
• Turning Ignition Off Causes Belt to Move to A-pillar.	• Damaged Door Ajar lamp assembly.	• Replace Door Ajar lamp assembly.
• Fasten Belt Indicator Remains On, Chime Sounds for Four to Eight Seconds.	• Connector to shoulder emergency release retractor switches not connected.	• Check for unconnected connectors in console near shoulder strap connector.
• Excessive Noise While Motor is Running.	• Motor adjustment knob on top of motor hits body.	• Loosen motor, side motor downward, then tighten motor.

Figure 36-2 Motorized seatbelt system troubleshooting chart. (Reprinted with permission of Ford Motor Company)

DRIVER AND PASSENGER OPERATIONAL LOGIC CHART

Vehicle Condition	EFI Fuel Switch (Inertia)		Door Switch		A Pillar Limit Switch		B Pillar Limit Switch	
	Closed	Open	On	Off	On	Off	On	Off
Ignition ON (after entry)								
• Door Closed — Belt starts to move	X			X	X		X	
• Door Closed — Belt Moving from A to B-Pillar	X			X	X		X	
• Door Closed — Belt stopped at B-Pillar	X			X	X			X
• Open Door — Belt starts to move	X		X		X			X
• Door Opened — Belt moving from B to A-Pillar	X		X		X		X	
• Door Opened — Belt stopped at A-Pillar	X		X			X	X	
Ignition OFF (before exit)								
• Parked—Door Closed — Belt at B-Pillar	X			X	X			X
• Parked—Open Door — Belt starts to move	X		X		X			X
• Parked—Door Opened — Belt moving from B to A-Pillar	X		X		X		X	
• Parked—Door Opened — Belt stopped at A-Pillar	X		X			X	X	
• Parked—Close Door — Belt stays at A-Pillar	X			X		X	X	
Impact—Ignition On/Off								
• Ignition On—Door Opened (upon impact) — Belt stays at B-Pillar		X	X			X		X
• Ignition Off—Door Open (after impact) — Belt moves from B to A-Pillar		X	X			X	X	

Figure 36-3 Switch operational logic chart. (Reprinted with permission of Ford Motor Company)

operation of the system and to determine faults with the module (Figure 36-4).

Some systems are capable of storing diagnostic trouble codes that can be retrieved by a scan tester. The scanner is also used to perform a functional test of the system. To perform this test turn the ignition switch to the RUN position and connect the scanner to the diagnostic connector. Select the system test and follow the scan tool instructions. If there is a failure in the system, the scanner will display the code.

Drive Motor Assembly Replacement

A faulty drive motor can cause slow or no operation of the automatic seatbelts. The motor must be replaced if it has been determined it is the faulty component. A typ-

	TEST	RESULT	ACTION
A1.	Fasten both front safety belts. With a passenger in the RH front seat, operate vehicle at 5 mph. Apply brakes firmly.	Safety belts lock-up.	GO TO A2.
		Safety belts do not lock-up.	GO TO A4.
A2.	Close door firmly and immediately try to open it. Repeat this step several times and ask customer if safety belts have ever locked-up when attempting to open door.	Safety belts do not lock-up (and customer has experienced no problem with safety belts locking-up) when attempting to open door.	GO TO A3.
		Safety belts lock-up (or customer has experienced problem with safety belts locking-up) when attempting to open door.	GO TO A5.
A3.	Backprobe PASSIVE RESTRAINT MODULE connector with a test lamp from cavity 3 to chassis ground. Operate vehicle at 20 mph with either front door open.	Test lamp lights but goes out at 2 mph.	All systems diagnosed in this cell are functioning normally.
		Test lamp stays lit.	GO TO A11.
A4.	Disconnect PASSIVE RESTRAINT MODULE connector. Fasten both front safety belts. With a passenger in the RH front seat, operate vehicle at 5 mph. Apply brakes firmly.	Safety belts lock-up.	Replace PASSIVE RESTRAINT MODULE.
		Safety belts do not lock-up.	Repair short to voltage in YEL (528) wire.
A5.	Disconnect PASSIVE RESTRAINT MODULE connector. Connect a test lamp from PASSIVE RESTRAINT MODULE connector cavity 1 to chassis ground.	Test lamp does not light.	Repair open in ORN (40) wire.
		Test lamp lights.	GO TO A6.
A6.	Connect a digital multimeter from PASSIVE RESTRAINT MODULE connector cavity 5 to chassis ground. Measure resistance.	More than 0.3 ohms.	Repair BLK (151) ground wire.
		Less than 0.3 ohms.	GO TO A7.
A7.	Connect a test lamp from PASSIVE RESTRAINT MODULE connector cavity 4 to chassis ground. Open LH front door.	Test lamp does not light.	Repair open in PNK (527) wire or LH FRONT DOOR LATCH SWITCH.
		Test lamp lights	Leave test lamp connected, close LH front door, and GO TO A8.
A8.	Open RH front door.	Test lamp does not light.	Repair open in PNK (527) wire or RH FRONT DOOR LATCH SWITCH.
		Test lamp lights.	GO TO A9.
A9.	Connect a test lamp from PASSIVE RESTRAINT MODULE connector cavity 2 to chassis ground.	Test lamp does not light.	Replace PASSIVE RESTRAINT MODULE.
		Test lamp lights.	GO TO A10.
A10.	Disconnect VEHICLE SPEED SENSOR BUFFER connector C1. Connect a test lamp from VEHICLE SPEED SENSOR BUFFER connector C1 cavity C to chassis ground.	Test lamp lights.	Repair short to voltage in LT BLU/BLK (824) wire.
		Test lamp does not light.	Refer to Cell 33 Vehicle Speed Sensor Buffer System Diagnosis.
A11.	Backprobe PASSIVE RESTRAINT MODULE connector with digital multimeter from cavity 2 to chassis ground. Operate vehicle at 20 mph. Measure voltage.	4-6 volts.	Replace PASSIVE RESTRAINT MODULE.
		Less than 4 or more than 6 volts.	Refer to Cell 33 Vehicle Speed Sensor Buffer System Diagnosis.

Figure 36-4 Automatic seatbelt diagnostic charts are used in a systematic approach to troubleshooting. (Courtesy of General Motors Corporation, Service Technology Group)

ical procedure for motor assembly replacement is described here.

Remove the B-pillar trim to gain access to the drive motor. Then remove the screws attaching the drive belt track to the motor. Next remove the screw attaching the vertical guide to the pillar. Be sure to remove only the screw located next to the motor (Figure 36-5).

Slide the belt track down far enough to disengage the sprockets from the motor drive gear teeth. Disconnect the electrical connectors to the motor. Remove the hex nuts retaining the mounting bracket and remove the motor.

Reverse the order to install the motor.

Drive Belt Replacement

If the drive belt is twisted, kinked, or broken it will cause the seatbelt to operate improperly. A binding drive belt may prevent the seatbelt from moving its full travel or may strip the gear teeth on the motor.

To replace a defective drive belt, gain access to the assembly by removing the A- and B-pillar trim covers.

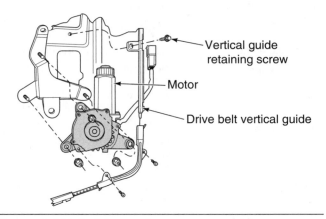

Figure 36-5 Guide and motor attaching screws. (Reprinted with permission of Ford Motor Company)

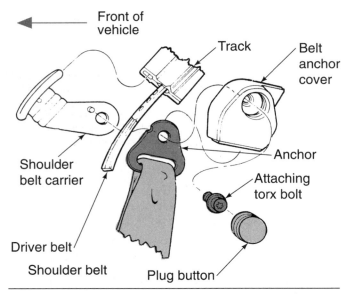

Figure 36-6 Shoulder belt to carrier attachment. (Reprinted with permission of Ford Motor Company)

Remove the plug button covering the torx attaching bolt (Figure 36-6). Cycle the shoulder belt to the A-pillar and remove the anchor bolt. Remove the belt anchor cover and the shoulder belt from the belt carrier.

Next remove the screws attaching the drive belt track to the drive motor. Then remove the vertical guide retaining screw located next to the motor. Disconnect the drive belt track from the motor. Allow the belt to pass by the gear teeth without engaging them (Figure 36-7). Remove limit switch A from the retaining bracket (Figure 36-8). Remove the bracket to the A-pillar attaching screw and the two screws attaching the bracket to the track. Slide the shoulder belt carrier forward off the track. The drive belt should slide out of the track with the shoulder belt. Remove the shoulder belt carrier from the drive belt.

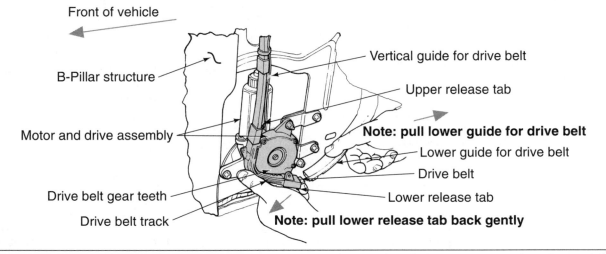

Figure 36-7 Disconnecting the drive belt from the motor. (Reprinted with permission of Ford Motor Company)

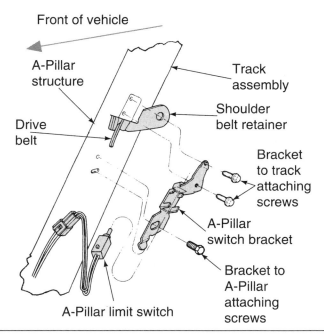

Figure 36-8 Limit switch A retaining bracket attachment to the track assembly. (Reprinted with permission of Ford Motor Company)

Before installing the new drive belt, lubricate the track with approved lubricant. Insert the drive belt into the front of the track at the A-pillar. Install the shoulder belt carrier into the large slot at the front end of the drive belt. Slide the drive belt rearward in the track until it is a half inch (12.7 mm) in the upper track slot.

Align the sprocket holes in the drive belt with the gear teeth of the motor (Figure 36-7). Install the attaching bolts to secure the vertical guide and track.

Engage the track locator on the A limit switch bracket into the slot in the A-pillar (Figure 36-8). Install the limit switch onto the bracket.

Cycle the belt anchor to the B-pillar. Pull the webbing out of the retractor located in the console (Figure 36-9). Lay the seatbelt webbing flat on the seat and check for twists. Align the notch in the shoulder belt anchor with the pin on the carrier. Install the anchor cover and bolt. Torque the bolt to specifications and replace the plug button. Replace the A- and B-pillar trim.

WARNING *The anchor bolt is usually a self-locking bolt. It must be replaced with a new one whenever it is removed.*

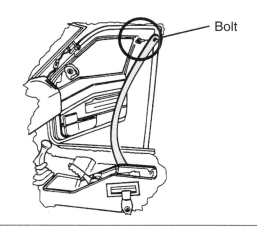

Figure 36-9 Install the shoulder belt to the carrier after checking it for twists and locating the carrier into the B-pillar. (Reprinted with permission of Ford Motor Company)

Air Bag Safety and Service Warnings

Whenever working on the air bag system, it is important to follow all safety warnings. There are safety concerns with both deployed and live air bag modules.

1. Wear safety glasses when handling an air bag module or servicing the bag system.

2. Always disconnect the battery negative cable, isolate the cable end, and wait for the amount of time specified by the vehicle manufacturer before proceeding with the necessary diagnosis or service. The average waiting period is 2 minutes, but some vehicle manufacturers specify up to 10 minutes. Failure to observe this precaution may cause accidental air bag deployment and personal injury.

3. Handle all sensors with care. Do not strike or jar a sensor in such a manner that deployment may occur.

4. Replacement air bag system parts must have the same part number as the original part. Replacement parts of lesser quality must not be used. Improper or inferior components may result in improper air bag deployment and injury to the vehicle occupants.

5. Do not strike or jar a sensor or an air bag system diagnostic monitor (ASDM). This may cause air bag deployment or make the sensor inoperative.

6. All sensors and mounting brackets must be properly torqued before an air bag system is powered up to ensure correct sensor operation. If sensor fasteners do not have the proper torque, improper air bag deployment may result in injury to the vehicle occupants.

7. When carrying a live air bag module, face the trim and bag away from your body.

8. Do not carry the module by its wires or connector.

9. When placing a live module on a bench, face the trim and air bag up.

10. Deployed air bags may have a powdery residue on them. Sodium hydroxide is produced by the deployment reaction and is converted to sodium carbonate when it comes into contact with atmospheric moisture. It is unlikely that sodium hydroxide will still be present. However, wear safety glasses and gloves when handling a deployed air bag. Wash hands after handling.

11. A live air bag module must be deployed before disposal. Because the deployment of an air bag is through an explosive process, improper disposal may result in injury and in fines. A deployed air bag should be disposed of in a manner consistent with EPA and manufacturer procedures.

12. Do not use a battery- or A/C-powered voltmeter, ohmmeter, or any other type of test equipment not specified in the service manual. Never use a test light to probe for voltage.

13. Never reach across the steering wheel to turn on the ignition switch.

Diagnostic System Check

A diagnostic system check must be performed to avoid diagnostic errors. The diagnostic system check involves observing the air bag warning light to determine if it is operating normally. A typical diagnostic system check follows:

1. Turn on the ignition switch and observe the air bag warning light. On some General Motors systems, this light should flash 7 to 9 times and then go out. On other vehicles, the air bag warning light should be illuminated continually for 6 to 8 seconds and then go out. If the air bag warning light does not operate properly, further system diagnosis is necessary.

2. Observe the air bag warning light while cranking the engine. On many General Motors vehicles, this light should be illuminated continually while cranking the engine. Always refer to the vehicle manufacturer's service manual. During engine cranking if the air bag warning light does not operate as specified by the vehicle manufacturer, complete system diagnosis is required.

3. Observe the air bag warning light after the engine starts. This light should flash 6 to 9 times and then remain off. If the air bag warning light remains off, there are no current diagnostic trouble codes (DTCs) in the air bag system diagnostic module (ASDM). When the air bag warning light remains on, obtain the DTCs with a scan tester or flash code method. History DTCs on General Motors vehicles can only be obtained with a scan tester. Some vehicles such as Chrysler products do not provide air bag flash codes. The scan tester must be used to diagnose these systems.

The following are examples of system diagnostic procedures. Be sure to follow the service manual procedures for diagnosing any particular SIR system.

General Motors SIR Diagnostics

The SIR diagnostic system check must be the first step of any SIR diagnosis. This check tests the warning light and the ability of the DERM to communicate through the data line. It also checks for trouble codes. The SIR diagnostic system check should be repeated whenever any repairs or diagnostic procedures have been performed. This will assure that repairs were done correctly and no other malfunctions exist.

See the SIR diagnostic system check chart (Figure 36-10). Perform the check as directed and it will instruct you to go to the correct diagnostic chart. Bypassing these procedures will result in prolonged diagnostic time, incorrect diagnosis, and costly parts replacement.

Retrieving Trouble Codes

Trouble codes can be retrieved through the use of a scan tool, flash codes, or the digital panel cluster (if equipped).

Scan Tool Diagnostics

Connect the scan tool to the data link connector. Turn the ignition switch to the RUN position and request "SIR Code Display" from the scan tool. Record all history and current codes.

A displayed code 52 indicates enough accident information is stored to fill memory. In most systems it takes four simultaneously closed arming and discriminating sensor events to fill memory. This code requires the DERM be replaced. A code 52 cannot be erased. Nor can any further diagnosis be performed until the DERM is replaced.

If a code 71 is set, then a DERM failure is detected. A code 71 requires the DERM be replaced before any more diagnostic procedures are performed. A code 71 cannot be erased.

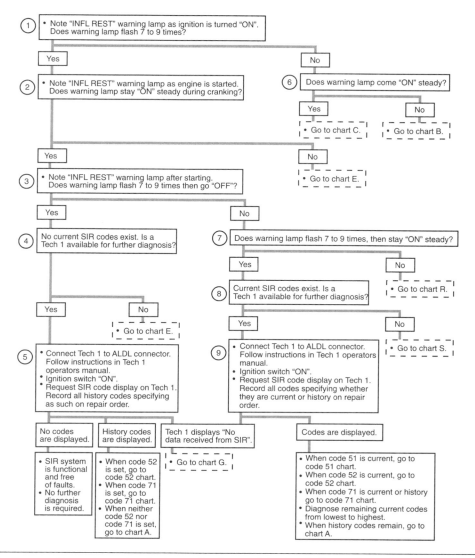

Figure 36-10 SIR diagnostic system check chart. Perform this check first when diagnosing the SIR system. (Courtesy of General Motors Corporation, Service Technology Group)

If code 52 or 71 are not displayed, and there are other history codes, go to chart A to locate the causes of the intermittent fault (Figure 36-11). If there are no history codes, diagnose remaining current codes from the lowest to the highest number.

Flash Code Diagnostics

SERVICE TIP *The air bag systems on some General Motors vehicles must be diagnosed with a scan tester. Always consult the proper vehicle manufacturer's service manual.*

Flash codes allow the technician to retrieve current trouble codes without the use of a scan tool. Flash codes will not provide any history codes. However, they will indicate if any are in memory.

With the ignition switch in the RUN position, connect a jumper wire from DLC terminal K to ground (Figure 36-12). The SIR warning lamp will display trouble codes through a series of flashes. Each displayed code will consist of a number of flashes that represent the first number. This number will be followed by a half-second delay. Then the second number of the code will be flashed. For example, a code 52 would be displayed by five flashes, a half-second pause, followed by two more flashes.

A code 12 will be displayed to indicate that the system is in flash code mode. This is not a fault code. Code 13 will be displayed if there are no history codes stored in memory.

Each code is displayed once, until all codes have been given. The codes will then repeat until the ground is lifted from DLC terminal K.

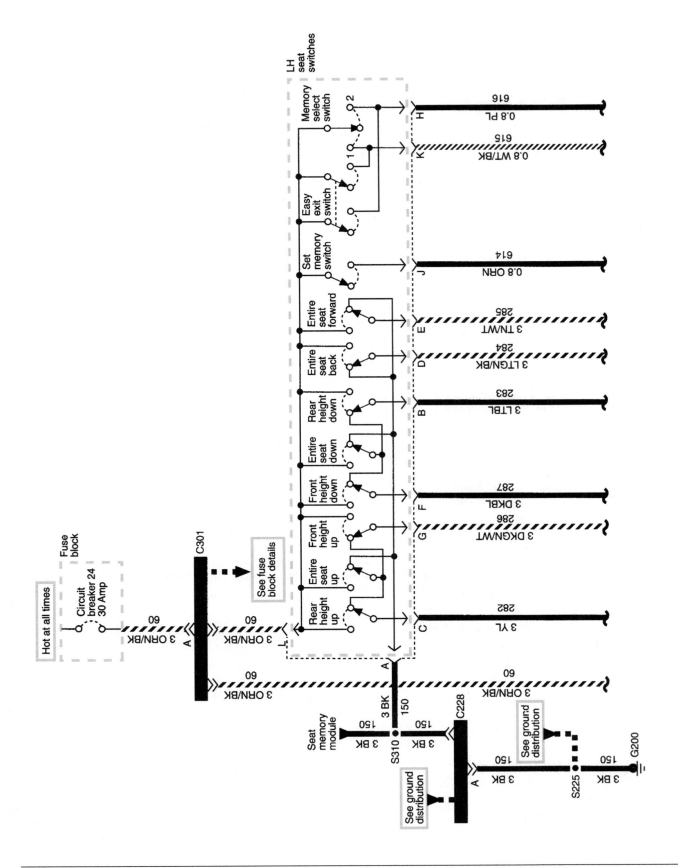

Figure 36-11 Chart A—history code diagnosis. This chart will lead the technician to intermittent faults. (Courtesy of General Motors Corporation, Service Technology Group)

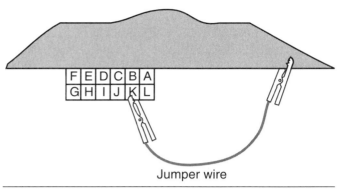

Figure 36-12 Jump the K terminal to ground to cause the module to enter self-diagnostics and to flash trouble codes.

Digital Instrument Panel Cluster Display

Late model Cadillac SIR system diagnosis can be entered by pressing the OFF and WARMER buttons on the electronic climate control panel (ECC) simultaneously. Press the LOW button four times to enter the SIR system. Then press HIGH to display any recorded codes (Figure 36-13).

Clearing Trouble Codes

All current trouble codes will be cleared automatically when the fault is repaired. The exception is code 51, which requires a scan tool to erase it from current code memory. History codes can only be cleared by using a scan tool. A code 51 cannot be erased from history code memory.

Ford Air Bag System Flash Code Diagnosis

SERVICE TIP *On some air bag systems such as those on Ford vehicles, if the air bag warning light is illuminated continually with the ignition switch on or the engine running, the ASDM is defective or disconnected. The air bag warning light may not be illuminated at full brilliance.*

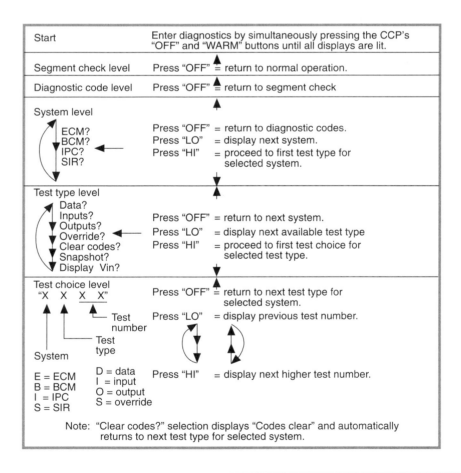

Figure 36-13 Onboard diagnostics using the ECC panel. (Courtesy of General Motors Corporation, Service Technology Group)

SERVICE TIP *On some Ford vehicles, the ASDM provides an audible tone if the air bag warning light is not operating properly. Under this condition, the ASDM provides 5 sets of 5 beeps every half-hour.*

SERVICE TIP *On Ford vehicles, the air bag deployment loop is connected to the battery positive terminal even with the ignition switch off. Therefore, air bag deployment is possible with the ignition switch off.*

On some air bag systems, the ASDM disarms the system if a fault exists that could result in unwarranted air bag deployment. The ASDM disarms the system by opening a thermal fuse inside the ASDM. This fuse is not replaceable.

On many Ford vehicles, the air bag warning light begins flashing a DTC when a defect occurs in the air bag system. On many Ford systems, the air bag warning light prioritizes the DTCs and flashes the highest priority DTC if there is more than one fault. When the fault represented by the flashing air bag warning light is corrected, the light flashes the DTC with the next highest priority. Since DTCs vary depending on the model year, the technician must have the DTC list for the model year being diagnosed.

Toyota Air Bag System Flash Code Diagnosis

SERVICE TIP *On Toyota vehicles, if the air bag warning light flashes a DTC that is not on the fault code list for that model year, the ASDM is defective.*

On Toyota vehicles, air bag flash codes may be obtained by cycling the ignition switch on and off five times. Each time the ignition switch is cycled on or off, the technician must wait 20 seconds. Toyota air bag DTCs may also be obtained by connecting a special jumper wire supplied by the vehicle manufacturer between terminals TC and E1 in the DLC2. Before connecting this jumper wire, make sure the ignition switch is in the ACC or ON position. After the ignition switch is in one of these positions, wait 30 seconds. If there are no DTCs in the ASDM memory, the air bag warning light flashes two times per second. When DTCs are present in the ASDM memory, the air bag warning light flashes these codes in numerical order. A DTC indicates a fault

in a certain area such as a specific air bag sensor. Voltmeter or ohmmeter tests recommended in the vehicle manufacturer's service manual are usually necessary to locate the exact cause of the problem.

Honda Air Bag System Voltmeter Diagnosis

CAUTION *Use only the vehicle manufacturer's recommended tools and equipment for air bag system service and diagnosis. Failure to observe this precaution may result in unwarranted air bag deployment and personal injury*

CAUTION *Do not use battery-powered or A/C-powered voltmeters or ohmmeters except those meters specified by the vehicle manufacturer. Failure to observe this precaution may result in unwarranted air bag deployment and personal injury.*

CAUTION *Do not use non-powered probe-type test lights or self-powered test lights to diagnose the air bag system. Unwarranted air bag deployment and personal injury may result from this procedure.*

CAUTION *Follow the vehicle manufacturer's service and diagnostic procedures. Failure to observe this precaution may cause inaccurate diagnosis and unnecessary repairs, or unwarranted air bag deployment, resulting in personal injury.*

Honda recommends testing the voltage at specific terminals on the ASDM, inflator modules, and sensor to diagnose the air bag system. When voltage tests are performed at these terminals, special jumper wires are connected to the terminals to allow the necessary voltmeter connections without damaging the terminals. Before connecting the special wiring harness, remove and isolate the battery negative cable. Then wait for the time period specified by the vehicle manufacturer.

Wiring harness A is connected to a terminal on the ASDM, and wiring harness B is connected in series between the large ASDM wiring connector and the matching terminals on the ASDM (Figure 36-14). Wiring harness C is connected in series at the inflator module connector, and harness D is connected in series at the dash sensor (Figure 36-15). After the special wiring harness is connected, the voltmeter tests provided in the vehicle manufacturer's service manual may be performed to diagnose the system.

Inspection After an Accident

Anytime the vehicle is involved in an accident, even if the air bag was not deployed, all air bag system com-

Test Harness A

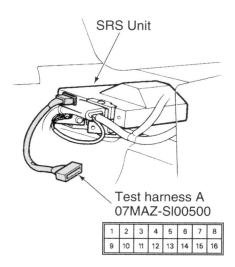

Test Harness B

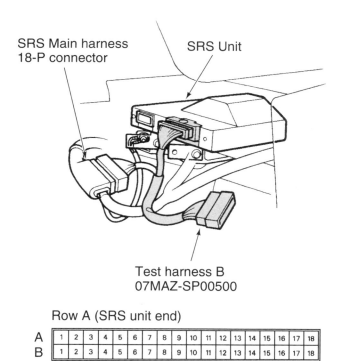

Figure 36-14 Wiring harness A and B connected at the ASDM terminals. (Courtesy of American Honda Motor Co., Inc.)

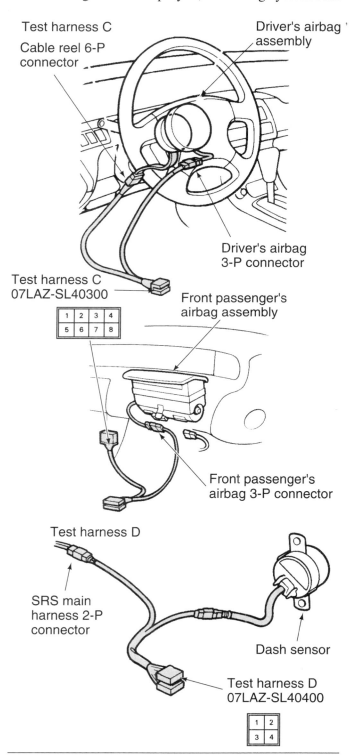

Figure 36-15 Wiring harness C and D connected at the inflator module and dash sensor terminals. (Courtesy of American Honda Motor Co, Inc.)

ponents should be inspected. The wiring harness must be inspected for damage and repaired or replaced as needed. Any damaged or dented components must also be replaced. Do not attempt to repair any of the sensors or modules. Service is by replacement only.

In the event of deployment, the service manual will provide a list of components requiring replacement. The list of components will vary depending on manufacturer.

Clean-up Procedure After Deployment

If the air bag has been deployed, the residue inside the passenger compartment must be removed before entering the vehicle. Tape the air bag exhaust vents closed to prevent additional powder from escaping. Use the shop vacuum cleaner to remove any powder from the vehicle's interior. Work from the outside to the center of the vehicle. Vacuum the heater and A/C vents. Run the heater fan blower motor on low speed and vacuum any powder that is blown from the plenum.

Component Replacement

Module Replacement

WARNING *Before replacing any component of the air bag system, follow the service manual procedure for disarming the system, even if the air bag is deployed.*

Follow Photo Sequence 17 as a guide to replacement of an air bag module on a 1993 Buick Roadmaster. Always follow the service manual for the vehicle you are working on.

Reverse the procedure to reinstall the module. To rearm the system, connect the yellow two-way electrical connector at the base of the steering wheel and install the CPA. Replace the SIR fuse and reconnect the battery.

CAUTION *Make sure the ignition switch is in the OFF position before connecting the battery.*

Turn the ignition switch to the RUN position while observing the SIR warning light. It should flash seven to nine times and then shut off. Perform the SIR diagnostic system check.

Clock Spring Replacement

The clock spring should be inspected any time the vehicle has been involved in an accident, even if the air bag was not deployed. In addition, the heat generated when an air bag is deployed may damage the clock spring. For this reason it should be replaced whenever the air bag is deployed. Exact procedures vary according to manufacturer. The following is typical for the Ford SRS system.

CAUTION *Wear safety glasses when servicing the air bag system.*

1. Place the front wheels in a straight ahead position and place the ignition in the LOCK position. Rotate the steering wheel 16 degrees counterclockwise until it locks.
2. Disarm the airbag system.

CAUTION *The procedure for disarming the system is different between years. Vehicles from 1992 and on require the use of an air bag simulator tool. Do not attempt to service the system if proper disarming is not possible.*

3. Remove the four nuts that attach the air bag module to the steering wheel.
4. Lift the module from the steering wheel and disconnect the wire connection to the clock spring.
5. Remove and discard the steering wheel attaching bolt.
6. Mark the shaft and steering wheel with index marks for re-installation.
7. Use a steering wheel puller to remove the steering wheel from the shaft.
8. Remove the upper and lower steering column shrouds.
9. Disconnect the clock spring connector from the steering column harness.
10. Tape the clock spring to prevent rotation of the rotor.
11. Remove the two retaining screws and the clock spring.

Replacement of the clock spring is in reverse order. The replacement clock spring will have a locking insert to prevent rotation of the rotor. Do not remove this insert until the clock spring is secured onto the column by the two retaining screws. Torque the new steering wheel attaching nut to specifications. When the steering wheel is replaced, rearm the system and perform the verification test.

PHOTO SEQUENCE 14 Typical Procedure for Removing the Air Bag Module

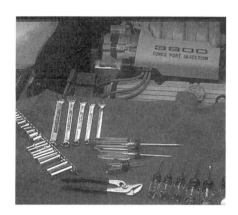

P14-1 Tools required to remove the air bag module: Safety glasses, seat covers, screw driver set, torx driver set, battery terminal pullers, battery pliers, assorted wrenches, ratchet and socket set, and service manual.

P14-2 Place the seat and fender covers on the vehicle.

P14-3 Place the front wheels in the straight ahead position and turn the ignition switch to the lock position.

P14-4 Disconnect the negative battery cable.

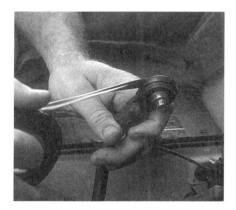

P14-5 Tape the cable terminal to prevent accidental connection with the battery post. Note: A piece of rubber hose can be substituted for the tape.

P14-6 Remove the SIR fuse from the fuse box. Wait 10 minutes to allow the reserve energy to dissipate.

PHOTO SEQUENCE 14 Typical Procedure for Removing the Air Bag Module

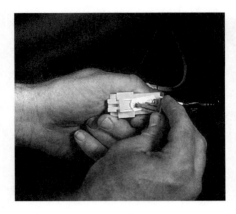

P14-7 Remove the connector position assurance (CPA) from the yellow electrical connector at the base of the steering column.

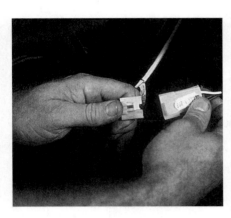

P14-8 Disconnect the yellow two-way electrical connector.

P149 Remove the four bolts that secure the module from the rear of the steering wheel.

P14-10 Rotate the horn lead a quarter turn and disconnect it.

P14-11 Disconnect the electrical connectors.

P14-12 Remove the module.

Summary

❏ Some passive seat belt systems are capable of storing diagnostic trouble codes that can be retrieved by a scan tester. The scanner is also used to perform a functional test of the system.

❏ A faulty drive motor can cause slow or no operation of the automatic seatbelts. The motor must be replaced if it has been determined it is the faulty component.

❏ If the drive belt is twisted, kinked, or broken it will cause the seatbelt to operate improperly. A binding drive belt may prevent the seatbelt from moving its full travel or may strip the gear teeth on the motor.

❏ Before installing the new drive belt, lubricate the track with approved lubricant.

❏ Whenever working on the air bag system, it is important to follow all safety warnings.

❏ A diagnostic system check must be performed on an air bag system to avoid diagnostic errors. The diagnostic system check involves observing the air bag warning light to determine if it is operating normally.

❏ Air bag system trouble codes can be retrieved through the use of a scan tool, flash codes, or the digital panel cluster (if equipped).

❏ Anytime the vehicle is involved in an accident, even if the air bag was not deployed, all air bag system components should be inspected.

❏ In the event of deployment, the service manual will provide a list of components requiring replacement. The list of components will vary depending on manufacturer.

ASE Style Review Questions

1. The automatic seatbelt system is being discussed. Technician A says to use standard test procedures for testing the circuits and components. Technician B says that some manufacturers have designed the system to store trouble codes. Who is correct?
   a. A only
   b. B only
   c. Both A and B
   d. Neither A nor B

2. General Motors SIR trouble codes are being discussed. Technician A says a code 52 indicates that the EEPROM memory is full. Technician B says that a code 52 is erased when the battery is disconnected. Who is correct?
   a. A only
   b. B only
   c. Both A and B
   d. Neither A nor B

3. Technician A says that trouble codes indicate the exact location of the fault. Technician B says trouble codes will lead you to the correct diagnostic chart to use. Who is correct?
   a. A only
   b. B only
   c. Both A and B
   d. Neither A nor B

4. Replacement of an air bag module is being discussed. Technician A says to follow the service manual procedure for disarming the system. Technician B says to wear safety glasses. Who is correct?
   a. A only
   b. B only
   c. Both A and B
   d. Neither A nor B

5. Technician A says the clock spring should be inspected any time the vehicle has been involved in an accident, even if the air bag was not deployed. Technician B says the heat that is generated when an air bag is deployed may damage the clock spring. Who is correct?
   a. A only
   b. B only
   c. Both A and B
   d. Neither A nor B

6. Technician A says air bag residue should be swept from the vehicle's interior using a whisk broom. Technician B says whenever a vehicle is involved in an accident the air bag control module must be replaced. Who is correct?
   a. A only
   b. B only
   c. Both A and B
   d. Neither A nor B

7. While discussing air bag system service: Technician A says before an air bag system component is replaced, the negative battery cable should be disconnected, and the technician should wait 2 minutes. Technician B says this waiting period is necessary to dissipate the reserve energy in the air bag system computer. Who is correct?
   a. A only
   b. B only
   c. Both A and B
   d. Neither A nor B

8. While discussing air bag sensor service: Technician A says incorrect torque on air bag sensor fasteners may cause improper air bag deployment. Technician B says the arrow on an air bag sensor must face toward the driver's side of the vehicle. Who is correct?
   a. A only
   b. B only
   c. Both A and B
   d. Neither A nor B

9. While discussing air bag system flash code diagnosis Technician A says on some Ford products, the air bag computer prioritizes faults and flashes the code representing the highest priority fault. Technician B says on some air bag systems, the air bag computer disarms the system if a fault occurs that could result in an unwarranted air bag deployment Who is correct?
   a. A only
   b. B only
   c. Both A and B
   d. Neither A nor B

10. While discussing air bag system flash code diagnosis: Technician A says on Toyota vehicles if there are no faults in the air bag system, the air bag warning light flashes 4 times per second in the diagnostic mode. Technician B says on Toyota vehicles a jumper wire must be connected between terminals TC and E1 in the DLC to obtain air bag system codes. Who is correct?
   a. A only
   b. B only
   c. Both A and B
   d. Neither A nor B

Glossary

Accumulator: A gas-filled pressure chamber that provides hydraulic pressure for ABS operation.

Acumulador: Cámara de presión que contiene gas y que proporciona gran presión hidráulica para una función ABS.

A circuit: A generator circuit that uses an external grounded field circuit. The regulator is on the ground side of the field coil.

Circuito A: Circuito regulador del generador que utiliza un circuito inductor externo puesto a tierra. En el circuito A, el regulador se encuentra en el lado a tierra de la bobina inductora.

Actuators: Devices that perform the actual work commanded by the computer. They can be in the form of a motor, relay, switch, or solenoid.

Accionadores: Dispositivos que realizan el trabajo efectivo que ordena la computadora. Dichos dispositivos pueden ser un motor, un relé, un conmutador o un solenoide.

Adaptive suspension systems: Suspension systems that are able to change ride characteristics by continuously altering shock damping and ride height.

Sistemas adaptadores de suspensión: Sistemas de suspensión que pueden cambiar las características del viaje al alterar continuamente el amortiguamiento y la altura del viaje.

Air bag module: Composed of the air bag and inflator assembly that is packaged into a single module.

Unidad del Airbag: Formada por el conjunto del Airbag y el inflador. Este conjunto se empaqueta en una sola unidad.

Air bag system: A supplemental restraint that will deploy a bag out of the steering wheel or passenger side dash panel to provide additional protection against head and face injuries during an accident.

Sistema de Airbag: Resguardo complementario que expulsa una bolsa del volante o del panel de instrumentos del lado del pasajero para proveer protección adicional contra lesiones a la cabeza y a la cara en caso de un accidente.

Air core gauge: Gauge design that uses the interaction of two electromagnets and the total field effect upon a permanent magnet to cause needle movement.

Calibrador de núcleo de aire Calibrador diseñado para utilizar la interacción de dos electroimanes y el efecto inductor total sobre un imán permanente para generar el movimiento de la aguja.

Alternating current: Electrical current that changes direction between positive and negative.

Corriente alterna: Corriente eléctrica que recorre un circuito ya sea en dirección positiva o negativa.

Ambient temperature: The temperature of the outside air.

Temperatura ambiente: Temperatura del aire ambiente.

Ambient temperature sensor: Thermistor used to measure the temperature of the air entering the vehicle.

Sensor de temperatura ambiente: Termistor utilizado para medir la temperatura del aire que entra al vehículo.

American wire gauge (AWG): System used to determine wire sizes based on the cross-sectional area of the conductor.

Calibrador americano de alambres Sistema utilizado para determinar el tamaño de los alambres, basado en el área transversal del conductor.

Ammeter: A test meter used to measure current draw.

Amperímetro: Instrumento de prueba utilizado para medir la intensidad de una corriente.

Amperes: See current.

Amperios: Véase corriente.

Analog: A voltage signal that is infinitely variable or can be changed within a given range.

Señal analógica Señal continua y variable que debe traducirse a valores numéricos discontinuos para poder ser trataba por una computadora.

Anode: The positive charge electrode in a voltage cell.

Ánodo Electrodo de carga positiva de un generador de electricidad.

Antilock brakes (ABS): A brake system that automatically pulsates the brakes to prevent wheel lock-up under panic stop and poor tractión conditions.

Frenos antibloqueo: Sistema de frenos que pulsa los frenos automáticamente para impedir el bloqueo de las ruedas en casos de emergencia y de tracción pobre.

Antitheft device A device or system that prevents illegal entry or driving of a vehicle. Most are designed to deterentry.

Dispositivo a prueba de hurto Un dispositivo o sistema quepreviene la entrada o conducción ilícita de un vehículo. Lamayoría se diseñan para detener la entrada.

A-pillar: The pillar in front of the driver or passenger that supports the windshield.

Soporte A: Soporte enfrente del conductor o del pasajero que sostiene el parabrisas.

Arming sensor: A device that places an alarm system into "ready" to detect an illegal entry.

Sensor de armado: Un dispositivo que pone "listo" un sistema dealarma para detectar una entrada ilícita.

Aspirator: Tubular device that uses a venturi effect to draw air from the passenger compartment over the in-car sensor. Some manufacturers use a suction motor to draw the air over the sensor.

Aspirador: Dispositivo tubular que utiliza un efecto venturi para extraer aire del compartimiento del pasajero sobre el sensor dentro del vehículo. Algunos fabricantes utilizan un motor de succion para extraer el aire sobre el sensor.

Atom: The smallest part of a chemical element that still has all the characteristics of that element.

Átomo: Partícula más pequeña de un elemento químico que conserva las cualidades íntegras del mismo.

Audio system: The sound system for a vehicle; can include radio, cassette player, CD player, amplifier, and speakers.

Sistema de audio: El sistema de sonido de un vehículo; puedeincluir el radio, el tocacaset, el toca discos compactos, el amplificador, y las bocinas.

Automatic door locks: A system that automatically locks all doorsthrough the activation of one switch.

Cerraduras automáticas de puerta: Un sistema que cierra todas laspuertas automaticamente al activar un solo conmutador.

Automatic headlight dimming: An electronic feature that automatically switches the headlights from high beam to low beam under two different conditions: light from oncoming vehicles strikes the photocell-amplifier; or light from the taillights of a vehicle that is being passed strikes the photocell-amplifier.

Reducción automática de intensidad luminosa de los faros delanteros: Característica electrónica que conmuta los faros delanteros automáticamente de luz larga a luz corta dadas las siguientes circunstancias: la luz de los vehículos que se aproximan alcanza el amplificador de fotocélula, o la luz de los faros traseros de un vehículo que se ha rebasado alcanza el amplificador de fotocélula.

Automatic Traction Control: A system that prevents slippage of one of the drive wheels. This is done by applying the brake at that wheel and/or decreasing the engine's power output.

Control Automático de Tracción: Un sistema que previene el patinaje de una de las ruedas de mando. Esto se efectúa aplicando el freno en esa rueda y/o disminuyendo la salida de potencia del motor.

Average Responding: A method used to read AC voltage.

Respuesta media Un método que se emplea para leer la tensión decorriente alterna.

Back probe: A term used to mean that a test is being performed on the circuit while the connector is still connected to the component. The test probes are inserted into the back of the wire connector.

Sonda exploradora de retorno: Término utilizado para expresar que se está llevando a cabo una prueba del circuito mientras el conectador sigue conectado al componente. Las sondas de prueba se insertan a la parte posterior del conectador de corriente.

Balanced atom : An atom that has an equal amount of protons and electrons.

Átomo equilibrado Átomo que tiene el mismo número de protones y de electrones.

Ballast resistor: A resistance put in series with a power lead to a component. Its purpose is to reduce the voltage applied to the component and to control the amount of current in the circuit.

Resistencia autorreguladora: Una regulación de serie con un conectador de alimentación a un componente. Su propósito es de reducir la tensión que se aplica al componente y controlar la cantidad del corriente en el circuito.

Base: The center layer of a bipolar transistor.

Base: Capa central de un transistor bipolar.

Battery cell: The active unit of a battery.

Acumulador de batería: Componente activo de una batería.

Battery holddowns: Brackets that secure the battery to the chassis of the vehicle.

Portabatería Los sostenes que fijan la batería al chasis del vehículo.

Battery leakage test: Used to determine if current is discharging across the top of the battery case.

Prueba de pérdida de corriente de la batería. Prueba utilizada para determinar si se está descargando corriente a través de la parte superior de la caja de la batería.

Battery terminal test: Checks for poor electrical connections between the battery cables and terminals. Use a voltmeter to measure voltage drop across the cables and terminals.

Prueba del borne de la batería: Verifica si existen conexiones eléctricas pobres entre los cables y los bornes de la batería. Utiliza un voltímetro para medir caídas de tensión entre los cables y los bornes.

Battery terminals: Terminals at the battery to which the positive and the negative battery cables are connected. The terminals may be posts or threaded inserts.

Bornes de la batería: Los bornes en la batería a los cuales se conectan los cables positivos y negativos. Los terminales pueden ser postes o piezas roscadas.

B circuit: A generator regulator circuit that is internally grounded. In the B circuit, the voltage regulator controls the power side of the field circuit.

Circuito B: Circuito regulador del generador puesto internamente a tierra. En el circuito B, el regulador de tensión controla el lado de potencia del circuito inductor.

Baud rate: The measure of computer data transmission speed in bits per second.

Razón de baúd: Medida de la velocidad de la transmisión de datos de una computadora en bits por segundo.

B circuit: A generator regulator circuit that is internally grounded. In the B circuit, the voltage regulator controls the power side of the field circuit.

Circuito B: Circuito regulador del generador puesto internamente a tierra. En el circuito B, el regulador de tensión controla el lado de potencia del circuito inductor.

B-pillar: The pillar located over the shoulder of the driver or passenger.

Soporte B: Soporte ubicado sobre el hombro del conductor o del pasajero.

Bench test: A term used to indicate that the unit is to be removed from the vehicle and tested.

Prueba de banco: Término utilizado para indicar que la unidad será removida del vehículo para ser examinada.

Bendix drive: A type of starter drive that uses the inertia of the spinning starter motor armature to engage the drive gear to the gears of the flywheel. This type starter drive was used on early models of vehicles and is rarely seen today.

Acoplamiento Bendix: Un tipo del acoplamiento del motor de arranque que usa la inercia de la armadura del motor de arranque giratorio para endentar el engranaje de mando con los engranajes del volante. Este tipo de acoplamiento del motor de arranque se usaba en los modelos vehículos antiguos y se ven raramente.

Bias voltage: Voltage applied across a diode.

Tensión polarizadora: Tensión aplicada a través de un diodo.

Bimetallic strip: A metal contact wiper consisting of two different types of metals. One strip will react quicker to heat than the other, causing the strip to flex in proportion to the amount of current flow.

Banda bimetálica: Contacto deslizante de metal compuesto de dos tipos de metales distintos. Una banda reaccionará más rápido al calor que la otra, haciendo que la banda se doble en proporción con la cantidad de flujo de corriente.

Binary code: A series of numbers represented by 1s and 0s. Any number and word can be translated into a combination of binary 1s and 0s.

Código binario: Serie de números representados por unos y ceros. Cualquier número y palabra puede traducirse en una combinación de unos y ceros binarios.

Bipolar: The name used for transistors because current flows through the materials of both polarities.

Bipolar: Nombre aplicado a los transistores porque la corriente fluye por conducto de materiales de ambas polaridades.

Bit: A binary digit.

Bit: Dígito binario.

Brushes: Electrically conductive sliding contacts, usually made of copper and carbon.

Escobillas: Contactos deslizantes de conducción eléctrica, por lo general hechos de cobre y de carbono.

Bucking coil: One of the coils in a three-coil gauge. It produces a magnetic field that bucks or opposes the low reading coil.

Bobina compensadora: Una de las bobinas de un calibre de tres bobinas. Produce un campo magnético que es contrario o en oposición a la bobina de baja lectura.

Buffer: A buffer cleans up a voltage signal. These are used with PM generator sensors to change the AC voltage to a digitalized signal.

Separador: Un separador aguza una señal del tensión. Estos se usan con los sensores generadores PM para cambiar la tensión de corriente alterna a una señal digitalizado.

Bulkhead connector: A large connector that is used when many wires pass through the bulkhead or firewall.

Conectador del tabique: Un conectador que se usa al pasar muchos alambres por el tabique o mamparo de encendios.

Bus bar: A common electrical connection to which all of the fuses in the fuse box are attached. The bus bar is connected to battery voltage.

Barra colectora: Conexión eléctrica común a la que se conectan todos los fusibles de la caja de fusibles. La barra colectora se conecta a la tensión de la batería.

Buzzer: An audible warning device that is used to warn the driver of possible safety hazards.

Zumbador: Dispositivo audible de advertencia utilizado para prevenir al conductor de posibles riesgos a la seguridad.

Capacitance: The ability of two conducting surfaces to store voltage.

Capacitancia: Propiedad que permite el almacenamiento de electricidad entre dos conductores aislados entre sí.

Capacity test: The part of the battery test series that checks the battery's ability to perform when loaded.

Prueba de capacidad: Parte de la serie de prueba de la batería que verifica la capacidad de funcionamiento de la batería cuando está cargada.

Carbon monoxide: An odorless, colorless, and toxic gas that is produced as a result of combustion.

Monóxido de carbono: Gas inodoro, incoloro y tóxico producido como resultado de la combustión.

Carbon tracking: A condition where paths of carbon will allow current to flow to points that are not intended. This condition is most commonly found inside distributor caps.

Rastreo de carbón: Una condición en la cual las trayectorias del carbón permiten fluir el corriente a los puntos no indicados. Esta condición se encuentra comunmente dentro de las tapas del distribuidor.

Cartridge fuses: See maxi-fuse.

Fusibles cartucho: Véase maxifusible.

Cathode: Negatively charged electrode of a voltage cell.

Cátodo: Electrodo de carga negativa de un generador de electricidad.

Cathode ray tube: Similar to a television picture tube. It contains a cathode that emits electrons and an anode that attracts them. The screen of the tube will glow at the points that are hit by the electrons.

Tubo de rayos catódicos: Parecidos a un tubo de pantalla de televisor. Contiene un cátodo que emite los electrones y un ánodo que los atrae. La pantalla del tubo iluminará en los puntos en donde pegan los electrones.

Cell element: The assembly of a positive and negative plate in a battery.

Elemento de pila: La asamblea de una placa positiva y negativa en una bateria.

Charging system requirement test: Diagnóstic test used to determine the total electrical demand of the vehicle's electrical system.

Prueba del requisito del sistema de carga: Prueba diagnóstica utilizada para determinar la exigencia eléctrica total del sistema eléctrico del vehículo.

CHMSL: The abbreviation for center high mounted stop light, often referred to as the third brake light.

CHMSL: La abreviación para el faro de parada montada alto en el centro que suele referirse como el faro de freno tercero.

Choke coil: Fine wire wound into a coil used to absorb oscillations in a switched circuit.

Bobina de inducción: Alambre fino devanado en una bobina, utilizado para absorber oscilaciones en un circuito conmutado.

Circuit: The path of electron flow consisting of the voltage source, conductors, load component, and return path to the voltage source.

Circuito: Trayectoria del flujo de electrones, compuesto de la fuente de tensión, los conductores, el componente de carga y la trayectoria de regreso a la fuente de tensión.

Clamping diode: A diode that is connected in parallel with a coil to prevent voltage spikes from the coil from reaching other components in the circuit.

Diodo de bloqueo: Un diodo que se conecta en paralelo con una bobina para prevenir que los impulsos de tensión lleguen a otros componentes en el circuito.

Clock circuit: A crystal that electrically vibrates when subjected to current at certain voltage levels. As a result, the chip produces very regular series of voltage pulses.

Circuito de reloj: Cristal que vibra electrónicamente cuando está sujeto a una corriente a ciertos niveles de tensión. Como resultado, el fragmento produce una serie sumamente regular de impulsos de tensión.

Clock spring: Maintains a continuous electrical contact between the wiring harness and the air bag module.

Muelle de reloj: Mantiene un contacto eléctrico continuo entre el cableado preformado y la unidad del Airbag.

Closed circuit: A circuit that has no breaks in the path and allows current to flow.

Circuito cerrado: Circuito de trayectoria ininterrumpida que permite un flujo continuo de corriente.

Cold cranking amps (CCA): Rating indicates the battery's ability to deliver a specified amount of current to start an engine at low ambient temperatures.

Amperios de arranque en frío: Tasa indicativa de la capacidad de la batería para producir una cantidad específica de corriente para arrancar un motor a bajas temperaturas ambiente.

Cold fouling: A condition of a spark plug in which deposits and other materials have not burned off the electrodes. The presence of the deposits will shorten the plug's gap and reduce firing voltages.

Engrase frío: Una condición de la bujía en la cual los depósitos y otras materiales no se han consumido de los electrodos. La presencia de los depósitos disminuirán el entrehierrro de la bujía y bajarán las tensiones del encendido.

Collector: The portion of a bipolar transistor that receives the majority of current carriers.

Dispositivo de toma de corriente: Parte del transistor bipolar que recibe la mayoría de los portadores de corriente.

Color codes: Used to assist in tracing the wires. In most color codes, the first group of letters designates the base color of the insulation and the second group of letters indicates the color of the tracer.

Códigos de colores: Utilizados para facilitar la identificación de los alambres. Típicamente, el primer alfabeto representa el color base del aislamiento y el segundo representa el color del indicador.

Common connector: A connector that is shared by more than one circuit and/or component.

Conector común: Un conector que se comparte entre más de un circuito y/o componente.

Commutator: A series of conducting segments located around one end of the armature.

Conmutador: Serie de segmentos conductores ubicados alrededor de un extremo de la armadura.

Component locator: Service manual used to find where a component is installed in the vehicle. The component locator uses both drawings and text to lead the technician to the desired component.

Manual para indicar los elementos componentes: Manual de servico utilizado para localizar dónde se ha instalado un componente en el vehículo. En dicho manual figuran dibujos y texto para guiar al mecánico al componente deseado.

Composite bulb: A headlight assembly that has a replaceable bulb in its housing.

Bombilla compuesta: Una asamblea de faros cuyo cárter tiene una bombilla reemplazable.

Compound motor: A motor that has the characteristics of a series-wound and a shunt-wound motor.

Motor compuesta: Un motor que tiene las características de un motor exitado en serie y uno en derivación.

Computer: An electronic device that stores and processes data and is capable of operating other devices.

Computadora: Dispositivo electrónico que almacena y procesa datos y que es capaz de ordenar a otros dispositivos.

Condenser: A capacitor made from two sheets of metal foil separated by an insulator.

Condensador: Capacitor hecho de dos láminas de metal separadas por un medio aislante.

Conduction: Bias voltage difference between the base and the emitter has increased to the point that the transistor is switched on. In this condition, the transistor is conducting. Output current is proportional to that of the current through the base.

Conducción: La diferencia de la tensión polarizadora entre la base y el emisor ha aumentado hasta el punto que el transistor es conectado. En estas circunstancias, el transistor está conduciendo. La corriente de salida está en proporción con la de la corriente conducida en la base.

Conductor: A material in which electrons flow or move easily.

Conductor: Una material en la cual los electrones circulen o se mueven fácilmente.

Continuity: Refers to the circuit being continuous with no opens.

Continuidad: Se refiere al circuito ininterrumpido, sin aberturas.

Conventional theory: Electrical theory that states current flows from a positive point to a more negative point.

Teoría convencional: Teoría de electricidad la cual enuncia que el corriente fluye desde un punto positivo a un punto más negativo.

Corner lights: Lamps that illuminate when the turn signals are activated. They burn steadily when the turn signal switch is in a turn position to provide additional illumination of the road in the direction of the turn.

Faros laterales: Lámparas que se encienden cuando se activan las luces indicadoras para virajes. Se encienden de manera continua cuando el conmutador de las luces indicadoras para virajes está conectado para proveer iluminación adicional de la carretera hacia la dirección del viraje.

Corona effect: A condition where high voltage leaks through a wire's insulation and produces a light or illumination; worn insulation on spark plug wires causes this.

Efecto corona: Una condición en la cual la alta tensión se escapa por la insulación del alambre y produce una luz o una iluminación; esto se causa por la insulación desgastada en los alambres de las bujías.

Counterelectromotive force (CEMF): An induced voltage that opposes the source voltage.

Fuerza cóntraelectromotriz: Tensión inducida en oposición a la tensión fuente.

Courtesy lights: Lamps that illuminate the vehicle's interior when the doors are open.

Luces interiores: Lámparas que iluminan el interior del vehículo cuando las puertas están abiertas.

Covalent bonding: When atoms share valence electrons with other atoms.

Enlace covalente: Cuando los átomos comparten electrones de valencia con otros átomos.

CRT: The common acronym for a cathode ray tube.

CRT: La sigla común de un tubo de rayos catódicos.

Crash sensor: Normally open electrical switch designed to close when subjected to a predetermined amount of jolting or impact.

Sensor de impacto: Un conmutador normalmente abierto diseñado a cerrarse al someterse a un sacudo de una fuerza predeterminada o un impacto.

Crimping: The process of bending, or deforming by pinching, a connector so that the wire connection is securely held in place.

Engarzado: Proceso a través del cual se curva o deforma un conectador mediante un pellizco para que la conexión de alambre se mantenga firme en su lugar.

Cross-fire: The undesired firing of a spark plug that results from the firing of another spark plug. This is caused by electromagnetic induction.

Encendido transversal: El encendido no deseable de una bujía que resulta del encendido de otra bujía. Esto se causa por la inducción electromagnética.

Crystal: A term used to describe a material that has a definite atom structure.

Cristal: Término utilizado para describir un material que tiene una estructura atómica definida.

Current: The aggregate flow of electrons through a wire. One ampere represents the movement of 6.25 billion billion electrons (or one coulomb) past one point in a conductor in one second.

Corriente: Flujo combinado de electrones a través de un alambre. Un amperio representa el movimiento de 6,25 mil millones de millones de electrones (o un colombio) que sobrepasa un punto en un conductor en un segundo.

Curb height: The height of the vehicle when it has no passengers or loads, and normal fluid levels and tire pressure.

Altura del contén: La altura del vehículo cuando no lleva pasajeros ni cargas, y los niveles de los fluidos y de la presión de las llantas son normales.

Current: The aggregate flow of electrons through a wire. One ampere represents the movement of 6.25 billion billion electrons (or one coulomb) past one point in a conductor in one second.

Corriente: Flujo combinado de electrones a través de un alambre. Un amperio representa el movimiento de 6,25 mil millones de mil millones de electrones (o un colombio) que sobrepasa un punto en un conductor en un segundo.

Current draw test: Diagnostic test used to measure the amount of current that the starter draws when actuated. It determines the electrical and mechanical condition of the starting system.

Prueba de la intensidad de una corriente: Prueba diagnóstica utilizada para medir la cantidad de corriente que el arrancador tira cuando es accionado. Determina las condiciones eléctricas y mecánicas del sistema de arranque.

Current limiting: A method used by some electronically controlled ignition systems to ensure enough time and current is available to the primary windings of a coil to saturate it.

Resistencia limitadora: Un método empleado por algunos sistemas de encendido controlados electronicamente que asegura que se dispone bastante tiempo y corriente para saturar los enrollados primarios de una bobina.

Current output testing: Diagnostic test used to determine the maximum output of the ac generator.

Prueba de la salida de una corriente: Prueba diagnóstica utilizada para determinar la salida máxima del generador de corriente alterna.

Cutoff: When reverse-bias voltage is applied to the base leg of the transistor. In this condition, the transistor is not conducting and no current will flow.

Corte: Cuando se aplica tensión polarizadora inversa a la base del transistor. En estas circunstancias, el transistor no está conduciendo y no fluirá ninguna corriente.

d'Arsonval gauge: A gauge design that uses the interaction of a permanent magnet and an electromagnet, and the total field effect to cause needle movement.

Calibrador d'Arsonval: Calibrador diseñado para utilizar la interacción de un imán permanente y de un electroimán, y el efecto inductor total para generar el movimiento de la aguja.

Darlington pair: An arrangement of transistors that amplifies current by one transistor acting as a preamplifier which creates a larger base current to the second transistor.

Par Darlington: Conjunto de transistores que amplifica la corriente. Un transistor actúa como preamplificador y produce una corriente base más ámplia para el segundo transistor.

Deep cycling: Discharging the battery completely before recharging it.

Operacion cíclica completa: La descarga completa de la batería previo al recargo.

DI: The J1930 acronym for an electronic controlled distributor ignition system.

DI: La sigla J1930 de un sistema de encendido de distribuidor controlado electronicamente.

Delta stator: A three-winding ac generator stator with the ends of each winding connected to each other.

Estátor Delta: Estátor generador de corriente alterna de devanado triple, con los extremos de cada devanado conectados entre sí.

Diagnostic module: Part of an electronic control system that provides self-diagnostics and/or a testing interface.

Módulo de diagnóstico: Parte de un sistema controlado electronicamente que provee autodiagnóstico y/o una interfase de pruebas.

Dielectric: An insulator material.

Dieléctrico: Material aislante.

Digital: A voltage signal is either on-off, yes-no, or high-low.

Digital: Una señal de tensión está Encendida-Apagada, es Sí-No o Alta-Baja.

Dimmer switch: A switch in the headlight circuit that provides the means for the driver to select either high beam or low beam operation, and to switch between the two. The dimmer switch is connected in series within the headlight circuit and controls the current path for high and low beams.

Conmutador reductor: Conmutador en el circuito para faros delanteros que le permite al conductor elegir la luz larga o la luz corta, y conmutar entre las dos. El conmutador reductor se conecta en serie dentro del circuito para faros delanteros y controla la trayectoria de la corriente para la luz larga y la luz corta.

Diode: An electrical one-way check valve that will allow current to flow in one direction only.

Diodo: Válvula eléctrica de retención, de una vía, que permite que la corriente fluya en una sola dirección.

Diode rectifier bridge: A series of diodes that are used to provide a reasonably constant dc voltage to the vehicle's electrical system and battery.

Puente rectificador de diodo: Serie de diodos utilizados para proveerles una tensión de corriente continua bastante constante al sistema eléctrico y a la batería del vehículo.

Diode trio: Used by some manufacturers to rectify the stator of an AC generator current so that it can be used to create the magnetic field in the field coil of the rotor.

Trío de diodos: Utilizado por algunos fabricantes para rectificar el estátor de la corriente de un generador de corriente alterna y poder así utilizarlo para crear el campo magnético en la bobina inductora del rotor.

Direct current (dc): Electric current that flows in one direction.

Corriente continua: Corriente eléctrica que fluye en una dirección.

Direct drive: A situation where the drive power is the same as the power exerted by the device that is driven.

Transmisión directa: Una situación en la cual el poder de mando es lo mismo que la potencia empleada por el dispositivo arrastrado.

Discrete devices: Electrical components that are made separately and have wire leads for connections to an integrated circuit.

Dispositivos discretos: Componentes eléctricos hechos uno a uno; tienen conductores de alambre para hacer conexiones a un circuito integrado.

Discriminating sensors: Part of the air bag circuitry; these sensors are calibrated to close with speed changes that are great enough to warrant air bag deployment. These sensors are also referred to as crash sensors.

Sensores discriminadores: Una parte del conjunto de circuitos de Airbag; estos sensores se calibran para cerrar con los cambios de la velocidad que son bastante severas para justificar el despliegue del Airbag. Estos sensores también se llaman los sensores de impacto.

Display pattern: A pattern of an oscilloscope in which the cylinders of an engine appear one after another according to the firing order.

Patrón visualizador: Un patrón en un osciloscopo en el cual los cilindros de un motor aparecen uno tras otro según el orden del encendido.

Doping: The addition of another element with three or five valence electrons to a pure semiconductor.

Impurificación: La adición de otro elemento con tres o cinco electrones de valencia a un semiconductor puro.

Double filament lamp: A lamp designed to execute more than one function. It can be used in the stoplight circuit, taillight circuit, and the turn signal circuit combined.

Lámpara con filamento doble: Lámpara diseñada para llevar a cabo más de una función. Puede utilizarse en una combinación de los circuitos de faros de freno, de faros traseros y de luces indicadoras para virajes.

Drain: The portion of a field-effect transistor that receives the holes or electrons.

Drenador: Parte de un transistor de efecto de campo que recibe los agujeros o electrones.

Drive coil: A hollowed field coil used in a positive—engagement starter to attract the movable pole shoe of the starter.

Bobina de excitación: Una bobina inductora hueca empleada en un encendedor de acoplamiento directo para atraer la pieza polar móvil del encendedor.

DSO: A common acronym for a digital storage oscilloscope.

DSO: Una sigla común del osciloscopio de almacenamiento digital.

Duty-cycle: The percentage of on time to total cycle time.

Ciclo de trabajo: Porcentaje del trabajo efectivo a tiempo total del ciclo.

Eddy currents: Small induced currents.

Corriente de Foucault: Pequeñas corrientes inducidas.

EI: The J1930 acronym for a distributorless ignition system.

EI: La sigla J1930 de un sistema de encendido sin distrubuidor.

Electrical load: The working device of the circuit.

Carga eléctrica: Dispositivo de trabajo del circuito.

Electrically Erasable PROM (EEPROM): Memory chip that allows for electrically changing the information one bit at a time.

Capacidad de borrado electrónico PROM: Fragmento de memoria que permite el cambio eléctrico de la información un bit a la vez.

Electrochemical: The chemical action of two dissimilar materials in a chemical solution.

Electroquímico: Acción química de dos materiales distintos en una solución química.

Electrolysis: The producing of chemical changes by passing electrical current through an electrolyte.

Electrólisis: La producción de los cambios químicos al pasar un corriente eléctrico por un electrolito.

Electrolyte: A solution of 64% water and 36% sulfuric acid.

Electrolito: Solucion de un 64% de agua y un 36% de ácido sulfúrico.

Electromagnetic gauge: Gauge that produces needle movement by magnetic forces.

Calibrador electromagnético: Calibrador que genera el movimiento de la aguja mediante fuerzas magnéticas.

Electromagnetic induction: The production of voltage and current within a conductor as a result of relative motion within a magnetic field.

Inducción electrómagnética: Producción de tension y de corriente dentro de un conductor como resultado del movimiento relativo dentro de un campo magnético.

Electromagnetic interference (EMI): An undesirable creation of electromagnetism whenever current is switched on and off.

Interferencia electromagnética: Fenómeno de electromagnetismo no deseable que resulta cuando se conecta y se desconecta la corriente.

Electromagnetism: A form of magnetism that occurs when current flows through a conductor.

Electromagnetismo: Forma de magnetismo que ocurre cuando la corriente fluye a través de un conductor.

Electromechanical: A device that uses electricity and magnetism to cause a mechanical action.

Electromecánico: Un dispositivo que causa una acción mecánica por medio de la electricidad y el magnetismo.

Electromotive force (EMF): See voltage.

Fuerza electromotriz: Véase tensión.

Electron: Negative-charged particles of an atom.

Electrón: Partículas de carga negativa de un átomo.

Electron theory: Defines electrical movement as from negative to positive.

Teoría del electrón: Define el movimiento eléctrico como el movimiento de lo negativo a lo positivo.

Electrostatic field: The field that is between the two oppositely charged plates.

Campo electrostático: Campo que se encuentra entre las placas de carga opuesta.

EMI: Electro-magnetic interference.

EMI: La interferencia electromagnética.

Emitter: The outer layer of the transistor, which supplies the majority of current carriers.

Emisor: Capa exterior del transistor que suministra la mayor parte de los portadores de corriente.

Equivalent series load (equivalent resistance): The total resistance of a parallel circuit. It is equivalent to the resistance of a single load in series with the voltage source.

Carga en serie equivalente (resistencia equivalente): Resistencia total de un circuito en paralelo, equivalente a la resistencia de una sola carga en serie con la fuente de tensión.

Erasable PROM (EPROM): Similar to PROM except that its contents can be erased to allow for new data to be installed. A piece of Mylar tape covers a window. If the tape is removed, the microcircuit is exposed to ultraviolet light and erases its memory.

Capacidad de borrado PROM Parecido al PROM, pero su contenido puede borrarse para permitir la instalación de nuevos datos. Un trozo de cinta Mylar cubre una ventana; si se remueve la cinta, el microcircuito queda expuesto a la luz ultravioleta y borra la memoria.

Excitation current: Current that magnetically excites the field circuit of the ac generator.

Corriente de excitación: Corriente que excita magnéticamente al circuito inductor del generador de corriente alterna.

Failsoft: Computer substitution of a fixed input value if a sensor circuit should fail. This provides for system operation, but at a limited function. Also referred to as the "Limp-In" mode.

Falla activa: Sustitución por la computadora de un valor fijo de entrada en caso de que ocurra una falla en el circuito de un sensor. Esto asegura el funcionamiento del sistema, pero a una capacidad limitada.

Fast charging: Battery charging using a high amperage for a short period of time.

Carga rápida: Carga de la batería que utiliza un amperaje máximo por un corto espacio de tiempo.

Feedback: 1. Data concerning the effects of the computer's commands are fed back to the computer as an input signal. Used to determine if the desired result has been achieved. 2. A condition that can occur when electricity seeks a path of lower resistance, but the alternate path operates another component than that intended. Feedback can be classified as a short.

Realimentación: 1. Datos referentes a los efectos de las órdenes de la computadora se suministran a la misma como señal de entrada. La realimentación se utiliza para determinar si se ha logrado el resultado deseado. 2. Condición que puede ocurrir cuando la electricidad busca una trayectoria de menos resistencia, pero la trayectoria alterna opera otro componente que aquel deseado. La realimentación puede clasificarse como un cortocircuito.

Fiber optics: A medium of transmitting for the transmission of light through polymethylmethacrylate plastic that keeps the light rays parallel even if there are extreme bends in the plastic.

Transmisión por fibra óptica: Técnica de transmisión de luz por medio de un plástico de polimetacrilato de metilo que mantiene los rayos de luz paralelos aunque el plástico esté sumamente torcido.

Field current: The current going to the field windings of a motor or generator.

Corriente inductora: El corriente que va a los devanados inductores de un motor o generador.

Field current draw test: Diagnostic test that determines if there is current available to the field windings.

Prueba de la intensidad de una corriente inductora: Prueba diagnóstica que determina si se está generando corriente a los devanados inductores.

Field-effect transistor (FET): A unipolar transistor in which current flow is controlled by voltage in a capacitance field.

Transistor de efecto de campo: Transistor unipolar en el cual la tensión en un campo de capacitancia controla el flujo de corriente.

Field relay: The relay that controls the amount of current going to the field windings of a generator. This is the main output control unit for a charging system.

Relé inductor: El relé que controla la cantidad del corriente a los devanados inductores de un generador. Es la unedad principal de potencia de salida de un sistema de carga.

Fire extinguisher: A portable apparatus that contains chemicals, water, foam, or special gas that can be discharged to extinguish a small fire.

Extinctor de incendios: Aparato portátil que contiene elementos químicos, agua, espuma o gas especial que pueden descargarse para extinguir un incendio pequeño.

Firing line: That section of a scope waveform that appears as a tall spike and represents the amount of voltage necessary to overcome the resistance in the secondary circuit.

Linea de activación: Esa sección de una forma de onda de un osciloscopio que aparece como un punto de impulso y representa la cantidad del tensión que se requiere para sobrepasar la resistencia en el circuito secundario.

Firing voltage: The amount of voltage needed to overcome the resistance in the secondary and to establish a spark across the electrodes of a spark plug.

Tensión de activación: La cantidad de la tensión que se requiere para sobrepasar la resistencia del circuito secundario y para establecer una chispa a través de los electrodos de una bujía.

Floor jack: A portable hydraulic tool used to raise and lower a vehicle.

Gato de pie: Herramienta hidráulica portátil utilizada para levantar y bajar un vehículo.

Flux density: The number of flux lines per square centimeter.

Densidad de flujo: Número de líneas de flujo por centímetro cuadrado.

Flux lines: Magnetic lines of force.

Líneas de flujo: Líneas de fuerza magnética.

Forward-bias: A positive voltage that is applied to the P-type material and negative voltage to the N-type material of a semiconductor.

Polarización directa: Tensión positiva aplicada al material P y tensión negativa aplicada al material N de un semiconductor.

Free speed test: Diagnostic test that determines the free rotational speed of the armature. This test is also referred to as the no-load test.

Prueba de velocidad libre: Prueba diagnóstica que determina la velocidad giratoria libre de la armadura. A dicha prueba se le llama prueba sin carga.

Frequency: The number of complete oscillations that occur during a specific time, measured in Hertz.

Frecuencia: El número de oscilaciones completas que ocurren durante un tiempo específico, medidas en Hertz.

Full field: Field windings that are constantly energized with full battery current. Full fielding will produce maximum ac generator output.

Campo completo: Devanados inductores que se excitan constantemente con corriente total de la batería. EL campo completo producirá la salida máxima de un generador de corriente alterna.

Full field test: Diagnostic test used to isolate if the detected problem lies in the ac generator or the regulator.

Prueba de campo completo: Prueba diagnóstica utilizada para determinar si el problema descubierto se encuentra en el generador de corriente alterna o en el regulador.

Fuse: A replaceable circuit protection device that will melt should the current passing through it exceed its rating.

Fusible: Dispositivo reemplazable de protección del circuito que se fundirá si la corriente que fluye por el mismo excede su valor determinado.

Fuse box: A term used that indicates the central location of the fuses contained in a single holding fixture.

Caja de fusibles: Término utilizado para indicar la ubicación central de los fusibles contenidos en un solo elemento permanente.

Fusible link: A wire made of meltable material with a special heat-resistant insulation. When there is an overload in the circuit, the link melts and opens the circuit.

Cartucho de fusible: Alambre hecho de material fusible con aislamiento especial resistente al calor. Cuando ocurre una sobrecarga en el circuito, el cartucho se funde y abre el circuito.

Gain: The ratio of amplification in an electronic device.

Ganancia: Razón de amplificación en un dispositivo electrónico.

Ganged: Refers to a type of switch in which all wipers of the switch move together.

Acoplado en tándem : Se refiere a un tipo de conmutador en el cual todos los contactos deslizantes del mismo se mueven juntos.

Gap bridging: A condition of a spark plug in which deposits connect the center electrode with the side electrode.

Puente: Una condición de la bujía en la cual los depósitos conectan el electrodo central con el electrodo lateral.

Gassing: The conversion of a battery's electrolyte into hydrogen and oxygen gas.

Burbujeo: La conversión del electrolito de una bateria al gas de hidrógeno y oxígeno.

Gate: The portion of a field-effect transistor that controls the capacitive field and current flow.

Compuerta: Parte de un transistor de efecto de campo que controla el campo capacitivo y el flujo de corriente.

Gauge: 1. A device that displays the measurement of a monitored system by the use of a needle or pointer that moves along a calibrated scale. 2. The number that is assigned to a wire to indicate its size. The larger the number the smaller the diameter of the conductor.

Calibrador: 1. Dispositivo que muestra la medida de un sistema regulado por medio de una aguja o indicador que se mueve a través de una escala calibrada. 2. El número asignado a un alambre indica su tamaño. Mientras mayor sea el número, más pequeño será el diámetro del conductor.

Gauss gauge: A meter that is sensitive to the magnetic field surrounding a wire conducting current. The gauge needle will fluctuate over the portion of the circuit that has current flowing through it. Once the ground has been passed, the needle will stop fluctuating.

Calibrador gauss: Instrumento sensible al campo magnético que rodea un alambre conductor de corriente. La aguja del calibrador se moverá sobre la parte del circuito a través del cual fluye la corriente. Una vez se pasa a tierra, la aguja dejará de moverse.

Gear reduction: Occurs when two different sized gears are in mesh and the driven gear rotates at a lower speed than the drive gear but with greater torque.

Desmultiplicación: Ocurre cuando dos engranajes de distinctos tamaños se endentan y el engranaje arrastrado gira con una velocidad más baja que el engranaje de mando pero con más par.

Glazing: A condition where deposits on a spark plug melt and form a glaze on the insulator and electrode.

Vidriado: Una condición en la cual los depósitos sobre una bujía se derritan y forman un barníz en el insulador y el electrodo.

Glitches: Unwanted voltage spikes that are seen on a voltage trace. These are normally caused by intermittent opens or shorts.

Irregularidades espontáneos: Impulsos de tensión no deseables que se ven en una traza de tensión. Estos se causan normalmente por las aberturas o cortos intermitentes.

Grid growth: A condition where the grid grows little metallic fingers that extend through the separators and short out the plates.

Expansión de la rejilla: Una condición en la cual la rejilla produce protrusiones metálicas que se extienden por los separadores y causan cortocircuitos en las placas.

Grids: The frame structure of a battery that normally has connector tabs at the top. It is generally made of lead alloys.

Rejillas: La estructura encuadrador de una batería que normalmente tiene orejas de conexión en la parte superior. Generalmente se fabrica de aleaciones de plomo.

Ground: The common negative connection of the electrical system that is the point of lowest voltage.

Tierra: Conexión negativa común del sistema eléctrico. Es el punto de tensión más baja.

Grounded circuit: An electrical defect that allows current to return to ground before it has reached the intended load component.

Circuito puesto a tierra: Falla eléctrica que permite el regreso de la corriente a tierra antes de alcanzar el componente de carga deseado.

Ground circuit test: A diagnostic test performed to measure the voltage drop in the ground side of the circuit.

Prueba del circuito a tierra: Prueba diagnóstica llevada a cabo para medir la caída de tensión en el lado a tierra del circuito.

Ground side: The portion of the circuit that is from the load component to the negative side of the source.

Lado a tierra: Parte del circuito que va del componente de carga al lado negativo de la fuente.

Growler: Test equipment used to test starter armatures for shorts and grounds. It produces a very strong magnetic field that is capable of inducing a current flow and magnetism in a conductor.

Indicador de cortocircuitos: Equipo de prueba utilizado para localizar cortociruitos y tierra en armaduras de arranque. Genera un campo magnético sumamente fuerte, capaz de inducir flujo de corriente y magnetismo en un conductor.

Half-wave rectification: Rectification of one-half of an ac voltage.

Rectificación de media onda Rectificación en la que la corriente fluye únicamente durante semiciclos alternados.

Hall-effect switch: A sensor that operates on the principle that if a current is allowed to flow through thin conducting material being exposed to a magnetic field, another voltage is produced.

Conmutador de efecto Hall: Sensor que funciona basado en el principio de que si se permite el flujo de corriente a través de un material conductor delgado que ha sido expuesto a un campo magnético, se produce otra tensión.

Halogen: The term used to identify a group of chemically related nonmetallic elements. These elements include chlorine, fluorine, and iodine.

Halógeno: Término utilizado para identificar un grupo de elementos no metálicos relacionados químicamente. Dichos elementos incluyen el cloro, el flúor y el yodo.

Hand tools: Tools that use only the force generated from the body to operate. They multiply the force received through leverage to accomplish the work.

Herramientas manuales Herramientas que para funcionar sólo necesitan la fuerza generada por el cuerpo. Para llevar a cabo el trabajo, las herramientas multiplican la fuerza que reciben por medio de la palancada.

Hard-shell connector: An electrical connector that has a hard plastic shell that holds the connecting terminals of separate wires.

Conectador de casco duro: Conectador eléctrico con casco duro de plástico que sostiene separados los bornes conectadores de alambres individuales.

Heat sink: An object that absorbs and dissipates heat from another object.

Dispersador térmico: Objeto que absorbe y disipa el calor de otro objeto.

Heat-shrink tubing: A hollow insulation material that shrinks to an airtight fit over a connection when exposed to heat.

Tubería contraída térmicamente: Material aislante hueco que se contrae para acomodarse herméticamente sobre una conexión cuando se encuentra expuesto al calor.

H-gate: A set of four transistors that can reverse current.

Compuerta H: Juego de cuatro transistores que pueden invertir la corriente.

HID: High Intensity Discharge; a lighting system that uses an arc across electrodes instead of a filament.

HID: Descarga de Alta Intensidad; un sistema de iluminación que utiliza un arco por dos electrodos en vez de un filamento.

Hoist: A lift that is used to raise the entire vehicle.

Elevador: Montacargas utilizado para elevar el vehículo en su totalidad.

Hold-in winding: A winding that holds the plunger of a solenoid in place after it moves to engage the starter drive.

Devanado de retención: Un devanado que posiciona el núcleo móvil de un solenoide después de que mueva para accionar el acoplamiento del motor de arranque.

Hole: The absence of an electron in an element's atom. These holes are said to be positively charged since they have a tendency to attract free electrons into the hole.

Agujero: Ausencia de un electrón en el átomo de un elemento. Se dice que dichos agujeros tienen una carga positiva puesto que tienden a atraer electrones libres hacia el agujero.

Hybrid battery: A battery that combines the advantages of low maintenance and maintenance-free batteries.

Batería híbrida: Una batería que combina las ventajas de las baterías de bajo mantenimiento y de no mantenimiento.

Hydrometer: A test instrument used to check the specific gravity of the electrolyte to determine the battery's state of charge.

Hidrómetro: Instrumento de prueba utilizado para verificar la gravedad específica del electrolito y así determinar el estado de carga de la batería.

Hydrostatic lock: Liquid entering the cylinder and preventing the piston from moving upward.

Cierre hidrostático: La entrada de líquido en el cilindro que impide el movimiento ascendente del pistón.

Igniter: A combustible device that converts electric energy into thermal energy to ignite the inflator propellant in an air bag system.

Ignitor: Un dispositivo combustible que convierte la energía eléctrica a la energía termal para encender el propelente inflador en un sistema Airbag.

Impedance: The combined opposition to current created by the resistance, capacitance, and inductance of a test meter or circuit.

Impedancia: Oposición combinada a la corriente generada por la resistencia, la capacitancia y la inductancia de un instrumento de prueba o de un circuito.

Incandescence: The process of changing energy forms to produce light.

Incandescencia: Proceso a través del cual se cambian las formas de energía para producir luz.

Induced voltage: Voltage that is produced in a conductor as a result of relative motion within magnetic flux lines.

Tensión inducida: Tensión producida en un conductor como resultado del movimiento relativo dentro de líneas de flujo magnético.

Induction: The magnetic process of producing a current flow in a wire without any actual contact to the wire. To induce 1 volt, 100 million magnetic lines of force must be cut per second.

Inducción: Proceso magnético a través del cual se produce un flujo de corriente en un alambre sin contacto real alguno con el alambre. Para inducir 1 voltio, deben producirse 100 millones de líneas de fuerza magnética por segundo.

Inductive reactance: The result of current flowing through a conductor and the resultant magnetic field around the conductor that opposes the normal flow of current.

Reactancia inductiva: El resultado de un corriente que circule por un conductor y que resulta en un campo magnético alrededor del conductor que opone el flujo normal del corriente.

Inertia engagement: A type of starter motor that uses rotating inertia to engage the drive pinion with the engine flywheel.

Conexión por inercia Tipo de motor de arranque que utiliza inercia giratoria para engranar el piñón de mando con el volante de la máquina.

Instrument voltage regulator (IVR): Provides a constant voltage to the gauge regardless of the voltage output of the charging system.

Instrumento regulador de tensión. Le provee tensión constante al calibrador, sin importar cual sea la salida de tensión del sistema de carga.

Insulated circuit resistance test: A voltage drop test that is used to locate high resistance in the starter circuit.

Prueba de la resistencia de un circuito aislado: Prueba de la caída de tensión utilizada para localizar alta resistencia en el circuito de arranque.

Insulated side: The portion of the circuit from the positive side of the source to the load component.

Lado aislado: Parte del circuito que va del lado positivo de la fuente al componente de carga.

Insulator: A material that does not allow electrons to flow easily through it.

Aislador: Una material que no permite circular fácilmente los electrones.

Integrated circuit (IC chip): A complex circuit of thousands of transistors, diodes, resistors, capacitors, and other electronic devices that are formed onto a small silicon chip. As many as 30,000 transistors can be placed on a chip that is 1/4 inch (6.35 mm) square.

Circuito integrado (Fragmento CI): Circuito complejo de miles de transistores, diodos, resistores, condensadores, y otros dispositivos electrónicos formados en un fragmento pequeño de silicio. En un fragmento de 1/4 de pulgada (6,35 mm) cuadrada, pueden colocarse hasta 30.000 transistores.

Interface: Used to protect the computer from excessive voltage levels and to translate input and output signals.

Interfase: Utilizada para proteger la computadora de niveles excesivos de tensión y traducir señales de entrada y salida.

Intermediate section: The part of an ignition trace on a scope that displays the coil's action after the firing of the plug.

Sección intermedia: La parte de un razgo de encendido en un osiloscopio que manifiesta la acción de la bobina después de encenderse la bujía.

Ion: An atom or group of atoms that has an electrical charge.

Ion: Átomo o grupo de átomos que poseen una carga eléctrica.

ISO: An abbreviation for International Standards Organizations.

ISO: Una abreviación de las Organizaciones de Normas Internacionales.

Jack stands: Support devices used to hold the vehicle off the floor after it has been raised by the floor jack.

Soportes de gato. Dispositivos de soporte utilizados para sostener el vehículo sobre el suelo después de haber sido levantado con el gato de pie.

Jumper wire: A wire used in diagnostics that is made up of a length of wire with a fuse or circuit breaker and has alligator clips on both ends.

Cable conector: Una alambre empleado en los diagnósticos que se comprende de un trozo de alambre con un fusible o un interruptor y que tiene una pinza de conexión en ambos lados.

Keyless entry: A lock system that allows for locking and unlocking of a vehicle with a touch keypad instead of a key.

Entrada sin llave: Un sistema de cerradura que permite cerrar y abrir un vehículo por medio de un teclado en vez de utilizar una llave.

KV: Kilovolt or 1000 volts.

KV: Kilovolito o 1000 voltios.

Lamination: The process of constructing something with layers of materials that are firmly connected.

Laminación: El proceso de construir algo de capas de materiales unidas con mucha fuerza.

Lamp: A device that produces light as a result of current flow through a filament. The filament is enclosed within a glass envelope and is a type of resistance wire that is generally made from tungsten.

Lámpara: Dispositivo que produce luz como resultado del flujo de corriente a través de un filamento. El filamento es un tipo de alambre de resistencia hecho por lo general de tungsteno, que es encerrado dentro de una bombilla.

Lamp outage module: A current-measuring sensor that contains a set of resistors, wired in series with the power supply to the headlights, taillights, and stop lights. If the sensor indicates that a lamp is burned out, the module will alert the driver.

Unidad de avería de la lámpara: Sensor para medir corriente que incluye un juego de resistores, alambrado en serie con la fuente de alimentación a los faros delanteros, traseros y a las luces de freno. Si el sensor indica que se ha apagado una lámpara, la unidad le avisará al conductor.

Light-emitting diode (LED): A gallium-arsenide diode that converts the energy developed when holes and electrons collide during normal diode operation into light.

Diodo emisor de luz: Diodo semiconductor de galio y arseniuro que convierte en luz la energía producida por la colisión de agujeros y electrones durante el funcionamiento normal del diodo.

Limit switch: A switch used to open a circuit when a predetermined value is reached. Limit switches are normally responsive to a mechanical movement or temperature changes.

Disyuntor de seguridad: Un conmutador que se emplea para abrir un circuito al alcanzar un valor predeterminado. Los disyuntores de seguridad suelen ser responsivos a un movimiento mecánico o a los cambios de temperatura.

Linearity: Refers to the sensor signal being as constantly proportional to the measured value as possible. It is an expression of the sensor's accuracy.

Linealidad: Significa que la variación del valor de una magnitud es lo más proporcional posible a la variación del valor de otra magnitud. Expresa la precisión del sensor.

Liquid crystal display (LCD): A display that sandwiches electrodes and polarized fluid between layers of glass. When voltage is applied to the electrodes, the light slots of the fluid are rearranged to allow light to pass through.

Visualizador de cristal líquido: Visualizador digital que consta de dos láminas de vidrio selladas, entre las cuales se encuentran los electrodos y el fluido polarizado. Cuando se aplica tensión a los electrodos, se rompe la disposición de las moléculas para permitir la formación de carácteres visibles.

Logic gates: Electronic circuits that act as gates to output voltage signals depending on different combinations of input signals.

Compuertas lógicas: Circuitos electrónicos que gobiernan señales de tensión de salida, dependiendo de las diferentes combinaciones de señales de entrada.

Logic probe: A test instrument used to detect a pulsing signal.

Sonda lógica: Un instrumento de prueba que se emplea para detectar una señal pulsante.

Magnetic field: The area surrounding a magnet where energy is exerted due to the atoms aligning in the material.

Campo magnético: Espacio que rodea un imán donde se emplea la energía debido a la alineación de los átomos en el material.

Magnetic flux density: The concentration of the magnetic lines of force.

Densidad de flujo magnético: Número de líneas de fuerza magnética.

Magnetic pulse generator: Sensor that uses the principle of magnetic induction to produce a voltage signal. Magnetic pulse generators are commonly used to send data concerning the speed of the monitored component to the computer.

Generador de impulsos magnéticos: Sensor que funciona según el principio de inducción magnética para producir una señal de tensión. Los generadores de impulsos magnéticos se utilizan comúnmente para transmitir datos a la computadora relacionados a la velocidad del componente regulado.

Magnetism: An energy form resulting from atoms aligning within certain materials, giving the materials the ability to attract other metals.

Magnetismo Forma de energía que resulta de la alineación de átomos dentro de ciertos materiales y que le da a éstos la capacidad de atraer otros metales.

Material expanders: Fillers that can be used in place of the active materials in a battery. They are used to keep the cost of manufacturing low.

Expansores de materias: Los rellenos que se pueden usar en vez de las materiales activas de una bateria. Se emplean para mantener bajos los costos de la fabricación.

Matrix: A rectangular array of grids.

Matriz: Red lógica en una rejilla de forma rectangular.

Maxi-fuse: A circuit protection device that looks similar to blade-type fuses except they are larger and have a higher amperage capacity. Maxi-fuses are used because they are less likely to cause an underhood fire when there is an overload in the circuit. If the fusible link burns in two, it is possible that the "hot" side of the fuse could come into contact with the vehicle frame and the wire could catch on fire.

Maxifusible: Dispositivo de protección del circuito parecido a un fusible de tipo de cuchilla, pero más grande y con mayor capacidad de amperaje. Se utilizan maxifusibles porque existen menos probabilidades de que ocasionen un incendio debajo de la capota cuando ocurra una sobrecarga en el circuito. Si el cartucho de fusible se quemase en dos partes, es posible que el lado "cargado" del fusible entre en contacto con el armazón del vehículo y que el alambre se encienda.

Memory seats: Power seats that can be programmed to return or adjust to a point designated by the driver.

Asientos con memoria: Los asientos automáticos que se pueden programar a regresar o ajustarse a un punto indicado por el conductor.

Metri-pack connector: Special wire connectors used in some computer circuits. They seal the wire terminals from the atmosphere, thereby preventing corrosion and other damage.

Conector metri-pack: Los conectores de alambres especiales que se emplean en algunos circuitos de computadoras. Impermealizan los bornes de los alambres, así previniendo la corrosión y otros daños.

Molded connector: An electrical connector that usually has one to four wires that are molded into a one-piece component.

Conectador moldeado: Conectador eléctrico que por lo general tiene hasta un máximo cuatro alambres que se moldean en un componente de una sola pieza.

Momentary contact: A switch type that operates only when held in position.

Contacto momentáneo: Tipo de conmutador que funciona solamente cuando se mantiene en su posición.

MSDS: Material Safety Data Sheet.

MSDS Hojas de Dato de Seguridad de los Materiales.

Multimeter: A test instrument that measures more than one electrical property.

Multímetro: Un instrumento diagnóstico que mide más de una propiedad eléctrica.

Multiplexing: A means of transmitting information between computers. It is a system in which electrical signals are transmitted by a peripheral serial bus instead of conventional wires, allowing several devices to share signals on a common conductor.

Multiplexaje: Medio de transmitir información entre computadoras. Es un sistema en el cual las señales eléctricas son transmitidas por una colectora periférica en serie en vez de por líneas convencionales. Esto permite que varios dispositivos compartan señales en un conductor común.

Mutual induction: An induction of voltage in an adjacent coil by changing current in a primary coil.

Inducción mutua: Una inducción de la tensión en una bobina adyacente que se efectúa al cambiar la tensión en una bobina primaria.

N-type material: When there are free electrons, the material is called an N-type material. The N means negative and indicates that it is the negative side of the circuit that pushes electrons through the semiconductor and the positive side that attracts the free electrons.

Material tipo N: Al material se le llama material tipo N cuando hay electrones libres. La N significa negativo e indica que el lado negativo del circuito empuja los electrones a través del semiconductor y el lado positivo atrae los electrones libres.

Negative logic: Defines the most negative voltage as a logical 1 in the binary code.

Lógica negativa: Define la tensión más negativa como un 1 lógico en el código binario.

Negative temperature coefficient (NTC) thermistors: Thermistors that reduce their resistance as the temperature increases.

Termistores con coeficiente negativo de temperatura: Termistores que disminuyen su resistencia según aumenta la temperatura.

Neon lights: A light that contains a colorless, odorless inert gas called neon. These lamps are discharge lamps.

Luces de neón: Una luz que contiene un gas inerto sin color, inodoro llamado neón. Estas lámparas son lámparas de descarga.

Neutral atom: See balanced atom.

Átomo neutro: Véase átomo equilibrado.

Neutral junction: The center connection to which the common ends of a Y-type stator winding are connected.

Empalme neutro: Conexión central a la cual se conectan los extremos comunes de un devanado del estátor de tipo Y.

Neutral safety switch: A switch used to prevent the starting of an engine unless the transmission is in PARK or Neutral.

Disyuntor de seguridad en neutral: Un conmutador que se emplea para prevenir que arranque un motor al menos de que la transmisión esté en posición PARK o Neutral.

Neutrons: Particles of an atom that have no charge.

Neutrones: Partículas de un átomo desprovistas de carga.

No-crank: A term used to mean that when the ignition switch is placed in the START position, the starter does not turn the engine.

Sin arranque: Término utilizado para expresar que cuando el botón conmutador de encendido está en la posición START, el arrancador no enciende el motor.

No-crank test: Diagnostic test performed to locate any opens in the starter or control circuits.

Prueba sin arranque: Prueba diagnóstica llevada a cabo para localizar aberturas en los circuitos de arranque o de mando.

Nonvolatile RAM: RAM memory that will retain its memory if battery voltage is disconnected. NVRAM is a combination of RAM and EEPROM into the same chip. During normal operation, data is written to and read from the RAM portion of the chip. If the power is removed from the chip, or at programmed timed intervals, the data is transferred from RAM to the EEPROM portion of the chip. When the power is restored to the chip, the EEPROM will write the data back to the RAM.

Memoria de acceso aleatorio no volátil [NV RAM] Memoria de acceso aleatorio (RAM) que retiene su memoria si se desconecta la carga de la batería. La NV RAM es una combinación de RAM y EEPROM en el mismo fragmento. Durante el funcionamiento normal, los datos se escriben en y se leen de la parte RAM del fragmento. Si se remueve la alimentación del fragmento, o si se remueve ésta a intervalos programados, se transfieren los datos de la RAM a la parte del EEPROM del fragmento. Cuando se restaura la alimentación en el fragmento, el EEPROM volverá a escribir los datos en la RAM.

Normally closed (NC) switch: A switch designation denoting that the contacts are closed until acted upon by an outside force.

Conmutador normalmente cerrado: Nombre aplicado a un conmutador cuyos contactos permanecerán cerrados hasta que sean accionados por una fuerza exterior.

Normally open (NO) switch: A switch designation denoting that the contacts are open until acted upon by an outside force.

Conmutador normalmente abierto: Nombre aplicado a un conmutador cuyos contactos permanecerán abiertos hasta que sean accionados por una fuerza exterior.

Nucleus: The core of an atom that contains the protons and neutrons.

Núcleo: Parte central de un átomo que contiene los protones y los neutrones.

Occupational safety glasses: Eye protection that is designed with special high-impact lens and frames, and provides for side protection.

Gafas de protección para el trabajo: Gafas diseñadas con cristales y monturas especiales resistentes y provistas de protección lateral.

Odometer: A mechanical counter in the speedometer unit that indicates total miles accumulated on the vehicle.

Odómetro: Aparato mecánico en la unidad del velocímetro con el que se cuentan las millas totales recorridas por el vehículo.

Offset Placed off center. Refers to the number of degrees a timing light or meter should be set to provide accurate ignition timing readings.

Desviación Ubicado fuera de lo central. Se refiere al número de grados que se debe ajustar una luz de temporización o un medidor para proveer las lecturas exactas del tiempo de encendido.

Ohm: Unit of measure for resistance. One ohm is the resistance of a conductor such that a constant current of one ampere in it produces a voltage of one volt between its ends.

Ohmio: Unidad de resistencia eléctrica. Un ohmio es la resistencia de un conductor si una corriente constante de 1 amperio en el conductor produce una tensión de 1 voltio entre los dos extremos.

Ohmmeter: A test meter used to measure resistance and continuity in a circuit.

Ohmiómetro: Instrumento de prueba utilizado para medir la resistencia y la continuidad en un circuito.

Ohm's law: Defines the relationship between current, voltage, and resistance.

Ley de Ohm: Define la relación entre la corriente, la tensión y la resistencia.

Open circuit: A term used to indicate that current flow is stopped. By opening the circuit, the path for electron flow is broken.

Circuito abierto: Término utilizado para indicar que el flujo de corriente ha sido detenido. Al abrirse el circuito, se interrumpe la trayectoria para el flujo de electrones.

Open circuit voltage test: Used to determine the battery's state of charge. It is used when a hydrometer is not available or cannot be used.

Prueba de la tensión en un circuito abierto: Sirve para determinar el estado de carga de la batería. Esta prueba se lleva a cabo cuando no se dispone de un hidrómetro o cuando el mismo no puede utilizarse.

Optical horn: A name Chrysler uses to describe their "flash-to-pass" headlamp system.

Claxón óptico: Un nombre que usa Chrysler para describir su sistema de faros "relampaguea para rebasar."

Oscillate: Fast back and forth movement.

Oscilar: Moverse rápidamente de atrás para adelante.

Overload: Excess current flow in a circuit.

Sobrecarga: Flujo de corriente superior a la que tiene asignada un circuito.

Overrunning clutch: A clutch assembly on a starter drive used to prevent the engine's flywheel from turning the armature of the starter motor.

Embrague de sobremarcha: Una asamblea de embrague en un acoplamiento del motor de arranque que se emplea para prevenir que el volante del motor dé vueltas al armazón del motor de arranque.

Oxygen sensor: A voltage generating sensor that measures the amount of oxygen present in an engine's exhaust.

Sensor de oxígeno: Un sensor generador de tensión que mide la cantidad del oxígeno presente en el gas de escape de un motor.

P-material: Silicon or germanium that is doped with boron or gallium to create a shortage of electrons.

Material-P: Boro o galio añadidos al silicio o al germanio para crear una insuficiencia de electrones.

Parallel circuit: A circuit that provides two or more paths for electricity to flow.

Circuito en paralelo: Circuito que provee dos o más trayectorias para que circule la electricidad.

Parasitic loads: Electrical loads that are still present when the ignition switch is in the OFF position.

Cargas parásitas: Cargas eléctricas que todavía se encuentran presentes cuando el botón conmutador de encendido está en la posición OFF.

Park switch: Contact points located inside the wiper motor assembly that supply current to the motor after the wiper control switch has been turned to the PARK position. This allows the motor to continue operating until the wipers have reached their PARK position.

Conmutador PARK: Puntos de contacto ubicados dentro del conjunto del motor del frotador que le suministran corriente al motor después de que el conmutador para el control de los frotadores haya sido colocado en la posición PARK. Esto permite que el motor continue su funcionamiento hasta que los frotadores hayan alcanzado la posición original.

Pass key: A specially designed vehicle key with a coded resistance value. The term pass is derived from Personal Automotive Security System.

Llave maestra: Una llave vehícular de diseño especial que tiene un valor de resistencia codificado. El termino pass se derive de las palabras Personal Automotive Security System (sistema personal de seguridad automotriz).

Passive restraints: A passenger restraint system that automatically operates to confine the movement of a vehicle's passengers.

Correas passivas: Un sistema de resguardo del pasajero que opera automaticamente para limitar el movimiento de los pasajeros en el vehículo.

Passive seatbelt system: Seatbelt operation that automatically puts the shoulder and/or lap belt around the driver or occupant. The automatic seatbelt is moved by dc motors that move the belts by means of carriers on tracks.

Sistema pasivo de cinturones de seguridad: Función de los cinturones de seguridad que automáticamente coloca el cinturón superior y/o inferior sobre el conductor o pasajero. Motores de corriente continua accionan los cinturones automáticos mediante el uso de portadores en pistas.

Passive suspension systems: Use fixed spring rates and shock valving.

Sistemas pasivos de suspensión: Utilizan elasticidad de muelle constante y dotación con válvulas amortigadoras.

Permeability: Term used to indicate the magnetic conductivity of a substance compared with the conductivity of air. The greater the permeability, the greater the magnetic conductivity and the easier a substance can be magnetized.

Permeabilidad: Término utilizado para indicar la aptitud de una sustancia en relación con la del aire, de dar paso a las líneas de fuerza magnética. Mientras mayor sea la permeabilidad, mayor será la conductividad magnética y más fácilmente se comunicará a un cuerpo propiedades magnéticas.

Photocell: A variable resistor that uses light to change resistance.

Fotocélula: Resistor variable que utiliza luz para cambiar la resistencia.

Phototransistor: A transistor that is sensitive to light.

Fototransistor: Transistor sensible a la luz.

Photovoltaic diodes: Diodes capable of producing a voltage when exposed to radiant energy.

Diodos fotovoltaicos: Diodos capaces de generar una tensión cuando se encuentran expuestos a la energía de radiación.

Pick-up coil: The stationary component of the magnetic pulse generator consisting of a weak permanent magnet that is wound around by fine wire. As the timing disc rotates in front of it, the changes of magnetic lines of force generate a small voltage signal in the coil.

Bobina captadora: Componente fijo del generador de impulsos magnéticos compuesta de un imán permanente débil devanado con alambre fino. Mientras gira el disco sincronizador enfrente de él, los cambios de las líneas de fuerza magnética generan una pequeña señal de tensión en la bobina.

Piezoelectricity: Voltage produced by the application of pressure to certain crystals.

Piezoelectricidad: Generación de polarización eléctrica en ciertos cristales a consecuencia de la aplicación de tensiones mecánicas.

Piezoresistive sensor: A sensor that is sensitive to pressure changes.

Sensor piezoresistivo: Sensor susceptible a los cambios de presión.

Pinion gear: A small gear; typically refers to the drive gear of a starter drive assembly or the small drive gear in a differential assembly.

Engranaje de piñón: Un engranaje pequeño; tipicamente se refiere al engranaje de arranque de una asamblea de motor de arranque o al engranaje de mando pequeño de la asamblea del diferencial.

Plate straps: Metal connectors used to connect the positive or negative plates in a battery.

Abrazaderas de la placa: Los conectores metálicos que sirven para conectar las placas positivas o negativas de una batería.

Plates: The basic structure of a battery cell; each cell has at least one positive plate and one negative plate.

Placas: La estructura básica de una celula de batería; cada celula tiene al menos una placa positiva y una placa negativa.

PMGR: An abbreviation for permanent magnet gear reduction.

PMGR: Una abreviación de desmultiplicación del engrenaje del imán permanente.

PN junction: The point at which two opposite kinds of semiconductor materials are joined together.

Unión pn: Zona de unión en la que se conectan dos tipos opuestos de materiales semiconductores.

Pneumatic tools: Power tools that are powered by compressed air.

Herrimientas neumáticas: Herramientas mecánicas accionadas por aire comprimido.

Polarizers: Glass sheets that make light waves vibrate in only one direction. This converts light into polarized light.

Polarizadores: Las láminas de vidrio que hacen vibrar las ondas de luz en un sólo sentido. Esto convierte la luz en luz polarizada.

Polarizing: The process of light polarization or of setting one end of a field as a positive or negative point.

Polarizadora: El proceso de polarización de la luz o de establecer un lado de un campo como un punto positivo o negativo.

Pole shoes: The components of an electric motor that are made of high-magnetic permeability material to help concentrate and direct the lines of force in the field assembly.

Expansión polar: Componentes de un motor eléctrico hechos de material magnético de gran permeabilidad para ayudar a concentrar y dirigir las líneas de fuerza en el conjunto inductor.

Positive engagement starter: A type of starter that uses the magnetic field strength of a field winding to engage the starter drive into the flywheel.

Acoplamiento de arranque positivo: Un tipo de arrancador que utilisa la fuerza del campo magnético del devanado inductor para accionar el acoplamiento del arrancador en el volante.

Positive temperature coefficient (PTC) thermistors: Thermistors that increase their resistance as the temperature increases.

Termistores con coeficiente positivo de temperatura: Termistores que aumentan su resistencia según aumenta la temperatura.

Potential: The ability to do something; typically voltage is referred to as the potential. If you have voltage, you have the potential for electricity.

Potencial: La capacidad de efectuar el trabajo; típicamente se refiere a la tensión como el potencial. Si tiene tensión, tiene la potencial para la electricidad.

Potentiometer: A variable resistor that acts as a circuit divider, providing accurate voltage drop readings proportional to movement.

Potenciómetro: Resistor variable que actúa como un divisor de circuito para obtener lecturas de perdidas de tensión precisas en proporción con el movimiento.

Power formula: A formula used to calculate the amount of electrical power a component uses. The formula is P = I x E, whereas P stands for power (measured in watts), I stands for current, and E stands for voltage.

Formula de potencia: Una formula que se emplea para calcular la cantidad de potencia eléctrica utilizada por un componente. La formula es P = I x E, en el que el P quiere decir potencia (medida en wats), I representa el corriente y el E representa la tensión.

Power tools: Tools that use forces other than those generated from the body. They can use compressed air, electricity, or hydraulic pressure to generate and multiply force.

Herramientas mecánicas: Herramientas que utilizan fuerzas distintas a las generadas por el cuerpo. Dichas fuerzas pueden ser el aire comprimido, la electricidad, o la presión hidráulica para generar y multiplicar la fuerza.

Pressure control solenoid: A solenoid used to control the pressure of a fluid, commonly found in electronically controlled transmissions.

Solenoide de control de la presión: Un solenoide que controla la presión de un fluido, suele encontrarse en las transmisiones controladas electronicamente.

Primary wiring: Conductors that carry low voltage and low current. The insulation of primary wires is usually thin.

Hilos primarios: Hilos conductores de tensión y corriente bajas. El aislamiento de hilos primarios es normalmente delgado.

Printed circuit: Made of thin phenolic or fiberglass board with copper deposited on it to create current paths. These are used to simplify the wiring of circuits.

Circuito impreso: Un circuito hecho de un tablero de fenólico delgado o de fibra de vidrio el cual tiene depósitos del cobre para crear los trayectorios para el corriente. Estos se emplean para simplificar el cableado de los circuitos.

Prism lens: A light lens designed with crystal-like patterns, which distort, slant, direct, or color the light that passes through it.

Lente prismático: Un lente de luz con diseños cristalinos que distorcionan, inclinan, dirigen o coloran la luz que lo atraviesa.

Program: A set of instructions that the computer must follow to achieve desired results.

Programa: Conjunto de instrucciones que la computadora debe seguir para lograr los resultados deseados.

PROM (programmable read only memory): Memory chip that contains specific data that pertains to the exact vehicle that the computer is installed in. This information may be used to inform the CPU of the accessories that are equipped on the vehicle.

PROM (memoria de sólo lectura programable). Fragmento de memoria que contiene datos específicos referentes al vehículo particular en el que se instala la computadora. Esta información puede utilizarse para informar a la UCP sobre los accesorios de los cuales el vehículo está dotado.

Protection device: Circuit protector that is designed to "turn off" the system that it protects. This is done by creating an open to prevent a complete circuit.

Dispositivo de protección: Protector de circuito diseñado para "desconectar" el sistema al que provee protección. Esto se hace abriendo el circuito para impedir un circuito completo.

Proton: Positively charged particles contained in the nucleus of an atom.

Protón: Partículas con carga positiva que se encuentran en el núcleo de todo átomo.

Prove-out circuit: A function of the ignition switch that completes the warning light circuit to ground through the ignition switch when it is in the START position. The warning light is on during engine cranking to indicate to the driver that the bulb is working properly.

Circuito de prueba: Función del boton conmutador de encendido que completa el circuito de la luz de aviso para que se ponga a tierra a través del botón conmutador de encendido cuando éste se encuentra en la posición START. La luz de aviso se encenderá durante el arranque del motor para avisarle al conductor que la bombilla funciona correctamente.

Pulse width: The length of time in milliseconds that an actuator is energized.

Duración de impulsos: Espacio de tiempo en milisegundos en el que se excita un accionador.

Pulse width modulation: On/off cycling of a component. The period of time for each cycle does not change, only the amount of on time in each cycle changes.

Modulación de duración de impulsos: Modulación de impulsos de un componente. El espacio de tiempo de cada ciclo no varía; lo que varía es la cantidad de trabajo efectivo de cada ciclo.

Radial grid: A type of battery grid that has its patterns branching out from a common center.

Rejilla radial: Un tipo de rejilla de bateria cuyos diseños extienden de un centro común.

Radio choke: Absorbs voltage spikes and prevents static in the vehicle's radio.

Impedancia del radio: Absorba los impulsos de la tensión y previene la presencia del estático en el radio del vehículo.

Radiofrequency interference (RFI): Radio and television interference caused by electromagnetic energy.

Interferencia de frecuencia radioeléctrica: Interferencia en la radio y en la televisión producida por energía electromagnética.

RAM (random access memory): Stores temporary information that can be read from or written to by the CPU. RAM can be designed as volatile or nonvolatile.

RAM (memoria de acceso aleatorio): Almacena datos temporales que la UCP puede leer o escribir. La RAM puede ser volátil o no volátil.

Ratio: A mathematical relationship between two or more things.

Razón: Una relación matemática entre dos cosas o más.

Recombination battery: A type of battery that is sometimes called a dry-cell battery because it does not use a liquid electrolyte solution.

Batería de recombinación: Un tipo de batería que a veces se llama una pila seca porque no requiere una solución líquida de electrolita.

Rectification: The converting of ac current to dc current.

Rectificación: Proceso a través del cual la corriente alterna es transformada en una corriente continua.

Reflectors: A device whose surface reflects or radiates light.

Reflectores: Un dispositivo cuyo superficie refleja o irradia la luz.

Relay: A device that uses low current to control a high current circuit. Low current is used to energize the electromagnetic coil, while high current is able to pass over the relay contacts.

Relé: Dispositivo que utiliza corriente baja para controlar un circuito de corriente alta. La corriente baja se utiliza para excitar la bobina electromagnética, mientras que la corriente alta puede transmitirse a través de los contactos del relé.

Reluctance: A term used to indicate a material's resistance to the passage of flux lines.

Reluctancia: Término utilizado para señalar la resistencia ofrecida por un circuito al paso del flujo magnético.

Reserve-capacity rating: An indicator, in minutes, of how long the vehicle can be driven with the headlights on, if the charging system should fail. The reserve-capacity rating is determined by the length of time, in minutes, that a fully charged battery can be discharged at 25 amperes before battery cell voltage drops below 1.75 volts per cell.

Clasificación de capacidad en reserva: Indicación, en minutos, de cuánto tiempo un vehículo puede continuar siendo conducido, con los faros delanteros encendidos, en caso de que ocurriese una falla en el sistema de carga. La clasificación de capacidad en reserva se determina por el espacio de tiempo, en minutos, en el que una batería completamente cargada puede descargarse a 25 amperios antes de que la tensión del acumulador de la batería disminuya a un nivel inferior de 1,75 amperios por acumulador.

Resistance: Opposition to current flow.

Resistencia: Oposición que presenta un conductor al paso de la corriente eléctrica.

Resistance wire: A special type of wire that has some resistance built into it. These typically are rated by ohms per foot.

Alambre de resistencia: Un tipo de alambre especial que por diseño tiene algo de resistencia. Estos tipicamente tienen un valor nominal de ohm por pie.

Resistive shorts: Shorts to ground that pass through a form of resistance first.

Cortocircuitos resistivos: Cortocircuitos a tierra que primero pasan por una forma de resistencia.

Resistor block: A series of resistors with different values.

Bloque resistor: Serie de resistores que tienen valores diferentes.

Reversed-bias: A positive voltage is applied to the N-type material and negative voltage is applied to the P-type material of a semiconductor.

Polarización inversa: Tensión positiva aplicada al material N y tensión negativa aplicada al material P de un semiconductor.

RFI: Common acronym for radio frequency interference.

RFI: Una sigla común de la interferencia de radiofrecuencia.

Rheostat: A two-terminal variable resistor used to regulate the strength of an electrical current.

Reóstato: Resistor variable de dos bornes utilizado para regular la resistencia de una corriente eléctrica.

RMS Root-mean-square; a method for measuring AC voltage.

RMS Raíz de la media de los cuadrados; un método para medir la tensión del corriente alterna.

ROM (read only memory): Memory chip that stores permanent information. This information is used to instruct the computer on what to do in response to input data. The CPU reads the information contained in ROM, but it cannot write to it or change it.

ROM (memoria de sólo lectura): Fragmento de memoria que almacena datos en forma permanente. Dichos datos se utilizan para darle instrucciones a la computadora sobre cómo dirigir la ejecución de una operación de entrada. La UCP lee los datos que contiene la ROM, pero no puede escribir en ella o puede cambiarla.

Rotor: The component of the ac generator that is rotated by the drive belt and creates the rotating magnetic field of the ac generator.

Rotor: Parte rotativa del generador de corriente alterna accionada por la correa de transmisión y que produce el campo magnético rotativo del generador de corriente alterna.

Safety goggles: Eye protection device that fits against the face and forehead to seal off the eyes from outside elements.

Gafas de seguridad: Dispositivo protector que se coloca delante de los ojos para preservarlos de elementos extraños.

Safety stands: See Jack stands.

Soportes de seguridad: Véase soportes de gato.

Saturation: 1. The point at which the magnetic strength eventually levels off, and where an additional increase of the magnetizing force current no longer increases the magnetic field strength. 2. The point where forward-bias voltage to the base leg is at a maximum. With bias voltage at the high limits, output current is also at its maximum.

Saturación: 1. Máxima potencia posible de un campo magnético, donde un aumento adicional de la corriente de fuerza magnética no logra aumentar la potencia del campo magnético. 2. La tensión de polarización directa a la base está en su máximo. Ya que polarización directa ha alcanzado su límite máximo, la corriente de salida también alcanza éste.

Scanner: A diagnostic test tool that is designed to communicate with the vehicle's on-board computer.

Dispositivo de exploración: Herramienta de prueba diagnóstica diseñada para comunicarse con la computadora instalada en el vehículo.

Schmitt trigger: An electronic circuit used to convert analog signals to digital signals or vice versa.

Disparador de Schmitt: Un circuito electrónico que se emplea para convertir las señales análogas en señales digitales o vice versa.

Sealed-beam headlight: A self-contained glass unit that consists of a filament, an inner reflector, and an outer glass lens.

Faro delantero sellado: Unidad de vidrio que contiene un filamento, un reflector interior y una lente exterior de vidrio.

Secondary wiring: Conductors, such as battery cables and ignition spark plug wires, that are used to carry high voltage or high current. Secondary wires have extra thick insulation.

Hilos secundarios: Conductores, tales como cables de batería e hilos de bujías del encendido, utilizados para transmitir tensión o corriente alta. Los hilos secundarios poseen un aislamiento sumamente grueso.

Semiconductor: An element that is neither a conductor nor an insulator. Semiconductors are materials that conduct electric current under certain conditions, yet will not conduct under other conditions.

Semiconductores: Elemento que no es ni conductor ni aislante. Los semiconductores son materiales que transmiten corriente eléctrica bajo ciertas circunstancias, pero no la transmiten bajo otras.

Sender unit: The sensor for the gauge. It is a variable resistor that changes resistance values with changing monitored conditions.

Unidad emisora: Sensor para el calibrador. Es un resistor variable que cambia los valores de resistencia según cambian las condiciones reguladas.

Sensitivity controls: A potentiometer that allows the driver to adjust the sensitivity of the automatic dimmer system to surrounding ambient light conditions.

Controles de sensibilidad: Un potenciómetro que permite que el conductor ajusta la sensibilidad del sistema de intensidad de iluminación automático a las condiciones de luz ambientales.

Sensor: Any device that provides an input to the computer.

Sensor: Cualquier dispositivo que le transmite información a la computadora.

Separators: Normally constructed of glass with a resin coating. These battery plates offer low resistance to electrical flow but high resistance to chemical contamination.

Separadores: Normalmente se construyen del vidrio con una capa de resina. Estas placas de la batería ofrecen baja resistencia al flujo de la electricidad pero alta resistencia a la contaminación química.

Series circuit: A circuit that provides a single path for current flow from the electrical source through all the circuit's components, and back to the source.

Circuito en serie: Circuito que provee una trayectoria única para el flujo de corriente de la fuente eléctrica a través de todos los componentes del circuito, y de nuevo hacia la fuente.

Series-parallel circuit: A circuit that has some loads in series and some in parallel.

Circuito en series paralelas: Circuito que tiene unas cargas en serie y otras en paralelo.

Series-wound motor: A type of motor that has its field windings connected in series with the armature. This type of motor develops its maximum torque output at the time of initial start. Torque decreases as motor speed increases.

Motor con devanados en serie: Un tipo de motor cuyos devanados inductores se conectan en serie con la armadura. Este tipo de motor desarrolla la salida máxima de par de torsión en el momento inicial de ponerse en marcha. El par de torsión disminuye al aumentar la velocidad del motor.

Servomotor: An electrical motor that produces rotation of less than a full turn. A feedback mechanism is used to position itself to the exact degree of rotation required.

Servomotor: Motor eléctrico que genera rotación de menos de una revolución completa. Utiliza un mecanismo de realimentación para ubicarse al grado exacto de la rotación requerida.

Shell: The electron orbit around the nucleus of an atom.

Corteza: Órbita de electrones alrededor del núcleo del átomo.

Short: An unwanted electrical path; sometimes this path goes directly to ground.

Corto: Una trayectoria eléctrica no deseable; a veces este trayectoria viaja directamente a tierra.

Shunt circuits: The branches of the parallel circuit.

Circuitos en derivación: Las ramas del circuito en paralelo.

Shunt-wound motor: A type of motor whose field windings are wired in parallel to the armature. This type of motor does not decrease its torque as speed increases.

Motor con devanados en derivación: Un tipo de motor cuyos devanados inductores se cablean paralelos a la armadura. Este tipo de motor no disminuya su par de torsión al aumentar la velocidad.

Shutter wheel: A metal wheel consisting of a series of alternating windows and vanes. It creates a magnetic shunt that changes the strength of the magnetic field from the permanent magnet of the Hall-effect switch or magnetic pulse generator.

Rueda obturadora: Rueda metálica compuesta de una serie de ventanas y aspas alternas. Genera una derivación magnética que cambia la potencia del campo magnético, del imán permanente del conmutador de efecto Hall o del generador de impulsos magnéticos.

Sine wave: A waveform that shows voltage changing polarity.

Onda senoidal: Una forma de onda que muestra un cambio de polaridad en la tensión.

Single phase voltage: The sine wave voltage induced in one conductor of the stator during one revolution of the rotor.

Tensión monofásica: La tensión en forma de onda senoidal inducida en un conductor del estator durante una revolución del rotor.

Sinusoidal: A waveform that is a true sine wave.

Senoidal: Una forma de onda que es una onda senoidal verdadera.

Slow charging: Battery charging rate between 3 and 15 amps for a long period of time.

Carga lenta: Indice de carga de la batería de entre 3 y 15 amperios por un largo espacio de tiempo.

Slow cranking: A term used to mean that the starter drive engages the ring gear, but the engine turns too slowly to start.

Arranque lento: Término utilizado para expresar que el mecanismo de transmisión de arranque engrana la corona, pero que el motor se enciende de forma demasiado lenta para arrancar.

Soft codes: Codes are those that have occurred in the past, but were not present during the last BCM test of the circuit.

Códigos suaves: Códigos que han ocurrido en el pasado, pero que no estaban presentes durante la última prueba BCM del circuito.

Soldering: The process of using heat and solder (a mixture of lead and tin) to make a splice or connection.

Soldadura: Proceso a través del cual se utiliza calor y soldadura (una mezcla de plomo y de estaño) para hacer un empalme o una conexión.

Solderless connectors: Hollow metal tubes that are covered with insulating plastic. They can be butt connectors or terminal ends.

Conectadores sin soldadura: Tubos huecos de metal cubiertos de plástico aislante. Pueden ser extremos de conectadores o de bornes.

Solenoid: An electromagnetic device that uses movement of a plunger to exert a pulling or holding force.

Solenoide: Dispositivo electromagnético que utiliza el movimiento de un pulsador para ejercer una fuerza de arrastre o de retención.

Solenoid circuit resistance test: Diagnostic test used to determine the electrical condition of the solenoid and the control circuit of the starting system.

Prueba de la resistencia de un circuito solenoide: Prueba diagnóstica utilizada para determinar la condición eléctrica del solenoide y del circuito de mando del sistema de arranque.

Source: The portion of a field-effect transistor that supplies the current-carrying holes or electrons.

Fuente: Terminal de un transistor de efecto de campo que provee los agujeros o electrones portadores de corriente.

Spark duration: The length in time that the spark continues across the electrodes to burn the air/fuel mixture.

Duración de la chispa: La cantidad del tiempo que dura la chispa viajando entre los electrodos para consumir la mezcla de aire/combustible.

Spark line: The section of an ignition's waveform that displays the length of time and the amount of voltage needed to maintain a spark across the electrodes of a spark plug.

Linea de encendido: La sección de la forma de onda del encendido que muestra la duración del tiempo y la cantidad de la tensión requerido para mantener una chispa entre los electrodos de una bujía.

Specific gravity: The weight of a given volume of a liquid divided by the weight of an equal volume of water.

Gravedad específica: El peso de un volumen dado de líquido dividido por el peso de un volumen igual de agua.

Speedometer: An instrument panel gauge that indicates the speed of the vehicle.

Velocímetro: Calibrador en el panel de instrumentos que marca la velocidad del vehículo.

Splash fouling: A condition of a spark plug that is caused by melted carbon in the combustion chamber.

Engrase por salpicadura: Una condición de la bujía causado por el carbón derritido en la cámara de combustión.

Splice: The joining of single wire ends or the joining of two or more electrical conductors at a single point.

Empalme: La unión de los extremos de un alambre o la unión de dos o más conductores eléctricos en un solo punto.

Splice clip: A special connector used along with solder to assure a good connection. The splice clip is different from solderless connectors in that it does not have insulation.

Grapa para empalme: Conectador especial utilizado junto con la soldadura para garantizar una conexión perfecta. La grapa para empalme se diferencia de los conectadores sin soldadura porque no está provista de aislamiento.

Starter drive: The part of the starter motor that engages the armature to the engine flywheel ring gear.

Transmisión de arranque: Parte del motor de arranque que engrana la armadura a la corona del volante de la máquina.

State of charge: The condition of a battery's electrolyte and plate materials at any given time.

Estado de carga: Condición del electrolito y de los materiales de la placa de una batería en cualquier momento dado.

Static electricity: Electricity that is not in motion.

Electricidad estática: Electricidad que no está en movimiento.

Static neutral point: The point at which the fields of a motor are in balance.

Punto neutral estático: El punto en que los campos de un motor estan equilibrados.

Stator: The stationary coil of the ac generator in which current is produced.

Estátor: Bobina fija del generador de corriente alterna donde se genera corriente.

Stator neutral junction: The common junction of wye stator windings.

Unión de estátor neutral: La unión común de los devanados de un estátor Y.

Stepped resistor: A resistor that has two or more fixed resistor values.

Resistor de secciones escalonadas: Resistor que tiene dos o más valores de resistencia fija.

Stepper motor: An electrical motor that contains a permanent magnet armature with two or four field coils. Can be used to move the controlled device to whatever location is desired. By applying voltage pulses to selected coils of the motor, the armature will turn a specific number of degrees. When the same voltage pulses are applied to the opposite coils, the armature will rotate the same number of degrees in the opposite direction.

Motor paso a paso: Motor eléctrico que contiene una armadura magnética fija con dos o cuatro bobinas inductoras. Puede utilizarse para mover el dispositivo regulado a cualquier lugar deseado. Al aplicárseles impulsos de tensión a ciertas bobinas del motor, la armadura girará un número específico de grados. Cuando estos mismos impulsos de tensión se aplican a las bobinas opuestas, la armadura girará el mismo número de grados en la dirección opuesta.

Stranded wire: A conductor comprised of many small solid wires twisted together. This type conductor is used to allow the wire to flex without breaking.

Cable trenzado: Un conductor que comprende muchos cables sólidos pequeños trenzados. Este tipo de conductor se emplea para permitir que el cable se tuerza sin quebrar.

Sulfation: A chemical action within the battery that interferes with the ability of the cells to deliver current and accept a charge.

Sulfatado: Acción química dentro de la batería que interfiere con la capacidad de los acumuladores de transmitir corriente y recibir una carga.

Superimposed pattern: The type of scope display that takes the traces of individual ignition patterns and stacks them on top of each other.

Imagen sobrepuesta: Un tipo de visualizador del osciloscopio que toma las trayectorias de los patrones individuales de los encendidos y los amontona.

Tachometer: An instrument that measures the speed of the engine in revolutions per minute (rpm).

Tacómetro: Instrumento que mide la velocidad del motor en revoluciones por minuto (rpm).

Thermistor: A solid-state variable resistor made from a semiconductor material that changes resistance in relation to temperature changes.

Termistor: Resistor variable de estado sólido hecho de un material semiconductor que cambia su resistencia en relación con los cambios de temperatura.

Three-coil gauge: A gauge design that uses the interaction of three electromagnets and the total field effect upon a permanent magnet to cause needle movement.

Calibrador de tres bobinas: Calibrador diseñado para utilizar la interacción de tres electroimanes y el efecto inductor total sobre un imán permanente para producir el movimiento de la aguja.

Three-minute charge test: A reasonably accurate method for diagnosing a sulfated battery on conventional batteries.

Prueba de carga de tres minutos: Método bastante preciso en baterías convencionales para diagnosticar una batería sulfatada.

Throw: Term used in reference to electrical switches or relays referring to the number of output circuits from the switch.

Posición activa: Término utilizado para conmutadores o relés eléctricos en relación con el número de circuitos de salida del conmutador.

Thyristor: A semiconductor switching device composed of alternating N and P layers. It can also be used to rectify current from ac to dc.

Tiristor: Dispositivo de conmutación del semiconductor compuesto de capas alternas de N y P. Puede utilizarse también para rectificar la corriente de corriente alterna a corriente continua.

Timer control: A potentiometer that is part of the headlight switch in some systems. It controls the amount of time the headlights stay on after the ignition switch is turned off.

Control temporizador: Un potenciómetro que es parte del conmutador de los faros en algunos sistemas. Controla la cantidad del tiempo que quedan prendidos los faros después de apagarse la llave del encendido.

Torque converter: A hydraulic device found on automatic transmissions. It is responsible for controlling the power flow from the engine to the transmission; works like a clutch to engage and disengage the engine's power to the drive line.

Convertidor de par: Un dispositivo hidráulico en las transmisiones automáticas. Se encarga de controlar el flujo de la potencia del motor a la transmisión; funciona como un embrague para embragar y desembragar la potencia del motor con la flecha motríz.

Transducer: A device that changes energy from one form into another.

Transductor: Dispositivo que cambia la energía de una forma a otra.

Transistor: A three-layer semiconductor used as a very fast switching device.

Transistor: Semiconductor de tres capas utilizado como dispositivo de conmutación sumamente rápido.

Trouble codes: Output of the self-diagnostics program in the form of a numbered code that indicates faulty circuits or components. Trouble codes are two or three digital characters that are displayed in the diagnostic display if the testing and failure requirements are both met.

Códigos indicadores de fallas: Datos del programa autodiagnóstico en forma de código numerado que indica los circuitos o los componentes defectuosos. Dichos códigos se componen de dos o tres caracteres digitales que se muestran en el visualizador diagnóstico si se llenan los requisitos de prueba y de falla.

Troubleshooting: The diagnostic procedure of locating and identifying the cause of the fault. It is a step-by-step process of elimination by use of cause-and-effect.

Detección de fallas: Procedimiento diagnóstico a través del cual se localiza e identifica la falla. Es un proceso de eliminación que se lleva a cabo paso a paso por medio de causa y efecto.

TVRS: An abbreviation for television-radio-suppression cable.

TVRS: Una abreviación del cable de supresíon del televisión y radio.

Two-coil gauge: A gauge design that uses the interaction of two electromagnets and the total field effect upon an armature to cause needle movement.

Calibrador de dos bobinas: Calibrador diseñado para utilizar la interacción de dos electroimanes y el efecto inductor total sobre una armadura para generar el movimiento de la aguja.

Vacuum distribution valve: A valve used in vacuum-controlled concealed headlight systems. It controls the direction of vacuum to various vacuum motors or to vent.

Válvula de distribución al vacío: Válvula utilizada en el sistema de faros delanteros ocultos controlado al vacío. Regula la dirección del vacío a varios motores al vacío o sirve para dar salida del sistema.

Vacuum fluorescent display (VFD): A display type that uses anode segments coated with phosphor and bombarded with tungsten electrons to cause the segments to glow.

Visualización de fluorescencia al vacío: Tipo de visualización que utiliza segmentos ánodos cubiertos de fósforo y bombardeados de electrones de tungsteno para producir la luminiscencia de los segmentos.

Valence ring: The outermost orbit of the atom.

Anillo de valencia: Órbita más exterior del átomo.

Valve body: A unit that consists of many valves and hydraulic circuits. This unit is the central control point for gear shifting in an automatic transmission.

Cuerpo de la válvula: Una unedad que consiste de muchas válvulas y circuitos hidráulicos. Esta unedad es el punto central de mando para los cambios de velocidad en una transmisión automática.

Variable resistor: A resistor that provides for an infinite number of resistance values within a range.

Resistor variable: Resistor que provee un número infinito de valores de resistencia dentro de un margen.

Vehicle Identification Number (VIN): A number that is assigned to a vehicle for identification purposes. The identification plate is usually located on the cowl, next to the left upper instrument panel.

Número de identificación del vehículo: Número asignado a cada vehículo para fines de identificación. Por lo general, la placa de identificación se ubica en la bóveda, al lado del panel de instrumentos superior de la izquierda.

Vehicle lift points: The areas that the manufacturer recommends for safe vehicle lifting. They are the areas that are structurally strong enough to sustain the stress of lifting.

Puntos para elevar el vehículo: Áreas específicas que el fabricante recomienda para sujetar el vehículo a fin de lograr una elevación segura. Son las áreas del vehículo con una estructura suficientemente fuerte para sostener la presión de la elevación.

Volatile RAM: RAM memory that is erased when it is disconnected from its power source. Also known as Keep Alive Memory.

RAM volátil: Memoria RAM cuyos datos se perderán cuando se la desconecta de la fuente de alimentación. Conocida también como memoria de entretenimiento.

Volt: The unit used to measure the amount of electrical force.

Voltio: Unidad práctica de tensión para medir la cantidad de fuerza eléctrica.

Volatility: The tendency for a fluid to evaporate quickly or pass off in the form of a vapor.

Volatilidad: La tendencia de un fluido a evaporarse rápidamente o disiparse en forma de vapor.

Voltage: The difference or potential that indicates an excess of electrons at the end of the circuit the farthest from the electromotive force. It is the electrical pressure that causes electrons to move through a circuit. One volt is the amount of pressure required to move one amp of current through one ohm of resistance.

Tensión: Diferencia o potencial que indica un exceso de electrones al punto del circuito que se encuentra más alejado de la fuerza electromotriz. La presión eléctrica genera el movimiento de electrones a través de un circuito. Un voltio equivale a la cantidad de presión requerida para mover un amperio de corriente a través de un ohmio de resistencia.

Voltage drop: A resistance in the circuit that reduces the electrical pressure available after the resistance. The resistance can be either the load component, the conductors, any connections, or unwanted resistance.

Caída de tensión: Resistencia en el circuito que disminuye la presión eléctrica disponible después de la resistencia. La resistencia puede ser el componente de carga, los conductores, cualquier conexión o resistencia no deseada.

Voltage limiter: Connected through the resistor network of a voltage regulator. It determines whether the field will receive high, low, or no voltage. It controls the field voltage for the required amount of charging.

Limitador de tensión: Conectado por el red de resistores de un regulador de tensión. Determina si el campo recibirá alta, baja o ninguna tensión. Controla la tensión de campo durante el tiempo indicado de carga.

Voltage regulator: Used to control the output voltage of the ac generator, based on charging system demands, by controlling field current.

Regulador de tensión: Dispositivo cuya función es mantener la tensión de salida del generador de corriente alterna, de acuerdo a las variaciones en la corriente de carga, controlando la corriente inductora.

Voltmeter: A test meter used to read the pressure behind the flow of electrons.

Voltímetro: Instrumento de prueba utilizado para medir la presión del flujo de electrones.

Wake-up signal: An input signal used to notify the body computer that an engine start and operation of accessories is going to be initiated soon. This signal is used to warm up the circuits that will be processing information.

Señal despertadora: Señal de entrada para avisarle a la computadora del vehículo que el arranque del motor y el funcionamiento de los accesorios se iniciarán dentro de poco. Dicha señal se utiliza para calentar los circuitos que procesarán los datos.

Warning light: A lamp that is illuminated to warn the driver of a possible problem or hazardous condition.

Luz de aviso: Lámpara que se enciende para avisarle al conductor sobre posibles problemas o condiciones peligrosas.

Watt: The unit of measure of electrical power, which is the equivalent of horsepower. One horsepower is equal to 746 watts.

Watio: Unidad de potencia eléctrica, equivalente a un caballo de vapor. 746 watios equivalen a un caballo de vapor (CV).

Wattage: A measure of the total electrical work being performed per unit of time.

Vataje: Medida del trabàjo eléctrico total realizado por unidad de tiempo.

Waveform: The electronic trace that appears on a scope; it represents voltage over time.

Forma de onda: La trayectoria electrónica que aparece en un osciloscopio; representa la tensión a través del tiempo.

Weather-pack connector: An electrical connector that has rubber seals on the terminal ends and on the covers of the connector half to protect the circuit from corrosion.

Conectador resistente a la intemperie: Conectador que tiene sellos de caucho en los extremos de los bornes y en las cubiertas de la parte del conectador para proteger el circuito contra la corrosión.

Wet fouling: A condition of a spark plug in which it is wet with oil.

Engrase húmedo: Una condición de la bujía en la cual se moja de aceite.

Wheatstone bridge: A series-parallel arrangement of resistors between an input terminal and ground.

Puente de Wheatstone: Conjunto de resistores en series paralelas entre un borne de entrada y tierra.

Wiring diagram: An electrical schematic that shows a representation of actual electrical or electronic components and the wiring of the vehicle's electrical systems.

Esquema de conexiones: Esquema en el que se muestran las conexiones internas de los componentes eléctricos o electronicos reales y las de los sistemas eléctricos del vehículo.

Wiring harness: A group of wires enclosed in a conduit and routed to specific areas of the vehicle.

Cableado preformado: Conjunto de alambres envueltos en un conducto y dirigidos hacia áreas específicas del vehículo.

Worm gear: A type of gear whose teeth wrap around the shaft. The action of the gear is much like that of a threaded bolt or screw.

Engranaje de tornillo sin fin: Un tipo de engranaje cuyos dientes se envuelven alrededor del vástago. El movimiento del engranaje es muy parecido a un perno enroscado o una tuerca.

Wye connection: A type of stator winding in which one end of the individual windings are connected at a common point. The structure resembles the letter "Y."

Conexión Y: Un tipo de devanado estátor en el cual una extremidad de los devanados individuales se conectan en un punto común. La estructura parece la letra "Y."

Y-type stator: A three-winding ac generator that has one end of each winding connected at the neutral junction.

Estátor de tipo Y: Generador de corriente alterna de devanado triple; un extremo de cada devanado se conecta al empalme neutro.

Zener diode: A diode that allows reverse current to flow above a set voltage limit.

Diodo Zener: Diodo que permite que el flujo de corriente en dirección inversa sobrepase el límite de tensión determinado.

Zener voltage: The voltage that is reached when a diode conducts in reverse direction.

Tensión de Zener: Tensión alcanzada cuando un diodo conduce en una dirección inversa.

Index

A

A circuit, 185, 187-88
ABS. (See Antilock brake system)
AC. (See Alternating current)
AC generator
  bench testing, 207-8
  circuits, 183
  component testing, 210-14
  components, 177
  disassembly, 209
  housings, 182
  noises, 197
  operation, 183-84
  reassembly, 215
  removal/replacement, 207
Acceleration pickup, 757
Accumulator, 387, 392. (See also Central processing unit)
ACT. (See Air charge temperature)
Active testing, 557
Actuators, 392, 396, 399-400, 413-14, 536
Adaptive
  memory, 478
  numerator, 554
  strategy, 389
Address, 387. (See also Memory)
ADL. (See Automatic door lock)
Advanced, 338
After top dead center, 331, 338-39
AIR. (See Secondary air injection)
Air
  charge temperature, 389-90, 470
  coil gauge, 269
  density systems, 476-77
Air bag
  clean-up procedure, 778
  deployment, 755
  diagnostic module, 753
  module, 749, 751, 779-80
  safety/service warnings, 771-72

  systems, 749-60
  warning lamp, 755
Alternate good trips, 551-52
Alternating current, 9
Alternator, 177
ALU. (See Arithmetic logic unit)
Ambient light sensor, 609
Ambient sensor, 647, 666
American wire gauge, 81-82
Ammeter, 49-50, 59, 190
Ampere-hour rating, 120
Amperes, 4
Amplification circuit, 386
Amplifier test, 622
Analog, 384
  meter, 46-47
  scope. (See Oscilloscope)
  signal, 74
  to digital (A/D) converter, 386, 389, 401
AND gate, 390
André Ampère, 4, 16
Antiflashover ribs, 337
Antilock brake system, 70, 544
Anti-theft systems, 695-700, 730, 732-34
Arithmetic logic unit, 387
Armature, 142-43
  commutation test, 169
  ground test, 166
  short test, 166
Armed, 695. (See also Vehicle theft security systems)
Arming sensor, 756
ASD. (See Automatic shutdown relay)
ASDM. (See Air bag diagnostic module)
Aspirator, 646, 666-68
ATC. (See Automatic temperature control)
ATDC. (See After top dead center)
Atom, 1-3, 15
Atomic structure, 1-2
Audio
  entertainment systems, 707-10, 742-43

warning systems, 273, 289-90
Auto range, 51
Automatic door lock, 700-1, 734-37
Automatic headlight
  dimming, 586
  on/off with time delay, 587-89
  system diagnosis, 620-24
Automatic shutdown relay, 438, 481-82
Automatic temperature control, 645
Average responding, 48
AWG. (See American wire gauge)

B

B circuit, 185
Backup lights, 235
Balanced atom, 2
Ballast resistor, 80
BARO. (See Barometric pressure sensor)
Barometric pressure sensor, 466
Barrel-type drive, 146. (See also Starter drive)
Base, 31
Base timing, 338
Battery
  cables, 121, 147
  capacity ratings, 120
  capacity test, 132-33
  charging, 127-129
  drain test, 134
  failure, 121-22
  inspection, 126-27
  jumping, 136-37
  leakage test, 130
  open circuit voltage test, 132
  removal and cleaning, 133-36
  selecting, 120-21
  state of charge, 131
  terminal test, 129-30
  three-minute charge test, 133
  vibration, 122
  working on precautions, 126
BCI. (See Battery Council International)
BCM. (See Body control module)
Before top dead center, 338, 343
Bench tested, 163
Bendix drive, 145-46. (See also Starter drive)
Bezel, 244-45
Big slope, 552
Bimetallic
  gauges, 266
  strip, 38, 232, 266

Binary code, 384-85
Bipolar transistor, 32
Blend air door actuator, 648-49
Block learn, 479, 532
Blower motor, 301, 323-25
Blower motor speed controller, 656, 660
BMSC. (See Blower motor speed controller)
Body computer, 583-86, 596, 627-28, 677
Body control module, 383, 407
  diagnostics, 410-12
  trouble code retrieval, 411
Brake light switch, 229-32, 257-58
Braking system combination valve, 271-72
Branches. (See Leg, parallel circuit)
Breaker
  plate, 341
  point, 341
Breakout box, 416-17, 532
Brushes, 142, 209
BTDC. (See Before top dead center)
Bucking coil, 267
Buffer, 607
Bulkhead connector. (See Connectors)
Bus, 400
  bar, 36, 325-26
  data links, 400-1
Butt connector. (See Connectors)
Buzzer, 26-27, 68, 289
Bypass valve, 226

C

C65 Tempmatic system, 668-69
CAFE. (See Corporate average fuel economy)
Calibrated frequency, 569
California Air Resources Board, 541-42
Camshaft position sensor, 422, 431, 462, 490-91
Capacitance, 14-15
Capacitor, 14-15
Capacity test, 132-33
CARB. (See California Air Resources Board)
Carbon tracking, 356
Cartridge fuses. (See Maxi-fuse)
Catalyst monitor, 545, 554, 562. (See also Onboard diagnostics second generation)
Catalyst monitor sensor, 568
Cathode ray tube, 52-53, 604, 607
CCA. (See Cold cranking amps)
CCP. (See Climate control panel)
Cell, 114
Cell input voltage test, 367-69

CEMF. (See Self-induction and Counter electromotive force)

Center
high mounted stop lamp, 228, 230
pole negative, 334

Central
port injection system, 493-95
processing unit, 386-87, 389-90

Centrifugal advance, 341-42

Charging, 127

Charging system
components, 175-76
indicators, 189-91
inspection, 196
lamp circuit testing, 287-89
operating principles, 176-82
purpose, 175
requirement test, 206-7
service precautions, 196-97
troubleshooting, 197-207
types, 175

Child safety latch, 307, 309

Chime module, 289. (See also Audio warning systems)

CHMSL. (See Center high mounted stop lamp)

Choke coil, 18

Chrysler full field testing, 203

Circuit
amplification, 386
breaker, 38, 61
composition, 5
determining amount of resistance in, 5
electrical, 9-14, 23
grounded, 44
open, 43, 537
opening relay, 482
parallel, 10-11
primary ignition, 332-34
printed, 83
secondary ignition, 332-34
series, 10
series-parallel, 11-12
shorted, 44

CIS-E. (See Electronic continuous injection system)

CKP. (See Crankshaft position sensor)

Clamping diode, 30. (See also Diodes)

Class 2 data links, 543-44

Clear flood, 474

Clearance lights, 235-36

Climate control panel, 652

Clock, 385

Clock spring, 752, 778

Clocked RS flip-flop circuits, 391

Closed loop, 475

CMP. (See Camshaft position sensor)

CMS. (See Catalyst monitor sensor)

Coil, pickup, 347-48

Coil-condenser zone. (See Intermediate section)

Coil
pack, 428
synchronizing, 427
winding Ohmmeter tests, 452

Cold
cranking amps, 120
fouling, 373
engine lock-out switch, 652
start injector, 487-88, 523-24
valve, 497

Collector, 31

Collision avoidance light. (See Center high mounted stop lamp)

Color codes, wire, 87-90, 94

Common connection, 83

Commutator, 142-43, 169

Comparators, 689

Component locator, 90

Composite headlights, 224
aiming, 247-48
replacing, 245-46

Compound motor, 145

Comprehensive components, 552, 561, 573

Computer, 383-84
service precautions, 409
visual inspection, 410

Concealed headlight system, 226, 249, 618-20

Condenser, 14

Conduction, 32

Conductor, 2-3

Conflict, 549

Connectors, 82-83, 108-10

Constant control relay module, 491-93

Continuity, 9

Continuity tester, 45-46

Continuous DTCs, 568

Control assembly, 646

Conventional theory, 5

Cornering lights, 234-35

Corona effect, 355

Corporate average fuel economy, 461

Counter electromotive force, 144, 295-96

Courtesy lights, 236, 258. (See also Interior lights)

Covalent bonding, 15

CPI. (See Central port injection system)

CPU. (See Central processing unit)

Cranking motor designs, 150-52

Crankshaft

position sensor, 422, 431, 452, 462, 490-91
pulley, 429
Crash sensors, 753
Crimping, 102
Cross-fire, 81
Crossfiring, 355
CRT. (See Cathode ray tube)
Cruise control, 685-91, 715-23
Crystals, 15
Curb height, 246. (See also Headlight, aiming)
Current, 3-5
draw test, 159-61
fault, 436
limiting, 363
output testing, 199-200
precautions, open, 364
types, 9
Cutoff, 32
Cycle, 9, 50, 74
Cylinder output test, 532

D

D'Arsonval
gauge, 266-67. (See also Electromagnetic gauges)
movement, 46-47
Darlington pair, 33
Dash-mount switches, 250-51
Data. (See Information)
Data
counter, 387. (See also Central processing unit)
link connector, 437, 577-78. (See also Universal diagnostic
test connector)
recordings, 532-35
Daytime running lamps, 589-92
DC. (See Direct current)
DC generator, 176-77
DDM. (See Driver door module)
Decoder circuit, 390
Deep cycling, 118
Defogger, electric. (See Electric defoggers)
Delco-Remy
CS system, 287
SI system, 287
Delta
connection, 179-80, 182
pressure feedback EGR sensor, 571-72
Demultiplexer, 390
DEMUX. (See Demultiplexer)
Density speed system, 477
Depletion-type FET, 33

Depressed-park wiper system, 299-300
DERM. (See Diagnostic energy reserve module)
Detonation. (See Ping)
Detonation sensor, 394
DI. (See Distributor ignition)
Diagnostic
energy reserve module, 756
executive, 560
management system, 557, 559
mode, 628
service mode, 717
switch, 566
system check, 772
trouble codes, 70, 529, 560-61, 568, 716-22
weight, 563-64
Diaphragm, 293-94
Dielectric, 14
Differential pressure regulator, 497
Diffuser, 751-52
Digital, 384
instrument clusters, 597-98, 775
multimeter, 47, 417
signal, 74
storage oscilloscope, 53, 72-75
to analog (D/A) converter, 386
volt/ohmmeter, 47, 50-52
Dimmer switch, 225, 251-55
Diode, 18, 28-30, 65-66
pattern testing, 215-17
photovoltaic diode, 647
rectifier bridge, 181
rectifier bridge testing, 211
stator test, 206
trio, 183, 211
Direct
current, 9, 113
drive, 150. (See also Cranking motor designs)
Discriminating sensors, 756-57
Display
check, 632
pattern. (See Oscilloscope)
Distributor, 333, 340-41, 371-72. (See also Ignition)
cap and rotor, 333, 336-37, 355-56
ignition, 333, 335, 346-49, 421, 445-46
DLC. (See Data link connector)
DMM. (See Digital multimeter)
DOM. (Digital volt/ohmmeter)
Doped, 15-16
Dot matrix display, 604-5
Double filament lamp, 222
DPFE. (See Delta pressure feedback EGR sensor)
DRBIII, 411, 416

Drive
  belt replacement, 770-71
  coil, 151-52
  cycle, 548
  motor assembly, 768-70
Driver
  circuits, 391
  door module, 544
  input circuits, 393
  seat module, 544
  test mode, 527
DRL. (See Daytime running lamps)
DSM. (See Driver seat module)
DSO. (See Digital storage oscilloscope)
DTCs. (See Diagnostic trouble codes)
Dual
  ramping ECT sensor circuit, 468-69
  plug systems, 428
  sensor, 756
  trace scope, 54
Dump valve test, 725
Duty
  cycle, 50, 74
  cycled, 399
DVOM. (See Digital volt/ohmmeter)
Dwell, 363

E

E coil, 334
EAP. (See Electric air pump system)
Easy exit, 691
EATC. (See Electronic automatic temperature control)
ECA. (See Electronic control assembly)
ECC. (See Electronic climate control panel)
ECM. (See Engine control module)
ECT. (See Engine coolant temperature)
ECU. (See Electronic control unit)
Eddy currents, 143, 264, 282
EEC-V system, 440
EEPROM. (See Electrically erasable programmable read only memory)
EFI. (See Electronic fuel injection)
EI. (See Electronic ignition)
EGO. (See Exhaust gas oxygen)
EGR. (See Exhaust gas recirculation)
Electric
  air pump system, 572-73
  defoggers, 302-3, 325-26
  grid, 302
Electrical

  behavior, 1
  circuit. (See Circuit)
  current, 3-5, 9
  power, 9
  pressure, 3, 6
  reserve, 333
  schematic, 87
Electrically erasable programmable read only memory, 389
Electricity
  definition, 3
  discovery, 3
  elements, 3-5
Electrochemical, 113, 686
Electrochromic mirrors, 305
Electrolyte solution, 115-16
Electromagnetic
  field wiper circuits, 296-98
  gauges, 266-69
  interference, 444
Electro-magnetic induction, 18-19
Electromagnetism, 16-17
Electromechanical
  cluster self test, 635
  gauge, 265
  regulator, 185-87
Electromotive force, 3-4, 6, 16
Electron
  drift, 4
  theory, 5
Electronic
  automatic temperature control, 645, 652-60, 670-81
  climate control panel, 411
  continuous injection system, 495-97
  control assembly, 439
  control unit, 345-49
  cruise control, 687-89
  fuel injection, 461-62, 477
  gauges, 639-40
  ignition, 333, 335, 427-30, 447-57
  instrument clusters, servicing, 628
  regulator, 187-88
  sunroof, 692, 694-95, 730
  touch climate control, 655-57, 672
  voltage monitor, 189
  voltage regulator system, 287-88
Electrons, 1-4
Electrostatic
  discharge, 409
  field, 14-15
Emergency
  release mechanism, 748-49
  tensioning retractor, 757

EMF. (See Electromotive force)

EMI. (See Electro-magnetic induction)

Emitter, 31

Energy accumulator, 757

Engine

  control module, 411, 599

  coolant temperature, 72, 504

  coolant temperature sensor, 468, 475, 511-12

  detonation, 359

  firing order, 339

  identification codes, 441

  load, 338-39

  misfiring, 359

  speed, 338

Enhanced EVAP, 548

Enhancement-type FET, 33

EPROM. (See Erasable programmable read only memory)

Equivalent

  resistance, 11-12

  series load, 11. (See also Equivalent resistance)

Erasable programmable read only memory, 389

ETCC. (See Electronic touch climate control)

ETR. (See Emergency tensioning retractor)

EVAP monitor, 548, 555, 566-68. (See also Onboard diagnostics second generation)

Evaporator temperature sensor, 647-48

EVP. (See Exhaust gas recirculation valve position)

EVR. (See Electronic voltage regulator system)

Excess vacuum test, 567

Exclusive-OR gate, 390

Exhaust gas

  oxygen, 470

  gas recirculation, 468

  gas recirculation monitor, 547, 556, 564-65. (See also Onboard diagnostics second generation)

  gas recirculation valve position, 472, 519-20

Exterior lights, 228-36

External grounded field circuit. (See A circuit)

F

Failsoft, 407

Farads, 15 (See also Capacitor)

Fast

  codes, 441

  charging, 128

Fault codes, erasing, 579-80

Federal test procedure, 543

Feedback, 238-39

  signal, 396

FET. (See Field-effect transistor)

Fiber optics, 592-93

Field, 142

  coils, 143-44, 166

  current, 178

  current draw test, 198-99

  effect transistor, 32-33, 390

  of force, 16

  open, 208

  relay, 185-87

  service mode, 529

Filament, 222-23, 229

FIN sensor. (See Evaporator temperature sensor)

Fingers, 177. (See also Rotor)

Firing

  line, 362. (See also Oscilloscope)

  voltage, 364-65

Fixed resistors, 27

Flash

  codes, 526-30, 773-76

  to pass, 228, 586

Flasher, 232. (See also Taillight assemblies)

Flashing, 389. (See also Electrically erasable programmable read only memory)

Flip-flop circuit. (See Sequential logic circuits)

Floor-mounted switches, 252

Folo-thru drive, 146. (See also Starter drive)

Ford

  full field testing, 202-3

  self-test diagnostics, 440

  servicing, 449-52

Forward-biased, 29. (See also Diodes)

Free speed test, 163-64

Freeze frame, 550, 552

Frequency, 50, 74

FTP. (See Federal test procedure)

Fuel

  distributor, 495-96

  injector, 473

    inspection, 502

    service precautions, 502

    testing, 520-23

  level sender unit testing, 286-87

  mileage reduction, 360

  pressure regulator, 482-84

  pressure testing, 505-8

  pump circuit, 480-82

  pump volume testing, 508

  system good trip, 551

  system monitor, 547. (See also Onboard diagnostics second generation)

  tank pressure sensor, 566

Full

field current, 186
fielding, 201
field test, 201-6
Full-wave rectification, 181-82
Function diagnostic mode, 628, 631-32
Functional test, 620
Fuse, 35-36, 43
box, 35-36
links, 61-62
ratings, 60
types, 60
Fusible links, 36

G

Ganged switch, 24
Gap bridging, 362
Gassing, 117, 119
Gauges
air core, 269
bimetallic, 266
definition, 265
electromagnetic, 266-69
electromechanical. 265
sending units, 269-70
three-coil, 267-68
troubleshooting
multiple gauge failure, 284
sending units, 285
single gauge failure, 283-84
two-coil, 268-69
Gauss gauge, 69
Gear reduction starter, 150-51. (See also Cranking motor designs)
General Motors full field testing, 201-2
Glazing, 374
Glitches, 53, 72
Global good trip, 551
Grid, 114
growth, 118
wire repair, 325
Ground, 5
Ground circuit
inspecting, 355
test, 161-62. (See also Starting system, testing)
Ground side of circuit, 10
Grounded circuit, 44. (See also Circuit)
Growler, 166

H

Half cycle counter, 552
Half field current, 186
Half-wave rectification, 181
Hall-effect
sensor, 422-24, 429
sensor testing, 416, 444
switch, 396, 427
Halogen lamp, 223, 234
Hard
code, 410
fault, 436
Hazard warning system, 232, 234. (See also Taillight assemblies)
HC. (See Hydrocarbon)
Headlight
aiming, 246-48
brighter than normal, 248-49
bypass valve, 226
circuit, 225-26
concealed system, 226, 249
composite, 224
dimmer switch, 225, 251-55
high intensity discharge, 224
limit switches, 227-28
low illumination, 248
replacement, 244-46
sealed-beam, 222-23
switch, 224-25, 250-51
types, 222
vacuum distribution valve, 226
warning systems, 273
Head-up display, 609-10, 641-42
Heat
of compression, 338
range, 336
shrink tubing, 102, 104-5
Heated
backlight. (See Electric defoggers)
exhaust gas oxygen, 470, 561, 571
Heater core flow valve, 649
HEGO. (See Heated exhaust gas oxygen)
HEI ignition module test, 371
Hertz, 50
H-gate, 599
HID. (See Headlight, high intensity discharge)
High-note horn, 294
High pressure TBI, 486
High-reading coil, 267
High resistance, 44-45, 47
High temperature bypass, 344

High voltage circuit. (See Circuit, secondary)
HO2S monitor. (See also Onboard diagnostics second generation)
Hold-downs, 121
Hold-in windings, 148. (See also Magnetic switches)
Hole-flow theory, 5
Holographic combiner, 609-10
Horn, 293-94
  circuits, 294
  switches, 294
  troubleshooting, 315-17
HPTBI. (See High pressure TBI)
HUD. (See Head-up display)
Hybrid
  battery, 119
  inflator module, 758, 760
Hydrometer, 118
Hydrocarbon, 345
Hydrostatic lock, 158

I

IAC. (See Idle air control stepper motor)
IAR. (See Integral alternator/regulator system)
IAT. (See Intake air temperature sensors)
IC. (See Integrated circuit)
Idle
  air control stepper motor, 474, 525-26, 573
  speed control motor, 474
  switch, 473
Igniter, 751
Ignition
  breaker-point system, 346
  cables, 337
  circuitry, 332-333
  coil, 333-34
  coil resistance test, 369
  component testing, 366-73
  computer-controlled system, 349
  distributor, 333
  electrical reserve, 333
  module
    removal/replacement, 370-71
    test, 370
  no-start, 358-59
  primary circuit, 334
  primary resistor, 366-67
  purpose, 421
  setting timing, 375-78
  stress testing, 372-73
  switch, 149-50, 366

  timing, 337-39, 341-45
  troubleshooting, 353-54, 436
  visual inspection, 354-58
  warning systems, 273
Illuminated
  entry actuator, 592
  entry system, 592, 624-27
Illumination lights, 237. (See also Interior lights)
I/M test. (See Inspection/maintenance test)
Impedance, 47
IMTV. (See Intake manifold tuning valve)
Incandescence, 222
Indicator lights, 189, 237. (See also Interior lights)
Induced voltage, 17-18
Induction, 17-18
Inductive reactance, 184
Inertia, 145
  lock retractors, 747
  switch, 481, 749
Inflator assembly, 751
Information, 383
  processing, 389-92
Initial timing, 338
Injector
  balance test, 520
  deposits, 473-74
  flow test, 523
  kill test, 521
  rail, 480
  sound test, 522
Input, 383, 392-99
Inspection/maintenance test, 551
Installation
  diagrams, 87
  register, 387. (See also Central processing unit)
Instrument
  cluster, 237-38, 260, 277-90, 599. (See also Interior lights)
  panel dimming system, 592
  voltage regulator, 265-66, 278
Insulated
  circuit resistance test, 161. (See also Starting system, testing)
  side of circuit, 10
Insulator, 2-3
Intake
  air temperature sensors, 469-70, 512
  manifold tuning valve, 494-95
Integral alternator/regulator system, 199, 202-4, 211-14, 288-89
Integrated
  circuit, 34
  cruise control, 690

vehicle speed control, 717
Integrator, 479, 532
Intelligent wiper systems, 706-7
Interface, 386
Interior lights, 236-38, 258-60, 270-73
Intermediate section, 362
Intermittent
  code, 410
  wiper, 298-99
Internally grounded circuit. (See B circuit)
International Standards Organization, 26, 544, 598
Intrusive testing, 557
In-vehicle sensor, 646-47, 666, 668
Inverter, 390
Ion, 3
Ionize, 333
ISC. (See Idle speed control motor)
ISO. (See International Standards Organization)
ISO relays, 26
Isolated field, 185
IVR. (See Instrument voltage regulator)
IVSC. (See Integrated vehicle speed control)

J

Jumper wire, 45, 64, 324

K

KAM. (See Keep alive memory)
Karman vortex phenomenon, 465-66
Keep alive memory, 389, 440
Kettering, Charles F., 141, 340
Keyless entry, 702-3, 737-39
Key on, engine off test, 527-28, 717, 721
Key on, engine running test, 528, 717, 721
Key-off loads, 114
Keypad, 702
Kirchhoff, Gustav, 6
Kirchhoff's law, 6
Knock sensor, 394, 425
KOEO. (See Key on, engine off test)
KOER. (See Key on, engine running test)

L

Lab scope, 53, 413-14
Lamp, 222
  outage indicator, 593-96

outage module, 593-96
Latch, child safety, 307, 309
LCD. (See Liquid crystal display)
LCM. (See Lighting control module)
LDP. (See Leak detection pump)
Leak detection pump, 555, 557
LED. (See Light-emitting diode)
Leg, parallel circuit, 10, 13
Lens, 222-23
Lenz's law, 18
Light-emitting diode, 30, 46, 289, 604. (See also Diodes)
Lighting circuits
  components, 221
  servicing, 244-60
Lighting control module, 544
Lights, exterior. (See Exterior lights)
Limit switches, 227-28
Limp-in value, 464, 468
Linearity, 393
Lines of force, 16, 142
Liquid crystal display, 47, 52, 604-6
LM. (See Logic module)
Load device, 27
Loaded canister test, 567
Locks, power door, 327-28
Logic
  gates, 390-92, 694
  module, 438
  probe, 46
Logical troubleshooting, 353-54
Long term adaptive, 479
Low maintenance battery, 118
Low-note horn, 294
Low pressure TBI, 486
Low-reading coil, 267
Low voltage circuit. (See Circuit, primary)
LPTBI. (See Low pressure TBI)

M

MAF. (See Mass air flow sensor)
Magnetic
  field. (See Field of force)
  flux density, 16
  induction, 17
  offset angle, 376
  pulse, 346-48
  sensor tests, 455-56
  switches, 147-49
  timing meter, 376
Magnetic pulse generators, 395-96, 416, 422-23

Magnetism, 16-17
Magnets, 16
Maintenance-free, 117-18
Make-before-break wiper, 24
Malfunction indicator light, 437-38, 526-29, 568-69
Manifold absolute pressure sensor, 462-64, 476, 514-15
MAP. (See Manifold absolute pressure sensor)
Mass air
  flow sensor, 394, 464-66, 476, 515-19, 531-32
  air systems, 476-77
Maturing code, 550
Material expanders, 114
Maxi-fuse, 36-38, 61
Mechanical advance units, checking, 378
Memory, 387-89
  cartridge, 70-71
  seat, 691-93, 728-29
Mercury switches, 24-25
Metal detection sensors, 422-23
Metri-pack connector. (See Connectors)
Microfarads, 15
Microprocessor. (See Central processing unit)
MIL. (See Malfunction indicator light)
Minimum
  idle speed, 524-25
  TPS, 491
Min Max function, 50
Mirror
  electrochromic, 305
  heaters, 303
  power, 303
  rear view, 303-4
Misfire, 545, 571
  good trip, 551
  monitor, 545-47, 562-63, 571. (See also Onboard diagnostics second generation)
Mode door actuator, 649-50
Molded connector. (See Connectors)
Momentary contact switch, 24
Movable-pole shoe starters. (See Positive-engagement starter)
MPI. (See Multipoint fuel injection)
Multimeter, 46-47, 50-52
Multiple-wire hard-shell connector. (See Connectors)
Multiplexer, 390
Multiplexing, 400-2
Multiplier. (See Resistor)
Multipoint fuel injection, 486, 491-92
Mutual induction, 334
MUX. (See Multiplexer)

N

NAND gate, 390
NDS. (See Neutral drive switch)
Negative
  logic, 694
  temperature coefficient, 393, 603
Nematic fluid, 605
Neon lights, 234
Neutral
  drive switch, 472
  safety switch, 150, 235
Neutrons, 1
NC. (See Normally closed switch)
NO. (See Normally open switch)
No-crank, 158
No-crank test, 162-63. (See also Starting system, testing)
Noid lights, 522
Noise, 50
No-load test. (See Free speed test)
Non-enhanced EVAP, 548
Nonintegrated cruise control, 690
Nonvolatile RAM, 388-89
NOR gate, 390
Normally
  closed switch, 23
  open switch, 23, 62-63
North seeking pole, 16
NOT gate, 390
No-touch starting, 474
NOx. (See Oxides of nitrogen)
NTC. (See Negative temperature coefficient)
N-type material, 16, 28-29, 31
Nucleus, 1-2
NVRAM. (See Nonvolatile RAM)

O

O2. (See Oxygen sensor)
OBD II. (See Onboard diagnostics second generation)
Odometer, 264-65, 280, 599-600, 636
Offset, 376
Ohm, Georg S., 7
Ohm's law, 6, 8
  using, 12-14
Ohmmeter, 48-49, 60
Ohms, 5
Onboard diagnostics, 436-38
Onboard diagnostics second generation, 71, 541-73
On-demand codes, 440
One trip monitors, 549

Open
  circuit, 43, 48, 67, 107. (See also Circuit)
  circuit voltage test, 132
  loop, 475
OR gate, 390
Orifice spark advance control, 344
OSAC. (See Orifice spark advance control)
Oscillator, 602
Oscilloscope, 52-55, 510, 522
  display pattern, 361-64
  raster pattern, 361
  superimposed pattern, 361-62
  scales, 360-61
  trace interpretation, 361
OTIS. (See Overhead travel information system)
Output, 384, 399-400
  driver, 399
Overhead travel information system, 640-41
Overload, 35
Overrunning clutch, 146-47. (See also Starter drive)
Oxides of nitrogen, 344
Oxygen sensor, 397-99, 470-71, 509-11

P

Panel lights, 237-38, 260. (See also Interior lights)
Parabolic reflector, 222
Parallel circuit. (See Circuit)
Parasitic loads, 114
Park contacts, 296
Park/neutral switch, 519-20
Passive
  restraint, 747, 765-80
  seatbelt system, 747-49
  testing, 557
Pass-key, 697-98
Pawl, 24
PBR. (See Potentio balance resistor)
PCM. (See Powertrain control module)
Peak, 73
Peak-and-hold current, 494
Peak-to-peak voltage, 73
Pended, 549
Pending DTC, 568
Permanent magnet
  gear reduction, 152
  motors, 145
  wiper circuits, 296
Permanent magnet generators. (See Magnetic pulse generators)
Permeability, 17
Personal automotive security system, 697

PFI. (See Port fuel injection)
PFS. (See Purge flow sensor)
Photo diode, 30. (See also Diodes)
Photocell, 586, 623
  resistance assembly, 624
  test, 620-22
Photoelectric sensor, 424-25
Phototransistor, 34
Photovoltaic diode, 647
Pickup
  air gap, 358
  coil, 347-48, 358, 369-70, 395
Picofarad, 15
Piezoelectric devices, 394
Piezoresistive, 394
  sensor, 270, 285, 603
Ping, 338
Pinion
  factor, 607
  gear, 145-46
Planetary gear assembly, 152
Plate, 114
  straps, 114-15
PM. (See Power module)
PMGR. (See Permanent magnet gear reduction)
PN junction, 28-29
Polarized, 176-77
Polarizers, 605-6
Pole, 24, 177-78. (See also Fingers)
  piece. (See Reluctor)
  shoes, 142-43
Polymethylmethacrylate plastic, 592-93
Poppet nozzles, 494
Port fuel injection, 461, 486-88
Ported vacuum, 342
Positive-engagement starter, 151-52. (See also Cranking motor designs)
Positive temperature coefficient, 38-39, 393-94
Post-catalyst oxygen sensor, 545
Potential difference, 3 (See also Voltage)
Potentio balance resistor, 653-54
Potentiometers, 28, 394-95, 415-16
Power. (See Electrical Power)
Powertrain control module, 71, 147, 150, 429, 437-39, 501
Power-up vacuum test, 567
Power-vacuum module, 660
Power loss diagnosis, 359
Power
  mirror, 303
  module, 438
  unit test, 633-34
  window system, 305-7, 326-27
PTC. (See Positive temperature coefficient)

P-type material, 16, 28-29
Preignition damage, 374
Preprogrammed signal check tests, 632-33
Pressure
  check mode, 555
  difference, 4
  regulators, 482-85, 493
Primary
  ignition circuit, 334
  wiring, 80, 101, 354-55
Printed circuit, 83
Processing, 383
Program, 387
  counter, 387. (See also Central processing unit)
  number, 654
Programmable read only memory, 389, 417-19
Programmer, 646, 649
PROM. (See Programmable read only memory)
Protons, 1-2
Prove-out
  circuit, 271
  display, 628
P-type material, 16
Pull-in windings, 148. (See also Magnetic switches)
Pull to seat, metri-pack connector, 110
Pulse
  transformer, 333
  width, 50, 74
  width modulation, 74-75
Pump mode, 555
Punch through, 356, 545
Purge
  flow sensor, 569-71
  free cells, 552
  solenoid leak test, 568
Push to seat, metri-pack connector, 110
PVM. (See Power-vacuum module)
PWM. (See Pulse width modulation)

Q

Quartz swing needle display, 607-9
Quick
  Checker, 717
  test, 158-59. (See also Starting system, testing)

R

Radio
  choke, 266
  frequency interference, 50, 444, 538, 743
Radio-stereo sound systems. (See Audio entertainment systems)
RAM. (See Random access memory)
Random access memory, 388
Rapid deceleration cutoff, 689
Raster pattern. (See Oscilloscope)
Ratio, 145
Reach, 335
Read only memory, 388
Readiness indicator, 551
Real-time scope. (See Oscilloscope)
Rear view mirror, 303-4
Recirc/air inlet door actuator, 650
Recombination battery, 119
Rectification, 181-82
Rectifier and stator ground test, 207-8
Rectifier bridge, 211
Rectifiers, silicon-controlled. (See Semiconductors)
Reference pickup, 427
Registers, 392
Regulators, 185-88
  pressure, 482-85
  test, 206
Relays, 25-26, 63-65, 294-95
Reluctance, 17
Reluctor, 357
Remote keyless entry, 702, 704
Reserve capacity, 120
Resistance, 5
  total, parallel circuit, 10-11
  total, series circuit, 10
  total, series-parallel circuit, 11-12
  wire, 80-81
Resistance assembly test, 624
Resistor, 27-28
  block, 301
  module, 757
  plugs, 336
Response time test, 561
Retarded timing, 338
Returnless fuel system, 484-85
Reverse-biased, 29. (See also Diodes)
RFI. (See Radio frequency interference)
Rheostats, 28, 237
Right-hand rule, 17
Rise time, 341
RMS. (See Root mean square)

ROM. (See Read only memory)
Root mean square, 48, 73
Rotational velocity fluctuations, 546
Rotor, 177-78, 210
  punch through, 356
RS circuits, 391

S

SAE. (See Society of Automotive Engineers)
SAE J2012, 578
Safing sensor, 753
SAM. (See Security alarm module)
SATC. (See Semiautomatic temperature control)
Saturation, 32
Scan testers, 55-56, 70-72, 530-32, 543, 551, 679-81, 772-73
Schmitt trigger, 396
Schrader valve, 505-6
Scope controls, 54-55
SCP. (See Standard corporate protocol)
SCR. (See Silicon controlled rectifier, 34)
SDV. (See Spark-delay valve)
Sealed-beam, 222
  aiming, 246-47
  replacing, 244-45
Seats, power, 307, 327
Secondary air injection
  monitor, 565, 572
  system, 548
Secondary
  circuit resistance, 365-66
  reserve voltage, 335
  wiring, 80, 354-55
Sector gear. (See Window)
Security alarm module, 698
Self-diagnostics, 716
Self-induction, 18. (See also Counter electromotive force)
Self-test, 632
  input, 720-21
Semiautomatic temperature control, 645-52, 666-70
Semiconductor, 3, 15-16
Sender unit, 265. (See also Gauge)
Sensing voltage, 184-85
Sensitivity control, 586
Sensors, 393, 414-16
  accelerometers, 754
  camshaft position, 422
  crankshaft position, 422
  Hall-effect, 422-24
  knock, 425
  mass-type, 753-54

metal detection, 422-23
photoelectric, 424-25
roller-type, 754
voltage test, 561
Separators, 114
  code, 440-41
Sequential
  fuel injection, 431, 486-90
  logic circuits, 391
  sampling, 390
Series circuit. (See Circuit)
Series-parallel circuit. (See Circuit)
Series-wound, 144
Service bulletins, 503
Servo, 686, 725, 728
Servomotor, 668, 687
SFI. (See Sequential fuel injection)
Short, 35, 60
  short to ground, 68-69 , 211
  testing, 67-68, 208
Short term adaptive, 478-79
Shorted circuit, 44. (See also Circuit)
Shorting bars, 755
Shunt
  circuit. (See Leg, parallel circuit)
  motors. 144
  resistor, 267
Shutter wheel, 396, 423
Side marker lights, 235-36
Signal conditioning, 386
Similar conditions window, 550-51
Simulated road test, 723-24
Single-phase voltage, 176
Silicon
  controlled rectifier, 34
  diaphragm, 462
Sine wave, 9, 48, 54, 176
Single-pole
  double-throw, 24
  single-throw, 23-24
Sinusoidal, 54
SIR. (See Air bag systems)
Slip rings, 178-79
Slow
  charging, 128
  cranking, 158
Small leak test, 568
Society of Automotive Engineers, 90, 120
Soft code, 410
Software look-up tables, 426
Soldering, 102
Solderless connections, 102